Advanced We

David Watson
Consultant Engineer
Southern International Inc.

Terry Brittenham
President
Southern International Inc.

Preston L. Moore
Consultant

Henry L. Doherty Memorial Fund of AIME
Society of Petroleum Engineers

Richardson, Texas
2003

SPE Textbook Series

The Textbook Series of the Society of Petroleum Engineers was established in 1972 by action of the SPE Board of Directors. The Series is intended to ensure availability of high-quality textbooks for use in undergraduate courses in areas clearly identified as being within the petroleum engineering field. The work is directed by the Society's Books Committee, one of the more than 40 Society-wide standing committees. Members of the Books Committee provide technical evaluation of the book. Below is a listing of those who have been most closely involved in the final preparation of this book.

Book Editors

Hans Juvkam-Wold, Texas A&M U., College Station, Texas
Susan Peterson, J. Murtha & Assocs., Houston

Books Committee (2003)

Waldo J. Borel, Devon Energy Production Co. LP, Youngsville, Louisiana, Chairman
Bernt S. Aadnoy, Stavanger U. College, Stavanger
Jamal J. Azar, U. of Tulsa, Tulsa
Ronald A. Behrens, ChevronTexaco Corp., San Ramon, California
Ali Ghalambor, U. of Louisiana-Lafayette, Lafayette, Louisiana
James E. Johnstone, WZI Inc., Plano, Texas
Gene E. Kouba, ChevronTexaco Corp., Houston
William R. Landrum, ConocoPhillips, Houston
Eric E. Maidla, Noble Engineering & Development Ltd., Sugar Land, Texas
Erik Skaugen, Stavanger U. College, Stavanger
Sally A. Thomas, ConocoPhillips, Houston

ISBN 1-55563-101-0

David Watson is a consultant petroleum engineer with Southern International Inc. in Oklahoma City, Oklahoma. He holds a BS degree in petroleum engineering from Texas Tech U.; after graduating, he went to work for Unocal Corp. and was with Unocal for 12 years. Watson's professional experience includes responsibilities in production and reservoir engineering, but most of his career has been in drilling. He has been an instructor in numerous industry schools on drilling practices and well control and has authored or coauthored papers and articles on pipe design, horizontal drilling, and well control. Watson is a registered professional engineer in Oklahoma.

Terry Brittenham is the president and owner of Southern International Inc. He holds a BS degree with honors from the U. of Wyoming and was employed by Continental Oil Co., Monsanto Co., and Grace, Shursen, Moore and Assocs. before cofounding SII in 1982. Although practiced in many aspects of petroleum engineering, Brittenham has considerable experience in petroleum, geothermal, and scientific drilling operations, including management, engineering, and extensive assignments as a wellsite supervisor. Brittenham has taught drilling practices and well-control short courses since 1979 and has authored or coauthored several technical papers, articles, books, and manuals on drilling. He is a registered engineer in Oklahoma and Wyoming.

Preston L. Moore has been active in the drilling business since 1949, including 14 years of teaching petroleum engineering at the U. of Oklahoma. Moore received the SPE Drilling Engineering Award in 1993 and was named a Distinguished Member in 1996. He is known throughout the world for his drilling practices schools, which he initiated in 1959. Moore has written three books on drilling practices and has authored more than 100 articles in various trade magazines. He is co-owner of three patents concerning well control and is a recognized authority on well control. Moore remains active in the oil industry and is currently helping to develop a school on deep-well drilling and associated costs.

Introduction

Well-control fundamentals have been understood and taught since at least the early 1960s. Accident statistics have demonstrated the merits of training, and most individuals involved in drilling or other well operations have received some well-control training. Yet, for various reasons, well-control problems and blowouts persist in the industry. The consequences of a blowout (personnel safety, environmental impact, and financial outlays) more than justify efforts to develop effective countermeasures. This book addresses almost all phases of well control, and we hope that its content will contribute to those efforts. We anticipate that the book will be used as a text to train young engineers and as a reference for working engineers and supervisors.

Acknowledgments and Dedications

David Watson

First, I would like to thank Terry Brittenham and Southern International Inc. for providing the commitment and resources necessary to get this book written and in its present form. A special thanks goes to our draftsman, Don Willis, for his fine work on the charts and illustrations. Many mentors, associates, and coworkers have been an inspiration to me over the years. There are too many to mention here, but those who have particularly sparked my interest in well control include Mac Laurie, Dennis Black, and Preston Moore. Thanks to Hans Juvkam-Wold for his valuable comments and to Juliana Brandys for her sharp editorial pencil. Finally, this project consumed a lot of time that would otherwise have been spent with my family, and I would like to thank them for their patience and understanding.

Terry Brittenham

To my children, for understanding why Dad spent half his life, and most of theirs, "at the rig"... and to Perry L. Moore, my lifelong friend and partner, who forgot more about drilling than most will ever know.... "happy trails," PL.

Preston L. Moore

I dedicate my portion of this book to my wife, Mary Jo Moore. Mary Jo has always supported my activities during our 53 years of marriage. Drilling operations place a substantial demand on a person's time; Mary Jo accepted these demands and was always a source of encouragement and support.

Contents

Chapter 1
Gas Behavior and Fluid Hydrostatics

1.1 Introduction

The nature of gas and how a gas behaves in response to changing wellbore conditions is the basis for all pressure control techniques. One feature common to the methods for handling an influx is that efforts are directed towards maintaining a relatively constant bottomhole pressure throughout the procedure. An influx of formation fluid into a wellbore may be in the form of gas, liquid, or any combination thereof. Applied correctly, each of the control techniques will succeed regardless of the influx attributes.

Some early clues indicating the type of influx will be discussed, but the only time the presence of a gassy influx can be eliminated from consideration is after the physical properties of the formation fluids have been ascertained at the surface. Hence, all well-control procedures are designed to move gas up a wellbore while maintaining the desired bottomhole pressure. To accomplish this task, the selected approach must account for the compressible nature of gas and allow gas, if present, to expand during the wellbore displacement.

1.2 Phase Behavior

The ability to predict hydrocarbon phase behavior is essential in reservoir engineering and other petroleum-related disciplines. Phase behavior principles are perhaps less important for those who design and drill wells. Even so, drilling engineers and those who supervise drilling operations should have some fundamental understanding of how reservoir fluids may react to changing wellbore conditions during a well-control event. Accurate predictions require knowledge or at least a reasonable estimation of the hydrocarbon composition along with the pressure and temperature. Generally, the composition and constituents of an influx are unknown and the temperature at any point in a well is, at best, an educated guess. However, the ability to accurately predict well fluid behavior is less important than understanding that influx phase changes can and do occur in the process of killing a well.

Fig. 1.1 portrays a typical pressure/temperature phase diagram for a pure substance. The line separating the gas phase from the liquid phase is defined as the vapor pressure curve while the separation between liquid and solid is the melting point curve. Our primary interest as petroleum engineers is the portion of the diagram that depicts the gas and liquid phases. For a pure substance, the critical temperature T_c at point C defines that temperature above which only gas can exist while the critical pressure p_c defines that pressure above which liquid and gas coexist in an undifferentiable state. Critical constants and molecular weights of various natural gas constituents are given in **Table 1.1.**

Rarely, if ever, do pure liquids or gases reside in a rock formation and produced fluids are usually a mixture. Any combination of methane and the heavier hydrocarbon components may be present in an influx plus, possibly, such benign or offensive gases as nitrogen, carbon dioxide, or hydrogen sulfide. **Fig. 1.2** depicts a typical pressure/temperature phase diagram for fluid mixtures.

The area within the envelope describes the combination of pressures and temperatures at which gas and liquid co-exist. Note that increasing liquid concentration within this region is seen at increasing pressure and at decreasing temperature. The 100% (by volume) liquid line defines the bubble point pressure at any given temperature while the 100% gas line gives the dew point pressure as a function of temperature. The critical point at C characterizes the unique pressure and temperature (p_c and T_c) at which the properties of the bubble point liquid are indistinguishable from the properties of the dew point gas. Thus the definitions of p_c and T_c for a mixture are markedly different than for a pure substance. A series of phase diagrams for various mixtures of methane and ethane is shown in **Fig. 1.3.** The critical points unique to each mixture concentration are shown as points 1 through 10 on the diagram.

Refer to the line A-A′ on Fig. 1.2 as an example of what may occur in removing an influx from a well. Under this scenario, point A depicts the reservoir pressure and temperature and is in the region of the phase diagram where the influx mixture is all gas. Traversing the dashed line to the surface conditions at A′, liquid or condensate begins to fall out of the gas at point B in the wellbore and the composition is almost 40% liquid by the time the influx surfaces. It can also be seen from this diagram that conditions may also exist such that the gas concentration increases as a fluid mixture approaches the surface.

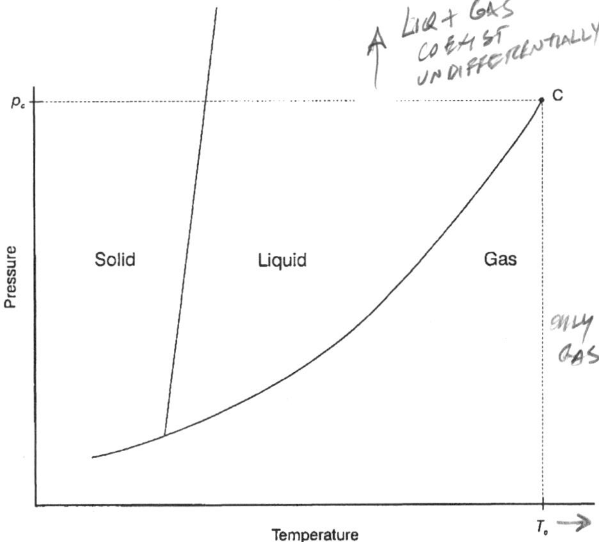

Fig. 1.1—Pressure/temperature phase diagram for a pure substance.

1.3 Gas Law Principles

An equation of state (EOS) describes the pressure/volume/temperature (PVT) relationships of that fluid. One of the simpler equations for gas was first described in the 17th century by Robert Boyle, who found by experiment that, at constant temperature, the volume of a quantity of gas is inversely proportional to its pressure. Boyle's law may then be expressed by

$$p_1V_1 = p_2V_2 = \text{constant}, \qquad (1.1)$$

where p and V are the pressure and volume of the gas at conditions 1 and 2.

Charles later discovered the direct proportionality between the temperature and volume of a given quantity of gas. Charles' law is given by

$$\frac{V_1}{T_1} = \frac{V_2}{T_2} = \text{constant}. \qquad (1.2)$$

All PVT relationships require the use of absolute pressure and temperature. The absolute pressure is simply the gauge pressure plus the atmospheric pressure. Given the imprecise nature of well control predictions, the use of unadjusted gauge pressures is probably acceptable in most cases. Exceptions to this generalization include those situations where pressures are low or approach atmospheric conditions.

Absolute temperatures are referenced to absolute zero and are determined in customary oilfield units by

$$°R = °F + 460, \qquad (1.3)$$

and in the SI metric system by

$$K = °C + 273, \qquad (1.4)$$

where °F and °R are temperatures in degrees Fahrenheit and Rankine. In the SI metric system, °C and K are degrees Celsius and Kelvin.

The volume of an ideal gas depends on the number of gas molecules, or moles, present as well as pressure and temperature. From Avogadro's law, the type of gas molecule or the presence of a mixture of different molecules is not a factor. Combining this principle with the observations of Boyle and Charles leads to the ideal gas law:

$$pV = nR_gT, \qquad (1.5)$$

where n is the number of moles (mass divided by molecular weight) and R_g is the universal gas constant, whose numerical value depends on the chosen unit system. Some common units and associated gas constant values are shown in **Table 1.2.**

In the case of a gas influx contained within a wellbore, n is constant and it follows that

$$\frac{p_1V_1}{T_1} = \frac{p_2V_2}{T_2}. \qquad (1.6)$$

The application of ideal gas concepts is demonstrated in the following problem.

Example 1.1. A 20-bbl gas influx has entered a well at a bottomhole pressure of 3,500 psia.

1. Determine the volume of this same influx when it exits the well if atmospheric pressure at the well location is 14.4 psia and the gas temperature does not change.

2. Recalculate the volume at atmospheric conditions assuming an initial gas temperature of 150°F and a surface temperature of 65°F.

Solution. 1. Using Boyle's law,

$$V_2 = \frac{(3,500)(20)}{(14.4)} = 4,861 \text{ bbl}.$$

2. For the second case, Eq. 1.6 yields

$$V_2 = \frac{(3,500)(20)(525)}{(14.4)(610)} = 4,183 \text{ bbl}.$$

The density of a gas or any other material is its mass per unit volume, or

$$\rho = \frac{m}{V} = \frac{nM}{V}, \qquad (1.7)$$

TABLE 1.1—PHYSICAL PROPERTIES OF NATURAL GAS CONSTITUENTS[1]

Compound	Formula	Molecular Weight	p_c psia (MPa)	T_c °F (°C)
Methane	CH_4	16.043	667.8 (4.60)	–116.7 (–82.7)
Ethane	C_2H_6	30.070	707.8 (4.88)	90.1 (32.3)
Propane	C_3H_8	44.097	616.3 (4.25)	206.0 (96.7)
n-Butane	C_4H_{10}	58.124	550.7 (3.80)	305.6 (152.0)
Isobutane	C_4H_{10}	58.124	529.1 (3.65)	275.0 (135.0)
n-Pentane	C_5H_{12}	72.151	488.6 (3.34)	385.6 (196.4)
Isopentane	C_5H_{12}	72.151	490.4 (3.38)	369.0 (187.2)
Carbon Dioxide	CO_2	44.010	1,071.0 (7.38)	87.8 (31.0)
Hydrogen Sulfide	H_2S	34.076	1,306.0 (9.00)	212.6 (100.3)
Nitrogen	N_2	28.013	493.0 (3.40)	–232.7 (–147.1)
Water	H_2O	18.015	3,207.9 (22.11)	705.5 (374.2)

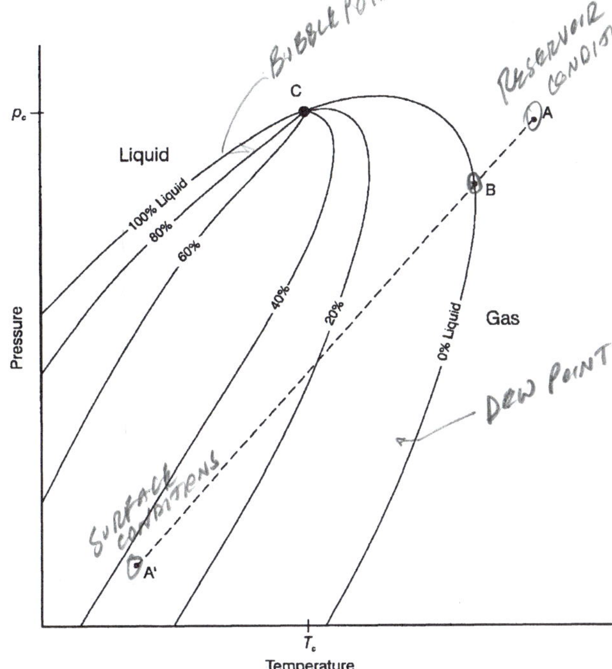

Fig. 1.2—Pressure/temperature phase diagram for a mixture.

where M is the molecular weight of the substance. Since

$$n = \frac{pV}{R_g T},$$

it follows that the density of an ideal gas may be determined by

$$\rho_g = \frac{pM}{R_g T}. \quad \text{(1.8)}$$

The specific gravity of a gas (γ_g) is the ratio of its molecular weight to that of air (M_a).

$$\gamma_g = \frac{M}{M_a} = \frac{M}{29}. \quad \text{(1.9)}$$

Rearranging Eq. 1.9 and substitution into Eq. 1.8 leads to a convenient relationship for gas density,

$$\rho_g = \frac{29\gamma_g p}{R_g T}. \quad \text{(1.10)}$$

The apparent molecular weight of a gas mixture can be obtained by

$$M = f_{g1}M_1 + f_{g2}M_2, \ldots, f_{gn}M_n, \quad \text{(1.11)}$$

where f_{gn} and M_n denote the mole fraction and molecular weight of the mixture components.

Example 1.2. A gas mixture consists of 95% methane, 3% ethane, and 2% of the heavier hydrocarbons. Determine the specific gravity of this mixture assuming an average heavy-end molecular weight of 47.

Solution. First, determine the apparent molecular weight of the mixture,

$$M = (0.95)(16.043) + (0.03)(30.070) + (0.02)(47.0)$$

$$= 17.083.$$

The gas specific gravity is determined as

$$\gamma_g = \frac{17.083}{29} = 0.59.$$

The gas specific gravity is an important variable in many of the well-control predictions that follow through the course of this text. The parameter can be readily obtained if the nature of the formation fluid is known and if a gas analysis is available for that fluid. Precise formation fluid constituent fractions in well-control problems, however, are generally unknown which means that some estimated value is required. Lacking specific knowledge, the use of a relatively low value is recommended. A common assumption in well control is for γ_g to range between 0.6 and 0.7.

Early investigators noted that gas behaved in an ideal fashion only under a limited range of pressure and temperature conditions. Compressibility factors, or z factors, were introduced as an empirical adjustment for non-ideal behavior. An EOS for a non-ideal, or real, gas is given by

$$pV = znR_g T. \quad \text{(1.12)}$$

Real gas adjustments for Eqs. 1.6, 1.8, and 1.10 follow as

$$\frac{p_1 V_1}{z_1 T_1} = \frac{p_2 V_2}{z_2 T_2}, \quad \text{(1.13)}$$

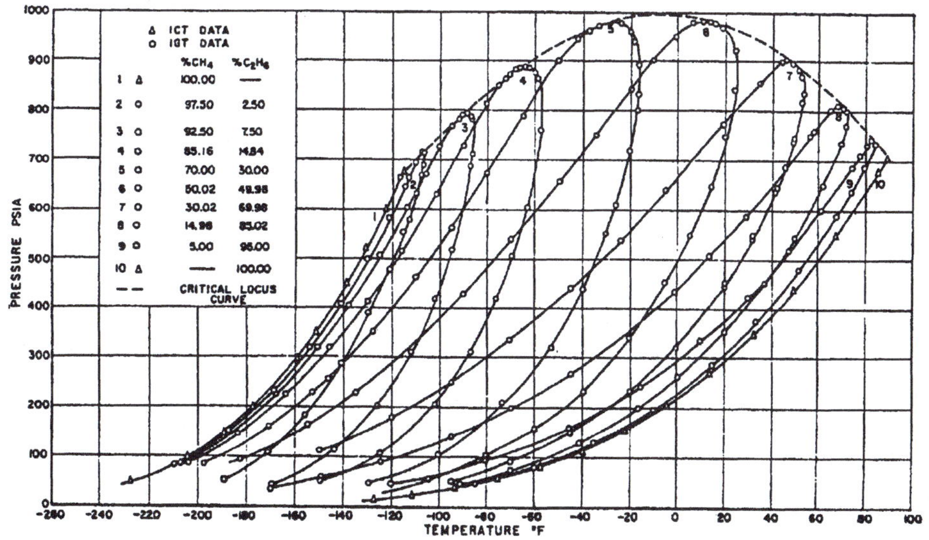

Fig. 1.3—Pressure/temperature phase diagram for various methane/ethane mixtures.[2]

TABLE 1.2—UNIVERSAL GAS CONSTANT VALUES

p	V	T	n	R_g
psia	cu ft	°R	lbm-mole	10.732
psia	gal	°R	lbm-mole	80.275
psia	bbl	°R	lmb-mole	1.911
psfa	cu ft	°R	lbm-mole	1,545.3
kPa	m^3	K	g·mole	0.0083145
kPa	m^3	K	kg·mole	8.3145

$$\rho_g = \frac{pM}{zR_gT}, \quad \text{.......... (1.14)}$$

and

$$\rho_g = \frac{29\gamma_g p}{zR_gT}. \quad \text{.......... (1.15)}$$

The magnitude of the z factor for a specific gas is dependent on both pressure and temperature. Compressibility factor curves have been obtained experimentally for a wide range of pure gases, one of which is depicted in **Fig. 1.4** for methane.

The z factor isotherm curves for all pure gases have the characteristic appearance shown in Fig. 1.4. This similarity naturally follows from the theorem of corresponding states, which says that two or more substances should have similar properties at corresponding conditions with reference to some basic property.[4-5] Another way of stating the theorem is that all pure gases should have the same z factor when the pressure and temperature of the gas are referenced to the critical pressure and temperature of the gas. The reduced pressure and reduced temperature (p_r and T_r) of a pure gas are the ratio of the gas pressure and temperature to the critical constants of the gas. Hence, all pure gases should have the same compressibility factor at equivalent p_r and T_r.

The technique for obtaining z factors must be modified if the gas is a mixture, as essentially all formation gases are. Pseudocritical pressure and temperature parameters (p_{pc} and T_{pc}) were devised by Kay[6] for gas mixtures and can be obtained by molal averaging the critical constants of the respective gas components.

$$p_{pc} = f_{g1}p_{c1} + f_{g2}p_{c2}, \ldots, f_{gn}p_{cn} \quad \text{.......... (1.16)}$$

and

$$T_{pc} = f_{g1}T_{c1} + f_{g2}T_{c2}, \ldots, f_{gn}T_{cn}. \quad \text{.......... (1.17)}$$

Pseudocritical properties can be correlated with specific gravity if the molecular structure of the gas components are similar. Sutton's[7] correlation, based on 264 gas samples, is presented as **Fig. 1.5.** Sutton also used regression analysis of the data to obtain

$$p_{pc} = 756.8 - 131.0\gamma_g - 3.6\gamma_g^2 \quad \text{.......... (1.18)}$$

and

$$T_{pc} = 169.2 + 349.5\gamma_g - 74.0\gamma_g^2. \quad \text{.......... (1.19)}$$

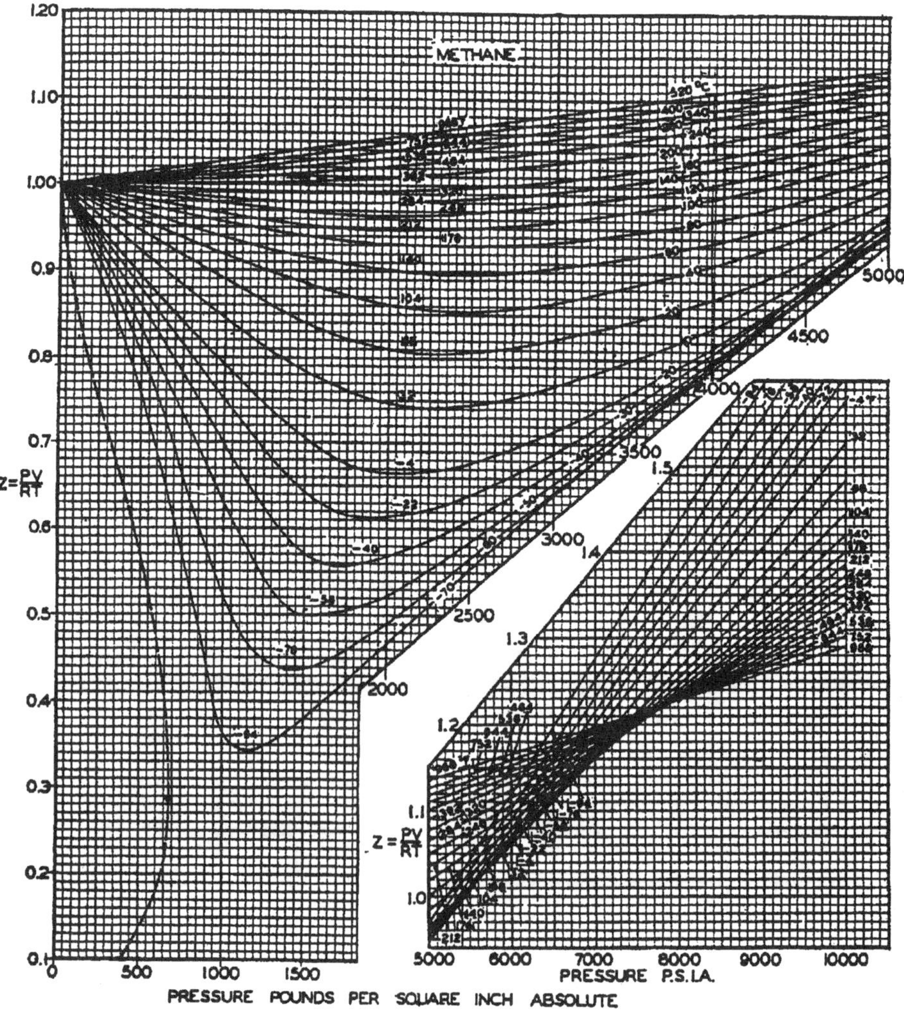

Fig. 1.4—Compressibility factors for methane.[3]

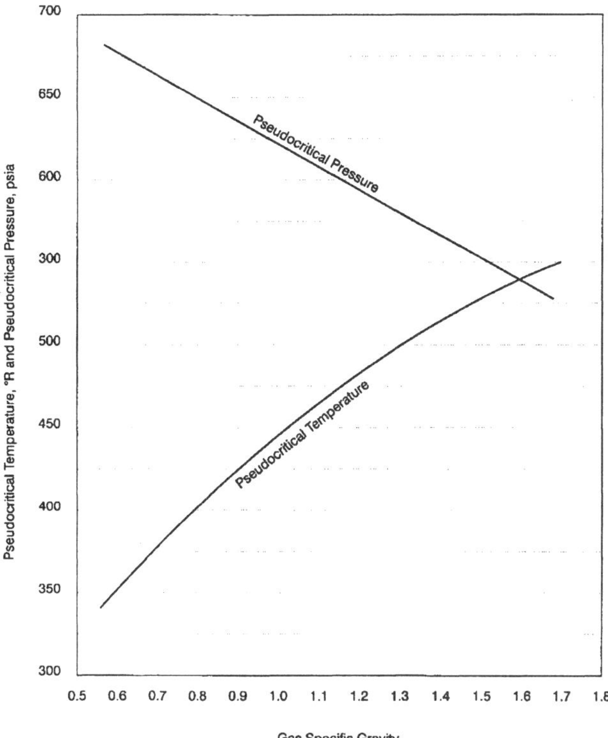

Fig. 1.5—Pseudocritical properties of natural gas (after Sutton[7]).

After calculating or otherwise acquiring p_{pc} and T_{pc}, the pseudoreduced pressure and temperature (p_{pr} and T_{pr}) are determined using Eqs. 1.20 and 1.21:

$$p_{pr} = p/p_{pc} \quad \text{.......... (1.20)}$$

and

$$T_{pr} = T/T_{pc}. \quad \text{.......... (1.21)}$$

Having the pseudoreduced properties, the z factor of any *hydrocarbon* gas can be obtained from the classic Standing and Katz[8] chart shown in **Fig. 1.6.** Based on the work of Kvalnes and Gaddy,[9] **Fig. 1.7** was later developed as an extension to the chart for higher pressures. Use of the pseudocritical property equations and compressibility factor chart is demonstrated in Example 1.3.

Example 1.3. Determine the initial z factor and gas density for the influx described in Example 1.1.

Solution. First we must assume a gas specific gravity. Taking γ_g to be 0.6, p_{pc} and T_{pc} are computed using Eqs. 1.18 and 1.19:

$$p_{pc} = 756.8 - (131.0)(0.60) - (3.6)(0.60)^2 = 677 \text{ psia}$$

and

$$T_{pc} = 169.2 + (349.5)(0.60) - (74.0)(0.60)^2 = 352°\text{R}.$$

Eqs. 1.20 and 1.21 yield the pseudoreduced properties,

$$p_{pr} = 3{,}500/677 = 5.170$$

and

$$T_{pr} = 610/352 = 1.733.$$

The z factor is found to be 0.892 in Fig. 1.6.

Substituting 80.275 for R_g in Eq. 1.15, the gas density in lbm/gal is

$$\rho_g = \frac{\gamma_g p}{2.77 z T}. \quad \text{.......... (1.22)}$$

Therefore,

$$\rho_g = \frac{(0.60)(3{,}500)}{(2.77)(0.892)(610)} = 1.39 \text{ lbm/gal}.$$

The arrival of computers to the oil industry in the early 1960's led to the introduction of equations for calculating z. At least 13 techniques have been published thus far. Some are more accurate than others for a given range of pseudoreduced values and some require more computing power than others.[10] Of these, the most common approach is to mathematically describe the empirical data presented by Standing and Katz.

Dranchuk and Abou-Kassem[11] used an EOS to develop a numerical model with coefficients to fit the Standing and Katz data. Their equation follows:

$$z = 1 + C_1(T_{pr})\rho_r + C_2(T_{pr})\rho_r^2 - C_3(T_{pr})\rho_r^5 + C_4(\rho_r, T_{pr}), \quad \text{.......... (1.23)}$$

The "reduced" density term ρ_r is obtained from the expression,

$$\rho_r = \frac{0.27 p_{pr}}{z T_{pr}}. \quad \text{.......... (1.24)}$$

The other functions are described by Eqs. 1.25 through 1.28:

$$C_1(T_{pr}) = 0.3265 - 1.07/T_{pr} - 0.5339/T_{pr}^3 + 0.01569/T_{pr}^4 - 0.05165/T_{pr}^5, \quad \text{.......... (1.25)}$$

$$C_2(T_{pr}) = 0.5475 - 0.7361/T_{pr} + 0.1844/T_{pr}^2, \quad \text{.......... (1.26)}$$

$$C_3(T_{pr}) = 0.1056(-0.7361/T_{pr} + 0.1844/T_{pr}^2), \quad \text{.......... (1.27)}$$

and

$$C_4(T_{pr}, \rho_r) = 0.6134(1 + 0.721\rho_r^3)(\rho_r^2/T_{pr}^3)\exp(-0.721\rho_r^2). \quad \text{.......... (1.28)}$$

Solving the Dranchuk and Abou-Kassem relation is an iterative process since the z factor depends on functions that contain the term. The Newton-Raphson iteration technique has the form,

$$z_{i+1} = z_i - f(z)/f'(z), \quad \text{.......... (1.29)}$$

where $f(z)$ is a function of z and $f'(z)$ is the first derivative of that function. The function for the z factor is obtained by rearranging Eq. 1.23,

$$f(z) = z - \left[1 + C_1(T_{pr})\rho_r + C_2(T_{pr})\rho_r^2 - C_3(T_{pr})\rho_r^5 + C_4(\rho_r, T_{pr})\right] = 0. \quad \text{.......... (1.30)}$$

Taking the derivative with respect to z yields

$$f'(z) = \frac{\partial f(z)}{\partial z} = 1 + C_1(T_{pr})\rho_r/z + 2C_2(T_{pr})\rho_r^2/z$$

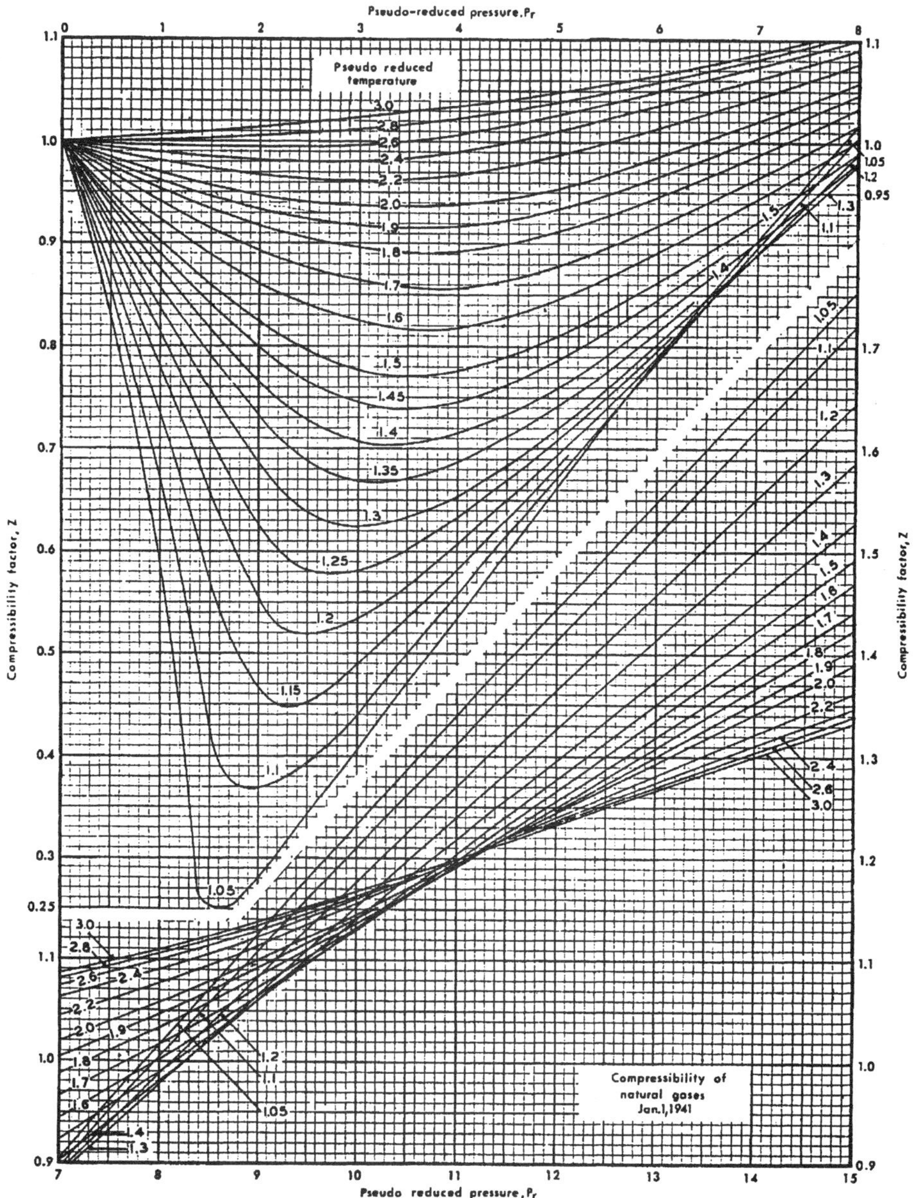

Fig. 1.6—Compressibility factors for hydrocarbon gases.[8]

$$- 5C_3(T_{pr})\rho_r^5/z + \frac{1.2268\rho_r^2}{T_{pr}^3 z}\left[1 + 0.721\rho_r^2 - (0.721\rho_r^2)\right]$$

$$\exp(-0.721\rho_r^3). \quad \ldots\ldots\ldots\ldots\ldots\ldots\ldots\ldots (1.31)$$

The technique is demonstrated in the next example.

Example 1.4. Determine the z factor for the previous example using the Dranchuk and Abou-Kassem method.

Solution. The pseudoreduced properties were calculated before. The parameters defined in Eqs. 1.25 through 1.27 depend only on the pseudoreduced temperature and will thus remain constant for a given gas specific gravity and temperature:

$$C_1(T_{pr}) = 0.3265 - (1.07/1.733) - (0.5339/1.733^3)$$

$$+ (0.01569/1.733^4) - (0.05165/1.733^5) = -0.3951,$$

$$C_2(T_{pr}) = 0.5475 - (0.7361/1.733)$$

$$+ (0.1844/1.733^2) = 0.1841,$$

and

$$C_3(T_{pr}) = (0.1056)$$

$$\left[(-0.7361/1.733) + (0.1844/1.733^2)\right] = -0.0384.$$

For the first iteration, assume $z_1 = 1.0000$ and compute the reduced density,

$$\rho_r = \frac{(0.27)(5.170)}{(1.0000)(1.733)} = 0.8055.$$

Now use Eq. 1.28 to determine $C_2(T_{pr}, \rho_r)$.

$$C_4(T_{pr}, \rho_r) = 0.6134\left[1 + (0.721)(0.8055^2)\right]$$

$$(0.8055^2/1.733^3)\exp\left[(-0.721)(0.8055^2)\right] = 0.0703.$$

$f(z)$ and $f'(z)$ are determined as

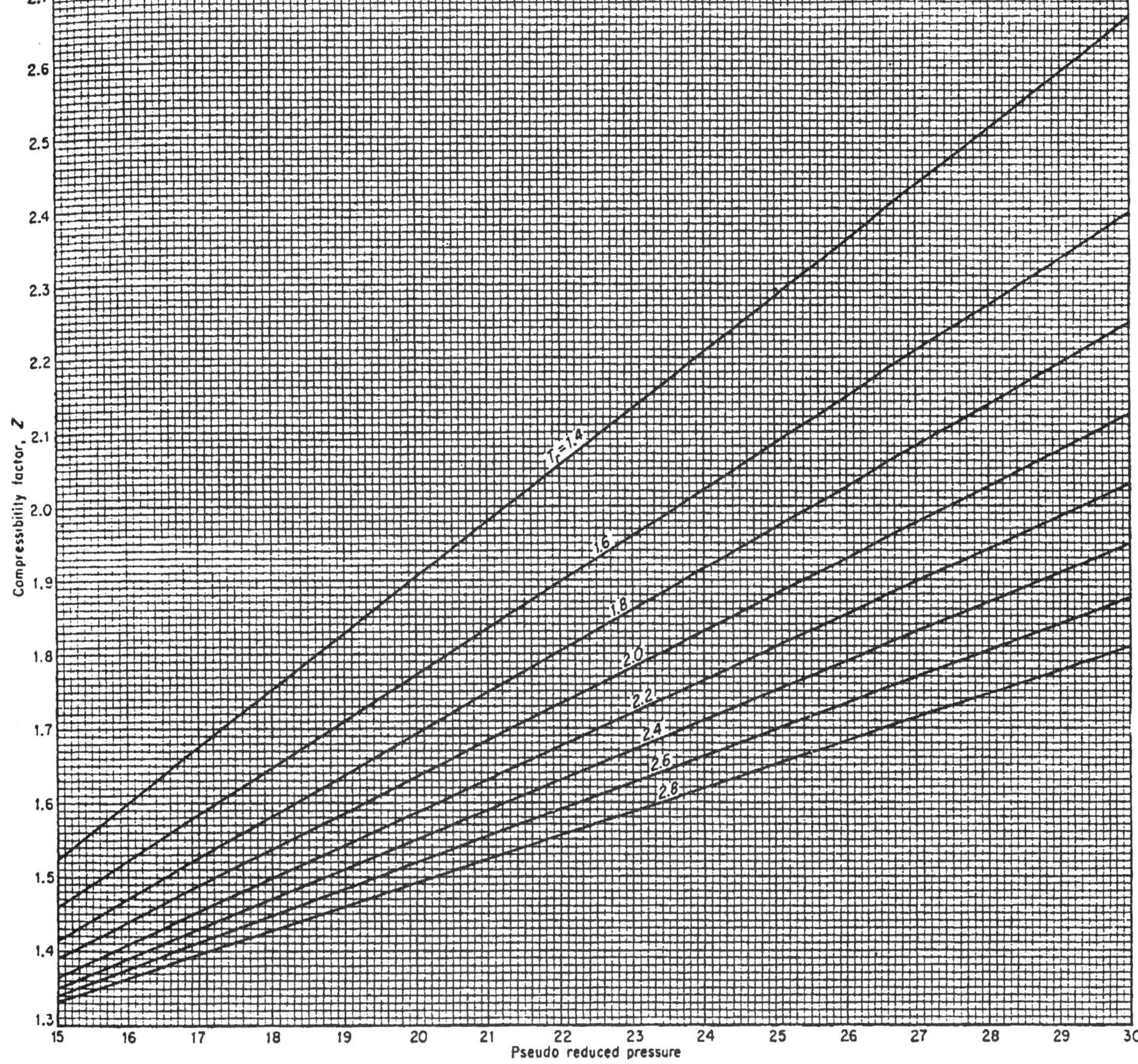

Fig. 1.7—Compressibility factors for hydrocarbon gases at high pressure.[4]

$$f(z) = 1.000 - [1 + (-0.3951)(0.8055)$$

$$+ (0.1841)(0.8055^2) - (-0.0384)(0.8055^5) + 0.0703]$$

$$= 0.1155$$

and

$$f'(z) = 1 + [(-0.3951)(0.8055)/1.0000]$$

$$+ [(2)(0.1841)(0.8055^2)/1.0000]$$

$$- [(5)(-0.0384)(0.8055^5)/1.0000]$$

$$+ \frac{(1.2268)(0.8055^2)}{(1.733^2)(1.0000)}$$

$$\times \left\{1 + (0.721)(0.8055^2) - [(0.721)(0.8055^2)]^2\right\}$$

$$\times \exp[(-0.721)(0.8055^2)] = 1.1054.$$

Eq. 1.29 yields the solution for z_{i+1}.

$$z_2 = 1.0000 - (0.1155/1.1054) = 0.8955.$$

This value is used for the next iteration and we ultimately obtain $z_3 = 0.8996$. Another iteration does not substantially change the result and the problem is solved.

Caution is advised if the gas is known or suspected to have non-hydrocarbon fractions. Use of the pseudocritical correlation and z factor charts in these cases, particularly if H_2S or CO_2 are present, can lead to an inaccurate result. Wichert and Aziz[12] offer a technique, not covered here, for correcting z factors when these gases are part of the mixture.

The techniques for predicting wellbore pressures with gas in the hole are generally accompanied by various simplifying assumptions. A typical assumption is that either the wellbore gas behaves according to Boyle's law (temperature and z factors are ignored) or as an ideal gas (wellbore temperature is included but z is taken as unity). Real gas computations re-

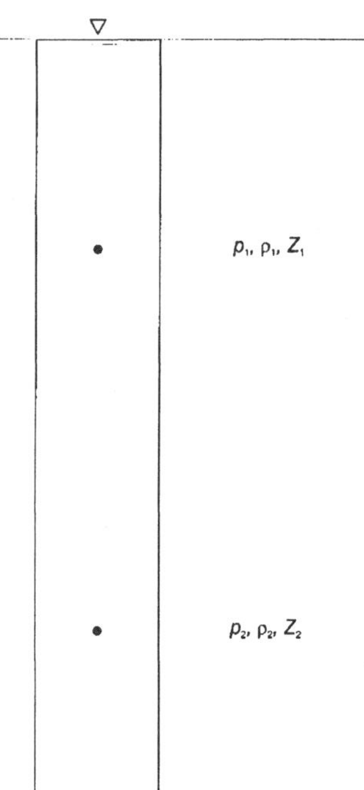

Fig. 1.8—Energy balance variables in a static wellbore.

quire iteration since the z factor is a function of the desired result, pressure. Even so, accurate modeling of what happens in a well during a control event demands that deviation from ideal behavior be considered.

1.4 Hydrostatics

An understanding of hydrostatics is basic to well-control concepts and a review of the fundamentals is therefore in order. We start with the mechanical energy balance relationship for fluids:

$$\int_{p_1}^{p_2} \frac{dp}{\rho_f} + \frac{g}{g_c}(Z_2 - Z_1) + \frac{\rho_f\left(v_2^2 - v_1^2\right)}{2g_c} + W + E_l = 0,$$

where

p_1, p_2 = pressure at positions 1 and 2,
ρ_f = fluid density,
g = acceleration of gravity,
g_c = gravitational system conversion constant,
Z_1, Z_2 = fluid elevation at positions 1 and 2,
v_1, v_2 = fluid velocity at positions 1 and 2,
W = work done by the fluid while in flow, and
E_l = irreversible energy loss between positions 1 and 2 (i.e., friction).

The last three terms drop out when the fluid is at rest, which leaves

$$\int_{p_1}^{p_2} \frac{dp}{\rho_f} = \frac{g}{g_c}(Z_1 - Z_2).$$

The variables at two arbitrary points in a wellbore are depicted in **Fig. 1.8.** In the strict sense, the elevation shown as Z_2 in the diagram is below Z_1, which means that Z_2 is smaller than Z_1 from any datum plane reference. Eq. 1.32 modifies the relation in keeping with customary usage—i.e., depth increases with distance from surface.

$$\int_{p_1}^{p_2} \frac{dp}{\rho_f} = \frac{g}{g_c}(D_2 - D_1). \quad \text{(1.32)}$$

1.4.1 Incompressible Fluids. Liquids such as oil, water, and drilling mud can be considered incompressible most of the time. Make the futher assumption that thermal expansion is negligible and the density at any point in a well will be constant. Integrating Eq. 1.32 and multiplying both sides by the density yields the change in hydrostatic pressure for a constant-density fluid:

$$p_2 - p_1 = \frac{g}{g_c}\rho_f(D_2 - D_1). \quad \text{(1.33)}$$

Inserting the acceleration of gravity and conversion constants gives

$$\Delta p(\text{psi}) = \frac{(32.17\ \text{ft/s}^2)}{(32.17\ \text{lbm-ft/lbf}-\text{s}^2)}\frac{\rho_f(\text{lbm/ft}^3)\Delta D(\text{ft})}{(144\text{in.}^2/\text{ft}^2)}.$$

The hydrostatic gradient (g_f) is the change in hydrostatic pressure with depth. For the preceding relation,

$$g_f(\text{psi/ft}) = \rho_f(\text{lbm/ft}^3)/144 = 0.00694\rho_f(\text{lbm/ft}^3). \quad \text{(1.34a)}$$

Inserting a conversion constant yields

$$g_f(\text{psi/ft}) = \rho_f(\text{lbm/gal})/19.25 = 0.00519\rho_f(\text{lbm/gal}). \quad \text{(1.34b)}$$

Consider the case where a hole is standing full of consistent fluid. The hydrostatic pressure at any depth can be determined using

$$p = g_f D. \quad \text{(1.35)}$$

The well depth to use in any hydrostatic pressure calculation is the well's true vertical depth (TVD), which is defined as the vertical distance from the kelly bushing (KB) datum plane to the point of interest in the wellbore. Another term, the measured depth (MD) is the length of the drilled hole from the KB datum. Refer to **Fig. 1.9** for the schematic difference between the two. The TVD and MD are the same in a vertical well.

The hydrostatic pressure in a stacked column of wellbore fluids is determined in additive fashion,

$$p = g_{f1}h_1 + g_{f2}h_2, \ldots, g_{fn}h_n, \quad \text{(1.36)}$$

where the subscripts denote the respective hydrostatic gradient and vertical heights of each fluid.

Finally, applying some pressure p_0 on top of the static fluid column will result in a wellbore pressure at any depth given by

$$p = p_0 + g_{f1}h_1 + g_{f2}h_2, \ldots, g_{fn}h_n. \quad \text{(1.37)}$$

The concept is illustrated in **Fig. 1.10.**

1.4.2 Gases. The density of a gas depends on the resident pressure and temperature, and is therefore dependent on its position in a wellbore. An acceptable practice for relatively short gas columns is to (1) determine the pressure and temperature at either the top or bottom of the gas column, (2) determine the gas density at the stated conditions using Eq. 1.15, and (3) assume that this density is constant throughout the gas column. The equations developed for incompressible fluids

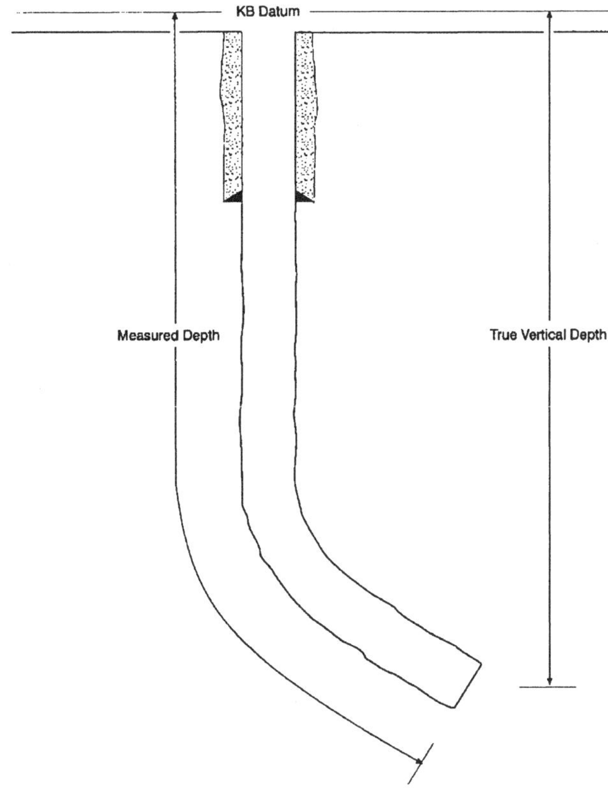

Fig. 1.9—Definition of true vertical and measured depths.

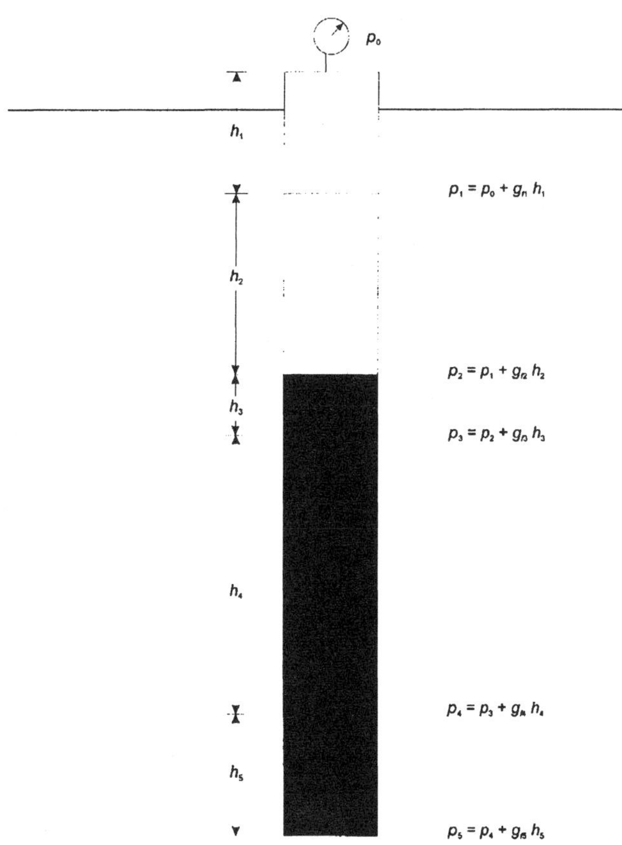

Fig. 1.10—Pressure in a static wellbore that contains stratified fluid layers.

can then be applied with reasonable accuracy. The simplified procedure is the usual approach when predicting gas kick behavior if the kick height is small relative to well depth.

Example 1.5. The 12,000-ft vertical well shown in **Fig. 1.11** is shut in with a 0.6 specific gravity gas influx on bottom. The initial shut-in casing pressure (SICP) is 500 psia. The initial influx height is estimated to be 400 ft and the annular mud density is 11.5 lbm/gal. Determine the bottomhole pressure if the bottomhole temperature is 205°F.

Solution. Eq. 1.37 is used to determine the pressure at the top of the influx.

$$p_{11,600} = 500 + (0.0519)(11.5)(12,000 - 400)$$

$$= 7,423 \text{ psia.}$$

The pseudocritical pressure and temperature for the specified gas gravity were determined previously and the pseudoreduced properties at the problem conditions are

$$p_{pr} = 7,423/677 = 10.96$$

$$\text{and } T_{pr} = 665/352 = 1.89.$$

From Fig. 1.6, the z factor at the influx top is 1.195 and the gas density is calculated using Eq. 1.22.

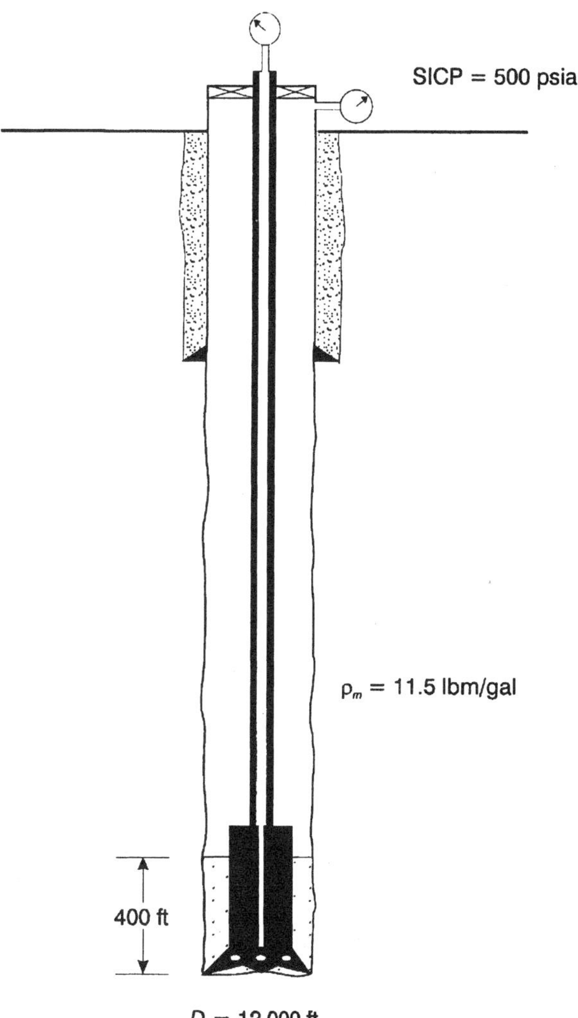

Fig. 1.11—Schematic of the wellbore described in Example 1.5.

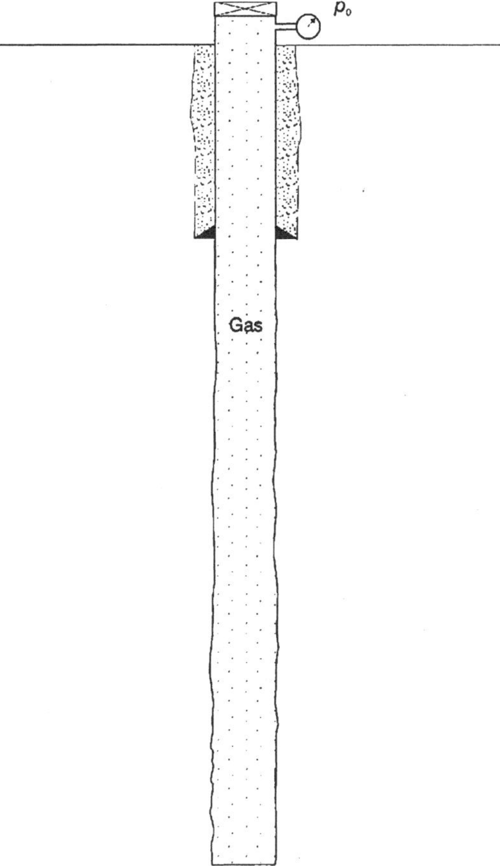

Fig. 1.12—Pressure in a gas column.

$$\rho_g = \frac{(0.60)(7,423)}{(2.77)(1.195)(665)} = 2.02 \text{ lbm/gal.}$$

The bottomhole pressure is

$$p_{bh} = 7,423 + (0.0519)(2.02)(400) = 7,465 \text{ psia.}$$

Assuming a constant density can lead to significant error for long gas columns or when gas pressure is small. For example, the well shown in **Fig. 1.12** has been shut in on a hole filled with dry gas and the density variation in the gas column must be considered if the intent is to determine the wellbore pressure at a given depth.

We substitute the gas density expression from Eq. 1.15 and obtain

$$\int_{p_1}^{p_2} \frac{zRT}{29\gamma_g p} \mathrm{d}p = \frac{g}{g_c}(D_2 - D_1). \quad \text{(1.38)}$$

Eq. 1.38 is the starting point for all derivations used to calculate the static bottomhole pressure in a gas well. Acceptable accuracy (at least for the drillers) can be realized by using an average wellbore temperature and z factor between the two depths of interest. Accordingly,

$$\int_{p_1}^{p_2} \frac{1}{p} \mathrm{d}p = \frac{g}{g_c} \frac{\gamma_g(D_2 - D_1)}{\bar{z}R\bar{T}}. \quad \text{(1.39)}$$

The applicable gas constant for customary units is 1,545.3. Substitution and integration yield

$$p_2 = p_1 \exp\left[\frac{\gamma_g(D_2 - D_1)}{53.3\bar{z}\bar{T}}\right]. \quad \text{(1.40)}$$

At least one iteration will be required because of the dependency of z on the average well pressure. An example demonstrates the calculation technique.

Example 1.6. Consider the well described in the last example. What would the shut-in casing pressure be if the drilling fluid had been unloaded from the hole prior to shut-in? Use the same bottomhole pressure as computed before and assume that the average wellbore temperature is 160°F [71°C].

Solution. We must solve Eq. 1.40 by trial-and-error. Assume for the first iteration that z is equal to unity and rearrange the relation to obtain

$$7,465 = p_{cs} \exp\left[\frac{(0.6)(12,000)}{(53.3)(1.0)(620)}\right]$$

$$p_{cs} = 6,004 \text{ psia.}$$

Now average the pressures and determine the average z factor.

$$p_{pr} = \frac{(7,465 + 6,004)}{(2)(677)} = 9.95,$$

$$T_{pr} = 620/352 = 1.76,$$

and $\bar{z} = 1.132$.

Substitute the value into Eq. 1.40,

$$7,465 = p_{cs} \exp\left[\frac{(0.6)(12,000)}{(53.3)(1.132)(620)}\right]$$

$$p_{cs} = 6,158 \text{ psia,}$$

and again determine the average z factor.

$$p_{pr} = \frac{(7,465 + 6,158)}{(2)(677)} = 10.06, \text{ so}$$

$$\bar{z} = 1.137.$$

Finally,

$$7,465 = p_{cs} \exp\left[\frac{(0.6)(12,000)}{(53.3)(1.137)(620)}\right]$$

$$p_{cs} = 6,163 \text{ psia.}$$

1.4.3 Equivalent Density. A useful concept in well control and any situation involving dissimilar wellbore fluids or applied surface pressure is the notion of equivalent density. The equivalent density (ρ_{eq}) or, more commonly, equivalent mud weight (EMW) at any point in a well is the wellbore fluid density that the hole "feels" from the standpoint of pressure. Given a wellbore pressure resulting from any combination of applied, hydrostatic, and dynamic pressures, the density in lbm/gal equivalent can be expressed as

$$\rho_{eq} = 19.25p/D. \quad \text{(1.41)}$$

The constant 19.25 becomes 102.0 when expressed in SI metric units.

Example 1.7. Take the hypothetical well from the last two examples and determine the equivalent density at total depth and at 6,000 ft [1828.8 m]. Assume the average temperature from surface to 6,000 ft is 120°F [49°C] for the case where the hole is filled with gas.

Solution. The bottomhole pressure is 7,465 psia, which is approximately 7,450 psig if the surface location is near sea level. The equivalent density at total depth is therefore

$$\rho_{eq} = (19.25)(7,450)/12,000 = 12.0 \text{ lbm/gal}.$$

The pressure at 6,000 ft is

$$p_{6,000} = 500 + (0.0519)(11.5)(6,000) = 4,081 \text{ psia}$$

$$= 4,066 \text{ psig},$$

which yields an equivalent density of

$$\rho_{eq} = (19.25)(4,066)/6,000 = 13.0 \text{ lbm/gal}.$$

The surface pressure of the gas-filled hole has been calculated as 6,163 psia. Assume that the average z factor from surface to 6,000 ft is 1.137 and compute

$$p_{6,000} = (6,163)\exp\left[\frac{(0.6)(6,000)}{(53.3)(1.137)(580)}\right] = 6,828 \text{ psia}.$$

As before,

$$p_{pr} = \frac{(6,828 + 6,163)}{(2)(677)} = 9.59,$$

$$T_{pr} = 580/352 = 1,65, \text{ and}$$

$$\bar{z} = 1.103.$$

Iterating again,

$$p_{6,000} = (6,163)\exp\left[\frac{(0.6)(6,000)}{(53.3)(1.103)(580)}\right] = 6,849 \text{ psia}.$$

No further iterations are necessary. The equivalent density at 6,000 ft is therefore

$$\rho_{eq} = (19.25)(6,849 - 14.7)/6,000 = 21.9 \text{ lbm/gal}.$$

An important point made by the example is that the EMW depends on the depth at which the determination is made, thus depth must be specified whenever the term is used. Also note that applied pressure causes the EMW to increase with shallower depth if the wellbore fluid density is consistent. This is demonstrated in the second shut-in condition where we see that the equivalent densities are the same for both cases at total depth. Up the hole, however, ρ_{eq} increases substantially.

1.5 Gas Migration

Gas, because of its lower density with respect to the drilling fluid medium, will tend to migrate upward in a well. Failure to expect and manage this fact of nature can lead to excessive wellbore pressures, possibly to the point that subsurface or surface control of the well is lost.

Refer to the wellbore schematics in **Fig. 1.13** for an illustration of the problem nature. Stage 1 shows the condition immediately after a well has been shut in on a gas influx. Drillpipe has been left out of the picture for demonstration purposes, but its presence is irrelevant to the discussion. The idealized single phase bubble has an initial volume V_1 at pressure p_1. Assume for now that the wellbore is sealed and that gas volume remains fixed as bubble migration occurs. Therefore, Boyle's law tells us that the bubble pressure at Stages 2 and 3 will be the same as at Stage 1:

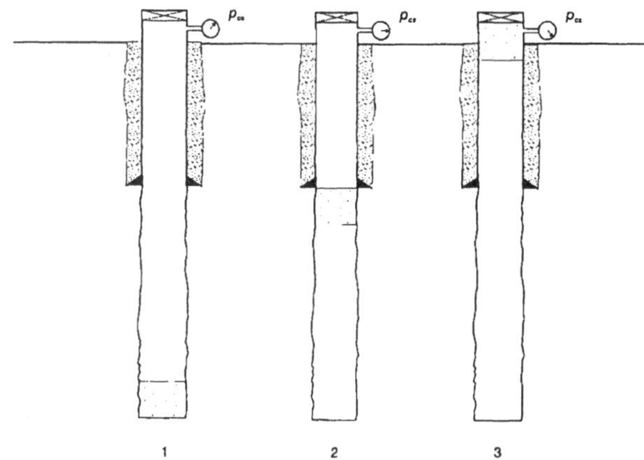

Fig. 1.13—Effect of gas migration on wellbore pressure.

$$p_1 = p_2 = p_3.$$

From Eq. 1.37, the constant gas pressure combined with the drilling fluid hydrostatic pressure will drive up the surface pressure and the pressure at every point in the well as migration occurs. The potential magnitude of the wellbore pressures associated with uncontrolled gas migration is demonstrated in Example 1.8.

Example 1.8. A 0.7 specific gravity gas bubble enters the bottom of a 9,000-ft vertical well when the drill collars are being pulled through the rotary table. Flow is noted with pipe out of the hole and the well is shut in with an initial recorded casing pressure of 50 psig. The casing pressure immediately begins to rise. Based on the pit gain volume, the influx height is estimated to be 350 ft. The mud density is 9.6 lbm/gal.

1. Assume no change in hole geometry (bubble height is constant) and determine the final casing pressure if the gas bubble is allowed to reach the surface without expanding.

2. Also determine the pressure and equivalent density at total depth for this final condition. Assume the temperature in the well is 70°F ambient plus 1.1°F/100 ft. Atmospheric pressure is 14 psia.

Solution. 1. The temperature and pressure at the top of the influx are

$$T_{8,650} = 70 + (0.011)(8,650) + 460 = 625°\text{R and}$$

$$p_{8,650} = 14 + 50 + (0.0519)(9.6)(8,650) = 4,374 \text{ psia}.$$

Eqs. 1.18 and 1.19 give the pseudocritical properties $T_{pc} = 378°\text{R}$ and $p_{pc} = 663$ psia. The pseudoreduced properties at bottomhole conditions are

$$T_{pr} = 625/378 = 1.65$$

$$\text{and } p_{pr} = 4,374/663 = 6.60.$$

The compressibility factor $z_{8,650}$ is determined as 0.934. The bubble pressure at surface temperature must be obtained by iteration. We first assume z_0 is 1.0 and solve for p_0 using Eq. 1.13.

$$\frac{4,374V}{(0.934)(625)} = \frac{p_0 V}{(1.0)(70+460)}.$$

$$p_0 = 3,971 \text{ psia}.$$

Now determine z_0 at this pressure.

$$T_{pr} = 530/378 = 1.40,$$

$$p_{pr} = 3,971/663 = 5.99,$$

and $z_0 = 0.829$.

Substitution yields

$$p_0 = \frac{(4,374)(530)(0.829)}{(625)(0.934)} = 3,292 \text{ psia}.$$

A few more iterative steps results in $z_0 = 0.729$ and $p_0 = 2,895$ psia.

2. The gas density at surface conditions is determined as

$$\rho_g = \frac{(0.7)(2,895)}{(2.77)(0.729)(530)} = 1.89 \text{ lbm/gal}.$$

The bottomhole pressure can now be obtained using Eq. 1.37.

$$p_{9,000} = 2,895 + (0.0519)[(1.89)(350) + (9.6)(8,650)]$$

$$= 7,239 \text{ psia}.$$

The equivalent density at this depth is

$$\rho_{eq} = (19.25)(7,239 - 14)/9,000 = 15.45 \text{ lbm/gal}.$$

Controlling gas migration simply means reducing the gas pressure as it rises in a well by allowing the gas to expand. This is a fairly simple procedure in most cases and operational details will be covered in a later chapter. Our discussion thus far has been focused on migration through a static borehole, but the same concepts apply if gas is circulated from a well. In fact, the basis for conventional kick displacement techniques is to maintain a constant bottomhole pressure by allowing gas to expand as it moves up the hole.

1.5.1 Predicting Migration Velocity. Having some estimate as to how fast gas is migrating through a static mud column may be desirable. It follows from a simple hydrostatics model that the rise in casing pressure reflects the drilling fluid hydrostatic pressure across the incremental hole section through which the gas has traveled. This assumption requires the gas pressure and volume remain constant across the traversed interval, which is true only if (1) the z factor and temperature do not change, (2) wellbore and fluid compressibility are zero, and (3) no fluid loss to the formation occurs. With these limiting assumptions, the gas migration velocity can be estimated using Eq. 1.42.

$$v_{sl} = \Delta p_{cs}/g_m \Delta t, \quad \ldots\ldots (1.42)$$

where v_{sl} is the slip velocity of the gas and Δt is the time over which the rise in casing pressure occurs.

Example 1.9. A well takes an influx and is shut in with an initial casing pressure of 500 psig. Thirty minutes later, the gauge pressure has increased to 800 psig. Estimate the slip velocity of the gas if the bubble length does not change during this period (hole geometry is constant). The mud density is 10.0 lbm/gal.

Solution. Eq. 1.42 yields

$$v_{sl} = \frac{(19.25)(800 - 500)}{(10.0)(0.5)} = 1,155 \text{ ft/hr}.$$

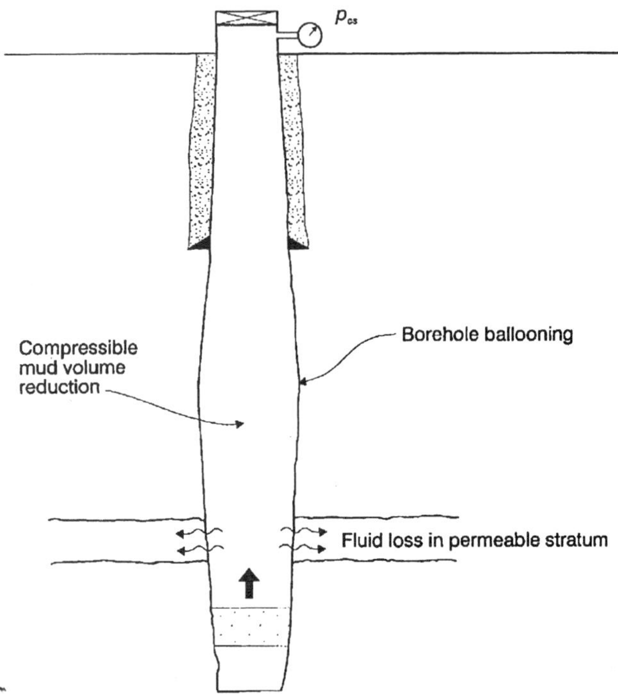

Fig. 1.14—Wellbore and mud volume changes resulting from an increase in pressure.

The calculated slip velocity from the previous problem falls in the range of what observers have calculated from field data. In fact, a rule of thumb that has been around for a long time is that gas will migrate between 600 and 1,000 ft/hr [180 and 300 m/hr]. Recent studies,[13,14] however, have shown that Eq. 1.42 can severely underestimate the actual slip velocity because migrating gas *does not* retain constant volume and pressure. Instead, the gas volume tends to increase as the wellbore and mud volumes change in response to an increase in the system pressure. Three associated processes are shown in **Fig. 1.14.**

One volume change follows from the elasticity of the casing and open hole, which circumferentially strains or balloons with internal pressure. Mud elasticity (i.e., compressibility) leads to a mud volume reduction with increasing pressure and further reduction in the form of increased filtrate loss if permeable stata are exposed. Quantifying the individual effects and the relative importance of each depends on such things as well depth, hole size, mud properties, formation characteristics, and other factors. The cited references discuss these parameters and offer a slip velocity equation that considers the combined effects.

1.5.2 Factors Affecting Slip Velocity. Johnson and White[15] categorized bubble slip behavior according to the size of the gas void fraction, which is defined as the ratio of the gas cross-sectional area to the total flow area. The concept is illustrated in **Fig. 1.15.** Larger bubbles, generally those that occupy a

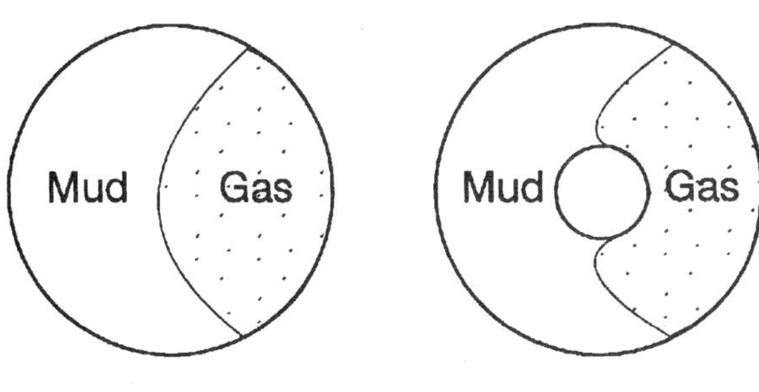

Fig. 1.15—Depiction of gas void fraction in openhole and annular flow.

void fraction larger than 25%, assume a bullet nose shape and, as shown in **Fig. 1.16,** migrate along the high side of the hole with concurrent liquid backflow down the opposing side. These so-called Taylor bubbles are influenced by the pipe and hole boundaries and will rise faster in liquid than smaller gas bubbles. Smaller bubbles are not influenced by the boundaries and are more dispersed in the drilling fluid medium. A transition from small, dispersed bubbles to larger Taylor[16] bubbles was noted to begin at a void fraction of about 12%.

Rader, Bourgoyne, and Ward[17] studied the migration of large Taylor bubbles and the factors that affect slip velocity in 1975. From laboratory observations, they concluded that hole geometry, mud viscosity, circulation rate, and hole inclination were most important. Based on their findings, we would expect gas slip velocities to increase with increased annular clearance (hole diameter relative to pipe diameter), increased velocity of the liquid medium, and reduced liquid viscosity. Hole inclination was also significant with maximum migration rates observed when when the test chamber attitude was close to 45°. Changing the gas and liquid densities did not have a major effect on migration rate as long as the gas density was small in comparison to that of the liquid.

More recently, Johnson and White[15] constructed a larger, improved model and measured gas rise velocities in both water and viscous mud. One of their conclusions which would seem to conflict with the earlier study was that gas migrated through a static column of viscous mud at approximately the same rate as measured in water. Another unexpected finding was that the migration rates through the viscous mud did not depend on the bubble void fraction. Gas bubbles in the thick mud tended to be longer at the small void fractions and had an observed size that was independent of the void fraction. Enhanced bubble stability deriving from the medium viscosity was believed to be the reason for this phenomenon.

Hovland and Rommetveit[18] measured gas migration rates through water and oil-base muds in a 7,317-ft [2230.2-m] deviated test well by injecting gas through coiled tubing and tracking its movement up the hole with a series of pressure transducers. Many of the previous conclusions based on laboratory work were validated by these full-scale tests. For example, the gas void fractions in viscous mud were found to have little impact on the migration rates. However, the test well results differed from the laboratory observations in one significant respect in that hole inclination did not seem to make much of a difference. This is a question that could use some additional research.

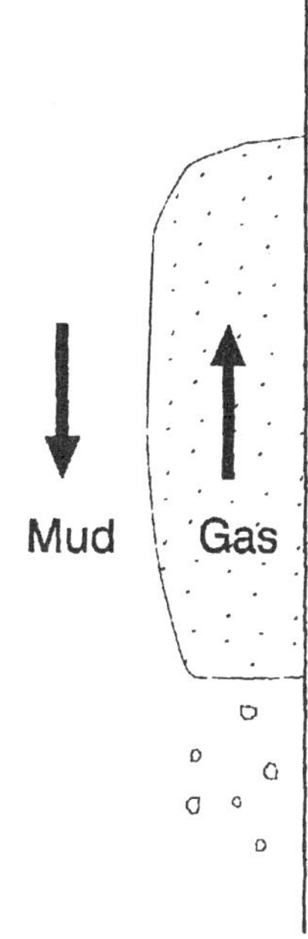

Fig. 1.16—Liquid backflow adjacent to migrating gas bubble.

One finding common to all of the investigations, and perhaps the most important contribution to the industry, is that gas tends to rise through a static or moving column of water or mud faster than we once believed. The experimental results have also been verified by measurements in the field. An operator who expects relatively rapid gas movement during a well-control event is in a position to make better decisions concerning job planning and execution, and will not be caught unaware in that critical period when the gas approaches and reaches the blowout preventers.

1.6 Gas Solubility

An assumption in most well-control problems is that an influx does not react to any degree with the drilling fluid and that the PVT properties of the formation fluid at wellbore conditions correspond to its surface properties. In other words, gas law predictions at any point in a well can usually be made as if the initial influx volume was the same as the volume gained in the surface pits. This reasoning does not hold true if the influx is gas and a significant proportion of the gas is dissolved in the drilling fluid. Hydrocarbon gas will dissolve to some extent in any drilling fluid, but the effect can generally be ignored with a water-base mud. A gas kick in an oil-base mud, however, is a different matter as are kicks that contain an appreciable quantity of CO_2 or H_2S into either mud type.

O'Brien[19] was one of the first to discuss the problems associated with well control and oil muds, which all derive from the fact that gas readily dissolves in the oil phase. Failing to consider gas solubility can lead to confusion, misapplied techniques, and potential disaster. An operator drilling with a diesel, mineral oil, or synthetic mud system must therefore be equipped with an understanding of how gas behaves in these fluids.

1.6.1 Solubility Limits and Bubblepoint Pressure. The solubility of a gas/liquid mixture is generally expressed as the amount of free gas in scf/bbl that can go into solution at a given temperature and pressure. The solubility of gas in liquids is a function of the gas and liquid composition, pressure, and

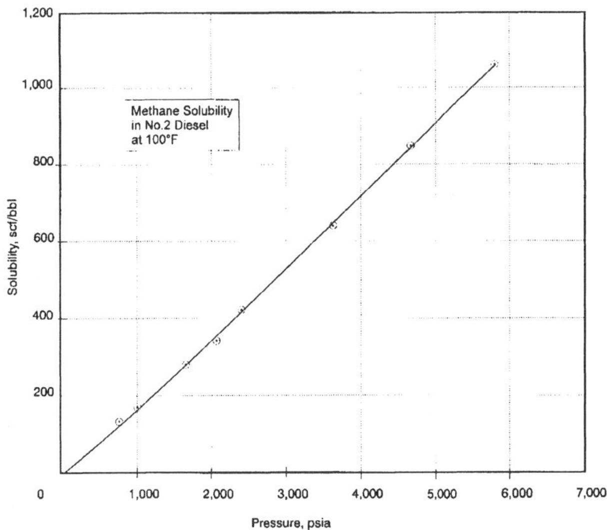

Fig. 1.17—Methane solubility in No. 2 diesel at 100°F (after Thomas *et al.*[20]).

temperature. Generally, solubility will increase as pressure increases, as temperature decreases, and as the molecular similarity between the gas and liquid components increases.

Closely related to solubility is the concept of bubblepoint pressure, which is defined as the pressure at which the first bubble of free gas breaks out of solution at a given solution gas/liquid ratio and temperature. Recalling the mixture phase diagram given in Fig. 1.2, the bubblepoint pressures are indicated along the 100% liquid line to the left of the critical point C. From the phase diagram, free gas cannot coexist with the liquid at pressures in excess of the bubblepoint pressure.

Thomas *et al.*[20] experimentally determined methane solubilities in No. 2 diesel at 100°F and presented the data plotted in **Fig. 1.17.** For a methane/diesel mixture, all of the gas will go into solution if the system gas/oil ratio (GOR) falls below the solubility curve. Free gas will be present with the diesel if sufficient methane is available to saturate the diesel. One use of solubility curves is demonstrated in Example 1.10.

Example 1.10. Determine the amount of free gas after blending 8,000 scf of methane with 10 bbl of diesel if the pressure and temperature are 3,000 psia and 100°F.

Solution. Fig. 1:17 indicates a solubility limit of 530 scf/bbl at the stated conditions. The free gas volume at standard conditions is therefore

$$V_{gsc} = \left[(8,000/10) - 530\right](10) = 2,700\,\text{scf}.$$

Thomas *et al.* also used the Redlich-Kwong[21] EOS to predict methane solubility in No. 2 diesel at temperatures ranging from 100°F to 600°F. The EOS-computed solubility curves are reproduced here as **Fig. 1.18.** Note that the gas solubility is approximately linear at low to moderate pressures, but rapidly increases and becomes infinite at some higher pressure. The vertical line for each isotherm corresponds to the critical pressure of the mixture, which means that the gas and diesel are completely miscible if the system pressure is any higher.

Fig. 1.18 seems to violate our previous statement concerning the relationship between gas solubility and temperature. Indeed, computer simulations using an EOS model suggest that temperature has little effect at low pressure and that temperature increases solubility as the pressure approaches critical. This direct proportionality between solubility and temperature may be true at high pressure but, to our knowledge, has not been experimentally confirmed at the pressure ranges investigated in the laboratory.

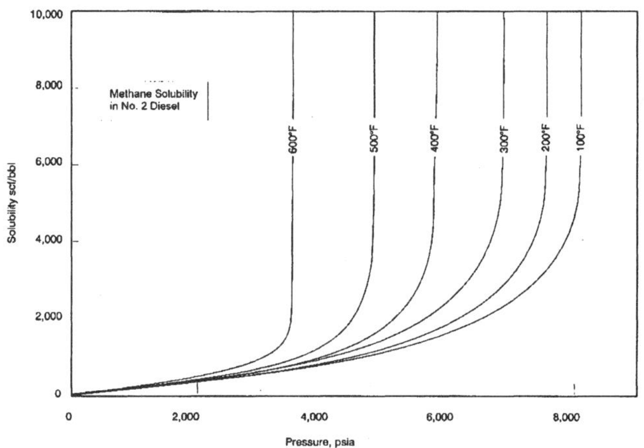

Fig. 1.18—EOS predictions of methane solubility in No. 2 diesel (after Thomas *et al.*[20]).

O'Bryan *et al.*[22] ran extensive solubility tests for various natural gas and base oil combinations. Much use is made of their findings as represented here in a series of solubility charts. **Fig. 1.19** shows the solubility of methane in diesel and two common mineral oils of 1988 vintage. As indicated, methane is most soluble in the Conoco LVT oil and least soluble in the Exxon Chemicals Mentor 28, with diesel intermediate. The disparities become less important at low pressure. It would be a worthwhile research effort to study the solubility characteristics of the new synthetic oil systems.

The solubility dependence on temperature is shown in **Fig. 1.20** where bubblepoint pressures were determined for

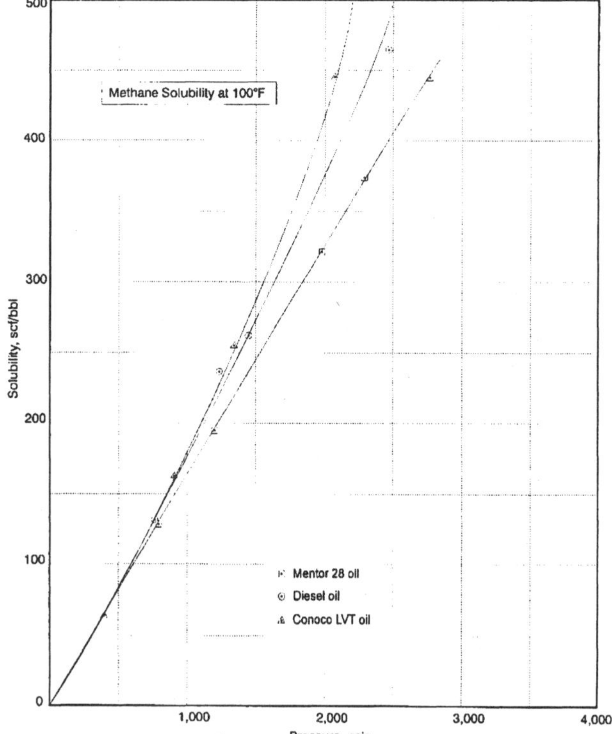

Fig. 1.19—Methane solubility in various base oils at 100°F (after O'Bryan *et al.*[22]).

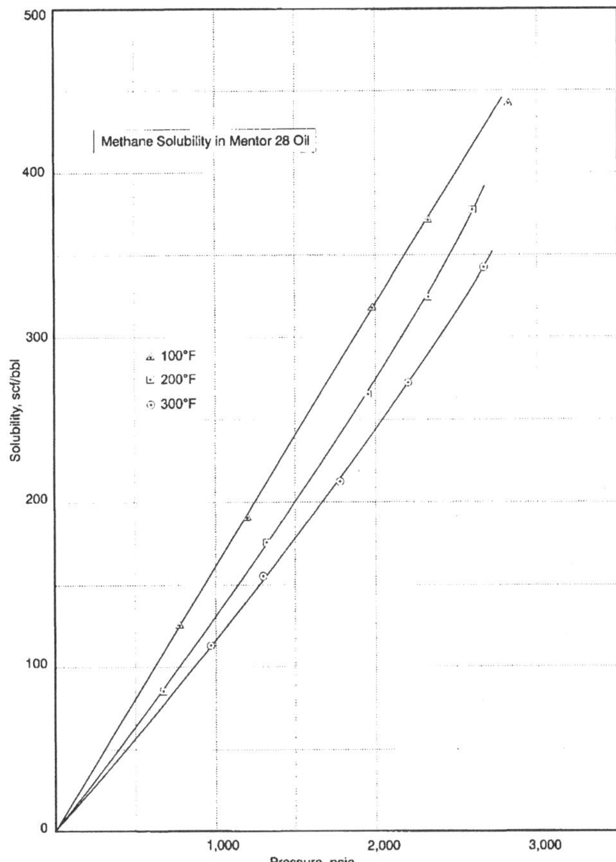

Fig. 1.20—Methane solubility in Mentor 28 oil at various temperatures (after O'Bryan *et al.*[22]).

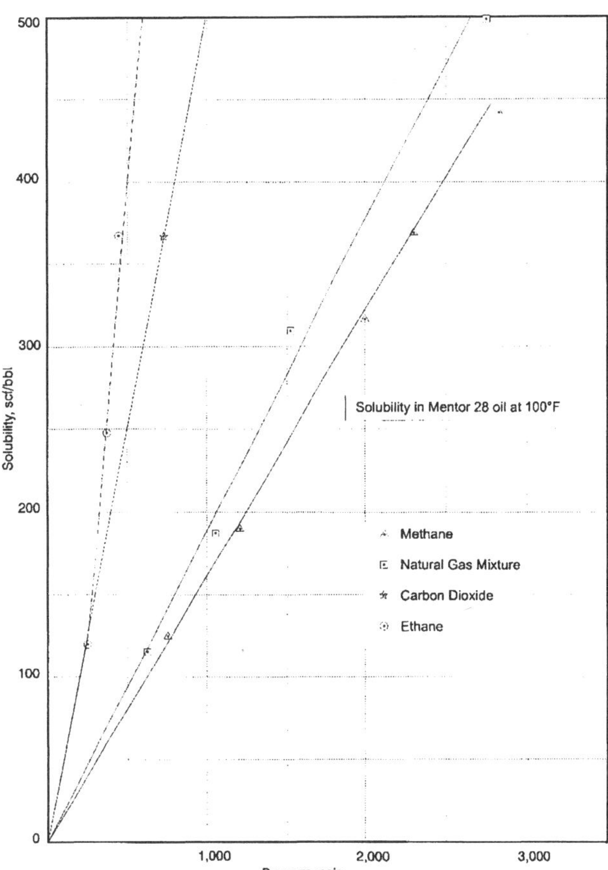

Fig. 1.21—Solubility of various gases in Mentor 28 oil at 100°F (after O'Bryan *et al.*[22]).

methane/Mentor 28 mixtures at 100°F, 200°F, and 300°F. As discussed, laboratory observations suggest an inverse relationship between temperature and solubility whereas the opposite holds true with some EOS predictions.

We mentioned that solubility of two substances, gas and oil in this case, should increase as the properties of the two become more similar. This is demonstrated in the **Fig. 1.21** solubility curves where ethane, which has the highest molecular weight of the tested hydrocarbon gases, is demonstrated to be much more soluble than methane in the Mentor 28 oil.

The solubility of methane in distilled water is depicted in **Fig. 1.22.** Though developed for methane, the chart can be used for any hydrocarbon gas mixture with little practical consequence. Gas solubility in water decreases with salinity and a chart for adjusting the results from Fig. 1.22 is included as **Fig. 1.23.**

CO_2 and H_2S are soluble in both water and oil, but with much higher solubility in the common base oils. Matthews[25] discussed well-control considerations for these gases, primarily for oil-base mud applications, and presented the diesel solubility data shown in **Fig. 1.24.** As indicated, H_2S is extremely soluble in diesel.

Fig. 1.25 gives the solubility of CO_2 in fresh water with salinity correction factors as presented in **Fig. 1.26.** Little experimental work has been done with H_2S solubility in water, mainly because the gas is so noxious to laboratory equipment and personnel. Instead, we offer the theoretical results shown in **Fig. 1.27,** which were computed using Henry's law and the gas constants derived by Selleck *et al.*[26] The isobars pertain to H_2S partial pressure, defined as the product of pressure and the H_2S mole fraction. Henry's law assumes ideal gas behavior so the chart loses accuracy in the higher partial pressure ranges. Even so, it may still be useful for comparative pur-

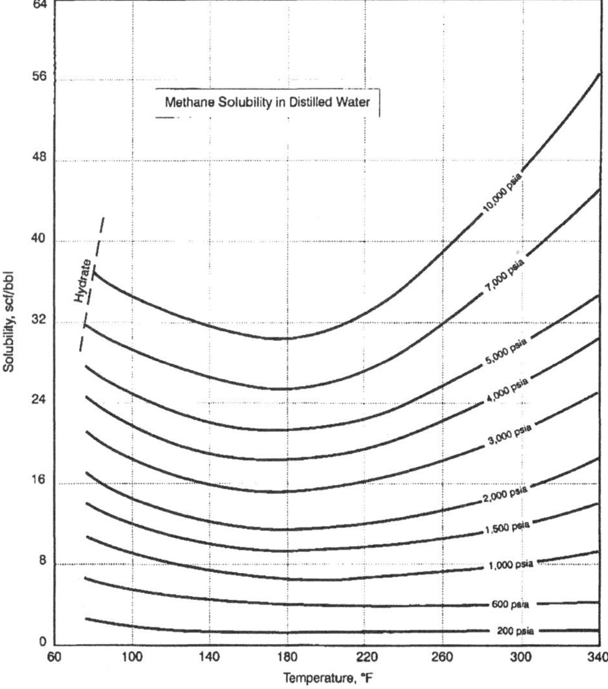

Fig. 1.22—Methane solubility in distilled water (after Culberson and McKetta[23]).

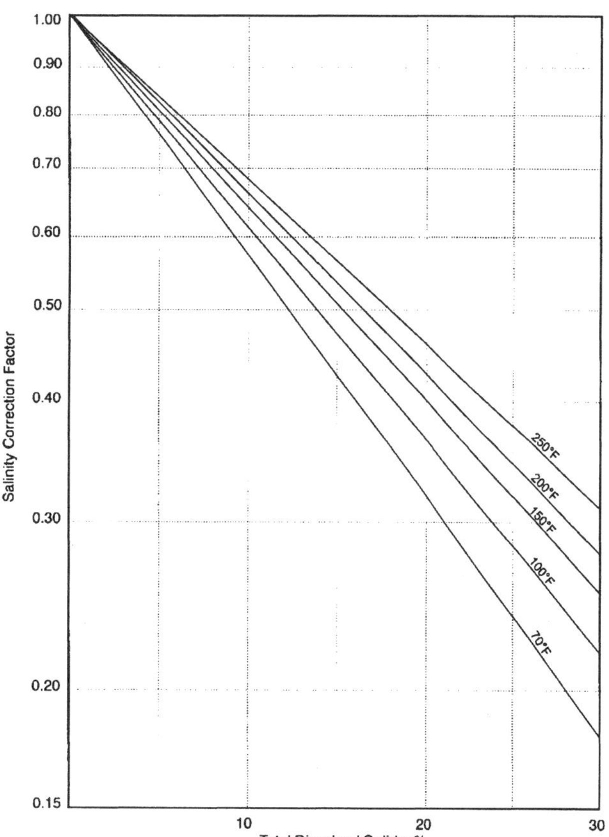

Fig. 1.23—Salinity correction factors for natural gas solubility in water (after McKetta and Wehe[24]).

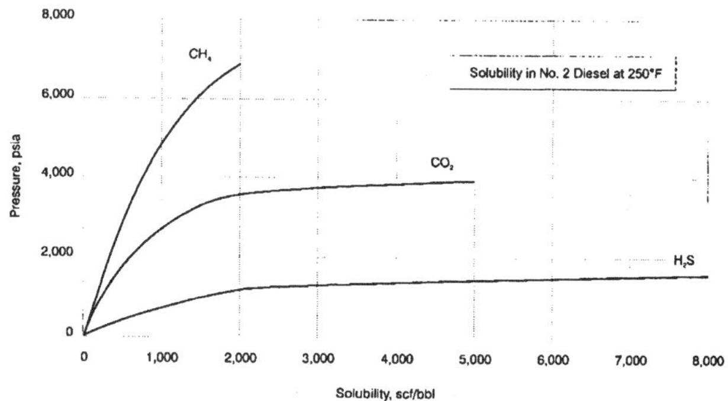

Fig. 1.24—Methane, CO_2, and H_2S solubility in diesel at 250°F (after Matthews[25]).

poses when we consider the inaccuracies inherent to most well-control predictions anyway.

At minimum, a weighted invert emulsion mud contains oil, emulsifiers, brine, and solids. Gas will not dissolve in solids and its solubilty in water is minor, so we would expect the gas solubility in an oil mud to be less than that of the base oil. This is demonstrated in **Fig. 1.28,** which compares methane solubility in Mentor 28 with two weighted muds made from the same base oil. Solubility is reduced with increasing mud weight because of the higher solids content.

The gas solubility in any mud system can be estimated by summing the respective solubilities of each component on a volume fraction basis.

$$r_{sm} = f_o r_{so} + f_w r_{sw} + f_e r_{se}. \qquad (1.43)$$

r_{sm} through r_{se} respectively describe the solution gas/component ratios of the mud, base oil, water, and emulsifier. The volume fraction of each component (f_o through f_e) is determined by material balance calculation or retort analysis. Any other mud additives that are capable of dissolving gas can also be incorporated in the relation.

The combined solubility of the hydrocarbon gas, CO_2, and H_2S in each of the mud components (oil, water, and emulsifier) can be estimated using Eq. 1.44.

$$r_{s(o,w,e)} = f_h r_{sh} + f_{CO_2} r_{sCO_2} + f_{H_2S} r_{sH_2S}, \qquad (1.44)$$

where the subscripted f terms represent the mole fractions of the natural gas constituents.

The solubility curves offered in this textbook or elsewhere may be used to estimate gas solubilities. As another tool, O'Bryan *et al.*[22] presented the following empirical relation to estimate hydrocarbon and CO_2 solubilities in oil and oil-mud emulsifiers.

$$\mathrm{d}r_{s(o,e)} = \left(\frac{p}{aT^b}\right)^c, \qquad (1.45)$$

where a, b, and c are constants that depend on the gas and liquid type. Values for a and b can be obtained from **Table 1.3.** The value for c is unity if the determination is made for CO_2. Otherwise, c is calculated using Eq. 1.46 or 1.47:

$$c = 0.3576 + 1.168\gamma_g + (0.0027 - 0.00492\gamma_g)T$$
$$- (4.51 \times 10^{-6} - 8.198 \times 10^{-6}\gamma_g)T^2, \qquad (1.46)$$

for hydrocarbon gases dissolved in oil, and

$$c = 0.40 + 1.65\gamma_g - 1.01\gamma_g^2, \qquad (1.47)$$

for hydrocarbon gases dissolved in the emulsifier. The temperature in these relations is in degrees Fahrenheit. The equations do not apply across the entire pressure spectrum and should only be used when the pressure is less than half of the mixture critical pressure.

Example 1.11. A 13.0-lbm/gal 70:30 invert emulsion oil mud consists of (by volume) 54% diesel, 23% $CaCl_2$ water, 4% emulsifiers, and 19% solids. Estimate the natural gas solubility in the mud at 150°F and 2,000 psia if the gas contains 95% hydrocarbons and 5% CO_2. Assume the brine phase has 200,000 ppm total-dissolved-solids (TDS) and use a gas specific gravity of 0.65.

Solution. We will use Eq. 1.45 to estimate the gas solubilities in the oil and emulsifiers. The a and b constants for CO_2 are obtained from Table 1.3 and c has a value of 1.00. Hence,

$$r_{s(CO_2/o)} = \left[\frac{2,000}{(0.059)(150)^{0.7134}}\right]^{1.0} = 950 \text{ scf/bbl}.$$

Substituting the CO_2 constants for the emulsifier yields

$$r_{s(CO_2/e)} = \left[\frac{2,000}{(0.135)(150)^{0.8217}}\right]^{1.0} = 241 \text{ scf/bbl}.$$

Now use Eq. 1.46 to determine c for hydrocarbon gas in oil.

$$c = 0.3576 + (1.168)(0.65)$$
$$+ [0.0027 - (0.00492)(0.65)](150)$$
$$- [4.51 \times 10^{-6} - (8.198 \times 10^{-6})(0.65)](150)^2$$
$$= 1.0605.$$

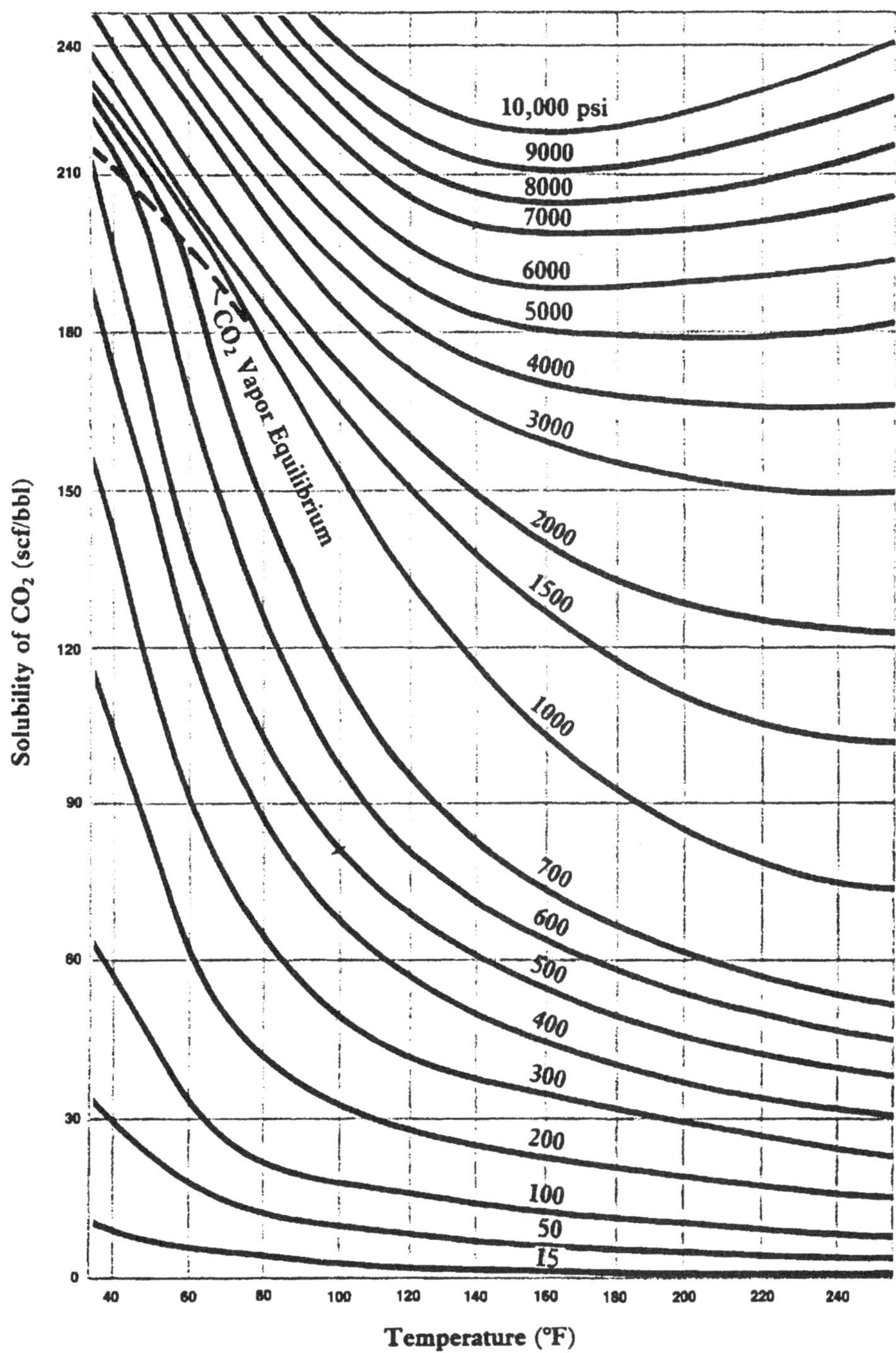

Fig. 1.25—CO_2 solubility in distilled water (courtesy of Halliburton Energy Services).

Use this result and the constants from Table 1.3 to predict hydrocarbon gas solubility in the base oil.

$$r_{s(h/o)} = \left[\frac{2,000}{(1.922)(150)^{0.2552}}\right]^{1.0605} = 408 \text{ scf/bbl.}$$

The hydrocarbon gas solubility in the emulsifiers is determined:

$$c = 0.40 + (1.65)(0.65) - (1.01)(0.65)^2 = 1.0458$$

and

$$r_{s(h/e)} = \left[\frac{2,000}{(4.162)(150)^{0.1770}}\right]^{1.0458} = 252 \text{ scf/bbl.}$$

Eq. 1.44 can now be used to determine the natural gas mixture's solubility in the oil and emulsifiers.

$$r_{so} = (0.95)(408) + (0.05)(950) = 435 \text{ scf/bbl}$$

and

$$r_{se} = (0.95)(252) + (0.05)(241) = 251 \text{ scf/bbl.}$$

The hydrocarbon gas solubility in fresh water is 12 scf/bbl from Fig. 1.22. Applying the salinity correction factor from Fig. 1.23 yields

$$r_{s(h/w)} = (12)(0.40) = 5 \text{ scf/bbl.}$$

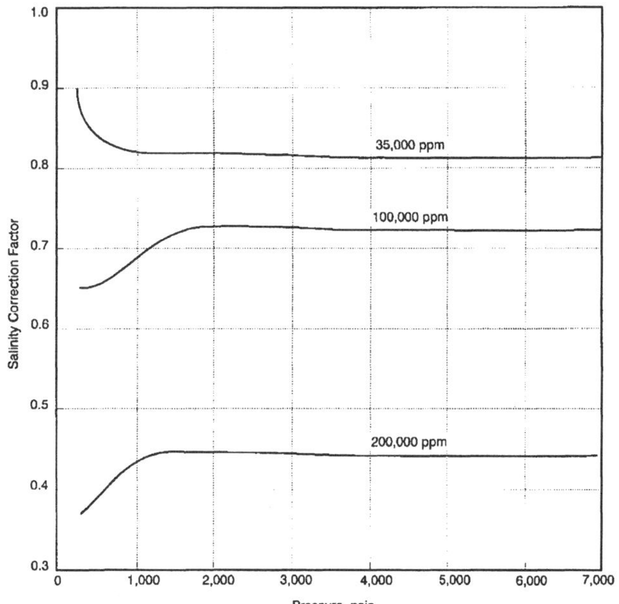

Fig. 1.26—Salinity correction factors for CO_2 solubility in water (after Crawford *et al.*[26]).

According to Fig. 1.25, the CO_2 solubility in fresh water is 145 scf/bbl. The salinity correction factor is obtained from Fig. 1.26 and the adjusted CO_2 solubility is computed as

$$r_{s(CO_2/w)} = (145)(0.45) = 65 \text{ scf/bbl.}$$

Again we apply Eq. 1.44 and determine the combined gas solubility in the brine phase.

$$r_{sw} = (0.95)(5) + (0.05)(65) = 8 \text{ scf/bbl.}$$

Finally, the gas solubility in the whole mud is calculated using Eq. 1.43.

$$r_{sm} = (0.54)(435) + (0.23)(8) + (0.04)(251)$$
$$= 247 \text{ scf/bbl.}$$

Example 1.12. A retort analysis on a fresh water mud indicates 6% solids and no oil. Estimate gas solubility in the mud at 180°F [82°C] and 5,200 psia [35.85 MPa] if the gas analysis shows mole fractions of 0.92 for methane, 0.06 for CO_2, and 0.02 for H_2S.

Solution. Fig. 1.22 and 1.25 give respective methane and CO_2 solubilities of 21 scf/bbl and 182 scf/bbl. The H_2S partial pressure is calculated as

$$p_{H_2S} = (0.02)(5{,}200) = 104 \text{ psia.}$$

The partial H_2S solubility from Fig. 1.27 is about 36 scf/bbl. Substituting terms into Eq. 1.44 gives

$$r_{sw} = (0.92)(21) + (0.06)(182) + 36 = 66 \text{ scf/bbl.}$$

The gas solubility in the whole mud is thus

$$r_{sm} = (0.94)(66) = 62 \text{ scf/bbl.}$$

Circulating a hole with an oil-base mud will provide a continuous supply of fresh oil available for taking gas into solution if an influx is taken while drilling. All of the gas may very well go into solution if the formation deliverability is low and/

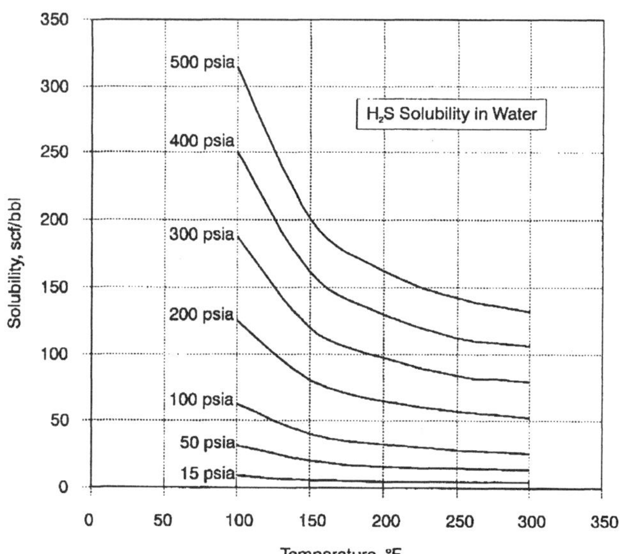

Fig. 1.27—H_2S solubility in distilled water as function of partial pressure and temperature.

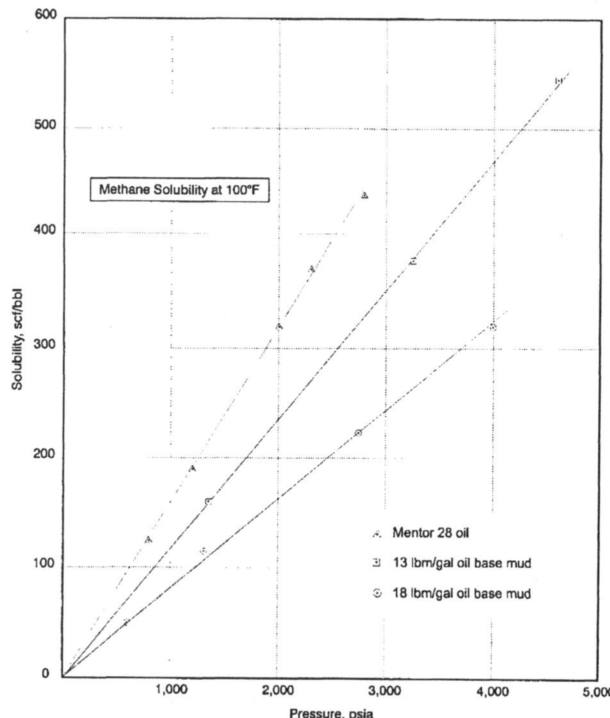

Fig. 1.28—Methane solubility in Mentor 28 and two oil-base muds at 100°F (after O'Bryan *et al.*[22]).

or circulation rate is high. Conversely, gas will eventually reach the solubility limit of the oil if a gas zone kicks into a static wellbore, say during a connection or trip. Once the oil is saturated, any additional entry will be in the form of free gas and thus occupy free gas volume. However, it will not take long for this gas to be dissolved if migration into the undersaturated mud takes place.

TABLE 1.3—CONSTANTS USED IN EQ. 1.45

Gas	Component	*a*	*b*
Hydrocarbon	Oil	1.922	0.2552
CO_2	Oil	0.059	0.7134
Hydrocarbon	Emulsifier	4.162	0.1770
CO_2	Emulsifier	0.135	0.8217

Drilled gas has caused some difficulties with oil muds. This is not a problem of being hydrostatically underbalanced, but involves what can transpire when gas removed by the bit is circulated up the annulus to a lower pressure. **Fig. 1.29** demonstrates what can happen and has happened. In Stage 1, two gas-bearing sands have been drilled with an oil mud and are being circulated up the hole with all of the associated gas in solution. At Stage 2, the bubblepoint pressure for the first gas package is attained at some critical point in the annulus and gas breaks out of solution. This can happen rather violently, with rapidly expanding gas displacing mud out the flowline. A domino effect, shown as Stage 3, can be created as the resulting drop in pressure releases more gas from the oil, perhaps from a deeper drilled sand, and more mud is expelled from the hole. As can be imagined, such a situation can develop and lead to an underbalanced hole if allowed to proceed.

The ratio of drilled gas to whole mud can be calculated knowing the penetration rate, bit diameter, and the circulation rate if some assumptions are made regarding the rock and gas characteristics. The rate at which bulk rock volume is removed by the bit is given by

$$q_r = \frac{\pi}{4}\frac{d_b^2 R(12)}{(1,728)(60)} = \frac{d_b^2 R}{11,000},$$

where

$q_r =$ rock removal rate (ft^3/min),
$d_b =$ bit diameter (in.), and
$R =$ penetration rate (ft/hr).

The gas portion of the bulk rock enters the mud at the rate of

$$q_g = q_r \phi S_g,$$

where ϕ is the formation porosity and S_g the gas saturation. Substitution and conversion to standard conditions yield the gas entry rate in scf/min:

$$q_{gsc} = \frac{d_b^2 R \phi S_g p(520)}{(11,000)(14.65)zT}, \text{ or}$$

$$q_{gsc} = \frac{d_b^2 R \phi S_g p}{310zT}.$$

The pressure, temperature, and compressibility factors are at circulating bottomhole conditions. Finally, the gas/mud ratio in scf/bbl is obtained when we divide the gas entry rate by the mud circulation rate q_m (bbl/min).

$$r_m = \frac{d_b^2 R \phi S_g p}{310zTq_m}. \quad \text{(1.48a)}$$

The equivalent expression in SI metric units is

$$r_m(\text{std m}^3/\text{m}^3) = \frac{d_b^2(\text{mm})^2 R(\text{mm/s})\phi S_g p(\text{MPa})}{445zT(\text{K})q_m(\text{L/s})}. \quad \text{(1.48b)}$$

Example 1.13. A 50-ft gas sand is drilled with a 12¼-in. bit at 250 ft/hr. Circulating conditions at the present total depth of 6,000 ft are 3,000 psia and 140°F. An oil-base mud is in use and its density is 10.5 lbm/gal.

1. Assuming the sand's porosity is 25% with a gas saturation of 80%, determine the drilled gas concentration in scf/bbl if the circulation rate is 8.0 bbl/min.

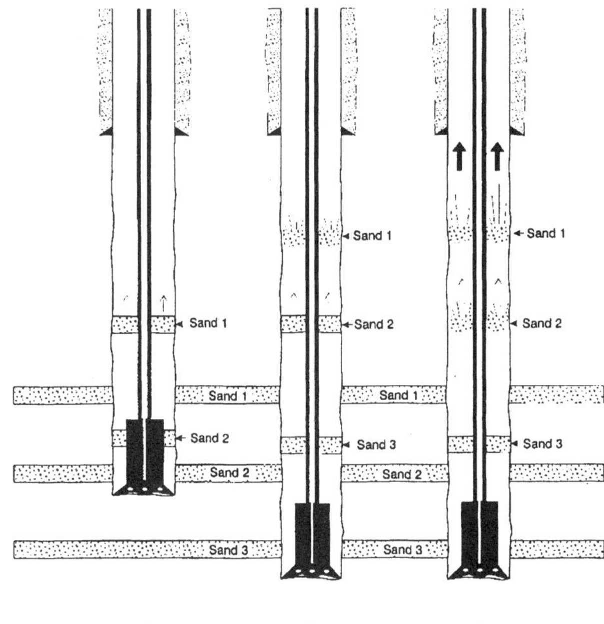

Fig. 1.29—Effect of drilled gas in oil-base mud when bubblepoint pressure is reached in the annulus.

2. Determine the expansion of the drilled gas if the bubblepoint is reached at annular conditions of 70 psia and 90°F.

Solution. 1. A 0.6 gas specific gravity is assumed and the z factor at bottomhole conditions is determined as 0.855. Substituting terms into Eq. 1.48a yields,

$$r_m = \frac{(12.25)^2(250)(0.25)(0.80)(3,000)}{(310)(8.0)(0.855)(600)} = 17.7 \text{ scf/bbl}.$$

This is a fairly low gas concentration and we can conclude that all of the gas is initially dissolved in the mud.

2. The downhole gas volume in bbl is

$$V_{g1} = \frac{\pi}{4}(12.25 \text{ in.})^2\left(\frac{12 \text{ in.}}{\text{ft}}\right)\left(\frac{\text{bbl}}{9,702 \text{ cu in.}}\right)(50 \text{ ft})(0.25)$$

$$(0.80),$$

$$V_{g1} = 1.5 \text{ bbl}.$$

Using the gas law, the free gas volume when released from the mud at bubblepoint pressure is

$$V_{g2} = \frac{(1.5)(3,000)(0.995)(550)}{(70)(0.855)(600)} = 68.6 \text{ bbl}.$$

The bubblepoint depth for this hypothetical situation would have been at approximately 100 ft [30 m] and all of the mud above this point would likely have been ejected from the hole when gas broke out of the mud. Disregarding the loss in hydrostatic pressure, such an event would cause a mess and a create a hazard to the crews if mud and gas were allowed to be blown onto the floor. For this reason, rotating heads are a necessary piece of equipment when drilling with oil-base muds.

O'Bryan and Bourgoyne[28] discussed the the drilled gas hazard potential for oil muds and presented a technique for predicting the cumulative loss in bottomhole circulating pressure under situations such as we have described. Their iterative procedure is somewhat involved for hand calculation, but

Insoluble Mixtures

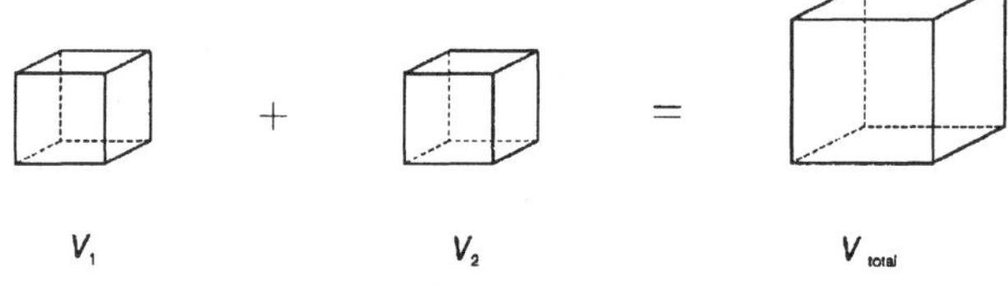

Soluble Mixtures

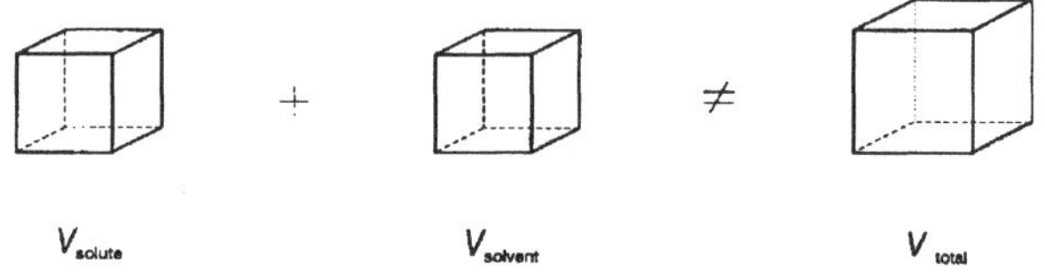

Fig. 1.30—The volumetric nature of solubility.

could easily be translated to a software spreadsheet. This would certainly be a worthwhile exercise when planning a well where the possibility of underbalancing a well in this fashion exists. Corrective action could then be taken before the problem potential becomes a problem actuality.

1.6.2 Solution Volume Factors. Fig. 1.30 demonstrates the nature of solubility. A volume balance principle applies when two insoluble substances are mixed together. That is, the final mixture volume is simply the sum of the two independent volumes. This is not the case when the one substance, the solute, is wholly or partially dissolved into the other, the solvent. For solutions, the final mixture volume is less than the sum of the two separate component volumes.

A primary indicator of an kick is a pit gain, which is an increase in the surface mud volume that results when the formation fluid volume displaces mud from the wellbore into the pits. The influx volume at bottomhole conditions is close to the observed gain if the formation fluid does not go into solution. If the entry is soluble, however, the resulting pit gain will be smaller. The implication to well control is that gas that goes into solution partially "hides" in the mud and is harder to detect than gas that remains in the free state.

O'Bryan and Bourgoyne[29] experimentally determined solution volumes of No. 2 diesel for various dissolved methane concentrations at 100°F. Their data are shown in **Table 1.4.** The solution gas/oil ratio (R_{so}) gives the standard cubic feet of dissolved methane per barrel of diesel at the test pressures listed in the second column. The measured volume factors (B_o) in the last column show the ratios of the diesel volume at wellbore conditions, including any dissolved gas, to the stock tank volume. The PVT properties of the diesel alone are indicated in the gas-free volume factors.

These authors also used a gas solubility correlation and the Peng-Robinson[30] EOS to develop a series of volume factor charts for methane/diesel combinations at temperatures ranging from 100°F to 400°F. The chart applicable to a temperature of 200°F, reproduced here as **Fig. 1.31,** will be used in the following example to demonstrate how a surface pit gain can be predicted for a given gas influx volume.

TABLE 1.4—VOLUME FACTORS FOR METHANE DISSOLVED IN NO. 2 DIESEL AT 100°F[2]

r_{so} scf/bbl	p psia	B_o bbl/STB
0	14.7	1.005
0	3,320	0.993
0	3,775	0.991
0	4,705	0.987
0	4,940	0.986
234	1,225	1.070
234	1,585	1.060
234	2,205	1.053
259	1,475	1.069
259	2,125	1.054
259	2,690	1.049
259	3,365	1.045
467	2,545	1.137
467	2,625	1.127
467	3,710	1.117
695	3,825	1.197
695	4,120	1.191
695	4,660	1.186
695	5,305	1.182
821	4,075	1.254
821	4,265	1.243
821	4,490	1.233
821	5,070	1.225

Example 1.14. The circulating bottomhole pressure and temperature in a well drilled with an oil-base mud are 5,000 psia and 200°F. The bit encounters a gas sand, which begins to flow methane into the wellbore.

1. The gas zone is flowing at a rate sufficient to give a concentration in the diesel phase of 400 scf/bbl. Use Fig. 1.31 to estimate the pit gain when 10 bbl of gas enters the mud.

2. Determine the pit gain for the same 10-bbl influx if the circulation rate leads to a gas concentration in the diesel of 600 scf/bbl.

Assume the compressibility and thermal expansion characteristics of the water phase are negligible compared to the diesel properties.

Solution. 1. From Fig. 1.31, the predicted bubblepoint pressure at the given bottomhole temperature and gas concentration is 3,000 psia. Alternatively, the solubility limit at 5,000 psia and 200°F is about 660 scf/bbl, so we conclude that all of the gas goes into solution. The dissolved gas causes the diesel to swell by an amount equal to the difference between the volume factor with gas (B_{og}) and the gas-free volume factor (B_{ong}). The two values are obtained from the chart, giving

$$B_{og} - B_{ong} = 1.128 - 1.012 = 0.116 \text{ bbl/STB}.$$

We need to convert the gas concentration to a downhole GOR in bbl gas/bbl diesel (r'_{so}). The bottomhole z factor for methane is obtained from Fig. 1.4 and the real gas law (Eq. 1.13) gives

$$r'_{so} = \frac{(400)(14.65)(1.029)(660)}{(520)(5,000)(5.6146)} = 0.273 \text{ bbl/bbl}.$$

Hence the pit gain is 0.116 bbl for each 0.273 bbl of free gas dissolved in the diesel. The pit gain for the 10-bbl influx is therefore

$$G = 0.116\left(\frac{10}{0.273}\right) = 4.2 \text{ bbl}.$$

2. At 600 scf/bbl dissolved gas,

$$B_{og} - B_{ong} = 1.205 - 1.012 = 0.193 \text{ bbl/STB}.$$

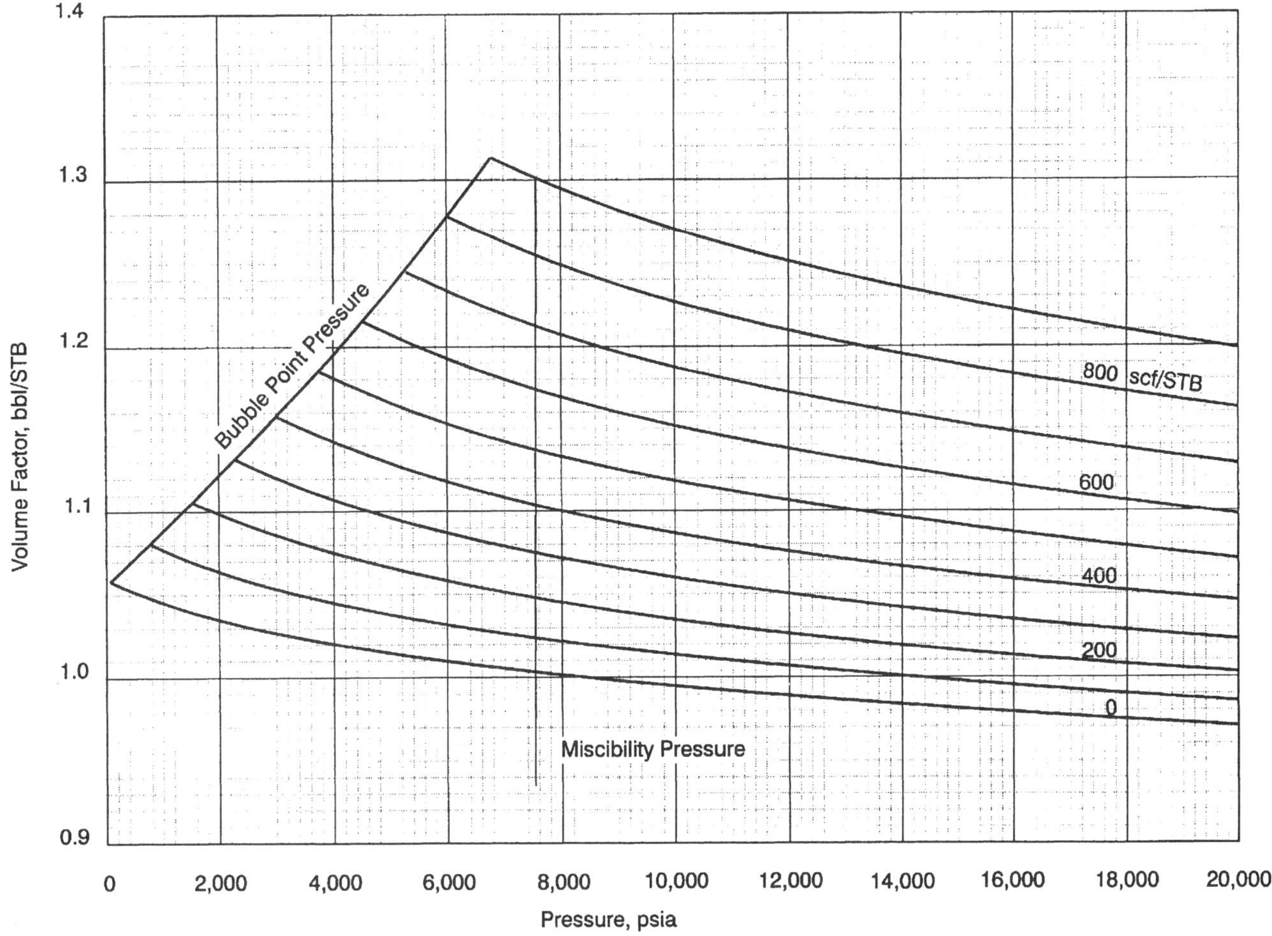

Fig. 1.31—Volume factors and solubility characteristics of methane/diesel mixtures at 200°F (after O'Bryan and Bourgoyne[29]).

As before,

$$r'_{so} = (0.273)(600/400) = 0.409 \text{ bbl/bbl}$$

and

$$G = 0.193\left(\frac{10}{0.409}\right) = 4.7 \text{ bbl.}$$

1.6.3 Oil Mud Recommendations. Gas wells can be drilled safely with oil-base drilling fluids. Actually, the solubility characteristics of these muds lead to some distinct advantages in well control. These will be discussed in a later chapter, but certain precautions should be taken with respect to minimizing operational risks whenever these mud types are being used. One prudent step, mentioned previously, is to equip the well with a rotating head to direct any evolved gas and expelled mud away from the rig floor.

Drilled gas hazards can be managed to a great extent by controlling how much gas enters the mud. As indicated by Eq. 1.36, the GOR is directly proportional to the penetration rate, so controlling the drill rate while in a gas sand limits the free gas volume when bubblepoint pressure is attained in the annulus. Alternatively, placing some minimum on the total amount of drilled gas in the annulus has become policy for some operators in areas that exhibit fast penetration rates. In practice, a limit is placed on the number of sand "packages" in the well at the same time. Once this limit is reached, drilling ceases and the bottoms-up hole volume is circulated through the choke manifold.

O'Bryan and Bourgoyne[28] also discussed some methods for minimizing or eliminating the detrimental effects of drilled gas evolution near the surface. One proposed solution was to drill with an annular backpressure higher than the predicted bubblepoint pressure, the objective being to have the gas break out of the mud in the surface equipment. Of course, this would have some effect on penetration rate and openhole fracture integrity is always a consideration when pressure is placed on the backside.

The mud/gas separator (MGS) and related equipment must be sized and designed properly for the potential mud and gas rates during a well-control procedure. This statement holds true for any drilling fluid, but the specifications become more critical with an oil mud. Any gas taken into the mud will break out relatively shallow, hence the effects of migration and general dispersion are less than in an equivalent operation with a water-base mud. Higher maximum gas flow rates through the choke and downstream equipment, though of shorter duration, are the end result.

A pit gain is one the best kick indicators, regardless of the drilling fluid.[20] As we have demonstrated, however, the surface indications of a kick may be much less gas with an oil mud. Pit level alarms should be set at a lower level or use of one of the more advanced kick detection systems should be considered. It follows that more frequent flow checks may be necessary with these muds.

A flow check, or shutting down the pump and observing for flow, is the standard procedure when an influx is suspected.

Note, however, that a gas influx into an oil mud may not provide an immediate flow indication until the saturation limit has been reached on bottom and more time may be necessary before deciding that the situation is safe. In the interest of minimizing the potential kick volume, an operator may instead choose to shut the well in and use the drillpipe pressure gauge as the kick confirmation tool.

Finally, the crews should be educated on the basics of gas solubility and how gas behaves differently in an oil mud. The rig personnel generally have the initial responsibility for detecting an influx and shutting a well in. It is imperative then that everyone understands that a gas intrusion into an oil mud will not have a dramatic effect at surface and that all be especially alert to the warning signs of an influx.

Problems

1.1 Refer to Fig. 1.2. Under what well-control event would you most likely see gas concentrations increase as formation fluids approach surface?

1.2 An influx consisting of 0.05 mole fraction CH_4 and 0.95 mole fraction C_2H_6 is circulated from a well. The pressure and temperature upstream of the choke are 700 psia and 80°F, respectively.

a. According to Fig. 1.3, what fluid phase or phases would you expect under these conditions?

b. What phases would you expect in the choke line if the fluids cool to 40°F upon expansion through the choke?

1.3 An influx consisting of 90% CH_4 and 10% H_2S enters a deep well. How would this mixture affect the wellbore phase behavior as compared to pure methane?

1.4 A pipeline is transporting gas under a pressure of 900 psig and at a temperature of 90°F. Atmospheric pressure is 13 psia. Determine the absolute pressure and temperature of the gas stream.

1.5 An accumulator unit serves to store hydraulic fluid under pressure for purposes of operating selected blowout preventer (BOP) equipment. A rubber bladder within a bottle or canister is typically charged with nitrogen to 1,000 psig before the control fluid is introduced. Referring to **Fig. 1.32,** determine the hydraulic fluid volume required to pressure the N_2 in an 11-gal nominal bottle (actual fluid capacity = 10 gal) to 3,000 psig. Assume isothermal compression and ideal gas behavior.

1.6 A 10-gal capacity accumulator bottle in Houston, Texas (sea level elevation) is charged with nitrogen to a pressure of 1,000 psig. The temperature that day in Houston was 85°F. The accumulator unit bottle is then transported to an above-sea-level (ASL) elevation of 5,280 ft in Denver, Colorado. The atmospheric pressure in Houston is 14.7 psia and declines linearly at a gradient of 0.49 psi per 1,000 ft of ground elevation. What will be the pressure gauge reading in Denver if the temperature dropped to 40°F in transit? Assume ideal behavior.

1.7 Determine the specific gravity of methane gas. Do the same for water vapor.

1.8 Determine the density of methane in lbm/gal at conditions of 4,500 psia and 160°F. Assume ideal behavior.

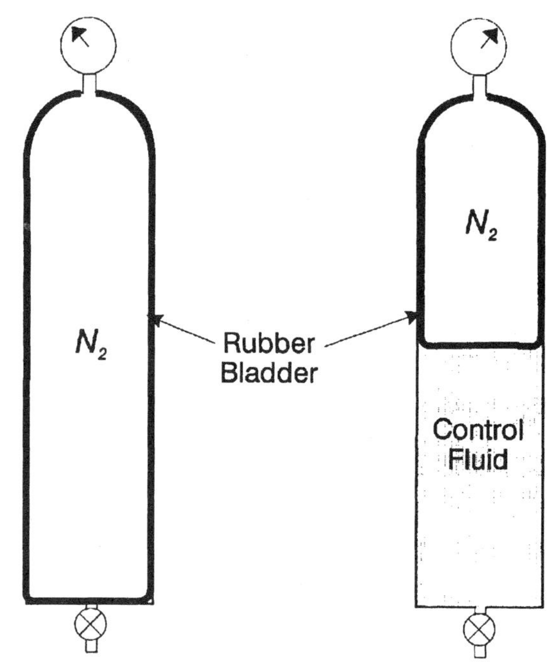

Fig. 1.32—Accumulator bottle described in Problem 1.5.

1.9 Air is composed of approximately 78% nitrogen (M=28.013) and 22% oxygen (M=31.999). Determine the apparent molecular weight of this gas mixture.

1.10 A gas mixture contains methane, ethane, and propane at respective mole fractions of 0.918, 0.063, and 0.019. Calculate the pseudocritical constants for this mixture and compare your results with the calculated values from Eqs. 1.18 and 1.19.

1.11 Write a spreadsheet program for calculating z factors using the method described by Dranchuk and Abou-Kassem.

1.12 A drillstem test of a gas reservoir at 8,200 ft indicates a pore pressure of 3,800 psia. Ambient surface temperature is 70°F and the undisturbed temperature gradient for the area is 0.9°F/100 ft. A subsequent analysis gives the gas specific gravity as 0.65. Determine the compressibility factor of the gas at initial reservoir conditions.

1.13 Work Problem 1.5 using the real gas law. Compressibility factors for N_2 are given in **Fig. 1.33.** The assumption of isothermal compression still applies.

1.14 A 0.6 specific gravity gas influx has entered a well at a depth of 21,000 ft. The bottomhole pressure and temperature are 19,700 psia and 300°F. The initial influx volume is 50 bbl.

a. Estimate, using real gas principles, the pressure at the time the gas reaches the choke if the bubble expands by a factor of 30. Assume the wellbore temperature immediately upstream of the choke is 100°F.

b. Do the problem again, but use an ideal gas.

1.15 Refer to Eq. 1.34c and derive the constant term relating mud density to hydrostatic gradient.

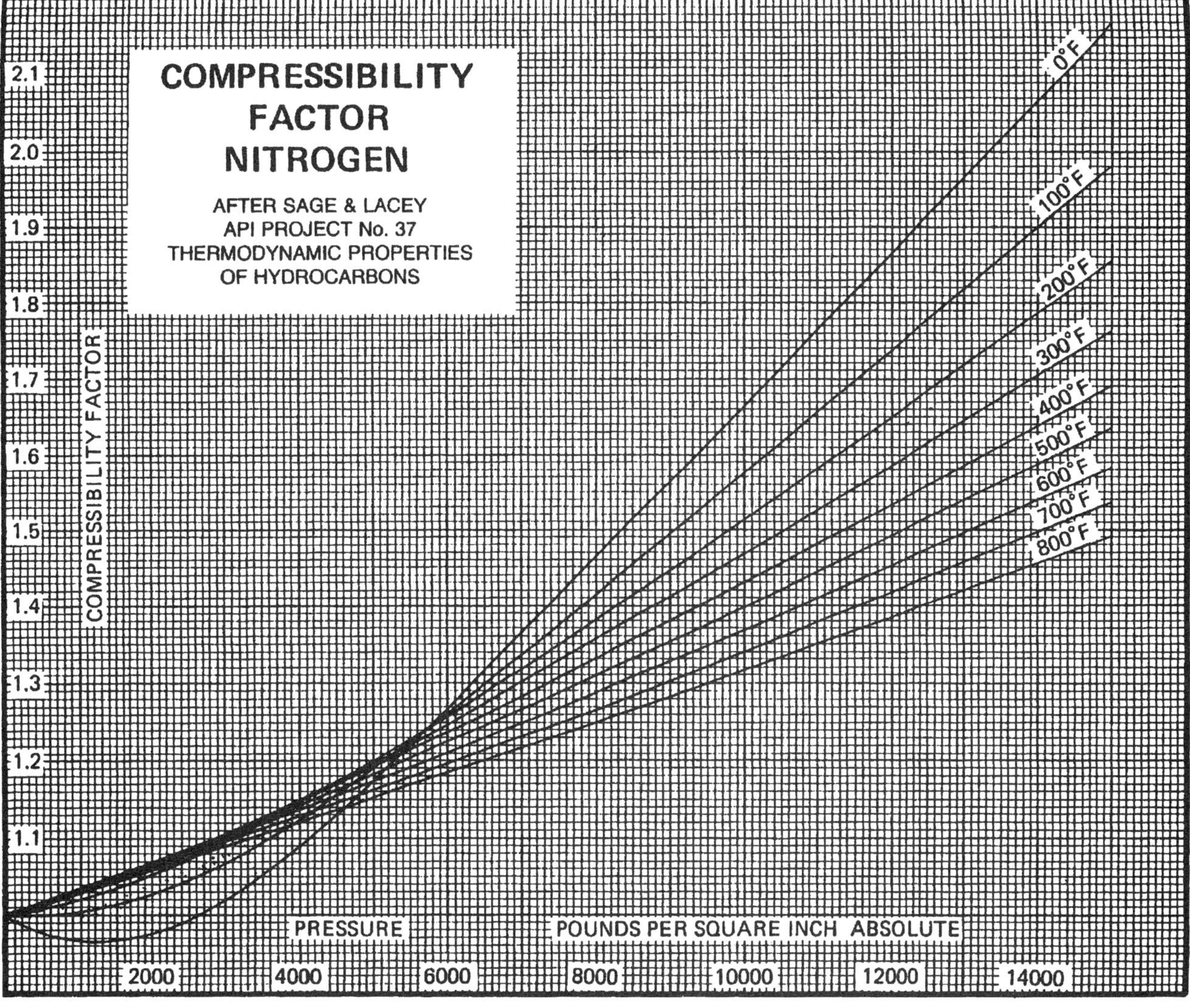

Fig. 1.33—Nitrogen compressibility factors.

1.16 The mud density in your horizontal well is 11.1 lbm/gal. Determine the bottomhole pressure if the present depths are 10,500 ft MD and 7,700 ft TVD.

1.17 A cementing program calls for 500 ft of 12-lbm/gal spacer followed by 2,000 ft of 15.6-lbm/gal cement. The mud density is 9.9 lbm/gal and the top plug will be displaced with fresh water.

a. Ignore friction losses and determine the pressure at the float collar depth of 9,900 ft immediately prior to bumping the plug.

b. Calculate the surface pump pressure at this point in the job.

1.18 Your 8,000-ft gas well has been shut-in for several months because of a marketing problem. The well is completed without a packer and the shut-in casing pressure is 1,300 psig. A fluid level survey is shot on the annulus and 500 ft of water is discovered over the perforations. Estimate the reservoir pressure if the wellbore temperature is 70°F + 1.2°F/100 ft. Assume that the well produces a 1.1 specific gravity water and that the gas specific gravity is 0.7.

1.19 Determine the equivalent density at total depth and at the cement top for the well described in Problem 1.17. Also determine the equivalent density at the perforations for the gas well in Problem 1.18.

1.20 You need to pull the well described in Problem 1.18 and the well is to be killed with 2% KCl water, which has a density of 8.43 lbm/gal. Estimate the KCl water level after the pressures have equalized assuming open perforations and high-permeability rock.

1.21 You have taken a gas kick on a 15,000-ft well and a 300-psig initial shut-in casing pressure is recorded. The mud density is 10.2 lbm/gal and the gas bubble occupies a height of 500 ft. Assume the bottomhole temperature is 210°F.

a. Estimate the equivalent mud weight at the kicking formation.

b. Determine the equivalent density at the last casing depth of 3,500 ft. What hazard could this situation pose?

1.22 Rework Example 1.8 except consider the effect of the volumetric changes that occur as result of the mud's compressibility. Assume the average borehole capacity is 0.1458 bbl/ft and the mud has a compressibility of 6.00×10^{-6} psi^{-1}. Specify any other assumptions you need to make to solve this problem.

1.23 Many well-control calculations assume that a gas influx remains in a discrete bubble and that no mixing or intermingling occurs with the drilling fluid. What effect would gas dispersion in the drilling fluid have on actual wellbore pressures as compared to the predicted values?

1.24 You are drilling a 9½-in. hole at a depth of 11,000 ft with a 12.0-lb/gal mud. The well kicks and an initial shut-in casing pressure of 300 psig is recorded. You leave to notify your supervisor of the problem. Upon returning 15 minutes later, you discover that the casing pressure has increased to 600 psig. At what approximate rate is gas migrating up the wellbore? Assume hole geometry is consistent across the depth of interest.

1.25 Rework Example 1.9 except consider the effect of the volumetric changes that occur as result of the the gas PVT properties and compressibility of the mud. Use the following information and specify any other assumptions you need to make to solve this problem.

gas specific gravity = 0.70,
wellbore temperature = 70°F + 1.1°F/100 ft,
atmospheric pressure = 14 psia,
initial SICP = 500 psig,
final SICP = 800 psig,
time elapsed between readings = 0.5 hr,
well depth = 9,000 ft,
initial kick height = 350 ft,
mud density = 10.0 lbm/gal,
average borehole capacity = 0.1458 bbl/ft, and
mud compressibility = 6.00×10^{-6} psi^{-1}.

1.26 Under which of the following comparative conditions would you expect Eq. 1.42 to more accurately predict gas migration velocity? Explain the reasoning behind your answers.

a. Deep well or shallow well?
b. Shallow casing or deep casing?
c. Cemented casing or uncemented casing?
d. Large hole or small hole?
e. Clear water or drilling mud?
f. Water-base or oil-base drilling fluid?
g. Tight rock or permeable rock?
h. Gas-cut mud or gas-free mud?

1.27 Based on the findings of Rader *et al.*, would you expect a gas bubble to migrate faster in a 10.0-lbm/gal mud or a 12.0-lbm/gal mud?

1.28 How much gas migration would you anticipate in a horizontal wellbore?

1.29 In Fig. 1.28, why is the methane more soluble in the 13.0-lbm/gal oil mud than in the 18.0-lbm/gal mud?

1.30 Use the correlation described by Eq. 1.45 to predict the solubility of methane in No. 2 diesel at 100°F in 500-psi increments from 1,000 to 5,000 psi. Plot your results and compare to the experimental data reflected in Fig. 1.17.

1.31 A 10-lbm/gal 85:15 invert emulsion drilling fluid has respective component volume fractions of: 0.724 diesel, 0.133 $CaCl_2$ brine (200,000 ppm TDS), 0.037 emulsifiers, and 0.106 solids. Prepare a gas solubility calculation spreadsheet for the pressures 50, 100, 500, 1,000, and 2,000 psia. Assume the gas specific gravity is 0.65 and that it contains 97% hydrocarbons and 3% CO_2. Present your results in the form of plotted 100°F, 200°F, and 300°F isotherms.

1.32 Derive the constant shown in Eq. 1.48b.

1.33 The following conditions are given for a well drilled with an oil mud.

Mud density = 11.2 lbm/gal,
circulating bottomhole temperature = 160°F,
circulation rate = 6.0 bbl/min,
bit diameter = 8.5 in., and
drillpipe outer diameter = 5.0 in.

Three sandstone drilling breaks have been noted on the penetration rate recorder: from 95 to 150 ft/hr between 10,100 and 10,120 ft, from 90 to 140 ft/hr between 10,150 and 10,180 ft, and from 90 to 150 ft/hr between 10,200 and 10,250 ft. Assume each sand has a porosity of 20% and that a 0.60 specific gravity gas occupies 75% of the pore space.

a. Determine the average drilled gas concentration in the whole mud for each sand. Express your answers in scf/bbl.
b. Determine the average drilled gas concentration in the diesel phase for each sand. Express your answers in scf/bbl.
c. Determine the average drilled gas concentration in the whole mud for each sand in terms of the downhole concentration. Express your answers in bbl/bbl.
d. What potential hazards does this situation pose?

1.34 Refer to the gas-free volume factors shown in Table 1.4. Plot these factors as function of pressure and construct a curve to fit the data points.

1.35 Refer to the volume factors with gas in Table 1.4.

a. Prepare another table that includes columns showing calculated values for $B_{og} - B_{ong}$ and r'_{so}. Use the plot constructed in the last problem to interpolate the needed B_{ong} values.
b. Plot $B_{og} - B_{ong}$ for each GOR as function of pressure.
c. Plot r'_{so} for each GOR as function of pressure.
d. Compute the pit gain per bbl of methane influx for each entry.

1.36 The following conditions apply to a gas kick on a well being drilled with an oil-base mud.

Vertical depth = 8,000 ft,
mud density = 12.0 lbm/gal,
base oil type = No. 2 diesel,
oil volume fraction in the mud = 0.64,
circulation rate = 10.0 bbl/min,
circulating bottomhole temperature = 200°F,
circulating bottomhole pressure = 5,400 psia,
bit diameter = 12.25 in.,
gas type = methane, and
gas entry rate = 3,500 scf/min.

Assume the compressibility and thermal expansion characteristics of the mud's water phase are negligible compared to the diesel. The solubility characteristics and volume factors can be obtained from Fig. 1.31. Answer the following.

a. Does all of the gas go into solution?
b. Determine the pit gain volume per 1,000 scf gas entry.
c. The pit level monitors are set to give an audible alarm at a pit gain of 10 bbl. Determine the total influx volume in bbl when the alarm is heard.

1.37 Refer to Fig. 1.31 and assume that the bottomhole pressure exceeds the miscibility pressure of the methane/diesel mixture. How might this situation affect the observed pit gain for a kick taken while drilling? What if the influx occurs on a trip?

Nomenclature

a = solubility equation constant
b = solubility equation constant
c = solubility equation constant
B_o = oil volume factor, bbl/STB
B_{og} = oil volume factor including dissolved gas, bbl/STB
B_{ong} = oil volume factor absent dissolved gas, bbl/STB
c = solubility equation constant
$C_1(T_{pr})$ = function in the Dranchuk and Abou-Kassem equation
$C_2(T_{pr})$ = function in the Dranchuk and Abou-Kassem equation
$C_3(T_{pr})$ = function in the Dranchuk and Abou-Kassem equation
$C_4(\rho_r, T_{pr})$ = function in the Dranchuk and Abou-Kassem equation
d_b = bit diameter, in.
D = depth, ft
E_l = energy loss per unit mass, (ft-lbf)/lbm [J/kg]
$f(z)$ = function for the z factor
$f'(z)$ = first derivative of the z function
f_{CO_2} = CO_2 mole fraction, dimensionless
f_e = emulsifier volume fraction, dimensionless
f_g = gas mole fraction, dimensionless
f_h = hydrocarbon mole fraction, dimensionless
f_{H_2S} = H_2S mole fraction, dimensionless
f_o = oil volume fraction, dimensionless
f_w = water volume fraction, dimensionless
g = acceleration of gravity, 32.17 ft/s^2
g_c = gravitational system conversion constant, 32.17 (lbm-ft)/(lbf-s^2)
g_f = fluid hydrostatic gradient, psi/ft
g_m = mud hydrostatic gradient, psi/ft
G = pit gain, bbl
h = height, ft
m = mass, lbm
M = molecular weight, lbm/(lbm-mole)
M_a = molecular weight of air, lbm/(lbm-mole)
n = number of moles, lbm-mole
p = pressure, psi
p_{bh} = bottomhole pressure, psi
p_c = critical pressure, psia
p_{cs} = shut-in casing pressure, psi
p_{pc} = pseudocritical pressure, psia
p_{pr} = pseudoreduced pressure, dimensionless
p_r = reduced pressure, dimensionless
q_m = mud circulation rate, bbl/min
q_{gsc} = drilled gas entry rate, scf/min
q_r = rock removal rate, ft^3/min
r_m = gas/mud ratio, scf/bbl
r_{sCO_2} = CO_2 gas/component ratio, scf/bbl
r_{sH_2S} = H_2S gas/component ratio, scf/bbl
r_{sh} = hydrocarbon gas/component ratio, scf/bb
r_{se} = solution gas/emulsifier ratio, scf/bbl
r_{sm} = solution gas/mud ratio, scf/bbl
r_{so} = solution gas/oil ratio, scf/bbl
r'_{so} = downhole gas/oil ratio, bbl/bbl
r_{sw} = solution gas/water ratio, scf/bbl
R = penetration rate, ft/hr
R_g = universal gas constant, (psia-gal)/(°R-lbm-mole)
S_g = gas saturation, dimensionless
T = temperature, °F or °R [°C or K]
T_c = critical temperature, °F or °R [°C or K]
T_{pc} = pseudocritical temperature, °R [K]
T_{pr} = pseudoreduced temperature, dimensionless
T_r = reduced temperature, dimensionless
t = time, hr
V = volume, bbl, cu ft, or gal
V_g = gas volume, bbl
V_{gsc} = gas volume at standard conditions, scf
v = velocity, ft/s
v_{sl} = gas slip or migration velocity, ft/hr
W = work per unit mass, (ft-lbf)/lbm
z = gas compressibility factor, dimensionless
Z = elevation, ft
γ_g = gas specific gravity, dimensionless
ρ = density, lbm/gal
ρ_{eq} = equivalent density, lbm/gal
ρ_f = fluid density, lbm/gal or lbm/ft^3
ρ_g = gas density, lbm/gal
ρ_r = reduced density, dimensionless
ϕ = formation porosity, dimensionless

References

1. Metcalfe, R.S.: "Gas Properties and Correlations," *Petroleum Engineering Handbook*, H.B. Bradley (ed), SPE, Richardson, TX (1987) **20,** 3.
2. Bloomer, O.T., Gami, D.C., and Parent, J.D: *Physical-Chemical Properties of Methane-Ethane Mixtures*, Institute of Gas Technology (1952) 3–8.
3. Brown, G.G., Katz, D.L., Oberfell, G.G., and Alden, R.C.: *Natural Gasoline and the Volatile Hydrocarbons*, Natural Gas Assn. of America, Tulsa (1948).
4. Katz, D.L. *et al.*: *Handbook of Natural Gas Engineering*, McGraw-Hill Book Co. Inc., New York City (1959) 103–106.
5. McCain, W.D. Jr.: *The Properties of Petroleum Fluids*, second edition, PennWell Publishing Co., Tulsa (1990) 108.
6. Kay, W.B.: "Density of Hydrocarbon Gases and Vapors at High Temperature and Pressure," *Ind. Eng. Chem.* (September 1932) **28,** 1,014–1,016.
7. Sutton, R.P.: "Compressibility Factors for High-Molecular-Weight Reservoir Gases," paper SPE 14265 presented at the 1985 SPE Annual Technical Conference and Exhibition, Las Vegas Nevada, 22–25 September.
8. Standing, M.B. and Katz, D.L.: "Density of Natural Gases," *Trans.*, AIME (1942) **146,** 140.
9. Kvalnes, H.M. and Gaddy, V.L.: "The Compressibility Isotherms of Methane at Pressures to 1,000 Atmospheres and Temperatures -70 °C to 200 °C," *J. Am. Chem. Soc.* (1931) **53,** 394.
10. Takacs, G.: "Comparing Methods for Calculating Z-factor," *Oil & Gas J.* (26 May 1989) 43–46.
11. Dranchuk, P.M. and Abou-Kassem, J.H.: "Calculations of Z-Factors for Natural Gases Using Equations of State," *JCPT* (July–September 1975) 34–36.
12. Wichert, E. and Aziz, K.: "Calculate Z's for Sour Gases," *Hyd. Proc.* (May 1972) **51,** 119–122.
13. Johnson, A.B. and Cooper, S.: "Gas Migration Velocities During Gas Kicks in Deviated Wells," paper SPE 26331 presented at the 1993 SPE Annual Technical Conference and Exhibition, Houston, 3–6 October.

14. Johnson, A.B. and Tarvin, J.: "New Model Improves Gas Migration Velocity Estimates in Shut-in Wells," *Oil & Gas J.* (15 May 1993) 55–60.
15. Johnson, A.B. and White, D.B.: "Gas-Rise Velocities During Kicks," *SPEDE* (December 1991) 257–263.
16. Davies, R.M. and Taylor, G.I.: "The Mechanics of Large Bubbles Rising Through Extended Liquids and Through Liquids in Tubes," *Proc.*, Royal Soc. London (1950) **A200**, 387–388.
17. Rader, D.W., Bourgoyne, A.T. Jr., and Ward, R.H.: "Factors Affecting Bubble-Rise Velocity of Gas Kicks," *JPT* (May 1975) 571–584.
18. Hovland, F. and Rommetveit, R.: "Analysis of Gas-Rise Velocities From Full-Scale Kick Experiments," paper SPE 24580 presented at the 1992 SPE Annual Technical Conference and Exhibition, Washington D.C., 4–7 October.
19. O'Brien, T.B.: "Handling Gas in an Oil Mud Takes Special Precautions," *World Oil* (January 1981) 83–46.
20. Thomas, D.C., Lea, J.F. Jr., and Turek, E.A.: "Gas Solubility in Oil-Based Drilling Fluids: Effects on Kick Detection," *JPT* (June 1984) 959–968.
21. Redlich, O. and Kwong, J.N.S.: "On the Thermodynamics of Solutions—V. An Equation of State. Fugacities of Gaseous Solutions," *Chemical Reviews* (1949) **44**, 233–244.
22. O'Bryan, P.L. *et al.*: "An Experimental Study of Gas Solubility in Oil-Based Drilling Fluids," *SPEDE* (March 1988) 33–42.
23. Culberson, O.L. and McKetta, J.J. Jr.: "Phase Equilibria in Hydrocarbon-Water Systems. III. The Solubility of Methane in Water at Pressures to 10,000 psia," *Trans.*, AIME (1951) **192** 223–226.
24. McKetta, J.J. Jr. and Wehe, A.H.: "Hydrocarbon-Water and Formation Water Correlations," *Petroleum Production Handbook*, Vol.II, T.C Frick and R.W. Taylor (eds), SPE, Richardson, Texas (1962) **22,** 13.
25. Matthews, W.R.: "How to Handle Acid Gas HS and CO Kicks," *Pet. Eng. Intl.* (15 November 1984) 22–29.
26. Crawford, H.R. *et al.*: "Carbon Dioxide—A Multipurpose Additive for Effective Well Stimulation," *JPT* (March 1963) 237–242.
27. Selleck, F.T., Carmichael, L.T., and Sage, B.H.: "Phase Behavior in the Hydrogen Sulphide-Water System," *Ind. Eng. Chem.* (1932) **44,** 2219.
28. O'Bryan, P.L. and Bourgoyne, A.T. Jr.: "Methods of Handling Drilled Gas in Oil-Based Drilling Fluids," *SPEDE* (September 1989) 237–246.
29. O'Bryan, P.L. and Bourgoyne, A.T. Jr.: "Swelling of Oil-Based Drilling Fluids Resulting Fron Dissolved Gas," *SPEDE* (June 1990) 149–155.
30. Peng, D.Y. and Robinson, D.B.: "A New Two Constant Equation of State," *Ind. & Eng. Chem. Fund.* (1976) **15**, No. 1, 59–64.

SI Metric Conversion Factors

bbl	× 1.589 873	E − 01	= m^3
bbl/ft	× 5.216 119	E − 01	= m^3/m
bbl/min	× 2.649 788	E + 00	= L/s
ft	× 3.048*	E − 01	= m
ft/hr	× 8.466 667	E − 02	= mm/s
cu ft	× 2.831 685	E − 02	= m^3
scf/bbl	× 1.801 175	E − 01	= std m^3/m^3
scf/min	× 4.719 475	E − 04	= std m^3/s
(ft-lbf)/lbm	× 2.989 067	E − 03	= kJ/kg
°F	(°F − 32)/1.8		= °C
°F/100 ft	× 1.822 689	E + 01	= mK/m
gal	× 3.785 412	E + 00	= L
in.	× 2.54*	E + 01	= mm
lbm	× 4.535 924	E − 01	= kg
lbm/ft^3	× 1.601 846	E + 01	= kg/m^3
lbm/gal	× 1.198 264	E + 02	= kg/m^3
psi	× 6.894 757	E − 03	= MPa
psi/ft	× 2.262 059	E + 01	= kPa/m
°R	°R/1.8		= K

* Conversion factor is exact.

Chapter 2
Pore Pressure

2.1 Introduction

Pore pressure and wellbore-fracture pressure substantially affect, indeed control, a drilling operation. The driller usually attempts to offset formation pressures with some minimum hydrostatic pressure supplied by the drilling fluid. A maximum wellbore-pressure limitation, however, is dictated by the fracture integrity of the rock. The allowable mud density or combination of applied and hydrostatic pressure across any hole interval has an upper and lower bound. Knowing or having some reasonable prediction of these limits is essential to well planning and subsequent plan execution.

These topics have received much attention over the past 40 years. Theories and explanations have been proposed and predictive techniques have been presented in the literature. Some have been accepted and applied universally, while others have not. Many procedures work well in a given area or depositional environment but less so or not at all in others. This chapter focuses on the more common or accepted concepts. Chapter 3 discusses fracture-pressure prediction and measurement.

2.2 Pore-Pressure Origins

Pore pressure, sometimes called formation or formation-fluid pressure, is defined as the pressure contained in the pore space of subsurface rock. Pore pressures can be classified by the magnitude of the corresponding pressure gradient in a given area as normal, subnormal, and abnormal.

2.2.1 Normal Pore Pressures. Normal pressure gradients correspond to the hydrostatic gradient of fresh or saline water. **Fig. 2.1** shows a normally pressured rock where a formation stratum was deposited in a marine environment. The bulk rock includes the grain framework or matrix plus interstitial water within the pore space. Assuming that the porosity is interconnected and extends back to the ground surface through the overlying sediments, the pore pressure at a point in the rock element is the product of the vertical depth D and the hydrostatic gradient of the pore water,

$$p_n = g_n D, \quad \text{(2.1)}$$

where p_n and g_n are normal pore pressure and normal pressure gradient, respectively.

What is considered normal pressure depends on the geographic area or depositional basin. For example, a normal pressure gradient is considered to be 0.465 psi/ft in the gulf coast region of the U.S. Midcontinent regions of North America and other continents often exhibit a 0.433-psi/ft normal pore-pressure gradient, which is equivalent to that of fresh water. **Table 2.1** lists normal pore-pressure gradients for some of these areas. Because normal gradients may vary, the values given should not be considered absolute.

Example 2.1. Determine the pore pressure of a normally pressured formation in the Gulf of Mexico at a depth of 9,000 ft. What would be considered normal at the same depth for a well drilled in the North Sea?

Solution. In the Gulf of Mexico, the normal pore pressure at the designated depth is

$$p_n = (0.465)(9,000) = 4,185 \text{ psig},$$

whereas in the North Sea,

$$p_n = (0.452)(9,000) = 4,068 \text{ psig}.$$

2.2.2 Subnormal Pore Pressures. A subnormal pore pressure is less than what would be considered normal for the area. Hence, the pore pressure in a normally pressured rock in Oklahoma would be considered subnormal at the same depth in south Louisiana. One reason for subnormal pressures is seen in areas displaying uneven surface terrain characteristics. **Fig. 2.2** depicts an aquifer outcropping below the surface drilling location that results in a piezometric water table at some depth below the kelly bushing (KB) datum of the rig. The pore pressure above the water level is near atmospheric, which lowers the pore pressure gradient of the rock to virtually nothing. These occurrences are common in mountain regions and create severe lost-circulation problems at shallow depths.

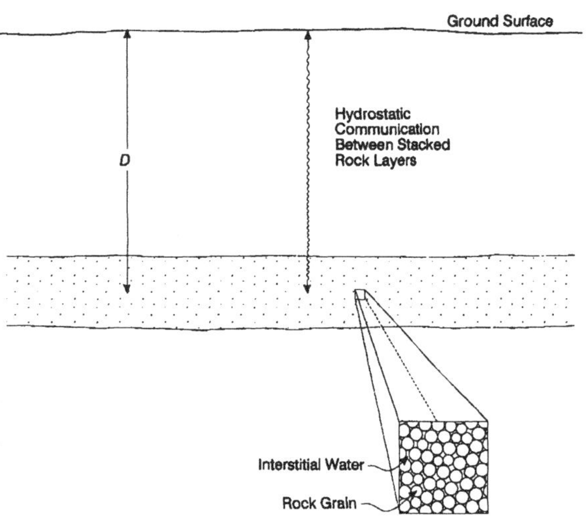

Fig. 2.1—Demonstration of a normal-pore-pressure environment.

TABLE 2.1—NORMAL-PORE-PRESSURE GRADIENTS IN SPECIFIC GEOGRAPHIC AREAS[1]		
	Pre-Pressure Gradient	
Region	psi/ft	kPa/m
Anadarko basin	0.433	9.64
California	0.439	9.77
Gulf of Mexico	0.465	10.35
Mackenzie delta	0.442	9.84
Malaysia	0.442	9.84
North Sea	0.452	10.06
Rocky Mountains	0.436	9.71
Santa Barbara channel	0.452	10.06
West Africa	0.442	9.84
West Texas	0.433	9.64

One cause of subnormal pressures is gross earth movements. **Fig. 2.3** shows a series of rock strata on each side of a sealing tension or normal fault. Sands A and B retain their original pressure at the greater depth if the fault plane does not leak and other means of thermal or hydraulic repressuring are not provided to the downthrown side. The sands to the left of the fault will be subnormal if those on the right are normally pressured.

Not all subnormal pore pressures occur naturally. Some are manmade, specifically those resulting from production. Subnormal pressures resulting from reservoir depletion are increasingly common in mature development areas. **Fig. 2.4** shows a deep prospect being drilled in a field that has produced for some period from the more shallow horizon. The productive formation may have had a normal or even an abnormal pore pressure when discovered, but production has drawn down the average reservoir pressure to a level that can create problems in the later development program. Well plans may need to incorporate another casing string to reach objective depth.

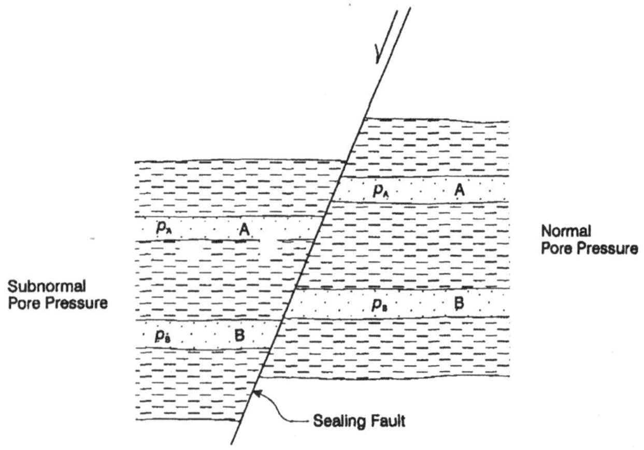

Fig. 2.3—Subnormal pore pressures resulting from a sealed normal fault.

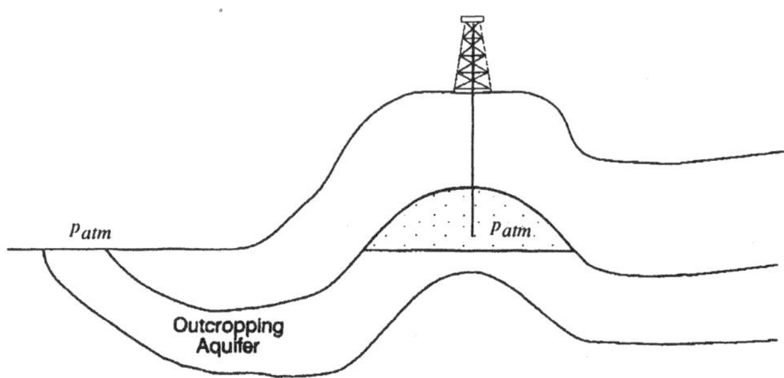

Fig. 2.2—Subnormally pressured aquifer resulting from a depressed outcrop.

A few explanations for the occurrence of subnormal pressures are covered here; Ref. 2 describes several others. Subnormal pressures can lead to drilling hazards, such as differential sticking and lost circulation. It is essential that drilling engineers make every effort to identify potentially troublesome zones and take corrective action in the planning stages of a well.

2.2.3 Abnormal Pore Pressure. Abnormal pore pressures are formation pressures that are higher than normal for an area. The term could apply to either abnormally high or abnormally low pressures; this chapter uses the expression as it applies to the higher pressure gradients. Other common terms for this phenomenon are "geopressures" and "overpressures."

Drilling problems and well costs generally increase in overpressured rock. Abnormally pressured shales create hole-instability problems that usually can be remedied with higher mud densities. Abnormally pressured formations with good

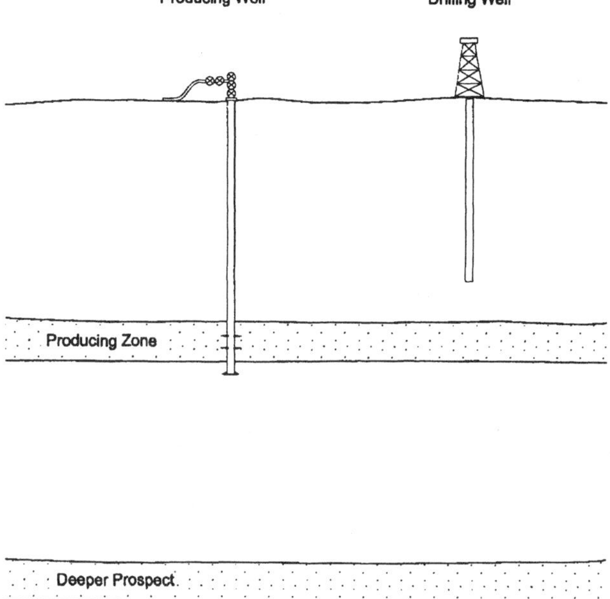

Fig. 2.4—Reservoir pressure depletion as a reason for subnormal pore pressures.

TABLE 2.2—POSSIBLE SOURCES OF ABNORMAL PORE PRESSURES
Artesian systems
Structural reasons
Tectonics
Faults
Salt or shale diapirs
Other
Surface erosion
Rock diagenesis
Sulfates
Precipitation
Clays
Thermal effects
Osmosis through shale
Biochemical effects
Undercompacted sediments
External pressure sources
Natural
Manmade

permeability require mud densities equivalent to or slightly in excess of the pore pressure if the intent is to keep formation fluids from entering the wellbore.

Many theories have been proposed to explain why abnormal pore pressures occur. Fertl[3] gives a comprehensive discussion of possible overpressuring mechanisms. These are listed in **Table 2.2** and are discussed in differing degrees of detail in this chapter. Many of the explanations given within the scope of this text are by necessity oversimplifications of a complex and not totally explained phenomenon. Abnormal-pore-pressure incidents may have more than one origin and may be the end result of many causes.

Regardless of the source, all abnormal pressures require some means of sealing or trapping the pressure within the rock body. Otherwise, hydrostatic equilibrium back to a normal gradient would be restored eventually. **Fig. 2.5** demonstrates some possible sealing mechanisms.

Figs. 2.5a and 2.5b show sand bodies isolated laterally by stratigraphic pinchouts and vertically by massive shale sections. Although shale has some permeability and Darcy's law must be obeyed, these sands need a long time to achieve a normal pressure gradient. Many believe that this time requirement is exceedingly long relative to the age of the younger rocks. Thick Tertiary shales, for example, have been attributed as abnormal pressure seals in the Gulf of Mexico and other geologically young areas. **Table 2.3** provides an abbreviated geologic time scale as a convenient reference for discussing geologic age.

Fig. 2.5c shows rock that contacts an impermeable salt bed or salt dome may be sealed effectively in the lateral or vertical direction. Fig. 2.5d shows a dense caprock seal found in many areas. An example of this are the anhydrites and tight carbonates associated with the Buckner formation that have been attributed as the primary seal mechanism for the deep, abnormally pressured Jurassic sands in Mississippi.[4] Another example is the Anadarko basin of western Oklahoma where the Pennsylvanian Morrow shales and sands are usually overpressured. The pore-pressure gradient in the Morrow varies considerably and seems to be dependent on the number and thicknesses of dolomite and hard-rock stringers within the transition section.

Abnormal pressures also are associated with massive salt beds. Rather than serving as a geopressure source, the impermeable nature of rock salt provides a highly effective seal for any subsequent processes, such as diagenesis and thermal expansion of pore fluids. In the Michigan basin, for example, the Silurian A-1 and A-2 carbonate formations are sandwiched within a massive salt section. Because of the large overburden gradients unique to Michigan, these formations may require mud densities in excess of 22 lbm/gal if sufficient carbonate permeability has been developed.

Dense rock encountered in a drilling operation is a warning to an operator that something may be changing with respect to the current pore-pressure gradient. Hard, tight rock can isolate pressure from above as well as below and may be an indication of rapid transition from a high-pressure environment to a lower or even normal pressure regime.

Mississippian-age limestones underlie the abnormally pressured Morrow in the Anadarko basin and signal the onset of much lower pore pressures. Other hard-rock areas subject to pore-pressure reversals include southern Iran and the Delaware basin of west Texas. Soft-rock areas are not immune to this occurrence; an example is the Tuscaloosa Trend in south Louisiana. **Fig. 2.6** show how pore-pressure reversals can make drilling a difficult task. Each pore-pressure reversal introduces significant pressure-control problems and several casing strings may be required to reach the objective depth.

Many abnormal-pore-pressure processes are simply the reverse of those that generate subnormal pressures. The converse to a low piezometric water level is abnormal pressure resulting from an artesian source. **Fig. 2.7** illustrates an aquifer outcropping above a surface drilling location. The high water table relative to the rig site results in an abnormal pore pressure in the aquifer. These occurences are common in mountainous regions and other areas such as the Permian Basin.

Example 2.2 demonstrates the problematic nature of an anticlinal or dipping bed structure with associated low-density pore fluids leading to abnormal pore pressure.

Example 2.2. Fig. 2.8 represents a 300-ft-thick gulf coast sand. The anticline crests at a depth of 1,000 ft and has a gas/water contact at 1,300 ft. Gas with a hydrostatic-pressure gradient of 0.05 psi/ft occupies the pore space down to 1,300 ft. Determine the pore-pressure gradient at the top of the structure if the sand is normally pressured at the gas/water contact.

Solution. The normal-pore-pressure gradient in the gulf coast is 0.465 psi/ft, giving a pore pressure at 1,300 ft of

$$p_{p1,300} = (0.465)(1,300) = 605 \text{ psig}.$$

The pressure at the top of the sand is reduced by an amount equal to the hydrostatic pressure of the gas,

$$p_{p1,000} = 605 - (0.05)(300) = 605 \text{ psig}.$$

Therefore, the pressure gradient at the top of the structure is

$$g_{p1,000} = 590/1,000 = 0.590 \text{ psi/ft}.$$

The shallow gas zone in Example 2.2 would kick if the mud weight was any lower than 11.4 lbm/gal. In some areas, shallow gas presents a real problem whose severity increases at shallower depths and as gas column height increases relative to where the bit penetrates the formation. Chapter 8 covers the hazards to a drilling operation and methods for dealing with the problem.

Large-scale earth movements, primarily those that involve compressive folding and faulting, play many roles in the creation of abnormal pore pressures. An immediate effect of tec-

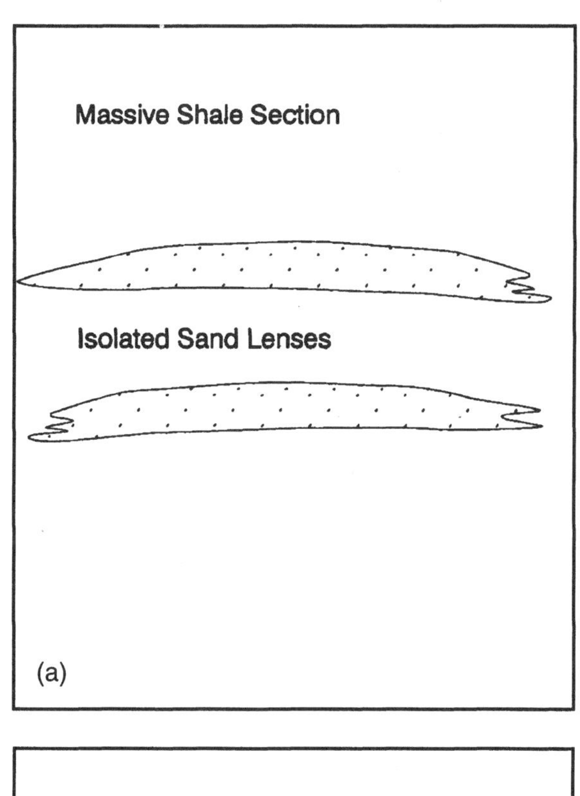

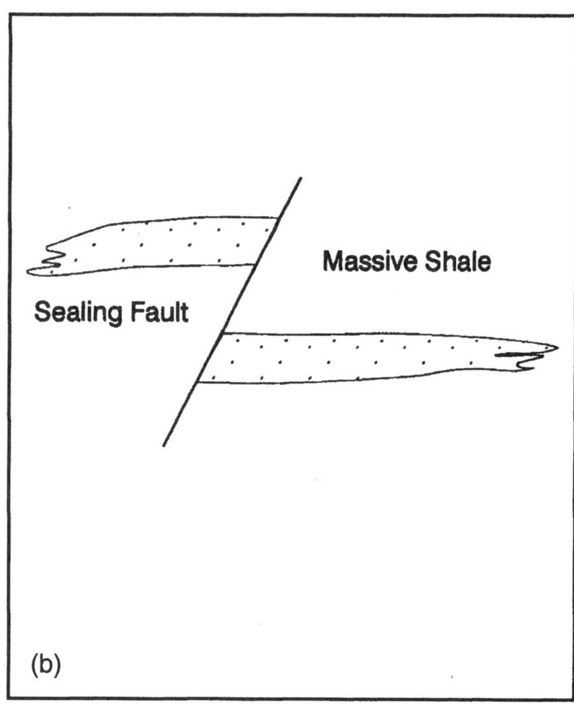

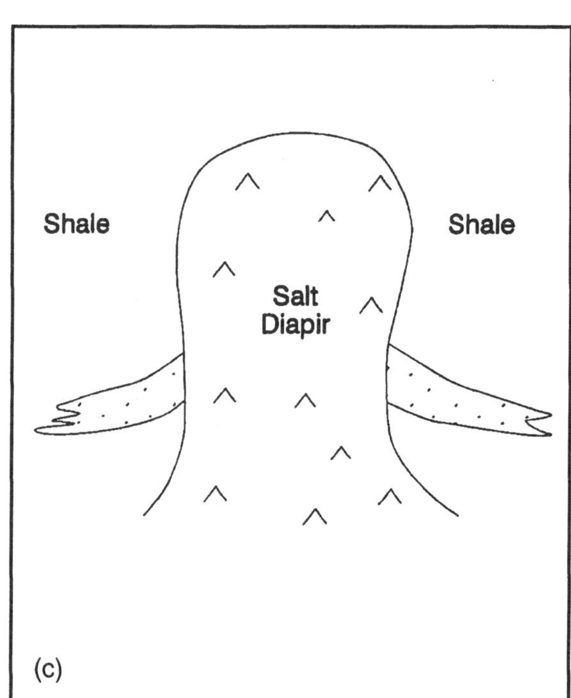

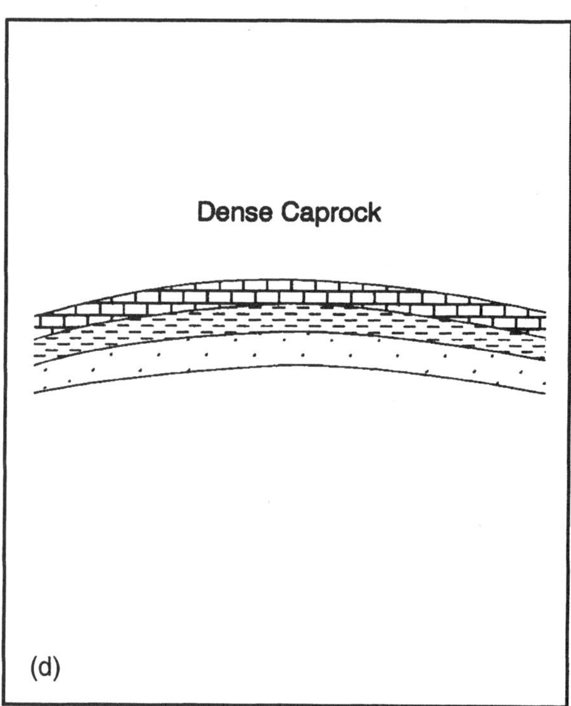

Fig. 2.5—Examples of some abnormal pressure seals.

tonic activity is redistribution of rock stress and rock deformation. Compression associated with rock deformation serves as a geopressuring agent if pressures do not dissipate through any created fractures or "leaky" fault planes.

The thrust fault depicted in **Fig. 2.9** shows two sands shifted upward relative to their previous positions in the normal fault shown in Fig. 2.3. Being elevated, Sands A and B would be abnormally pressured if they retained their original pore pressure. However, tectonic uplift, as an origin of abnormal pressures termed paleopressures, has some serious weaknesses. An uplift likely would be associated with fracturing and the means to dissipate pore pressures back to hydrostatic equilibrium. The effect of reducing the system temperature is to reduce pore-water volume, thereby lowering the pore pressure unless gas existed in the pore space before uplift.

Fig. 2.10 depicts a geopressure source associated with normal growth faulting. The dashed line indicates the top of the transition zone before the fault movement. After the block on the left shifts downward, strata that were normally pressured before the tectonic event are exposed to abnormal pressures on the other side of the fault. Thus, a lateral flow gradient into the downthrown side is created. Abnormal pressure occurrences in the Niger delta appear to have been created in this manner.[5]

A common tectonic event is a salt diapir that plastically "flows" or extrudes into the younger sediments. **Fig. 2.11** shows the overlying strata deformed and perhaps failed in

TABLE 2.3—GEOLOGIC TIME SCALE IN NORTH AMERICA

Era	Period	Epoch	Approximate Age (millions of years)
Cenozoic	Quatenary	Recent Pleistocene	1
		Pliocene Miocene	25
	Tertiary	Oligocene Eocene Paleocene	63
Mesozoic	Cretaceous	Upper and Lower	135
	Jurassic	Upper, Middle, and Lower	181
	Triassic	Upper, Middle, and Lower	
Paleozoic	Permian	Upper, Middle, and Lower	280
	Pennsylvanian	Upper, Middle, and Lower	
	Mississippian	Upper, Middle, and Lower	345
	Devonian	Upper, Middle, and Lower	405
	Silurian	Upper, Middle, and Lower	
	Ordivician	Upper, Middle, and Lower	500
	Cambrian	Upper, Middle, and Lower	600
Precambrian	Proterozoic Archeozoic		

shear by the intrusion of the salt dome. The resulting compression of the rock and pore fluid results in tremendous overpressures if a means of regaining hydrostatic equilibrium is not provided. Plastic shale diapirs that follow the same density inversion process are found in many areas. Besides the compression effects, as a rule, shale domes are abnormally pressured and may serve as a lateral pressure source for the penetrated strata.

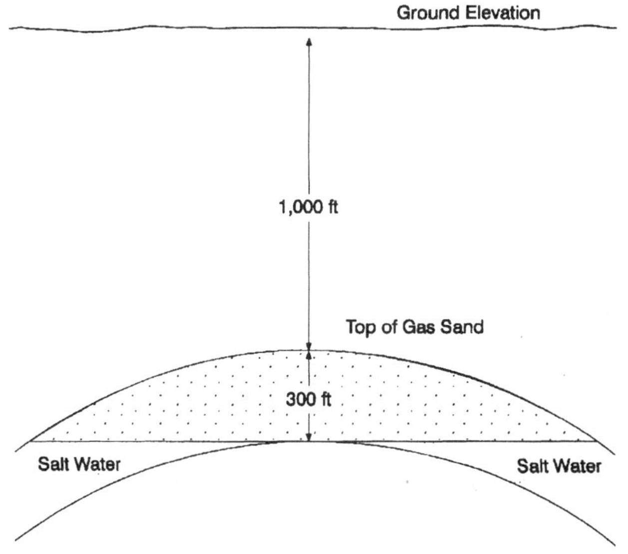

Fig. 2.8—Shallow gas structure described in Example 2.2.

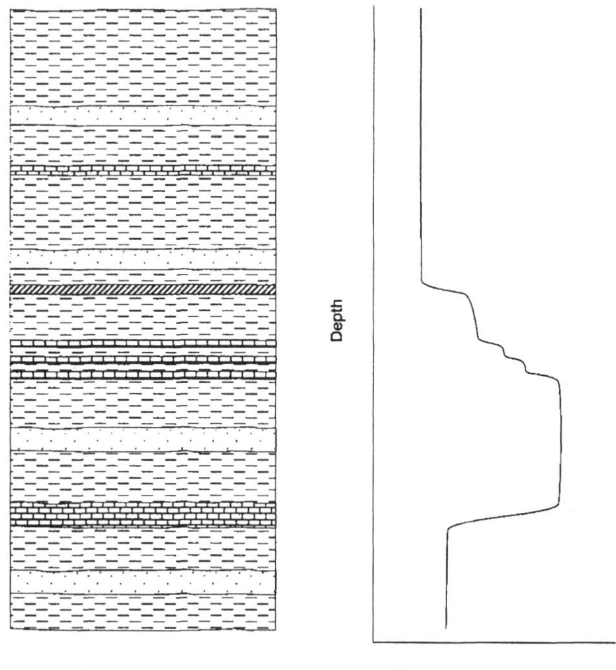

Fig. 2.6—A lithological sequence that sometimes results in a pore pressure reversal.

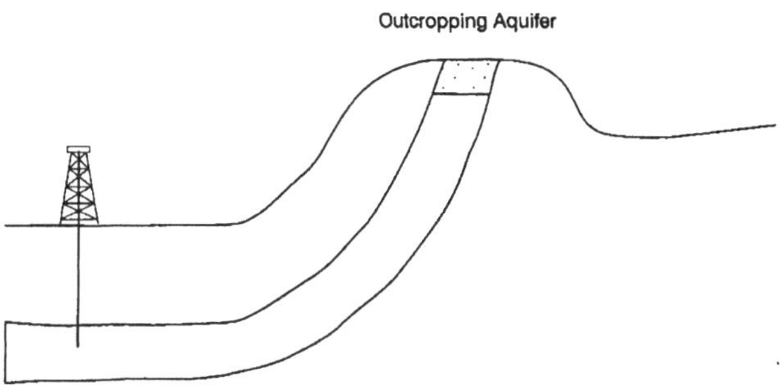

Fig. 2.7—Abnormal pore pressures arising from an artesian source.

Fig. 2.12 illustrates another potential cause for abnormal pore pressure. The isolated sand body on the left contains fluid under some pressure. On the right, the ground surface has eroded over time. For a sealed reservoir, the sand retains its original pore pressure and exhibits a higher gradient at depth D_2.

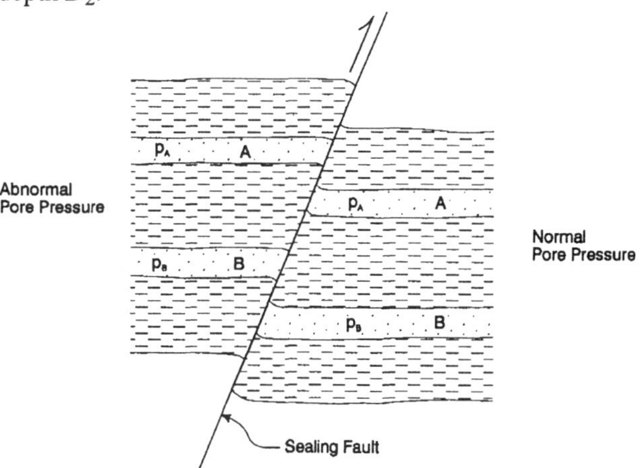

Fig. 2.9—Abnormal pressures resulting from a sealed thrust fault.

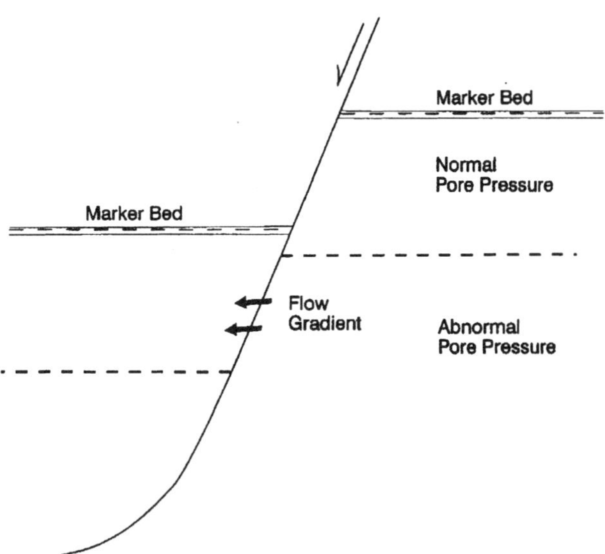

Fig. 2.10—Abnormal pressure source across a growth fault.

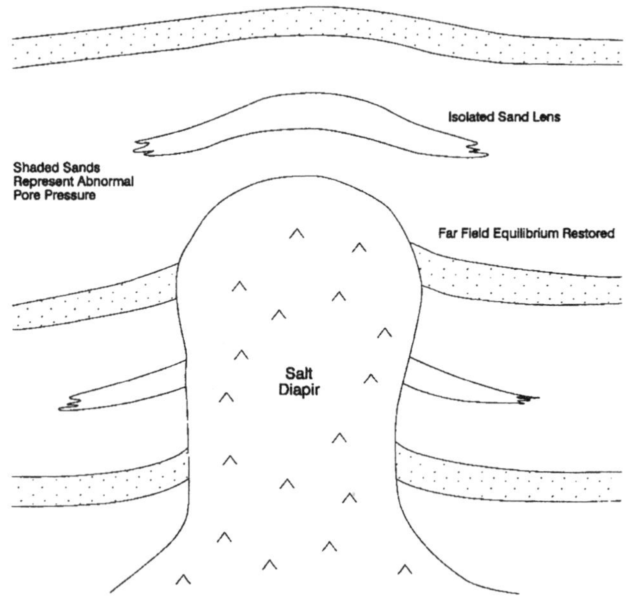

Fig. 2.11—Rock deformation and compression adjacent to a salt dome.

Rock diagenesis often has been cited as a source for abnormal pore pressures. As defined by Pettijohn,[6] "Diagenesis denotes the processes leading to the lithification of a rock, or the conversion of newly deposited sediments into an indurated rock." Post-depositional diagenetic processes that either increase pore water or reduce pore volume can lead to geopressures. Processes attributed as geopressure sources are the release of water from gypsum during the conversion to anhydrite and the precipitation of cementing materials from pore-water solution.

Powers[7] and Burst[8] discussed the role of clay diagenesis in causing the abnormal pressures seen in deep Gulf of Mexico Tertiary sediments. At elevated temperatures and with an available potassium ion source, montmorillonite converts to illite and releases its tightly held interlayer water into the pore space. In the process, the water undergoes a volume increase and thereby increases pore pressure. However, Magara[9] convincingly argued that clay diagenesis alone could not account for these pressures and that the mechanism was secondary to undercompaction.

Barker[10] proposed that thermal expansion of pore water was a viable and substantial source of abnormal pressures. This is a controversial topic with regard to its significance in shale geopressures. Thermal expansion must fulfill several requirements to be a major source of abnormal pressures, including completely impervious beds and for the heating to occur after the beds have been sealed.[11,12] Its importance, while probably real, is more likely secondary to other processes.[13]

Osmosis refers to the flow potential of low-salinity to high-salinity water across a semipermeable membrane. Young and Low[14] demonstrated experimentally that naturally occurring clays or shales serve as a semipermeable membrane by allowing water molecules to pass but blocking salt ions. The flow potential could result in overpressuring a shale and has been attributed as a source for abnormal pressures in the San Juan basin.[15]

Shale as a semipermeable membrane is thought to be a reason for the dense caprock seals often seen beneath a shale section. **Fig. 2.13** shows that a possible mechanism for caprock formation results from the pressure gradient driving the upward flow of water. Ion exclusion from the mobile water leads to the precipitation of carbonate and silica minerals at the base of the shale membrane.

Biochemical processes refer to the formation of gas and graphite by thermal cracking of kerogen or oil.[16] The com-

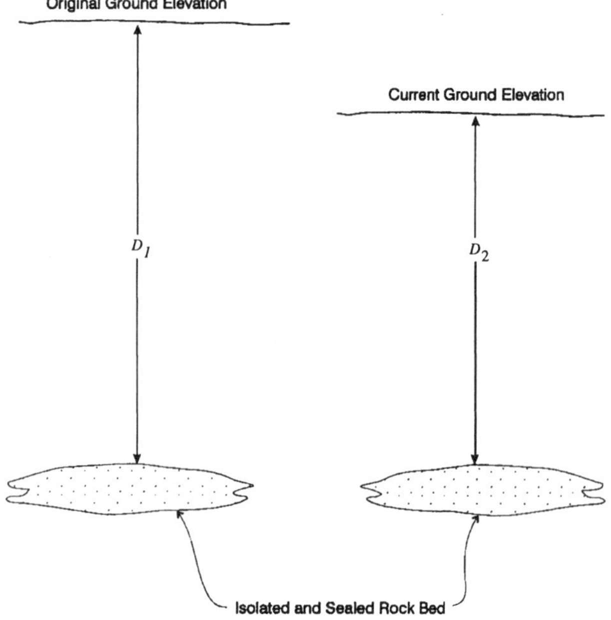

Fig. 2.12—Abnormal pore pressure resulting from erosion processes.

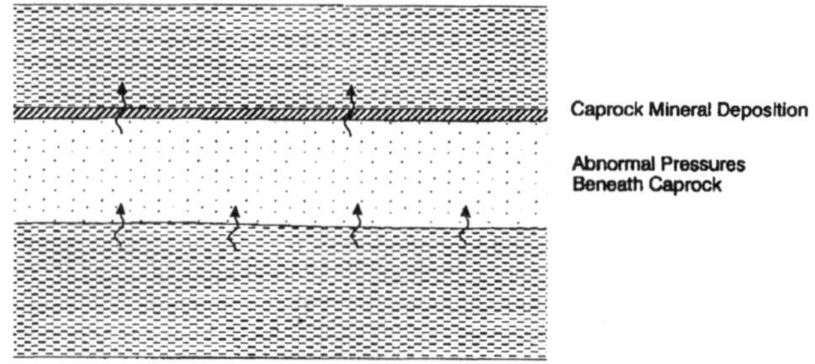

Fig. 2.13—Caprock mineral growth resulting from water flow across a semipermeable membrane into a shale.

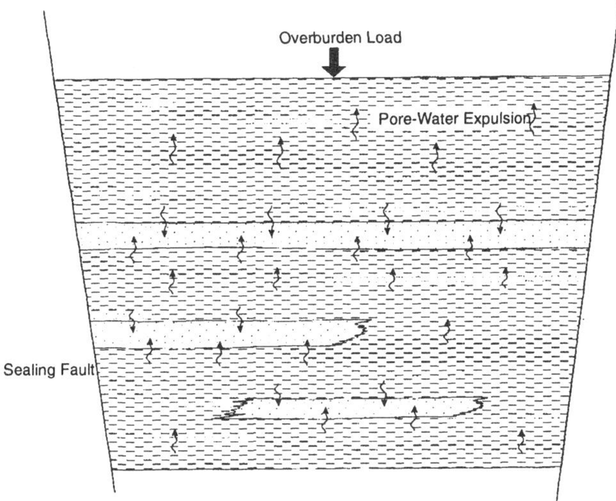

Fig. 2.14—Restricted pore-water expulsion as rock compacts in response to overburden load.

bined oil, gas, and graphite volume after the cracking reaction is larger than the volume of the uncracked oil, thus leading to an increase in pore pressure. This process is thought to be a significant driving agent behind hydrocarbon migration from source rocks.

During deposition, sediments are compacted by the overburden load and are subjected to greater temperature with increasing burial depth. Porosity decreases as water is expelled from the mud or rock by the increasing weight of the overburden and thermal expansion of the water. Hydrostatic equilibrium within the compacted layers is retained as long as the expelled water is free to migrate vertically up through the overlying strata or by other, less direct drainage routes.

Fig. 2.14 shows a deposition model. Pore water expelled from thick shales into interbedded sands is restricted from flowing in the lateral direction by stratigraphic barriers or sealing faults. In the vertical direction, the shales restrict flow and other permeability barriers in the form of caprock material may be present. Thus, a state of hydrostatic disequilibrium is achieved if and when the sediment deposition rate exceeds the rate of water expulsion. Porosity is maintained and the pore water begins to support more of the increasing overburden. Consequently, pore pressures in excess of hydrostatic develop in the shales and adjacent sands and the sediments are undercompacted for their burial depth and overburden load.

The compaction theory best fits most naturally occurring abnormal pressures. Most pore-pressure prediction techniques are based on this theory. This does not imply that other explanations are invalid. However, few methods have been proposed to predict or detect abnormal pressures caused by other sources.

Waterflooding and other secondary-recovery methods increase the pore pressure of the flooded zone if fluid input exceeds reservoir withdrawals. One example of this is seen in a 7,900-ft Pennsylvanian-age sandstone in the Texas panhandle. A waterflood was initiated at a time when primary production had reduced the average reservoir pressure to subnormal levels. Since that time, the average sand pressure has increased to the point where mud densities in excess of 12 lbm/gal are required for infill-drilling projects.

Another particularly dangerous manmade event occurs where shallow formations inadvertently or unknowingly are charged with deeper gas. **Fig. 2.15** shows three of these situations. Fig. 2.15a shows an underground blowout where a

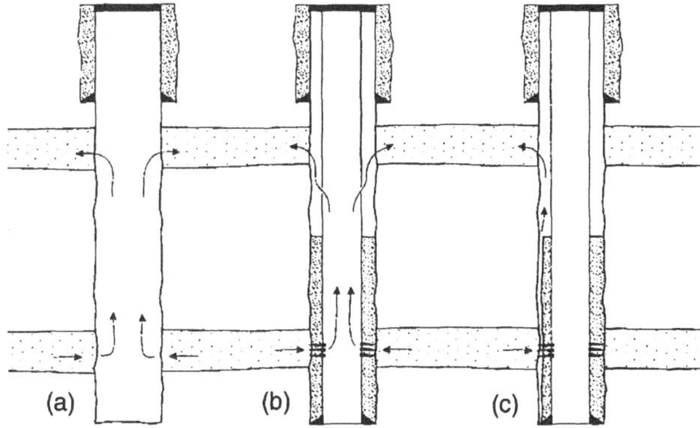

Fig. 2.15—Three examples of shallow formations being charged with deeper gas.

flow circuit has been established from the deeper interval to the shallow horizon. Figs. 2.15b and 2.15c portray flow from some deeper formation into a shallow zone through a casing leak and defective primary cement job. Many of these incidents are documented, and rigs have burned down while drilling in old fields. Typically, the pressure communication is unknown and the shallow overpressures are not detected until later infill or deeper exploration drilling takes place.

2.3 Overburden and Effective Stress Concepts

Normally compacted or undercompacted rock at any burial depth must support the weight of overlying rocks and pore fluids. The overburden load expressed in terms of gradient is the geostatic or overburden-stress gradient, g_{ob}. **Fig. 2.16** shows a sedimentary rock element containing mineral grains and pore fluid in the matrix interstices. The overburden stress imposed on the element, σ_{ob}, is opposed by vertical stress in the matrix framework, σ_{Ve}, and by pore pressure, giving

$$\sigma_{ob} = \sigma_{Ve} + p_p. \quad \quad (2.2)$$

Eq. 2.2 demonstrates that the maximum theoretical pore pressure is equivalent to the overburden stress.

Terzaghi[17] proved Eq. 2.2 in his classical soil mechanics work, where the relationship was expressed as

$$\sigma_e = \sigma - p_p. \quad \quad (2.3)$$

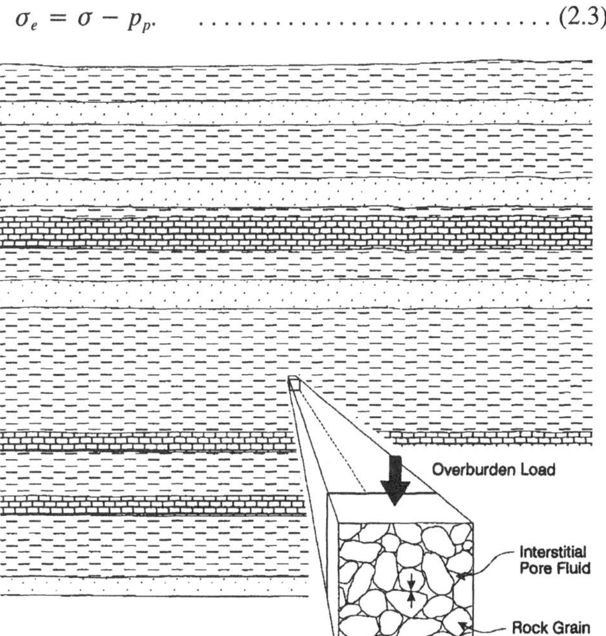

Fig. 2.16—Overburden support in a sedimentary-rock element.

The matrix stress presented is designated as σ_e and refers to the effective stress reaction in the direction of the applied compressive stress, σ. The latter term is the overburden stress if applied to subsurface rock in the vertical direction (if the overburden is the maximum principal stress). Terzaghi demonstrated that effective stress rather than total principal stress controls matrix behavior with regard to the effect on rock properties and strain deformation.

Terzaghi worked in unconsolidated soils, not rock. Eq. 2.3 is modified by the Biot[18] relationship to account for the observed effect of grain cementation on the ability of rock pore pressure to counteract the overburden or other loading. Eq. 2.4 is a more accurate expression for effective vertical stress in consolidated rock.

$$\sigma_e = \sigma - sp_p, \quad \text{(2.4)}$$

where s = the poroelasticity constant. The poroelasticity constant is a rock property that theoretically can vary between zero and one but is commonly taken to be one. It has been shown to be near unity for shales. Because pore-pressure predictions generally use shale measurements, Eq. 2.3 is the working expression for effective stress.

Chap. 4 discusses the use of effective stress to characterize fracture gradients, however, effective stress is important in several other respects. The overburden at any depth is unchanging within our portion of the geologic time scale, so matrix stress remains constant unless something happens to decrease or increase the pore pressure. Eq. 2.3 states that reducing pore pressure by production leads to an increased matrix stress. This is not a problem where rock grains are well-cemented and the matrix has adequate compressive strength. However, loosely consolidated or weak formations are subject to compressive failure and consequent problems such as permeability reduction, perforation collapse, sand production, and surface subsidence.[19]

Eq. 2.5 relates the bulk density, ρ_b, of a rock to the constituent grain and fluid densities.

$$\rho_b = \rho_{ma}(1 - \phi) + \rho_f\phi, \quad \text{(2.5)}$$

where ρ_{ma} = grain or matrix density, ρ_f = pore-fluid density, and ϕ = porosity. The composite overburden-stress gradient at any depth is obtained by integrating the relation,

$$g_{ob} = \frac{1}{D}\frac{g}{g_c}\int_0^D \rho_b dD, \quad \text{(2.6)}$$

where g = the acceleration of gravity and g_c = the proportionality constant necessary to preserve dimensional consistency. Eq. 2.6 reduces to Eq. 2.7.

$$g_{ob}(\text{psi/ft}) = \frac{1}{19.25D}\int_0^D \rho_b(\text{lbm/gal})dD(\text{ft}). \quad \text{(2.7)}$$

Eq. 2.8 is obtained if the bulk density of the rock is constant to the depth of interest.

$$g_{ob}(\text{psi/ft}) = \rho_b(\text{lbm/gal})/19.25. \quad \text{(2.8)}$$

Example 2.3. Determine the overburden-stress gradient of a sandstone having 20% porosity and fresh water in the pore space. Repeat the calculation for a shale with 30% porosity and a 1.07-specific-gravity pore fluid. Assume the sand and shale matrix specific gravities are 2.65 and 2.60, respectively.

Solution. Combining Eqs. 2.5 and 2.8 for the sandstone obtains

$$g_{ob} = [(2.65)(8.33)(0.80) + (8.33)(0.20)]/19.25$$
$$= 1.004 \text{ psi/ft}.$$

Combining Eqs. 2.5 and 2.8 for the shale obtains

$$g_{ob} = [(2.60)(8.33)(0.70) + (8.33)(1.07)(0.30)]/19.25$$
$$= 0.926 \text{ psi/ft}.$$

Example 2.3 shows that the overburden stress at any depth depends on the bulk rock constituents and porosity. A composite overburden gradient of 1 psi/ft often is assumed for sediments and works well in many older hard-rock areas. But it should be obvious intuitively that this assumption can be greatly in error. The best way to obtain the overburden-stress gradient in an area is to measure and integrate the bulk densities from a density log. Most major wireline companies provide this service.

Eaton[20] determined composite bulk densities from numerous density logs along the gulf coast and Santa Barbara channel and published the two curves shown as **Figs. 2.17 and 2.18,** respectively. The effect of compaction on the young Tertiary sediments is clear. Near the surface, the most recent sediments have low bulk densities deriving from their high porosity. Porosity reduction with depth is evidenced thereafter by the increasing density values.

In shales, water may exist in the pore space as free water or be held tightly between clay layers by electrostatic forces. This bound water constitutes part of the porosity; its complete removal from a montmorillonite clay lattice can be accomplished but only at extreme pressures or through diagenesis. Free water, on the other hand, is relatively mobile and can be expelled readily during compaction. This partially explains the asymptotic character of the Eaton density curves.

Eaton's curves were based on the composite or combined bulk densities of the different rock strata. However, not all rocks exhibit the same degree of compactibility (i.e., porosity reduction under compressive loading). For example, shales are more compactible than sandstones and young shales are more compactible than older shales. Limestones and dolomites typically have little or no intergranular porosity and are only slightly compactible. **Fig. 2.19** demonstrates the relative difference between representative shales and sandstones.

In the same work, Eaton averaged the density-log data over 1,000-ft increments and developed overburden-stress gradient correlations for the two areas. The curves in **Figs. 2.20 and 2.21** reflect total overburden-stress gradient rather than incremental values so that the desired overburden gradient at the depth of interest can be read directly from the selected chart.

Mitchell[22] approximated Eaton's overburden relationship for the gulf coast with the curve-fitting equation,

$$g_{ob} = 0.84753 + 0.01494\left(\frac{D}{1,000}\right) - 0.0006\left(\frac{D}{1,000}\right)^2 + 1.199 \times 10^{-5}\left(\frac{D}{1,000}\right)^3. \quad \text{(2.9a)}$$

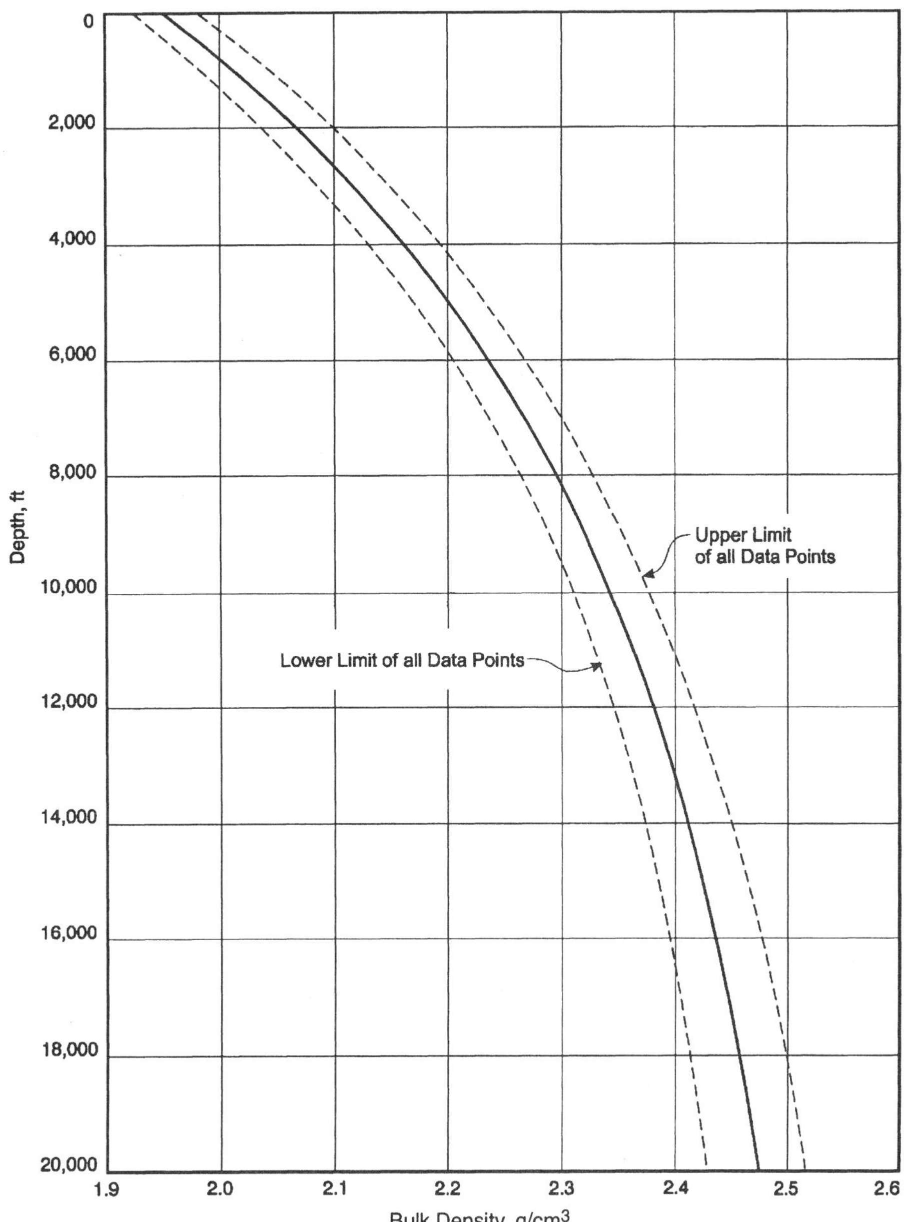

Fig. 2.17—Composite bulk-density curve for the U.S. gulf coast.[20]

The expression in SI metric units is given by

$$g_{ob} = 19.172 + 1.109\left(\frac{D}{1,000}\right) - 0.146\left(\frac{D}{1,000}\right)^2 + 0.009578\left(\frac{D}{1,000}\right)^3. \quad \text{(2.9b)}$$

These equations are more useful for programming applications than a chart.

Eaton's density curves and Fig. 2.19 suggest that an exponential or power-law mathematical model can be used to describe the relationship between shale porosity and depth. **Fig. 2.22** was prepared with data given by Magara[23] and portrays the porosities of assorted shales from various geographic areas as functions of depth on semilogarithmic graph paper. Some of the shales can be modeled with an exponential equation because the porosity points plot as an approximate straight line. This approximation is less suitable for others, such as the Oklahoma Permian shales, except over a relatively small spread of the data. Nevertheless, an exponential relationship between shale porosity and depth is a common assumption that leads to some useful relationships.

In 1959, Rubey and Hubbert[24] started with an earlier treatise[25] and developed Eq. 2.10 as an expression of sediment porosity with depth.

$$\phi = \phi_0 \exp\left(-K_\phi D\right), \quad \text{(2.10)}$$

where ϕ_0 = initial or surface porosity and K_ϕ = porosity decline constant. This is an equation of a straight line on semilogarithmic graph paper and the ϕ_0 and K_ϕ terms may be obtained from a plot of the data. Bourgoyne *et al.*[1] used Eaton's bulk-density data from Fig. 2.17 and obtained ϕ_0 and K_ϕ values of 0.41 and 8.5×10^{-5} ft^{-1}, respectively, for the gulf coast. Example 2.4 uses Eaton's data from the Santa Barbara channel.

Example 2.4. Use the bulk-density curve in Fig. 2.18 and estimate ϕ_0 and K_ϕ for the Santa Barbara channel. Assume an average grain specific gravity (SG) of 2.60.

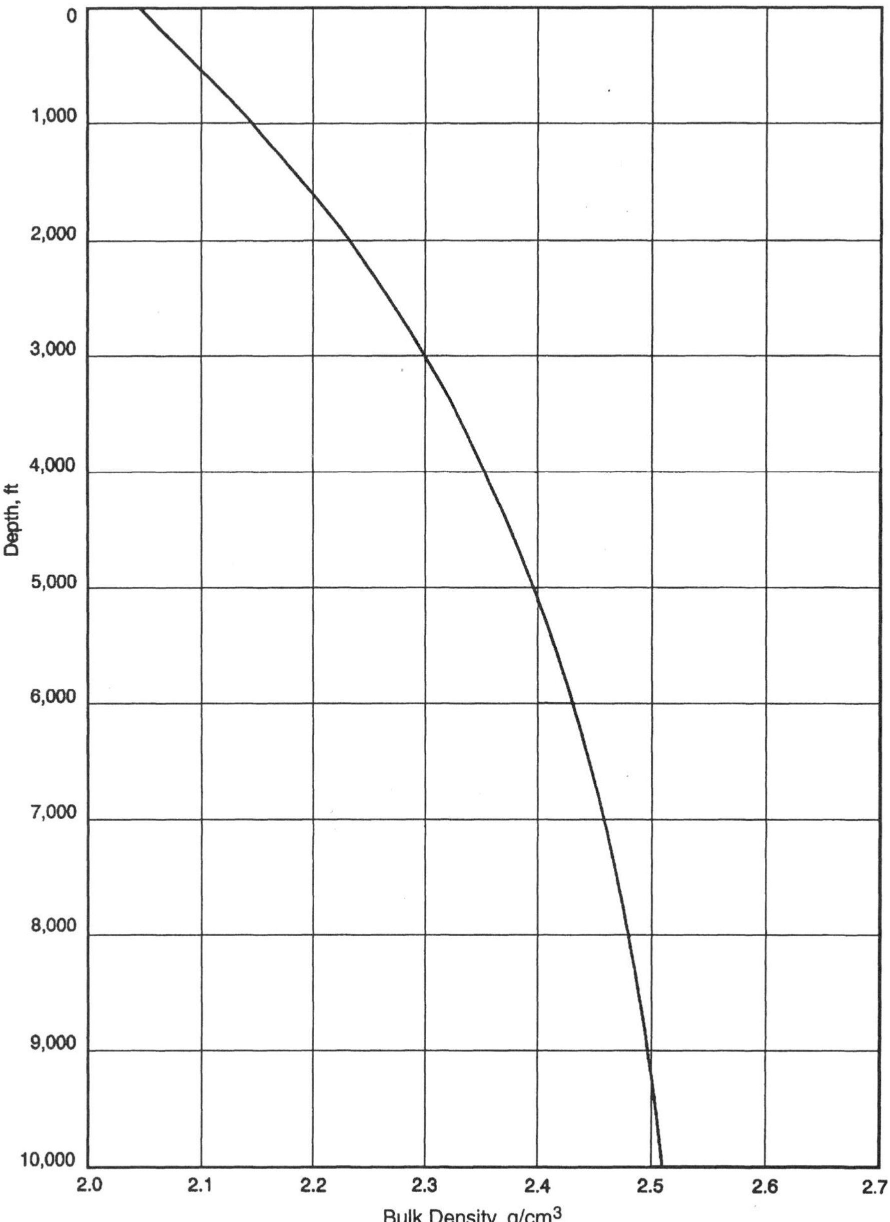

Fig. 2.18—Composite bulk-density curve for the Santa Barbara channel.[20]

Solution. Eq. 2.5 is rearranged to solve for porosity.

$$\phi = \frac{\rho_{ma} - \rho_b}{\rho_{ma} - \rho_f}. \quad \text{(2.11)}$$

The unit system is arbitrary but, for convenience, convert terms to densities in g/cm^3. Table 2.1 gives the normal-pressure gradient for the area as 0.452 psi/ft. The pore-fluid and matrix densities are, respectively,

$$\rho_f = (0.452/0.433)(1.0) = 1.044 \text{ cm}^3$$

and $\rho_{ma} = (2.60)(1.0) = 2.60 \text{ cm}^3$.

Fig. 2.18 shows Eaton's bulk densities in 500-ft increments. **Table 2.4** lists porosities computed with Eq. 2.11. The results are plotted in **Fig. 2.23.** Fitting a straight line to the data shows that the surface porosity, ϕ_0, is 0.37 from the line intercept at surface. The line slope K_ϕ is determined with the porosity at 10,000 ft.

$$K_\phi = \frac{\ln\frac{\phi_0}{\phi}}{D - 0} = \frac{\ln\frac{0.370}{0.074}}{10,000 - 0} = 1.61 \times 10^{-4}\text{ft}^{-1}.$$

Sediment porosity in the Santa Barbara channel declines much faster with depth than in the gulf coast.

Using Eqs. 2.5, 2.7, and 2.10, Constant and Bourgoyne[26,27] derived the relation for overburden stress as

$$\sigma_{ob} = 0.0519\left\{\rho_{ma}D - \frac{(\rho_{ma} - \rho_f)\phi_0}{K_\phi}\left[1 - \exp(-K_\phi D)\right]\right\}. \quad \text{(2.12)}$$

The constant 0.0519 expressed in SI metric units is 9.81×10^{-3}. Example 2.5 shows one application of Eq. 2.12.

Example 2.5. Calculate the overburden at 7,200 ft in the Santa Barbara channel. Compare this result to Eaton's prediction.

Solution. The surface porosity and porosity decline constant were found to be 0.37 and 1.61×10^{-4} ft^{-1} in Example 2.4. Substituting variables into Eq. 2.12 yields

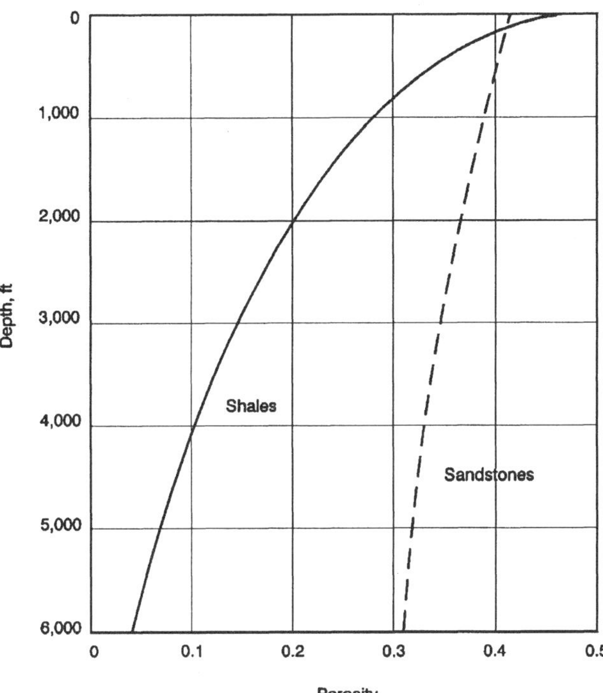

Fig. 2.19—The effect of compaction on shale and sandstone porosity.[21]

$$\sigma_{ob} = 0.0519\{(2.60)(8.33)(7,200)$$

$$-\frac{(2.60 - 1.044)(8.33)(0.37)}{0.000161}$$

$$\times\left[1 - \exp(-0.000161 \times 7,200)\right]\}$$

$$= 7,032\,\text{psig}.$$

Eaton's predicted overburden-stress gradient is obtained from Fig. 2.21 as 0.995, giving

$$\sigma_{ob} = (0.995)(7,200) = 7,164\,\text{psig}.$$

The difference between the two results lies in the straight-line fit of the calculated porosity data and to a lesser extent, the matrix-density assumption.

2.4 Conventional Pressure-Prediction Concepts

Other than direct readings, most pore-pressure-prediction or -detection techniques rely on measured or inferred shale porosity. The compaction theory is the basis for most predictions, and abnormal pressures arising from other sources generally are more difficult to identify or determine. Actually, the degree of shale compaction depends on factors other than burial depth and pore pressure. The soft shales, Pliocene and Miocene for example, compact more than the older, harder shales like the Pennsylvanian. As expected, pore pressures are easier to predict in the more compactible shales.

The approach common to most compaction methods is to measure porosity indicators in normally pressured shales and to establish a normal compaction trend with depth on a graph similar to the one shown in **Fig. 2.24.** For accuracy, it is important that measurements be taken in "clean" or "pure" shales (i.e., those shales with a minimum of other rock constituents). Application of these procedures becomes more difficult, if not impossible, in areas having a scarcity of clean shales.

After establishing a normal trend line, any measured data that deviate into higher porosity indicate a transition into abnormal pore pressure. The parameter trend depicted in Fig. 2.24 is a straight line on semilogarithmic graph paper. Logarithmic and Cartesian relationships proposed for certain correlations work well in some cases; however, most standard approaches are based on an exponential relationship.

Fig. 2.25 shows a shift in normal-pressure-trend lines, which usually indicates a change in the geologic age of the shale. The older shales have compacted and therefore exhibit a lower porosity for the applied geostatic load. Going from Tertiary into Cretaceous, for example, would be reflected by a new normal-compaction-trend line. The newly established trend line, which becomes the basis for the deeper predictions, may or may not be parallel to the previous line.

Fig. 2.26 demonstrates the equivalent-depth method for quantifying abnormal pore pressure. Every data point in the undercompacted region has a counterpart in the normally pressured section. For example, the abnormally pressured shale at Depth D is under the same state of compaction as its counterpart at D_{eq}. It is a reasonable conclusion that the shales at D and D_{eq} have the same matrix stress.

$$\sigma_{Ve} = \sigma_{Ve(eq)},$$

where $\sigma_{Ve(eq)}$ = effective (matrix) vertical stress at the equivalent depth. From Terzaghi's relationship, we obtain

$$\sigma_{ob} - p_p = \sigma_{ob(eq)} - p_{n(eq)},$$

where $\sigma_{ob(eq)}$ and $p_{n(eq)}$ = overburden stress and pore pressure at the equivalent depth, respectively. Rearranging terms yields

$$p_p = p_{n(eq)} + \left[\sigma_{ob} - \sigma_{ob(eq)}\right]. \quad \text{.............. (2.13)}$$

Example 2.6 demonstrates the the equivalent-depth method for quantifying abnormal pore pressure.

Example 2.6. Shale porosity indicator data are obtained and plotted on a graph similar to Fig. 2.24. Estimate the pore pressure at 10,200 ft if the vertical extrapolation from this depth intersects the normal trend line at 9,100 ft. The normal-pore-pressure gradient for the area is 0.433 psi/ft. Assume that the overburden gradient is a constant 1.000 psi/ft.

Solution. At 9,100 ft, the overburden stress is 9,100 psig and the pore pressure is

$$p_{n(eq)} = (0.433)(9,100) = 3,940\,\text{psig}.$$

The overburden at 10,200 ft is 10,200 psig. Substitution in Eq. 2.13 yields

$$p_p = 3,940 + (10,200 - 9,100) = 5,040\,\text{psig}.$$

In terms of gradient,

$$g_p = 5,040/10,200 = 0.494\,\text{psi/ft}.$$

Fig. 2.27 illustrates another standard approach that uses an empirical correlation that relates pore pressure to some function involving the observed parameter in the abnormally pressured interval and the value taken from the normal-trend-line extrapolation. The observed and normal porosity indicator

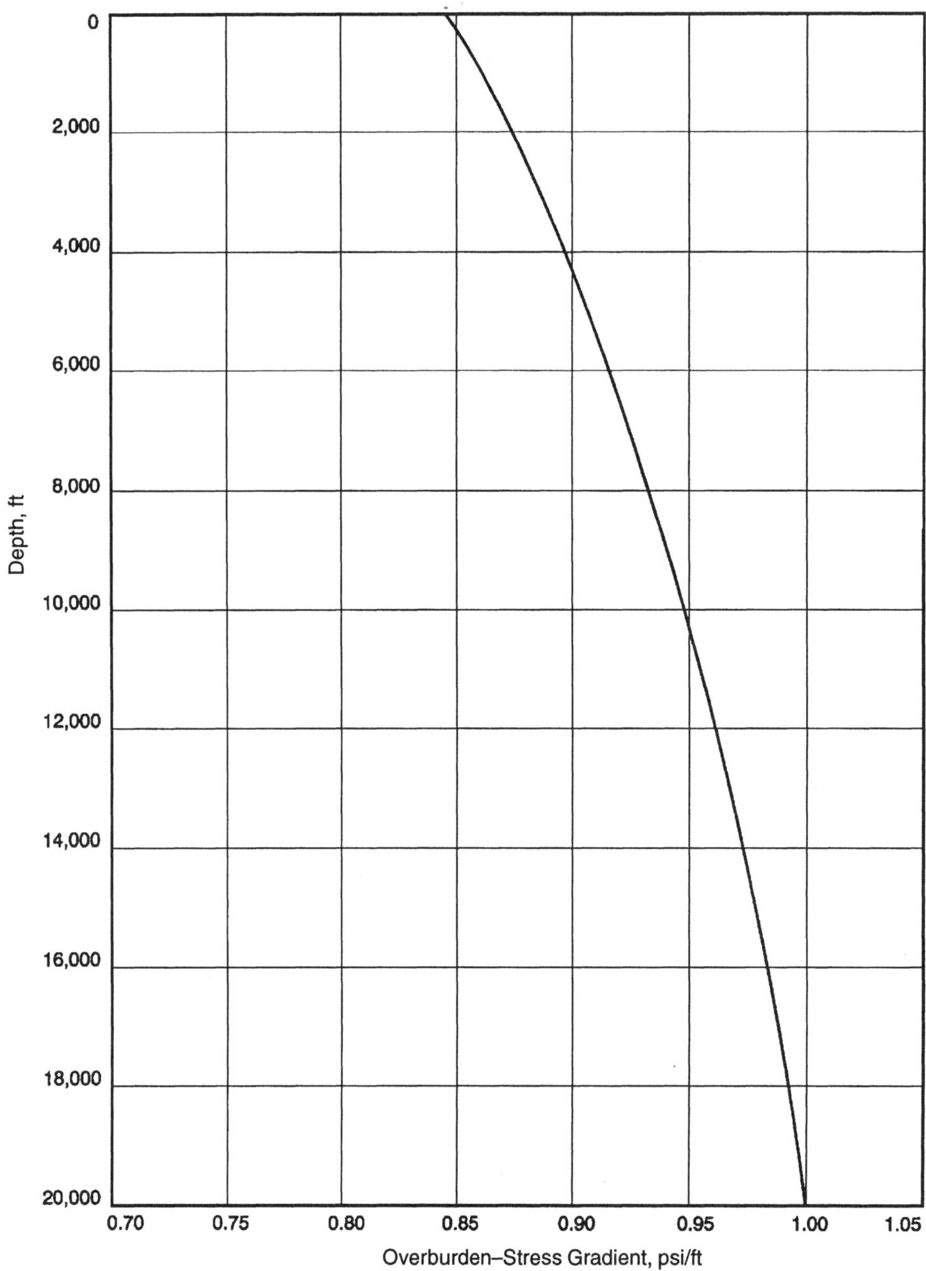

Fig. 2.20—Overburden-stress gradients for the U.S. gulf coast.[20]

values, X_o and X_n, respectively, are obtained from the plot at the depth of interest and are used in an empirical equation to predict the pore pressure. The limitation to any empirical method is that the correlation is developed for a specific area and lacks universal application. Operators who attempt to extend empirical relationships beyond their intended application create problems.

Pore-pressure predictions may be grouped into three broad classifications: (1) those relied on in planning a well, (2) those that can be applied while drilling, and (3) after-the-fact techniques. Methods falling into the first two categories are most beneficial to the design and operation of a drilling project. During the well-planning stage, an operator is limited to using information from offset wells, geological analogy, and seismic data. Several manifestations of abnormal pore pressure may be available after drilling operations begin. **Table 2.5** provides a partial list of quantitative and qualitative indicators. Many of these are discussed in detail, including their theoretical underpinnings, when they should be applied, and any inherent weaknesses in or limitations to their use.

2.5 Pressure Prediction by Analogy

Pore pressures and fracture gradients usually control well design and impact well costs. The number and setting depth of the casing strings, hole diameters, equipment pressure ratings, mud-density requirements, and other elements constituting a well plan rely on an accurate assessment of the expected pore pressure and fracture gradients. Every effort should be made to gather all available information and to use sound engineering judgment in applying the relevant information to the well plan. Anything less can lead to major difficulties or, in the extreme situation, a blowout or junked hole.

Table 2.6 lists a few of the numerous information sources available for planning and drilling oil and gas wells. Direct pressure measurements are superior to a correlation that in-

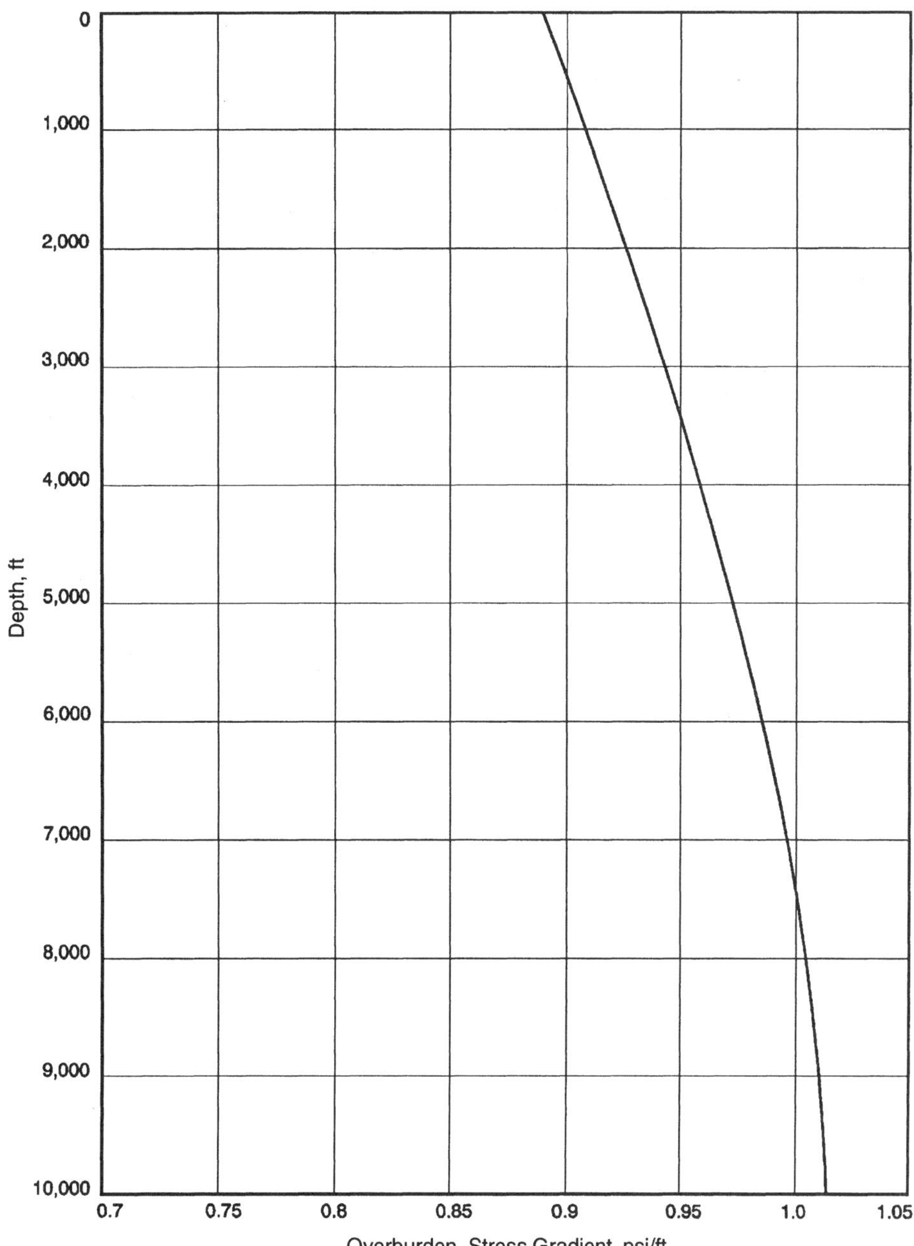

Fig. 2.21—Overburden-stress gradients for the Santa Barbara channel.[20]

directly infers pore pressure. Drillstem tests, shut-in producing well pressures, and recorded pressures during a well-control operation help to establish the known pore pressures in a given prospect provided that the information is timely and the lithology correlates to the proposed drilling location. Known data points can be used to fine tune other, more indirect methods.

Offset mud densities are available from a variety of sources, but the information should be used cautiously. Mud weights depend on several factors other than hydrostatic-pressure balance, including hole stability and operating practices. The information can be valuable, however, in helping to establish the range of allowable mud densities across a given hole section. Accurate lithological correlation is essential for mud densities or any other offset data to be beneficial.

Without any direct offsets, an operator may need to rely on region or basinwide analogy to anticipate conditions in a new hole. Open lines of communication between the drilling department and the exploration or geophysical group is always important, but even more so when planning these types of wells. The conscientious drilling engineer learns as much as possible about the expected lithology, potential pressure seals, tectonic features that might impact rock stress, and other variables. However, there are likely to be unknown or at least questionable data, even in the most scientific of prospects. Prudence dictates a rank exploration well be planned so that one or more additional casing strings can be set if actual well conditions so dictate.

2.6 Abnormal-Pressure Prediction From Seismic Data

A valuable exploration tool for predicting pore pressures and other potential drilling problems is a seismic survey. Seismic surveys, as used in conventional geophysical prospecting, take known or computed velocities of sound through rock media to determine depths to subsurface reflector beds. Structural characteristics then may be delineated across the line of shot points. Structural information alone can be useful in anticipating potential pore-pressure anomalies. Salt domes,

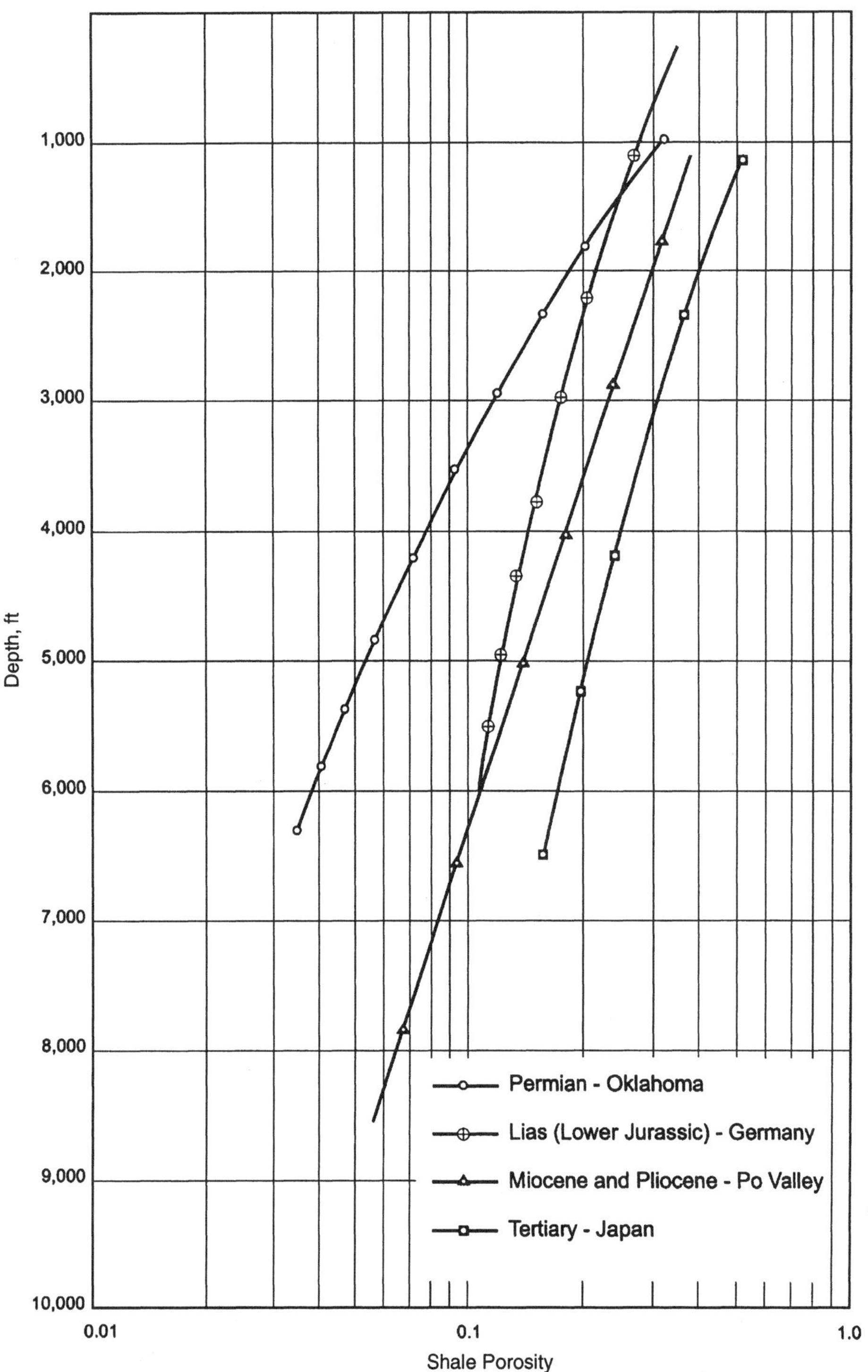

Fig. 2.22—Assorted shale porosities as functions of burial depth (after Magara[23]).

shale diapirs, faults, and other tectonic features often can be identified clearly by the geophysicist and can be factored into the well plan as depths where abrupt changes in pore pressure are possible.

The use of seismic as a method for detecting and quantifying abnormal pressure involves the relationship between computed sound velocity and the degree of sediment compaction. Pennebaker[28] first described the approach in 1968. The velocity of sound in a medium increases with the density of that medium. For instance, the sonic velocity is approximately 1,100 ft/sec in atmospheric air and approximately 4,600 ft/sec in distilled water. For sedimentary formations, the sonic velocity in a low-density rock like highly porous shale may be as low as 6,000 ft/sec whereas the velocity in a dense dolomite may exceed 20,000 ft/sec. Seismic velocity, therefore, can be considered to be an indirect measure of porosity and used to predict pore pressure.

The desired sonic velocity information is the velocity induced in a specific interval. Stacked or root-mean-squared (RMS) average velocities (see **Fig. 2.28**) can be converted to interval velocities with the Dix[29] equation if some assumptions are made concerning the lithological sequence.[30] Sediment densities and average interval velocities increase with burial depth under normal compaction. **Fig. 2.29** plots a normal-compaction trend for average interval velocity.

TABLE 2.4—AVERAGE SEDIMENT POROSITY FOR THE SANTA BARBARA CHANNEL

Depth (ft)	Bulk Density (g/cm^3)	Average Porosity (fraction)
0	2.050	0.374
500*	2.100	0.342
1,000	2.140	0.318
1,500	2.180	0.293
2,000	2.210	0.274
2,500	2.260	0.243
3,000	2.290	0.224
3,500	2.320	0.205
4,000	2.340	0.193
4,500	2.370	0.174
5,000	2.400	0.156
5,500	2.420	0.143
6,000	2.440	0.131
6,500	2.460	0.118
7,000	2.470	0.112
7,500	2.480	0.106
8,000	2.490	0.100
8,500	2.495	0.097
9,000	2.500	0.093
9,500	2.505	0.090
10,000	2.510	0.087

In practice, it is more convenient to use interval travel or transit time. Transit time is the reciprocal of velocity and usually is expressed in units of microseconds (μsec) per foot or meter (10^{-6} sec/ft or 10^{-6} s/m). The equation relating porosity to the transit time in rock media, Δt, is given by

$$\Delta t = \Delta t_{ma}(1 - \phi) + \Delta t_f \phi, \qquad (2.14)$$

where Δt_{ma} and Δt_f = transit times of the rock matrix and pore fluid, respectively.

The technique for predicting abnormal pore pressures involves first computing the average interval-transit times underlying the point on the seismic line closest to the drilling location and then plotting the data vs. depth. Pennebaker assumed a power-law relationship between interval-transit time and depth. Power-law functions plot as a straight line on logarithmic paper and all the Pennebaker trend lines were presented in this format. Actually, a derivation of the expression relating transit time to depth does not represent a straight line on logarithmic or any other type of graph paper.[1] However, the common procedure in use today assumes that compaction causes normal interval-transit times to decrease exponentially with depth. Hence, the data are plotted on a semilogarithmic plot.

Given quality data, a normal trend line (or lines if significant geological age shifts underly the prospect) should be apparent from the semilog plot. Deviation from the trend to higher transit times indicates transition into undercompacted rock. The pore pressure within any interval can be estimated with the equivalent-depth method or an empirical correlation specific to the area. Pennebaker's correlation, shown in **Fig. 2.30,** was developed for gulf coast sediments and relates pore-pressure gradient to the ratio of the observed interval-transit time to the normal trend extrapolation.

Example 2.7. Table 2.7 shows average interval-transit times for a Miocene prospect in south Louisiana.

1. Determine the top of the transition zone. The interval between 9,000 and 11,000 ft is known to be a highly calcerous sequence and should be disregarded in fitting the normal trend line.

2. Estimate the pore pressure at 19,000 ft using the equivalent-depth method.

3. Use Pennebaker's empirical correlation to predict pore pressure at this depth.

Solution.

1. **Fig. 2.31** shows interval-time data plotted on semilog paper. A normal trend line is constructed by ignoring the calcerous sediment data and concentrating on the interval between 6,000 and 9,000 ft. From the plot, the abnormal pressure transition appears to occur somewhere in the vicinity of 11,000 ft.

2. The overburden-stress gradient at 19,000 ft for this gulf coast prospect is obtained as 0.997 from Fig. 2.20. The overburden is thus

$$\sigma_{ob} = (0.997)(19,000) = 18,943 \text{ psig}.$$

A vertical line drawn from the interval travel time intersects the normal trend line at 2,000 ft. The overburden and pore pressures at this equivalent depth are, respectively,

$$\sigma_{ob(eq)} = (0.875)(2,000) = 1,750 \text{ psig}.$$

and $p_{n(eq)} = (0.465)(2,000) = 930$ psig.

Eq. 2.13 yields

$$p_p = 930 + (18,943 - 1,750) = 18,123 \text{ psig},$$

which gives a pore-pressure gradient of

$$g_p = \frac{18,123}{19,000} = 0.954 \text{ psi/ft}.$$

3. For the second approach, we find from the normal trend line extrapolation that Δt_n is about 65 μsec/ft. The ratio of normal to observed transit time at 19,000 ft is

$$\Delta t_o/\Delta t_n = 95/65 = 1.46.$$

The pore-pressure gradient from Fig. 2.30 is approximately 0.95 psi/ft, which leads to the estimated pore pressure,

$$p_p = (0.95)(19,000) = 18,050 \text{ psig}.$$

Note that the close agreement between the two methods was pure happenstance. A comparison between empirical results and values obtained by the equivalent-depth method generally show more disparity. Example 2.19 demonstrates and discusses why this frequently is the case.

One of the weaknesses of seismic predictions is that the interval lithology must be known to some extent. Thick sequences containing a high ratio of shale to sand are most suitable for establishing a normal-compaction trend and for applying the technique in general. The lithological sequence in hard-rock country typically contains more carbonates and the thick shale beds, when present, are much harder than the clays and shales in the coastal areas. Seismic information in these areas may not permit an operator to find a transition, much less quantify pore pressure.

Another point that Example 2.7 makes is that judgment must be exercised in selecting the normal-compaction trend. Seismic results are highly sensitive to the interpretation of the chart reader, and the data interpretation must be valid to obtain meaningful results. Also, a mathematical model for tran-

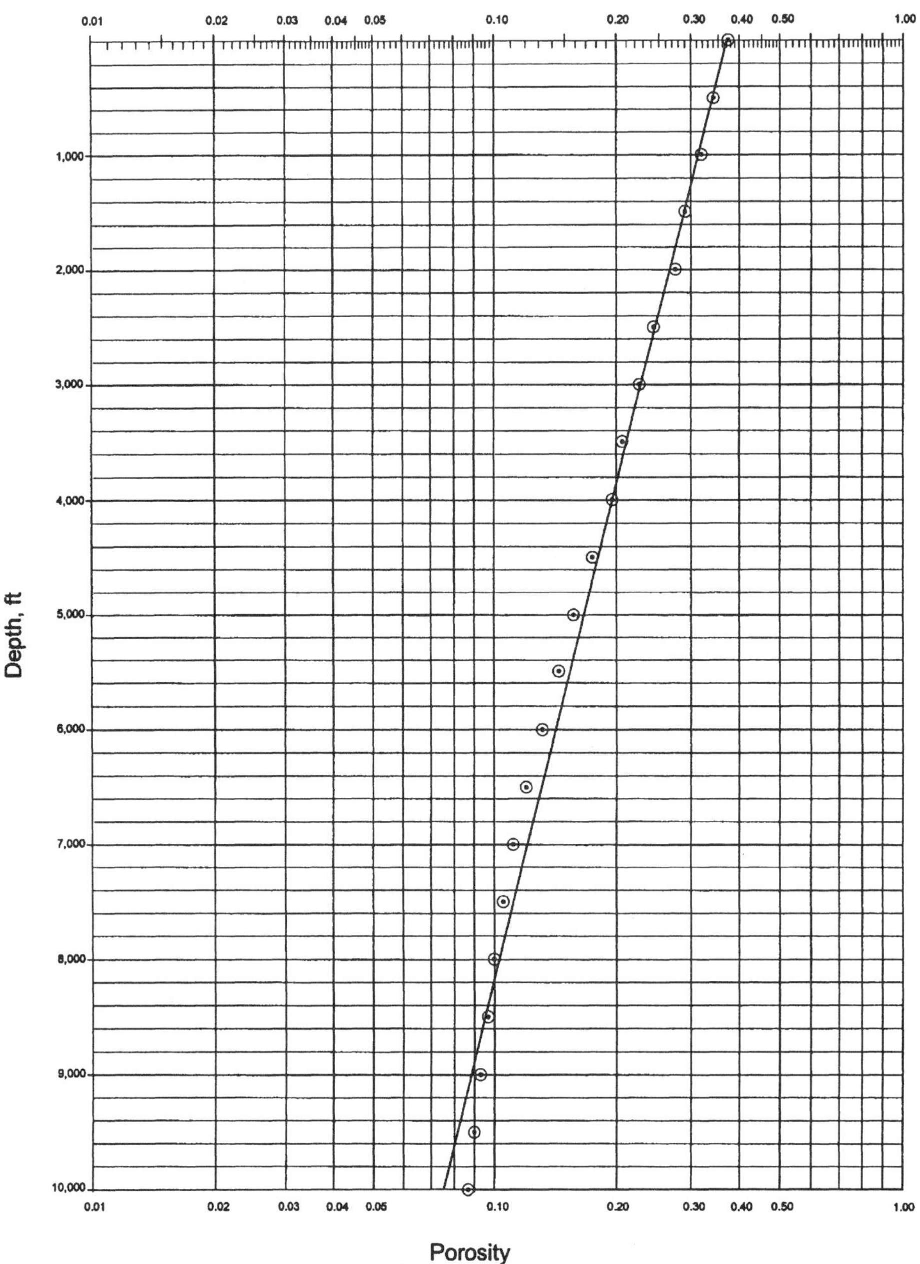

Fig. 2.23—Calculated sediment-porosity data for the Santa Barbara channel.

sit time does not support a semilog-straight-line extrapolation except across relatively short intervals. This technique is more successful when greater emphasis is placed on the data points closest to the transition.

Structural complexity adds to the difficulty in obtaining useful pore pressures from seismic. The basic assumption in the Dix conversion from RMS velocity to average interval velocity is that beds are flat and of uniform velocity throughout the lateral investigative distance. Compressive rock stress induced by tectonic events produces anomalously low travel times for the burial depth and introduces complications to the procedure. Lateral facies changes and high bed dips also violate the Dix assumption.

Seismic pore-pressure-detection techniques are most applicable in younger deltaic areas where normal faults predominate and where thick, horizontal shaly layers are prevalent. However, recent developments in both acquisition and processing technology have extended seismic capabilities beyond their traditional application.[31] Depth-migration-before-stack (DMBS) processing can compute interval velocities directly from the seismic data as opposed to using the Dix conversion and can deal with lateral velocity variations better. Other advances, such as 3D DMBS processing, will provide further capabilities in more geologically complex areas.

Given all of the current limitations to seismic predictions, DMBS may be the only tool available to the drilling engineer for an exploration prospect. In most cases, accurately predicting the magnitude of the pore pressure is less important than establishing the likelihood and probable depth of undercompaction. More precise measurements normally become available in the process of drilling the well, and operations can proceed safely when these potential transitions are incorporated into the well plan.

2.7 Penetration Rate

All other factors being equal, bits drill faster through overpressured rock than through normally or subnormally pressured rock. The most applied pore-pressure-prediction meth-

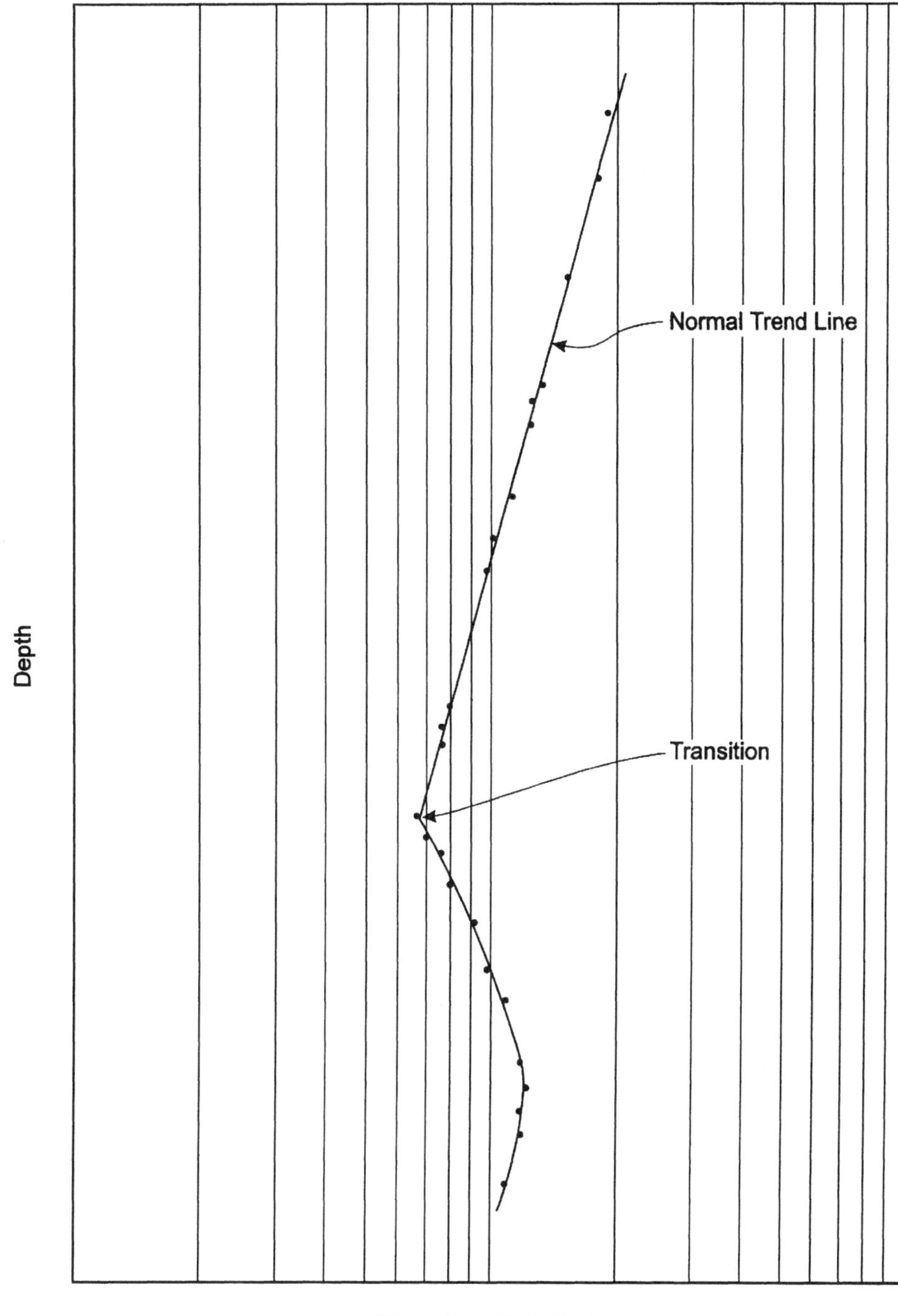

Fig. 2.24—Normal-compaction trend and indication of transition into abnormal pore pressures.

ods involve correlations related to penetration rate. One advantage to using drilling rate is that any changes are immediately apparent to the driller, while most other methods involve lag time. For example, shale-cuttings density measurements can be a useful tool if done correctly, but there is considerable delay because of the time it takes to circulate the cuttings to surface, physically prepare the sample, and measure the density.

Table 2.8 lists factors governing how fast a bit will drill through rock. Some of the factors are at least partially controlled by the operator, while others are strictly a function of the rock and burial depth. Important factors from the standpoint of pressure prediction are the differential pressure across the bit face, the state of rock compaction, and rock strength. Effectively ruling out or normalizing other variables as contributors to drill-rate variations is an important aspect of correlations based on penetration rate.

Another need that must be met is sufficient bottomhole cleaning at the selected weight-on-bit (WOB) and rotating speed. The bit must be continually in contact with fresh or undrilled formation to realize accurate predictions. Otherwise, changing formation conditions can be masked if much of the energy expended by the bit is involved in regrinding old cuttings.

A simple field technique demonstrates whether current hydraulics are providing adequate bottomhole cleaning. At constant rotating speed and lithology, the drilling rate can be considered to be directly proportional to the bit weight over

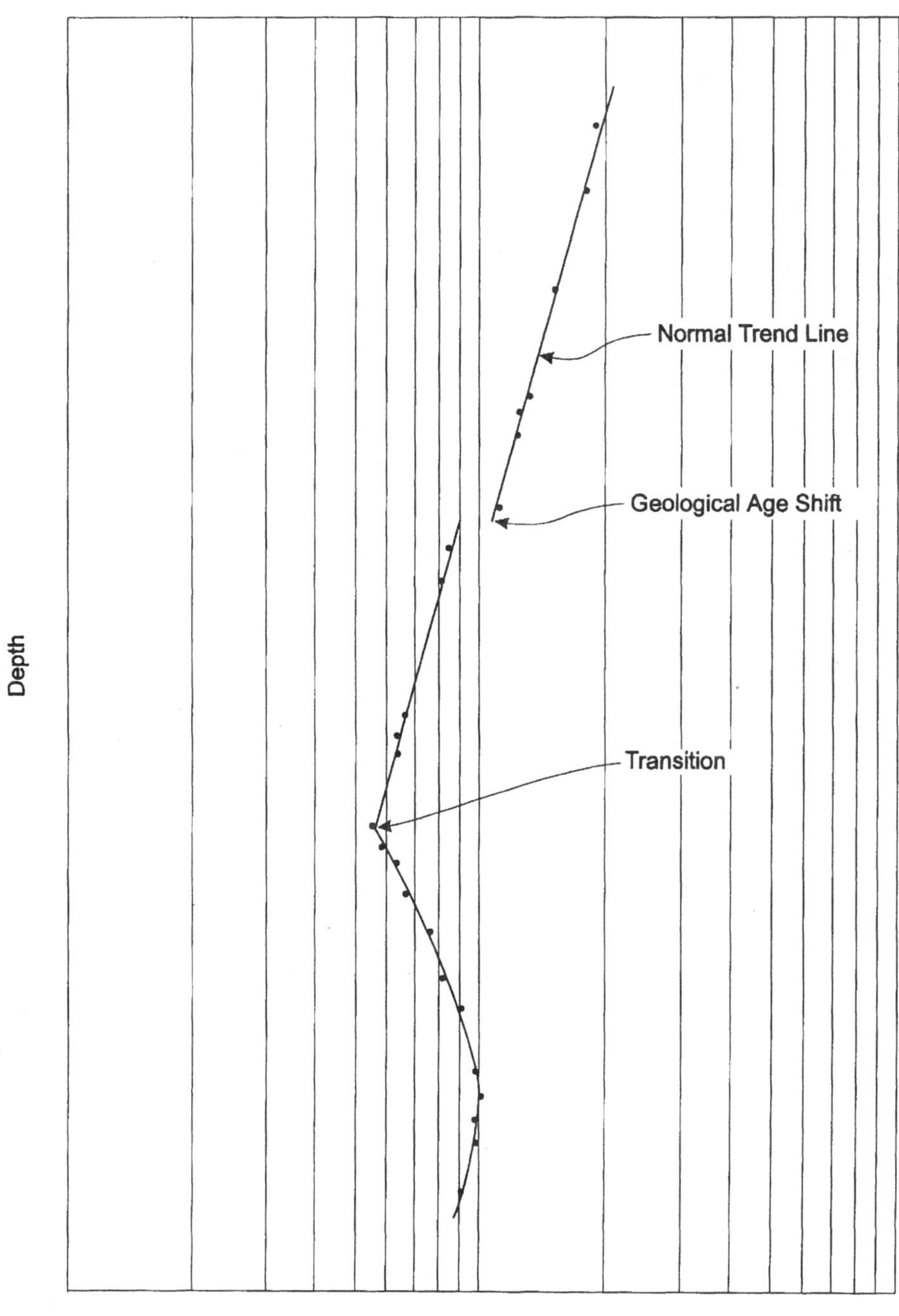

Fig. 2.25—Shift in a normal-pressure-trend line because of a change in geologic age.

a specific range of conditions. Starting with low values and at the desired rotating speed, the bit weight increases incrementally and the penetration rate is recorded for each step until the desired WOB is achieved. The data then are plotted similarly to the chart shown in **Fig. 2.32.** A flattening of the curve indicates one of the following conditions: more cuttings are being generated than can be swept away by the mud stream or the bit cutting structures are embedded fully in the rock. Operators should avoid bit flounder and maintain drilling parameters within the straight-line portion of the curve.

The obvious question at this point is how pore pressure affects penetration rate. There is no single simple reason and a combination of causes has been demonstrated. However, the mechanics break down into three or four major elements. One is related to the inverse relationship between rock drillability and compaction. Rock porosity decreases as drilling proceeds in a normal compaction trend and, as a result, penetration rate suffers. Undercompacted rock has higher porosity than normally pressured rock under the same overburden stress and therefore exhibits higher drillability. Another factor becomes apparent when the effective or matrix stress of an abnormally pressured rock is comparedwith a normally pressured rock at the same overburden. From Terzaghi's law, higher pore pressure results in lower matrix stress and, hence, reduced strength.

The differential (or overbalance) pressure between the wellbore and pore fluid is one of the more significant factors listed in Table 2.8. Rock drillability decreases with increasing overbalance for two primary reasons: the so-called chip hold-down phenomenon and the effect that wellbore pressure has on the rock strength immediately ahead of the bit. This subject

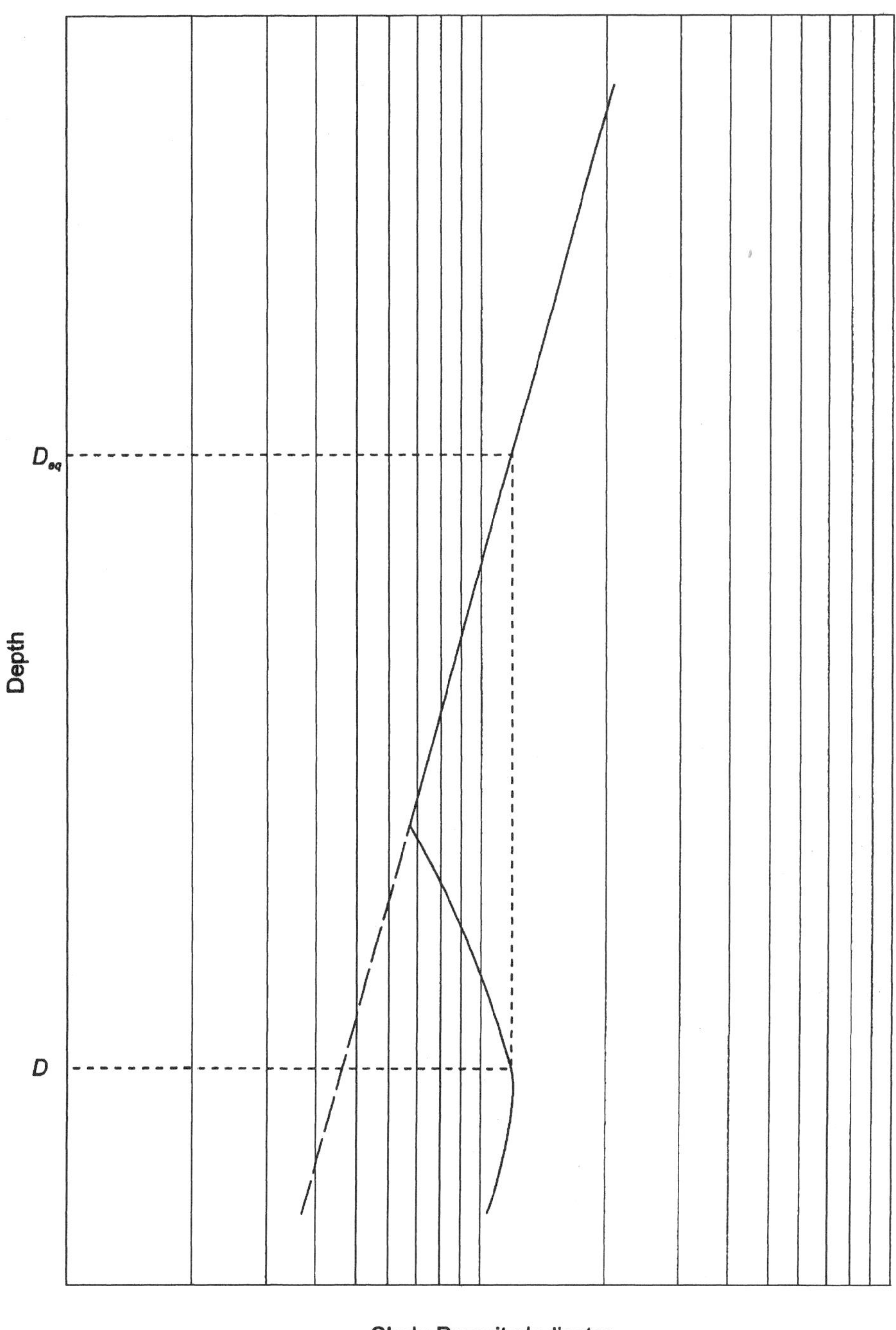

Fig. 2.26—Equivalent-depth method for quantifying abnormal pore pressure.

requires some explanation because of its relative importance in influencing penetration rates.

Investigators began studying how rocks drill under pressure in the laboratory in the mid-1950s. Murray and Cunningham[32] conducted one of the earliest series of microbit drilling experiments and found that drilling rates decreased with increasing hydrostatic pressure under constant-pore-pressure conditions. They came close to discovering the true role of wellbore pressure, but it was left to Eckel[33] to demonstrate experimentally that drilling rate was influenced more by differential pressure than by hydrostatic pressure in the borehole.

Three independent papers presented greatly advanced the understanding of how overbalance influences rock failure and removal mechanics.[34-36] Cunningham and Eenik[34] concluded that the differential pressure is the only pressure that influences penetration rate; **Fig. 2.33** shows that the effect becomes more pronounced with decreasing overbalance. In addition, they were among the first to offer chip holddown as a theory for reduced penetration rate. Garnier and van Lingen[35] and Robinson[36] discussed the secondary effect of how rock strength is enhanced by overbalance pressure.

Vidrine and Benit[37] corroborated earlier laboratory work with field observations. They made drilling-rate measurements in shale at variable differential pressures on eight south Louisiana wells and found that the penetration rates generally followed an exponential decline similar to that depicted in Fig. 2.33. **Fig. 2.34** shows one of their examples, normalized for tooth wear, bit weight, and rotary speed. Their data suggest that penetration rates continue to increase when the differential pressure becomes negative (i.e., pore pressure > wellbore pressure).

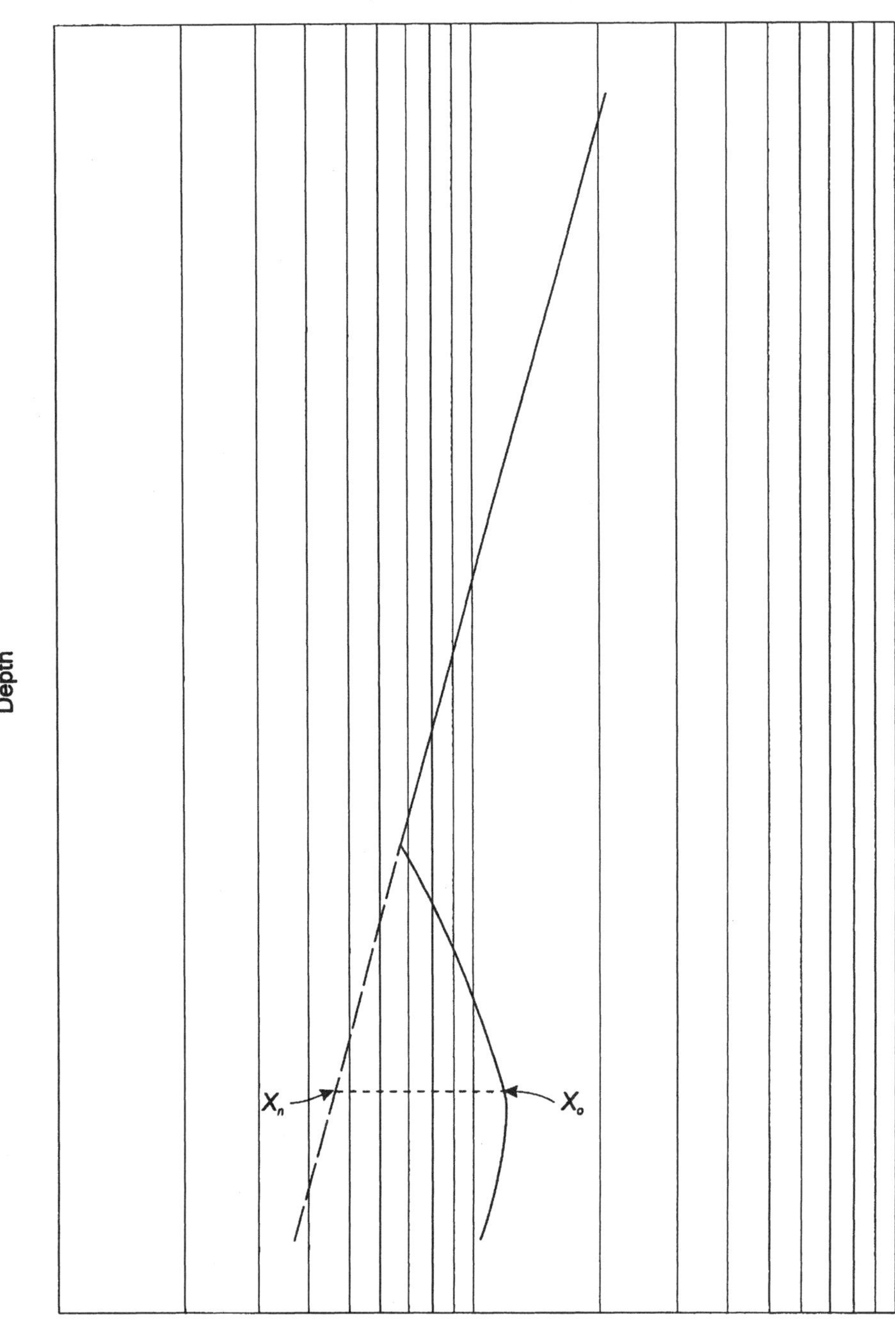

Fig. 2.27—Empirical approach for quantifying abnormal pore pressure.

Fig. 2.35 shows a rock chip created by the indentation of a roller-cone-bit tooth. Some means of equalizing the pressure below the chip to the wellbore pressure must be provided before the chip can be dislodged easily and removed by the mud stream. If it is not, the rock fragment is effectively "held down" by the pressure from above. It should be apparent that the degree of overbalance is a significant component of chip holddown. Other important variables include formation permeability, mud-filtration properties, and whether the bit removes the rock predominantly by shear or by crushing.

Recall that the rock below a drill bit actually gets stronger when the overbalance pressure is increased. A review of some rock mechanics fundamentals is in order. These concepts will be used in a discussion of fracture theory in Chap. 3. Take a piece of rock (or any other solid) and apply loads in an arbitrary fashion. Recall that we can resolve the loads on an element within or on the surface of the body into normal and shear stresses acting on its orthogonal planes. The element can be oriented so that the plane shear stresses vanish. The normal stresses acting on these two planes are defined as the principal stresses and represent maximum and minimum normal stress, σ_{max} and σ_{min}, respectively.

Fig. 2.36 graphically depicts with a Mohr's circle the normal and shear stresses on an arbitrary plane given by the angle a, σ_a and τ_a, respectively. The plane angle on the element corresponds to angle $2a$ on the circle. The maximum and minimum principal stresses on the circle have zero shear and are at relative positions 180° apart (90° apart on the element). The maximum shear stress, τ_{max}, acts on the plane that is positioned 45° from the principal stresses on the element (90° on the circle).

TABLE 2.5—PORE PRESSURE INDICATIONS WHILE DRILLING

Indicator	Correlations
Penetration rate	*d* exponent
	Modified *d* exponent
	Combs'[43] method
	Bourgoyne and Young's[46] method
	A exponent
	Sigmalog
	Prentice's[50] method
	Moore's[52, 53] method
	Other correlations
Cutting characteristics	Boatman's[81] density correlation
	Qualitative indicators
Hole conditions	Qualitative
Gas-cut mud	Qualitative
Change in mud properties	Qualitative
Flowline temperature	Qualitative
MWD/LWD	Openhole log correlations
	Pseudoporosity or strength correlations
Direct Measurements	

A brittle rock specimen typically fails in a laboratory uniaxial compression test by breaking along a shear plane similar to that illustrated in **Fig. 2.37**. The Mohr-Coulomb failure criterion is often used in rock mechanics to describe the behavior of a rock under compression and to predict its compressive strength at a given confining pressure. In practice, the compressive strength of a rock is determined under at least two confining stresses and a Mohr's circle is drawn at the point of failure for each condition. In Fig. 2.37, the specimen failed at the vertical stress σ_{max} when the circumferential confining pressure was σ_{min}. The smaller circle represents failure of an unconfined specimen.

Fig. 2.37 shows that the compressive strength of rock is highly dependent on the confining stress. As a corollary statement, the confining stress must be specified whenever the strength of a rock is given. **Table 2.9** lists tensile, shear, and compressive strength ranges for various rocks.[38] Note that the Mohr-Coulomb criterion for buried sediments is not applicable to total stress. Pore pressure cannot produce shear, nor can it deform rock. Hence, the Mohr-Coulomb behavior in rock penetrated by a bit is controlled by the effective or matrix stresses.

Fig. 2.38 depicts a buried rock element. The maximum in-situ principal stress is the overburden and confining stresses in the horizontal direction, $\sigma_{H max}$ and $\sigma_{H min}$, respectively, are provided by the surrounding rock. The confining stresses are related to the overburden and increase with burial depth. The compressive strength of rock also increases with depth because confining stress controls the strength.

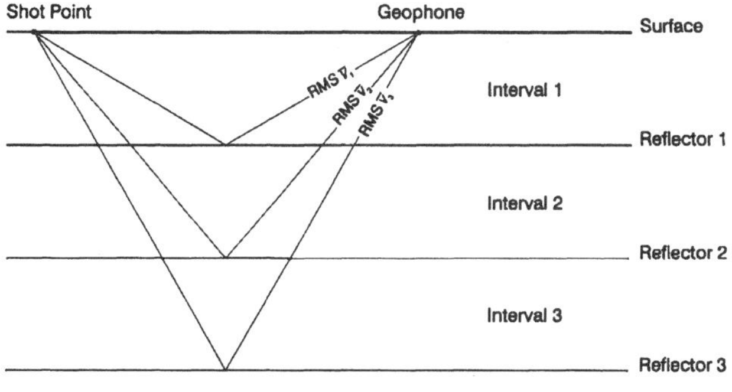

Fig. 2.28—RMS velocity in seismic surveys.

TABLE 2.6—PORE PRESSURE AND FRACTURE GRADIENT INFORMATION FROM OFFSET WELLS

Data Source	Information Provided
Mud logs	Instantaneous penetration rates
	Lithological sequence
	Measured shale densities
	Gas concentration in drilling fluid
	Well-control events
Openhole logs	Shale-compaction parameters
	Lithological sequence
	Wireline pressure tests
	Casing points
	Mud density at casing point
	Postcirculation temperatures
	Sonic log-derived rock properties
Mud recaps	Mud densities
	Well-control events
	Casing points
Bit records	Mud densities
	Casing points
Scout tickets	Drillstem-test pressures
	Casing points
	Initial test pressures
Tour sheets	Mud densities
	Casing points
	Well-control events
	Leakoff-test data
Public record sources	Annual test pressures
	Casing points
	Production and injection data
Technical papers and articles	Case histories
Service company database records and studies	Miscellaneous pore-pressure and rock-property data and correlations
Daily drilling reports	Mud densities
	Casing points
	Well-control events
	Leakoff-test data

Fig. 2.39 shows a rock element at the bottom of a wellbore. A column of drilling fluid replaces the overburden, and the minimum principal stress acting on an element near the bottom of the hole is now the wellbore pressure. On the basis of the Mohr-Coulomb model, the wellbore pressure can be considered as the confining pressure, which implies that the degree of overbalance controls the strength of the rock immediately ahead of the bit.

Maurer's[39] experimental study of rock failure by roller-cone bits provides additional insight into the rock-strengthen-

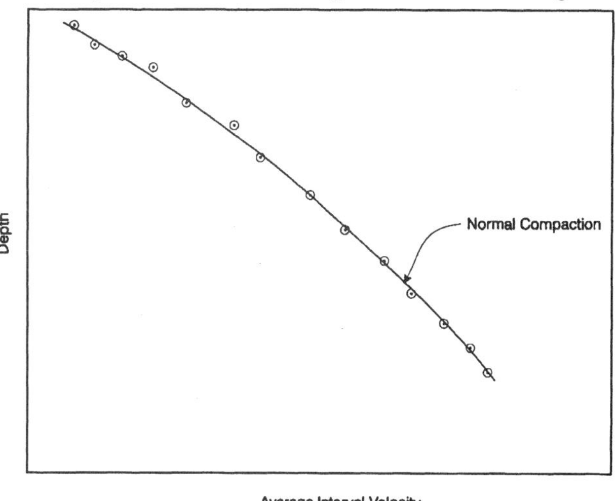

Fig. 2.29—Normal-compaction trend as evidenced by seismic interval velocities.

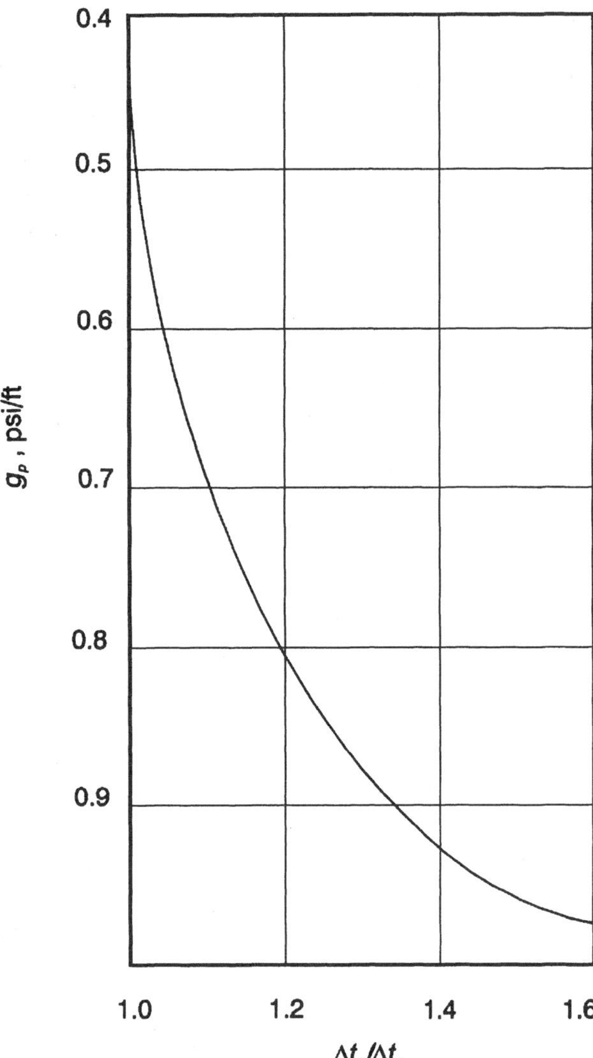

Fig. 2.30—Pennebaker's[28] gulf coast correlation between interval-transit time and pore-pressure gradient.

TABLE 2.7—AVERAGE INTERVAL-TRANSIT TIMES FOR A SOUTH LOUISIANA MIOCENE PROSPECT[28]

Interval (ft)	Midpoint (ft)	Average Transit Time (μsec/ft)
4,000 to 5,000	4,500	98
5,000 to 6,000	5,500	93
6,000 to 7,000	6,500	86
7,000 to 8,000	7,500	84
8,000 to 9,000	8,500	84
9,000 to 10,000	9,500	78
10,000 to 11,000	10,500	75
11,000 to 12,000	11,500	80
12,000 to 13,000	12,500	81
13,000 to 14,000	13,500	84
14,000 to 15,000	14,500	82
15,000 to 16,000	15,500	95
16,000 to 18,000	17,000	95
18,000 to 20,000	19,000	95
20,000 to 21,000	20,500	93
21,000 to 22,000	21,500	93

ing mechanism. **Fig. 2.40** shows a bit-tooth indentation into a rock just before the creation of a fracture along the dashed shear plane. The differential pressure from above provides a normal stress, σ_a, along the potentially failed shear plane. Fracture formation is resisted by the shear stress, τ_a, which is a function of the rock cohesion and friction between the top and bottom planes. The cohesion is a constant rock property, but the friction depends on the magnitude of σ_a, which, in turn, depends on the overbalance pressure. The same strengthening concepts apply to drag bits, such as the polycrystalline-diamond-compact (PDC) types.

Warren and Smith[40] drew some interesting conclusions from their analysis of stresses at the bottom of a wellbore. After wellbore pressure replaces the overburden, the rock immediately ahead of a bit undergoes an increase in pore volume because of the bulk rock compressibility. If the rock is a shale or otherwise relatively impermeable, the pore-fluid mass in the affected region is fixed leading to a localized area of reduced pore pressure. The effective stress in this region increases and results in a stronger rock. In effect, a differential pressure is induced in the rock that may be higher than the difference between the wellbore pressure and far-field pore pressure.

The finite-element method (FEM) was used to predict induced differential pressures at a distance 0.1 in. below the hole surface under various conditions. **Fig. 2.41** is their curve for the condition where the far-field pore pressure is equivalent to the wellbore pressure (i.e., a balanced situation). The induced differential pressure remains fairly constant at 1,400 psi from the center of the hole ($r/r_w = 0$) out to approximately half the distance toward the wellbore wall. Near the corners, deviator stresses also begin to influence the effective stress. The induced overbalance postulated by Warren and Smith may be a contributing factor to why weak shales often drill slower than strong, but permeable, sandstones.

In summary, the two fundamental pore-pressure-related factors that affect penetration rate are compaction and differential pressure. Of the two, differential pressure generally is conceded to be more significant. However, recent laboratory work suggests penetration rate is independent of differential pressure in low-permeability rock with low compactibility (hard shales).[41] The two factors are interrelated in soft shales, and distinguishing the predominant mechanism may not be important if the selected drill-rate correlation works for the area.

2.7.1 Drill-Rate Models. Given all the factors listed in Table 2.8, some means of relating penetration rate to a common reference must be used if penetration rate is to be a useful pressure-prediction tool. A normalized drilling rate based on a mathematical model provides the key. A penetration-rate model expresses the relationship between penetration rate, R, and the controlling variables in a general form by

$$R = K(f_1)(f_2)(f_3)\ \ldots\ (f_n), \qquad (2.15)$$

where f_i = functions of the variable parameters. K = a constant of proportionality and includes the effect of all the variables that are not accounted for in the selected model.

Numerous penetration-rate equations of varying complexity have been presented. One of the simpler models, expressed as Eq. 2.16, considers only the effect of bit weight and rotating speed.

$$R = K\left(\frac{W}{d_b}\right)^{a_W} N^{a_N}, \qquad (2.16)$$

where W = applied bit weight, d_b = bit diameter, a_W = bit weight exponent, N = bit rotating speed, and a_N = rotating speed exponent.

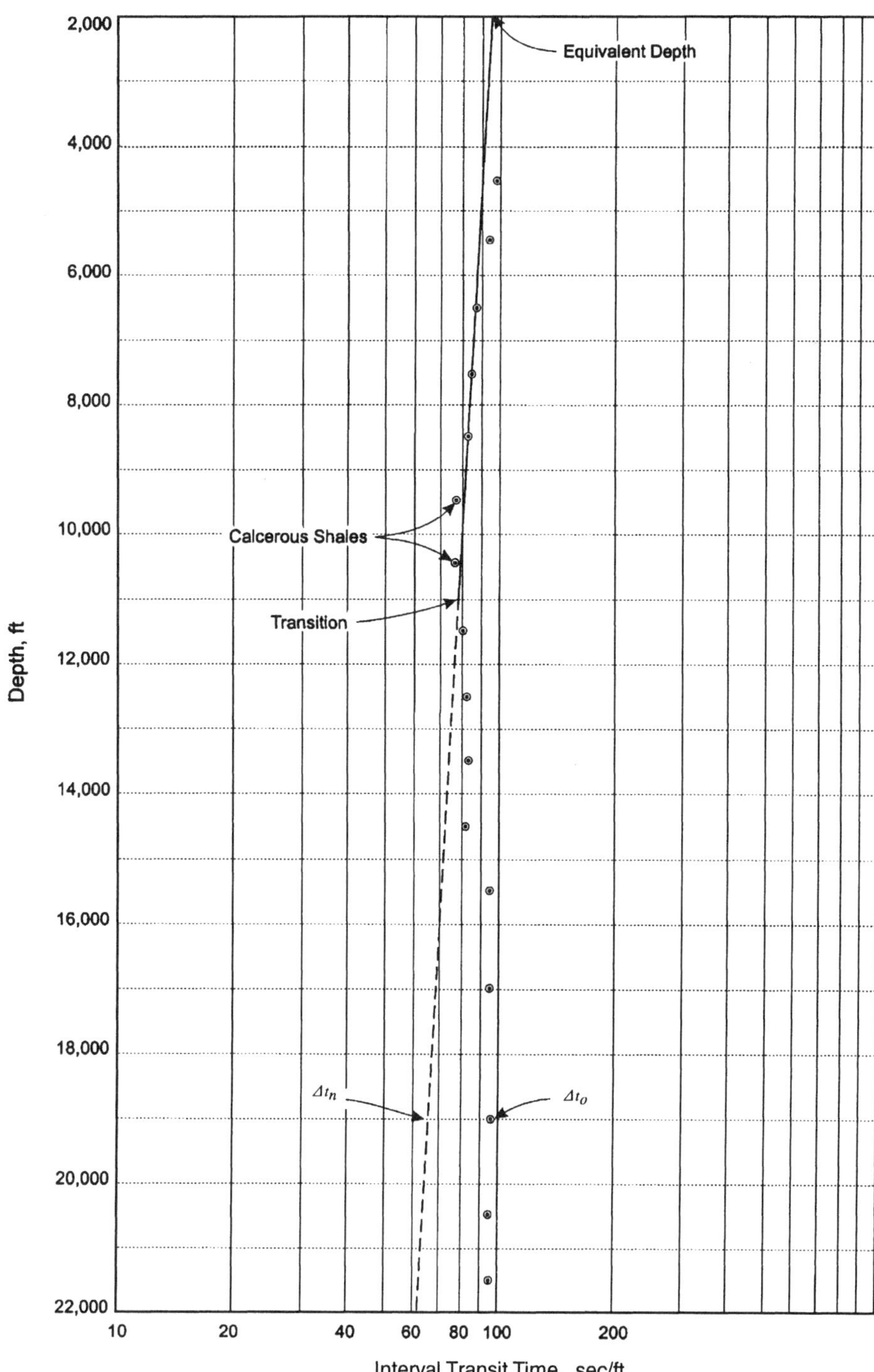

Fig. 2.31—Interval-transit times for the drilling prospect described in Example 2.7.

A graphical procedure can be used to evaluate the bit-weight and rotating-speed exponents in Eq. 2.16. For example, suppose that the objective is to determine the rotating-speed exponent. A penetration rate in consistent lithology, typically shale, is measured over a short interval. Additional measurements are made at other rotating speeds while holding the bit weight and other rate-related variables constant. These other factors are lumped into the proportionality constant and Eq. 2.16 is then expressed as

$$R = K'(N)^{a_N},$$

where K' = constant. An alternative form of the equation,

$$\log(R) = \log(K') + a_N \log(N),$$

gives a straight line on logarithmic graph paper. Hence, the value for a_N can be obtained by plotting R and N on log-log paper and measuring the slope of the line. In practice, obtaining reasonable values for the drilling-rate exponents is not as difficult as it might appear. In many cases, exponent values for shales of the same geologic age and at similar depths can be assumed equivalent.

Pore-pressure-prediction methods that rely on changes in normalized penetration rate have been introduced or modified from the early 1960s forward.[42-54] Most of the correlations were developed in a given rock hardness or geographic region and an engineer must be judicious when assessing their applicability. A standard relating to almost all drilling performance procedures is that measurements are made and trends

TABLE 2.8—FACTORS AFFECTING PENETRATION RATE	
Controlled by Operator	Out of Operator's Control
Hole diameter	Lithology
Bit type, design, and wear	Rock strength
Applied WOB	Confining stress
Rotating speed, rev/min	Formation permeability
Bit hydraulics	Porosity (degree of compaction)
Wellbore differential pressure	
Drilling-fluid properties	
Personnel and equipment	

noted in clean shales. Space does not allow discussion of each of these in detail, and we encourage those who want to pursue the subject to review the cited references. Ref. 1 describes Bourgoyne and Young's method and provides examples.

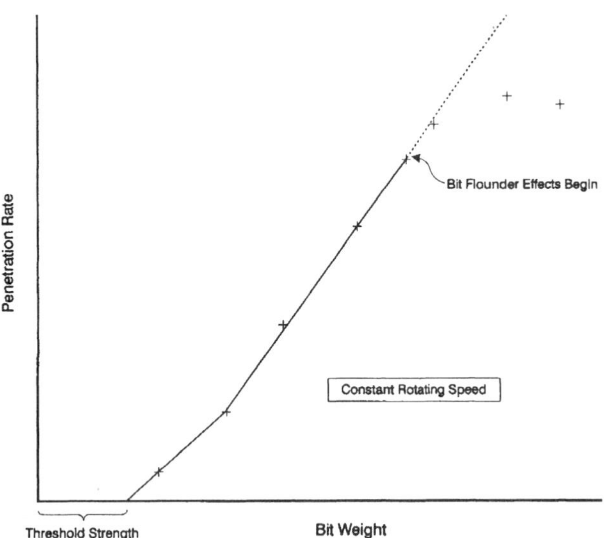

Fig. 2.32—Graphical field method for detecting the onset of bit flounder.

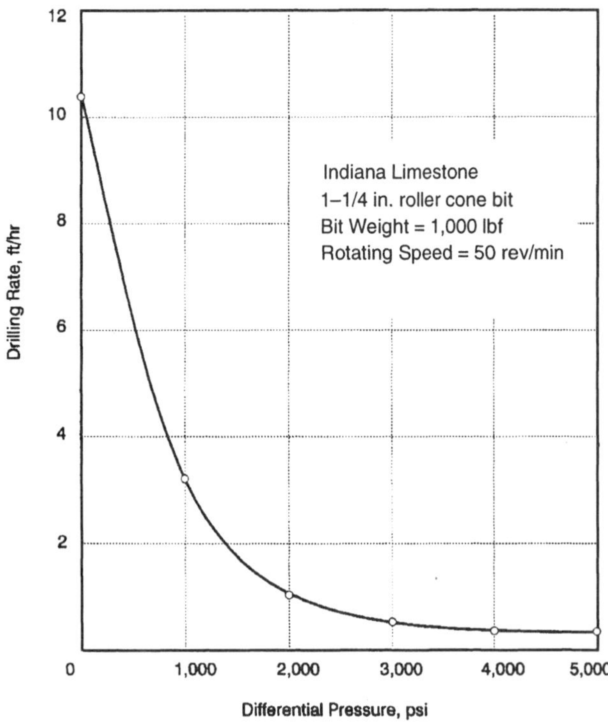

Fig. 2.33—Effect of differential pressure on penetration rate in Indiana limestone.[34]

2.7.2 *d* Exponent and Modified *d* Exponent.

The *d* exponent introduced by Jorden and Shirley[42] in the 1960s is the most widely used (and misused) drilling-rate prediction method. The technique was developed initially as an empirical relationship intended for application in the soft-rock areas of the gulf coast. Its use, however, has spread to all areas. The widespread application of *d* exponents is a mistake, however, particularly when the attempt is made to use the technique in hard-rock areas.

Jorden and Shirley started with the Bingham[55] model,

$$R = KN\left(\frac{W}{d_b}\right)^d, \quad \text{(2.17a)}$$

where d = bit weight exponent. The constant was described as being a function of the formation characteristics, although it should be apparent that other factors (e.g., bit type, bit wear, and hydraulics) are involved.

Bingham's model assumes that drilling rate is directly proportional to rotating speed, which may be a reasonable approximation in soft rock. However, the drilling-rate relationship in harder rock is expressed more accurately by Eq. 2.16 with an a_N value of less than one. Therefore, using the conventional *d* exponent in hard shales is inaccurate for

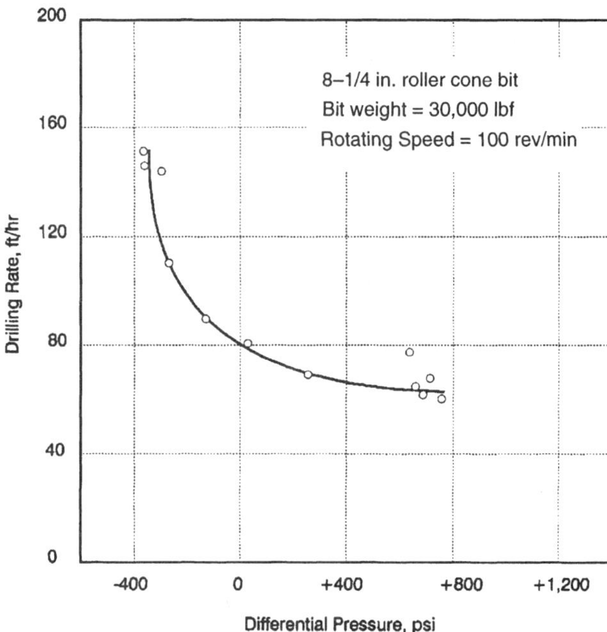

Fig. 2.34—The effect of differential pressure on normalized drilling rates in a south Louisiana shale.[37]

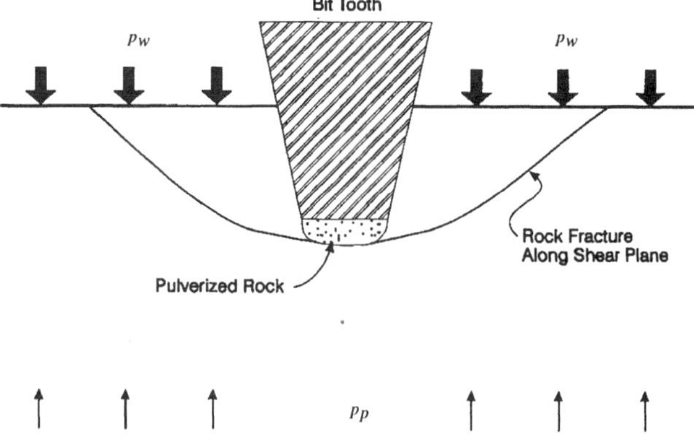

Fig. 2.35—Chip holddown resulting from a positive differential pressure.

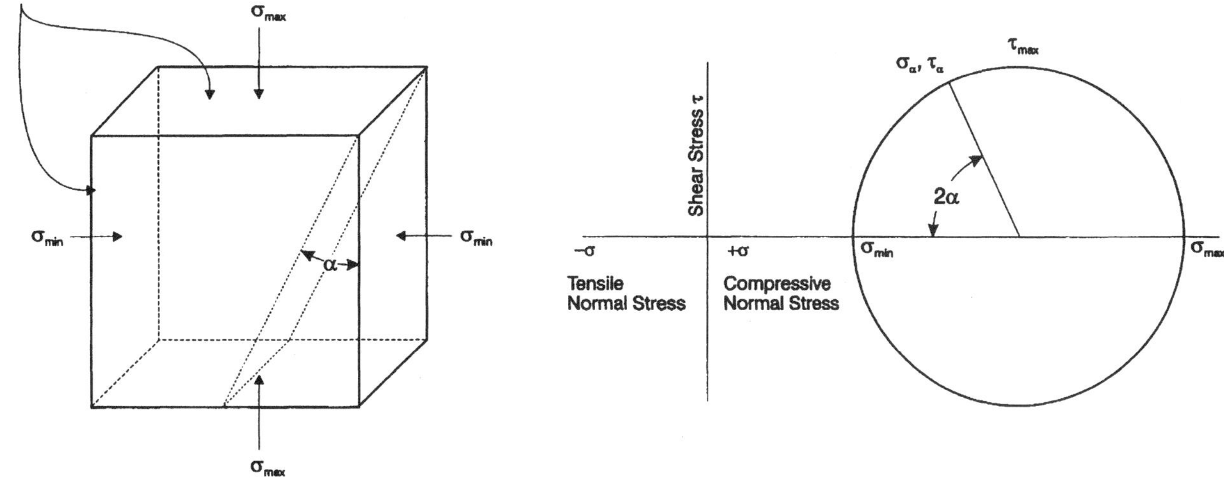

Fig. 2.36—Use of a Mohr's circle to define the stress states of an element.

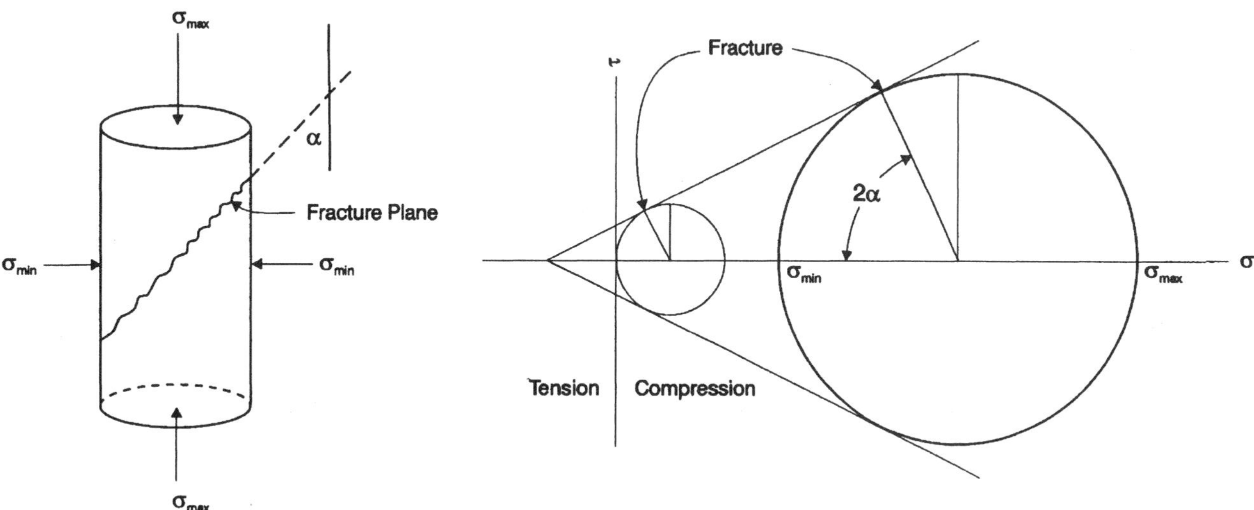

Fig. 2.37—Mohr-Coulomb failure envelope for rock in compression.

normalizing changes in the rotating speed unless modifications are made to the fundamental equation.

Other inaccuracies are inherent to the d exponent. Eq. 2.17a can be written as

$$\log\frac{R}{N} = \log K + d\log\frac{W}{d_b}. \quad (2.17b)$$

A plot of R/N vs. W/d_b represents a straight line on logarithmic paper where d=line slope and K=the intercept. Jorden and Shirley[42] set K equal to unity and justified doing so with the assumption that shale properties in the gulf coast study area were consistent and affected only by the degree of compaction.

After making this assumption and converting for customary units and magnitude, Eq. 2.17b is rearranged as

$$d = \frac{\log(R/60N)}{\log(12W/10^6 d_b)}. \quad (2.18)$$

The multiplying factor 10^6 is simply a constant inserted to ensure that the d exponent is not a negative number.

The d exponent normalizes R for the W and N variables and changes in response to a change in drilling rate. With normal compaction, R should decrease with depth, which results in an increasing d exponent trend. A decrease or flattening of the plotted trend implies transition into abnormal pore pressure if the mud density does not change across the interval of interest.

TABLE 2.9—STRENGTH PROPERTIES OF ROCK[38]

Rock Type	Unconfined Tensile Strength (psi)	Unconfined Compressive Strength (psi)	Shear Strength (psi)
Granite	600 to 3,600	14,000 to 40,000	700 to 7,100
Dolomite	400 to 3,600	2,100 to 36,000	—
Limestone	100 to 3,600	600 to 36,000	200 to 7,100
Sandstone	300 to 3,600	2,800 to 24,000	—
Shale	300 to 1,400	1,400 to 23,000	400 to 4,300

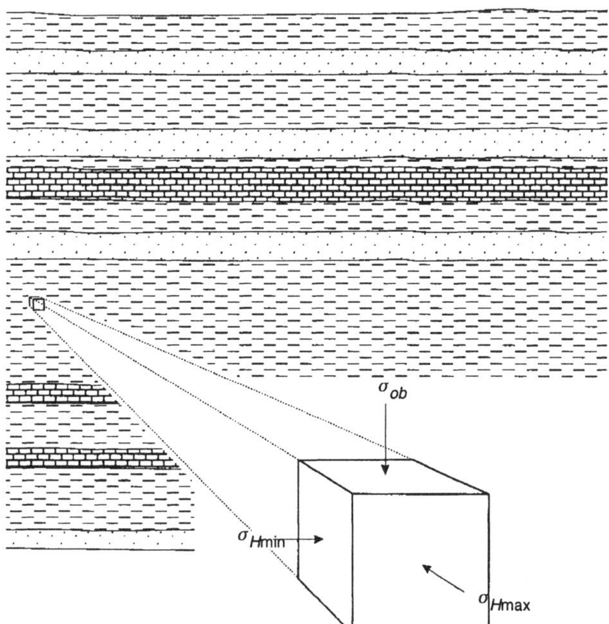

Fig. 2.38—Principal stresses acting on a subsurface rock element.

On the basis of empirical data, Rehm and McClendon[44] later proposed using a modified d exponent (d_c) to normalize mud density. Eq. 2.19 followed from their observations.

$$d_c = d(\rho_n/\rho_{eq}), \quad \text{(2.19)}$$

where ρ_n = density equivalent of the normal pore pressure and ρ_{eq} = equivalent circulating density (ECD) at bit depth. The ECD includes the hydrostatic pressure of the fluids in the annulus, the annular friction losses above the depth of interest, and any applied backpressure at the return outlet.

Example 2.8. A penetration rate of 50 ft/hr is recorded in a gulf coast shale with an applied bit weight and rotary speed of 20,000 lbf and 100 rev/min, respectively. Calculate the d exponent and modified d exponent if the ECD is 10.1 lbm/gal and the bit diameter is 8½ in.

Solution. The d exponent is obtained from Eq. 2.18a as

$$d = \frac{\log[50/(60)(100)]}{\log[(12)(20,000)/(10^6)(8.5)]} = 1.34.$$

The normal pore-pressure gradient in the gulf coast is 0.465 psi/ft. Thus,

$$\rho_n = (19.25)(0.465) = 8.95 \text{ lbm/gal}.$$

Eq. 2.19 gives

$$d_c = (1.34)(8.95)/(10.1) = 1.19.$$

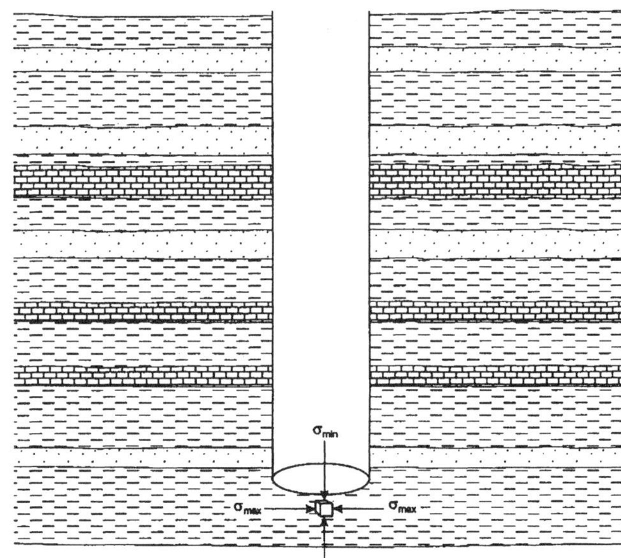

Fig. 2.39—Principal stresses acting on an element near the bottom of a wellbore.

In addition to identifying a gradient transition, empirical equations have been developed that use the normal and observed d_c exponents to quantify abnormal pressure. The Rehm and McClendon relationship (Eq. 2.20) is based on plotting the modified d exponent data on Cartesian coordinate graph paper.

$$g_p = 0.398\log(d_{cn} - d_{co}) + 0.86, \quad \text{(2.20)}$$

where d_{co} and d_{cn} = the observed and extrapolated normal d_c exponents, respectively.

Zamora[45] presented a different relationship.

$$g_p = g_n(d_{cn}/d_{co}). \quad \text{(2.21)}$$

Eaton[47,48] included the effect of a variable overburden in his equation.

$$g_p = g_{ob} - (g_{ob} - g_n)(d_{co}/d_{cn})^{1.2}. \quad \text{(2.22)}$$

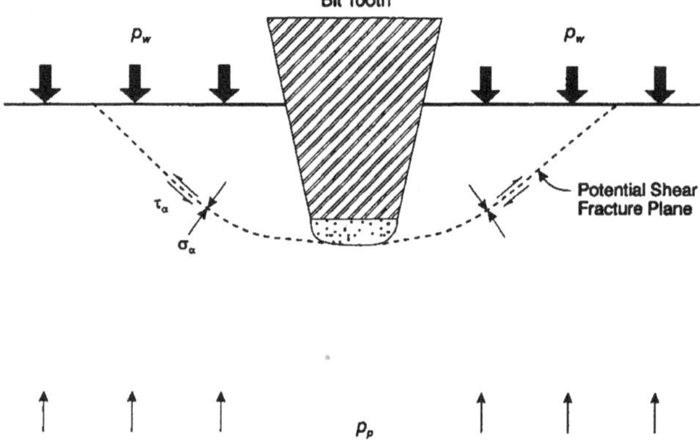

Fig. 2.40—The effect of overbalance pressure on crater formation forces beneath a bit tooth.

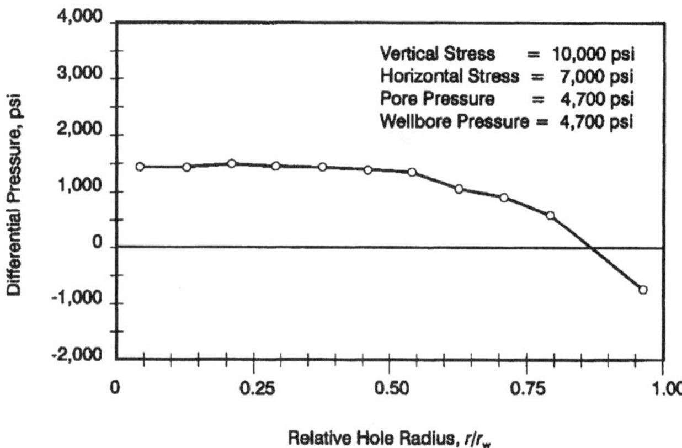

Fig. 2.41—FEM prediction of the induced differential pressure in impermeable rock for a balanced wellbore condition.[40]

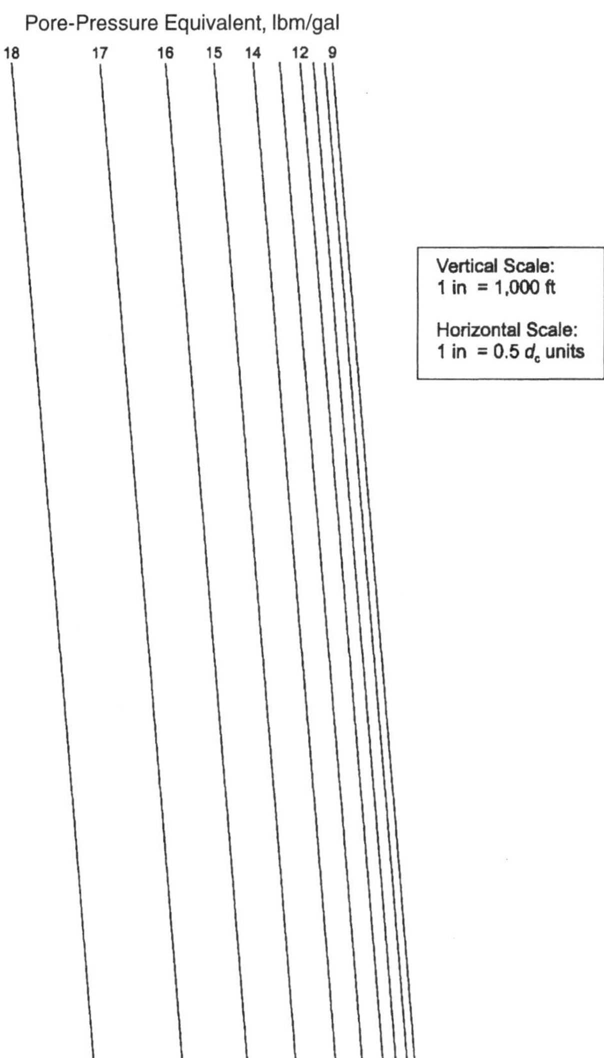

Fig. 2.42—Overlay for pore-pressure prediction based on the Rehm and McClendon[44] modified d_c exponent equation.

Both the Zamora and Eaton techniques are based on an exponential model, which means that the data are plotted on semilogarithmic graph paper.

A transparent overlay similar to that shown in **Fig. 2.42** can save a great deal of time and computation effort. The inscribed pore-pressure gradient or equivalent-density lines allow for direct pressure predictions by placing the normal gradient line on the plotted d_c exponent data corresponding to the normally compacted interval. Rehm and McClendon[44] found that the slope of the normal d_c exponent trend for Miocene-age rock from different geographic locations was 0.000038 ft^{-1}. Zamora's data showed that the semilog slope was consistently close to 0.000039 ft^{-1}. These slopes can be used to assist in curve-fitting the normal-pressure-trend line and as a guide in preparing an overlay.

Example 2.9. The *d* exponent data shown in **Table 2.10** were obtained in Miocene shales on a well off the Louisiana coast in South Marsh Island. Seismic data suggested a possible transition into abnormal pressure at approximately 5,400 ft. Cuttings analysis indicates a consistent Miocene sequence to present depth. Predict the pore pressure in equivalent density at 6,050 ft using (1) Rehm and McClendon's method, (2) Zamora's equation, and (3) Eaton's equation.

TABLE 2.10—*d*-EXPONENT AND MUD DENSITY DATA FOR A WELL LOCATED OFFSHORE LOUISIANA

Depth (ft)	Mud Density (lbm/gal)	*d* exponent	Modified *d* exponent
3,750	10.4	1.17	1.01
3,760	10.3	1.21	1.05
3,820	10.3	1.21	1.05
3,890	10.3	1.21	1.05
3,910	10.2	1.24	1.09
4,000	10.2	1.25	1.10
4,210	10.4	1.32	1.14
4,260	10.3	1.28	1.11
4,300	10.3	1.33	1.16
4,350	10.2	1.28	1.12
4,490	10.2	1.30	1.14
4,520	10.2	1.32	1.16
4,600	10.2	1.36	1.19
4,770	10.2	1.23	1.08
4,840	10.3	1.21	1.05
4,880	10.4	1.28	1.10
4,930	10.4	1.28	1.10
5,010	10.5	1.24	1.06
5,140	10.5	1.26	1.07
5,240	10.5	1.28	1.09
5,270	10.5	1.29	1.10
5,320	10.5	1.30	1.11
5,450	10.5	1.35	1.15
5,580	10.5	1.38	1.18
5,620	10.5	1.41	1.20
5,670	10.5	1.35	1.15
5,710	10.5	1.35	1.15
5,760	10.6	1.30	1.10
5,800	10.6	1.29	1.09
5,850	11.0	1.30	1.06
5,890	11.3	1.33	1.05
5,930	11.5	1.28	1.00
5,970	11.5	1.27	0.99
6,020	11.8	1.27	0.96
6,050	11.8	1.25	0.95

Solution.

1. The first step is to plot the modified *d* exponents on Cartesian graph paper. **Fig. 2.43** shows a perplexing interpretation problem. A possible normal-compaction-trend line is evident through the upper data points, which would indicate the transition depth to be approximately 4,700 ft. Another possibility is that an unconformity was crossed that established a new, parallel trend line below 4,700 ft. However, both of these interpretations conflict with the seismic data and geological indicators while drilling.

One clue is that the two possible trend lines have a much higher slope than would be expected by Rehm and McClendon's work. Fitting a line with a slope of 0.000038 ft^{-1} to establish a "best estimate" normal-trend line shows data scatter resulting from incorrect lithology picks or other reasons. This procedure leads to a transition depth closer to 5,700 ft.

The modified *d* exponent from the normal-trend-line extrapolation at 6,050 ft is 1.18. On the basis of the observed value of 0.95, the pore-pressure gradient is estimated with Eq. 2.20a as

$$g_p = 0.398\log(1.18 - 0.95) + 0.86 = 0.606 \text{ psi/ft}.$$

The equivalent density corresponding to this gradient is

$$\rho_p = (19.25)(0.606) = 11.7 \text{ lbm/gal}.$$

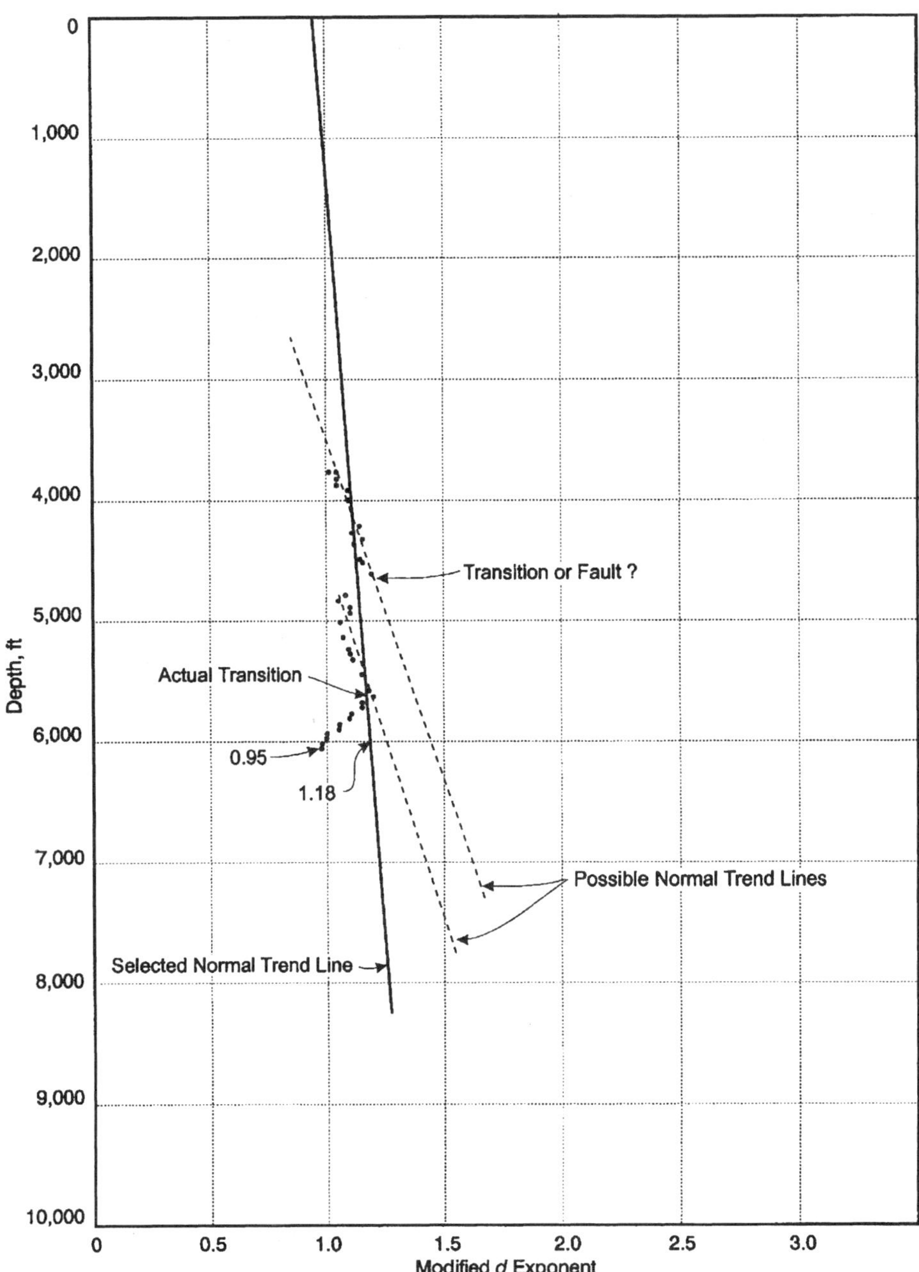

Fig. 2.43—Modified d_c-exponent data obtained from a well in South Marsh Island plotted on Cartesian graph paper.

2. **Fig. 2.44** shows the data as plotted on semilogarithmic graph paper. Constructing a line with Zamora's slope as a guide shows d_{cn} to be 1.18. Applying Eq. 2.21, the pore-pressure-gradient prediction at 6,050 ft is determined as

$$g_p = (0.465)(1.18/0.95) = 0.578 \text{ psi/ft}$$

and $\rho_p = (19.25)(0.578) = 11.1$ lbm/gal.

3. Eq. 2.9a yields an overburden-stress gradient of 0.919 psi/ft at 6,050 ft. Thus Eaton's equation gives

$$g_p = 0.919 - (0.919 - 0.465)(0.95/1.18)^{1.2} = 0.569 \text{ psi/ft}$$

and $\rho_p = (19.25)(0.578) = 11.1$ lbm/gal.

Example 2.9 shows that pore-pressure predictions in a real-world problem are subject to much interpretation. Experience and sound engineering judgment are valuable, of course, but these qualities alone may not be sufficient to solve pore-pressure-prediction problems. Some knowledge of the geological sequence is essential for accurate pressure-prediction work. Also, any serious pore-pressure-prediction effort should include more than one or a combination of indicators to enhance the interpretation. This is true particularly for *d* exponent interpretations because minor lithology changes and many other variables influence penetration rate.

The normalizing variables in the *d* exponent model may be difficult to ascertain by surface measurements. For example, the applied bit weight as read from the weight indicator may not correspond to the downhole WOB in deep or directional wells because of hole drag. Measurements from a floating drilling vessel add more complications. Measurement-while-drilling (MWD) capabilities offer some promise here. Beato[56] discussed some modified *d* exponent interpretation difficulties in deepwater Gulf of Mexico operations and how these

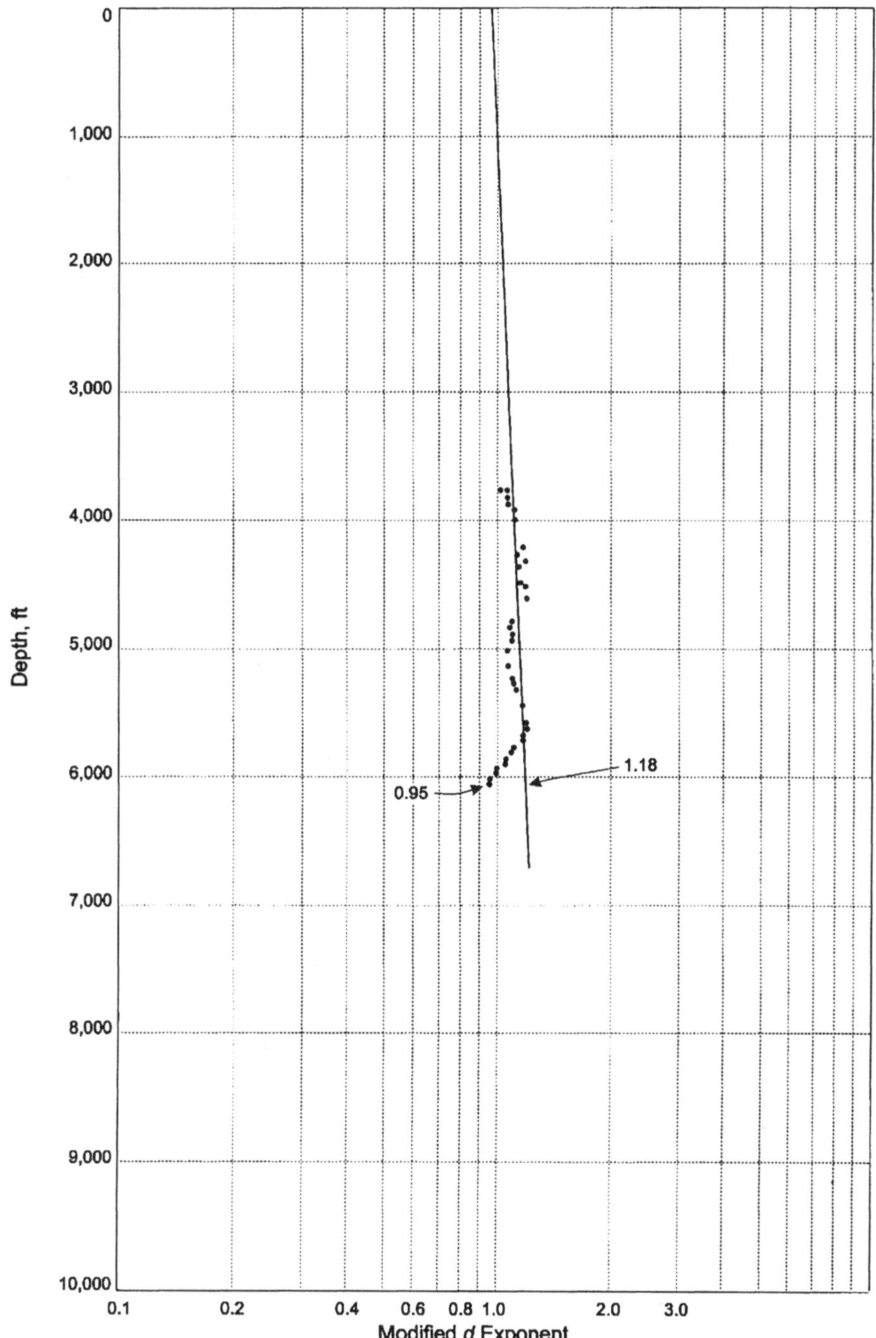

Fig. 2.44—Modified d_c-exponent data obtained from a well in South Marsh Island plotted on semilogarithmic paper.

were alleviated by measuring actual WOB with an MWD tool. MWD also allows for more accurate lithology picks through the use of logging-while-drilling (LWD) data and a lithology correlation involving downhole torque.

Drilling in extremely soft or unconsolidated rock can lead to interpretation problems. Making a hole in these formations can be achieved partially by the jetting action of the mud, a factor not considered in the base equation. For mill tooth bits, another factor that is difficult to describe mathematically is the effect of tooth wear on penetration rate. Some penetration-rate models include a means to normalize for tooth wear, and one of these functions could be used to adjust R before applying Eq. 2.18. However, published tooth-wear relationships may be difficult to apply in actual drilling conditions unless the operator has previous empirical evidence pertaining to the rock and bit types used in the area.

Finally, the shale compaction or differential pressure are not included in the base equation because these are, in effect, the indirect parameters measured with d exponents. Jorden and Shirley's[42] equation is based on the assumption that all shales compact in the same manner when exposed to an increase in differential pressure. It has been established, however, that not all shales compact the same for a given change in differential pressure, even in the coastal regions.

Despite the flaws in the base equation and possible interpretation problems, modified d exponents have worked quite well in the gulf coast and other areas where young sediments predominate. A few guidelines are suggested for improving technique accuracy. Although the d exponent theoretically

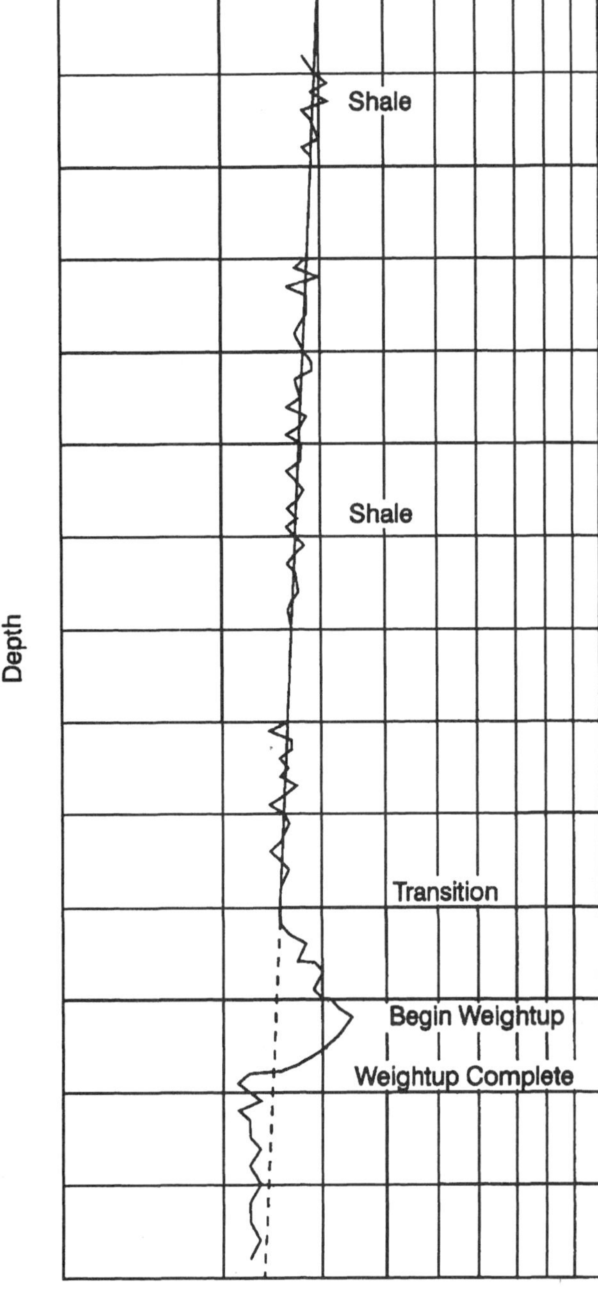

Fig. 2.45—Graphical method for maintaining an overbalance condition in a pressure transition.

normalizes for bit weight and rotating speed, better results are achieved if an attempt is made to hold these parameters relatively constant. This is especially true for rotating speed, because an exponent was not incorporated to the term in the Bingham[55] equation. Extreme overbalance pressures also lead to inaccuracies. Figs. 2.33 and 2.34 show that penetration rate is not very sensitive to overbalance at high differential pressure. Hence, a modified *d* exponent calculation in this region of the curve gives an erroneously high pressure prediction.

2.7.3 Moore's Technique. Moore[52,53] proposed a practical method for maintaining a pore-pressure overbalance while drilling into a transition. **Fig. 2.45** depicts shale penetration rates plotted vs. depth on semilog graph paper. Because of increasing differential pressure and compaction, a drill-rate trend of constant slope can be expected as long as the operational parameters do not change. The transition depth into increasing pore pressure is indicated on the diagram by the faster drill rates. The differential pressure continues to drop as drilling progresses into the transition, perhaps to a level that invites an influx if a permeable section is encountered.

A means of regaining the lost differential pressure and restoring the overbalance condition simply entails weighting up the drilling fluid to a density sufficient to push drill rates back onto the trend extrapolation. Fig. 2.45 shows the weight-up process and its effect on drill rates. A drawback to this uncomplicated approach, however, is that pore pressure at any depth in the transition is unknown. Lacking this knowledge, an operator can only guess the mud weight and the speculation probably will be incorrect. In this figure, the final mud weight exceeds the desired mud weight because the resulting penetration rates fall to the left of the normal extrapolation. Note, however, that the penetration rates with the excess mud density ultimately return to the normal trend line as drilling proceeds deeper into the transition.

The penetration rates shown on Fig. 2.45 represent actual rates; however, pretransition normalized rates from an appropriate model may be used to compensate for changes in the parameters. For instance, a model will have to be used to predict the target penetration rate if the operator control drills (reduces bit weight and/or rotating speed) while increasing mud density. The drill-rate equation adopted by Moore is

$$R = K\left(\frac{W}{d_b}\right)N^{a_N}. \qquad (2.23)$$

Note that this expression is equivalent to Eq. 2.16 if the bit weight is directly proportional to penetration rate. The selected model, however, is irrelevant to the technique. A bit-weight exponent and normalizing functions for other parameters, such as tooth wear and hydraulics, can be incorporated if needed.

Example 2.10. Shale penetration rates for a well in the mid-continent U.S. are listed in **Table 2.11** and plotted in **Fig. 2.46.** Bit-operating parameters before the transition were 4,700 lbf/in. and 80 rev/min. Transition was detected at 9,100 ft and the operator immediately reduced the bit weight to 2,900 lbf/in. and began to increase mud density. Determine the target penetration rate at 9,250 ft at the reduced WOB. A J-22 tungsten carbide insert (TCI) bit was used throughout the plotted intervals, and bit hydraulics did not change appreciably past 8,000 ft.

Solution. The extrapolated normal-penetration rate at 9,250 ft is 15.7 ft/hr, which would have been the target rate had the bit weight remained constant. Eq. 2.23 yields the target penetration rate at the reduced bit weight.

$$R = 15.7\left(\frac{2,900}{4,700}\right)\left(\frac{80}{80}\right)^{a_N} = 9.7 \text{ ft/hr.}$$

The target rate reverts back to 15.7 ft/hr if the operator resumes drilling at 4,700 lbf/in.

Obviously, some means of estimating the mud-weight requirement would be valuable in implementing this technique. Moore proposed a pore-pressure-prediction method that recognizes the difference in shale compactibility and offers promise for all sedimentary basins. Eq. 2.24 was developed

TABLE 2.11—SHALE PENETRATION RATES MEASURED IN A WELL IN THE MID-CONTINENT U.S.

Depth (ft)	Drilling Rate (ft/hr)
8,200	21.0
8,220	22.0
8,240	20.0
8,260	20.0
8,270	21.0
8,280	20.0
8,300	20.5
8,320	20.0
8,330	21.0
8,350	20.0
8,360	18.5
8,610	18.0
8,630	19.5
8,640	19.0
8,660	18.5
8,680	18.5
8,690	18.0
8,700	19.0
8,720	18.0
8,740	17.0
8,760	18.0
8,780	18.0
8,790	17.5
8,850	17.5
8,870	17.0
8,890	17.0
8,900	18.5
8,910	18.0
8,920	16.5
8,930	17.0
8,980	17.0
9,010	17.0
9,020	17.5
9,040	18.0
9,050	17.5
9,060	18.0
9,090	18.0
9,100	12.0
9,120	11.5

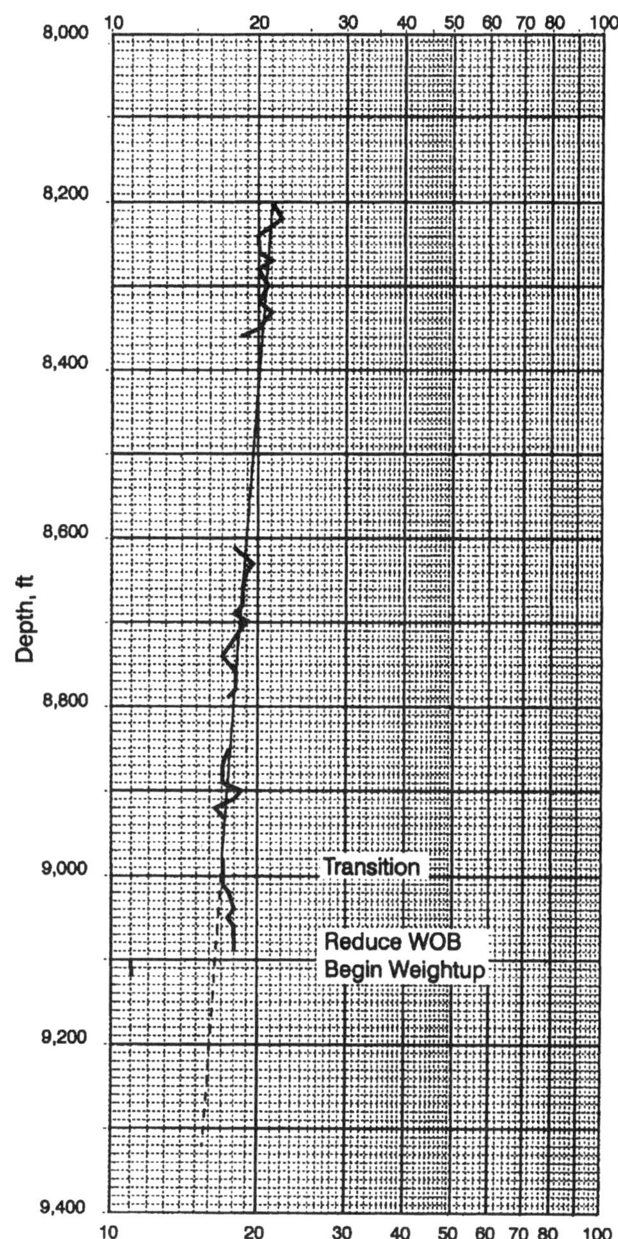

Fig. 2.46—Shale-penetration rates for the well described in Example 2.10.

from drilling data taken in a variety of shale types and degrees of hardness.

$$\rho^c_{eq1}\log(R_1) = \rho^c_{eq2}\log(R_2). \quad \text{(2.24)}$$

This equation relates the change in penetration rate between conditions 1 and 2 to a change in differential pressure expressed as a density equivalent. The manner in which shale compacts below the bit in response to a change in differential pressure is accommodated by the shale-compactibility coefficient, c. Thus, c depends on the shale hardness or compactibility and can be determined on the basis of a simple field experiment. Example 2.11 demonstrates the procedure.

Example 2.11. Refer to the 8,300-ft measurement plotted in Fig. 2.46. Assume that the penetration rate drops from 20.5 to 18.5 ft/hr in response to changing the ECD from 9.6 to 10.1 lbm/gal. Determine the shale-compactibility coefficient.

Solution. Substituting the known values into Eq. 2.24 yields

$$(9.6)^c\log(20.5) = (10.1)^c\log(18.5).$$

Rearrange and solve for c.

$$\left(\frac{10.1}{9.6}\right)^c = \frac{\log(20.5)}{\log(18.5)} = 1.035,$$

$$c\log(1.052) = \log(1.035),$$

and $c = 0.68$.

Increasing mud density is expensive in terms of barite consumption and reduced penetration rate and should be done only for reasons of hole conditions. A valid means for increasing the ECD and differential pressure without changing mud properties is to drill with backpressure supplied by the choke. Because applied pressure is reflected at all points in the wellbore, the wellbore-pressure integrity is a consideration if this procedure is used.

Once c has been determined for the shale being drilled, the change in pore pressure within the transition can be estimated with the same equation. Conceptually, the problem solution lies in calculating the mud density that reduces the enhanced drilling rate back to the normal rate.

Example 2.12. Using the well data given in Example 2.11, estimate the pore pressure at 9,090 ft. The ECD at the time of measurement is a 9.6-lbm/gal equivalent. Assume the normal pore-pressure gradient for the area is that of fresh water.

Solution. The extrapolated normal drilling rate at 9,090 ft is 16.5 ft/hr. First solve for ρ_{eq2} in the Moore equation.

$$(9.6)^{0.68}\log(18.0) = \rho_{eq2}^{0.68}\log(16.5)$$

and $\rho_{eq2} = 4.80^{\frac{1}{0.68}} = 10.0$ lbm/gal.

This represents an increase in 0.4 lbm/gal over the normal ECD of 9.6 lbm/gal. The pore-pressure gradient in density equivalent is thus

$$\rho_p = 8.3 + 0.4 = 8.7 \text{ lbm/gal}.$$

It is apparent that the well is still significantly overbalanced.

Pore-pressure predictions across an interval where operating parameters have changed requires normalizing the base penetration rate to the existing conditions. Example 2.13 demonstrates the procedure.

Example 2.13. The penetration rate on the subject well drops to 12 ft/hr at 9,100 ft after reducing bit weight from 4,700 to 2,900 lbf/in. The rotary speed did not change and weighted mud has not yet entered the well. Estimate the pore pressure at 9,100 ft.

Solution. The extrapolated normal-drilling rate at 9,100 ft is approximately 16.4 ft/hr. Normalizing to the current conditions yields

$$R_1 = 16.4\left(\frac{2,900}{4,700}\right)\left(\frac{80}{80}\right)^{a_N} = 10.1 \text{ ft/hr}.$$

Substitution in Eq. 2.24 leads to

$$(9.6)^{0.68}\log(12.0) = \rho_{eq2}^{0.68}\log(10.1)$$

and $\rho_{eq2} = 5.00^{\frac{1}{0.68}} = 10.7$ lbm/gal.

A 1.1-lbm/gal increase in the pore-pressure gradient is indicated. The predicted gradient at 9,100 ft in density equivalent then is

$$\rho_p = 8.3 + 1.1 = 9.4 \text{ lbm/gal}.$$

Accurate assessment of the compaction exponent requires that the pore-pressure gradient remain constant during data collection. If an operator increases mud density in anticipation of a pressure transition and evaluates the compactibility coefficient on the basis of reduced drilling rate, the coefficient will be in error if the pore-pressure gradient changes during this time. Example 2.14 presents a means of later correcting the coefficient value.

Example 2.14. While drilling in a 0.465-psi/ft shale at 11,800 ft, an operator increases mud density and raises the ECD from 9.5 to 10.5 lbm/gal. The penetration rate at 12,000 ft drops to 17.5 ft/hr from an extrapolated normal rate of 23 ft/hr. On this basis, a 0.91 compactibility coefficient is calculated. A drill-stem test (DST) in an adjacent sand indicates that the pore-pressure gradient at this depth is 0.480 psi/ft. Determine the actual value for c.

Solution. The pore-pressure increase over the 200 ft of drilled hole is

$$\Delta\rho_{eq} = (19.25)(0.480 - 0.465) = 0.29 \text{ lbm/gal}.$$

Rather than a 1.0-lbm/gal increase in density, the formation effectively "feels" a density of

$$\rho_{eq2} = 10.5 - 0.29 = 10.21 \text{ lbm/gal}.$$

Hence, the corrected value for c is

$$\left(\frac{10.21}{9.5}\right)^c = \frac{\log(23.0)}{\log(17.5)} = 1.095,$$

$$c\log(1.075) = \log(1.095),$$

and $c = 1.26$.

Example 2.14 illustrates one of the practical limitations of the technique. A fairly minor adjustment in effective mud density can have a major impact on the calculated coefficient value. This effect is particularly pronounced in the low-density ranges and emphasizes the need for valid data during the experimental procedure and in the subsequent pore-pressure calculation. Accurate annular friction-loss estimates, wellbore-mud densities, lithology picks, drill rates, and normal-trend-line extrapolations are essential.

2.8 Other Drilling Indications

The remaining qualitative and quantitative indicators listed in Table 2.5 have one disadvantage compared with penetration rate in that the data are not as timely. The indicators that rely on the drill-cuttings characteristics and mud properties (gas-cutting, contamination, and temperature) involve a lag period that consists of the time it takes to circulate the cuttings or mud from the abnormal-pressure source to surface.

Openhole-log correlations often provide the most accurate pore-pressure predictions, but historically were limited by the requirement that the tools be conveyed on wireline. Since the advent of and continuing advances in LWD technology, this is no longer a problem. Information obtained by an LWD tool transmits to surface at the speed of sound. Nonetheless, there is still a delay because the sensor reading the formation is an estimated 3 to 60 ft above the bit. Section 2.9 discusses the historical perspective and relative importance of conventional methods based on log analysis.

2.8.1 Drill-Cuttings Characteristics. Formation sample returns from a well can provide information related to the drilled or anticipated pore-pressure environment. Indications generally are obtained from visual observation or analytical measurements of shale cuttings because of their relative compactibility and other unique petrophysical characteristics. **Table 2.12** summarizes shale properties used to assess pore pressure.

Measuring and tracking the bulk density of shale cuttings is one of the earlier pore-pressure-detection procedures. Recalling the relationship between bulk density and porosity, the density of a shale sample gives an indication of the drilled rock's compaction state and pore pressure. **Fig. 2.47** shows a shale-cutting-density plot taken from a well offshore Nigeria.

TABLE 2.12—SHALE CUTTING INDICATORS OF ABNORMAL PORE PRESSURE	
Shale Property	Information Provided
Bulk density	Compaction state
Moisture content	Compaction state
Resistivity	Compaction state
CEC	Montmorillonite content
	Marker bed identification
Paleontology	Marker bed identification
Lithology	Marker bed identification
	Pressure seal indicators
	Density measurement quality control
Mineralogy	Montmorillonite content
	Marker bed identification
NMR	Compaction state
Size, shape, and volume	Matrix strength
	Differential pressure

The top of the pressure transition is indicated clearly at approximately 8,600 ft.

Boatman[57] used data from south Louisiana to develop the correlation shown as **Fig. 2.48.** He recommended plotting shale-cutting densities vs. depth of origin on Cartesian coordinate graph paper and fitting a normal-compaction-trend line through the pretransition points. Pore pressures within the transition are estimated by taking the difference between the extrapolated normal density, ρ_{shn}, and measured density, ρ_{sho}, then using Fig. 2.48 to obtain the gradient. In drawing the line, greater emphasis is placed on the deeper measurements because sediment compaction with depth is described better by an exponential relationship.

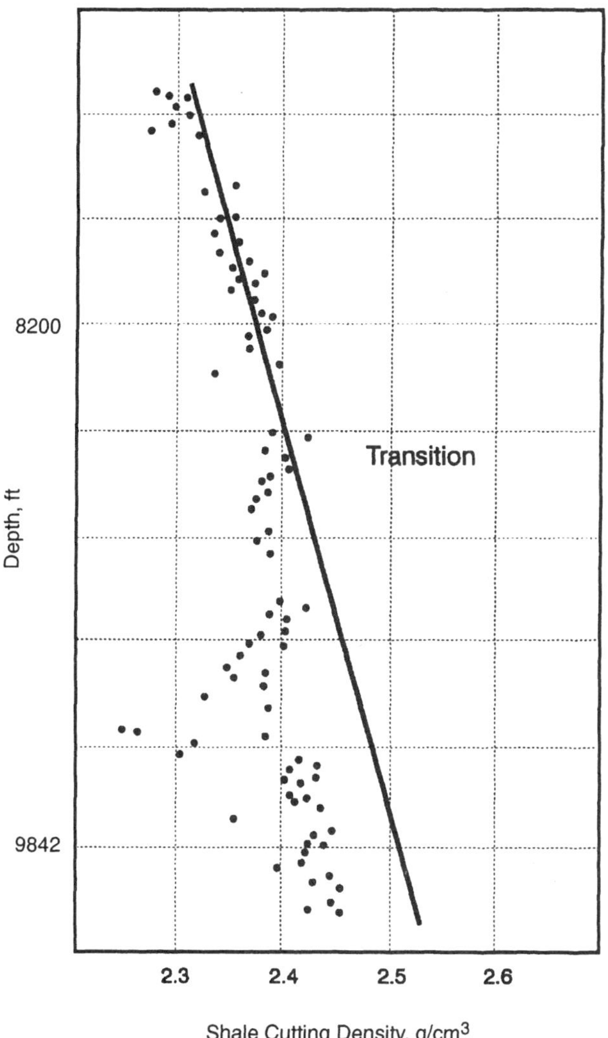

Fig. 2.47—Shale cutting densities with depth from offshore Nigeria.[5]

Example 2.15. Table 2.13 gives shale-cutting-density measurements obtained from a south Louisiana well. Estimate the pore pressure at 14,000 ft using Boatman's chart.

Solution. **Fig. 2.49** shows the plot for the shale-density data. Transition into abnormal pore pressure is indicated at approximately 13,500 ft. The extrapolated normal shale density at 14,000 ft is 2.54 g/cm^3, while the observed density is 2.44 g/cm^3. Taking the difference,

$$\rho_{shn} - \rho_{sho} = 2.54 - 2.44 = 0.10\ \text{g/cm}^3.$$

Fig. 2.48 gives a pore-pressure gradient of 14.6 lbm/gal. Hence, the predicted pressure at 14,000 ft is

$$p_p = (0.0519)(14.6)(14,000) = 10,608\ \text{psig}.$$

The use of shale-cutting densities to predict pore pressure has some drawbacks, and certain requirements must be met for the technique to be effective. For instance, samples obtained from the flowline must be representative of the rock generated beneath the bit at the depth of interest. This simple precaution applies to all techniques that rely on flowline samples but sometimes is difficult to achieve in actual practice. One potential problem arises from velocity variations across the annular cross section that occur when flow is laminar. A distribution of cuttings from different depths is a possible consequence. Uphole cavings also can be a source of error and some means of identifying and eliminating the extraneous rock from consideration is needed.

Ideally, the shale properties do not change during the circulation or sample preparation process. Some chemical or mechanical alteration probably is inevitable under the best of circumstances. Inhibitive drilling fluids, those systems that limit water absorption and swelling of reactive clays, yield better results. Normal sample preparation procedure is to

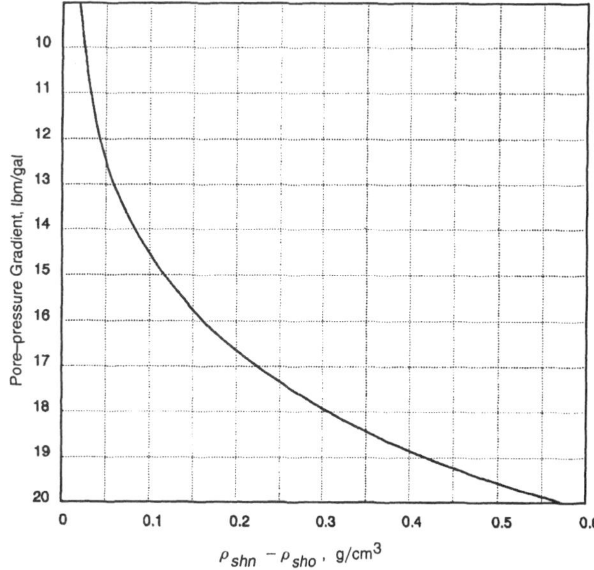

Fig. 2.48—Empirical gulf coast correlation between pore-pressure gradient and shale-cutting bulk density.[57]

TABLE 2.13—SHALE-CUTTING DENSITY MEASUREMENTS FROM A WELL IN SOUTH LOUISIANA[57]

Depth (ft)	Shale Density (g/cm³)
11,100	2.39
11,200	2.44
11,420	2.42
11,550	2.44
11,680	2.41
11,700	2.44
11,770	2.46
11,820	2.43
11,890	2.46
12,140	2.47
12,190	2.42
12,220	2.48
12,350	2.45
12,410	2.48
12,500	2.49
12,520	2.47
12,620	2.47
12,680	2.49
12,810	2.47
12,900	2.48
12,960	2.49
13,080	2.50
13,190	2.49
13,200	2.52
13,250	2.49
13,300	2.51
13,400	2.50
13,450	2.52
13,560	2.51
13,610	2.49
13,690	2.47
13,810	2.46
13,910	2.44
14,000	2.44
14,120	2.46
14,250	2.45
14,350	2.46
14,470	2.45
14,580	2.46
14,700	2.47
14,890	2.48
14,920	2.46

wash the shale cuttings with fresh water and to dry with a towel. Drying the shale in an oven liberates some of the bound water and is to be discouraged.

Small amounts of carbonate or heavy minerals within the shale matrix can influence bulk density dramatically. Some scatter of the points can be expected when considering the effect of lithology and other possible errors in the procedure. It follows that more accurate trend lines result if shale densities are plotted from a large data base.

The most accurate of the four different methods for measuring shale-cutting densities described involves using a mercury pump to measure the bulk volume of a known mass of cuttings. The variable-density column is another commonly applied procedure. A column of liquid that exhibits increasing density with depth is placed in a graduated cylinder. Calibration beads of known density are put into the cylinder and float at a liquid column height corresponding to the bead density. The density of a cutting chip placed in the cylinder is determined by interpolating the chip position to the height of the nearest calibration beads.

A less precise, though relatively fast, method does not require any specialized equipment or chemicals. First, the rider of a standard American Petroleum Inst. (API) mud balance is positioned at the density of fresh water and shale is added to the cup until the freshwater-density equivalence is obtained. Then, the cup is filled with fresh water and the combined density is recorded. The specific gravity of the shale originally placed in the cup is obtained with

$$\gamma_{sh} = \frac{\rho_w}{2\rho_w - \rho_{shw}} \quad \text{. . . . . (2.25)}$$

The subscripts w and shw = the density of fresh water and the measured shale/water mixture, respectively. Note that specific gravity is equivalent to density if the latter is defined in g/cm^3 units.

Example 2.16. The rider of an API mud balance is positioned at 8.33 lbm/gal, and dry shale cuttings are placed in the cup until the level is balanced. The cup is then filled with fresh water, and the mixture density is read as 13.3 lbm/gal. Determine the shale density.

Solution. Eq. 2.25 yields

$$\rho_{sh} = 8.33/(16.66 - 13.3) = 2.48 \text{ g/cm}^3.$$

A means of directly measuring the porosity of a shale sample is available now through application of nuclear magnetic resonance principles. Prepared shale cuttings are exposed to a magnetic field, and the electromotive force originating from the induced spin of any associated hydrogen atom nuclei is measured. The energy released from the shale corresponds to the amount of hydrogen in the rock, thus allowing porosity to be determined.

Another way to estimate shale porosity is to measure the moisture content of a dry sample. A prepared sample is weighed carefully, then heated to release the pore water. The sample's weight loss gives the moisture content of the shale and porosity can be determined. This procedure assumes that all the water, both bound and free, leaves the sample and ignores the effect of any retained salt ions.

Montmorillonite clays carry a predominant negative charge on the basal surfaces of the individual crystalline layers. In the natural state, sodium, calcium, or magnesium cations compensate for this charge deficiency and are adsorbed within the interlayer structure. Water molecules readily penetrate the space between the layers thus giving these clays their hydration or swelling characteristics in the presence of water.

In an aqueous environment, the interlayer cations may be displaced by other cations present in solution. A measure of the cation substitution potential of a given clay is a property defined as the cation exchange capacity (CEC), expressed in cation meq/100 of dry clay. Comparatively, montmorillonite has a much higher CEC than the nonswelling clays, such as illite or kaolinite. The CEC of a shale sample can be approximated in the field from the amount of methylene blue dye absorbed by a clay suspension. This measurement, called the shale factor, is expressed in mL methylene blue/100 crushed clay.

A shale-factor trend with depth may be a useful tool in overall pore-pressure prediction. Generally, shale factors in Tertiary basins decrease with increasing temperature as the clays gradually convert from montmorillonite to illite. A reversal

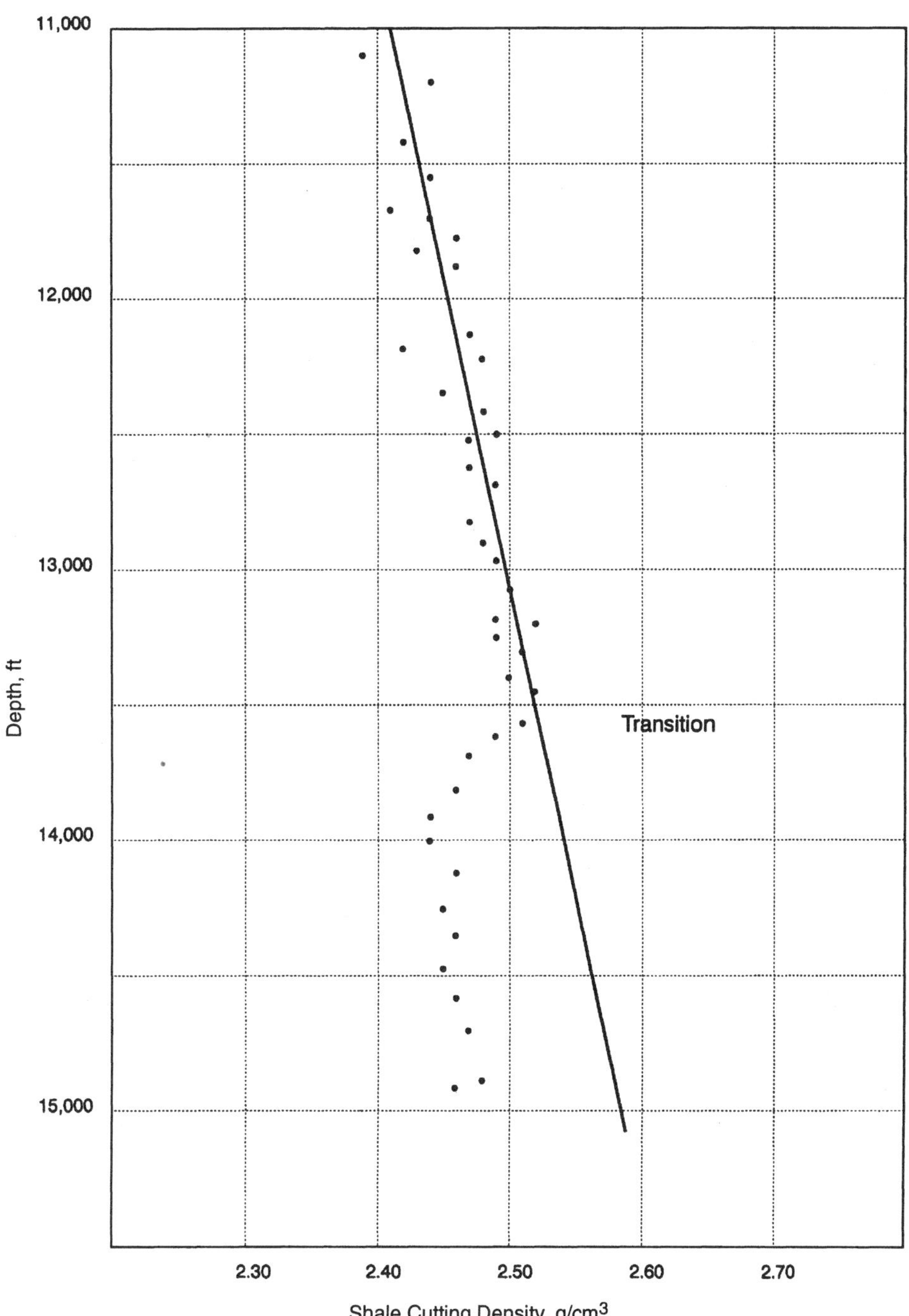

Fig. 2.49—Shale density plot corresponding to Example 2.15.

of the observed trend into higher montmorillonite concentration implies a loss of dehydration efficiency often associated with undercompaction. Conversely, a relatively sudden increase in shale factors indicates rapid conversion to illite and the possibility of diagenetic abnormal pressures. Shale factors, although helpful, are not positive indicators and should be considered only in conjunction with other signals given by the well.

A large volume of shale at the flowline may indicate a borehole-stability problem. One form of instability is chemical in nature and arises from montmorillonite-rich-shale hydration at the walls of the hole. A swelling stress builds up over time and can cause sloughing at the wellbore if the swelling pressure exceeds the matrix strength of the rock. Using an inhibitive mud system generally solves or at least controls any chemically induced instability problems.

Shale cavings also may signal an increase in pore pressure or, alternatively, a reduction in differential pressure. The associated instability mechanism may, in fact, be a combination of processes. Chap. 3 covers the phenomenon in more detail. Creating a hole in stressed rock causes a compressive-stress concentration at the wellbore and a hole collapse relieves stress concentration exceeding the rock strength. Plastic rock, like soft clays and salt, collapses by squeezing into the wellbore. Brittle rocks, on the other hand, fail by breaking loose from the wall of the hole as chunks (breakouts). Abnormal pressure tends to weaken rock so it is reasonable to conclude that an abnormally pressured shale will be more prone to collapse than a normally pressured shale.

If the wellbore pressure is underbalanced, a pressure gradient from the formation to the well develops that may create a high tensile stress normal to the wall of the hole in low-permeability rock. Instability is promoted, and failure of the rock may be explosive, (sometimes called "popping" shale). Typically, underbalanced shale spallings are long and splintery in appearance and have sharp edges. Conchoidal tension fractures often can be seen in shales that fail in this fashion. Breakouts and intact sloughings, on the other hand, generally have a block-like appearance.

Evidence usually is available for determining whether an unstable hole is the result of a chemical problem. Hydrating clays tend to disintegrate in the drilling fluid, thereby increasing solids content, and reactivity tests can be run on the intact shale samples. Also, chemical instability problems take some time, perhaps days, before symptoms arise, whereas stress-related breakouts can occur almost instantaneously. However, it may be difficult to differentiate between the other possible causes by appearance alone. This highlights the need for considering all available evidence before making conclusions. As an example, an increase in drilling rate should accompany or precede cavings generated by abnormal pore pressure.

The lithological information obtained by the wellsite geologist or mud logger often can suggest a change in pore pressure. For example, the possibility of pressure seals should always be regarded when drilling through tight carbonates, calcerous shales, anhydrites, or other potential caprock materials. Undercompacted rock typically is associated with thick shales, so advancing a hole in a massive shale section should be done with caution. Finally, marker beds that correlate to known pressure seals or transition zones may be identified by the minerals or fossils present in the rock.

2.8.2 Gas in the Drilling Fluid. Fig. 2.50 shows several avenues for gas to enter a circulating or static column of mud. Drilled gas, also called cuttings or liberated gas, refers to the gas released from rock cuttings generated by the bit. The

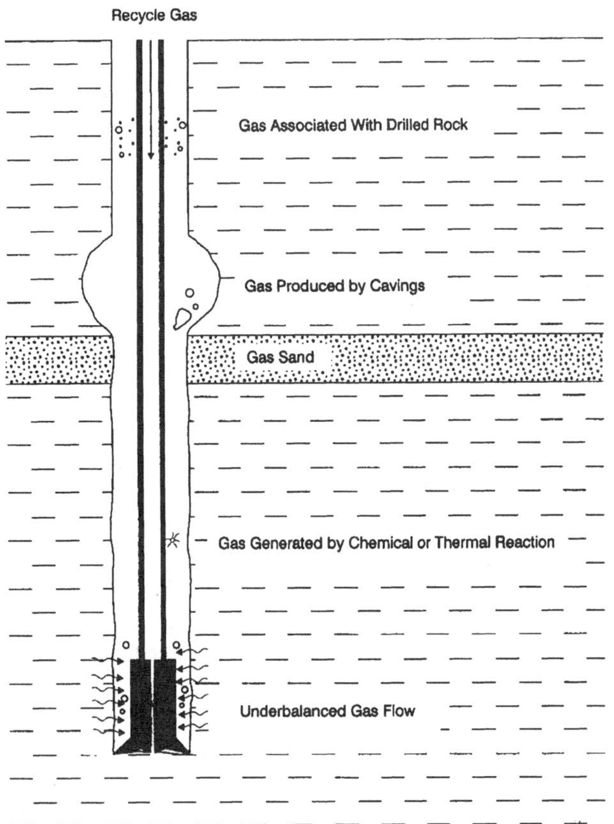

Fig. 2.50—Possible sources of gas in a drilling fluid.

associated gas volumes typically are small, and are a function of hole diameter, circulation rate, pore pressure, gas pore volume, and penetration rate. Drilled gas normally does not constitute a drilling hazard, and its presence is not an indication of abnormal pressure. Note that gas formation cuttings do not always give a gas "show," and the absence of drilled gas does not necessarily eliminate the possibility of gas production, be it desired or not. For example, in a permeable formation being drilled with high-filtrate mud at significant overbalance, the show is suppressed, if not elminated, by a near-complete flushing of the hydrocarbon pore volume ahead of the bit.

Produced gas refers to any gas that enters the wellbore from the walls of the hole. Gas may be produced into a well because of a pressure underbalance (i.e., negative differential pressure). Given sufficient permeability, an underbalanced zone may produce gas at a rate sufficient to cause a well-control emergency. On the other hand, low-permeability rock, such as a shale or tight sand, may continuously produce gas under negative differential pressure without causing a measurable increase in the mud-return rate. The ability to manage the associated gas depends on the flow rate and on the design of the gas-separation equipment.

By definition, gas liberated from uphole cavings is also produced gas. Other possible sources include recycle gas and contamination gas. Recycle gas is any wellbore gas, regardless of its origin, that remains in the mud after at least one pass through the pits. Contamination gas refers to gas released from any volatile hydrocarbons intentionally added to the system. Thermal breakdown of organic mud additives and other downhole reactions also results in the release of combustible gas.

Gas-cut mud typically has a fluffy or grainy appearance in the shale-shaker possum belly or pits, and gas bubbles may be seen breaking out of the mud. Measuring the return mud weight at the flowline gives a relative indication of the problem severity. Measuring the mud density under atmospheric conditions may lead to apparently alarming results and some concern about a reduction in the annular hydrostatic pressure. Severe gas cutting may be a warning sign, but its effect on the total equivalent mud weight at the bottom of the hole is probably negligible.

From a material balance of the system components (excluding drilled cuttings), the density of a gas/mud mixture is given by

$$\rho_{mg} = \rho_m(1 - f_{Vg}) + \rho_g f_{Vg}, \quad \ldots\ldots\ldots\ldots\ldots\ldots (2.26)$$

where ρ_{mg} = density of the gas/mud mixture, ρ_m = uncut mud density, ρ_g = gas density, and f_{Vg} = the volume fraction of the gas phase.

Eq. 2.26 holds true at the prevailing surface conditions when the mud is weighed and at any point in the annulus. Eq. 1.15 states that gas density increases in response to higher pressure. It follows that the density of a gas-cut mud escalates with increasing depth in a wellbore. More importantly, the gas-volume fraction also changes with increasing pressure. Combining a given quantity of gas with a specified volume of drilling fluid yields a gas-volume fraction of

$$f_{Vg} = V_g/(V_m + V_g).$$

The volume associated with n moles of gas is obtained by rearranging Eq. 1.12.

$$V_g = znR_gT/p.$$

Take V_m to be one gallon (or any other convenient quantity) of clean mud. Substitute terms and obtain

$$f_{Vg} = \frac{znR_gT/p}{1 + znR_gT/p}. \quad \text{.....................} (2.27)$$

Example 2.17 demonstrates the combined effect of the gas volume and density changes with depth in a well.

Example 2.17. A thick gas sand has been drilled at constant penetration rate and circulation rate. Gas cutting of the drilling fluid is noted on bottoms up, and a sample taken from the flowline weighs 7.0 lbm/gal. Determine the mixture density in the annulus 2 ft below the flowline outlet if the clean mud density is 12.0 lbm/gal. The sample temperature is 100°F and atmospheric pressure is 14.7 psia.

Solution. Assume a gas specific gravity and calculate the gas density in the cup using Eq. 1.22.

$$\rho_g = \frac{(0.6)(14.7)}{(2.77)(1.0)(560)} = 0.00569 \text{ lbm/gal.}$$

Eq. 2.26 is rearranged to yield the gas-volume fraction.

$$f_{Vg} = (12.0 - 7.0)/(12.0 - 0.00569) = 0.417.$$

Now Eq. 2.27 is rearranged to give the number of gas moles in 1 gal of the mud.

$$n = \frac{0.417}{\dfrac{(1.0)(80.275)(560)}{14.7} - \dfrac{(0.417)(1.0)(80.275)(560)}{14.7}}$$

$$= 0.0002338 \text{ lbm mol.}$$

This concentration is constant throughout the annulus if the downhole entry rate did not change. To simplify the problem, we assume that the mixture density is constant down to the depth of interest. Thus the absolute pressure at 2 ft is

$$p_2 = 14.7 + (0.0519)(7.0)(2) = 15.43 \text{ psia.}$$

At this pressure, the gas density and volume fraction, respectively, are

$$\rho_g = \frac{(0.6)(15.43)}{(2.77)(1.0)(560)} = 0.00597 \text{ lbm/gal}$$

$$\text{and } f_{Vg} = \frac{(1.0)(0.0002338)(80.275)(560)/15.43}{1 + [(1.0)(0.0002338)(80.275)(560)/15.43]}$$

$$= 0.405.$$

Therefore, mixture density at this depth is

$$\rho_{mg2} = (12.0)(1 - 0.405) + (0.00597)(0.405)$$

$$= 7.14 \text{ lbm/gal.}$$

Fig. 2.51 is a plot of the computed annular mud densities from surface to a depth of 10,000 ft for the conditions described in the Example 2.17. We see that most of the gas expansion and mud-density reduction occurs in the very top portion of the wellbore. In fact, the density variation between 1,500 ft and total depth is less than 0.1 lbm/gal.

It should be apparent from the curve that the cumulative effect on the hydrostatic pressure is likely to be small. Eq. 2.28 follows from White's[58] original derivation for an ideal gas and can be used to approximate the hydrostatic pressure loss for real gases.

$$p_m - p_{mg} = \frac{f_{Vgs}p_s\bar{z}\bar{T}}{(1 - f_{Vgs})z_sT_s}\ln\left(\frac{p_{mg} + p_s}{p_s}\right), \quad \text{.....} (2.28)$$

where p_m = the hydrostatic pressure of a clean mud column, p_{mg} = the hydrostatic pressure of the gas-cut mud, f_{Vgs} = the gas volume fraction at the top of the hole, p_s = surface pressure, T_s = surface temperature, z_s = the surface compressibility factor, $\bar{T}$ = the average temperature in the annulus, and $\bar{z}$ = the average compressibility factor in the annulus.

Eq. 2.28 presumes a constant gas concentration throughout the annulus. The hydrostatic pressure of a clean mud column can be used to determine the average z factor. However, the problem solution still requires iteration because p_{mg} is in the logarithmic function. A recommended approach is to use a p_{mg} value on the right side of the equation equivalent to the hydrostatic pressure of the clean mud and solve for the pressure difference on the left. Subsequent iterations then use calculated values in the logarithm until the results agree.

Example 2.18. Estimate the change in hydrostatic pressure at total depth for the well described in Example 2.17. Assume the circulating mud has an average temperature of 150°F.

Solution. The average annulus pressure given by a column of 12.0-lbm/gal mud is

$$\bar{p} = (14.7 + 6,248)/2 = 3,131 \text{ psia.}$$

The z factor at the average pressure and temperature is obtained as 0.868 from Fig. 1.6. For the first iteration, assume that p_{mg} is 6,248 psia and solve for the pressure change using Eq. 2.28.

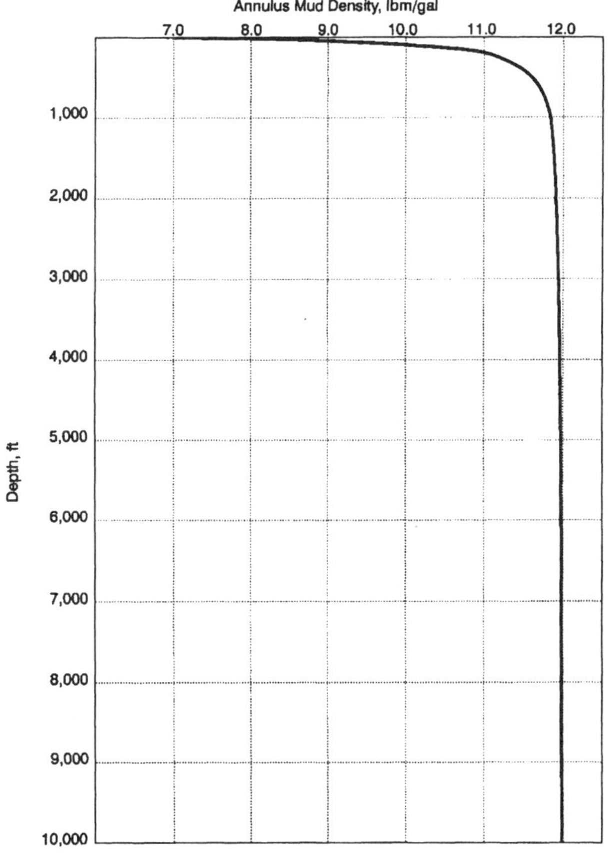

Fig. 2.51—Gas-cut mud density as function of depth for the well described in Example 2.17.

$$p_m - p_{mg} = \frac{(0.417)(14.7)(0.868)(610)}{(1 - 0.417)(1.0)(560)} \times \ln\left(\frac{6,248 + 14.7}{14.7}\right) = 60 \text{ psi.}$$

Solve for p_{mg}.

$$p_{mg} = 6,248 - 60 = 6,188 \text{ psia.}$$

Another iteration yields

$$p_m - p_{mg} = \frac{(0.417)(14.7)(0.868)(610)}{(1 - 0.417)(1.0)(560)} \times \ln\left(\frac{6,188 + 14.7}{14.7}\right) = 60 \text{ psi.}$$

A second iteration is unnecessary if the first assumed value for p_{mg} is reasonably accurate.

Adding gas to a drilling fluid increases the system viscosity, and any hydrostatic pressure reduction is countered to some extent by an increase in the annular friction loss. Rather than predicting two-phase-flow behavior, the net change in bottomhole pressure could be noted by observing the standpipe pressure. For example, a 50-psi reduction in pump pressure (with no change in circulation rate or base mud properties) indicates that the pressure in the annulus had decreased by the same amount. As a practical consideration, however, the sensitivity of most oilfield gauges plus the effect of pump pulsation may not allow for detection of a small pressure change.

An operator usually has the option of shutting down the operation and closing in the well if there is concern that the well is underbalanced. Checking for flow, the normal precursor to shut-in, may falsely indicate formation entry if the wellbore gas is shallow. Such a flow indication may only be mud displacement resulting from gas expansion. By shutting in the well, any negative pressure imbalance should be reflected on the standpipe pressure gauge regardless of the conditions in the annulus. One point is worth considering, however. The ability to circulate gas to the surface when the mud density is insufficient to control pore pressure implies that the producing zone is tight. Consequently, some time may be required before a pressure buildup can be detected by the gauge.

At times, a well cannot be safely shut in. Under these circumstances, circulation should continue while the pump pressure and flowline returns are observed closely. The gas concentration ultimately decreases if drilled gas was the source. If the concentration remains the same after the annular volume has been displaced, gas is feeding into the borehole. Any drop in pump pressure exceeding 5% of the normal circulating pressure should lead the operator to direct returns down the choke manifold. Weighting up the system can begin while holding backpressure on the well equivalent to the loss in pump pressure.

One of the responsibilities of a mud logger is to monitor continuously for the presence of combustible gas in the mud. A gas trap placed in the possum belly of the shale shaker samples the mud as it exits the flowline. An agitator churns the mud, facilitating breakout, and gas is pulled from the trap by a vacuum pump. **Fig. 2.52** shows a gas trap in operation.

Fig. 2.52—A gas trap in operation.

Detection systems use a catalytic-filament or flame-ionization device to detect combustible gas in the vacuum stream. **Fig. 2.53** shows a portion of an example mud log. The recording instrument and log output for this particular system gives the total gas concentration as a fraction (percent or parts per million) of the vacuum stream. Alternatively, the concentration may be given as gas units. A gas unit is an arbitrary quantity that relates to the combustible gas fraction in the instrument throughput. Fifty gas units are equivalent to a 1% methane equivalent in many devices, but a gas unit may mean something different in other devices. It is important to use the information in relative terms rather than trying to attach some physical connotation to the measurements.

A gas chromatograph breaks down combustible gas into respective methane (C_1) through pentane (C_5) components. The recorded information typically is displayed on a fractional or percentage basis, giving an indication of the reservoir fluids and commercial potential of the drilled horizon. Chromatograph measurements may be used as a qualitative pore-pressure-prediction tool by tracking the ethane to propane ratio with depth. In normally compacted rock, the ethane (C_2) concentration is usually higher than propane (C_3) concentration. The reverse is often true within a pressure transition.

Some definitions are in order. The background-gas (BGG) level is the baseline concentration in the mud and usually is on the order of a few units under normal drilling conditions. Gas shows are composed primarily of drilled gas and are associated with a gas-bearing formation. During a drillstring connection, the backpressure provided by circulating losses in the annulus is removed and some swabbing occurs in a kelly-drilled hole. As result, a small amount of formation fluids

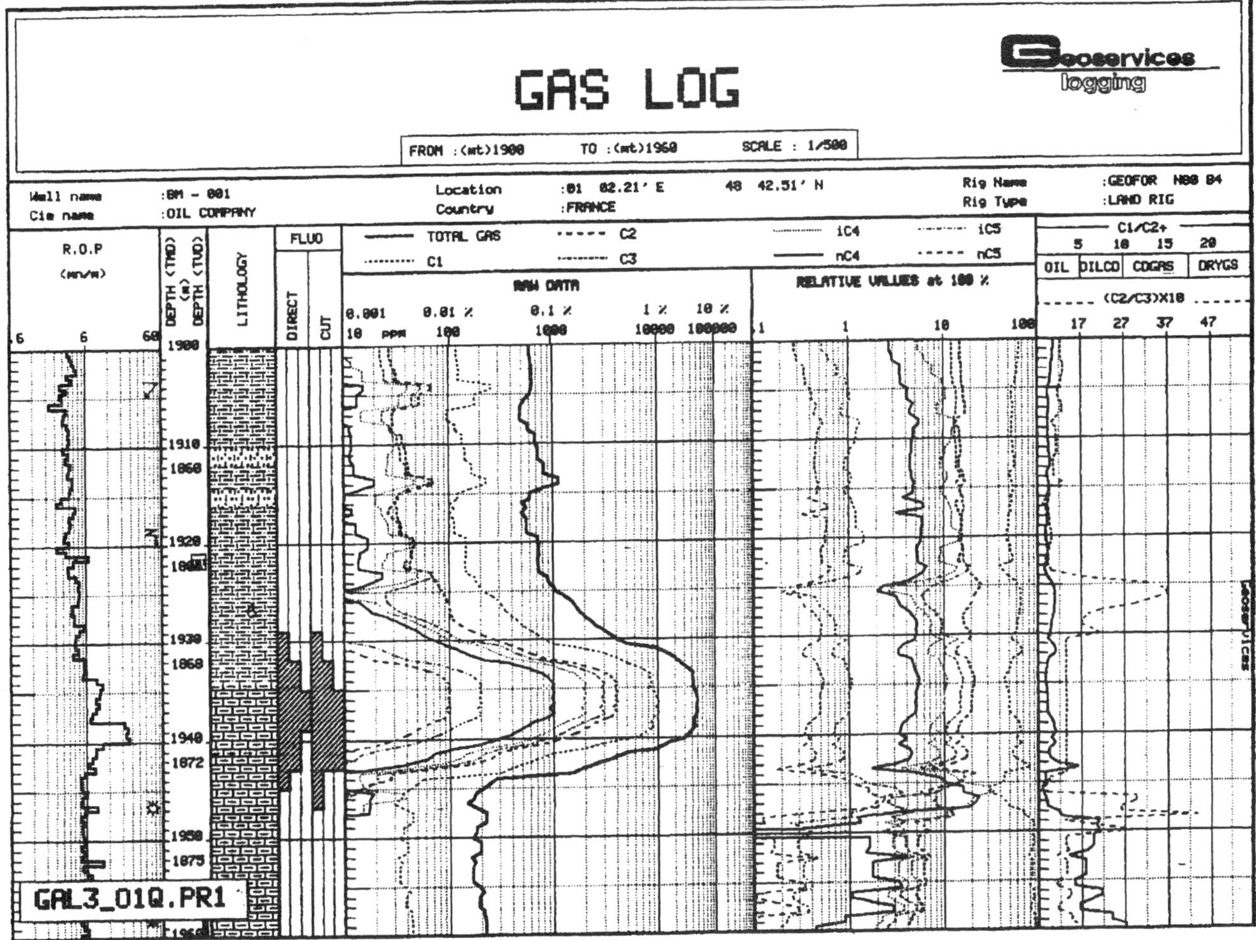

Fig. 2.53—Example mud log, courtesy of Geoservices.

may seep into a well during a connection and a connection-gas (CG) spike occurs on bottoms up. Trip-gas (TG) readings are normally higher than the nearby CG peaks because of the time factor and the greater potential for reducing wellbore pressure. The relative magnitude of the measured gas units is a direct function of the wellbore differential pressure. In fact, a highly overbalanced situation may suppress all the described gas indications.

Fig. 2.54 depicts characteristic log responses to some generalized downhole environments. Fig. 2.54a indicates a gas formation drilled at an overbalanced condition. BGG readings remain constant before and after the show as do the size of the CG peaks. Compare this response with that in Fig. 2.54b where the BGG attains a higher, although stable, level after drilling the show. Note that subsequent CG readings also are intensified. Continuing flow into the wellbore from the penetrated sand is the most logical conclusion and, although the inflow rate is small, an underbalanced condition is evident.

The track in Fig. 2.54c is more difficult to interpret. One possible inference that can be drawn from the size of the second CG response is that the annulus friction provided the overbalance margin across the show sand. In other words, the well becomes underbalanced when the pump is shut down. Alternatively, the static hydrostatic may be high enough but the newly exposed sand may be more permeable than the other exposed rock or the hole simply may have been swabbed harder during the next two connections.

The track in Fig. 2.54d shows a steady increase in both BGG and CG readings with depth. This is typical behavior for drilling through a thick shale transition. The increase in the amount of mud gas indicates a negative differential pressure and the need to increase mud density. The relative strength of the gas measurements will be suppressed back to approximately the original readings after the wellbore achieves a balanced or overbalanced condition.

2.8.3 Changes in Drilling-Fluid Properties. A change in mud properties may indicate contamination and an underbalanced wellbore if formation fluids are the contamination source. Gas cutting, a form of contamination, can influence drilling-fluid density and viscosity dramatically. Minor saltwater flows or acid-gas entries also affect various mud attributes. These changes may be sudden and noticed immediately or more subtle in nature and detected only by closely monitoring the mud properties over time.

Small saltwater additions will increase the chloride content of a freshwater mud progressively and flocculate bentonite clays. Flocculation is evidenced by an increased yield point, high gel strengths, increased water loss, and a reduction in filter-cake quality. Also, the system pH may be reduced in some cases. These signals are not definitive flow indications, however. Mud salinity can arise from other sources (makeup water or drilled rock salt) and muds can flocculate for reasons other than salt contamination.

Acid-gas (CO_2 and H_2S) contamination may result from slow seepage into a wellbore or from other nonformation sources. CO_2 detrimentally affects water-based muds by reducing the pH and providing a source of soluble carbonates,

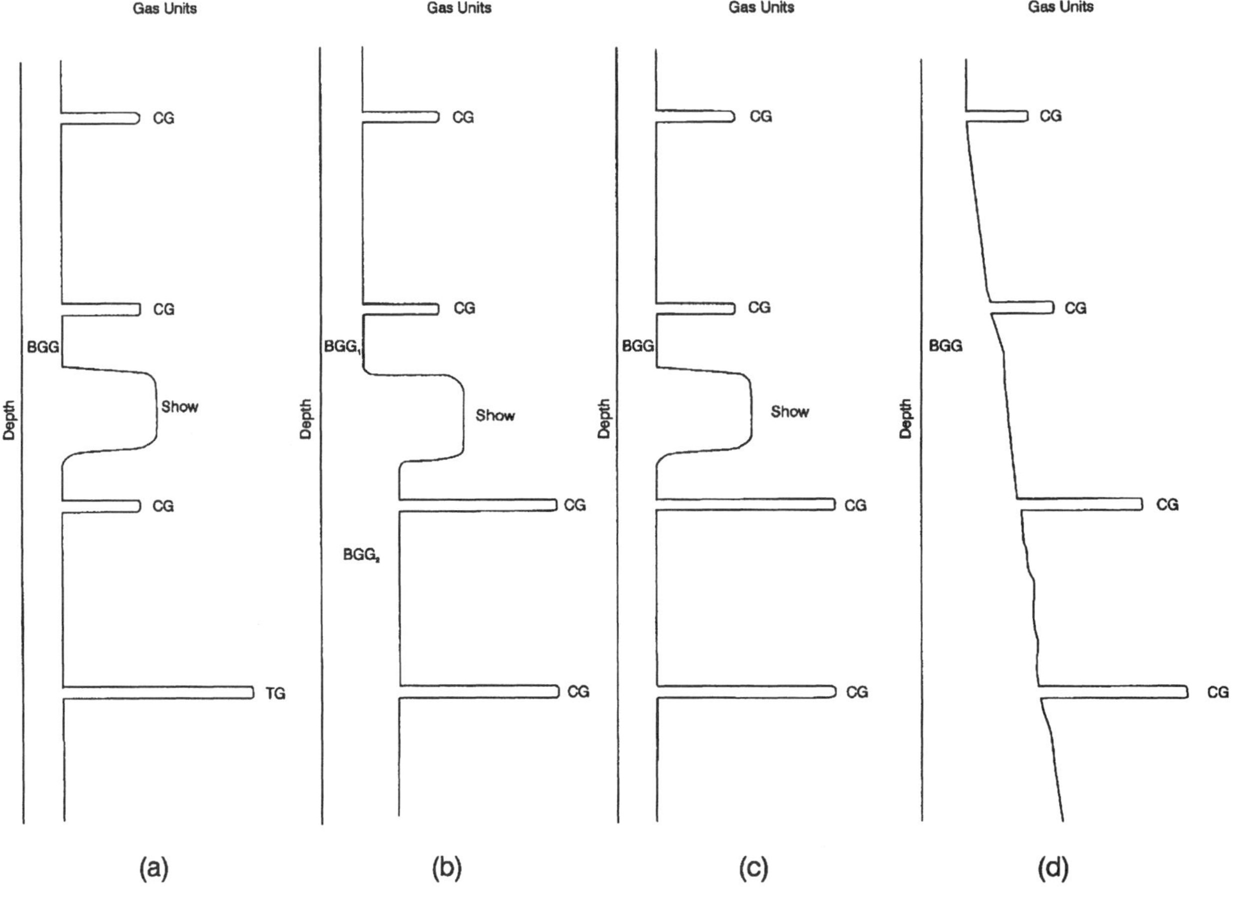

Fig. 2.54—Characteristic mud-logged gas-concentration response to various conditions.

another clay flocculant. Soluble carbonates may be detected by running a Garrett gas train analysis of the filtrate.

H_2S contamination is evidenced by pH reduction, a foul odor, and a black appearance in weighted muds because of its reaction with the iron minerals in commercial barite to form iron sulfide. The presence of H_2S can be confirmed by separating the gas with a Garrett gas train and analyzing the sample with a Dräger tube. H_2S is a deadly substance and highly corrosive to steel goods when it is in solution form. It is imperative that immediate steps be taken to raise the pH and precipitate any soluble sulfides with scavengers. Following or coincident to the mud treatment, efforts should focus on identifying and eliminating the problem source.

2.8.4 Flowline Temperature. Consider heat conduction outward from the earth's core. The heat flux u_H is described by the relation

$$u_H = \lambda g_G, \quad \text{(2.29)}$$

where λ = the thermal conductivity of the rock and g_G = the geothermal gradient. The thermal conductivity of bulk rock depends on the matrix material, porosity, and pore fluid. We can assume that the heat flux is constant for a given area. Hence, the temperature gradient in a rock stratum is dependent on its thermal conductivity.

Rock grains have a much higher thermal conductivity than pore fluids; and, under normal sediment compaction (porosity reduction), we would expect to see greater capacity to transmit heat with increasing depth. Lewis and Rose[60] proposed that undercompacted beds act as an insulating layer because of the lower thermal conductivity. Thus, heat is retained in the rock and a temperature gradient anomaly is evidenced across the abnormally pressured interval. **Fig. 2.55** illustrates this effect. Earth isotherms normally are perpendicular to the

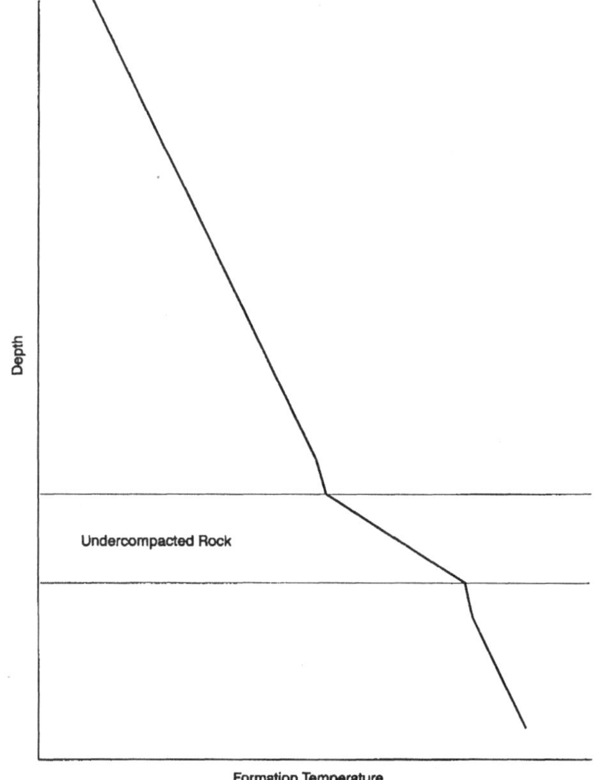

Fig. 2.55—Effect of undercompaction on formation temperatures (after Lewis and Rose[60]).

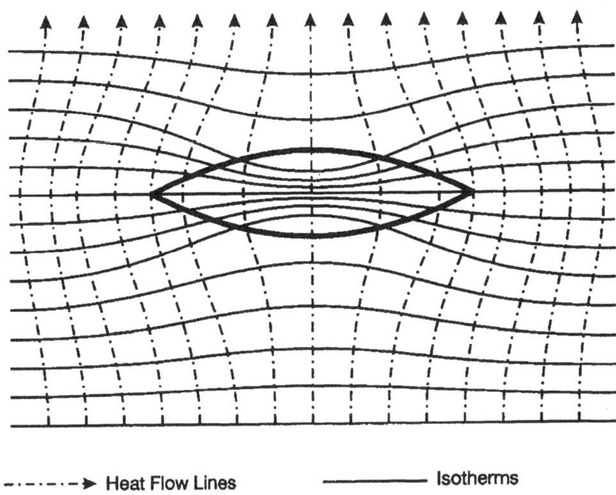

Fig. 2.56—Distribution of heat flow lines and isotherms in the vicinity of an insulating body (after Lewis and Rose[60]).

lines of heat flow. **Fig. 2.56** shows isotherms more widely spaced immediately above and below the insulating body, which is the reason for the reduced temperature gradients shown in Fig. 2.55.

Given sufficient data points, well-planning predictions may be assisted by downhole temperature measurements in offsetting wells. Wellbore temperature measurements, however, should be corrected for the effect of prior well events, such as circulation or production, to approximate the undisturbed earth temperature. Empirical correlations and simulation models are available for this purpose.

Measuring the temperature of mud as it exits a well also gives an indication of the subsurface temperature environment and can be a useful pore-pressure-detection method while drilling. The returning drilling-fluid temperature increases naturally as drilling proceeds and should follow a predictable trend with depth. As expected, a change in the subsurface rock temperature gradient leads to a change in the heat-transfer rate and to an anomaly in the established trend. In practice, the measured temperatures are plotted vs. depth after making the appropriate adjustment for bottoms-up lag time. A deviation from the normal temperature trend signals an increasing pore-pressure gradient. **Fig. 2.57** shows flowline temperature measurements from a North Sea well. A negative gradient anomaly is observed above the geopressured rock, followed by a rapid temperature increase within the transition.

Monitoring flowline temperature may be the only viable while-drilling prediction method in hard-rock country and areas lacking clean shales. The technique applies in soft rock as well, although more accurate methods are available for operations in these areas. Offshore drilling, particularly in cool or deep water, may pose some difficulty because the marine riser or conductor serves as an effective heat exchanger with the surrounding sea.

Several operational variables result in a temperature change at the flowline and should be considered when selecting the data. Circulation time is one of the more significant factors. Stopping circulation for a trip results in rapid cooling of the mud in the surface pits and in the upper part of the hole. Circulation time is necessary before an approximate equilibrium is achieved again. Raymond,[62] using an early simulation model, found that one or two full circulations were required before wellbore temperatures achieved fairly stable values.

The temperature of the mud in the suction pit is another important variable. Transferring mud between pits, liquid mud additions, and surface climate changes all affect the inlet mud temperature and, hence, the validity of the flowline temperature trend. **Fig. 2.58** shows data from frequent changes in outlet temperature plotted end-to-end without regard to actual temperature values.

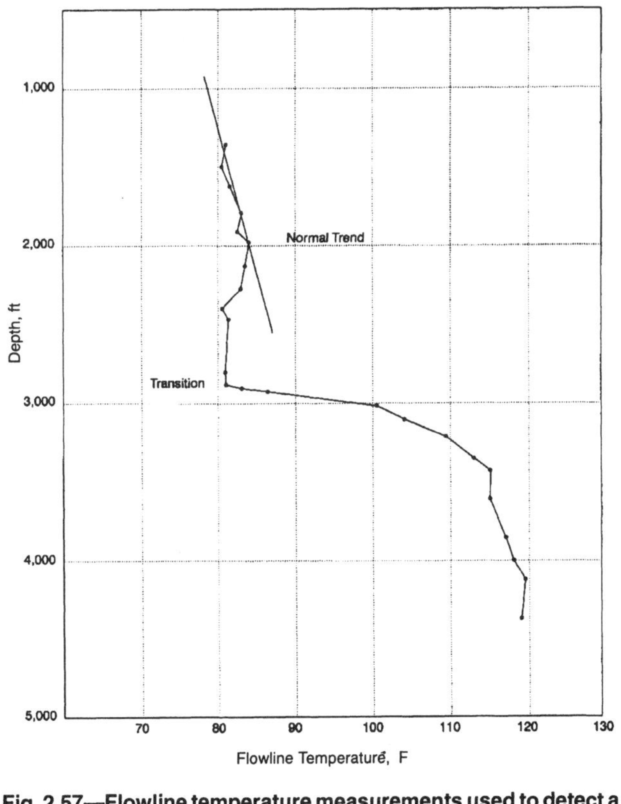

Fig. 2.57—Flowline temperature measurements used to detect a pore-pressure transition on a North Sea well.[61]

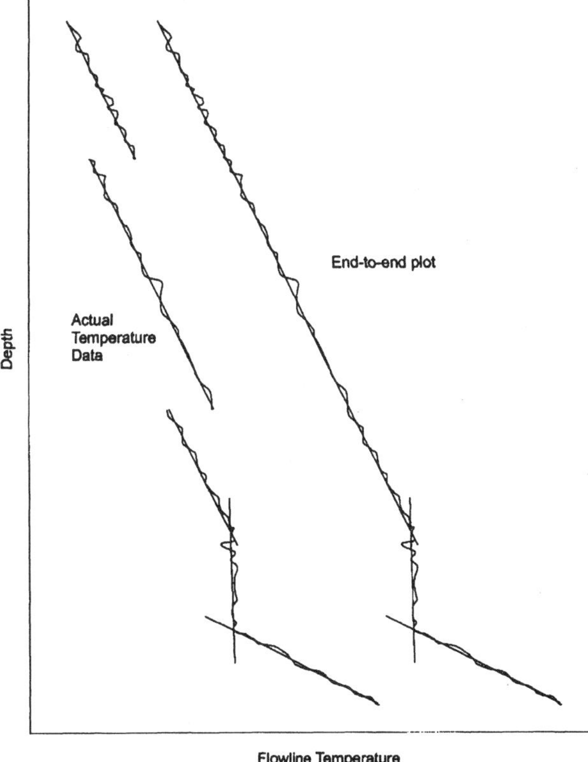

Fig. 2.58—Using an end-to-end plotting technique to compensate for flowline temperature scatter.

2.8.5 Hole Conditions. Drilling torque and drag during trips or connections result from friction between the drillstring or bit and the walls of the hole. Torque and drag increase naturally with depth, but a relatively sudden increase must have an underlying reason and a prudent operator investigates to determine the cause. Hole instability is just one possible reason for excessive torque and drag, though a rapid onset of hole drag is fairly substantive evidence of an unstable hole. Circulating bottoms up and observing the samples furnish a more direct indication of hole instability and may suggest a pore-pressure source. Torque and drag should be considered a secondary tool for predicting pore pressures and only when observed in conjunction with a drilling-rate increase or other primary indications.

2.9 Conventional Log Correlations

Several empirical correlations that relate logged parameters to pore pressure have been developed since the mid-1960s. Most have fairly narrow application and are limited to the areas where the techniques were developed unless modifications are made to the original procedure. Historically, the techniques have been devised in shales where undercompaction is the predominant geopressure mechanism and applied successfully. Work in recent years, however, has extended the potential for predicting pore pressures to rock other than shales and to areas where other overpressuring mechanisms play an important role.

Any log that implies shale porosity can indicate the compaction state of the rock and, hence, abnormal pore pressures associated with undercompaction. However, most of the published correlations are based on sonic- and electric-log data. Density logs, for instance, have been largely ignored in the literature. This may be because of the common practice of recording bulk densities only through potentially productive pay horizons, thus leaving a scarcity of shallow measurements required to establish a normal-compaction trend.

Pioneering work in this field was done in the U.S. gulf coast region. Most of the empirical log correlations discussed in the scope of this chapter originated in and are applicable to this area. Pore-pressure relationships for other regions have been developed and several are available in petroleum literature.[63-65] Unfortunately, many others have received scant publication and may be difficult to locate in the public domain.

2.9.1 Sonic-Log Correlations. In their most basic form, sonic- or acoustic-logging tools generate acoustic energy from transmitters located at the top and bottom of the tool. A series of receivers located in the center of the tool record the vibrational waveforms sent out by the two transmitters. The first waves to arrive at the receivers originate as pressure pulses in the borehole fluid, traverse the near-wellbore rock as compression waves, and are picked up by the receivers again as pressure pulses. Acoustic velocity through rock medium is calculated from tool geometry and recorded on log tracks in units of transit time (reciprocal of velocity).

The seismic discussion in Sec. 2.6 covered the relationship between transit time and porosity. However, seismic predictions, must use transit times averaged over a lithological sequence, whereas conventional acoustic-log methods concentrate on clean shales. Eq. 2.14 can be used to compute porosity if the matrix and fluid transit times are known. Most sonic-log procedures, however, use direct Δt measurements and estimate pore pressure with an empirical correlation or the equivalent-depth method.

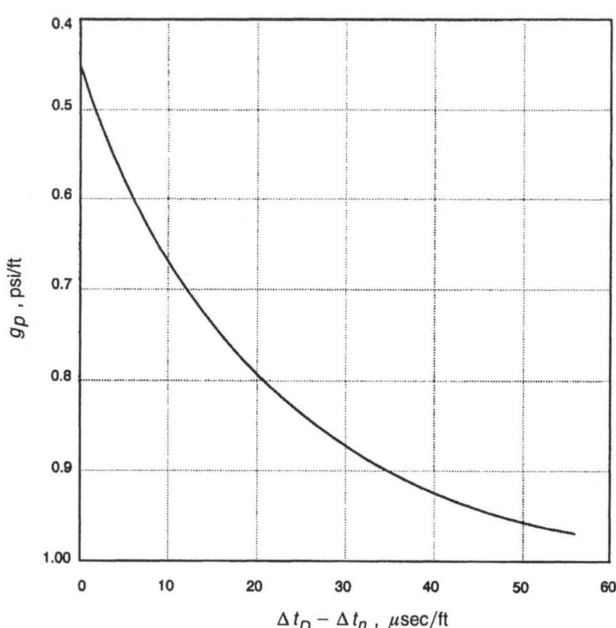

Fig. 2.59—Hottman and Johnson[66] correlation between shale transit time and pore pressure for gulf coast Miocene/Oligocene shales.

Fig. 2.59 shows the pore-pressure relationship developed by Hottman and Johnson[66] from transit-time data gathered in the Lousiana/upper Texas gulf coast. To use the correlation, travel times obtained in shales are plotted vs. depth on semilogarithmic graph paper. Sufficient measurements in normally compacted rock are obtained and a normal compaction trend line is established. Deviation from the normal trend into higher transit times indicates the onset of higher porosities, thus undercompaction for the burial depth. A quantitative pore-pressure estimate is obtained by taking the difference between the observed and normal transit times ($\Delta t_o - \Delta t_n$) and reading the pore-pressure gradient from the curve.

Matthews and Kelly[67] later presented empirical charts applicable to three Tertiary formations in the south Texas gulf coast. **Fig. 2.60** presents their sonic-log correlation, used in the same manner as Hottman and Johnson's curve. **Fig. 2.61** presents correlations specific to the North Sea and South China Sea Tertiary basins. Comparing the variety of established sonic-log relationships points out the often discussed requirement for applying empirical procedures only in the appropriate area and geologic age.

Example 2.19. Table 2.14 lists shale transit times for a well in Jefferson County, Texas.

1. Plot the data and determine the top of the pore-pressure transition.

2. Determine the pore-pressure gradient at 11,190 ft using the Hottman and Johnson correlation.

3. Use the equivalent-depth method to estimate the pore-pressure gradient at this depth and compare the result with that obtained by the empirical chart.

Solution.

1. **Fig. 2.62** plots the data. Transition depth is indicated at about 9,300 ft.

2. The observed transit time at 11,190 ft is 146 μsec/ft, while the extrapolated normal transit time is 109 μsec/ft. Taking the difference,

$$\Delta t_o - \Delta t_n = 146 - 109 = 37\,\mu\text{sec/ft}.$$

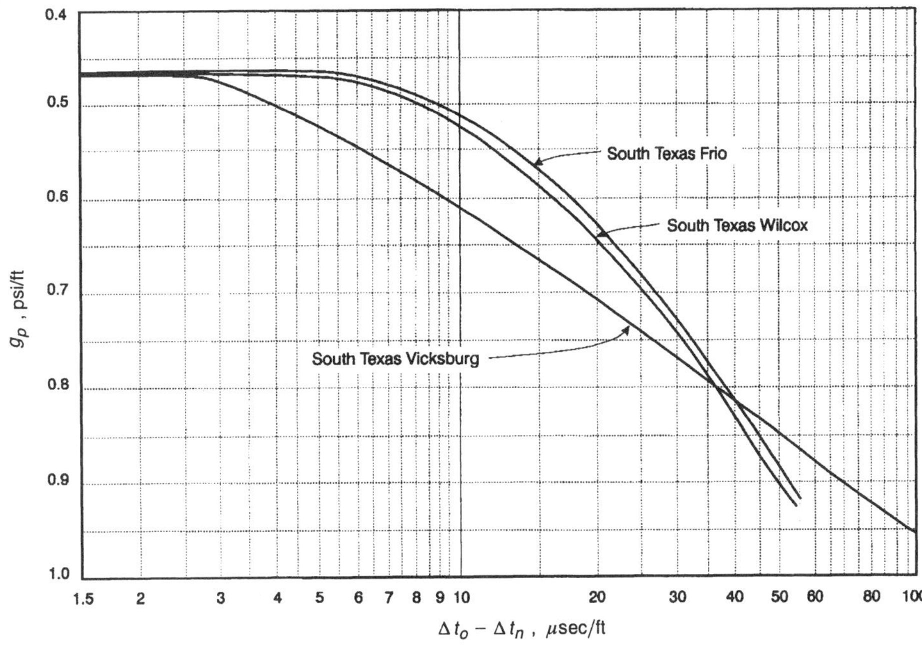

Fig. 2.60—Matthews and Kelly[67] transit-time correlation for Oligocene-Eocene shales in the south Texas gulf coast.

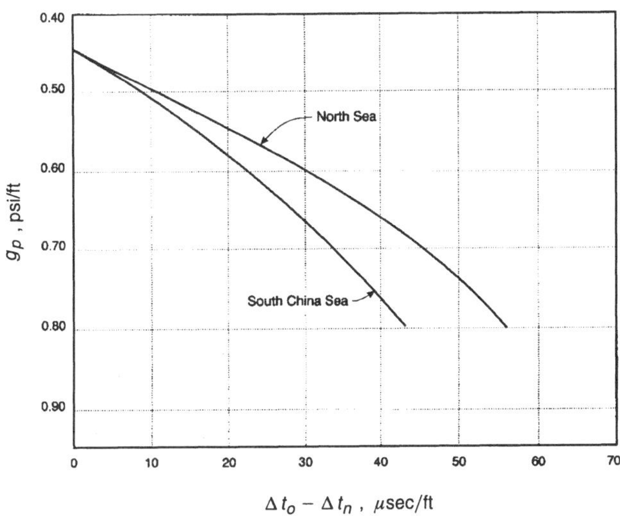

Fig. 2.61—Transit-time correlations for North Sea and South China Sea Tertiary sediments.[68]

The pore-pressure gradient from Fig. 2.59 is 0.91 psi/ft.

3. Use Eaton's gulf coast correlation from Fig. 2.20 to determine the overburden-stress gradient at 11,190 ft and compute

$$\sigma_{ob} = (0.956)(11,190) = 10,698 \text{ psig.}$$

The equivalent depth in terms of compaction and grain stress is found to be 4,520 ft from the semilog plot. The overburden and pore pressures at 4,520 ft are, respectively,

$$\sigma_{ob(eq)} = (0.902)(4,520) = 4,077 \text{ psig}$$

and $p_{n(eq)} = (0.465)(4,520) = 2,102$ psig.

Eq. 2.13 yields

$$p_p = 2,102 + (10,698 - 4,077) = 8,723 \text{ psig,}$$

which gives a gradient of

$$g_p = 8,723/11,190 = 0.780 \text{ psi/ft.}$$

TABLE 2.14—SHALE ACOUSTIC-TRANSIT TIMES FOR A WELL IN JEFFERSON COUNTY, TEXAS[66]

Depth (ft)	Transit Time (μsec/ft)
2,820	158
3,210	153
4,000	150
4,170	152
4,520	146
5,210	141
6,000	138
6,210	135
6,970	135
7,500	130
7,810	124
8,000	120
8,320	122
8,410	121
9,000	120
9,010	117
9,220	118
9,300	119
9,390	121
9,410	124
9,580	123
9,620	127
9,710	131
9,810	134
9,900	139
10,010	141
10,100	145
10,200	148
10,370	147
10,540	146
11,190	146
11,380	142
11,720	147
12,300	141
13,000	139

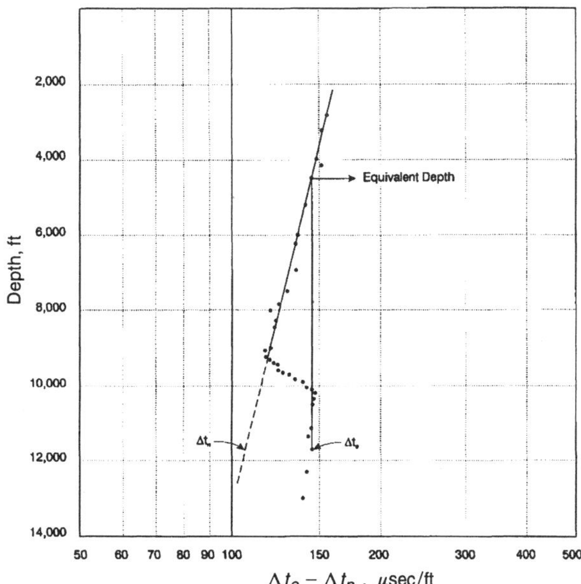

Fig. 2.62—Shale transit-time plot for a well in Jefferson County, Texas.

The equivalent-depth method with sonic-log data should be a valid approach to pore-pressure prediction if undercompaction is the abnormal pressure source. Yet the pore-pressure prediction with the Hottman and Johnson[66] curve differs from the equivalent-depth result by 1,460 psi. In other words, the predicted mud weight needed to balance the pore pressure exactly at 11,190 ft varies by 2.5 lbm/gal, depending on the selected technique. Obviously, something is wrong with the assumptions underlying one, if not both, of the two approaches.

The answer may lie in the basis of any pore-pressure-prediction curve generated from composite well data. Recall that pore pressure is a function of overburden stress, yet no provision is made in the Hottman and Johnson or similar correlations for a variable overburden-stress gradient. Instead, trend line divergence is related to pore pressure from a data base consisting of measurements generally taken from the same approximate depth range. Thus, a potential source of error exists if the depth of interest in the subject well is under an overburden stress different from what would be reflected by the average curve.

However, this is not a satisfactory explanation in this particular case because the example well was one of the 18 wells used to construct the Hottman and Johnson curve. Indeed, the actual pressure measured from a sand at 10,890 ft lends credence to our 0.91 psi/ft result. As an alternative explanation, geopressure mechanisms other than compaction appear to play a role in this area and the equivalent-depth method is not a suitable approach. We will return to this question later.

2.9.2 Resistivity and Conductivity Logs. The ability of rock to conduct electric current is another property used to infer porosity. Earlier electric-logging-tool designs (called normal devices) pass an electric current from the tool into the mud and rock and measure the voltage potential created by the current. The resistivity of the rock mass in units of ohm meters ($\Omega \cdot$ m) is proportional to this voltage. Conductivity is the reciprocal of resistivity and generally is displayed in units of 10^{-3} m℧/m.

With some exceptions the grains that make up a rock matrix generally are nonconductive. The fact that saline pore water is a conductor provides the basis for inferring porosity from bulk resistivity or conductivity measurements. Under normal shale compaction, a trend of increasing resistivity and decreasing conductivity should be noted with burial depth. Deviation from the trend into lower resistivities with corresponding higher conductivities signals a pore-pressure transition. However, factors other than porosity impact shale resistivity. Detecting a transition or predicting pore pressure with electric-log measurements requires due consideration of other effects.

The porosity of water-saturated rock can be determined with the classic Archie[69] relationships. The formation resistivity factor, F_R, is defined by

$$F_R = R_0/R_w, \quad \ldots\ldots (2.30)$$

where R_0 = the resistivity of a formation fully saturated with water and R_w = the resistivity of the pore water. R_w reflects the dissolved salt content in the water, but is also a function of temperature. For the same NaCl concentration, R_w decreases with temperature according to the Arps[70] equation.

$$R_{w2} = R_{w1}\left(\frac{T_1 + 6.77}{T_2 + 6.77}\right), \quad \ldots\ldots (2.31)$$

where T = the temperature, °F. The constant 6.77 becomes 21.5 if the temperature is expressed in °C. Hence, R_w of a given solution decreases naturally with increasing burial depth.

The porosity of water-saturated rock is calculated by

$$\phi = aF_R^{-1/m}, \quad \ldots\ldots (2.32)$$

where a and m = constants defined as the formation-resistivity factor coefficient and cementation exponent, respectively. These constants can be determined accurately in the laboratory, but acceptable results in clastics can be obtained by assigning a and m values of 1.0 and 2.0, respectively. Substitution leads to

$$\phi = \left(R_0/R_w\right)^{-0.5}. \quad \ldots\ldots (2.33)$$

Normally, shales are considered to be saturated with water, although hydrocarbons or organic materials may be present. Thus, the shale resistivity measured by the logging tool is taken to be R_0. The value for R_w in shales cannot be measured directly, but a common assumption is that shale pore water has the same salinity as that found in a nearby sand. R_w in a sand may be deduced by use of a spontaneous-potential (SP) -log response or by directly measuring a water sample and adjusting for the temperature.

Archie's equations were developed for clean, nonshaly rock. Clays are unique minerals in that the loosely held cations adjacent to the adsorbed water layer supplement the conductivity supplied by interstitial water. Hence, the R_0 measured in a shale is influenced by the pore-water resistivity, the CEC of the clay, and the relative proportion of bound to free water. For this reason, dual-water models such as that proposed by Waxman and Smits[71] describe formation factors in dirty sands or shales better. Nonetheless, Archie's equations are used in many cases to estimate shale porosity.

Pore-water salinity is expected to increase with burial depth in normally compacted sediments because shales, acting as a semipermeable membrane, sieve ions from the expelled water. A decrease in pore-water salinity and an increase in R_w should accompany any undercompacted rock. In fact, it is

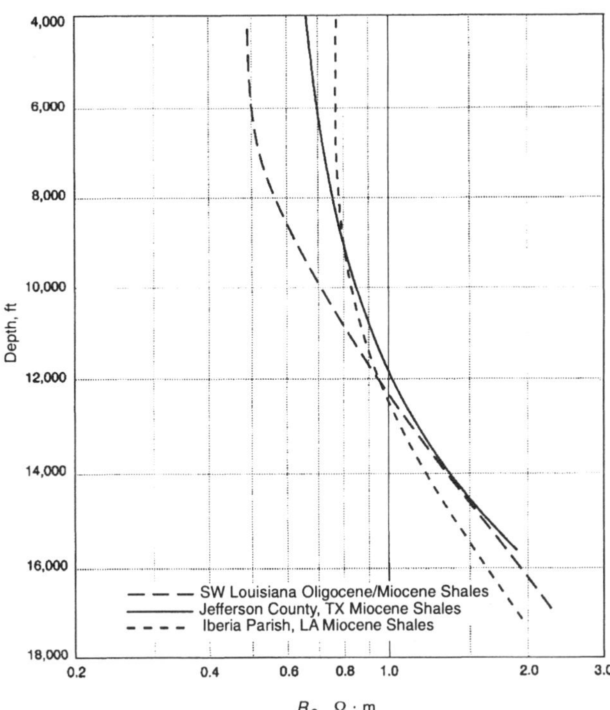

Fig. 2.63—Normal-compaction resistivity trends observed in gulf coast Miocene/Oligocene shales.[66]

common for some abnormally pressured sands to produce near-fresh water.

Because R_w over a given depth range is not constant, the proportionality between R_0 and porosity is not the same at every depth. Therefore, a plot of normally compacted shale resistivities with depth would not be expected to fit a straight line on semilogarithmic graph paper. This is evidenced by the Hottman and Johnson[66] composite data for the gulf coast shown in **Fig. 2.63.** It should be apparent that shale formation factors are more reliable porosity indicators than conductivity or resistivity measurements.

Foster and Whalen's[72] Method. Foster and Whalen discussed an effective-stress approach for predicting pore pressures with computed F_R data. R_0 measurements are obtained in shales, and R_w values for these shales are estimated from the spontaneous potential (SP) response in adjacent sands. F_R for each shale resistivity reading is determined by use of Eq. 2.30 and plotted vs. depth on semilog graph paper. Transition into abnormal pressure is evidenced by deviation from the normal compaction trend into lower F_R values, and the equivalent depth procedure is used to estimate pressures below the transition depth.

TABLE 2.15—SHALE RESISTIVITY AND ESTIMATED PORE-WATER RESISTIVITY DATA FROM AN OFFSHORE LOUISIANA WELL[72]

Depth (ft)	R_0 (Ω·m)	R_w (Ω·m)	R_w Depth (ft)	F_R
3,110	0.55	0.072	3,611	7.64
3,538	0.55	0.072	3,611	7.64
4,135	0.55	0.066	4,310	8.33
4,544	0.50	0.051	4,625	9.80
4,890	0.50	0.049	4,950	10.20
5,175	0.55	0.049	4,950	11.22
5,363	0.50	0.045	5,475	11.11
5,867	0.50	0.041	6,100	12.20
6,041	0.50	0.041	6,100	12.20
6,167	0.54	0.041	6,100	13.17
6,482	0.55	0.045	6,540	12.22
6,577	0.55	0.045	6,540	12.22
6,955	0.70	0.039	6,910	17.95
7,113	0.70	0.038	7,280	18.42
7,255	0.70	0.038	7,280	18.42
7,696	0.71	0.030	7,900	23.67
8,200	0.76	0.028	8,400	27.14
8,342	0.85	0.028	8,400	30.36
8,767	0.80	0.029	8,600	27.59
9,113	0.85	0.025	9,460	34.00
9,492	0.91	0.025	9,460	36.40
9,665	0.86	0.025	9,460	34.40
9,996	0.80	0.025	9,460	32.00
10,217	0.85	0.024	10,700	35.42
10,485	0.92	0.024	10,700	38.33
10,659	0.91	0.024	10,700	37.92
10,989	0.90	0.024	10,700	37.50
11,162	0.91	0.016	11,400	56.88
11,487	0.90	0.016	11,400	56.25
11,588	1.20	0.016	11,400	75.00
11,776	1.16	0.018	11,800	64.44
11,966	1.10	0.019	12,020	57.89
12,265	1.11	0.019	12,350	58.42
12,470	0.96	0.019	12,350	50.53
12,550	0.90	0.019	12,350	47.37
12,785	1.06	0.019	12,880	55.79
13,069	0.91	0.019	12,880	47.89
13,385	1.10	0.019	13,290	57.89
13,573	1.05	0.024	13,700	43.75
13,778	1.06	0.024	13,700	44.17
13,983	0.96	0.024	13,700	40.00
14,188	0.96	0.034	14,300	28.24
14,487	0.71	0.030	14,500	23.67
14,566	0.80	0.030	14,500	26.67
14,833	0.80	0.037	14,680	21.62
14,960	0.90	0.065	15,090	13.85
15,275	1.06	0.065	15,090	16.31

Example 2.20. Table 2.15 lists shale resistivities from an offshore Louisiana well. The R_w for each shale reading is estimated from an SP measurement in a nearby sand; Cols. 3 and 4 of Table 2.15 give R_w values and sand depths, respectively. Estimate the pore pressure at 14,188 ft using Foster and Whalen's technique.

Solution. The last column of Table 2.15 lists formation factors computed with Eq. 2.30. The formation factor at 14,188 ft is

$$F_R = 0.96/0.034 = 28.24.$$

Fig. 2.64 plots the data. Deviation from normal compaction appears to begin at approximately 11,800 ft. With Eaton's gulf coast correlation, g_{ob} at 14,188 ft is 0.974 psi/ft and the overburden stress is determined as

$$\sigma_{ob} = (0.974)(14,188) = 13,819 \text{ psig}.$$

The equivalent depth is found to be 8,720 ft, where the overburden and pore pressure, respectively, are

$$\sigma_{ob(eq)} = (0.937)(8,720) = 8,171 \text{ psig}$$

and $p_{n(eq)} = (0.465)(8,720) = 4,055$ psig.

The predicted pore pressure and pressure gradient (density equivalent), respectively, are

$$p_p = 4,055 + (13,819 - 8,171) = 9,703 \text{ psig}$$

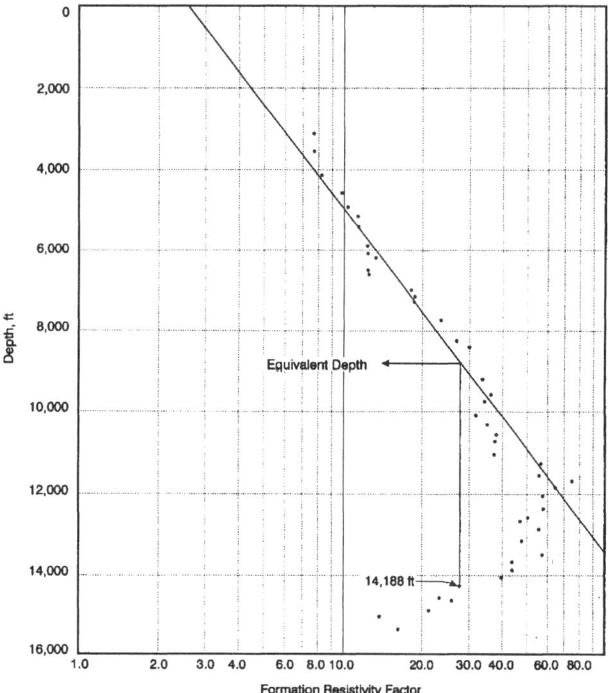

Fig. 2.64—Shale formation resistivity factor plot from an offshore Louisiana well.

and $\rho_p = (19.25)(9,703)/14,188 = 13.16$ lbm/gal.

Empirical Correlations. Lacking complete R_w information, quantitative pore-pressure predictions from electric logs must rely on site-specific empirical correlations. **Fig. 2.65** shows Hottman and Johnson's[66] upper gulf coast relationship between shale resistivity and pore pressure while **Fig. 2.66** shows the Matthews and Kelly[67] south Texas curves. These correlations use a procedure similar to the travel-time technique, except that the ratio between the normal and observed parameters (R_n/R_o or C_o/C_n) is entered as the abscissa.

Considering Fig. 2.63, shale resistivities do not obey an exponential relationship with depth. Some curve-fitting is recommended to yield the best normal-compaction trend in the subject well, particularly if the pore-pressure transition is shallow. Even so, the most common procedure is to extrapolate a straight normal-compaction line into the abnormally-pressured sediments.

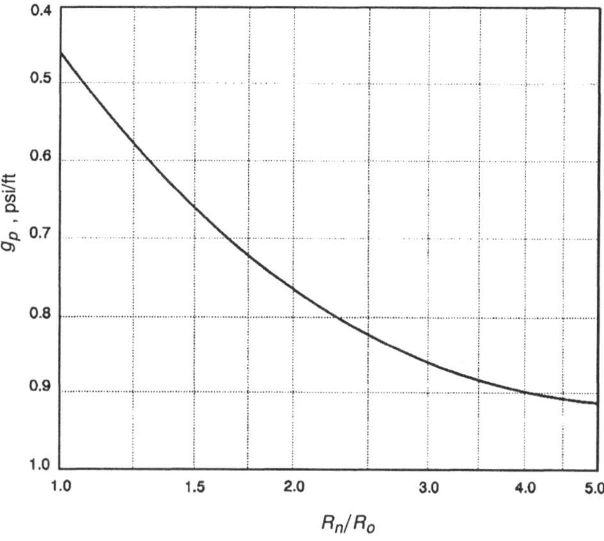

Fig. 2.65—Hottman and Johnson[66] correlation between shale resistivity and pore pressure for gulf coast Miocene/Oligocene shales.

Example 2.21. Table 2.16 lists Frio shale conductivities from a well in Nueces County, Texas. Determine the transition depth and estimate the pore-pressure gradient at 11,500 ft using the Matthews and Kelly correlation.

Solution. **Fig. 2.67** plots the conductivity values and a normal-compaction trend line is fit through the shallow points. Deviation from the normal-compaction trend begins at 9,600 ft, defining the transition depth. The observed conductivity at 11,500 ft is 1920 m℧/m while the extrapolated normal conductivity is 440 m℧/m. Taking the ratio between the observed and normal conductivities,

$$C_o/C_n = 1920/440 = 4.36.$$

The pore-pressure gradient is 0.81 psi/ft from the Frio curve in Fig. 2.66.

2.9.3 Eaton's[47,48] Equations. None of the log relationships discussed thus far account for the effect that a variable overburden gradient has on the effective stress and pore pressure. Hence, pore-pressure predictions from these correlations may be in significant error if the depth (overburden gradient) differs from the composite well average depth. Eaton made an important contribution to the industry when he developed a series of empirical equations that do incorporate the overburden gradient. Now more than 20 years old, these relationships are the most widely used of the log-derived methods. For acoustic transit time, resistivity, and conductivity data plots on semilog paper, respectively:

$$g_p = g_{ob} - (g_{ob} - g_n)(\Delta t_n/\Delta t_o)^3, \quad \text{(2.34)}$$

$$g_p = g_{ob} - (g_{ob} - g_n)(R_o/R_n)^{1.2}, \quad \text{(2.35)}$$

and $$g_p = g_{ob} - (g_{ob} - g_n)(C_n/C_o)^{1.2}. \quad \text{(2.36)}$$

Example 2.22. An offshore Louisiana well in West Cameron Block 192 has an observed to normal resistivity ratio (R_o/R_n) of 0.264 in a Miocene shale at 11,494 ft.[73] An integrated density log on the well indicates an overburden-stress gradi-

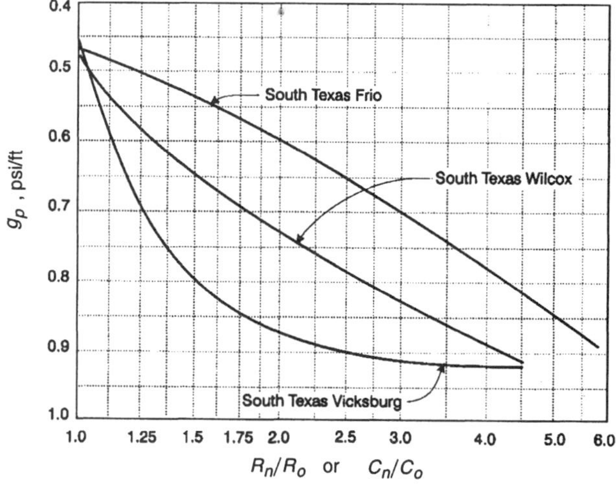

Fig. 2.66—Matthews and Kelly[67] correlation between shale resistivity or conductivity and pore pressure for gulf coast Tertiary shales.

TABLE 2.16—FRIO SHALE CONDUCTIVITY VALUES FOR A WELL IN NUECES COUNTY, TEXAS[67]	
Depth (ft)	Conductivity (m℧/m)
7,400	710
7,550	790
8,300	710
8,350	690
8,400	680
8,500	690
9,200	600
9,300	590
9,550	570
9,600	590
9,700	610
9,750	620
9,900	700
9,950	830
10,000	950
10,050	1,100
10,150	1,200
10,200	1,240
10,300	1,310
10,500	1,250
10,600	1,320
10,650	1,370
10,850	1,500
11,000	1,280
11,050	1,400
11,200	1,650
11,300	1,840
11,500	1,920

ent of 0.920 psi/ft at this depth. Estimate the pore-pressure gradient using (1) Eaton's technique and (2) Hottman and Johnson's correlation.

Solution.

1. Substituting terms into Eq. 2.35 gives

$$g_p = 0.920 - (0.920 - 0.465)(0.264)^{1.2}$$

$$= 0.827 \text{ psi/ft}.$$

2. Determine the R_n/R_o ratio using the Hottman and Johnson curve,

$$R_n/R_o = 1.0/0.264 = 3.79.$$

Fig. 2.65 gives a pore-pressure gradient of 0.894 psi/ft.

The pressure predictions from the two techniques differ by 770 psi and 1.3-lbm/gal equivalent density. However, the actual gradient in the subject well was determined to be 0.818 psi/ft, a result that supports Eaton's approach.

Eaton originally proposed that the equations as published were suitable for any area and offered three examples to support this conclusion. Subsequent usage, however, suggests that slight modifications to the exponent terms are necessary in older, less compactible shales. For example, it is common procedure to reduce the resistivity exponent to 1.0 when changing from Miocene to Oligocene rock. Modifications to the base equations should be derived by comparing calculated pore pressures with the actual measured values in adjacent permeable sections. Service companies working in a specific area may have more accurate exponent numbers for use in the Eaton equations.

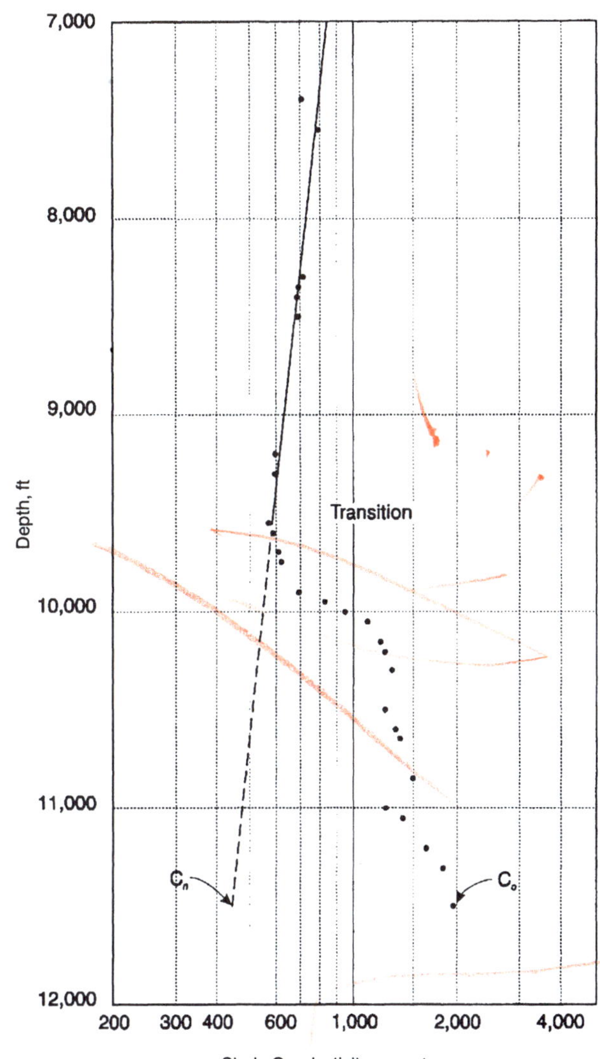

Fig. 2.67—Frio shale conductivity plot for a well in Nueces County, Texas.

2.9.4 Quality Control. Sonic logs are more versatile and offer better accuracy than electric-log correlations. However, almost every well has resistivity or conductivity information at least through the surface casing depth, while acoustic logs generally are scarcer across shallow strata. Regardless of which logs are available, some comments pertaining to quality control of the log data and trend-line construction are worthwhile.

Hydration of reactive clays in the vicinity of the borehole affect both the measured-sonic and electric-log readings. Inhibitive muds and deep-investigation tools mitigate the effects, but emphasis still should be placed on the more recent (deeper) data when constructing a normal trend line. Obviously, shale hydration should not be a major concern when obtaining resistivity data from an LWD tool.

Constructing a normal-compaction trend line may be difficult in many cases, particularly if clean shales are lacking or if some of the data are poor quality. In addition, the truism pertaining to trend-line shifts and slope changes with geologic age also applies to openhole-log methods. This presents a major interpretation problem in some cases, particularly if numerous fault blocks have been encountered or the geology is complicated in any way. Most wells require more than one trend-line fit of the normally compacted data.

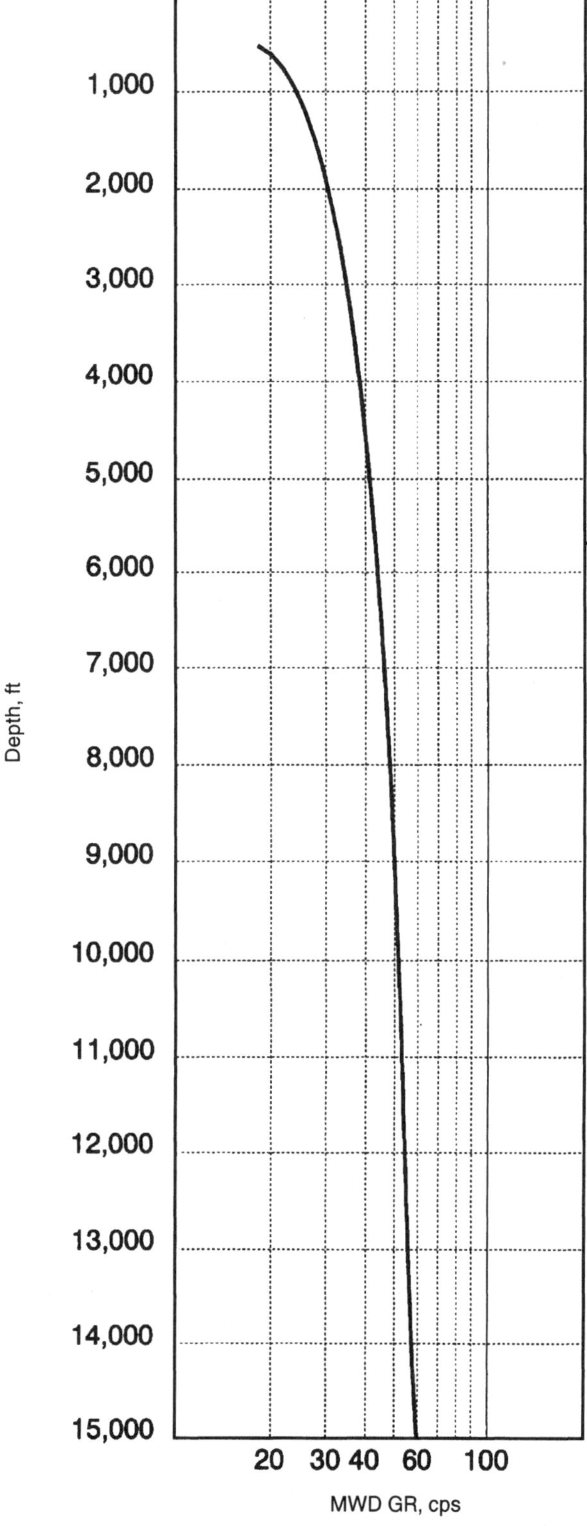

Fig. 2.68—MWD GR intensity with depth for normally compacted gulf coast shales.[75]

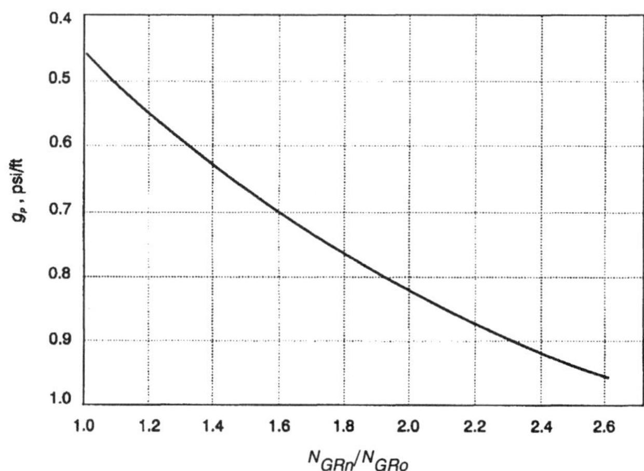

Fig. 2.69—Zoeller's empirical gulf coast correlation between MWD GR intensity ratio and pore pressure.[75]

Foster[74] discussed the preparation and use of multiple overlays, covering Pleistocene through Wilcox (late Paleocene) rock, to assist pore-pressure-prediction efforts in the gulf coast. Similar overlays, or at least the expected trend slope for a given geologic age, may be available in other areas from local service companies or operators. Characterizing the geologic sequence is an important consideration and input from the prospect geologist is a definite asset in predicting pressure.

2.9.5 Natural Gamma Ray (GR). In concluding the discussion of the trend-line methods, GR logs have seen some recent application. GR tools measure the natural radioactive emissions of rock. The three most common radioactive elements found in sediments are the potassium K^{40} isotope, uranium, and thorium. K^{40} tends to concentrate in shale minerals, which leads to the traditional use of GR tools to determine the shaliness of a stratum. GR intensity may be used to infer porosity in shales of consistent mineralogy.

Zoeller[75] developed a pore-pressure-prediction method for gulf coast shales using an MWD GR correlation. **Fig. 2.68** depicts GR intensity, measured in counts-per-second (cps), in this area increased with normal compaction along the composite profile. Deviation from an established trend into lower cps intensity indicates less shale matrix volume (i.e., greater porosity, and transition into abnormal pressure).

Readings transmitted by an MWD scintillation detector are affected by the size of the nonmagnetic housing, tool position, hole diameter, and mud density. Thus, some means of normalizing the observed count-rate data to some standard is necessary before applying a quantitative correlation. All the curves presented in this text are based on measurements obtained in an 8.0-in. collar while drilling a 12¼-in. hole with an 8.8-lbm/gal drilling fluid. The original article gives Zoeller's normalizing procedure to compensate for other conditions.

In practice, corrected GR count rates are selected at suitable depth increments (e.g., every 100 ft) and plotted on semilogarithmic graph paper. The slope of the normal-compaction trend line depends on burial depth and should correspond to the curve shown in Fig. 2.68. After detecting the transition, the normal-compaction curve is extrapolated to the depth of interest. **Fig. 2.69** uses the ratio between normal and observed count rates (N_{GRn}/N_{GRo}) to predict the pressure gradient.

Example 2.23. Table 2.17 lists GR count rates measured while drilling an offshore Louisiana well that have been corrected for borehole conditions. Estimate the pore-pressure gradient at 11,100 ft using Zoeller's correlation. Use the first three data points to establish the normal-compaction trend.

Solution. **Fig. 2.70** shows GR data plotted on semilog graph paper. A normal-compaction trend line having the same approximate slope as the equivalent section in Fig. 2.68 is constructed through the last two normal-compaction data

TABLE 2.17—CORRECTED GAMMA RAY COUNT RATES OBTAINED WHILE DRILLING A WELL LOCATED OFFSHORE LOUISIANA[75]	
Depth (ft)	GR Count Rate (cps)
7,900	48
8,400	50
8,900	51
9,200	50
9,300	50
9,400	50
9,500	50
9,600	48
9,700	45
9,800	48
9,900	47
10,000	50
10,100	50
10,200	44
10,300	45
10,400	44
10,500	47
10,600	45
10,700	44
10,800	45
10,900	44
11,000	43
11,100	42

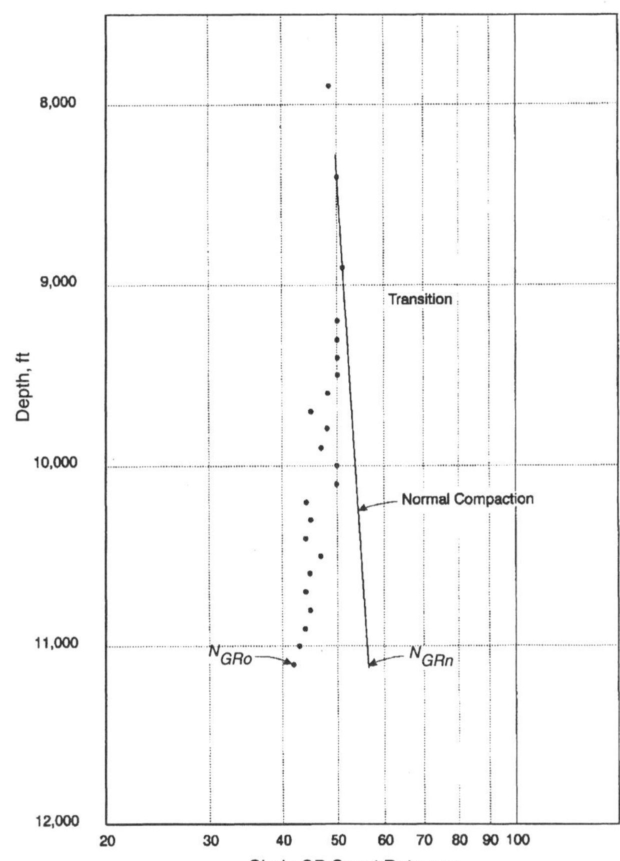

Fig. 2.70—MWD GR count rates for a well in south Louisiana.

points. The observed GR count rate at 11,100 ft is 42 cps and the extrapolated normal rate is 57 cps.

$$N_{GRn}/N_{GRo} = 57/42 = 1.36.$$

The predicted pore-pressure gradient from Fig. 2.69 is 0.61 psi/ft (11.7 lbm/gal).

2.10 Effective-Stress Models

Investigations since the late 1980s have done much to extend log pore-pressure predictions beyond their historical limitations.[76-82] The impetus for much of the recent work has been the developing LWD/MWD technology and the consequent possibilities for log predictions on a real-time basis. Timely data acquisition and processing capabilities are an obvious benefit when drilling into a transition zone. Other advantages over the traditional empirical methods also have been realized, including (1) elimination of the requirement for normal-compaction-trend extrapolation, (2) more widespread application in different geographic regions, (3) more widespread application in different lithologies and geologic age rock, and (4) ability to account for different geopressure mechanisms.

Space does not permit separate discussion of each referenced technique. A representative sample is offered in this text consisting of four different approaches to the problem, each of which is unique in some way. All rely on the effective-stress principle as the basis for empirical or analytical prediction. The common objective is to apply log-derived petrophysical parameters of the rock to a compaction model to quantify effective stress. Knowing the overburden and poroelastic constant (generally assumed to be unity), Terzaghi's[17] equation is resolved into one unknown (i.e., pore pressure).

In all cases, the effective stress needed to characterize pore pressure is the maximum effective stress. Usually, the maximum stress is controlled by the overburden and acts in the vertical direction. However, the maximum stress may act along a horizontal or inclined plane in those areas affected by active or fairly recent tectonic activity. Terzaghi's principle still applies in these cases, but quantifying the magnitude and direction of the maximum stress requires in-situ testing of the rock.

2.10.1 Model Based on Excess-Porosity Characterization. The first technique is a sophisticated variation of the equivalent-depth method in which abnormal pore pressures are quantified on the basis of the excess porosity observed in the undercompacted zone. Rasmus and Gray-Stephens[79] observed that sediment-porosity decline with depth could be modeled by

$$D = D_0 10^{-K_D \phi}, \quad \text{(2.37)}$$

where D_0 = the depth at which porosity is zero and and K_D = a constant reflecting the depth/compaction relationship for the area. Eq. 2.37 is similar to Eq. 2.10 except for the transposition of the depth and porosity terms. The direct expression for porosity is obtained as

$$\phi = \frac{1}{K_D}[\log(D_0) - \log(D)]. \quad \text{(2.38)}$$

The constants D_0 and K_D can be evaluated by a semilog plot of porosity vs. depth.

Example 2.24. Average porosities for North Sea Tertiary and Upper Cretaceous shales are listed in **Table 2.18.** Estimate the values of D_0 and K_D for this area.

TABLE 2.18—AVERAGE SHALE POROSITIES IN THE NORTH SEA (Data Furnished by Schlumberger Anadrill)

Depth (ft)	Porosity
1,000	0.475
2,000	0.380
3,000	0.320
4,000	0.280
5,000	0.250
6,000	0.225
7,000	0.205
8,000	0.185
9,000	0.170
10,000	0.150
11,000	0.135
12,000	0.125
13,000	0.115
14,000	0.105
15,000	0.095
16,000	0.085
17,000	0.080
18,000	0.070
19,000	0.060
20,000	0.055

Solution. **Fig. 2.71** shows a semilog plot reflecting an excellent straight-line fit of the data. The line slope, K_D, can be determined by use of any two points that fall on the line. Selecting porosity values from 5,000 ft and 18,000 ft gives

$$K_D = \frac{\log(D_2/D_1)}{\phi_1 - \phi_2} = \frac{\log(18,000/5,000)}{0.250 - 0.070} = 3.09.$$

D_0 is obtained by extrapolating the line to zero porosity. Alternatively, use the same equation to obtain

$$3.09 = \frac{\log(D_0/5,000)}{0.250 - 0} \text{ and}$$

$$D_0 = 10^{(3.09)(0.25)} \times 5,000 = 29,612 \text{ ft.}$$

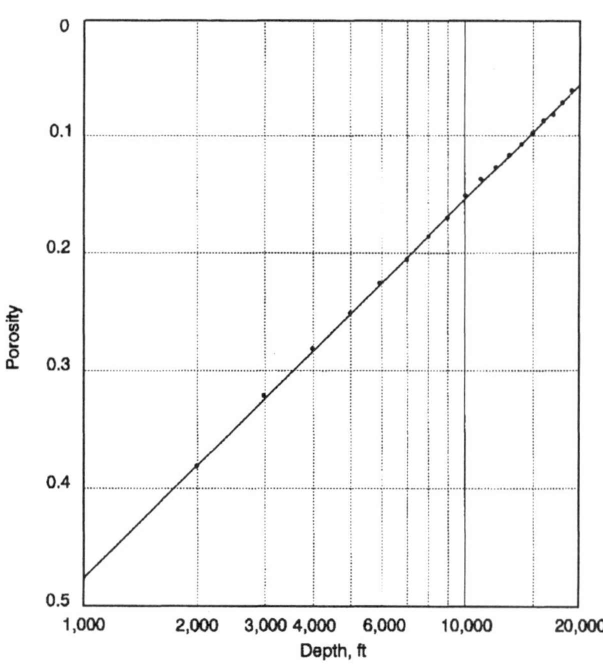

Fig. 2.71—Porosity vs. depth relationship for Tertiary and Upper Cretaceous shales in the North Sea.

Eq. 2.37 is represented as the curve shown in **Fig. 2.72.** The undercompacted rock at D exhibits abnormally high porosity for the burial depth; this porosity is expressed as the sum of the normal and excess or overpressure porosity,

$$\phi = \phi_n + \phi_{op} . \quad \text{(2.39a)}$$

The authors assumed a direct proportionality between depth and effective stress. Accordingly, effective stress replaces depth in Eq. 2.38 and the resulting expressions for normal and overpressure porosity are substituted into Eq. 2.39. Rearranging terms then yields

$$\phi_{op} = -\frac{1}{K_D}\log\left(\frac{\sigma_{Ve}}{\sigma_{Ven}}\right), \quad \text{(2.39b)}$$

where σ_{Ve} and σ_{Ven} = the abnormally pressured and normal effective stresses, respectively, for the burial depth. Eq. 2.2 is substituted for the effective stress terms and obtains the expression for pore pressure,

$$p_p = \sigma_{ob} - (\sigma_{ob} - p_n)10^{-K_D\phi_{op}}. \quad \text{(2.40)}$$

The decline constant for the area can be obtained graphically, but the overpressure porosity requires proper measurement and interpretation of the porosity and lithological indicators if the estimate is to have any validity.

In the described application, LWD tools furnish measurements of the formation GR count rate and any combination of porosity indicators, such as resistivity, travel time, or bulk density. MWD provides downhole bit weight and torque readings, while the drill rate and rotating speed are obtained from surface measurements. The data are fed into a wellsite computer, and the software characterizes the measurements (including a drilling strength function) in terms of the volume fractions occupied by the free water, bound water, overpressure water, and grain constituents. A mathematical minimization technique[83] is used to compute the "most likely" value for these parameters and the pore pressure is calculated with Eq. 2.40.

Example 2.25. From resistivity-log data, an overpressure porosity of 0.075 is predicted at 7,500 ft for a well drilled in the Central Graben of the North Sea. Calculate the pore pressure if the normal-pressure gradient is 0.452 psi/ft. The overburden stress at this depth is 6,953 psig.

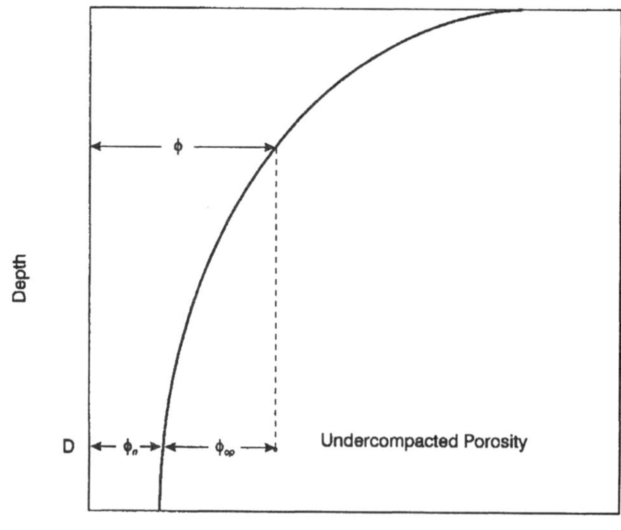

Fig. 2.72—Overpressure porosity in an undercompacted shale.

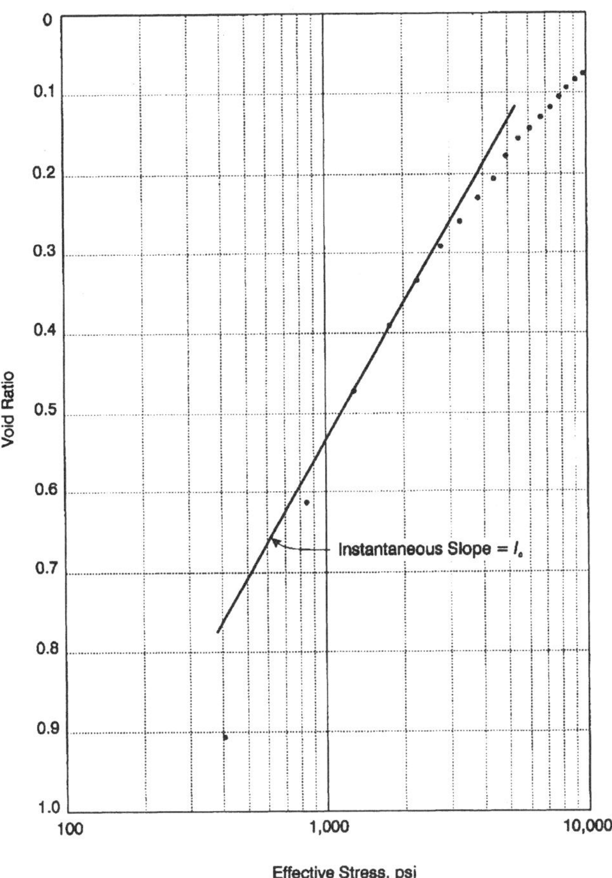

Fig. 2.73—Typical relationship between void ratio and effective stress for cohesive rock.

Solution. The normal pore pressure at 7,500 ft is

$$p_n = (7,500)(0.452) = 3,390 \text{ psig}.$$

Eq. 2.40 yields

$$p_p = 6,953 - (6,953 - 3,390)10^{-(3.09)(0.075)}$$

$$= 4,863 \text{ psig}.$$

The result is equivalent to a 12.5-lbm/gal density at 7,500 ft.

There is an alternative approach. Effective stress is not proportional to depth if the overburden stress gradient is variable. A more appropriate decline constant can be obtained by measuring the slope of the effective stress vs. porosity curve for the area. This procedure leads to a slope value of 3.46 for the North Sea and, in Example 2.25, a computed pore pressure of 4,992 psig. The same pore-pressure estimate is obtained when the equivalent-depth method is used on a porosity vs. depth plot.

2.10.2 Shale-Compaction Model. Alixant and Desbrandes[78] described another method for calculating effective stress in shales from log-derived porosity estimates. They discussed characterizing porosity in terms of shale resistivity but travel time or density porosities would be equally appropriate. They applied the Perez-Rosales[84] relationship between formation resistivity factor and porosity,

$$F_R = \frac{R_0}{R_w} = 1 + 1.85\left(\frac{1-\phi}{\phi - 0.1}\right). \qquad (2.41)$$

The equation constant 1.85 is the grain geometry factor, and 0.1 is the residual porosity (i.e., portion of the porosity which does not conduct electricity). Actually, the two terms are not constants but depend on the lithology. Nonetheless, the given values are appropriate in most sandstones and are assumed to apply in shales as well. Rearranging Eq. 2.41 yields a direct expression for porosity.

$$\phi = \frac{1.75 + 0.1F_R}{0.85 + F_R}. \qquad (2.42)$$

It was further assumed that the shale is buried sufficiently deep so that essentially all the pore water is electrostatically bound to the clays. The resistivity of the water bound to sodium clays may be considered as constant at a given temperature, with dependence approximated by

$$R_{wb} = 297.6T^{-1.76}, \qquad (2.43)$$

where R_{wb} = the bound-water resistivity and T = °F. Thus, shale porosity is estimated with only the measured resistivity and temperature.

Alixant and Desbrandes[78] selected a relationship from soil mechanics to decribe effective stress as a function of the compaction state. The void ratio, r_v, is introduced and defined as the ratio of sediment pore volume to matrix volume.

$$r_v = \phi/(1 - \phi). \qquad (2.44)$$

Triaxial compression testing of soils or cohesive rock, such as a shale, leads to determination of the compression index, I_c.

$$I_c = \frac{dr_v}{d(\log \sigma_{Ve})}. \qquad (2.45)$$

In **Fig. 2.73,** I_c defines the instantaneous slope of the void ratio vs. log effective-stress data. I_c is not a constant over the entire range of plotted effective-stress values because of the observed stiffening effect with increasing stress. Even so, a constant I_c can be approximated over a specified effective-stress range similar to the line drawn through the midrange data. Integrating Eq. 2.45 with constant I_c and choosing a reference effective-stress value of unity gives

$$\sigma_{Ve} = 10^{(r_v - r_{vi})/-I_c}, \qquad (2.46)$$

where r_{vi} = the void ratio at the reference effective stress.

Obtaining an estimated effective stress in this fashion and having knowledge of the overburden stress leads to a direct pore-pressure solution with Eq. 2.2. Ideally, I_c and r_{vi} are determined from laboratory testing on reconstituted shale samples. Lacking this information, reasonable values probably can be obtained with a log-derived compaction model for the area. Example 2.26, drawn from Ref. 78, demonstrates the method.

Example 2.26. Triaxial compression tests on a North Sea shale sample indicate that an I_c value of 1.1 psi^{-1} is represented in the effective-stress range between 1,100 and 2,300 psig. The experimentally determined r_{vi} constant for the shale is 3.84. Estimate the pore pressure at 5,000 ft if the measured shale resistivity is 0.48 Ω·m, the undisturbed formation temperature is 121°F, and the overburden stress is 4,570 psig.

Solution. Determine R_{wb} using Eq. 2.43.

$$R_{wb} = (297.6)(121)^{-1.76} = 0.064\ \Omega\text{·m}.$$

The formation resistivity factor is

$$F_R = 0.48/0.064 = 7.5,$$

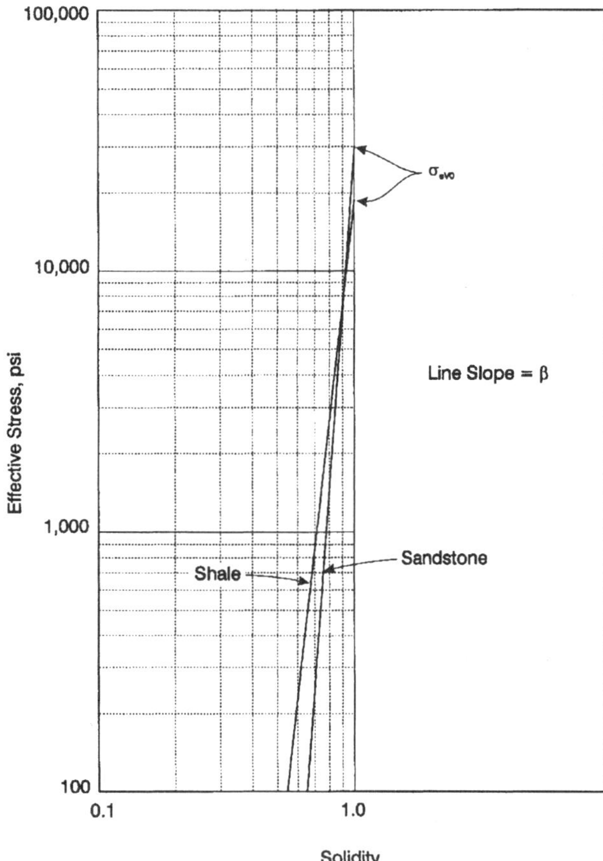

Fig. 2.74—Relationship between solidity and effective stress for a shale and a clean sandstone.

which leads to the porosity estimate,

$$\phi = \frac{1.75 + (0.1)(7.5)}{0.85 + 7.5} = 0.299.$$

The void ratio and effective stress are computed, respectively, as

$$r_v = 0.299/(1 - 0.299) = 0.43$$

and $\sigma_{Ve} = 10^{(0.43-3.84)/-1.1} = 1,259$ psig.

Eq. 2.2 yields the pore pressure,

$$p_p = 4,570 - 1,259 = 3,311 \text{ psig}.$$

This approach is well-suited to continuous LWD pressure predictions. The first requirement is to identify clean shale intervals with GR measurements. The resistivities measured in these shales then may be converted directly to pore pressure if valid input for the formation temperature gradient, overburden stress, and compaction-model constants are available. Despite many assumptions, some of which conflict with other established techniques, Alixant and Desbrandes[78] reported accurate pore-pressure predictions in three different geological basins.

2.10.3 Variable-Lithology Compaction Model. Holbrook *et al.*[81] discussed the problem of evaluating pore pressures in the variable Cretaceous/Jurassic lithology associated with the North Sea Central Graben. Conventional shale-compaction methods do not apply to this particular environment, and rock-specific compaction parameters were developed for a LWD prediction technique.

TABLE 2.19—POWER-LAW COMPACTION CONSTANTS FOR CERTAIN LITHOLOGIES[81]

Formation	σ_{Ve0} (psi)	β
Quartz sandstone	30,000	13.219
Average shale	18,461	8.728
Limestone	12,000	13.000
Anhydrite	1,585	20.000
Halite	85	31.909

Shale porosity is proportional to the matrix stress by an inverse function. This truism applies in other lithologies as well, but in a different way. Applying a suitable compaction model should result in the capability to determine effective stress and, hence, pore pressure from measured porosity regardless of the rock type. This reasoning is the basis for the procedure.

Holbrook *et al.*[81] selected a power-law compaction relationship of the form

$$\sigma_{Ve} = \sigma_{Ve0} \times (1 - \phi)^{\beta}, \quad \text{.................. (2.47)}$$

where σ_{Ve0} = a characteristic rock property defined as the effective stress that reduces porosity to zero, β = the natural compaction-strain-hardening coefficient, and the term $(1-\phi)$ = the solidity and represents the bulk-rock fraction occupied by solids. Eq. 2.47 shows that a plot of solidity vs. effective stress on logarithmic graph paper gives a straight line of slope β and zero-porosity intercept at σ_{Ve0}.

Fig. 2.74 shows the logarithmic relationships for a clean quartz sandstone from Louisiana[85] and a typical shale. **Table 2.19** gives the compaction-model constants for these two formations plus those measured for a Po Valley limestone,[86] anhydrite, and halite. These constants yield acceptable results in the North Sea study area but may not apply elsewhere.

The procedure for estimating σ_{Ve0} in mixed-grain lithology first takes the logarithm of each component value and computes a volume-weighted average of the individual coefficients. Then, the maximum effective stress is obtained by raising the logarithm base to the average coefficient power. The β exponent in mixed lithology is simply the weighted average of the individual β coefficients.

Example 2.27. LWD data from a North Sea well indicate 24.1% porosity in a mixed sand/shale sequence at 4,500 ft. Estimate the pore pressure if 29% sandstone and 71% shale make up the grain matrix. Assume that the constants given in Table 2.19 apply and use an overburden stress of 3,875 psig.

Solution. First, determine the applicable σ_{Ve0} term as

$$\log(\sigma_{Ve0}) = (0.29)\log(30,000) + (0.71)\log(18,461)$$

and $\sigma_{Ve0} = 10^{4.3274} = 21,252$ psig.

In similar fashion, the weighted-average β coefficient is

$$\beta = (0.29)(13.219) + (0.71)(8.728) = 10.03.$$

The rock solidity and computed effective stress, respectively, are

$$1 - 0.241 = 0.759$$

and $\sigma_{Ve} = (21,252) \times 0.759^{10.03} = 1,337$ psig.

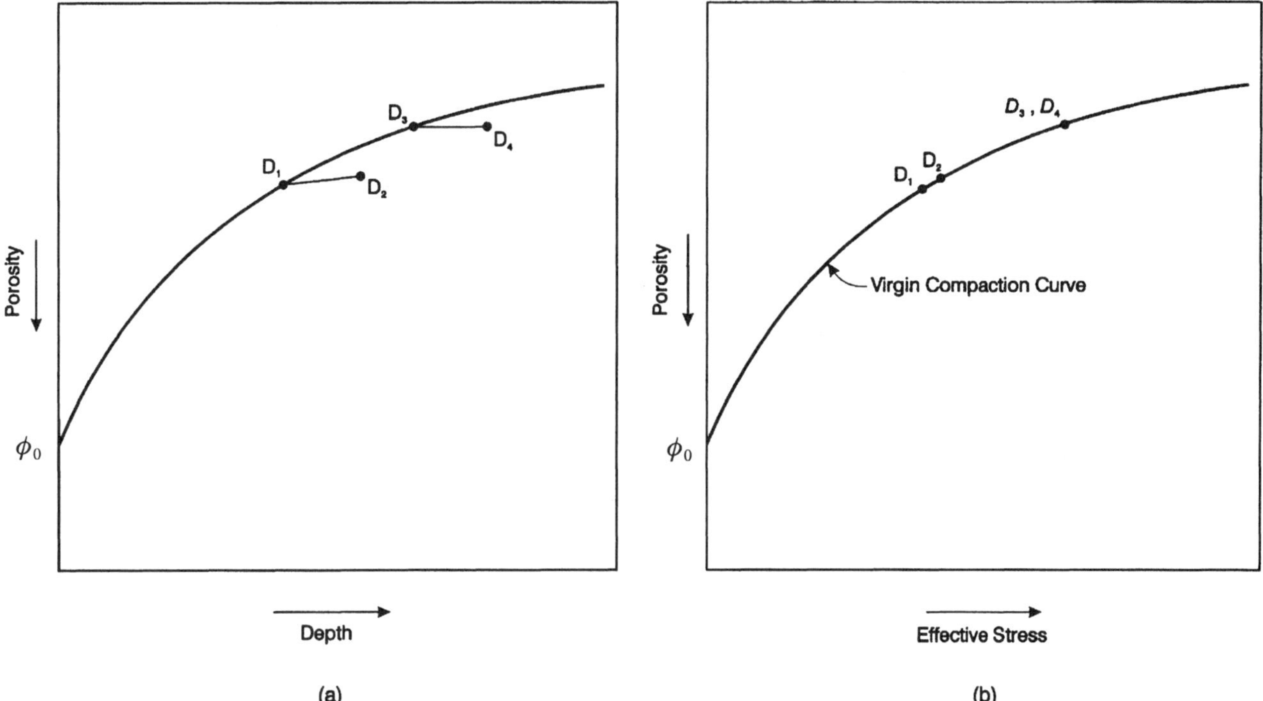

Fig. 2.75—Sediment compaction and undercompaction behavior as a function of depth and effective stress.

Terzaghi's[17] equation gives

$$p_p = 3,875 - 1,337 = 2,538 \text{ psig}$$

$$\text{or } g_p = 2,538/4,500 = 0.563 \text{ psi/ft.}$$

In practice, GR measurements from an LWD tool are used to compute the bulk-rock constituents and shale fractions. Porosities are determined from resistivity data by use of Archie's F_R relationships from a user-supplied C_w profile, and the compaction model is used to compute pore pressure. Archie's equations, while simple to apply, do not recognize the dual-water nature of clays and, in the strictest sense, are inadequate for characterizing porosity in shaly rock. This and other potential sources of error are managed by a calibration procedure in which the input C_w function is adjusted so that predictions match measured pore-pressure data. A modified C_w profile (actually a pseudo-C_w), determined in the calibration well, should serve to increase the accuracy in subsequent projects if these are drilled in a similar geological setting.

2.10.4 Effective-Stress Reversal Concepts. The effective-stress models (including the equivalent-depth method) all assume that shale undercompaction is the predominant geopressure mechanism. The sediment-loading diagrams in **Fig. 2.75** show compaction and undercompaction behavior. The bottom left of Fig. 2.75a shows surface conditions at maximum porosity; a decrease in porosity is seen with increasing burial depth. Point D_1 represents the transition depth into abnormal pore pressure, as noted by the maintenance of porosity through the depth D_2. Curve extension into greater burial depth corresponds to the normal-compaction trend for the sediment.

Fig. 2.75b depicts the same compaction process in an effective-stress diagram. This relationship between porosity and effective stress is defined from soil-mechanics principles as the virgin curve of the rock. The effective stress increases slightly in the previously defined transition zone (D_1 through D_2) and continues to track the virgin curve. This is an important point. Effective stresses in a normal-compaction trend or in pure undercompaction always fall on or near the established virgin curve. In the same lithological sequence, the transition interval between D_3 and D_4 also implies undercompaction because the effective stress remains constant.

The compaction strains shown in Fig. 2.75 are not reversible in sedimentary-basin rock. In other words, some permanent porosity loss occurs if the effective stress subsequently decreases. Consider the compaction state depicted at D_1 in **Fig. 2.76.** Reducing the effective stress leads to a corresponding increase in the porosity, but along a different track than shown by the virgin curve. The observed behavior during an effective-stress reversal is defined as the unloading curve for

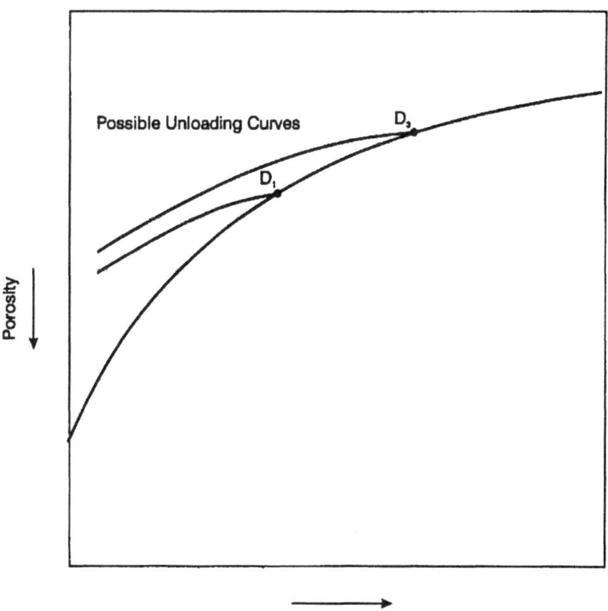

Fig. 2.76—Unloading curves associated with reductions in effective stress.

TABLE 2.20—ABNORMAL PORE PRESSURE SOURCES THAT CAN LEAD TO AN EFFECTIVE STRESS REVERSAL IN SHALES

Source	Mechanism
Surface erosion	Overburden reduction
Clay diagenesis	Fluid expansion
Gypsum diagenesis	Fluid expansion
Grain cementation	Pore volume reduction
Aquathermal pressuring	Fluid expansion
Biochemical processes	Fluid expansion

the sediment. The maximum effective stress experienced by the sediment during compaction must be specified when defining an unloading curve. For example, reducing the effective stress from D_3 or any other point on the virgin curve leads to a different unloading curve.

The importance of all this to pore-pressure prediction is that effective stress can be reduced by fluid expansion or other mechanisms and the consequent stress reversal follows an unloading curve. **Table 2.20** lists most of the geopressure processes that reduce effective stress. **Fig. 2.77** suggests that conventional compaction models overestimate effective stress and underestimate pore pressure if these other pressure sources play a significant role.

Bowers[82] presented a prediction method that uses sonic-log data that considers both virgin- and unloading-curve behavior. Recognizing the presence of an effective-stress reversal is a prerequisite step for the prediction method. One clue is offered by the appearance of a conventional travel time vs. depth plot.

Consider the plot shown in **Fig. 2.78a.** The transit time reaches a maximum and remains fairly constant throughout the transition. A constant transit time implies a constant effective stress and an undercompaction pressure source. On the other hand, a velocity reversal, such as the one shown in Fig. 2.78b, may imply an effective-stress reversal. Velocity reversals, however, do not give definitive proof of stress unloading

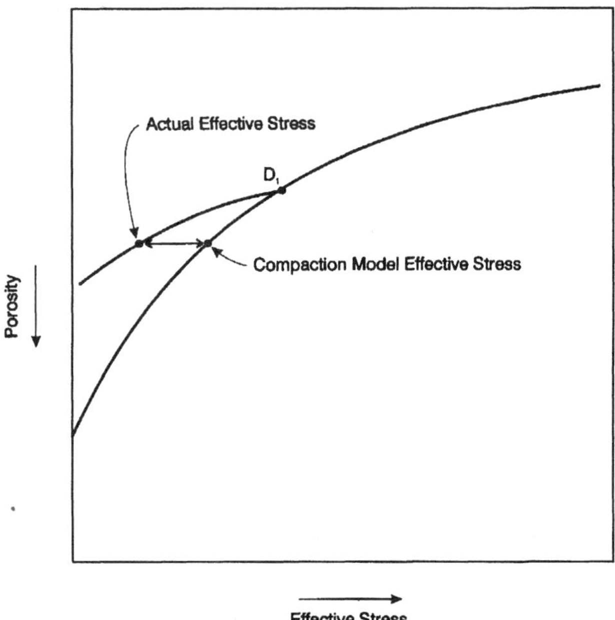

Fig. 2.77—Effect of a matrix-stress reversal on pressure predictions based on a compaction model.

because the same phenomenon can occur in massive undercompacted shales.

Bowers discussed two ways of determining whether undercompaction or stress reversal plays the more important role. One procedure compares a prediction made by one of the compaction models with a measured pressure in the transition. Stress unloading and a different, or at least contributing, geopressure source is indicated if the actual pressure reading exceeds the calculated value. The definitive method, however, generates a velocity/effective-stress diagram for a well or number of wells in the area. The virgin effective-stress curve is constructed with normal-compaction-trend data. Effective stresses in the transition are calculated from measured pore

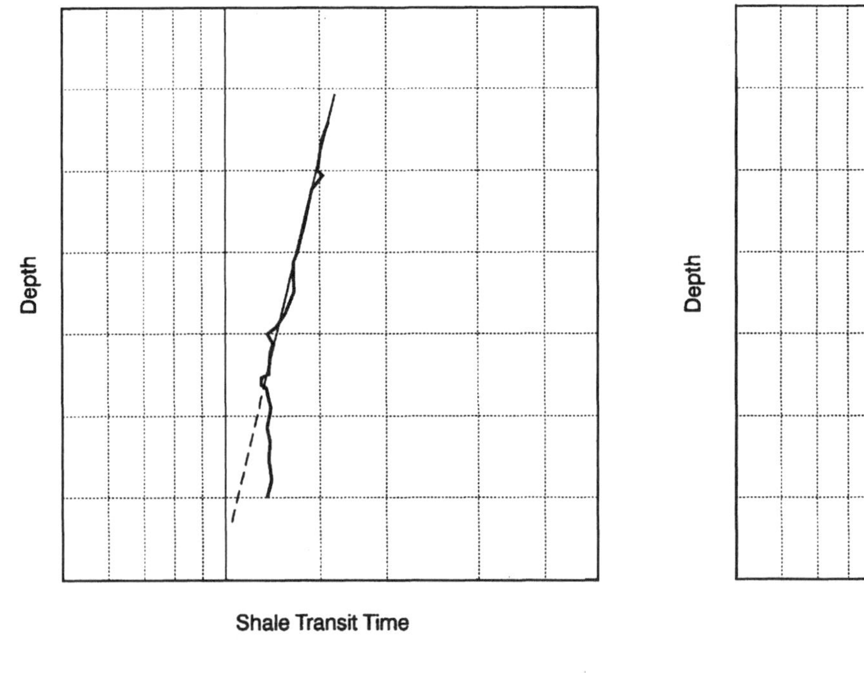

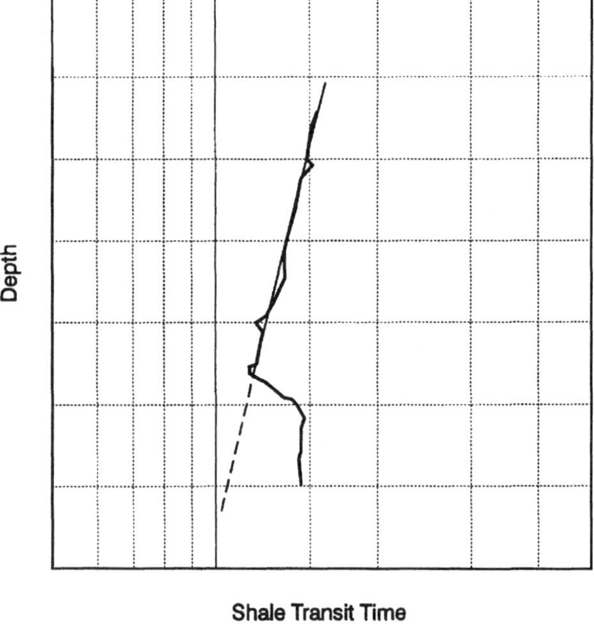

Fig. 2.78—Characteristic transit time plots for abnormal pressures generated by undercompaction and a stress reversal mechanism.

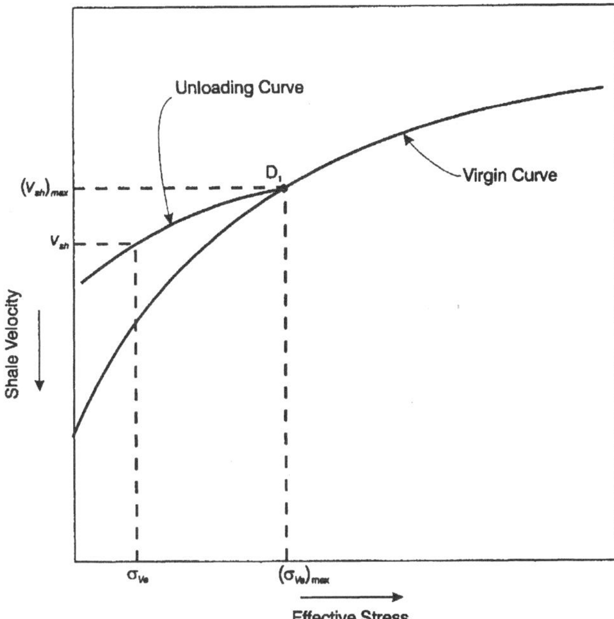

Fig. 2.79—Shale effective-stress curves and associated variables.

pressures. Stress unloading is observed if the effective stresses fall to the left of the virgin curve.

Both methods require at least one borehole in the area that has been drilled and logged through the transition and tested in an abnormally pressured permeable zone. Therefore, it is difficult to apply the technique in rank wildcats or in impermeable strata. Even so, a high/low range of possible pressures may be provided if suitable equation parameters are available for the area.

Over the practical effective-stress range, the virgin curve for shales can be approximated by

$$v_{sh} = 5{,}000 + A\sigma_{Ve}^{B}, \quad \text{(2.48)}$$

where v_{sh} = shale acoustic velocity (ft/sec) and A and B = constants unique to the rock.

Let $(\sigma_{Ve})_{max}$ be the vertical effective stress at the onset of unloading and $(v_{sh})_{max}$ the corresponding acoustic velocity. Substituting these terms into Eq. 2.48 gives

$$(\sigma_{Ve})_{max} = \left[\frac{(v_{sh})_{max} - 5{,}000}{A}\right]^{1/B}. \quad \text{(2.49)}$$

The unloading curve from $(\sigma_{Ve})_{max}$ is described by the empirical equation

$$v_{sh} = 5{,}000 + A\left\{(\sigma_{Ve})_{max}\left[\frac{\sigma_{Ve}}{(\sigma_{Ve})_{max}}\right]^{1/U}\right\}^{B}. \quad \text{(2.50)}$$

The constant U = a measure of the relative plasticity of the bulk rock and theoretically may vary in magnitude between one and infinity. The value $U = 1$ indicates a perfectly elastic system because the expression reduces back to Eq. 2.48, and an infinite value indicates permanent deformation because porosity (i.e., shale velocity) does not change during the unloading process. Typical values for U range between 3.0 and 8.0 in most sediments.

TABLE 2.21—NORMAL-COMPACTION SHALE ACOUSTIC VELOCITY AND EFFECTIVE STRESS DATA FOR A WELL IN JEFFERSON COUNTY, TEXAS

Depth (ft)	v_{sh} (ft/sec)	p_p (psig)	σ_{ob} (psig)	σ_{Ve} (psi)
2,820	6,329	1,311	2,496	1,184
3,210	6,536	1,493	2,850	1,358
4,000	6,667	1,860	3,584	1,724
4,170	6,579	1,939	3,745	1,806
4,520	6,849	2.102	4,077	1,975
5,210	7,092	2,423	4,731	2,308
6,000	7,246	2,790	5,490	2,700
6,210	7,407	2,888	5,695	2,807
6,970	7,407	3,241	6,440	3,199
7,500	7,692	3,488	6,960	3,473
7,810	8,065	3,632	7,271	3,639
8,000	8,333	3,720	7,464	3,744
8,320	8,197	3,869	7,779	3,910
8,410	8,264	3,911	7,872	3,961
9,000	8,333	4,185	8,460	4,275
9,010	8,547	4,190	8,469	4,280
9,220	8,475	4,287	8,676	4,389
9,300	8,403	4,325	8,761	4,436
9,390	8,264	4,366	8,855	4,488

An appropriate value for U can be obtained with a procedure for normalizing effective-stress reversal data from the area. Ref. 81 discusses the technique. Normally, $(\sigma_{Ve})_{max}$ can be selected as the computed effective stress at the transition depth, while $(v_{sh})_{max}$ is the corresponding acoustic velocity. **Fig. 2.79** depicts the equation variables.

Reconsider the question posed by Example 2.19. The Hottman and Johnson[66] correlation was used to predict the pore pressure in an upper gulf coast well. The equivalent-depth method also was applied, but the prediction with this procedure turned out to be 1,460 psi less than the empirical (and actual) result. The discrepancy is a strong clue that a geopressure mechanism other than undercompaction is at work. Also, refer to Fig. 2.62 and note the severe velocity reversal within the transition. Example 2.28 recasts this problem in the Bowers method.

Example 2.28. Table 2.21 lists acoustic-velocity data for the normally compacted interval in Example 2.19. The effective-stress values in the right column were computed with Eaton's overburden correlation and the normal pressure gradient for the gulf coast.

1. Estimate values for parameters A and B.
2. Estimate the pore pressure at 11,190 ft using the Bowers technique. The unloading parameter U for gulf coast and Gulf of Mexico shales is 3.13 based on regional normalization data.

Solution.

1. Use a graphical procedure to determine the virgin-curve parameters. Rearrange Eq. 2.48 as

$$\log(v_{sh} - 5{,}000) = \log(A) + B\log(\sigma_{Ve}).$$

Thus a plot of $(v_{sh} - 5{,}000)$ vs. σ_{Ve} on logarithmic graph paper should yield a straight line of Slope B. Accordingly, the data are plotted as shown in **Fig. 2.80** and a straight line is fit through the points. The line slope can be determined with

$$B = 3.32\log\left[\frac{(v_{sh} - 5{,}000)\ \text{at}\ 2\sigma_{Ve}}{(v_{sh} - 5{,}000)\ \text{at}\ \sigma_{Ve}}\right].$$

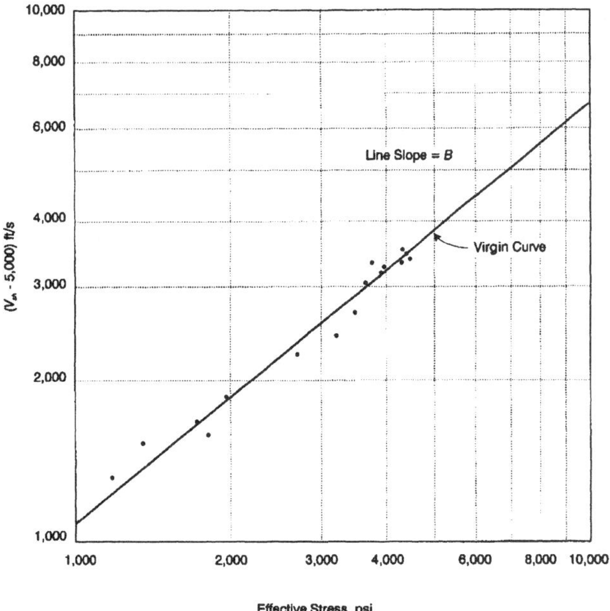

Fig. 2.80—Logarithmic relationship between normally compacted effective stress and shale velocity for a well in Jefferson County, Texas.

Choosing σ_{Ve} and $2\sigma_{Ve}$ as 2,000 and 4,000 psi gives

$$B = 3.32\log\left(\frac{3,205}{1,850}\right) = 0.7923.$$

Substitute values corresponding to an arbitrary σ_{Ve} and solve for A.

$$\log(1,850) = \log(A) + 0.7923\log(2,000)$$

and $A = 10^{0.6516} = 4.4837$.

2. From Fig. 2.62, the pressure transition begins at approximately 9,300 ft. Thus, $(\sigma_{Ve})_{\max}$ and $(v_{sh})_{\max}$ are 4,436 psig and 8,403 ft/sec, respectively. From Table 2.14, the measured transit time at 11,190 ft is 146 μsec/ft, which is equivalent to the velocity 6,849 ft/sec. Substituting terms into Eq. 2.50a yields the effective stress at this acoustic velocity.

$$6,849 = 5,000 + 4.4837\left[4,436\left(\frac{\sigma_{Ve}}{4,436}\right)^{1/3.13}\right]^{0.7923}$$

and $\sigma_{Ve} = 366$ psi.

Eaton's[20] overburden-stress gradient at 11,190 ft is 0.956 psi/ft. Terzaghi's[17] equation gives

$$p_p = (11,190)(0.956) - 366 = 10,332 \text{ psig}.$$

The result using Bower's[82] technique is in agreement with the Hottman and Johnson[66] correlation. A reasonable conclusion from Example 2.28 is that their correlation was based on measured pore pressures that had been affected significantly by fluid expansion. **Fig. 2.81** shows the virgin and unloading curves for this well. The open circle to the far left represents the effective stress calculated with Hottman and Johnson's 0.91-psi/ft prediction at 11,190 ft. The matching between the unloading curve and their correlation is apparent from the diagram. To illustrate this correspondence further, the other two

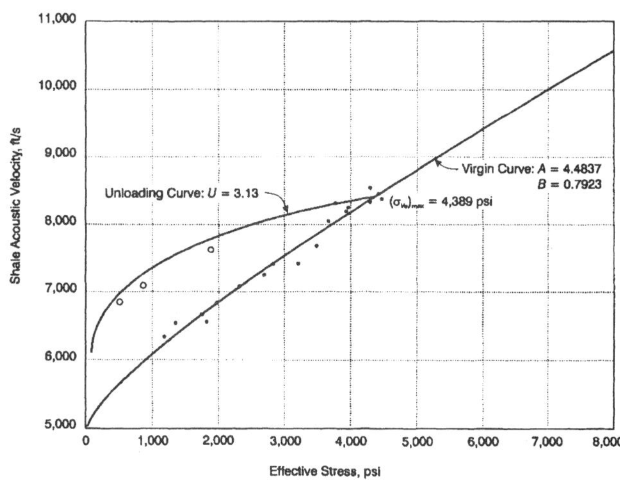

Fig. 2.81—Application of the Bowers[82] pressure-prediction method to the well in Jefferson County, Texas.

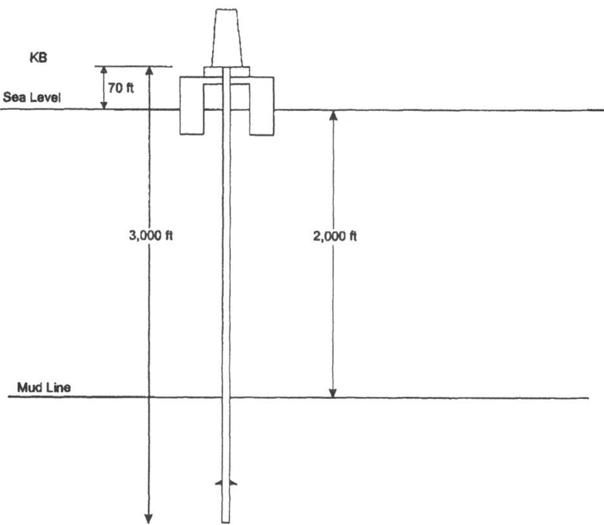

Fig. 2.82—Drilling operation described in Problem 2.2.

circles show the Hottman and Johnson effective stresses at 10,010 ft and 9,710 ft.

Problems

2.1 What formation pressure would be expected in a normally pressured sandstone at 5,000 ft in a well drilled along the gulf coast? What is the normal formation pressure at the same depth in Nigeria?

2.2 Consider the offshore drilling operation shown in **Fig. 2.82.** Assume seawater density is 8.65 lbm/gal and that the normal-pore-pressure gradient for this area is 0.465 psi/ft.

1. Determine normal pore pressure at present total depth.
2. What is the pore-pressure gradient at this depth as referenced from the KB?

2.3 A well encounters an aquifer at a depth of 500 ft that has a static water table at 400 ft relative to the rig's KB. What is the pore-pressure gradient at the top of this sand? Assume the pore fluid has a freshwater gradient.

2.4 Twenty years ago, a Pennsylvanian Morrow gas field underlying your drilling prospect was discovered at 10,100 ft and had an initial pore pressure equivalent to a 15.6-lbm/gal fluid.

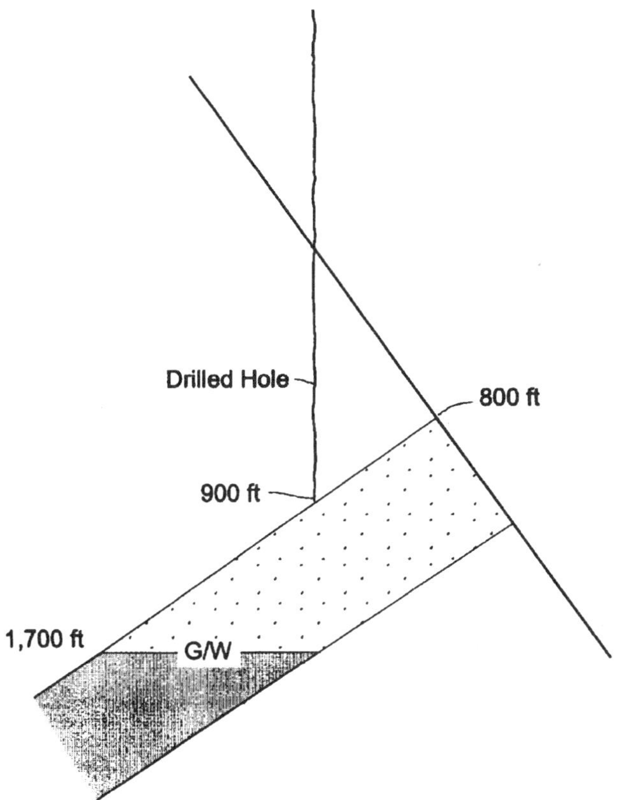

Fig. 2.83—Dipping geologic structure described in Problem 2.7.

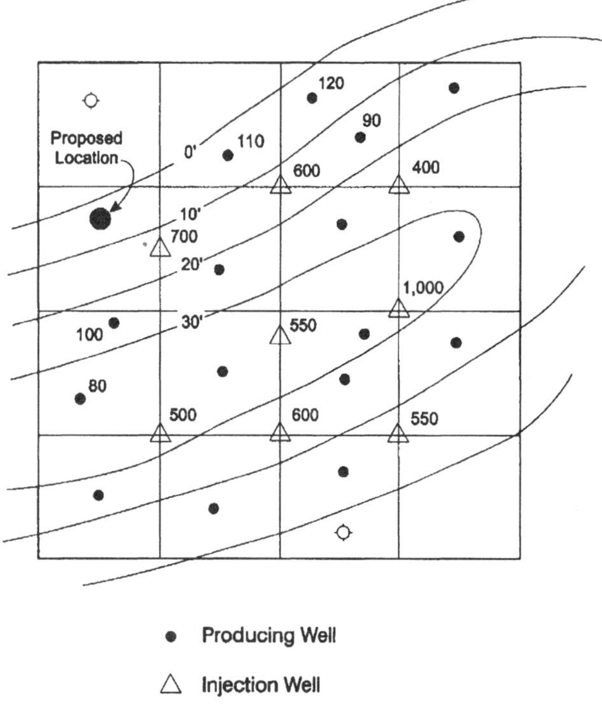

Fig. 2.84—Surface-pressure readings overlying the isopach map described in Problem 2.11.

The prospect objective is the Springer, which is expected at 10,600 ft and should have the same virgin pressure gradient as the Morrow. The reservoir engineering group tells you that the average Morrow pressure has declined to 2,500 psia.

1. Determine the current Morrow pressure gradient in terms of equivalent density.

2. What effect might this situation have on your pipe program?

2.5 Consider a massive Tertiary shale along the gulf coast that overlies an abnormally pressured sandstone at 8,500 ft. The pore-pressure gradient in this sand is 0.779 psi/ft. Above the shale is a normally pressured sand at 7,500 ft. A study[87] showed that a compacted shale at this depth should have a permeability on the order of 0.002 md (2.0×10^{-9} darcy). Use Darcy's law to answer the following.

1. How many years will it take for gas to migrate from the lower sand to the upper sand if the average gas viscosity is 0.021 cp? Assume that the pore-pressure differential is the only driving mechanism.

2. Repeat the calculation assuming that the pore pressure in the lower sand has been bled to a gradient of 0.500 psi/ft.

3. Compare your results to the age of the Tertiary.

2.6 Consider Problem 2.3 again. Determine the pressure gradient at 500 ft if the aquifer is full of fresh water and it outcrops 300 ft above the drilling location.

2.7 The dipping structure shown in **Fig. 2.83** is encountered at a depth of 900 ft. The downdip gas/water contact exhibits a pore-pressure gradient of 0.465 psi/ft at 1,700 ft. Assume that the hydrostatic-pressure gradient of the gas is 0.06 psi/ft.

1. What pore-pressure gradient can be expected upon drilling into this structure?

2. What pore-pressure gradient is present 100 ft updip at the fault contact?

2.8 Refer to Example 2.2 and assume the same conditions except that the top of the sand is at 500 ft instead of 1,000 ft. Determine the pore-pressure gradient in mud-weight equivalent at the top of the structure.

2.9 A thrust fault has isolated and sealed a small sand body. As a result, the sand was folded and compressed to the point that the gas pore volume was reduced by 25%. Quantify the effect this tectonic event has on the pore pressure.

2.10 Refer to Fig. 2.12 and assume that the sand lens on the left was at 2,000 ft before 300 ft of ground surface eroded. Determine the current pore-pressure gradient if the original gradient was 0.433 psi/ft and the sand retained its original pressure.

2.11 You are writing a drilling plan for a 5,000-ft prospect to develop banked oil in a waterflood. Over time, the producing wells in the field developed the capability to flow. You prepare the map shown in **Fig. 2.84,** which depicts the shut-in casing-tubing pressures for some of the producing well offsets and the injection pressures for the water input wells. What minimum mud density should you specify in the prognosis? Assume that the specific gravity (SG) of the produced water is 1.07 and that gas production is zero.

2.12 In an old field, gas has been leaking over time from a sand at 4,500 ft into a conglomerate at 2,900 ft. The current pore pressure of the deeper horizon is equivalent to a 7.0-lbm/gal fluid. Determine the maximum theoretical pore-pressure gradient in the conglomerate. Assume a 0.7 SG gas and an average temperature of 115°F.

TABLE 2.22—SHALE BULK DENSITY DATA FOR PROBLEM 2.19	
Depth (ft)	Bulk Density (g/cm^3)
550	2.12
700	2.14
1,250	2.13
1,800	2.20
1,950	2.20
2,200	2.18
2,700	2.21
3,250	2.23
3,600	2.26
3,900	2.25
4,500	2.26
4,600	2.28
5,000	2.29
5,500	2.31
5,650	2.31
6,000	2.32
6,400	2.34
6,600	2.33
7,000	2.35
7,200	2.35
7,600	2.36
7,900	2.36

2.13 An abnormally pressured formation at 19,000 ft has an initial pore pressure of 17,750 psig.

1. Determine the rock-matrix stress if the overburden gradient is 1.00 psi/ft.
2. Determine the matrix stress after production depletes the reservoir pressure to 2,500 psig.

2.14 A 1,000-ft oil zone in California has a virgin-pore-pressure gradient of 0.439 psi/ft. The overburden gradient at this depth is assumed to be 0.98 psi/ft.

1. Determine the rock-matrix stress at this initial condition.
2. At what pore pressure would you expect problems to develop if the compressive strength of the rock is 700 psi?

2.15 Determine the overburden-stress gradient for a rock mass with an average grain density of 2.60 g/m^3, 18% porosity, and fresh water as the pore fluid.

2.16 Write a spreadsheet program for computing overburden-stress gradients using Eq. 2.9a. Assume matrix and pore fluid densities of 2.60 and 1.074 g/cm^3, respectively, and generate a plot for the depth range of 0 to 20,000 ft. Compare the curve to Eaton's[47,48] gulf coast correlation in Fig. 2.20.

2.17 Use Eaton's overburden correlations to answer the following questions. Assume that the formations are normally pressured.

1. Estimate the overburden and matrix stress for a gulf coast shale buried at 6,000 ft.
2. Do the same for a rock layer in the Santa Barbara channel at this depth.

2.18 What conclusions can you draw from Fig. 2.22 regarding the relationship between a shale's age and its porosity?

2.19 Your company has made a major Miocene discovery in a new area. The reservoir drillstem test at 8,000 ft indicates a pore pressure of 3,670 psig, which, on the basis of all indications, is presumed to be normal. You obtain the logs and determine the bulk densities shown in **Table 2.22.** Estimate the surface porosity and the porosity-decline constant for the area. Assume an average sediment-matrix density of 2.60 g/cm^3.

2.20 Derive Eq. 2.12.

2.21 Use the surface porosity and decline constant for the gulf coast and determine the overburden stress in 1,000-ft depth increments using Eq. 2.12. Plot the calculated data from surface to 20,000 ft in terms of gradient and compare your curve to Fig. 2.20.

2.22 You have been predicting pore pressures from shale-porosity indicators during the process of drilling a well and have prepared the plot shown as **Fig. 2.85.**

1. Explain what might be affecting the data beginning at approximately 9,500 ft.
2. Indicate the transition depth into abnormal pore pressures.
3. Which normal-compaction-trend-line extrapolation would you select to predict the pore pressure at 14,000 ft?
4. What is the equivalent depth for the abnormal pressure at 14,000 ft?

2.23 Assume interval-transit times decrease with depth according to a power-law relationship. Plot the data shown in Table 2.7 on logarithmic graph paper and rework Example 2.7.

2.24 Table 2.23 shows average interval times for a prospect in Malaysia. The normal pore-pressure gradient is 0.442 psi/ft. Density-log data from the area is unavailable, and the overburden-stress gradient is assumed to be a constant 0.95 psi/ft.

1. Determine the transition depth.
2. Approximate the pore pressure at 7,500 ft.
3. Is it appropriate to use Pennebaker's[28] correlation for this problem?

2.25 Based on the chip hold-down theory, explain how the following conditions should affect penetration rate.

1. Mud density.
2. Mud viscosity.
3. Solids content of the mud.
4. Circulation rate.
5. Filtration rate.
6. Formation permeability.
7. Bit type (roller cone vs. drag bits).

2.26 Considering the Mohr-Coulomb failure criteria for rock, is it theoretically possible for a rock to fail in compression simply from the weight of the overburden?

2.27 On the basis of Fig. 2.41, would you expect rock at the bottom of a wellbore to break more readily at the center of the hole or toward the perimeter?

2.28 Give two reasons other than induced differential pressure for the observed relationship between penetration rate and rock permeability.

2.29 You conduct a drilloff test with an 8½-in. bit at a rotating speed of 70 rev/min and obtain the following penetration rates: 36 ft/hr at a bit weight of 40,000 lbf; 31 ft/hr at 34,000 lbf; and 23 ft/hr at 25,000 lbf.

1. Estimate the threshold bit weight.
2. What value would you assign to the bit-weight exponent?

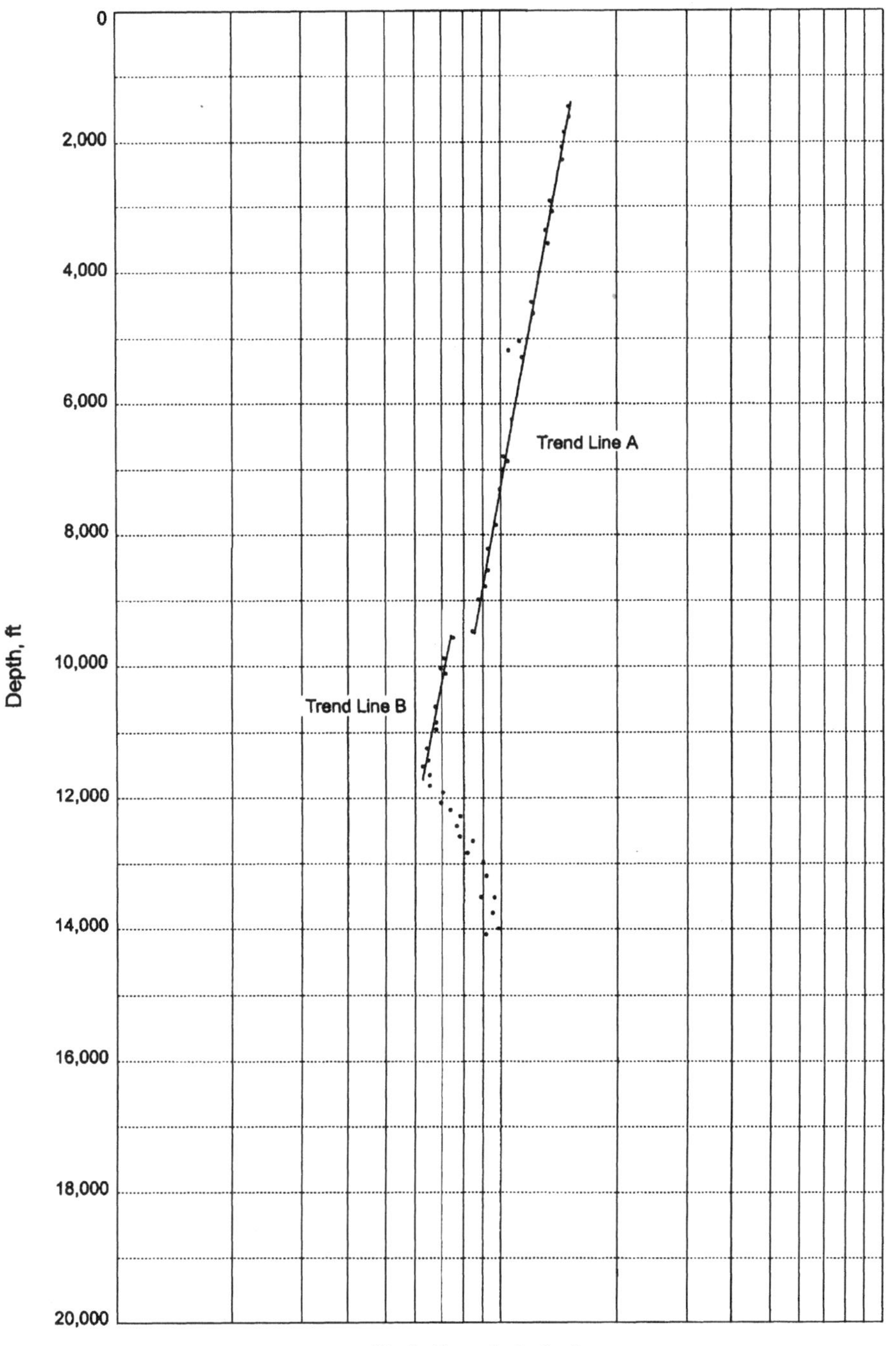

Fig. 2.85—Shale-porosity indicator plot for the well described in Problem 2.22.

2.30 You are drilling at 25 ft/hr with a $7^7/_8$-in. bit. The bit weight is 30,000 lbf and the rotating speed is 150 rev/min.

1. Determine the *d* exponent.
2. Determine the modified *d* exponent if the normal-pore-pressure gradient is 0.465 psi/ft and the mud weight is 10.0 lbm/gal.

2.31 Select appropriately scaled Cartesian and semilogarithmic graph paper. Using the d_c exponent vs. depth slopes discussed in Sec. 2.7.2, prepare transparent overlays for the Rehm and McClendon method and for Zamora's method. Draw pore-pressure lines in increments of 1.0-lbm/gal equivalent density.

2.32 The modified *d* exponents in **Table 2.24** were computed with shale-drilling data in a well drilled along the gulf coast. Predict the pore-pressure gradient and equivalent density at 14,000 and 16,000 ft using the three correlations discussed in Sec. 2.7.2.

2.33 Go back to Example 2.9 and use the equivalent-depth method to predict the pore-pressure gradient at 6,050 ft. How does your answer compare with the results obtained from the empirical correlations? In which method do you have the most faith?

2.34 While drilling at 10,000 ft in a normally pressured shale, the bottomhole ECD is increased from 9.2 to 9.8 lbm/gal. The

TABLE 2.23—AVERAGE INTERVAL TRANSIT TIMES FOR THE MALAYSIA PROSPECT DESCRIBED IN PROBLEM 2.24[5]

Interval (ft)	Midpoint (ft)	Average Transit Time (μsec/ft)	Lithology
2,500 to 3,000	2,750	160	Shale/sand
3,000 to 3,500	3,250	150	Shale/sand
3,500 to 4,000	3,750	141	Shale/sand
4,000 to 4,500	4,250	132	Shale/sand
4,500 to 5,000	4,750	121	Shale/sand
5,000 to 5,500	5,250	110	Limey shale
5,500 to 6,000	5,750	95	Limey shale
6,000 to 6,500	6,250	92	Limey shale
6,500 to 7,000	6,750	120	Shale/sand
7,000 to 7,500	7,250	150	Shale/sand
7,500 to 8,000	7,750	150	Shale/sand

penetration rate drops from 40 to 32 ft/hr as result. Determine the shale-compactibility coefficient.

2.35 Determine the choke backpressure that yields the ECD change described in Example 2.11. What impact does this procedure have on the equivalent density at the last shoe if the casing is set at 3,000 ft?

2.36 The following drilling data are obtained in massive shale above transition depth: depth, 10,500 ft; drill rate, 27 ft/hr; and ECD, 9.4 lbm/gal. The penetration rate at 10,800 ft increases to 35 ft/hr from an extrapolated normal rate of 24.5 ft/hr with no change in the drilling parameters. Determine the pore-pressure increase if $c = 1.10$ for this shale.

2.37 The following information pertains to a shale: depth, 13,000 ft; drill rate, 11 ft/hr; ECD, 10.1 lbm/gal; pore-pressure gradient, 9.5 lbm/gal; bit weight, 5,000 lbf/in.; rotary speed, 80 rev/min; rotary speed exponent, 0.8; and compactibility coefficient, 0.95. The penetration rate over the next 400 ft increases to 12 ft/hr. During this period, the bit weight and rotary speed are reduced to 4,500 lbf/in. and 70 rev/min. The extrapolated normal rate at 13,400 ft is 9.7 ft/hr. Estimate the pore-pressure gradient at this depth.

2.38 Refer to the data in Table 2.13. Estimate the pore pressure at 14,920 ft using both the Boatman[57] correlation and the equivalent-depth method.

2.39 Compute the bulk density of a shale if the grain density is 2.55 g/cm^3 and the porosity is 20%. Assume fresh water in the pore space. Recalculate the bulk density if the rock matrix contains 20% limestone (matrix density = 2.71 g/cm^3).

2.40 Work Example 2.16 again except use a measured mixture density of 13.0 lbm/gal.

2.41 Duplicate the curve shown in Fig. 2.51 using the calculation method and data from Example 2.17. Assume that the circulating-temperature gradient is a constant 1°F/100 ft and apply the real-gas law. Small depth iterations are necessary in the upper portion of the hole. The problem solution is assisted greatly by use of a spreadsheet computer program.

2.42 The following drilling conditions apply to a hypothetical well: depth, 8,100 ft; flowline temperature, 80°F; bottomhole temperature, 135°F; mud density, 9.7 lbm/gal; and atmospheric pressure, 13.5 psia. A surface sample of the mud

TABLE 2.24—MODIFIED *d* EXPONENT DATA FOR THE GULF COAST WELL DESCRIBED IN PROBLEM 2.32[44]

Depth (ft)	Modified *d* Exponent
8,150	1.51
9,000	1.55
9,600	1.58
10,150	1.51
10,400	1.60
10,650	1.61
10,900	1.62
11,100	1.58
11,300	1.66
11,600	1.50
11,700	1.61
11,850	1.58
12,100	1.70
12,200	1.45
12,300	1.30
12,450	1.22
12,750	1.21
12,900	1.26
13,000	1.19
13,300	1.18
13,450	1.12
13,550	1.06
13,750	1.02
13,900	1.05
13,950	0.96
14,050	1.00
14,200	0.91
14,300	0.88
14,400	0.91
14,600	0.89
14,700	0.97
14,800	0.90
14,950	0.94
15,050	0.98
15,200	0.93
15,500	0.87
16,300	0.85
16,800	0.67

weighs 5.1 lbm/gal because of entrained gas. Estimate the bottomhole pressure if a 50 psig backpressure is held on the well by a rotating head. Ignore the annular-friction loss and the effect of drilled solids on mud weight.

2.43 Table 2.25 gives flowline temperatures for a well in the South China Sea. Plot the data and determine the onset of abnormal pore pressure.

2.44 List five reasons for seeing a surface increase in drilling torque. List five reasons for increased hole drag.

2.45 Would it be appropriate to use Boatman's[57] pore-pressure correlation with wireline density measurements from a well located along the U.S. gulf coast? Defend your answer.

2.46 Plot the predicted pore pressures in density equivalent vs. depth for the well described in Table 2.14. Use the Hottman and Johnson[65] correlation.

2.47 The shale acoustic-travel times in **Table 2.26** were measured on a well located in the Mackenzie delta. Determine the

TABLE 2.25—DRILLING-FLUID TEMPERATURES FOR A WELL IN THE SOUTH CHINA SEA[61]

Lagged Depth (ft)	Flowline Temperature (°F)
3,020	106.0
3,180	107.0
3,390	107.0
3,600	104.5
3,770	109.5
4,000	110.5
4,210	112.0
4,240	113.5
4,500	112.5
4,710	116.0
4,760	120.5
4,890	120.0
5,050	115.5
5,120	116.5
5,230	116.0
5,300	117.0
5,390	123.0
5,510	123.5
5,520	125.5
5,600	128.0
5,700	121.0
5,780	122.5
5,890	121.5
5,910	127.0
6,010	128.5
6,050	134.0
6,060	140.5
6,100	143.0
6,210	138.0
6,300	138.0
6,430	141.0
6,460	144.0
6,720	141.0
6,810	136.0
6,870	132.5
6,890	131.5

transition depth and estimate the pore pressure at 11,800 ft using the equivalent-depth method.

2.48 Estimate the porosity of a shale if the logged resistivity is 1.50 $\Omega \cdot$ m and R_w in an adjacent sand is 0.17 $\Omega \cdot$ m. Assume a and m values of 1.0 and 2.0, respectively.

2.49 Plot the R_w values given in Table 2.15 vs. depth on semilogarithmic graph paper. Could R_w be considered as a tool for detecting abnormal pore pressure?

2.50 Plot predicted pore pressures in density equivalent vs. depth for the well described in Table 2.15. Use Foster and Whalen's[72] technique.

2.51 Table 2.27 gives Miocene shale-resistivity measurements from a U.S. gulf coast well. Estimate the pore pressure at 12,910 ft using the Hottman and Johnson[66] correlation.

2.52 Work Example 2.20 using Hottman and Johnson's resistivity correlation. How does your answer compare to that obtained from Foster and Whalen's[72] technique?

TABLE 2.26—SHALE TRANSIT TIMES FOR A WELL LOCATED IN THE MACKENZIE DELTA

Depth (ft)	Transit Time (μsec/ft)
4,500	100
4,910	96
5,480	85
5,620	89
5,910	90
6,490	85
6,900	79
7,090	85
7,730	84
7,910	99
8,290	90
8,690	90
9,010	95
9,680	89
9,900	83
10,200	84
10,580	87
10,770	90
11,800	90

2.53 Plot predicted pore pressures in density equivalent vs. depth for the well described in Table 2.16. Use the Matthews and Kelly[67] correlation.

2.54 Table 2.28 gives the shale transit times that accompanied the conductivity data in Table 2.16. Estimate the pore pressure at 11,500 ft using the Matthews and Kelly Frio shale correlation. How does your answer compare with the conductivity log result?

2.55 Demonstrate equivalence between Eaton's[47,48] pore pressure prediction equations and Terzhagi's[17] effective-stress principle. Hint: consider the relationships in normally compacted rock.

2.56 Solve Problems 2.51, 2.52, and 2.54 using Eaton's pore-pressure prediction equations.

2.57 A pore pressure of 11,900 psig was calculated with Eq. 2.34 in a Cretaceous age shale at 16,100 ft. Later completion tests of a sand at 16,200 ft indicate a true pore pressure of 11,700 psig. Determine the appropriate value for Eaton's travel-time ratio exponent.

2.58 Prepare overlays suitable for U.S. gulf coast use that are based on the three Eaton pore-pressure-prediction equations.

2.59 The following conditions may impact acoustic- and electric-log parameters to some degree. Discuss the effect (increase, decrease, or no change) and the relative importance that each of the following conditions has on shale travel-time and resistivity measurements.

1. Hole washouts.
2. High pore-water salinity.
3. Higher temperature.
4. Shale hydration.
5. Calcerous shales.
6. Pyritic shales.
7. Gas in the pore space.

TABLE 2.27—MIOCENE SHALE RESISTIVITY VALUES FOR A WELL IN THE U.S. GULF COAST AREA[68]	
Depth (ft)	Resistivity (Ω·m)
5,220	5.5
5,300	5.1
5,410	5.5
5,500	5.2
6,350	5.7
6,450	5.5
6,590	5.9
6,950	6.0
7,730	6.5
8,040	6.1
8,400	6.5
9,260	6.6
9,450	6.8
9,800	7.1
10,460	7.2
10,530	6.9
10,660	6.0
10,750	6.4
10,790	5.0
11,080	5.1
11,340	5.3
11,470	4.9
11,520	4.0
11,600	3.3
11,710	2.5
11,900	2.0
12,060	2.1
12,380	2.2
12,450	2.0
12,630	1.9
12,910	2.0

2.60 How would an unnoticed trend-line shift into older, more dense rock affect subsequent pore-pressure predictions?

2.61 Table 2.29 gives corrected GR count rates measured while drilling a U.S. gulf coast well. Estimate the pore pressure at 3,800 ft using Zoeller's[75] pore-pressure correlation. Extrapolate the normal-compaction trend using Fig. 2.68 as a guide.

2.62 Estimate K_D and D_0 values appropriate for use in the U.S. gulf coast.

2.63 Derive Eq. 2.40. Show all your steps.

2.64 The following questions pertain to the technique presented by Rasmus and Gray-Stephens.[79]

1. Estimate the surface porosity and porosity-decline constants from the North Sea porosity data in Table 2.18. Use the graphical procedure discussed in Example 2.4.
2. Develop an equation relating overburden stress as function of depth if the matrix density is 2.60 g/cm^3 and the pore-fluid density is 1.044 g/cm^3.
3. Compute the effective stress at each depth for the normal-pressure gradient of 0.452 psi/ft. Plot the results vs. linear porosity on semilogarithmic graph paper.
4. Determine the slope K_D from the effective-stress porosity plot.
5. Solve Example 2.25 using the K_D value determined from the effective-stress plot.

TABLE 2.28—FRIO SHALE TRANSIT TIMES FOR A WELL IN NUECES COUNTY, TEXAS[67]	
Depth (ft)	Transit Time (μsec/ft)
7,400	100
7,550	93
8,300	90
8,350	89
8,400	89
8,500	88
9,200	75
9,300	74
9,550	72
9,600	71
9,700	70
9,750	70
9,900	69
9,950	73
10,000	85
10,050	100
10,150	110
10,200	116
10,300	119
10,500	117
10,600	120
10,650	121
10,850	120
11,000	113
11,050	110
11,200	108
11,300	120
11,500	128

6. Solve Example 2.25 using the equivalent-depth method.

2.65 Table 2.30 lists sediment void ratios and overburden stresses derived from Eaton's[47,48] gulf coast bulk density and overburden gradient curves (Figs. 2.17 and 2.20). The last column of the table shows the calculated effective stress at each 1,000-ft depth increment. Using the shallower data, graphically determine suitable values for I_c and r_{vi}.

2.66 Determine the predicted normal-compaction porosity at 10,000 ft for the rocks listed in Table 2.19. Use an overburden-stress gradient of 0.95 psi/ft.

2.67 Subsequent LWD measurements on the well described in Example 2.27 indicate 24.1% porosity in a mixed sand/shale bed at 11,220 ft. Estimate the pore pressure if the rock is 35.3% sandstone and 64.7% shale. Assume the constants given in Table 2.19 and use an overburden-stress gradient of 0.935 psi/ft.

2.68 Make suitable assumptions and estimate the virgin-curve parameters for the North Sea using the data in Table 2.18.

2.69 On a central North Sea well, the logged shale velocity in a Jurassic formation at 15,500 ft is 12,000 ft/sec. Estimate the pore pressure at this depth if $A = 8.116$, $B = 0.8002$, $U = 4.48$, and $(v_{sh})_{max} = 5{,}200$ ft/sec.

Nomenclature

a = formation resistivity factor coefficient, dimensionless
a_N = rotating-speed exponent, dimensionless
a_W = bit-weight exponent, dimensionless

TABLE 2.29—CORRECTED GR COUNT RATES OBTAINED WHILE DRILLING A GULF COAST WELL[75]

Depth (ft)	GR Count Rate (cps)
1,500	29
1,600	30
1,700	30
1,800	31
1,900	30
2,000	32
2,100	33
2,200	34
2,300	32
2,400	32
2,500	33
2,600	33
2,700	34
2,800	35
2,900	39
3,000	41
3,100	42
3,200	40
3,300	38
3,400	40
3,500	39
3,600	35
3,700	34
3,800	34
3,900	36
4,000	38
4,100	40

A = virgin-curve compaction parameter, dimensionless
B = virgin-curve compaction exponent, dimensionless
c = shale-compactibility coefficient, dimensionless
C_n = normal conductivity, m℧ /m, q^2/mL^3
C_o = observed conductivity, m℧ /m, q^2/mL^3
C_w = water conductivity, m℧ /m, q^2/mL^3
d = bit-weight exponent in Bingham's equation, dimensionless
d_b = bit diameter, in.
d_c = corrected or modified d exponent, dimensionless
d_{cn} = normal d_c exponent, dimensionless
d_{co} = observed d_c exponent, dimensionless
D = depth, ft
D_{eq} = equivalent depth, ft
D_0 = depth of zero porosity, ft
f_1–f_n = penetration rate functions
f_{Vg} = gas volume fraction, dimensionless
f_{Vgs} = gas volume fraction at surface conditions in the wellbore, dimensionless
F_R = formation resistivity factor, dimensionless
g = acceleration of gravity, 32.17 ft/sec^2
g_c = gravitational system conversion constant, 32.17 (lbm-ft)/(lbf-sec^2)
g_G = geothermal gradient, °F/ft
g_n = normal-pore-pressure gradient, psi/ft
g_{ob} = overburden-stress gradient, psi/ft
g_p = pore-pressure gradient, psi/ft
I_c = compression index, psi^{-1}

TABLE 2.30—COMPUTED GULF COAST VOID RATIO AND EFFECTIVE-STRESS DATA FROM EATON'S[20,47,48] POROSITY AND OVERBURDEN CORRELATIONS

Depth ft	r_v fraction	σ_{ob} psig	p_n psig	σ_{Ve} psi
3,000	0.471	2,661	1,395	1,266
4,000	0.408	3,588	1,860	1,728
5,000	0.370	4,535	2,325	2,210
6,000	0.316	5,496	2,790	2,706
7,000	0.282	6,475	3,255	3,220
8,000	0.250	7,464	3,720	3,744
9,000	0.220	8,469	4,185	4,284
10,000	0.190	9,480	4,650	4,830
11,000	0.176	10,505	5,115	5,390
12,000	0.163	11,544	5,580	5,964
13,000	0.149	12,571	6,045	6,526
14,000	0.136	13,622	6,510	7,112
15,000	0.124	14,685	6,975	7,710
16,000	0.111	15,744	7,440	8,304
17,000	0.109	16,813	7,905	8,908
18,000	0.101	17,874	8,370	9,504
19,000	0.093	18,924	8,835	10,089
20,000	0.086	20,000	9,300	10,700

K = drill-rate model proportionality constant
K_D = depth-decline constant, dimensionless
K_ϕ = porosity-decline constant, ft^{-1}
m = cementation exponent, dimensionless
n = number of moles, lbm mol
N = bit rotating speed, rev/min
N_{GRn} = normal GR intensity, cps
N_{GRo} = observed GR intensity, cps
p = pressure, psi
p_m = mud-column hydrostatic pressure, psi
p_{mg} = gas-cut mud-column hydrostatic pressure, psi
p_n = normal pore pressure, psi
$p_{n(eq)}$ = normal pore pressure at the equivalent depth, psi
p_p = pore pressure, psi
p_s = surface pressure, psi
r = radius, in.
r_v = void ratio, dimensionless
r_{vi} = void ratio at an effective stress of unity, dimensionless
r_w = wellbore radius, in.
R = penetration rate, ft/hr
R_g = universal gas constant, (psia-gal)/(°R-lbm-mole)
R_0 = resistivity of water-saturated rock, mL^3/tq^2, $\Omega \cdot m$
R_n = normal resistivity, mL^3/tq^2, $\Omega \cdot m$
R_o = observed resistivity, mL^3/tq^2, $\Omega \cdot m$
R_w = water resistivity, mL^3/tq^2, $\Omega \cdot m$
R_{wb} = bound-water resistivity, mL^3/tq^2, $\Omega \cdot m$
s = poroelasticity constant, dimensionless
Δt = formation transit time, μs/ft
Δt_f = fluid transit time, μs/ft
Δt_{ma} = matrix transit time, μs/ft
Δt_n = normal formation travel time, μs/ft
Δt_o = observed formation travel time, μs/ft
T = temperature, T, °F or °R [°C or K]
T_s = surface temperature, T, °R
u_H = heat flux, Btu/(hr-ft^2)
U = unloading-curve exponent, dimensionless
v_{sh} = shale acoustic velocity, ft/sec

$(v_{sh})_{max}$ = shale acoustic velocity at the onset of unloading, ft/sec
V_g = gas volume, gal
V_m = mud volume, gal
W = applied bit weight, lbf
X_n = porosity indicator from the normal-trend-line extrapolation
X_o = observed porosity indicator
z = compressibility factor, dimensionless
z_s = compressibility factor at surface conditions in the wellbore, dimensionless
α = plane angle, degree
β = natural compaction strain-hardening coefficient, dimensionless
γ_{sh} = shale SG, dimensionless
λ = thermal conductivity, Btu/(hr-ft-°F/ft)
ρ_b = bulk density, lbm/gal, also g/cm^3
ρ_{eq} = equivalent density, lbm/gal
ρ_f = fluid density, lbm/gal, also g/cm^3
ρ_g = gas density, lbm/gal
ρ_m = mud density, lbm/gal
ρ_{ma} = matrix density, lbm/gal, also g/cm^3
ρ_{mg} = gas-cut mud density, lbm/gal
ρ_n = density equivalent of the normal-pore-pressure gradient, lbm/gal
ρ_p = pore-pressure equivalent density, lbm/gal
ρ_{sh} = shale bulk density, g/cm^3
ρ_{shn} = normal shale density, g/cm^3
ρ_{sho} = observed shale density, g/cm^3
ρ_{shw} = shale/water mixture density, lbm/gal
ρ_w = water density, lbm/gal
σ = stress applied normal to an element plane, psi
σ_e = effective stress, psi
σ_{Ve} = effective stress in the vertical direction, psi
σ_{Ve0} = effective vertical stress that gives zero porosity, psi
$\sigma_{Ve(eq)}$ = effective vertical stress at the equivalent depth, psi
$(\sigma_{Ve})_{max}$ = effective vertical stress at the onset of unloading, psi
σ_{Ven} = normal vertical effective stress, psi
$\sigma_{H\max}$ = maximum horizontal principal stress, psi
$\sigma_{H\min}$ = minimum horizontal principal stress, psi
σ_{max} = maximum principal stress, psi
σ_{min} = minimum principal stress, psi
σ_{ob} = overburden stress, psi
$\sigma_{ob(eq)}$ = overburden stress at the equivalent depth, psi
σ_a = normal stress on an arbitrary plane, psi
τ_{max} = maximum shear stress, psi
τ_a = shear stress on an arbitrary plane, psi
ϕ = formation porosity, dimensionless
ϕ_0 = porosity at zero depth, dimensionless
ϕ_n = normal porosity, dimensionless
ϕ_{op} = overpressure porosity, dimensionless

References

1. Bourgoyne, A.T. Jr. *et al.*: *Applied Drilling Engineering*, second printing, Textbook Series, SPE, Richardson, Texas (1991) 246–99.
2. Fertl, W.H. and Chilingarian, G.V.: "Importance of Abnormal Formation Pressures," *JPT* (April 1977) 347.
3. Fertl, W.H: *Abnormal Formation Pressures, Implications to Exploration, Drilling, and Production of Oil and Gas Resources*, Elsevier Scientific Publishing Co., New York City (1976).
4. Parker, C.A: "Geopressures in the Deep Smackover in Mississippi," *JPT* (August 1973) 971; *Trans.*, AIME **255**
5. Mouchet, J.P. and Mitchell, A.: *Abnormal Pressures While Drilling*, Elf Aquitane Manuels Techniques 2, Boussens, France (1989).
6. Pettijohn, F.J.: *Sedimentary Rocks*, Harper & Brothers, New York City (1949) 476.
7. Powers, M.C.: "Fluid-Release Mechanisms in Compacting Marine Mudrocks and Their Importance in Oil Exploration," *AAPG Bull.* (1967) **51,** 1240.
8. Burst, J.F.: "Diagenesis of Gulf Coast Clayey Sediments and Its Possible Relation to Petroleum Migration," *AAPG Bull.* (1969) **53,** 73.
9. Magara, K.: "Reevaluation of Montmorillonite Dehydration as Cause for Abnormal Pressures and Hydrocarbon Migration," *AAPG Bull.* (1975) **59,** 292.
10. Barker, C.: "Aquathermal Pressuring—Role of Temperature in Development of Abnormal-Pressure Zones," *AAPG Bull.* (1972) **56,** 2068.
11. Magara, K: "Importance of Aquathermal Pressuring Effect in the Gulf Coast," *AAPG Bulletin* (1975) **59,** 2037.
12. Chapman, R.E.: "Mechanical versus Thermal Cause of Abnormally High Pore Pressure in Shales," *AAPG Bull.* (1980) **64,** 2179.
13. Sharp, J.M. Jr.: "Permeability Controls on Aquathermal Pressuring," *AAPG Bull.* (1983) **67,** 2057.
14. Young, A. and Low, P.F.: "Osmosis in Argillaceous Rocks," *AAPG Bull.* (1965) **67,** 1004.
15. Berry, F.A.F.: "Hydrodynamics and Geochemistry of the Jurassic and Cretaceous Systems in the San Juan Basin, N.W. New Mexico and S.W. Colorado," PhD dissertation, Stanford U., Stanford, California (1959).
16. Barker, C: "Generation of Anomalous Internal Pressures in Source Rocks," *Migration of Hydrocarbons in Sedimentary Basins*, B. Doligez (ed.), Gulf Publishing Co., Houston (1987) 223–35.
17. Terzaghi, K.: *Theoretical Soil Mechanics*, John Wiley and Sons Inc., New York City (1943) 51.
18. Biot, M.A: "General Theory of Three-Dimensional Consolidation," *J. Appl. Phys.* (1941) **12,** 155.
19. Morita, N. *et al.*: "A Quick Method to Determine Subsidence, Reservoir Compaction, and In-Situ Stress Induced by Reservoir Depletion," *JPT* (January 1989) 71.
20. Eaton, B.A: "Fracture Gradient Prediction and its Application in Oilfield Operation," *JPT* (October 1969) 1353.
21. Bass, D.M. Jr.: "Properties of Reservoir Rocks," *Petroleum Engineering Handbook*, H.B. Bradley (ed.), SPE, Richardson, Texas (1987) 26, 7.
22. Mitchell, B.J.: *Advanced Oilwell Drilling Engineering Handbook*, ninth edition, SPE, Richardson, Texas (September 1992) 180.
23. Magara, K.: *Compaction and Fluid Migration*, Elsevier Scientific Publishing Co., New York City (1978).
24. Rubey, W.W. and Hubbert, M.K.: "Role of Fluid Pressure in Mechanics of Overthrust Faulting," *GSA Bull.* (1959) **70,** 115.
25. Athey, L.F.: "Density, Porosity, and Compaction of Sedimentary Rocks," *AAPG Bull.* (1930) **14,** 1.
26. Constant, W.D. and Bourgoyne, A.T. Jr.: "Method Predicts Frac Gradient for Abnormally Pressured Formations," *Pet. Eng. Intl.* (January 1986) 38.
27. Constant, W.D. and Bourgoyne, A.T. Jr.: "Fracture Gradient Prediction for Offshore Wells," *SPEDE* (June 1988) 136.
28. Pennebaker, E.S.: "An Engineering Interpretation of Seismic Data," paper SPE 2165 presented at the 1968 SPE Annual Fall Meeting, Houston, 29 September–2 October.
29. Dix, C.H.: "Seismic Velocities from Surface Measurements," *Geophysics* (1955) **20,** 68.

30. Keyser, W. *et al.*: "Pore Pressure Prediction from Surface Seismic," *World Oil* (September 1991) 115.
31. Davis, B. and Jones, T.: "Pore Pressure Prediction Cuts Exploratory Drilling Risk," *World Oil* (September 1994) 63.
32. Murray, A.S. and Cunningham, R.A.: "Effect of Mud Column Pressure on Drilling Rates," *Trans.*, AIME (1955) **204,** 196.
33. Eckel, J.R.: "Effect of Pressure on Rock Drillability," *Trans.*, AIME (1958) **213,** 1.
34. Cunningham, R.A. and Eenik, J.G.: "Laboratory Effect of Overburden, Formation, and Mud Column Pressures on Drilling Rate of Permeable Formations," *Trans.*, AIME (1959) **216,** 9.
35. Garnier, A.J. and van Lingen, N.H.: "Phenomena Affecting Drilling Rates at Depth," *Trans.*, AIME (1959) **216,** 232.
36. Robinson, L.H. Jr.: "Effects of Pore and Confining Pressure on Failure Characteristics of Sedimentary Rocks," *Trans.*, AIME (1959) **216,** 26.
37. Vidrine, D.J. and Benit, E.J.: "Field Verification of the Effect of Differential Pressure on Drilling Rate," *JPT* (July 1968) 676.
38. Jumikis, A.R.: *Rock Mechanics*, second edition, Gulf Publishing Co., Houston (1983) 202.
39. Maurer, W.C.: "Bit-Tooth Penetration Under Simulated Borehole Conditions," *JPT* (December 1965) 1433; *Trans.*, AIME, **234**.
40. Warren, T.M. and Smith, M.B.: "Bottomhole Stress Factors Affecting Drilling Rate at Depth," *JPT* (August 1985) 1523; *Trans.*, AIME, **279**.
41. Gray-Stephens, D., Cook, J.M., and Sheppard, M.C.: "Influence of Pore Pressure on Drilling Response in Hard Shales," *SPEDC* (December 1994) 263.
42. Jorden, J.R. and Shirley, O.J.: "Application of Drilling Performance Data to Overpressure Detection," *JPT* (November 1966) 1387.
43. Combs, G.D.: "Prediction of Pore Pressure From Penetration Rate," paper SPE 2162 presented at the 1968 SPE Annual Meeting, Houston, 29 September–2 October.
44. Rehm, W.A. and McClendon, M.T.: "Measurement of Formation Pressure From Drilling Data," paper SPE 3601 presented at the 1971 SPE Annual Meeting, New Orleans, 3–6 October.
45. Zamora, M.: "Slide-Rule Correlation Aids 'd' Exponent Use," *Oil & Gas J.* (18 December 1972).
46. Bourgoyne, A.T. Jr. and Young, F.S. Jr.: "A Multiple Regression Approach to Optimal Drilling and Abnormal Pressure Detection," *SPEJ* (August 1974) 371; *Trans.*, AIME, **257.**
47. Eaton, B.A.: "The Equation for Geopressure Prediction from Well Logs," paper SPE 5544 presented at the 1975 SPE Annual Techinal Conference and Exhibition, Dallas, 28 September–1 October.
48. Eaton, B.A.: "Graphical Method Predicts Geopressures Worldwide," *World Oil* (July 1976) 100.
49. Bellotti, P. and Gerard, R.E.: "Instantaneous Log Indicates Porosity and Pore Pressure," *World Oil* (October 1976) 90.
50. Prentice, C.M.: "Normalized Penetration Rate Predicts Formation Pressures," *Oil & Gas J.* (11 August 1980) 103.
51. Prentice, C.M.: "P-Rate Plot Interpretation Yields Pore Pressure," *Oil & Gas J.* (18 August 1980) 104.
52. Moore, P.L.: "How to Predict Pore Pressure," *Pet. Eng. Intl.* (March 1982) 144.
53. Moore, P.L.: *Drilling Practices Manual*, second edition, Pennwell Publishing Co., Tulsa, Oklahoma (1986) 468–70.
54. Cheli, E., Brancato, C., and Vagnoux, J.P.: "WOODIE: A Tool To Support Overpressure Detection," paper SPE 29346 presented at the 1995 SPE/IADC Drilling Conference, Amsterdam, 28 February–2 March.
55. Bingham, M.G.: "A New Approach to Interpreting Rock Drillability," *Oil & Gas J.* series, Petroleum Publishing Co., Tulsa, Oklahoma (April 1965).
56. Beato, C.: "New Rig Approach Corrects D-Exponent Values," *Pet. Eng. Intl.* (March 1991) 26.
57. Boatman, W.A. Jr.: "Measuring and Using Shale Density to Aid in Drilling Wells in High Pressure Areas," *Drill. & Prod. Prac.*, (1967) Dallas, 121.
58. White, R.J.: "Bottomhole Pressure Reduction Due to Gas Cut Mud," *JPT* (July 1957) 112, *Trans.*, AIME, **210**.
59. "Recommended Practice Standard Procedure for Field-Testing Water-Based Drilling Fluids," *RP 13B-1,* second edition, API, Washington, DC (September 1997).
60. Lewis, C.R. and Rose, S.C.: "A Theory Relating High Temperatures and Overpressures," *JPT* (January 1970) 11.
61. Adams, N.: *Well Control Problems and Solutions*, Petroleum Publishing Co., Tulsa (1980) 348.
62. Raymond, L.R.: "Temperature Distribution in a Circulating Drilling Fluid," *Well Completions*, Reprint Series, SPE, Richardson, Texas (1970) **5,** 98–106.
63. Herring, E.A.: "Estimating Abnormal Pressures From Log Data in the North Sea," paper SPE 4301 presented at the 1973 Annual European Meeting, London, 2–3 April.
64. Evers, J.F. and Ezeanyim, R.: "Prediction of Abnormal Pressures in Wyoming Sedimentary Basins Using Well Logs," paper SPE 11859 presented at the 1983 SPE Rocky Mountain Regional Meeting, Salt Lake City, Utah, 23–35 May.
65. Owolabi, O.O., Okpobiri, G.A., and Obomanu, I.A.: "Prediction of Abnormal Pressures in the Niger Delta Basin Using Well Logs," paper SPE 21575 presented at the SPE/CIM 1990 International Technical Meeting, Calgary, 10–13 June.
66. Hottman, C.E. and Johnson, R.K.: "Estimation of Formation Pressure From Log-Derived Shale Properties," *JPT* (June 1965) 717: *Trans.*, AIME, **234**.
67. Matthews, W.R. and Kelly, J.: "How to Predict Formation Pressure and Fracture Gradient from Electric and Sonic Logs," *Oil & Gas J.* (20 February 1967).
68. Reynolds, E.B., Timko, D.J., and Zanier, A.M.: "Potential Hazards of Acoustic Log Pressure Plots," *JPT* (September 1973) 1039.
69. Archie, G.E.: "Electric Resistivity Log as an Aid in Determining Some Reservoir Characteristics," *Trans.*, AIME (1942) **146,** 54.
70. Arps, J.J.: "The Effect of Temperature on the Density and Electrical Resistivity of Sodium Chloride Solutions," *JPT* (October 1953) 17; *Trans.*, AIME, **198.**
71. Waxman, M.H. and Smits, L.J.M.: "Electric Conductivities in Oil-Bearing Shaly Sands," *SPEJ* (June 1968) 107; *Trans.*, AIME, **243.**
72. Foster, J.B. and Whalen, H.E.: "Estimation of Formation Pressures From Electrical Surveys—Offshore Louisiana," *JPT* (February 1966) 165.
73. Lane, R.A. and Macpherson, L.A.: "A Review of Geopressure Evaluation From Well Logs—Louisiana Gulf Coast," *JPT* (September 1976) 963.
74. Foster, J. Jr.: "Pore-Pressure Plot Accuracy Increased by Multiple Trend Lines," *Oil & Gas J.* (7 May 1990) 100.
75. Zoeller, W.A.: "Determine Pore Pressures from MWD Gamma Ray Logs," *World Oil* (March 1984) 97.
76. Holbrook, P.W. and Hauck, M.L.: "A Petrophysical/Mechanical Math Model for Real-Time Wellsite Pore Pressure/Fracture Gradient Prediction," paper SPE 16666 presented at the 1987 SPE Annual Technical Conference and Exhibition, Dallas, 27–30 September.
77. Bryant, T.M.: "A Dual Shale Pore Pressure Prediction Technique," paper SPE 18714 presented at the 1989 SPE/IADC Drilling Conference, New Orleans, 28 February–3 March.
78. Alixant, J.L. and Desbrandes, R.: "Explicit Pore-Pressure Evaluation: Concept and Application," *SPEDE* (September 1991) 182.
79. Rasmus, J.C. and Gray-Stevens, D.M.R.: "Real-Time Pore-Pressure Evaluation From MWD/LWD Measurements and

Drilling-Derived Formation Strength," *SPEDE* (December 1991) 264.
80. Accarain, P. and Desbrandes, R.: "Neuro-Computing Helps Pore Pressure Determination," *Pet. Eng. Intl.* (February 1993) 39.
81. Holbrook, P.W., Maggiori, D.A., and Hensley, R.: "Real-Time Pore Pressure and Fracture Gradient Evaluation in All Sedimentary Lithologies," paper SPE 26791 presented at the 1993 Offshore European Conference, Aberdeen, 7–10 September.
82. Bowers, G.L.: "Pore Pressure Estimation From Velocity Data: Accounting for Overpressure Mechanisms Besides Undercompaction," *SPEDE* (June 1995) 89.
83. Mayer, C. and Sibbit, A.: "GLOBAL, A New Approach to Computer-Processed Log Interpretation," paper SPE 9341 presented at the 1980 SPE Annual Technical Conference and Exhibition, Dallas, 21–24 September.
84. Perez-Rosales, C.: "Generalization of Maxwell Equation for Formation Factor," paper SPE 5502 presented at the 1975 SPE Annual Technical Conference and Exhibition, Dallas, 28 September–1 October.
85. Atwater, G.L. and Miller, E.E.: "The Effect of Decrease in Porosity With Depth on Future Development of Oil and Gas Reserves in South Louisiana," *AAPG Bull.* (1965) **49,** 334.
86. Gandino, A. and Zenucchini, G.: "Density Depth Correlation in Po Valley Sediments," *Bollettino de Geofisica Teorica ed Applicata* (1987) **29,** 221.

SI Metric Conversion Factors

Btu	× 1.055 056	E + 00	= kJ
cp	× 1.0*	E + 03	= Pa · s
deg	× 1.745 329	E − 02	= rad
ft	× 3.048*	E − 01	= m
ft^2	× 9.290 304*	E − 02	= m^2
°F	(°F − 32)/1.8		= °C
gal	× 3.785 412	E + 00	= L
in.	× 2.54*	E + 01	= mm
lbf	× 4.448 222	E + 00	= N
lbm	× 4.535 924	E − 01	= Kg
psi	× 6.894 757	E − 03	= MPa
°R	°R/1.8		= K
μs/ft	× 3.280 840	E + 00	= μs/m

*Conversion factor is exact.

Chapter 3
Fracture Pressure

3.1 Introduction

In a drilling operation, the lower bound to the allowable wellbore pressure is controlled either by formation pressure or by wellbore collapse. In a general sense, pore pressure usually dictates the minimum pressure limit and most operations are conducted so that the mud-density hydrostatic pressure exceeds the pore pressure by some "safe" margin. An upper pressure limitation must be considered when planning a well and during all subsequent phases of operation. In some cases, the upper bound is controlled by the pressure rating of the casing or blowout-prevention equipment. More commonly, however, the pressure integrity of the exposed open hole dictates the maximum wellbore pressure allowed. This chapter focuses on fracture-pressure prediction and measurement.

Several methods for predicting fracture pressures have been developed and refined. We first discuss the governing rock mechanics principles, followed by analytical or empirical relationships. Sec. 3.5 covers techniques for measuring or approximating the true fracture gradient in an existing wellbore.

A fundamental understanding of the concepts discussed in Chap. 2 is prerequisite to this discussion. Pore pressure controls fracture pressure to a large extent, and a valid pressure vs. depth estimate is essential to predicting the fracture gradient. In addition, effective-stress, overburden-stress, and formation-strength concepts are essential parts of the foundation for understanding fracture pressure.

3.2 Basic Principles From Rock Mechanics

Rock mechanics, or how rock reacts to an imposed stress, plays an important role in many aspects of drilling, completion, and production design. As Chap. 2 discussed, rock strength and confining stress greatly influence formation drillability. Understanding rock behavior is important to production and reservoir engineering when it comes to evaluating perforating gun performance, controlling sand production, modeling the effect of compaction on reservoir performance, and many other considerations. Creating a fracture by applying pressure to a wellbore is another process controlled by the in-situ properties of the formation.

Hydraulic fracturing is a well-established procedure for enhancing reservoir performance. In fact, many of the fracture-gradient principles and rock-mechanics concepts discussed in this chapter were developed in response to this stimulation technology. This is not intended, nor does it need to be, a rigorous treatise on a complex phenomenon having many unknown or questionable variables. Simplifying assumptions must be made when planning or drilling a well, and these assumptions may carry severe limitations. The drilling engineer must realize when physical models are not applicable and be flexible in adjusting the plan for unknowns.

3.2.1 Elastic Rock Properties. Fig. 3.1 illustrates how a material under a tensile or compressive load deforms or strains in response to imposed stress. Certain useful material properties can be characterized in the laboratory by applying tension or compression to a material specimen and observing the response to the applied load. In the procedure, an increasing axial force, F_a, is applied to a sample of original length, L_1, and diameter, d_1, while strain gauges measure axial and lateral changes in dimension. The axial stress σ_a at any point in the test is determined by

$$\sigma_a = F_a/A_1, \quad \text{(3.1)}$$

where A_1 = the original cross-sectional area of the specimen. The axial and transverse strains, respectively, are

$$\epsilon_a = (L_1 - L_2)/L_1 \quad \text{(3.2)}$$

and $$\epsilon_{tr} = (d_1 - d_2)/d_1. \quad \text{(3.3)}$$

By convention, a negative strain denotes expansion.

Plotting ϵ_a as a function of the applied σ_a yields the familiar uniaxial-stress/-strain diagram for the material. **Fig. 3.2** is an example of a tension-stress/-strain diagram for a common aluminum alloy. It is worthwhile to review some definitions and distinguishing characteristics of the relationships depicted on the diagram. The straight-line portion from initial loading up to Point A represents the elastic-stress range. Elastic behavior is described by Hooke's law,

$$\sigma_a = E\epsilon_a, \quad \text{(3.4)}$$

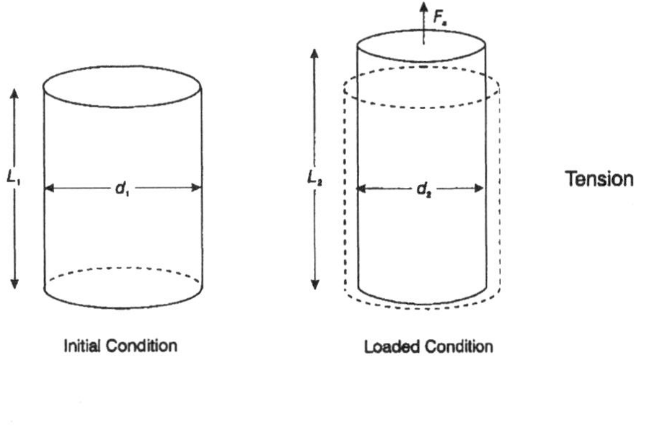

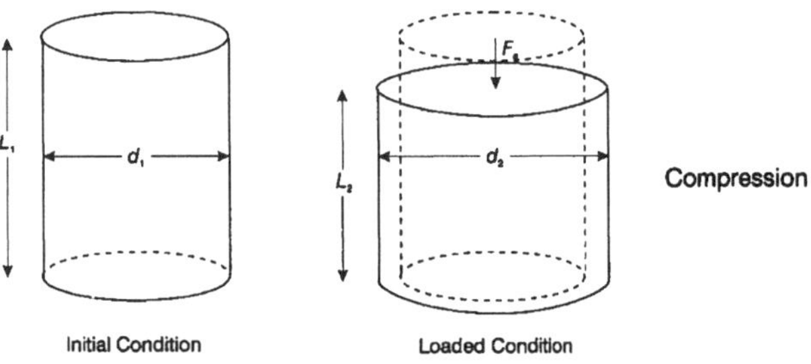

Fig. 3.1—Material deformation in reponse to an axial load.

where the proportionality constant E is Young's modulus of elasticity. Young's modulus is considered a reproducible parameter for materials having the same composition. Its value usually is independent of the material strength and axial loading direction (i.e., tension or compression).

True elastic behavior is defined as the capacity of a material to return to its original dimensions once the stress has been released. Exceeding the elastic limit to Point B results in a permanent deformation of the material shown by the strain ϵ_p. The nonlinear portion of the curve between Point A and the breaking point at Point C describes the plastic-behavior range. Materials that break or fracture at a stress below the elastic limit are defined as brittle, whereas those that rupture after significant plastic deformation are characterized as ductile.

The ratio between the transverse and axial strains is a material property defined as the Poisson's ratio μ.

$$\mu = -\epsilon_{tr}/\epsilon_a. \qquad (3.5)$$

The negative sign is necessary for the customary use of negative numbers to denote expansion. The Poisson's ratio for true elastic materials is constant within the elastic-stress range and is on the order of 0.3 for most metals. In the plastic range, however, the ratio begins to increase and may achieve the limiting value of 0.5.

Metallic alloys used as structural materials are polycrystalline materials refined to precise standards. The elastic constants of these well-ordered substances are predictable quantities for specific metallurgy and heat treatment. In contrast, rock is part of the disordered domain of nature, and the response of a rock element to stress depends on such things as its loading history, lithological constituents, cementing materials (or lack thereof), porosity, and any inherent defects. Even so, similar stress/strain behavior is observed and much of the same terminology has been adopted in the field of rock mechanics.

The most common rock principal-stress fields encountered in construction projects or within the earth's crust are compressive; hence, rock usually is tested for compression and shear strength. Triaxial behavior is analyzed in the laboratory by supplying a transverse (or confining) stress in addition to the axial-stress component. **Fig. 3.3** shows example triaxial-stress/-strain diagrams for a sandstone at confining stresses of 0 and 1,450 psi. Features of these two curves are of interest to the discussion.

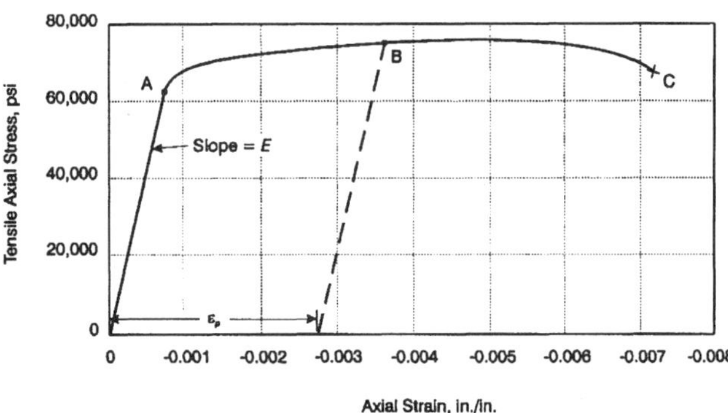

Fig. 3.2—Uniaxial-stress/-strain diagram for an aluminum alloy.[1]

Both curves are nonlinear from initial loading up to Point A. The high degree of initial compliance corresponds to the closing of any microcracks or other defects within the sample. The stress at Point B is the elastic limit marking the end of elastic behavior and the onset of plasticity.

Young's modulus for the sandstone and most rock is not a unique entity but depends instead on the magnitude of axial and confining stresses. For the nonlinear portion of the diagram (including the plastic range), E can be described one of three ways. The initial modulus E_i is the slope of the stress/strain curve at initial loading conditions, whereas the tangent modulus E_t is defined as the instantaneous slope at a specified stress. Generally, the tangent modulus at the in-situ-stress conditions is the most useful designation when modeling hydraulic fracture behavior. The secant modulus E_s is the ratio of the total stress to the total strain at a specified stress and is useful for describing behavior across a fairly large stress range. **Fig 3.4** illustrates the difference between the three moduli.

Fig. 3.5 shows transverse strains for the same sandstone. Eq. 3.5 states that Poisson's ratio is constant as long as the axial and transverse strains are linear with stress. However, this dual linearity is evidenced only within narrowly defined

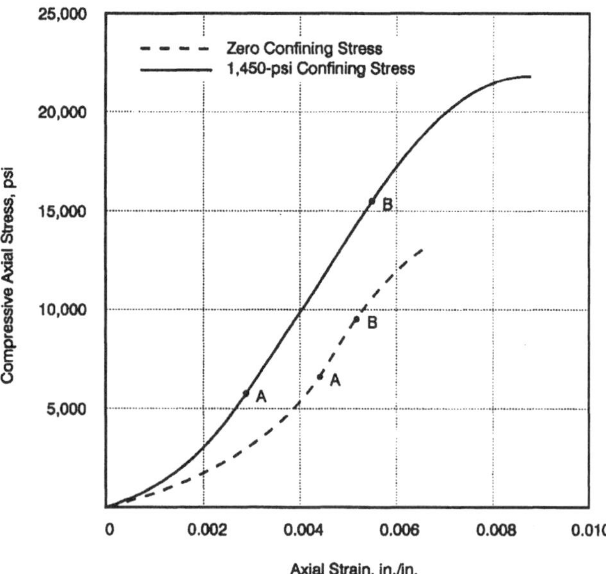

Fig. 3.3—Triaxial-stress/-strain diagram for a sandstone at two confining pressures.[2]

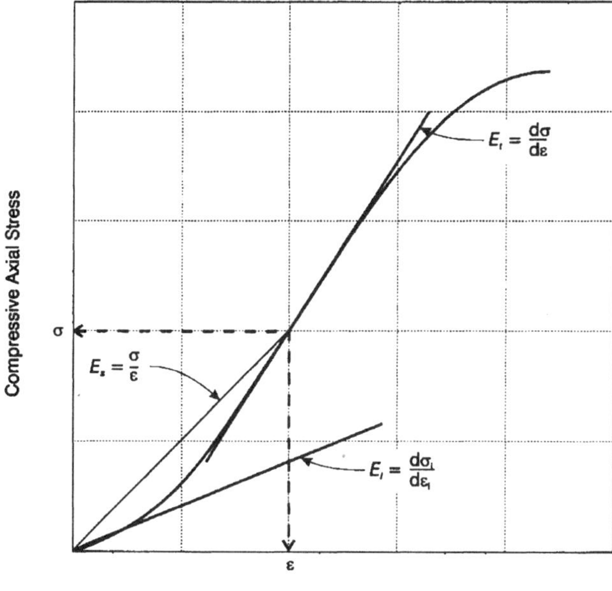

Fig. 3.4—Young's initial, tangent, and secant moduli defined.

stress ranges. Thus, Poisson's ratio for rock is also stress-dependent and the stress state must be specified to obtain representative values. The relative flattening of the transverse strains indicates the onset of plastic behavior and an increasing Poisson's ratio.

Example 3.1. Use the sandstone data shown in Fig. 3.5 and estimate the Young's modulus and Poisson's ratio at 10,000-psi axial stress and 1,450-psi confining stress.

Solution. It is apparent that the given stress conditions are within the elastic range of the material. The tangent modulus can be approximated as the line slope between 5,000 and 15,000 psi.

$$E_t \approx (15,000 - 5,000)/(0.00538 - 0.00266)$$
$$= 3.7 \times 10^6 \text{ psi}.$$

The axial and transverse strains at the specified stress are approximately 0.00404 and −0.00044, respectively. Eq. 3.5 yields

$$\mu = -(-0.00044/0.00404) = 0.109.$$

Rocks tend to be more ductile and plastic with increasing confining stress and temperature. Plastic behavior is more difficult to characterize, and a general simplifying assumption made in hydraulic-fracturing models is that the formation has not been stressed beyond its elastic limit. The relationships discussed in this chapter are based on linear-elastic mechanics unless stated otherwise.This is usually a reasonable assumption for rigid, well-consolidated rock at reasonable burial depths.

Other formations, however, react as plastic materials under low-stress conditions. For example, rock salt and many clays are fairly ductile and exhibit substantial deformation before failure. A reasonable conclusion is that these rocks have a relatively high Poisson's ratio compared with brittle rock at the same depth. It is not uncommon to see Poisson's ratio for shallow plastic formations attain the pure hydrostatic behavior limit of 0.5.

Another common assumption is that rock is a homogenous, isotropic material and that the properties of a laboratory specimen can be applied to its original stratum. Anyone who has observed a rock outcrop or road cut should realize that rock heterogeneity is more likely the norm. Also, rock tends to be anisotropic in nature—i.e., the elastic properties depend on the direction of the applied load. In other words, one should expect a different stress/strain diagram for loads applied in a direction parallel to the bedding plane than when the loads are normal to bedding.

Young's modulus is an important parameter in hydraulic-fracture design. It controls fracture width whereas Poisson's ratio is the property used to predict fracture gradient. **Table 3.1** gives ranges of Young's modulus and Poisson's ratio derived from laboratory measurements on various sedimentary and igneous rock.

In-situ elastic properties may be ascertained reasonably from triaxial testing if the rock and stress state accurately represent downhole conditions. This procedure is often used when gathering pretreatment data or when setting drawdown limits for highly compactible rock. Obtaining several cores and conducting extensive analyses on a reservoir rock and its boundary layers would be a prudent measure before designing a massive hydraulic-fracture treatment program. Obtain-

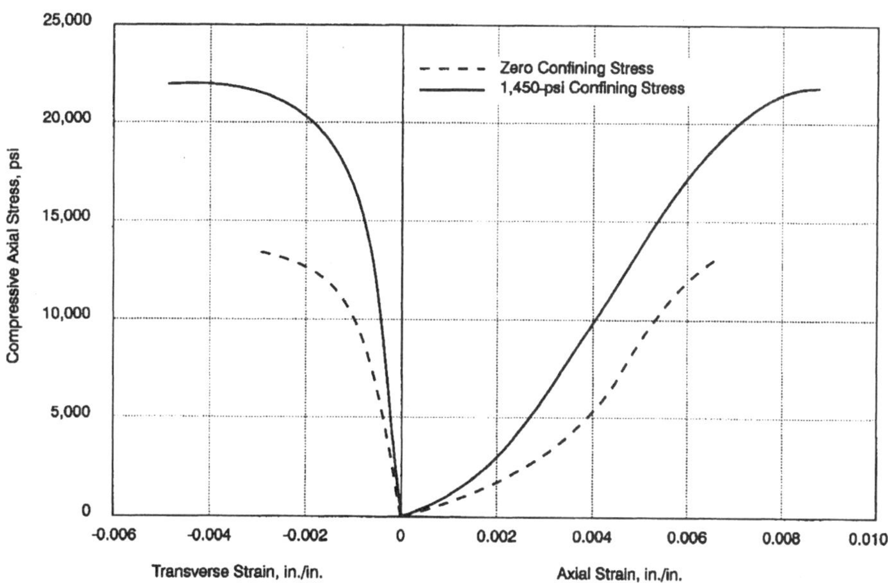

Fig. 3.5—Triaxial-stress/-strain diagram and transverse strains for the sandstone shown in Fig. 3.3.[2]

TABLE 3.1—TYPICAL ELASTIC PROPERTIES OF ROCK[3]		
Rock Type	Young's Modulus (10^6 psi)	Poisson's Ratio
Granite	3.7 to 10.0	0.125 to 0.250
Dolomite	2.8 to 11.9	0.08 to 0.20
Limestone	1.4 to 11.4	0.10 to 0.23
Sandstone	0.7 to 12.2	0.066 to 0.30
Shale	1.1 to 4.3	0.1 to 0.50

ing information in this fashion probably is not an option to the drilling well planner, and some other way of estimating the elastic constants is necessary.

Tables that give measured elastic constants for specific formations in an area may be available from the service sector.[4] However, these compilations typically are assembled for reservoir rock at unspecified stress conditions, so their usefulness is limited.

A potentially valuable information source is sonic logging tools that distinguish shear from compression waveforms or independently transmit and measure shear wave velocity. Shear and compression wave velocities depend on the bulk rock's compressive resistance; i.e., elasticity and several dynamic rock properties can be derived from the two acoustic velocities and the bulk density.

Clark[5] presented the following dynamic equations using sonic and density log data. The expression for Poisson's ratio depends only on the measured compression and shear velocities, v_c and v_s, respectively.

$$\mu = \frac{0.5(v_c/v_s)^2 - 1}{(v_c/v_s)^2 - 1}. \qquad (3.6)$$

Eq. 3.7 is an expression for the Young's modulus.

$$E = 0.0268\rho_b v_s^2(1 + \mu), \qquad (3.7)$$

where E is in psi, ρ_b is in g/cm^2, and v_s^2 is in (ft/sec)^2.

Example 3.2. Acoustic-log measurements across a sandstone at 8,695 ft indicate compression- and shear-wave transit times of 79 and 135 μsec/ft, respectively. The sand bulk density is 2.38 g/cm^3. Determine (1) the dynamic Poisson's ratio and (2) Young's modulus.

Solution.

1. An alternative form of Eq. 3.6 gives the dynamic Poisson's ratio:

$$\mu = \frac{0.5(\Delta t_s/\Delta t_c)^2 - 1}{(\Delta t_s/\Delta t_c)^2 - 1}$$

$$= \frac{0.5(135/79)^2 - 1}{(135/79)^2} = 0.240.$$

2. Use Eq. 3.7 to determine the dynamic Young's modulus.

$$E = (0.0268)(2.38)(7,407)^2(1 + 0.240)$$

$$= 4.34 \times 10^6 \text{ psi}.$$

Dynamic elastic constants are based on measurements taken at the speed of sound, whereas hydraulic fractures initiate and propagate at a much slower velocity. Thus, wellbore fracturing is considered a static or quasistatic process. Caution should be exercised when using log-derived elastic properties because there may be a discrepancy between the two. **Fig. 3.6** illustrates the difference between static and dynamic Young's moduli in the experimental correlation. The effect is believed to result from microfractures in the rock, which are essentially ignored by the acoustic waves but can be important in the quasistatic sense. However, log-derived properties do have appeal because they can be obtained economically across a long hole section and under actual stress conditions.

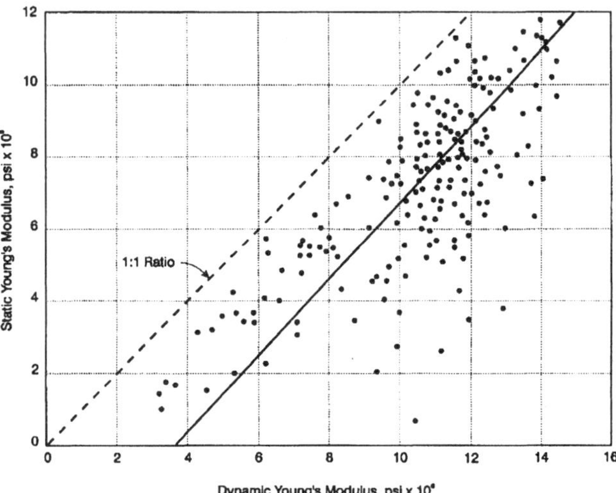

Fig. 3.6—Laboratory correlation between static and dynamic Young's moduli for rock.[6]

Elastic deformation of porous rock is complicated by the separate reaction of the matrix framework and the pore fluid, two components making the rock mass. Terzaghi's[7] effective-stress principle from Chap. 2 is modified by the Biot[8] equation,

$$\sigma_e = \sigma - sp_p, \qquad (2.4)$$

when applied to cemented rock. The correction factor, s, is called the poroelasticity constant or the Biot elastic constant. The poroelasticity constant has been described as a measure of the pore fluid's "efficiency" in counteracting an applied stress.[9] Its value can range between the rock porosity and unity, depending on pore geometry, degree of cementation, matrix constituents, and other factors. The constant can be estimated with the relation

$$s = 1 - (c_{ma}/c_b), \qquad (3.8)$$

where c_{ma} and c_b = compressibilities of the rock matrix and bulk rock, respectively.

Common practice uses Terzaghi's equation ($s = 1.0$) when estimating subsurface effective stresses. This is a reasonable approximation if the bulk compressibility of the rock greatly exceeds the matrix compressibility. Geertsma[10] reasoned that shales and sandstones having porosity greater than 15% meet this criterion. This condition also may be met in old and brittle rock, which tends to be fractured naturally and thus is highly compressible.[11] Another argument holds that the parameter is needed only when the stress field is undergoing change, a circumstance applying to a producing reservoir but not to relatively quiescent strata.[12] Adopting the Terzaghi effective-stress relation,

$$\sigma_e = \sigma - p_p, \qquad (2.3)$$

unless stated otherwise.

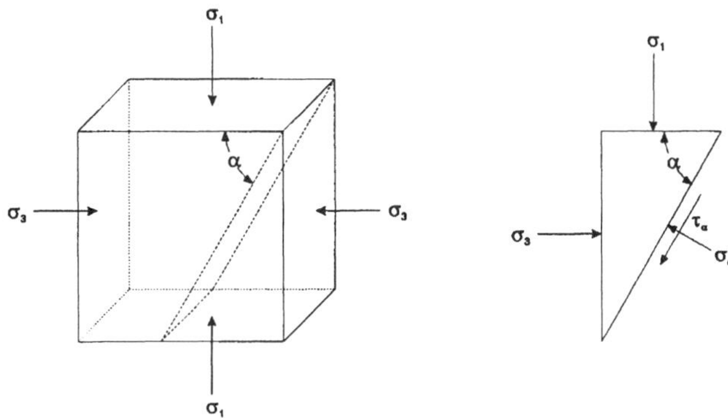

Fig. 3.7—Principal stresses acting on an element and the reaction shear and normal stress on an arbitrary plane.

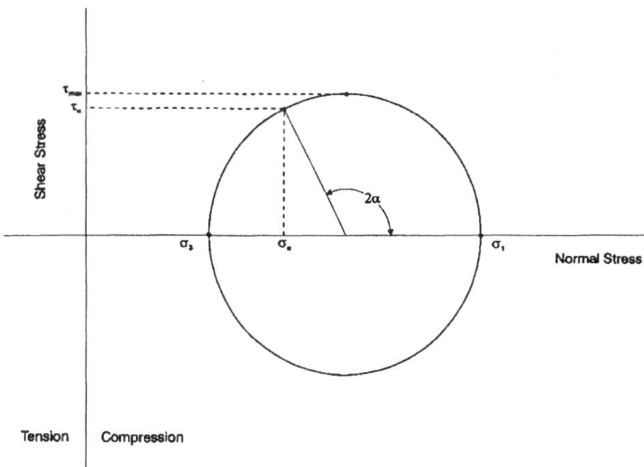

Fig. 3.8—Mohr's circle depiction of biaxial stress.

3.2.2 The Mohr-Coulomb Failure Criterion. A cubic element under any stress state can be oriented so that the shear stresses along the six orthogonal planes vanish. **Fig. 3.7** shows that this orientation results in three normal stresses and designates the maximum and minimum principal stresses as σ_1 and σ_3, respectively. The intermediate principal stress, σ_2, (not shown, but in a direction perpendicular to the page) is considered inconsequential in this failure theory.

Angle α describes an arbitrary plane. Present along any plane orientation other than principal are a shear stress, τ_a, and a stress normal to the plane surface, σ_a. The Mohr's[13] circle is a convenient method for defining these reaction shear and normal stresses and fully characterizing the biaxial-stress state at any desired plane angle.

Fig. 3.8 shows maximum and minimum principal stresses positioned on the circle at zero shear whereas the maximum shear stress is defined by the circle's radius,

$$\tau_{max} = (\sigma_1 - \sigma_3)/2. \quad \text{.......................} \quad (3.9)$$

The plane angle on the element corresponds to 2α on the circle. Thus the maximum shear stress always occurs on the plane 45° from the maximum principal stress. The quantities

$$\tau_a = \frac{\sigma_1 - \sigma_3}{2}\sin 2\alpha \quad \text{......................} \quad (3.10)$$

$$\text{and } \alpha_a = \frac{\sigma_1 + \sigma_3}{2} + \frac{\sigma_1 - \sigma_3}{2}\cos 2\alpha \quad \text{..........} \quad (3.11)$$

are evident from the diagram geometry.

Long before Mohr's invention, Coulomb[14] observed that rock under compression typically failed in shear but that the failed surface did not necessarily correspond to the plane of maximum shear stress. He found that the failure plane usually oriented at a lower plane angle relative to the maximum-stress direction and explained this apparently anomalous behavior by attributing the failure to a rock property called internal friction. Coulomb's law can be expressed as

$$\tau_f = \pm\left(c + \sigma_f \tan\omega\right), \quad \text{....................} \quad (3.12)$$

where τ_f and σ_f = the shear and normal stresses, respectively, acting on the incipient-failure plane; c = the cohesion, which is the shear strength of the rock at zero normal stress; and ω = the rock's angle of internal friction.

The stress conditions at compressive failure and the corresponding failure plane can be depicted on a Mohr's circle. As Chap. 2 discussed, increasing the confining stress, σ_3, increases the stress normal to the failure plane and the increase in friction results in a higher shear stress at failure. **Fig. 3.9** graphically depicts Coulomb's theory by drawing two of these circles and connecting the failure shear-stress points.

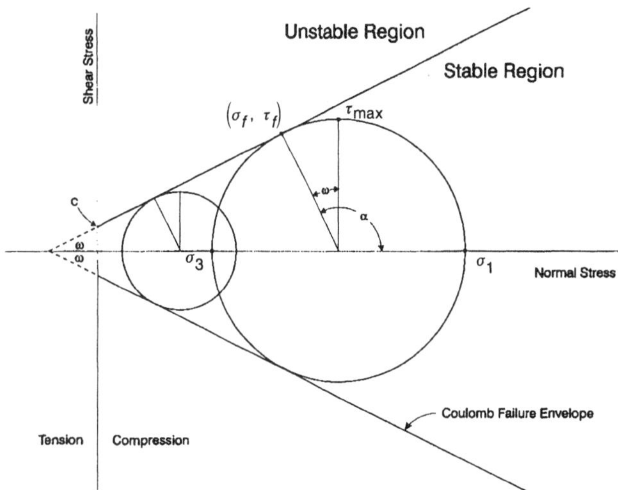

Fig. 3.9—Coulomb failure theory represented in graphical form.

The theory states that a rock under compression will fail in shear when its Mohr's circle intersects the line drawn tangent to other circles describing different principal-stress conditions. Eq. 3.12 describes the top and bottom tangent lines and thus defines the Coulomb failure envelope for the rock. In other words, stability dictates that any principal-stress combinations result in a circle that fits between the two tangent lines. Example 3.3 demonstrates one application of the Coulomb failure theory.

Example 3.3. A sandstone's properties include a 350-psi cohesion and a 35° angle of internal friction.

1. Estimate the compressive strength if 1,000-psi confining stress is applied to the specimen during a triaxial test.
2. On which plane is failure expected to occur?
3. What is the maximum shear stress?
4. What is the shear stress on the failure plane?

Solution. 1. A useful expression relates the two principal stresses at the point of failure.

$$\sigma_3 = \sigma_1\left(\frac{1 - \sin\omega}{1 + \sin\omega}\right) - \frac{2c\cos\omega}{1 + \sin\omega}. \quad \text{..........} \quad (3.13)$$

For this example,

$$1,000 = \sigma_1\left[\frac{1 - \sin(35)}{1 + \sin(35)}\right] - \frac{(2)(350)\cos(35)}{1 + \sin(35)} \text{ and}$$

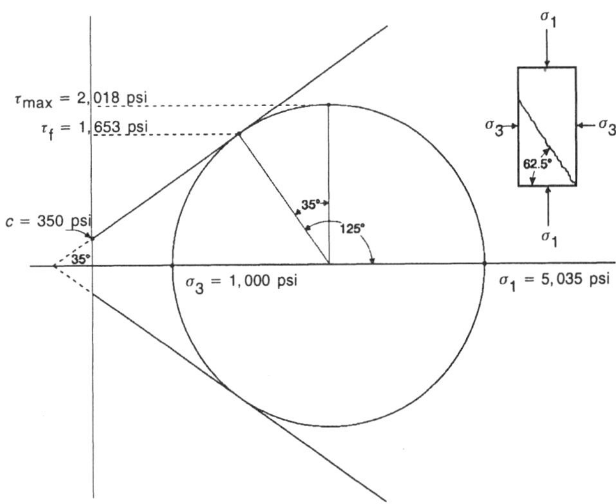

Fig. 3.10—Graphical solution to Example 3.3.

$\sigma_1 = 5,035$ psi.

2. From Fig. 3.9,

$2\alpha - 35° = 90°$

$\alpha = 62.5°$.

Thus, the expected failure plane is oriented 27.5° (90° – 62.5°) counterclockwise from the σ_1 direction.

3. Maximum shear stress occurs on the 45° plane. Eq. 3.9 gives its magnitude as

$$\tau_{max} = (5,035 - 1,000)/2 = 2,018 \text{ psi}.$$

4. Shear stress on the rupture plane is determined with an equivalent form of Eq. 3.10.

$$\tau_f = \tau_{max} \sin 2\alpha$$

$$= (2,018)\sin(125) = 1,653 \text{ psi}.$$

Fig. 3.10 shows an alternative graphical solution.

Coulomb's theory is useful in predicting brittle rock failure. However, all rocks evidence increasing ductility with increasing stress and many rocks under low-stress conditions behave like plastic materials. Coulomb's equation is inadequate in these circumstances, and a more general relationship is needed to predict shear failure. Mohr's criterion, expressed as

$$\tau_f = \pm f(\sigma_f), \quad \text{.......................... (3.14)}$$

better represents true rock behavior and is suited to all rock types and stress conditions. Eq. 3.14 recognizes that the shear failure of a compressed rock is controlled by some function of the normal stress, but makes no attempt to quantify this relationship. In practice, the function takes the form of a symmetrical failure envelope which must be determined in the laboratory. Consider the Coulomb equation as the special case where the function reduces to that of a straight line.

Fig. 3.11 shows a characteristic Mohr failure envelope. The curvilinear nature and flattening of the envelope indicate increased ductility with increasing stress. Note that the failure plane approaches 45° with increasing confining stress. Hence, the failure shear stress approaches, and would be ultimately the same as, the maximum shear stress under perfectly ductile or plastic behavior.

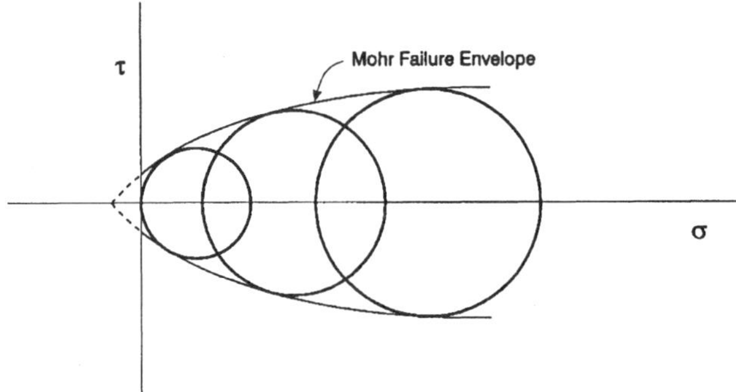

Fig. 3.11—Mohr failure envelope for a ductile rock.

Finally, a Mohr-Coulomb failure envelope must be constructed with Terzaghi's effective stress when dealing with porous rocks encountered by a drill bit. The poroelasticity constant is not considered in the analysis because Biot's relationship applies only to rock deformation, not its ultimate failure.

One weakness of the Mohr-Coulomb theory lies in the assumption that intermediate stresses are irrelevant. Although it has been shown that these stresses do play a role, the Mohr-Coulomb criterion continues to be a popular technique. Many wellbore-stability models and predictions pertaining to formation collapse rely on the method. Also, conditions resulting in such geological processes as faulting and fracturing can be anticipated by use of a Mohr-Coulomb diagram. Though not suitable for modeling hydraulic fractures directly (essentially a tensile mechanism), the criterion has been applied to the problem of estimating in-situ stresses.

3.3 Stress and Fracture-Pressure Relationships

Lost circulation is an expensive and potentially hazardous situation when drilling any well. Whole mud can enter a formation through two fundamentally different mechanisms. Mud losses may result when a positive pressure differential is imposed across high-permeability avenues, such as open natural fractures, solution channels, or coarse-grained porosity. These thief zones frequently are associated with relatively shallow rock in a normal- or subnormal-pressure environment. Hydraulic fracturing is a fundamentally different process where rock grains part along a preferred plane in response to excessive wellbore pressure.

Three conditions must be met before a fracture will initiate and extend out into the surrounding rock mass. First, wellbore pressure must exceed the tensile strength of the rock. Rocks have little tensile strength, however, and any intrinsic strength may be effectively zero if pre-existing joints or closed natural fractures are exposed to the wellbore. Thus tensile strength is ignored or assumed to be nonexistent.

The second requirement is that wellbore pressure be sufficient to overcome any compressive-stress concentration at the walls of the hole. Third, the pressure must exceed the minimum rock stress before a fracture can propagate away from the wellbore. Sec. 3.3 focuses on factors controlling the stress field beneath the earth's surface and how these stresses react to the presence of a wellbore. These concepts are essential to understanding hydraulic fracturing as it applies to a drilling operation.

3.3.1 In-Situ Rock Stress. The simplest fracturing model assumes that the subsurface stress field is governed solely by the rock's linear-elastic response to the overburden load. **Fig.**

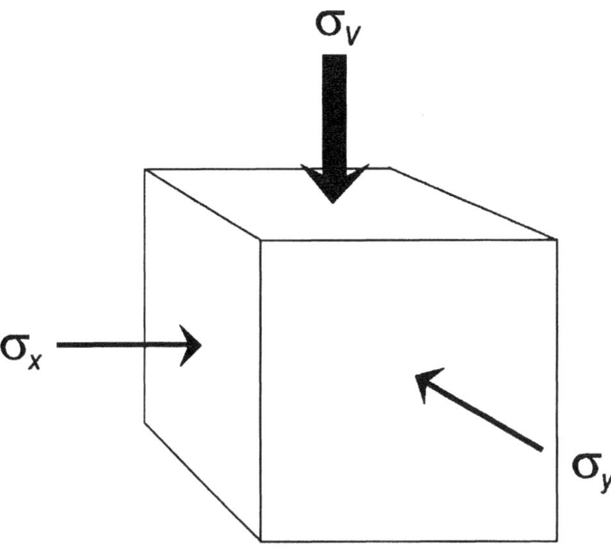

Fig. 3.12—Transverse-reaction stresses for a confined linear-elastic material.

3.12 shows how an elastic block, when loaded by an applied vertical stress, σ_V, strains in the x and y transverse directions according to Hooke's law.

$$\epsilon_x = \frac{\sigma_x}{E} - \mu\frac{\sigma_y}{E} - \mu\frac{\sigma_V}{E}, \quad (3.15)$$

$$\text{and } \epsilon_y = \frac{\sigma_y}{E} - \mu\frac{\sigma_x}{E} - \mu\frac{\sigma_V}{E}. \quad (3.16)$$

If the material is isotropic,

$$\epsilon_x = \epsilon_y = \epsilon_H, \quad (3.17)$$

where the subscript H = the horizontal direction. Thus,

$$\epsilon_H = \frac{\sigma_H}{E} - \mu\frac{\sigma_H}{E} - \mu\frac{\sigma_V}{E}. \quad (3.18)$$

Constraining the block on all sides prevents lateral strain. Setting $\epsilon_H = 0$ gives

$$\frac{\sigma_H}{E}(1 - \mu) = \mu\frac{\sigma_V}{E}. \quad (3.19)$$

Eliminating E and rearranging yields the fundamental relationship,

$$\sigma_H = \left(\frac{\mu}{1 - \mu}\right)\sigma_V. \quad (3.20)$$

For a confined linear-elastic and isotropic material, the horizontal stress is a function only of the Poisson's ratio and vertical stress. The block stress is analogous to the effective stress in a buried rock element if the material assumptions remain valid. Using overburden as the vertical stress and substituting Terzaghi's equation leads to

$$\sigma_H - p_p = \left(\frac{\mu}{1 - \mu}\right)(\sigma_{ob} - p_p). \quad (3.21a)$$

Finally,

$$\sigma_H = \left(\frac{\mu}{1 - \mu}\right)(\sigma_{ob} - p_p) + p_p. \quad (3.21b)$$

The poroelasticity constant may be applied to the pore pressure terms if so desired.

As Poisson's ratio is always less than or equal to 0.5, Eq. 3.21b demands that the horizontal stress be less than or equal to the overburden stress. Heim[15] proposed the latter in 1912, suggesting that rock stresses at great burial depths are the same in all directions. It follows that Poisson's ratio must be 0.5 to meet this hydrostatic-stress environment. Thus far, the effect of stress and temperature on plasticity suggests that Heim's hypothesis is reasonable although his conditions are unlikely for most rock at those depths currently penetrated by the bit.

Eq. 3.21b generalizes a condition where the horizontal stresses are less than the overburden stress. A hydraulic fracture takes the path of least resistance, parting rock along the plane easiest to open. This corresponds to a fracture plane perpendicular to the minimum principal stress. Hence, the equation predicts that fractures will be vertical in orientation.

However, the simplifying asssumptions needed to derive the expression are unsupportable in most, if not all, actual situations. Even a minor degree of bed anisotropy, for example, favors one of the principal horizontal stress directions. In addition, other processes besides sedimentation loading can lead to a significant change in horizontal stress over the deposition and burial history.

Diagenesis, grain cementation, viscoelastic effects (creep and relaxation), and thermal expansion alter the in-situ stresses over geologic time. Compressive earth movements associated with thrust faulting, folding, and diapirism provide one of the more important sources of horizontal stress. The minimum principal stress may in fact be the overburden in active tectonic regions.

Prats[16] discussed some of the parameters affecting horizontal stress and supplied a working model that incorporates the additional components. However, enough information rarely is known about the material properties and geological processes to predict the true stress field within a particular stratum accurately. Nonetheless, Eq. 3.21b serves as a generally conservative basis for estimating fracture pressures and provides insight into the predominant controlling factors.

A series of generic depth vs. stress profiles illustrate the effect of the variables that most strongly influence the subsurface stress environment and fracture characteristics. **Fig. 3.13**

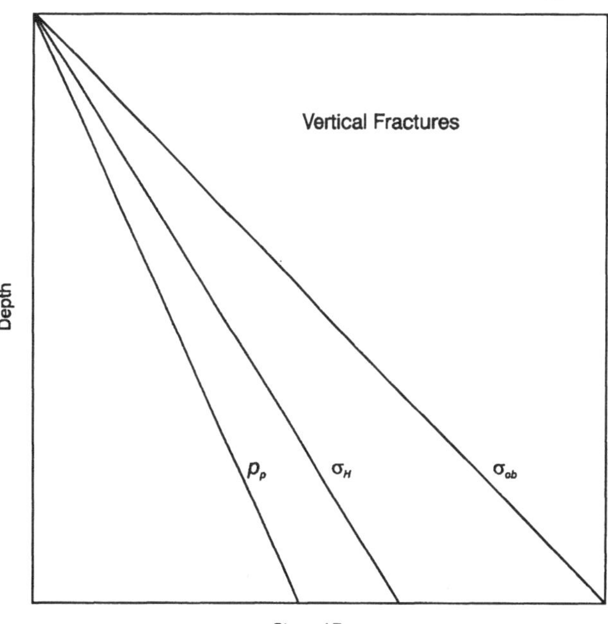

Fig. 3.13—Horizontal-stress profile for constant pore-pressure gradient, overburden gradient, and rock properties.

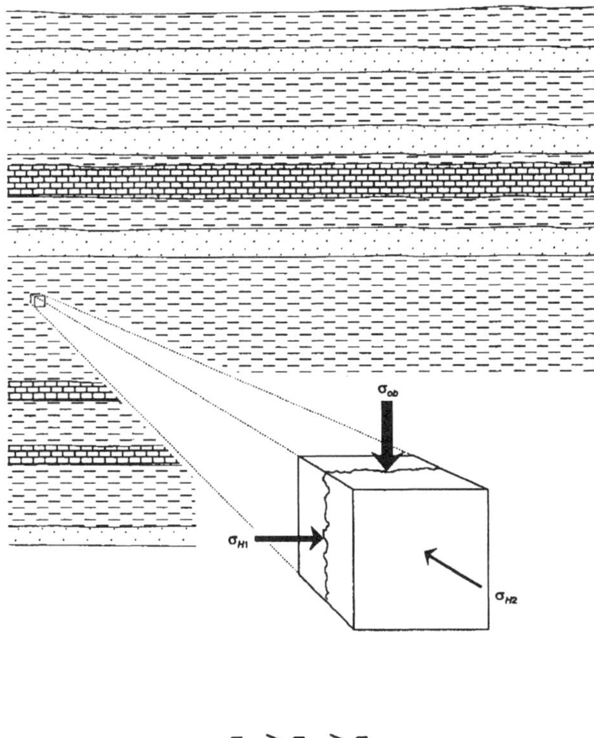

Fig. 3.14—Predicted hydraulic-fracture orientation if overburden is the maximum principal stress.

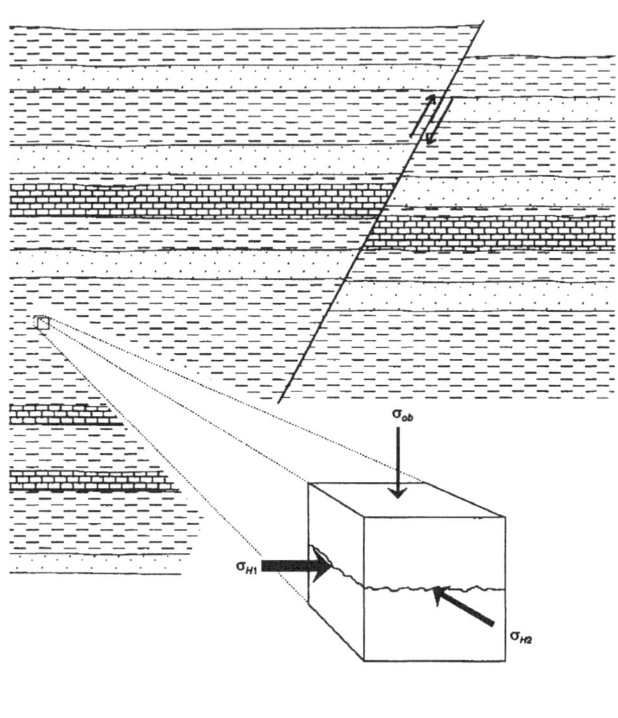

Fig. 3.15—Predicted hydraulic-fracture orientation if overburden is the least principal stress.

represents the hypothetical case where the rock properties and the overburden and pore-pressure gradients are consistent with depth. The horizontal stress is the least principal stress and vertical fractures are predicted at all depths. **Fig. 3.14** includes a minor horizontal-stress anisotropy that illustrates this concept.

Fig. 3.15 shows the effect of an incremental, consistent horizontal-stress component. The overburden stress in this case is less than the total horizontal stress down to a critical depth. **Fig. 3.16** illustrates anticipated horizontal fractures within the shallower strata and vertical fractures below the critical depth.

Tectonic forces can create a large horizontal-stress component. Other mechanisms, such as post-depositional erosion or glacial action, have been proposed and observed in the field. Consider the stress profile depicted in Fig. 3.13. Now erode away and lower the ground surface (or melt a glacier) over recent geological time. The overburden stress is reduced, but the horizontal stress may remain fixed because of inelastic deformation or recrystallization.

Topography can result in anomalously high horizontal stresses relative to the overburden. **Fig. 3.17** depicts a drilling location positioned in a valley or the foothills of a mountain range. Horizontal stresses in the rock beneath the rig are affected by the nearby overburden load, although vertical stress from the valley floor is a function only of well depth.

Fig. 3.18 illustrates changes in the Poisson's ratio or plasticity in different formations. Increasing or decreasing the Poisson's ratio can shift the horizontal stress profile to the

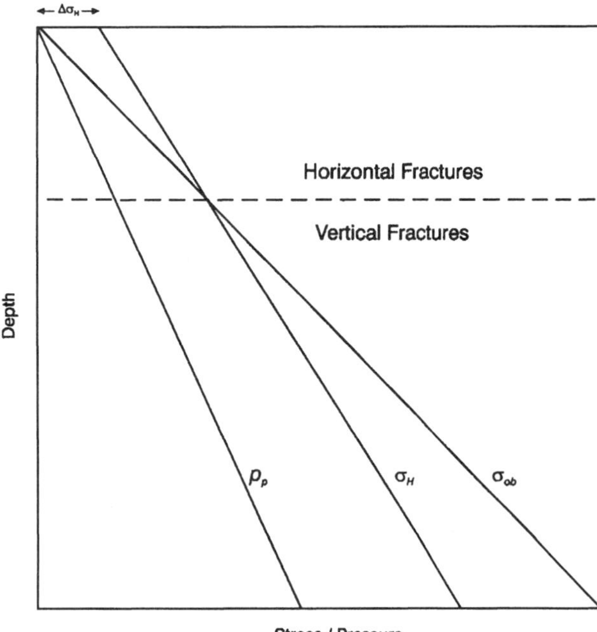

Fig. 3.16—Effect of an additional horizontal-stress component on the principal-stress regime and fracture orientation.

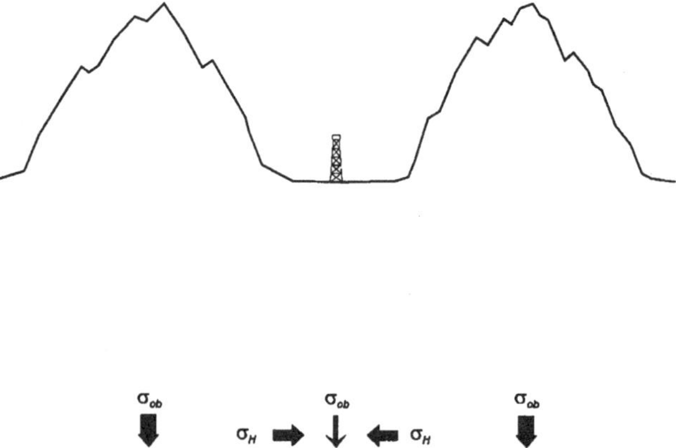

Fig. 3.17—Effect of topography changes on horizontal stress.

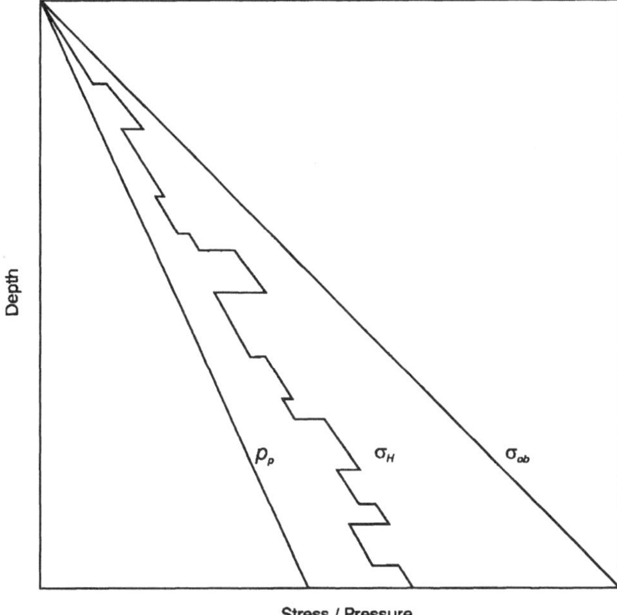

Fig. 3.18—Effect of variable Poisson's ratio on horizontal stress.

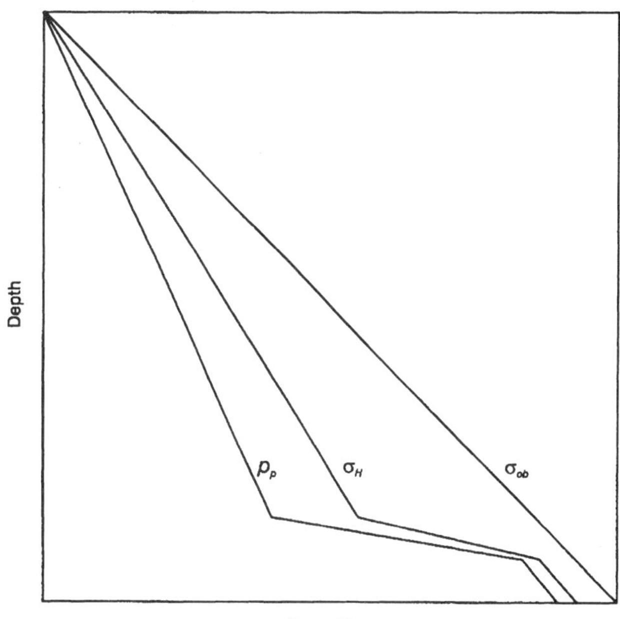

Fig. 3.19—Effect of abnormal pore pressure on horizontal stress.

right or left dramatically, which indicates the importance of having accurate material property data if applying Eq. 3.16. One consequence of these in-situ stress changes is of major importance in fracture-stimulation design in that high-stress layers above and below the reservoir are desirable for fracture-growth containment. To the drilling engineer, it should be apparent that the last casing seat is not necessarily the weakest part of the hole.

Fig. 3.19 shows the effect of abnormal pore pressures on horizontal stress. The disparity between horizontal stress and pore pressure greatly diminishes on entering the transition zone. A major consequence to drillers is that the tolerable-mud-density or equivalent-circulating-density (ECD) range is reduced when drilling abnormally pressured formations. Taking the theoretical extreme limit where the pore pressure equals the overburden stress leaves zero margin between the in-situ stress and pore pressure.

Fig. 3.20 shows a pressure-depleted horizon. This diagram indicates a substantial horizontal-stress reduction across the depleted sand. Consequently, the fracture gradient in the sand is less than what it had been originally, which may dictate a modified hole program to reach the objective depth safely.

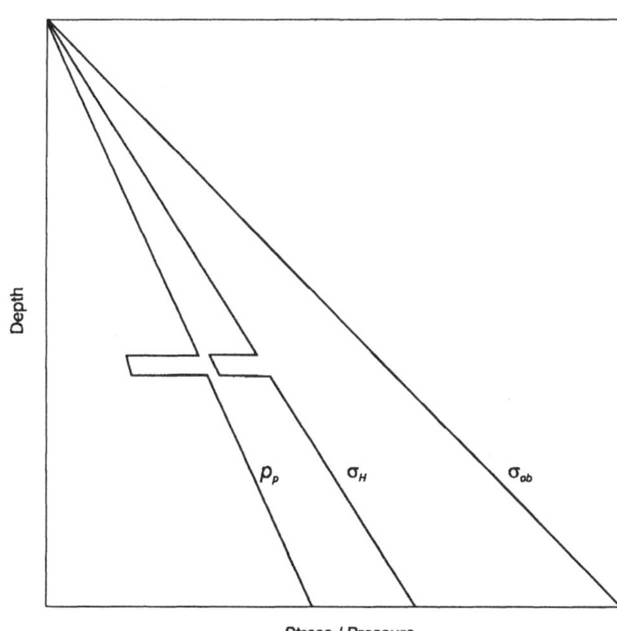

Fig. 3.20—Effect of pore-pressure depletion on horizontal stress.

3.3.2 Stresses Induced by a Wellbore. Fig. 3.21 depicts an infinite elastic plate under a uniaxial stress field. Removing a smooth cylindrical core normal to the stress field causes a realignment of the force streamlines and a stress concentration at the walls of the borehole. The top and bottom of the hole are placed in tension while additional compression is focused on the sides perpendicular to the stress direction.

Kirsch[17] originally derived the elastic plane-strain solution for the stresses acting on an elemental slice at some distance r from the hole center. A biaxial-stress analysis for the two principal stresses perpendicular to the hole is obtained by superposition, while the effects of a borehole pressure p_w is obtained by incorporating Lamé's[18] thick-walled cylinder equation.

Fig. 3.22 is a schematic of key variables for the biaxial-stress case. The polar coordinate system is used where the dis-

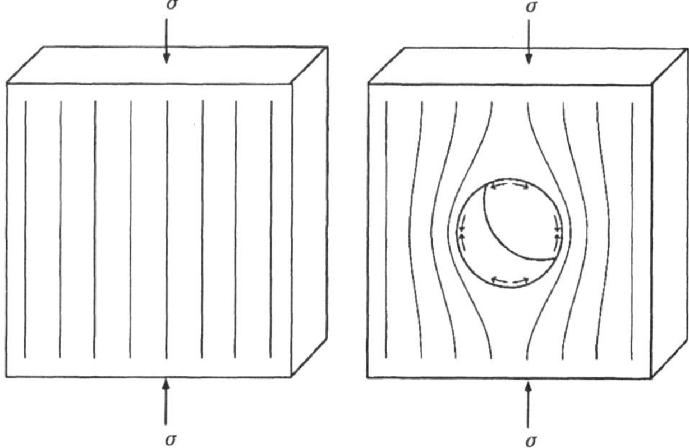

Fig. 3.21—Induced compression and tension at a borehole in the presence of a uniaxial-stress field.

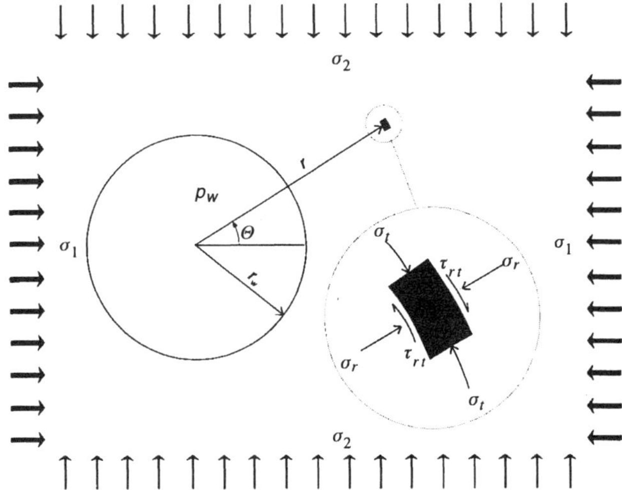

Fig. 3.22—Induced stresses in the presence of a biaxial-stress field.

tance r defines the distance from the borehole center to an arbitrary element and the angle θ is measured in degrees or radians from the plane parallel to the principal stress of interest. The radial stress, σ_r, acts on the element in the r direction, while the hoop or tangential stress, σ_t, acts circumferentially. The reaction shear stress on the surface normal to σ_r is defined by τ_{rt}. This shear vanishes when θ is a multiple of $\pi/2$ radians, which means that σ_t and σ_r are principal in these directions. Finally, p_w is the pressure in the borehole.

Eqs. 3.22 through 3.30 can be used to predict the induced normal stresses at distance r.[19-21] These relationships apply to a vertical hole where the principal earth stresses are orthogonal to the wellbore axis. A no-flow condition is dictated as well; thus, we must assume impermeable rock or filter cake if p_w exceeds p_p. Keep in mind that the calculated values are total rather than effective matrix stresses. The radial and tangential stresses, respectively, are given by

$$\sigma_r = \frac{\sigma_{H1} + \sigma_{H2}}{2}\left(1 - \frac{r_w^2}{r^2}\right) + \frac{\sigma_{H1} - \sigma_{H2}}{2}\left(1 + 3\frac{r_w^4}{r^4} - 4\frac{r_w^2}{r^2}\right) + \cos 2\theta + \frac{r_w^2}{r^2}p_w \quad \text{(3.22)}$$

and $$\sigma_t = \frac{\sigma_{H1} + \sigma_{H2}}{2}\left(1 + \frac{r_w^2}{r^2}\right) - \frac{\sigma_{H1} - \sigma_{H2}}{2}\left(1 + 3\frac{r_w^4}{r^4}\right) \times \cos 2\theta - \frac{r_w^2}{r^2}p_w. \quad \text{(3.23)}$$

Setting $r = r_w$ gives the stresses at the borehole (**Fig. 3.23**):

$$\sigma_r = p_w \quad \text{(3.24)}$$

and $$\sigma_t = \sigma_{H1} + \sigma_{H2} - p_w - 2(\sigma_{H1} - \sigma_{H2})\cos 2\theta. \quad \text{(3.25)}$$

The vertical stress along the borehole axis is approximately the same as the overburden stress.

The maximum tangential stress occurs at the borehole positions of $\pi/2$ and $3\pi/2$ radians as measured from the maximum stress direction.

$$(\sigma_t)_{\max} = 3\sigma_{H1} - \sigma_{H2} - p_w. \quad \text{(3.26)}$$

The minimum stress is seen at 0 and π radians,

$$(\sigma_t)_{\min} = 3\sigma_{H2} - \sigma_{H1} - p_w. \quad \text{(3.27)}$$

Eq. 3.28 is obtained if the horizontal stresses are equal.

$$\sigma_t = 2\sigma_H - p_w. \quad \text{(3.28)}$$

The radial and hoop stresses carried by the rock matrix are

$$\sigma_{re} = \sigma_r - p_p \quad \text{(3.29)}$$

and $$\sigma_{te} = \sigma_t - p_p, \quad \text{(3.30)}$$

where Subscript e = effective stress.

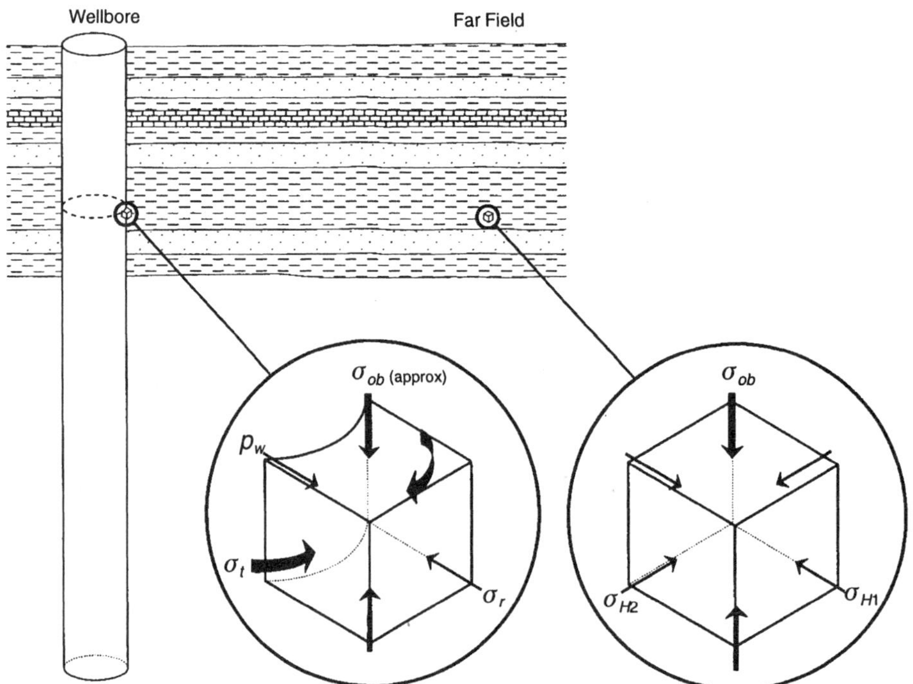

Fig. 3.23—Induced stresses at the borehole for a vertical well drilled orthogonal to the principal earth stresses.

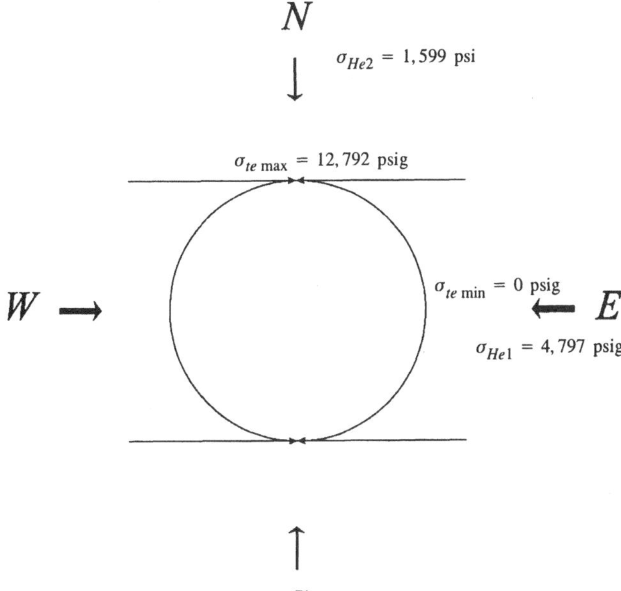

Fig. 3.24—Minimum and maximum effective tangential stresses for Question 5 of Example 3.4.

Example 3.4. A formation at 10,000 ft in a west Texas well has a Poisson's ratio of 0.22 and a 0.433-psi/ft pore-pressure gradient. The hole diameter is 9.0 in., and the mud density is equivalent to the pore-pressure gradient.

1. Estimate the in-situ horizontal stress if the rock behaves in a linear-elastic manner. Use a 1.0-psi/ft overburden gradient.
2. Estimate the tangential and radial stress at the borehole if the horizontal stresses are equal.
3. What are the normal stresses 2.0 ft from the wellbore centerline?
4. Estimate the tangential and radial stress at the borehole if a 0.5-lbm/gal overbalance is provided by the mud column.
5. Assume the maximum horizontal stress is oriented east/west while the minimum is north/south. Determine the minimum and maximum tangential stresses if the maximum effective horizontal stress exceeds the minimum by a factor of three. Use the result from Question 1 as the minimum horizontal stress.

Solution.

1. Eq. 3.21b yields the predicted isotropic horizontal stress for elastic rock deformation.

$$\sigma_{H1} - \sigma_{H2} = \left(\frac{0.22}{1 - 0.22}\right)(10,000 - 4,330)$$

$$+4,330 = 5,929 \text{ psig}.$$

The effective horizontal stress ($s = 1.0$) is

$$\sigma_{He} = 5,929 - 4,330 = 1,599 \text{ psi}.$$

2. Eq. 3.24 and 3.25 give the radial and tangential stress at the wellbore.

$$\sigma_r = 4,330 \text{ psig}$$

$$\text{and } \sigma_t = (2)(5,929) - 4,330 = 7,528 \text{ psig}.$$

The corresponding effective stresses are

$$\sigma_{re} = 4,330 - 4,330 = 0 \text{ psi}$$

$$\text{and } \sigma_t = 7,528 - 4,330 = 3,198 \text{ psi}.$$

3. Simplifying Eqs. 3.22 and 3.23 for the case of isotropic horizontal stress gives

$$\sigma_r = \sigma_H\left(1 - \frac{r_w^2}{r^2}\right) + \frac{r_w^2}{r^2}p_w \quad \text{.................. (3.31)}$$

$$\text{and } \sigma_t = \sigma_H\left(1 + \frac{r_w^2}{r^2}\right) - \frac{r_w^2}{r^2}p_w. \quad \text{.................. (3.32)}$$

At 2 ft from the borehole center,

$$\sigma_r = (5,929)\left(1 - \frac{0.375^2}{2.0^2}\right) + \frac{0.375^2}{2.0^2}(4,330)$$

$$= 5,873 \text{ psig}$$

$$\text{and } \sigma_t = (5,929)\left(1 - \frac{0.375^2}{2.0^2}\right) + \frac{0.375^2}{2.0^2}(4,330)$$

$$= 5,985 \text{ psig}.$$

The induced stresses rapidly approach the in-situ state within a short distance from the wellbore.

4. The wellbore pressure at 10,000 ft is

$$p_w = [0.433 + (0.5/19.25)](10,000) = 4,590 \text{ psig}.$$

Thus,

$$\sigma_r = 4,590 \text{ psig}$$

$$\text{and } \sigma_t = (2)(5,929) - 4,590 = 7,268 \text{ psig}.$$

The effective stresses are

$$\sigma_{re} = 4,590 - 4,330 = 260 \text{ psi}$$

$$\text{and } \sigma_{te} = 7,268 - 4,330 = 2,938 \text{ psi}.$$

5. The effective minimum and maximum horizontal stresses for this condition are, respectively,

$$\sigma_{He2} = 1,599 \text{ psi}$$

$$\text{and } \sigma_{He1} = 3(1,599) = 4,797 \text{ psi}.$$

The maximum horizontal stress is then

$$\sigma_{H1} = \sigma_{He1} + p_p$$

$$\sigma_{H1} = 4,797 + 4,330 = 9,127 \text{ psig}.$$

Maximum circumferential compression is seen at the north and south points on the borehole, and the minimum corresponds to the east and west positions. Apply Eqs. 3.21 and 3.22 and obtain

$$(\sigma_t)_{max} = (3)(9,127) - 5,929 - 4,330 = 17,122 \text{ psig}$$

$$\text{and } (\sigma_t)_{min} = (3)(5,929) - 9,127 - 4,330 = 4,330 \text{ psig}.$$

The corresponding effective stresses at these positions are

$$(\sigma_{te})_{max} = 17,122 - 4,330 = 12,792 \text{ psi}$$

$$\text{and } (\sigma_{te})_{min} = 4,330 - 4,330 = 0 \text{ psi}.$$

Fig. 3.24 illustrates the induced wellbore stresses.

Example 3.4 illustrates several important concepts. The effective radial stress is zero if the wellbore and formation pressures are balanced and increases in compression as wellbore pressure increases. An underbalanced situation, however, leads to an inward tensile stress that can destabilize the walls of the hole.

A positive differential pressure reduces the induced circumferential stress, while a negative differential pressure increases the compression. Considering the isotropic stress case in Question 2, a wellbore pressure in excess of 7,528 psi places the rock grains in tension and a fracture initiates if the tensile strength is zero. Because this is higher than the far-field stress, this fracture propagates out into the rock.

Question 5 shows an effective horizontal-stress ratio of three exactly reduces the minimum effective tangential stress to zero. In this case, any borehole pressure slightly greater than the pore pressure places the wellbore in tension along the $\theta = \pi$ plane and may create a fracture. Fracture extension, however, is inhibited within a short distance from the well because of the rapid stress transition.

Extremely high compression at the wellbore walls can develop because of the induced stresses. These stresses are a concern to hole stability because they may cause compressive failure, such as tight hole in plastic formations and breakouts in the brittle rock. The only practical means for mitigating or preventing compressive rock failure is to reduce the circumferential rock stress by increasing the mud weight. Pore pressure is not the only parameter to consider when selecting mud densities across a hole section.

3.3.3 Formation-Fracture Pressure. What does "formation-fracture pressure" actually mean? To drillers, the expression usually means the wellbore pressure or equivalent mud weight at which mud losses occur. To completion personnel, it may suggest the surface or bottomhole pressure required to place a stimulation treatment. We can differentiate at least two fracture pressures, called fracture-initiation and -propagation pressures.

If the rock is intact, initial separation requires exceeding its tensile strength, σ_{ts}. Conventional theories state that a fracture initiates when the effective tangential-stress concentration around the wellbore is reduced to zero. To recast Eq. 3.22 in terms of effective tangential stress, set σ_{te} equal to zero; replace p_w with the fracture-initiation pressure, p_{fi}; and incorporate an operative σ_{ts} to obtain

$$p_{fi} = 3\sigma_{H2} - \sigma_{H1} - p_p + \sigma_{ts}. \quad \text{.............. (3.33)}$$

Tensile strength is an undependable quantity, if present at all, and is usually ignored.

Example 3.4 demonstrates that fracture-initiation pressure is reduced when the ratio between the two effective horizontal principal stresses increases. Taking the maximum stress ratio to be three leads to an expression for the minimum fracture-initiation pressure.

$$\left(p_{fi}\right)_{\min} = p_p + \sigma_{ts}. \quad \text{...................... (3.34)}$$

Within a few wellbore diameters, however, the effective tangential stress rapidly approaches the in-situ state and any created fracture is limited by the far-field minimum-stress component. The working minimum fracture-initiation pressure is stated as

$$\left(p_{fi}\right)_{\min} = \sigma_{H2} + \sigma_{ts}. \quad \text{..................... (3.35)}$$

If the assumptions that led to Eq. 3.16 are valid,

$$\left(p_{fi}\right)_{\min} = \left(\frac{\mu}{1-\mu}\right)\left(\sigma_{ob} - p_p\right) + p_p + \sigma_{ts}. \quad \text{.... (3.36)}$$

The maximum initial parting pressure results when the horizontal stresses are equal. It follows that

$$\left(p_{fi}\right)_{\max} = 2\sigma_H - p_p + \sigma_{ts}. \quad \text{................ (3.37)}$$

Reasoning as before,

$$\left(p_{fi}\right)_{\max} = \left(\frac{2\mu}{1-\mu}\right)\left(\sigma_{ob} - p_p\right) + p_p + \sigma_{ts}. \quad \text{.... (3.38)}$$

Once a fracture has started, propagation pressure at the wellbore, p_{fp}, will exceed the minimum stress to some degree. Tensile strength no longer plays a role after a crack has formed, but crack extension still requires some excess pressure because of a related rock property known as fracture toughness.[22] Mud-gel strength and fluid-friction drop also come into play, but assuming equivalence between the the extension pressure and minimum stress is reasonable enough for our purpose. Accordingly, under the adopted elasticity assumptions,

$$p_{fp} = \left(\frac{\mu}{1-\mu}\right)\left(\sigma_{ob} - p_p\right) + p_p. \quad \text{............ (3.39)}$$

Eqs. 3.34, 3.35, and 3.36 can be expressed in terms of fracture gradient as follows.

$$\left(g_{fi}\right)_{\min} = \left(\frac{\mu}{1-\mu}\right)\left(g_{ob} - g_p\right) + g_p + \frac{\sigma_{ts}}{D}, \quad \text{.... (3.40)}$$

$$\left(g_{fi}\right)_{\max} = \left(\frac{2\mu}{1-\mu}\right)\left(g_{ob} - g_p\right) + g_p + \frac{\sigma_{ts}}{D}, \quad \text{.... (3.41)}$$

and $$g_{fp} = \left(\frac{\mu}{1-\mu}\right)\left(g_{ob} - g_p\right) + g_p. \quad \text{........... (3.42)}$$

Eqs. 3.36 through 3.42 apply to vertical fractures. However, the minimum principal stress may be vertical or even inclined in some cases. The initiation pressure for horizontal fractures is considered to be

$$p_{fi} = \sigma_{ob} + \sigma_{ts}, \quad \text{........................ (3.43)}$$

whereas the approximate propagation pressure is

$$p_{fp} = \sigma_{ob}. \quad \text{.............................. (3.44)}$$

In Fig. 3.15, the geotectonic conditions that can result in a situation where the minimum principal stress is vertical such as horizontal compression, surface erosion, and other such factors are usually found at shallow depths of less than 2,000 ft. This is one reason why we often see higher fracture gradients in shallower strata, which is most fortunate from the standpoint of kick tolerance and well safety.

Example 3.5. Consider a sandstone at 12,000 ft in a well located along the U.S. gulf coast. Assume isotropic horizontal stresses, linear-elastic behavior, and a Poisson's ratio of 0.19.

1. Determine fracture-initiation and -propagation gradients if the sand is normally pressured. Assume tensile strength is effectively zero.

2. What are the fracture gradients if this sand has a 15.0-lbm/gal equivalent pore pressure?

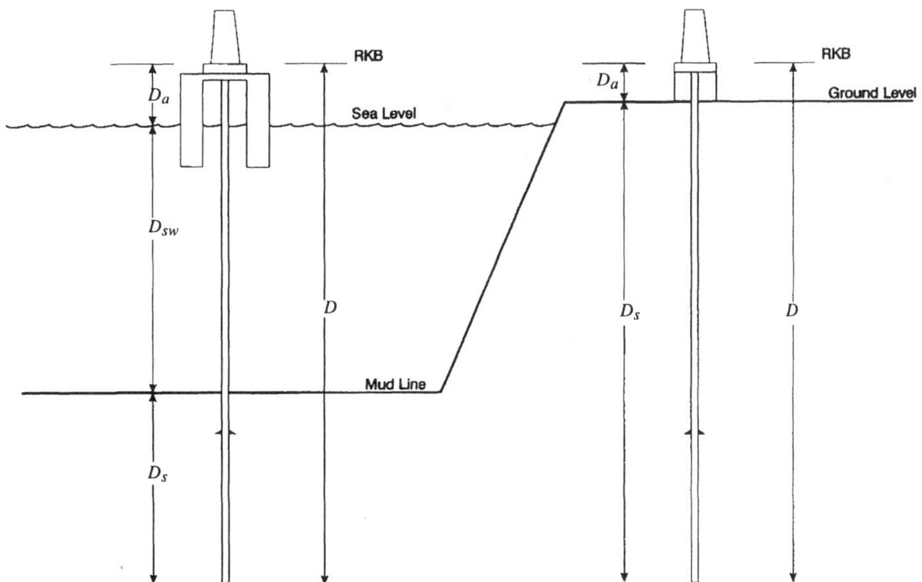

Fig. 3.25—Overburden-pressure components for an offshore well compared with a land well.

3. Determine the fracture gradient for a normally pressured sand at 500 ft if the minimum principal stress is vertical.

4. What impact would a 300-psig tensile strength have on the fracture-initiation gradients in the shallow and deep sands?

Solution. 1. Fig. 2.20 gives an overburden gradient at 12,000 ft of 0.961 psi/ft. Eq. 3.41 yields the fracture-initiation gradient.

$$g_{fi} = \left[\frac{(2)(0.19)}{1 - 0.19}\right](0.961 - 0.465) + 0.465$$

$$= 0.698 \text{ psi/ft.}$$

The fracture-extension gradient is

$$g_{fp} = \left(\frac{0.19}{1 - 0.19}\right)(0.961 - 0.465) + 0.465$$

$$= 0.581 \text{ psi/ft.}$$

2. A 15.0-lbm/gal pore pressure is equivalent to a 0.779-psi/ft gradient.

$$g_{fi} = \left(\frac{0.38}{1 - 0.19}\right)(0.961 - 0.779) + 0.779$$

$$= 0.864 \text{ psi/ft}$$

and $g_{fp} = \left(\frac{0.19}{1 - 0.19}\right)(0.961 - 0.779) + 0.779$

$$= 0.822 \text{ psi/ft.}$$

3. The shallow-fracture gradient in this case is simply that needed to lift the overburden. From Eaton's curve,

$$g_{fi} = g_{fp} = g_{ob} = 0.855 \text{ psi/ft.}$$

4. At 12,000 ft, the given tensile strength increases the fracture-initiation gradient by

$$\Delta g_{fi} = 300/12,000 = 0.025 \text{ psi/ft,}$$

whereas at 500 ft,

$$\Delta g_{fi} = 300/500 = 0.60 \text{ psi/ft.}$$

The latter results in a predicted breakdown pressure of almost 1.50 psi/ft. Thus, we have another possible reason why shallow-fracture gradients tend to be relatively high.

3.3.4 Factors That Affect Fracture Gradients. The great majority of wells are located on land and are classified as "vertical" holes. In a general sense, it is appropriate to focus on these types of wells and their associated problems. Fracture gradients, however, are affected once drilling locations move offshore and when wellbores begin to deviate from vertical. The effects of mud type and changes in wellbore temperature also need to be understood by drilling supervisors and engineers.

Offshore Locations. In 1997, offshore wells made up approximately 18% of the worldwide rig activity.[23] This statistic has been fairly consistent over the past few years and highlights the predominance of land operations compared with offshore. However, the gap between the two well categories narrows considerably when cost/reward is the measurement yardstick; risk and safety issues receive more attention on wells drilled from an offshore structure.

The impact of water depth on the fracture-initiation gradient is one of the factors that must be considered when planning and executing an offshore drilling project. Generally, an offshore well exhibits lower fracture-initiation gradients than an onshore well of similar depth and lithological sequence. A look at the fracture-initiation-gradient equations and reference to the two wellbore diagrams in **Fig. 3.25** tells us why this is so. The overburden stress is an important component to the predicted results, and an overburden component in the offshore location is seawater. Because water weighs less than rock, offshore overburden and fracture gradients are usually lower than onshore gradients.

The overburden-stress gradient imposed on the sediments underneath the land rig has two depth components. One is the sediment depth D_s, composed of the bulk rock (matrix and pore fluids) between ground surface and the depth of interest. Another minor contribution lies in the distance between ground level and the rotary kelly bushing (RKB) datum plane,

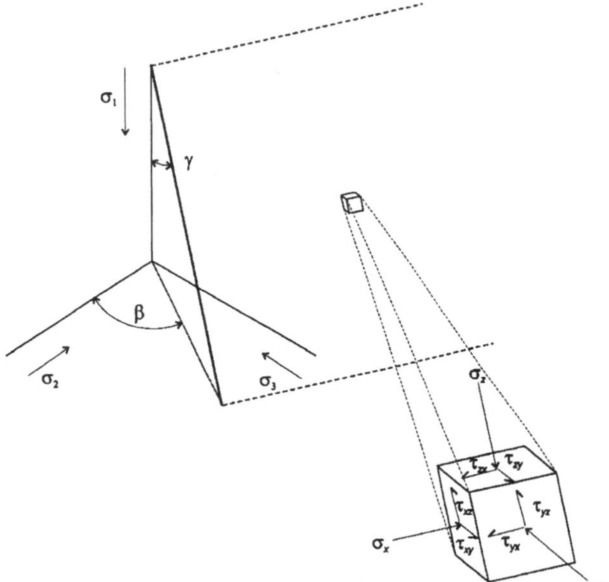

Fig. 3.26—Parameters used to transform earth principal stresses to the borehole orientation.

which can be called the air gap, D_a. The overburden gradient as referenced from the RKB is

$$g_{ob} = \sigma_s/D = \sigma_s/(D_a + D_s), \qquad (3.45)$$

where σ_S = total stress exerted by the bulk sediments and D_S is normally taken to be D in land operations.

The offshore stress relationship includes the seawater contribution.

$$g_{ob} = (p_{sw} + \sigma_s)/D = (p_{sw} + \sigma_s)/(D_a + D_{sw} + D_s), \qquad (3.46)$$

where p_{sw} and D_{sw} = water hydrostatic pressure and depth, respectively. Another influential factor on offshore rigs is the air gap. The height of the RKB above sea level may exceed 100 ft, a significant distance when compared with land rigs.

Eq. 3.38 suggests that these effects are most serious while drilling in deep water with shallow formations exposed to the wellbore. This is a consequence of major concern to operators and drilling contractors who are pushing the technology envelope by drilling and completing wells in the ultradeep waters offshore Brazil, the North Sea, the Gulf of Mexico, and other regions. New tools and techniques have been and will be devised to manage the lost circulation and well-control risks in these frontier areas better.

Inclined Wellbores. The induced stress equations presented thus far are valid if the wellbore axis is parallel to the maximum-principal-stress direction. This rather special case applies only to vertical wells at a sufficient depth so that the overburden is maximum. Directional (and some vertical) holes, however, are inclined with respect to the principal stress field, and a more general analytical approach must be used.

Inclined wells have lower fracture gradients and are more prone to compressive-stability problems than comparable near-vertical wells, and well plans should account for these. For example, a vertical discovery well is drilled offshore and the operator has assembled the leakoff pressures and allowed minimum mud weights. The fracture gradients and mud-weight data established may not transfer to subsequent directional wells. At the extreme, a highly deviated hole may require more casing to reach the depth objective safely.

Refs. 24 and 25 give the stress relationships that follow. The elastic solution to the problem first involves describing the earth stresses in the borehole frame of reference. The left side of **Fig. 3.26** illustrates the wellbore orientation with respect to the principal-stress field. The angle between the maximum-principal-stress, σ_1, direction and the borehole axis is γ, while β is the angle between the intermediate stress direction, σ_2, and the borehole projection onto the $\sigma_2-\sigma_3$ plane. To the right is a rock element oriented with the borehole axis and inscribed with the three normal and six shear stresses that act on the element surface.

Eqs. 3.47 through 3.59 express the normal stresses on the element in terms of the principal stress field and element orientation.

$$\sigma_x = \sigma_1 \sin^2\gamma + (\sigma_2 \cos^2\beta + \sigma_3 \sin^2\beta)\cos^2\gamma, \qquad (3.47)$$

$$\sigma_y = \sigma_2 \sin^2\beta + \sigma_3 \cos^2\beta, \qquad (3.48)$$

and $\sigma_z = \sigma_1 \cos^2\gamma + (\sigma_2 \cos^2\beta + \sigma_3 \sin^2\beta)\sin^2\gamma. \qquad (3.49)$

The three applicable shear stresses are defined by

$$\tau_{yz} = 0.5(\sigma_3 - \sigma_2)\sin(2\beta)\sin\gamma, \qquad (3.50)$$

$$\tau_{xz} = 0.5(\sigma_2 \cos^2\beta + \sigma_3 \sin^2\beta - \sigma_1)\sin(2\gamma), \qquad (3.51)$$

and $\tau_{xy} = 0.5(\sigma_3 - \sigma_2)\sin(2\beta)\cos^2\gamma. \qquad (3.52)$

The radial and tangential stresses at the walls of a penetrating borehole, respectively, are given by

$$\sigma_r = p_w \qquad (3.53)$$

and $\sigma_t = \sigma_x + \sigma_y - p_w - 2(\sigma_x - \sigma_y)\cos 2\theta - 4\tau_{xy}\sin 2\theta. \qquad (3.54)$

The induced normal stress acting along the wellbore axis is

$$\sigma_{zz} = \sigma_z - 2\mu(\sigma_x - \sigma_y)\cos 2\theta - 4\mu\tau_{xy}\sin 2\theta. \qquad (3.55)$$

One finite shear stress at the wellbore is given by

$$\tau_{tzz} = 2(-\tau_{xz}\sin\theta + \tau_{yz}\cos\theta). \qquad (3.56)$$

The least principal stress at the walls of the hole and tensile strength (if present) control fracture initiation. The three principal wellbore stresses are

$$\sigma_1' = \sigma_r = p_w, \qquad (3.57)$$

$$\sigma_2' = \frac{\sigma_{zz} + \sigma_t}{2} + \frac{\sqrt{(\sigma_t - \sigma_{zz})^2 + 4\tau_{tzz}^2}}{2}, \qquad (3.58)$$

and $\sigma_3' = \dfrac{\sigma_{zz} + \sigma_t}{2} - \dfrac{\sqrt{(\sigma_t - \sigma_{zz})^2 + 4\tau_{tzz}^2}}{2}. \qquad (3.59)$

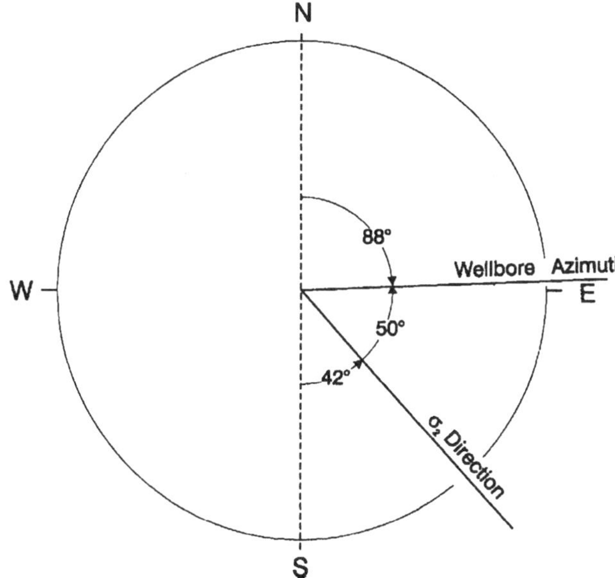

Fig. 3.27—Angle β between the σ_2 direction and wellbore azimuth described in Example 3.6.

We do not know which of these stresses is the smallest until the calculations are made. Once determined, the 1, 2, and 3 subscripts can be rearranged to fit the customary definition.

Elastic theory predicts that a fracture initiates away from the wellbore along the plane described by

$$\xi = \tan^{-1}\left(\frac{2\tau_{tzz}}{\sigma_t - \sigma_{zz}}\right), \quad \ldots\ldots\ldots\ldots (3.60)$$

where ξ = the angle between the wellbore axis and fracture plane. Experimental and analytical work, however, indicate that a created fracture tends to maintain along the hole axis.[26,27] In any event, eventually the fracture orients normal to the least principal earth stress.

Example 3.6. A gulf coast wellbore has a 70° hole inclination and a N88E azimuth at 14,000 ft true vertical depth. The maximum in-situ stress is vertical, and the minimum horizontal stress is 0.739 psi/ft. A salt-dome intrusion has created an incremental horizontal stress of 0.061 psi/ft in the S42W direction. Estimate the fracture-initiation pressure at this depth if the pore-pressure gradient is a 12.0-lbm/gal equivalent. Assume elastic behavior with Poisson's ratio equal to 0.25.

Solution. Eaton's overburden gradient is 0.972 psi/ft. The three principal earth stresses are

$$\sigma_1 = (0.972)(14,000) = 13,608 \text{ psig},$$

$$\sigma_2 = (0.739 + 0.061)(14,000) = 11,200 \text{ psig},$$

and $\sigma_3 = (0.739)(14,000) = 10,346$ psig.

The wellbore angle, γ, is 70°, and β is 50° from **Fig. 3.27.** Transform the in-situ normal stresses to the borehole reference frame with Eqs. 3.47 through 3.49.

$$\sigma_x = (13,608)\sin^2(70) + \left[(11,200)\cos^2(50) + (10,346)\sin^2(50)\right]\cos^2(70) = 13,268 \text{ psig},$$

$$\sigma_y = (11,200)\sin^2(50) + (10,346)\cos^2(50) = 10,847 \text{ psig},$$

and $\sigma_z = (13,608)\cos^2(70) + \left[(11,200)\cos^2(50) + (10,346)\sin^2(50)\right]\sin^2(70) = 11,039$ psig.

Eqs. 3.50 through 3.52 give the transformed shear stresses.

$$\tau_{yz} = 0.5(10,346 - 11,200)\sin(100)\sin(70) = -395 \text{ psig},$$

$$\tau_{xz} = 0.5[(11,200)\cos^2(50) + (10,346)\sin^2(50) - 13,608]\sin(140) = -935 \text{ psig},$$

and $\tau_{xy} = 0.5(10,346 - 11,200)\sin(100)\cos^2(70) = -49$ psig.

The problem solution requires a trial-and-error procedure, and we iterate by selecting values for p_w and solving for the least wellbore principal stress. The fracture-initiation pressure is found ultimately as that wellbore pressure resulting in an effective principal stress of zero. Note, however, that θ is also variable, so we must determine the borehole position where the minor stress is minimum. It can be shown that this condition is achieved when $\theta = 0$ if $\sigma_2 = \sigma_3$. We must determine the critical θ value numerically when this condition is not met.

For the first iteration, set p_w equal to the pore pressure, θ equal to zero radians, and solve Eqs. 3.50 through 3.56. The induced stresses orthogonal to the wellbore axis are

$$\sigma_r = p_w = p_p = (0.0519)(12)(14,000) = 8,722 \text{ psig},$$

$$\sigma_t = 13,268 + 10,847 - 8,722 - 2(13,268 - 10,847)\cos(0) - 0 = 10,552 \text{ psig},$$

$$\sigma_{zz} = 11,039 - (2)(0.25)(13,268 - 10,847)\cos(0) - 0 = 9,829 \text{ psig},$$

and $\tau_{tzz} = 2[-(-935)\sin(0) + (-395)\cos(0)] = -790$ psig.

The wellbore principal stresses (acting on the planes of zero shear) are calculated by

$$\sigma_1' = 8,722 \text{ psig},$$

$$\sigma_2' = 0.5(9,829 + 10,552) + 0.5\sqrt{(10,553 - 9,829)^2 + (4)(-790)^2} = 11,059 \text{ psig},$$

and $\sigma_3' = 10,190 - 869 = 9,321$ psig.

The minor principal stress is the wellbore pressure in this case. Nonetheless, we will maintain these subscripts for the time being. The calculation is repeated with different θ values until σ_3 reaches a minimum. **Table 3.2** shows the results of this exercise where the critical value for θ occurs at 2.950 ra-

θ (radians)	σ_t (psig)	σ_{zz} (psig)	τ_{tzz} (psig)	σ_2' (psig)	σ_3' (psig)
0.000	10,552	9,829	–790	11,059	9,321
0.393	12,109	10,218	–15	12,109	10,218
0.785	15,590	11,088	763	15,716	10,962
1.178	18,955	11,930	1,425	19,233	11,652
1.571	20,234	12,249	1,870	20,650	11,833
1.963	18,677	11,860	2,030	19,236	11,301
2.356	15,196	10,990	1,881	15,915	10,271
2.749	11,830	10,148	1,446	12,661	9,317
2.950	10,829	9,898	1,132	11,588	9,140
3.142	10,552	9,829	790	11,059	9,321

TABLE 3.2—ITERATIVE PROCEDURE TO DETERMINE THE BOREHOLE POSITION THAT MINIMIZES WELLBORE PRINCIPAL STRESS IN EXAMPLE 3.6

dians and its counterpart at 6.092 radians. **Fig. 3.28** identifies these positions on the hole cross section.

Having the critical θ, we are ready now to iterate to the fracture-initiation pressure by successively increasing wellbore pressure until the effective minimum principal stress at the wellbore vanishes. Changing the wellbore-pressure gradient to 0.650 psi/ft leads to

$$\sigma_r = p_w = (0.650)(14,000)$$
$$= 9,100 \text{ psig},$$
$$\sigma_t = 15,015 - 2(2,421)\cos(2 \times 169)$$
$$- (4)(-49)\sin(2 \times 169) = 10,451 \text{ psig},$$
$$\sigma_{zz} = 11,039 - (2)(0.25)(2,421)\cos(2 \times 169)$$
$$- (4)(0.25)(-49)\sin(2 \times 169)$$
$$\sigma_{zz} = 9,898 \text{ psig},$$

and $\tau_{tzz} = 2[-(-935)\sin(169) + (-395)\cos(69)]$

$$= 1,132 \text{ psig}.$$

Note that 2.950 radians has been converted to 169° for convenience.

$$\sigma_1' = 9,100 \text{ psig},$$
$$\sigma_2' = 0.5(9,898 + 10,451)$$
$$+ 0.5\sqrt{(10,451 - 9,898)^2 + (4)(1,132)^2}$$
$$= 11,340 \text{ psig},$$

and $\sigma_3' = 10,175 - 1,165$

$$= 9,010 \text{ psig}.$$

This is now the minor stress. The wellbore pressure is intermediate; 11,340 psig is the major stress. The controlling effective stress is obtained by subtracting pore pressure from least principal wellbore stress.

$$\sigma_{e3}' = 9,010 - 8,722 = 288 \text{ psi}.$$

Fracture initiation is not predicted at the 0.650-psi/ft wellbore-pressure gradient, and a higher value must be selected. Successive iterations, shown in **Table 3.3,** finally lead to a fracture-initiation gradient of approximately 0.695 psi/ft. The predicted fracture gradient for a comparable vertical hole is determined from Eq. 3.33 as

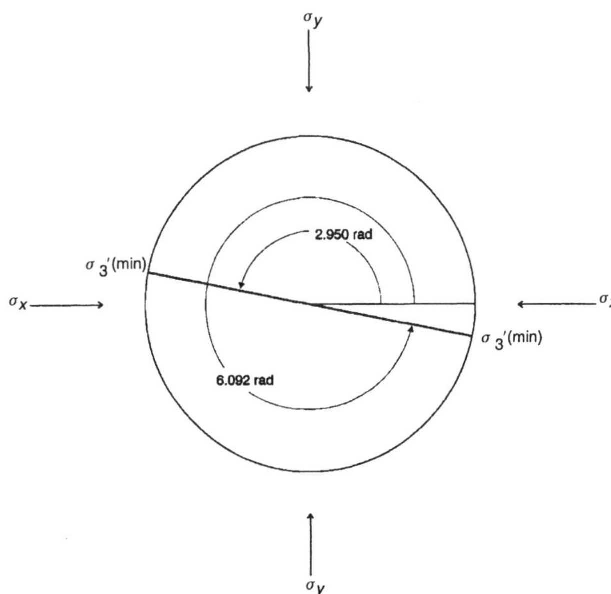

Fig. 3.28—Hole position θ that minimizes the least wellbore principal stress in Example 3.6.

TABLE 3.3—ITERATIVE PROCEDURE TO DETERMINE THE FRACTURE PRESSURE IN EXAMPLE 3.6

p_w/D (psi/ft)	p_w (psig)	σ_3' (psig)	σ_{e3}' (psi)
0.623	8,722	9,140	418
0.650	9,100	9,010	288
0.680	9,520	8,831	109
0.690	9,660	8,763	41
0.695	9,730	8,727	5
0.696	9,744	8,820	–2

$$p_{fi} = 3(10,346) - 11,200 - 8,722 = 11,116 \text{ psig}$$

and $g_{fi} = 11,116/14,000 = 0.794$ psi/ft.

Because it is lower than the minimum-horizontal-stress gradient, 0.695 psi/ft compared with 0.739 psi/ft, the fracture gradient in the 70° hole has the same value as the far-field stress.

Wellbore-collapse tendencies increase in highly deviated holes because of the increased vertical-stress component. For example, maximum grain stress in the well in Example 3.6 approaches 12,000 psig with balanced mud weight, whereas the predicted maximum stress in a similar vertical hole is 5,810 psig. The allowed mud density or ECD margin narrows as a function of hole inclination. The fracture gradient imposes a limit to higher mud densities, while compressive hole stability dictates the allowable minimum mud weight.

Bradley's[28] chart, shown as **Fig. 3.29,** illustrates this concept. He took a hypothetical well and used an extended von Mises yield criterion to determine the minimum mud densities for preventing hole collapse. Linear-elastic relationships discussed in this book then were used to predict fracture mud densities at various hole angles. As shown, the allowable mud-weight window on this particular well narrows considerably with inclination and reaches the point at approximately 65° where rock stability cannot be maintained. This is obviously a poor horizontal-well candidate if the diagram describes the wellbore behavior under stress.

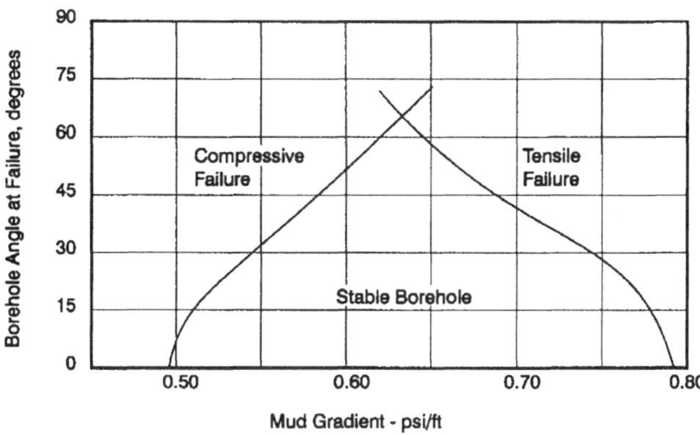

Fig. 3.29—Allowed mud weights for maintaining hole stability as functions of hole inclination.[28]

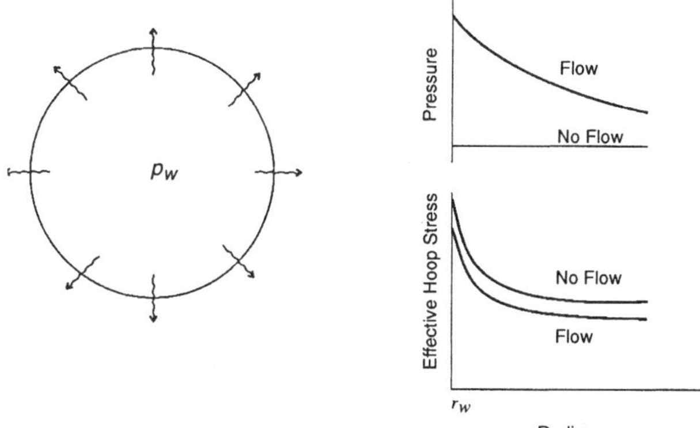

Fig. 3.30—Effect of fluid flow on the effective hoop stress around a wellbore.

Experience, however, has shown that elastic-mechanics principles yield exceedingly conservative results in predicting both compressive and tensile rock failure.[29-31] Indeed, many of today's horizontal and extended-reach wells would be impossible if these behavior assumptions held true. More complex models that consider plastic behavior,[32] elastoplasticity,[33] and poroelastic effects[34] better approximate the real world in most cases, although certain variables are difficult to quantify and require extensive laboratory testing to ascertain.

Drilling-Fluid Considerations. Fluid flow between the wellbore and surrounding rock alters the tangential stress around a borehole, an outcome not considered in the previous relationships. A penetrating fluid, resulting from an overbalance mud column against a permeable formation, serves to relieve the borehole-stress concentration. **Fig. 3.30** demonstrates one way of visualizing this effect. In a flowing condition, the pressure at any radial distance from the well is described by Darcy's law, whereas the pore pressure in the no-flow condition is essentially constant. As indicated, the increase in the local pore pressure during flow leads to a reduction in the effective tangential stress and a corresponding reduction in the fracture-initiation pressure.

Fig. 3.31 shows the effect of a filter cake. Total hoop stresses are transmitted through the filter cake to the sand face, but most of the pressure drop occurs across the low-permeability cake. The pressure at the sandface is less than would be seen had the cake been absent and the effective rock stress higher. A quality mud with good filtration properties can actually increase fracture gradients across permeable rock.

Mouchet and Mitchell[35] gave the following relation for modeling the incremental effective tangential stress arising from continuous flow through a porous, elastic medium.

$$\Delta\sigma_{te(q)} = \frac{1-2\mu}{1-\mu}(p_p - p_w). \qquad (3.61)$$

Wellbore pressure must be taken at the sandface if a filter cake is present.

Oil-based muds (OBMs) are more prone to lost circulation than water-based muds (WBMs).[36] Part of the difference may be attributed to the complex rheology of OBMs. WBMs generally observe the shear-thinning characteristics of a modified power-law fluid. A common procedure uses multi-speed Fann viscometer data to extrapolate viscosities at the shear rates found in the annulus. OBMs tend to be more viscous at lower shear rates than comparable WBMs, thus yielding higher ECD's and greater potential for excessive pressure. A variable-speed viscometer and a heat cup indicate oil-based-mud behavior better, although it should be noted that pressure thickens oil and downhole viscosities still may be underestimated greatly.

Another possible reason for lost-circulation problems with OBMs lies in the relative compressibility of mineral or diesel oil compared with the water and solid phases. Volume reduction of the relatively compressible oil increases downhole-mud densities from those measured at atmospheric pressure. However, this effect is offset by the relative thermal expansion properties of the base oils, and the net result may be an increase, reduction, or little change in the composite well-fluid density.

The determining factors in a static OBM column are wellbore-temperature gradient, water-volume fraction, and mud density (hydrostatic pressure and solids fraction effects). Static models that use geothermal-temperature gradients often predict the potential for underbalance in relatively warm holes.[37,38] One area of concern is tripping the drillstring when the mud weight is overbalanced only slightly compared with the pore pressure. Negative pressure surges cannot be avoided regardless of the mud type, but OBMs in

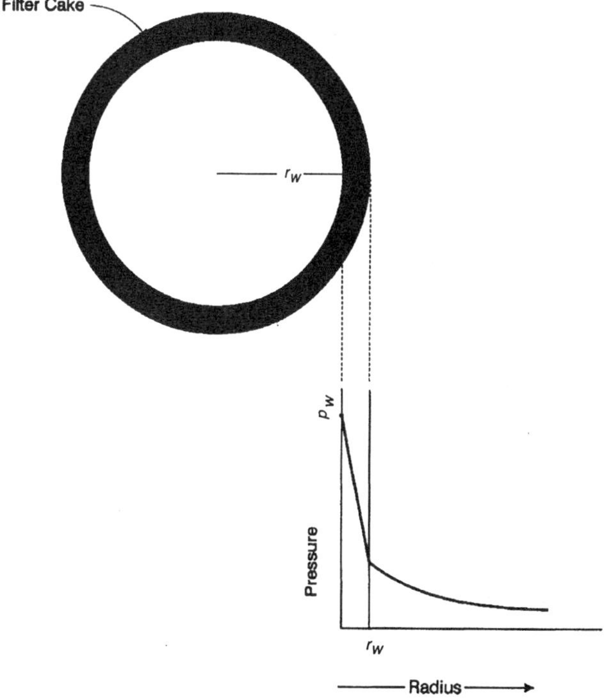

Fig. 3.31—Flowing pressure gradient into permeable rock for a mud with filtration properties.

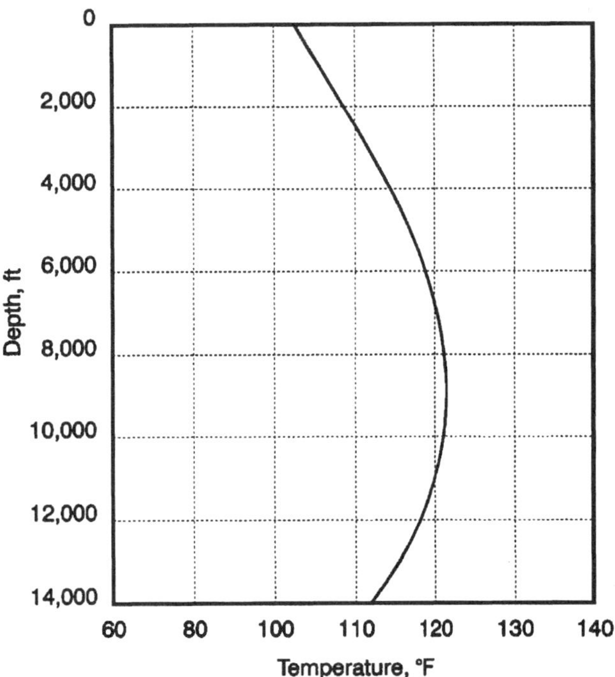

Fig. 3.32—Simulated annular-temperature profile after circulating a 15-lbm/gal OBM.[39]

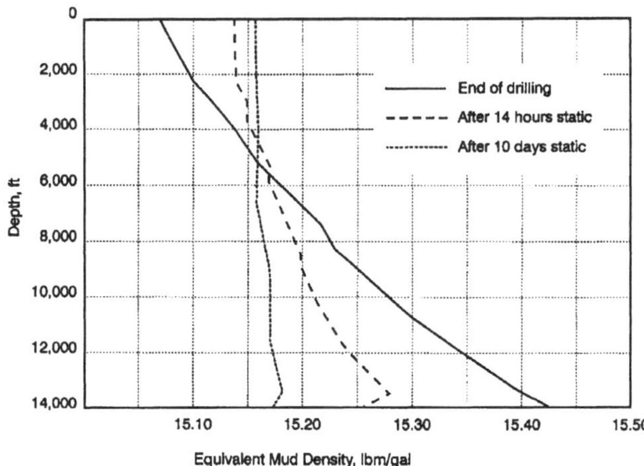

Fig. 3.33—Simulated annular-density profiles after circulating a 15-lbm/gal OBM.[39]

a hot hole lose even more pressure as quiescent wellbore temperatures increase.

Computer models that simulate circulating-temperature effects generally predict higher downhole pressures for OBMs than comparable WBMs.[39] **Fig. 3.32** depicts the predicted annular-temperature profile for a 15-lbm/gal OBM immediately after drilling to a depth of 14,000 ft. The mud is significantly cooler than the undisturbed earth temperatures below 4,500 ft. **Fig. 3.33** shows three simulated annulus density profiles for the same wellbore. The maximum densities along the curve to the right relate to the wellbore conditions immediately following circulation (no friction pressure). The mud temperature decays, and there is a corresponding reduction in composite-mud density for the 14-hour and 10-day shut-in periods.

Lost-circulation pressure estimates derived from linear elastic mechanics are often too conservative. Experimental work[40] under the aegis of the Drilling Engineering Assn. and analytical developments[41] demonstrate one reason why this is so. These studies also explained some reasons why an OBM may lose circulation when the same density WBM in the same rock maintains integrity.

Consider the borehole model shown in **Fig. 3.34a.** Assume that the rock was initially flawless and that excessive wellbore pressure created a fracture. Conventional theory holds that mud will enter the breach (i.e., lose circulation, once the tensile hoop stress exceeds the rock tensile strength). Laboratory investigation, however, showed that a created fracture must attain some minimum width before accepting any whole mud. The wellbore pressure required to achieve this critical width may far exceed the initial cracking pressure. The controlling rock property for fracture width is Young's modulus; hence, rock with a high Young's modulus should be more resistant to lost circulation than lower-moduli rock.

Initial lost-circulation pressure for intact rock did not vary significantly with mud type. Pre-existing cracks, on the other hand, reduced the initial lost-circulation pressure for all muds tested. Part of this can be attributed to the nullified tensile strength and the fact that fractured rock has no effective fracture toughness (resistance to propagation). Open-fracture width also has an effect because larger openings require less borehole pressure to achieve critical invasion width.

Breakdown pressure also was found to vary inversely with hole size, a factor not considered in conventional stress analysis. This is a reasonable conclusion if we consider the increased probability for encountering more cracks of greater width when wall surface area is increased. Other investigators have noted a similar correlation between hole size and fracture pressure.[42]

Relaxed-filtrate oil muds have long been promoted for their enhanced penetration rates.[43] Low-colloid mineral oil and diesel-based muds, however, breached the flawed rock specimens at significantly lower pressures than the higher-filtrate WBMs. **Fig. 3.35** gives some insight into why. Water-based muds primarily rely on bentonite gel or polymers for filtrate control and desired cake properties. Applying a positive differential pressure results in cake formation across permeable

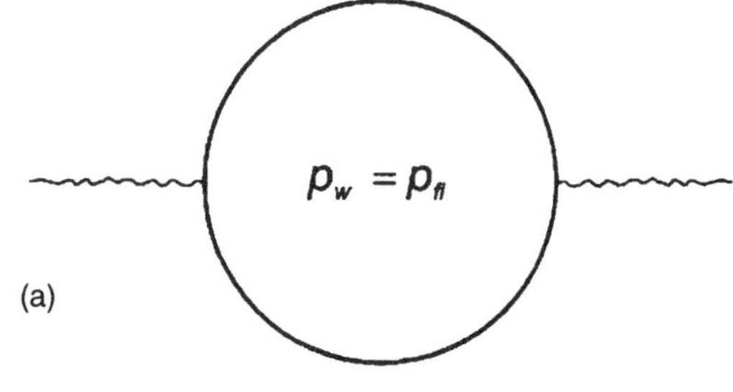

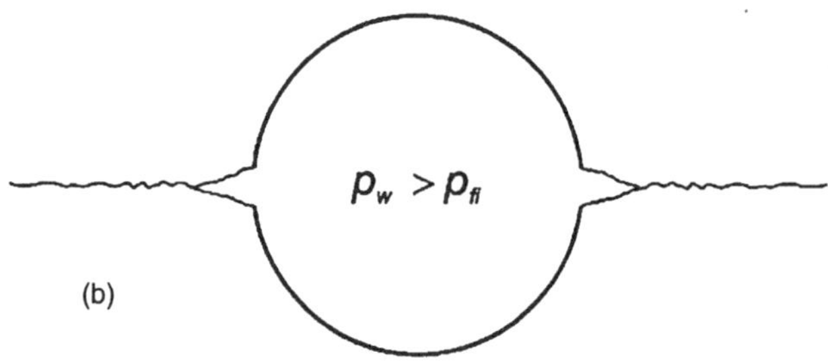

Fig. 3.34—Lost-circulation process for tensile rock failure when applying wellbore pressure in excess of fracture-initiation pressure.

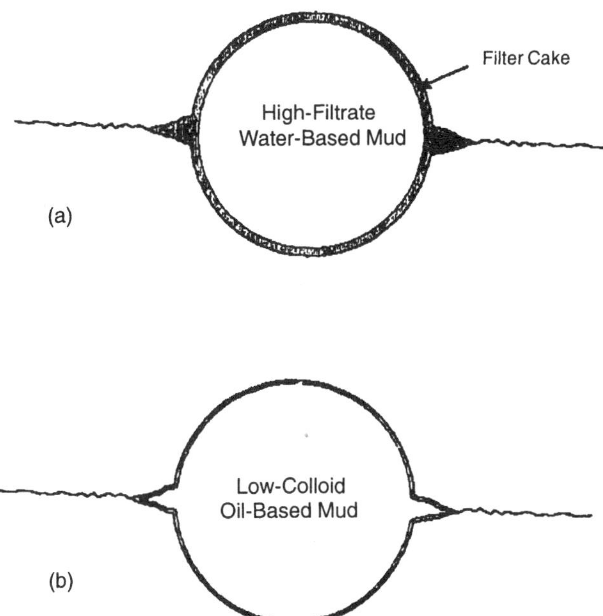

Fig. 3.35—Fracture-sealing characteristics of high-filtrate WBMs and low-colloid OBMs.

rock and solid plugging or bridging across natural fractures. The solids tend to seal the fracture-tip area from the wellbore and result in a higher pressure requirement to break through the dehydrated mud region.

In comparison, low-colloid invert muds depend less on commercial solids and more on emulsified brine water for filtration control. Thus, we can envision less fracture-sealing capacity from these drilling fluids and consequent reduction in breakdown pressure. By the same reasoning, the pressure at which a previously created fracture opens depends on how well the mud particles seal the closed fracture. Thin-wall cakes leave a smaller crack width on closure and, consequently, a lower reopening pressure. This has some implications for well planners and drilling supervisors with regard to leakoff or formation-integrity test procedures.

Fig. 3.36 compares measured fracture pressures for 16-lbm/gal WBMs and OBMs in Berea sandstone. Breakdown occurs at roughly the same pressure for the intact rock specimens, but note the spread between the ensuing propagation-pressure curves. A sawtooth pressure pattern occurs during water-based-mud propagation whereas the oil-based muds show much less variance.

The pressure behavior indicates a series of leak-seal steps during water-based-mud extension. Taking one of these increments, the fracture propagates some distance as shown by the downward pressure jag. Newly created fracture faces within the critical-width region are exposed to mud and filtrate spurt loss followed by nearly instantaneous cake deposition. A region consisting of dehydrated mud is created behind the narrow tip region. Pressure rises, the fracture inflates until sufficient width is achieved to allow further extension, and the process is repeated.

The WBM pressures shown in Fig. 3.36 reflect stable fracture-growth behavior. In comparison, the OBM fractures are propagating in an unstable manner because only minor wellbore-pressure disturbances result in further extension. The leak-seal steps are present to a much lesser degree, and less fracture width is required for the mud front to advance. In addition, allowing pressure to be applied closer to the tip region greatly increases growth tendencies in Griffith[44] cracks

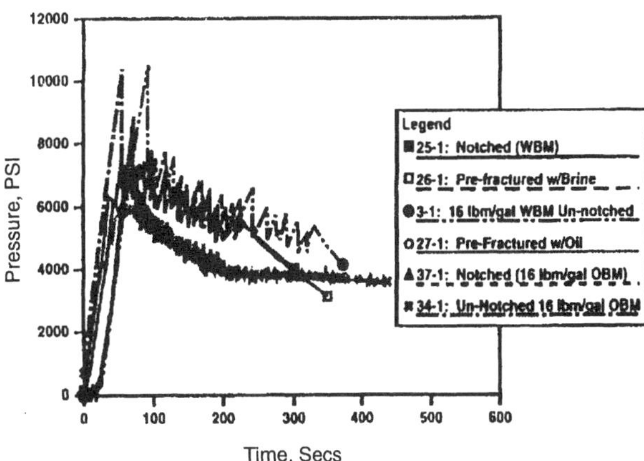

Fig. 3.36—Fracture-initiation and -propagation-behavior characteristics for 16-lbm/gal WBMs and OBMs in Berea sandstone.[40]

because of increased tensile stresses at the tip. The laboratory evidence is supported by field experience in fighting lost circulation with OBMs. Fractures, once initiated, can be very difficult to arrest, and ultimate control often requires much time and expense.

Obviously, a filter cake will not be deposited on completely impermeable rock. It follows that permeability should be an important parameter for designating stable or unstable growth behavior, regardless of the mud. Fracturing tests performed on Mancos shale indicate this to be the case. Both mud types evidenced unstable propagation, leading to the conclusion that lost circulation in shales and tight carbonates is more difficult to control. Shales are often, though not always, under higher stress than adjacent permeable strata and thus may have a higher fracture gradient.

Thermal Expansion or Contraction. There is substantial evidence that formations gain fracture resistance after being exposed to drilling for several days. Any natural fissures tend to plug with drill cuttings or fines over time benefiting hole integrity. But thermal contraction of relatively shallow rock is a more important reason for the observed increases in fracture integrity.

A fracture gradient is measured in newly exposed formation and the wellbore temperature is less than or equal to the geothermal earth temperature. Drilling proceeds and, as discussed in Chap. 2, the circulating-mud temperature in the upper portion of the wellbore increases. Rock is a relatively good insulator, but some heat transfers from the mud and is stored in the near-wellbore region. The effective tangential stress at the walls of an elastic, round hole changes in response to temperature according to

$$\Delta\sigma_{te(T)} = Eb\Delta T/(1 - \mu), \qquad (3.62)$$

where b = the material's coefficient of thermal expansion.

Example 3.7. The estimated fracture gradient for the 500-ft sand in Example 3.5 was 0.864 psi/ft. What fracture gradient is expected if circulating mud increases the sand temperature from 80 to 90°F. Assume E and b are 2.5×10^6 psi and 8.0×10^{-6} in./(in.-°F), respectively.

Solution. Eq. 3.54 yields the expected increase in the wellbore effective stress.

$$\Delta\sigma_{te(T)} = (2.5 \times 10^6)(8.0 \times 10^{-6})(90 - 80)/(1 - 0.19)$$

$$= 247 \text{ psi.}$$

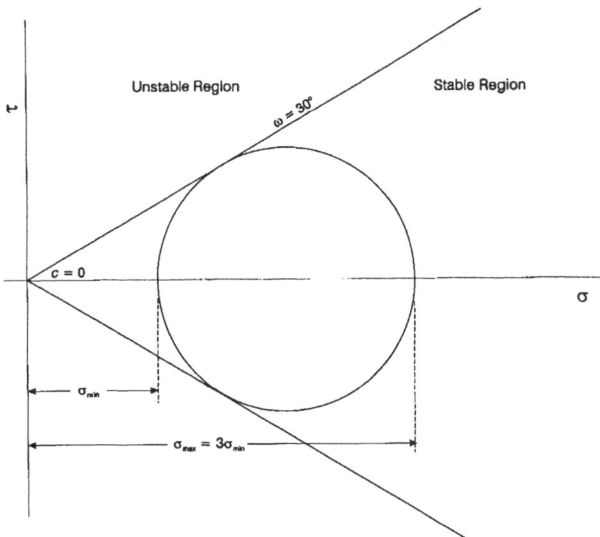

Fig. 3.37—Allowable ratio between minimum and maximum effective stresses for a rock with zero cohesion and 30° angle of internal friction.

The fracture gradient is now

$$g_{fi} = 0.864 + (247/500) = 1.358 \text{ psi/ft}.$$

The converse, of course, holds true in the deeper portion of a drilling well. Cooling a formation reduces its integrity. This is not of major concern, at least in the geologically young basins, because horizontal-stress gradients tend to increase with depth.

3.4 Prediction Methods

The ideal situation in well design is in development projects where pore pressures and fracture gradients have been measured and verified from top to bottom. Lacking such data, we must base our hole program and conduct the drilling operation using other, less certain criteria. Sec. 3.4 discusses some of the more important or widely used fracture-pressure-prediction techniques. Most are empirical correlations derived from regional fracture-gradient data and are subject to the same limitations as empirical pore-pressure-prediction methods. Others take a more physical/mechanical approach, although we attempted to consider only those that can be applied readily by the working engineer or drilling supervisor.

The majority of the published techniques were developed in relatively young and compactible sediments. Similar to pore-pressure predictions, hard-rock formations do not fit many of the established fracture-gradient techniques.

3.4.1 Hubbert and Willis'[45] Relation. When Hubbert and Willis published their classic paper on hydraulic fracturing in 1957, conventional thinking held that all fractures were created and extended away from a welbore like a horizontal "pancake." They demonstrated the link between fracture pressure and orientation to the stress state of the rock.

The authors assumed elastic behavior and reasoned that subsurface rock was stressed by tectonic action to the point of incipient failure by faulting. With zero cohesion and 30° angle of internal friction, a Coulomb stability envelope dictates shear failure when the maximum effective stress exceeds the minimum by a factor of three (**Fig. 3.37**). If the overburden is maximum, the assumed horizontal stress is

$$\sigma_H = \frac{1}{3}(\sigma_{ob} - p_p) + p_p. \quad \text{(3.63)}$$

Equating fracture-propagation pressure to minimum stress gives

$$p_{fp} = \frac{1}{3}(\sigma_{ob} + 2p_p). \quad \text{(3.64)}$$

This equation and its derivation are primarily of historical interest.

3.4.2 Matrix-Stress Ratio Correlations. The relationship between the effective horizontal and vertical stress can be expressed as

$$\sigma_{He} = F_\sigma \sigma_{Ve}, \quad \text{(3.65)}$$

where F_σ = the matrix-stress coefficient. Substituting Eq. 2.3 and equating the fracture-initiation pressure to the minimum horizontal stress gives

$$p_{fi} = F_\sigma(\sigma_{ob} - p_p) + p_p \quad \text{(3.66)}$$

and $$g_{fi} = F_\sigma(g_{ob} - g_p) + g_p. \quad \text{(3.67)}$$

Note that the matrix-stress ratio is some function of the Poisson's ratio if the rock is a linear-elastic material with zero tensile strength.

The most common prediction method uses an empirically derived correlation between the matrix-stress coefficient and depth. If the pore pressure is known, the fracture gradient can be estimated with Eq. 3.67. An implicit assumption made here regarding the horizontal-stress state at the borehole is that borehole hoop stress equals in-situ stress. Other factors such as plasticity effects and tensile strength also influence breakdown pressure; therefore, matrix-stress ratio is an unfortunate terminology. These correlations do not represent the true stress relationship but should be considered only as useful prediction tools.

Matthews and Kelly[46] Correlation. Matthews and Kelly were the first to use the matrix-stress-ratio concept to predict fracture gradients. **Fig. 3.38** shows the two correlations they developed from fracture-pressure data in normally pressured formations along the U.S. gulf coast. The authors assumed a 1.0-psi/ft overburden gradient in developing these curves.

The matrix stress in overpressured rock is abnormally low for the burial depth. In these cases, to calculate the effective vertical stress, determine the equivalent depth as defined in Chap. 2 and obtain F_σ using the equivalent depth instead of the actual depth. The equivalent depth can be calculated by

$$D_{eq} = \sigma_{Ve}/(g_{ob} - g_n) = (\sigma_{ob} - p_p)/(g_{ob} - g_n), \quad \text{(3.68)}$$

where σ_{Ve} denotes the effective vertical stress and g_n = normal pore-pressure gradient.

Example 3.8. Table 3.4 lists shale resistivities and computed pore pressures for a well located in the East Cameron Block, offshore Louisiana. Estimate the fracture-initiation gradient at 8,110 and 15,050 ft using the Matthews and Kelly correlation.

Solution. The first depth is in normally compacted rock and F_σ is obtained as 0.69 from the Louisiana curve in Fig. 3.38. The fracture gradient is predicted with Eq. 3.67.

$$g_{fi} = (0.69)(1 - 0.465) + 0.465 = 0.834 \text{ psi/ft}.$$

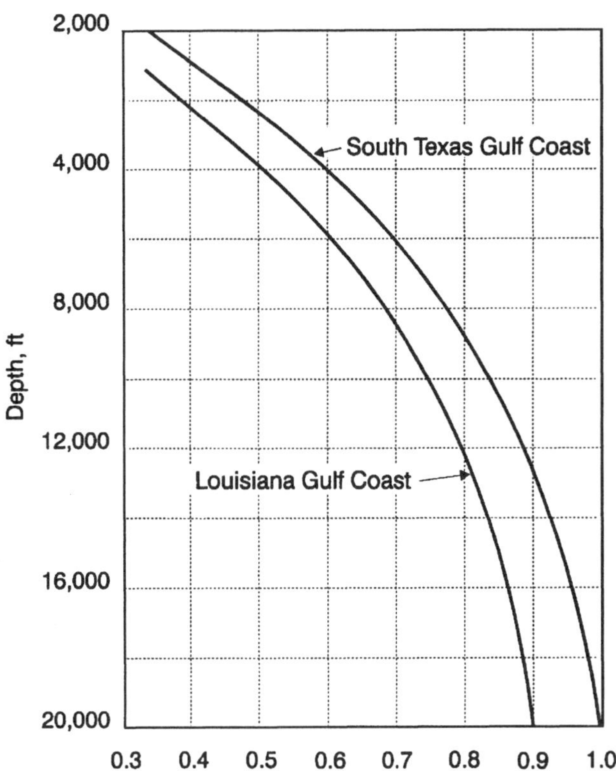

Fig. 3.38—Matthews and Kelly's[46] matrix-stress-ratio correlation for normally pressured formations along the U.S. gulf coast.

For the undercompacted interval at 15,050 ft, Eq. 3.58 determines the depth equivalent in terms of matrix compaction.

$$D_{eq} = [15,050 - (0.815)(15,050)]/(1.0 - 0.465)$$

$$= 5,204 \text{ ft}.$$

The matrix-stress coefficient at 5,204 ft is found to be 0.61. Hence,

$$g_{fi} = (0.61)(1 - 0.815) + 0.815 = 0.928 \text{ psi/ft}.$$

Pennebaker's[48] Gulf Coast Correlation. Pennebaker developed a similar correlation but improved on the Matthews and Kelly[46] approach by accounting for a variable gulf coast overburden. **Fig. 3.39** shows the effect of geologic age on the overburden gradients in the gulf coast region. Pennebaker used interval travel time obtained from seismic to define the degree of compaction and, thus, overburden stress at any depth. As indicated, the depth where 100 μsec/ft is achieved is related to the compactibility and geologic age of the sediments. This travel time in Pleistocene and young Tertiary, for example, would be found relatively deep while older, inland Cretaceous shows the same compaction level at a much shallower depth.

Fig. 3.40 shows that Pennebaker's effective-stress-ratio correlation and procedure differ from Matthews and Kelly's correlation in two respects. The initial shut-in pressure (ISIP) from stimulation treatments is used to define fracture pressure. An ISIP should be closer to the propagation pressure than initiation pressure, thus Pennebaker's stress ratio yields somewhat conservative results. Also, the correlation is based on actual vertical depth rather than equivalent depth when the rock is abnormally pressured.

TABLE 3.4—SHALE RESISTIVITIES AND PRESSURE GRADIENTS FOR EAST CAMERON WELL C[47]

Depth (ft)	Shale Resistivity (Ω·m)	Pore Pressure (psi/ft)
7,070	0.89	0.465
7,250	1.00	0.465
7,630	1.02	0.465
7,820	1.00	0.465
8,110	1.02	0.465
8,270	1.08	0.465
8,560	1.19	0.465
8,600	0.95	0.566
8,820	1.01	0.550
9,000	1.00	0.559
9,450	0.97	0.588
9,550	0.89	0.626
9,880	1.02	0.584
10,110	0.91	0.637
10,400	0.89	0.656
10,480	0.80	0.694
10,630	0.71	0.733
10,850	0.69	0.745
10,980	0.64	0.765
11,150	0.69	0.751
11,230	0.66	0.764
11,410	0.68	0.764
11,570	0.64	0.779
11,720	0.68	0.770
11,780	0.75	0.748
11,900	0.78	0.740
12,020	0.70	0.768
12,260	0.71	0.773
12,300	0.67	0.786
12,450	0.81	0.745
12,600	0.78	0.757
12,680	0.86	0.733
12,730	0.80	0.754
12,900	0.80	0.760
12,960	0.71	0.789
13,070	0.80	0.764
13,300	0.81	0.764
13,440	0.71	0.798
13,520	0.80	0.773
13,700	0.78	0.784
13,950	0.75	0.797
14,110	0.80	0.787
14,370	0.75	0.806
14,420	0.74	0.810
14,600	0.85	0.783
14,780	0.94	0.766
15,050	0.77	0.815
15,120	0.55	0.870
15,280	0.57	0.868

Example 3.9. Use the data given in Table 3.4 and estimate the fracture gradients at 8,110 and 15,050 ft using the Pennebaker correlation. Assume the young sediments offshore Louisiana exhibit a 100 μsec/ft interval travel time at 10,000 ft.

Solution. The overburden-stress gradient at 8,110 ft is approximately 0.945 psi/ft from a visual interpolation between the 9,000- and 12,000-ft curves in Fig. 3.39. The effective-stress ratio at 8,110 ft is found in Fig. 3.40, and Eq. 3.67 is applied to give

$$g_{fi} = (0.77)(0.945 - 0.465) + 0.465 = 0.835 \text{ psi/ft}.$$

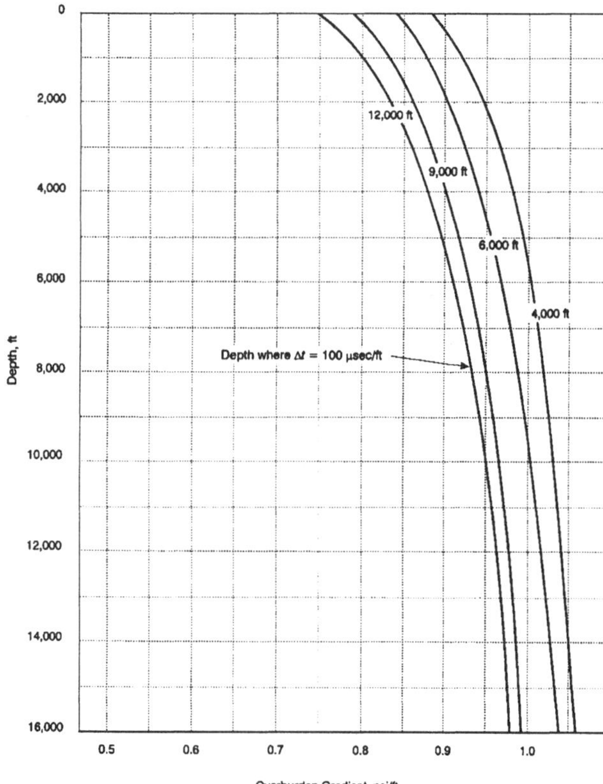

Fig. 3.39—Pennebaker's[48] overburden-gradient correlation for gulf coast sediments.

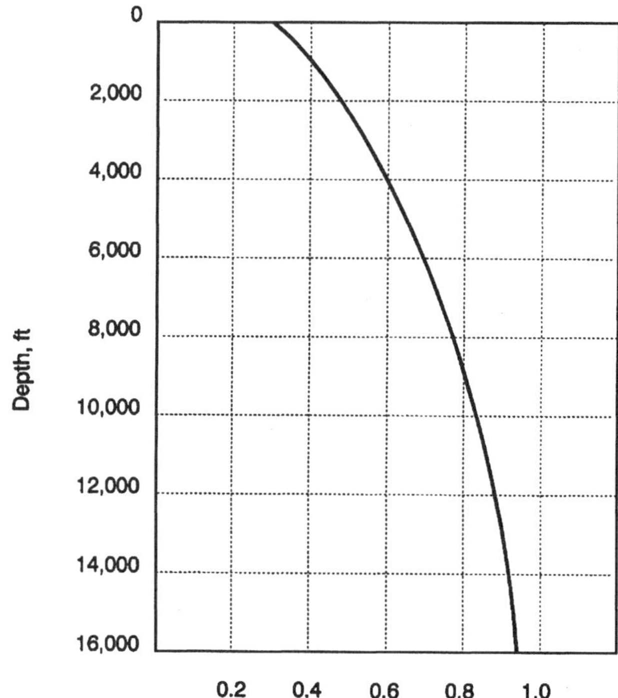

Fig. 3.40—Pennebaker's[48] matrix-stress-ratio correlation for the gulf coast.

The result is practically identical to the Matthews and Kelly[45] prediction. At 15,050 ft,

$$g_{fi} = (0.94)(0.984 - 0.815) + 0.815 = 0.974 \text{ psi/ft}.$$

Eaton's[47] Gulf Coast Correlation. Eaton's correlation, based on offshore Louisiana data in moderate water depths, is one of the more widely used prediction techniques. His relationship adopts the same form as Eqs. 3.37 and 3.39.

$$g_{fi} = \left(\frac{\mu_E}{1-\mu_E}\right)\left(g_{ob} - g_p\right) + g_p. \qquad (3.69)$$

The bracketed Poisson's ratio term is in fact a matrix-stress ratio. Eaton took fracture-initiation pressures in the subject area and backcalculated Poisson's ratios using measured pore pressures and the variable overburden curve presented as Fig. 2.20. **Fig. 3.41** shows the resulting correlation. The subscript E indicates that there is a difference between Poisson's ratio definition and Eaton's use of the term. The parameter as determined from field fracturing data also includes induced stress, poroelasticity, viscoelastic stress, and other such factors.

Mitchell[49] approximated Eaton's Poisson's ratio relationship with the following curve-fitting equation.

$$\mu_E = 0.23743 + 0.05945\left(\frac{D}{1,000}\right) - 0.00668\left(\frac{D}{1,000}\right)^2 + 0.00035\left(\frac{D}{1,000}\right)^3 - 6.71 \times 10^{-6}\left(\frac{D}{1,000}\right)^4. \qquad (3.70)$$

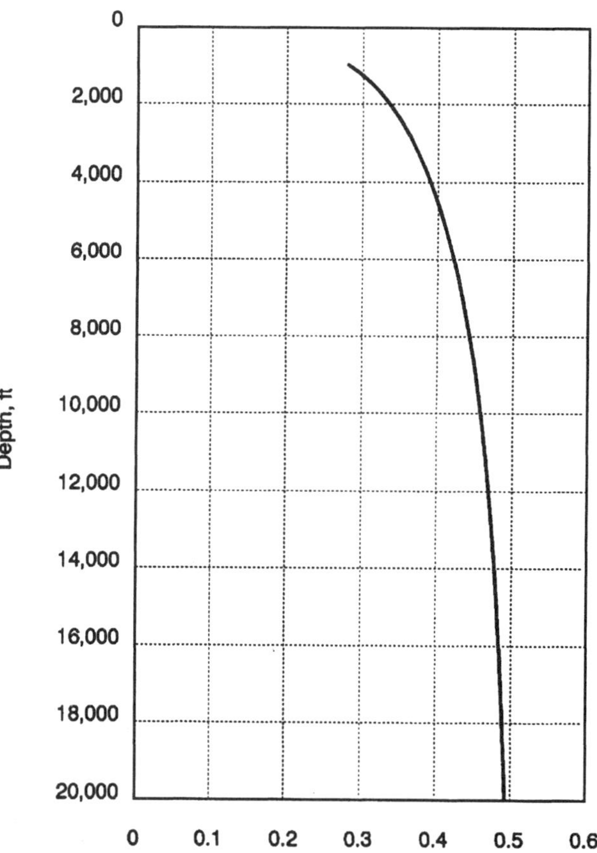

Fig. 3.41—Eaton's[47] Poisson's ratio correlation for variable overburden sediments in the gulf coast area.

Using nomographs or curves can be cumbersome when calculating a lot of data; these equations are more readily adapted to programming applications.

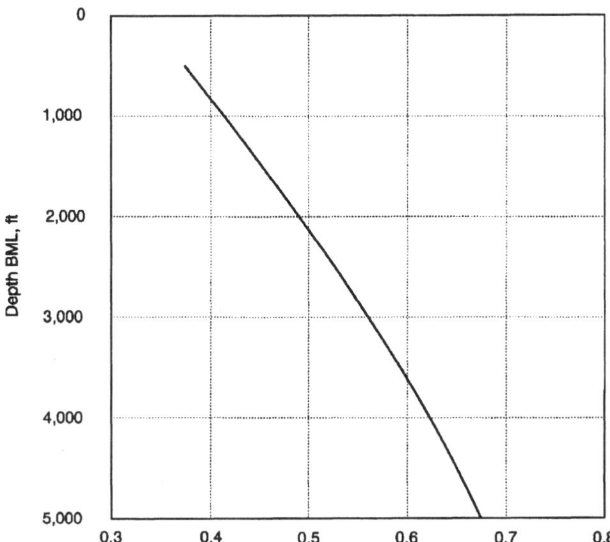

Fig. 3.42—Christman's[50] stress-ratio correlation for the Santa Barbara channel.

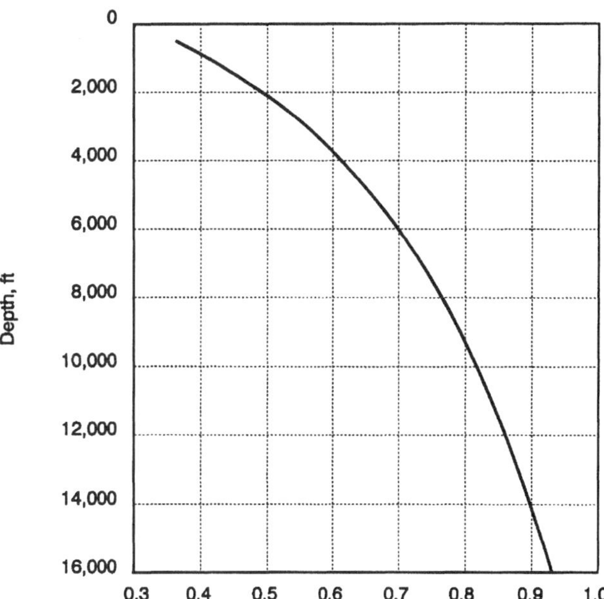

Fig. 3.43—Pilkington's[51] average-stress-ratio correlation based on data from the gulf coast and Santa Barbara channel.

Example 3.10. Estimate the fracture gradient at the two described depths in East Cameron Well C using Eaton's method.

Solution: Use Mitchell's equations to determine the overburden gradient (Eq. 2.9a) and Poisson's ratio (Eq. 3.70).

$$g_{ob} = 0.84753 + 0.01494(8.11) - 0.0006(8.11)^2$$

$$+ 1.199 \times 10^{-5}(8.11)^3 = 0.936 \text{ psi/ft}$$

$$\text{and } \mu_E = 0.23743 + 0.05945(8.11) - 0.00668(8.11)^2$$

$$+ 0.00035(8.11)^3 - 6.71 \times 10^{-6}(8.11)^4$$

$$= 0.438.$$

Eq. 3.69 predicts

$$g_{fi} = \left(\frac{0.438}{1 - 0.438}\right)(0.936 - 0.465) + 0.465$$

$$= 0.832 \text{ psi/ft}$$

for the zone at 8,110 ft. At 15,050 ft, Mitchell's approximation gives $g_{ob} = 0.977$ psi/ft and $\mu_E = 0.468$. Thus,

$$g_{fi} = \left(\frac{0.468}{1 - 0.468}\right)(0.977 - 0.815) + 0.815$$

$$= 0.958 \text{ psi/ft.}$$

Christman's[50] Santa Barbara Channel Correlation. Deepwater drilling locations require that the effect of seawater on overburden stress be considered. **Fig. 3.42** shows Christman's correlation for the Santa Barbara channel offshore California. The depth used in the chart is referenced from the mudline (seafloor).

Example 3.11. Estimate the fracture gradient for a normally pressured formation 1,490 ft below the mudline (BML) in the Santa Barbara channel if the water depth is 768 ft and the air gap is 75 ft. Use 0.44 psi/ft as the water hydrostatic gradient and assume that Eaton's overburden curve for the Santa Barbara channel from Fig. 2.21 applies to sediment depth.

Solution. Eaton's overburden-stress gradient for 1,490 ft of sediment is 0.917 psi/ft. Eq. 3.46 gives the combined overburden as referenced from the RKB datum plane.

$$g_{ob} = \frac{(768)(0.44) + (1,490)(0.917)}{75 + 768 + 1,490} = 0.730 \text{ psi/ft.}$$

Entering Fig. 3.42 with the 1,490-ft BML depth yields a 0.451 stress ratio.

$$g_{fi} = (0.451)(0.73 - 0.452) + 0.452 = 0.577 \text{ psi/ft.}$$

Pilkington's[51] Technique. The empirical relationships discussed thus far were derived from different geographic regions: south Texas, Louisiana, and offshore California. Despite the different locales, the stress-ratio curves that are based on a variable overburden show remarkable agreement, deviating at most by 0.05. Pilkington averaged the matrix-stress ratios developed by the previous authors and presented the curve shown as **Fig. 3.43.** Reasonable accuracy along the U.S. gulf coast and offshore California can be expected with ratios obtained from his chart.

The curve values can be approximated by

$$F_\sigma = 3.9g_{ob} - 2.88 \quad \text{.......................} \quad (3.71)$$

for $g_{ob} \leqq 0.94$ psi/ft and

$$F_\sigma = 3.2g_{ob} - 2.224 \quad \text{.....................} \quad (3.72)$$

at higher overburden gradients.

Pilkington[51] suggested that the similarity in the stress ratio vs. depth curves indicated overburden stress was the controlling variable, implicitly assuming the overburden gradient vs. depth functions to be approximately the same in the different study areas. **Fig. 3.44** shows the overlaid composite-overburden-gradient and stress-ratio curves. He proposed that the chart be used for preliminary fracture-gradient estimates in other Tertiary basins around the world.

The only parameters needed are the actual overburden gradient underlying the drilling location and pore pressure. The first requirement can be met with integrated density logs, while pore-pressure predictions can be made with one of the

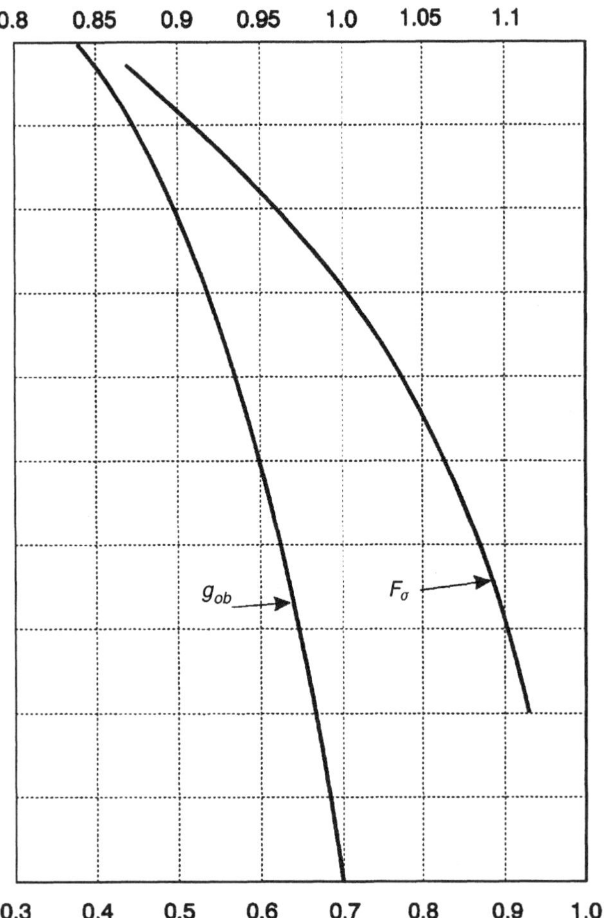

Fig. 3.44—Pilkington's[51] overburden-gradient and average-stress-ratio curves.

methods discussed in Chap. 2. To use the chart, the overburden gradient at the depth of interest is entered at the top and a line is constructed down to the g_{ob} curve and then over to read F_σ.

Example 3.12. Wireline-density-log data obtained from a wildcat well in the South China Sea indicates a 0.883-psi/ft overburden-stress gradient at the 3,970-ft casing seat. Estimate the fracture gradient at the shoe if the pore-pressure gradient is 0.442 psi/ft.

Solution. **Fig. 3.45** demonstrates how an effective-stress ratio of 0.55 is obtained from the chart represented in Fig. 3.44. The expected fracture gradient at 3,970 ft is then

$$g_{fi} = (0.55)(0.883 - 0.442) + 0.442 = 0.685 \text{ psi/ft}.$$

Constant and Bourgoyne's[52,53] Equation. Constant and Bourgoyne proposed the general model expressed by the following exponential equation.

$$F_\sigma = 1 - a[\exp(K_F D_s)], \quad \text{(3.73)}$$

where the a and K_F coefficients are derived empirically to fit local conditions or actual fracture-gradient measurements. The model curve approximates a plot of Eaton's $\mu_E/(1-\mu_E)$ data if a and K_F are set equal to 0.629 and -0.000128 ft^{-1}, respectively.

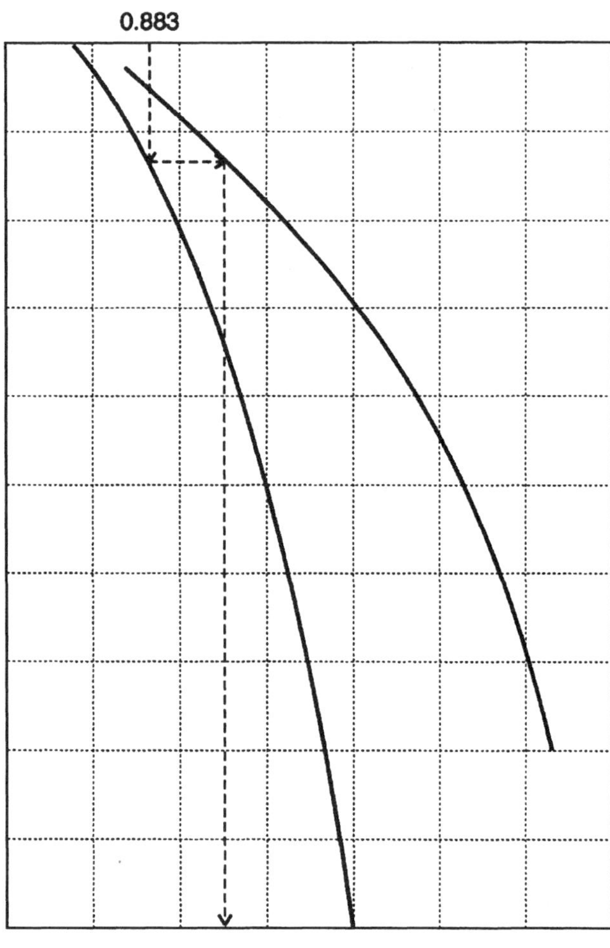

Fig. 3.45—Procedure used to determine the effective-stress ratio in Example 3.12.

Eq. 2.12 is Constant and Bourgoyne's exponential relationship for overburden stress. Modifying this relation for offshore use yields

$$\sigma_{ob} = 0.0519$$

$$\times\left\{\rho_{sw}D_{sw} + \rho_{ma}D_s - \frac{(\rho_{ma}-\rho_f)\phi_{ml}}{K_\phi}\left[1 - \exp(-K_\phi D_s)\right]\right\}. \quad \text{(3.74)}$$

The technique is demonstrated in the Example 3.13 with well data described in Ref. 53.

Example 3.13. The following conditions apply to a well in the Green Canyon area of the Gulf of Mexico: D = 10,000 ft, D_a = 85 ft, D_{sw} = 1,223 ft, ρ_p = 13.2 lbm/gal, ρ_{sw} = 8.5 lbm/gal, ρ_{ma} = 21.66 lbm/gal, ρ_f = 8.95 lbm/gal, ϕ_{ml} = 0.41 (sediment porosity at the mudline), and K_ϕ = 0.000085 ft^{-1}.

Assume that Eaton's Poisson's ratio correlation accurately represents the stress ratio with BML depth and determine the fracture gradient.

Solution. The combined overburden stress at 10,000 ft is determined with Eq. 3.74. For convenience, separate the overburden into its two components.

$$p_{sw} = (0.0519)(8.5)(1,223) = 540 \text{ psig}$$

$$\text{and } \sigma_s = 0.0519\left\{(21.66)(8,692) - \frac{(21.66-8.95)(0.41)}{0.000085}\right.$$

$$\left.\times\,[1-\exp(-\,0.000085\times 8,692)]\right\} = 8,109 \text{ psig.}$$

Thus,

$$\sigma_{ob} = 540 + 8,109 = 8,649 \text{ psig.}$$

Now use the Eaton-fit coefficients in Eq. 3.73 and solve for F_σ.

$$F_\sigma = 1 - 0.629\exp(-\,0.000128\times 8,692) = 0.793.$$

Convert the pore-pressure gradient to pressure,

$$p_p = (0.0519)(13.2)(10,000) = 6,851 \text{ psig,}$$

and determine the predicted fracture pressure as

$$p_{fi} = 0.793(8,649 - 6,851) + 6,851 = 8,276 \text{ psig.}$$

The fracture gradient as referenced from the RKB is

$$g_{fi} = 8,276/10,000 = 0.828 \text{ psi/ft.}$$

The authors obtained a better match with leakoff test measurements on this well by using $a = 0.1$.

Simmons and Rau's[54] Recommended Overburden and Stress-Ratio Modification for Offshore Use. An assumption in the preceding offshore prediction methods is that the stress ratio and sediment burden depend only on BML depth. In other words, we achieve the same F_σ and g_s whether sediment begins at ground surface or beneath 5,000 ft of water. Simmons and Rau,[54] however, argued that seawater pressure would alter both sediment compaction at the mudline and BML stress relationships.

They proposed converting water depth to an equivalent-sediment depth using the relation

$$D_{s(eq)} \approx D_{sw}/2. \qquad (3.75)$$

For a given BML depth, the depth to use in an empirical correlation (Eaton's in their case) is obtained in additive fashion.

$$D_{se} = D_{s(eq)} + D_s, \qquad (3.76)$$

where D_{se} = the effective sediment penetration.

Example 3.14. Estimate the fracture gradient for the Green Canyon well described in Example 3.13 using Eaton's correlations as modified by Simmons and Rau.[54]

Solution. Determine the equivalent-sediment depth and effective penetration using Eqs. 3.75 and 3.76, respectively.

$$D_{s(eq)} = 1,223/2 = 612 \text{ ft}$$

$$\text{and } D_{se} = 612 + 8,692 = 9,304 \text{ ft.}$$

Obtain the sediment gradient and Eaton's Poisson's ratio using D_{se} rather than D_s in Eaton's charts or curve-fit equations and obtain the values $g_s = 0.944$ psi/ft and $\mu_E = 0.444$. The composite overburden pressure computes as

$$\sigma_{ob} = (0.44)(1,223) + (0.944)(8,692) = 8,743 \text{ psig.}$$

The predicted fracture pressure and gradient, respectively, are

$$p_{fi} = \left(\frac{0.444}{1-0.444}\right)(8,743 - 6,851) + 6,851$$
$$= 8,362 \text{ psig}$$

$$\text{and } g_{fi} = 8,362/10,000 = 0.836 \text{ psi/ft.}$$

Zamora's[55] Method. Zamora[55] presented a prediction technique based on an empirical family of overburden gradient and stress-ratio curves, which are dependent on geological age and geographic setting. It is one of the few attempts to model fracture gradients in the older, hard-rock areas.

Zamora expressed the bulk-density relationship with depth as a power-law function. Substituting the function into Eq. 2.6 and integrating yield his overburden correlation,

$$\sigma_{ob} = (0.4171 + 0.01205B)D^{1.075}. \qquad (3.77)$$

The equation constants were obtained by matching field data to the integrated function and apply with customary units. Adapting Eq. 3.77 for offshore applications yields

$$\sigma_{ob} = g_{sw}D_{sw} + (0.4171 + 0.01205B)D_s^{1.075}. \qquad (3.78)$$

The B parameter is variable, depending on the geologic age of the formation and geographic area. Overburden gradients are expected to increase with geologic age because of lower porosity and compactibility, thus B also increases in older rock. Zamora solved Eq. 3.77 for parameter values ranging from 0 through 14 and generated the family of curves shown in **Fig. 3.46.** With a value for B, the sediment overburden gradient (in equivalent density) is obtained directly from the appropriate curve.

Zamora's stress-ratio equation in customary units is

$$F_\sigma = M[1 - 0.55\exp(-\,0.000134D_s)]. \qquad (3.79)$$

The M parameter also depends on the compaction environment and decreases with increasing geologic age. After a value for M is chosen, the stress ratio can be either calculated or read directly from the graphic solution shown as **Fig. 3.47. Tables 3.5 and 3.6** can be used as guides in selecting B and M values for a specific formation age or geographic region.

Example 3.15. Use Zamora's technique and estimate the fracture gradient at 19,900 ft in a Mississippi Smackover well if the pore-pressure equivalent is 16.9 lbm/gal.

Solution: Obtain B and M estimates of 10.5 and 0.34, respectively, from Table 3.6. Obtain the overburden and stress ratio by substituting the parameters into Eqs. 3.77 and 3.79, respectively,

$$\sigma_{ob} = [0.4171 + (0.01205)(10.5)](19,900)^{1.075}$$
$$= 22,728 \text{ psig}$$

$$\text{and } F_\sigma = 0.34\left(1 - 0.55e^{-0.000134\times 19,900}\right) = 0.327.$$

The pore pressure is

$$p_p = (0.0519)(16.9)(19,900) = 17,454 \text{ psig.}$$

Solving for the fracture gradient,

$$p_{fi} = 0.327(22,728 - 17,454) + 17,454$$

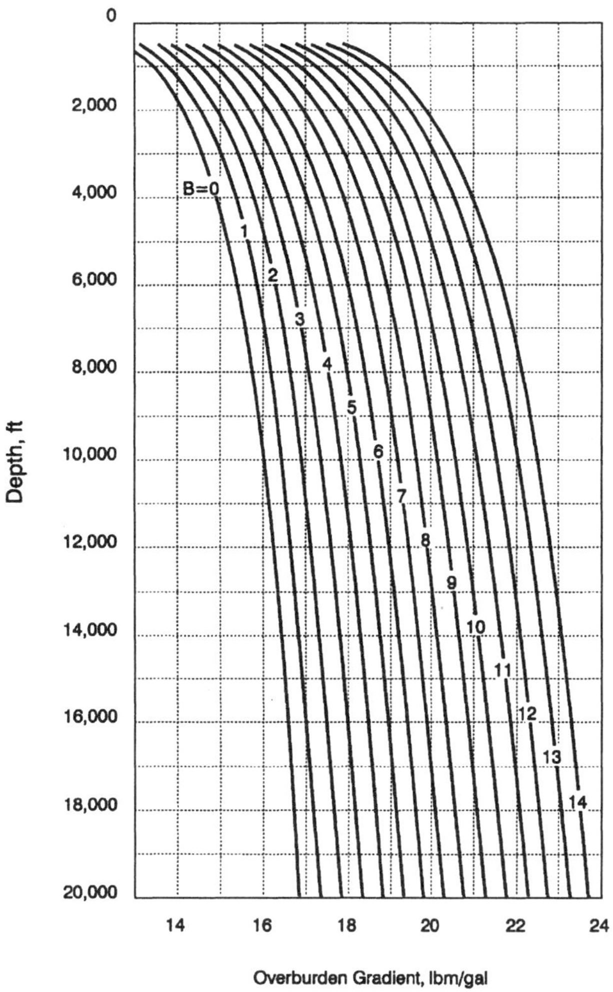

Fig. 3.46—Zamora's[55] overburden-gradient correlation.

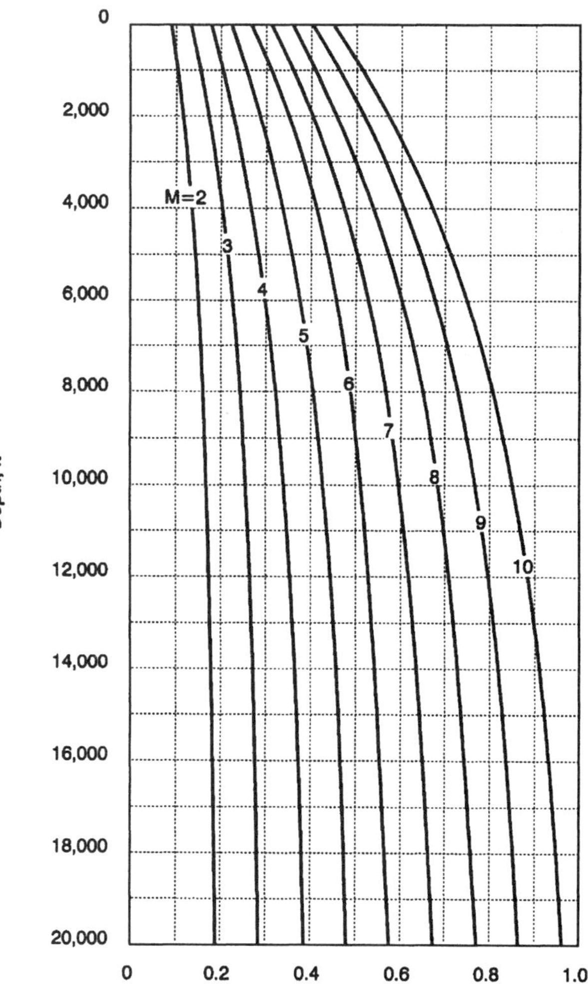

Fig. 3.47—Zamora's[55] matrix-stress-ratio correlation.

$$= 19,179 \text{ psig}$$

and $g_{fi} = 19,179/19,900 = 0.964$ psi/ft.

Brennan and Annis'[56] Gulf of Mexico Correlation. The effective-stress-ratio idea as applied by most investigators over the years uses sediment depth (or effective-sediment depth) as the determining criterion. There is little technical foundation for assessing stress ratio on the basis of depth, other than that it works in many cases. Brennan and Annis[56] introduced a more straightforward approach that has some philosophical appeal.

Brennan and Annis[56] met with limited success in applying conventional stress-ratio techniques in the central and western Gulf of Mexico, particularly in abnormally pressured intervals. They noted good correlation, however, when effective vertical stress from measured overburden and pore-pressure data were plotted directly vs. the $p_{fi} - p_p$ data obtained from

TABLE 3.5—ZAMORA'S[55] OVERBURDEN PARAMETER BY GEOLOGIC AGE

Sediment Age	B
Holocene—Pliocene	0 to 5
Miocene—Oligocene	5 to 9
Eocene—Paleocene	9 to 10
Cretaceous—Triassic	10 to 11
Permian and older	11 to 14

TABLE 3.6—EXAMPLE OVERBURDEN AND STRESS RATIO PARAMETERS APPLIED IN VARIOUS GEOGRAPHIC AREAS

Geographic Area	B	M
Mobile Bay	6 to 7	1.0
Alaska	9 to 10	0.9
Offshore Atlantic coast	9 to 10	0.7
Offshore California	6	1.0
Onshore California	7	1.0
California, Sacramento area	8 to 9	1.0
U.S. Gulf Coast (Eaton[47] fit)	4	1.0
Offshore Louisiana	3 to 5	1.0
Mississippi (Smackover)	10.5	0.34
Northwest New Mexico	10 to 11	0.45
North Sea (Gullfaks)	7 to 10	1.0
Oklahoma	7 to 10	0.39
Rocky Mountains	11	0.4
Texas (Austin chalk)	9	1.0
North Texas	12 to 13	0.4
Offshore south Texas	4 to 6	1.0
Onshore south Texas	6 to 8	1.0
West Texas	12 to 13	0.4

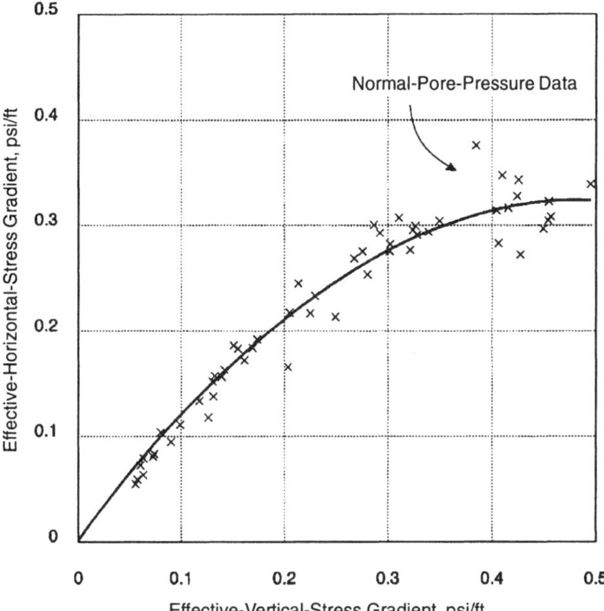

Fig. 3.48—Brennan and Annis'[56] Gulf of Mexico correlation between effective vertical and horizontal stress.

leakoff tests. Thus, results are depth independent. **Fig. 3.48** shows the least-squares-curve fit described by

$$g_{fi} - g_p = 1.35(g_{ob} - g_p) - 1.40(g_{ob} - g_p)^2.$$

Rearranging yields

$$g_{fi} = 1.35(g_{ob} - g_p) - 1.40(g_{ob} - g_p)^2 + g_p. \quad (3.80)$$

Pore pressure and overburden gradients used in the correlation were referenced from sea-level datum, and the fracture-gradient prediction must be corrected for the air gap.

The effective-stress ratio is defined by the slope of a line extending from the origin up to any point on the curve. The slope and stress ratio are fairly constant in the abnormal-pore-pressure data as represented by low effective-vertical-stress values. These points were obtained across a large depth range, so there would be a discrepancy between the curve results and those obtained using conventional stress-ratio curves.

Example 3.16. Estimate the fracture gradient for the Green Canyon well described in Example 3.13 using the Brennan and Annis[56] correlation.

Solution: Assume that Constant and Bourgoyne's[53] 8,649-psi overburden stress is correct. The overburden and pore-pressure gradient, respectively, with a subsea reference are

$$g_{ob} = 8,649/(10,000 - 85) = 0.872 \text{ psi/ft}$$

and $g_p = 6,851/9,915 = 0.691$ psi/ft.

Use Eq. 3.80 to determine the subsea fracture gradient and fracture pressure.

$$g_{fi} = 1.35(0.872 - 0.691) - 1.40(0.872 - 0.691)^2 + 0.691 = 0.889 \text{ psi/ft}$$

and $p_{fi} = (0.889)(9,915) = 8,814$ psi.

It follows that the RKB fracture gradient is

$$g_{fi} = 8,814/10,000 = 0.881 \text{ psi/ft}.$$

3.4.3 Empirical Equations. Breckels and van Eekelen[57] collected stimulation ISIP data in the gulf coast and other areas and presented a series of equations relating minimum horizontal stress (i.e., fracture-propagation pressure) with depth. They developed fracture-initiation-pressure equations by using leakoff test data in normally pressured rock. Vuckovic[58] later used the same method in waters off western Australia.

Table 3.7 presents the collective relationhips in customary units and limitations to their use. Water depth as a limiting criterion was not discussed in the Breckels and van Eekelen[57] paper. Vuckovic's[58] data, however, suggested that reasonable accuracy could be achieved if water depth did not exceed 600 ft. In deeper waters, he proposed modifying the result by reducing the calculated fracture pressure by 250 psi for every 1,000 ft of water.

TABLE 3.7—EMPIRICAL DEPTH/FRACTURE-PRESSURE RELATIONS DEVELOPED IN VARIOUS GEOGRAPHIC REGIONS

Geographic Region	Equation	Limitations
U.S. gulf coast[57]	$p_{fp} = 0.197D^{1.145} + 0.46(p_p - p_n)$	$D \leqq 11,500$ ft
	$p_{fp} = 1.167D^{1.145} - 4,596 + 0.46(p_p - p_n)$	$D > 11,500$ ft
	$p_{fi} = 0.219D^{1.145}$	$D \leqq 11,500$ ft Normal pore pressure
Lake Maracaibo (Venezuela)[57]	$p_{fp} = 0.210D^{1.145} + 0.56(p_p - p_n)$	5,900 $D < 9,200$ ft Eocene Formation
Brunei[57]	$p_{fp} = 0.227D^{1.145} + 0.49(p_p - p_n)$	$D < 10,000$ ft
	$p_{fi} = 0.252D^{1.145}$	$D < 10,000$ ft Normal pore pressure
Onshore Netherlands[57]	$p_{fi} = 0.332D^{1.104}$	Normal pore pressure Carboniferous excluded
North Sea[57]	$p_{fi} = 0.353D^{1.091}$	Normal pore pressure
Offshore west Australia[58]	$p_{fp} = 0.185D^{1.145}$	$D_{sw} < 600$ ft Normal pore pressure
	$p_{fi} = 0.197D^{1.145}$	$D_{sw} < 600$ ft Normal pore pressure

3.4.4 Methods Based on Measured or Inferred Rock Properties. Techniques discussed thus far ignore the effect of lithology on fracture pressure. Horizontal stress does change with elastic rock properties, thus introducing potential error to many conventional empirical methods. Most stress-ratio correlations were derived in Tertiary strata characterized by shale/sand sequences, and most leakoff tests used in these correlations were obtained in shales, which may not be representative of other rock types. It is advantageous to have a method that recognizes the effect of lithology on horizontal stress.

Predictions With Log-Derived Elastic Properties. The Poisson's ratio and poroelasticity constant, if applicable, as inferred from sonic-log measurements may be used directly to calculate fracture pressure. Major wireline companies offer this service, taking pore-pressure and tensile-strength data from the user and overburden pressure from an integrated density log. A continuous fracture-gradient prediction with depth is available in areas where later-generation sonic logs have been run.

Sec. 3.2.1 discussed limitations to dynamic rock-property measurements and the assumption that near-wellbore rock behaves as an isotropic linear-elastic material. The inherent weaknesses to this technique can be countered to some extent by calibrating the output to accommodate leakoff test or stress-measurement information from the field. Multiplication and additive scaling factors may be derived and incorporated into computation software to enhance accuracy.

***Correlation Proposed by Anderson* et al.[59]** Anderson *et al.*[59] introduced a method specifically for the gulf coast area that uses openhole log-porosity measurements and calibrated leakoff test data to estimate fracture gradient in sandstone. Eq. 3.33 represents fracture-initiation pressure for linear-elastic rock if the horizontal stresses are equal. Incorporating Biot's[8] poroelasticity constant and eliminating the tensile strength yields

$$g_{fi} = \left(\frac{2\mu}{1-\mu}\right)g_{ob} + \left(\frac{1-3\mu}{1-\mu}\right)sg_p. \quad \text{(3.81)}$$

This relation is the basis for the technique. If pore pressure and overburden can be obtained, all that remains is a means for determining μ and s. Their technique was published before sonic shear-wave-detection technology had matured, and the authors backed into these elastic parameters in a much more indirect and empirical manner.

In Eq. 3.8, the reduction in s corresponds to a reduction in bulk compressibility. Its limiting lower value is porosity, and the authors argued that this condition would be approached as increasing confining stress reduces bulk compressibility with depth. Their approximate relation for s was obtained as

$$s \approx \phi_D, \quad \text{(3.82)}$$

where ϕ_D = the porosity obtained from a density log.

Rearranging Eq. 3.81 to solve for Poisson's ratio yields

$$\mu_A = \frac{g_{fi} - sg_p}{g_{fi} + 2g_{ob} - 3sg_p}. \quad \text{(3.83)}$$

The Poisson's ratios in 29 sandy intervals were estimated with measured p_{fi} data and log-derived values for s, σ_{ob}, and p_p. Similar to Eaton's[47] correlation, the subscript A is used to distinguish the empirical Poisson's ratio from the true rock property.

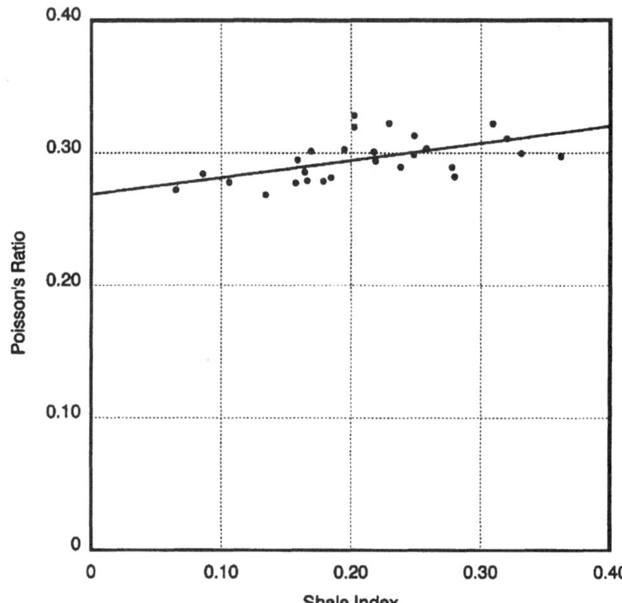

Fig. 3.49—Anderson *et al.*'s[59] correlation between shale index and Poisson's ratio.

The authors reasoned that sand shaliness is related to Poisson's ratio, citing laboratory and leakoff test evidence, and developed a correlation between Poisson's ratio and shale index. The shale index, defined as the ratio between dispersed-shale-volume fraction and intermatrix porosity, can be approximated using log data if the pore space is filled with water.

$$I_{sh} \approx (\phi_S - \phi_D)/\phi_D, \quad \text{(3.84)}$$

where I_{sh} = shale index and ϕ_S = sonic-log porosity. **Fig. 3.49** shows their data and correlation.

Example 3.17. Estimate the fracture gradient using the Anderson *et al.*[59] correlation from the following data obtained in a gulf coast well[58]: D = 10,431 ft; g_p = 0.550 psi/ft; g_{ob} = 0.950 psi/ft; ϕ_D = 0.30; and I_{sh} = 0.32.

Solution. Entering Fig. 3.49 with the given shale index yields a Poisson's ratio of 0.31. Taking s to be 0.30, substitute values into Eq. 3.81 and obtain

$$g_{fi} = 0.950\left[\frac{(2)(0.31)}{1-0.31}\right] + (0.30)(0.550)$$

$$\times \left[\frac{1-(3)(0.31)}{1-0.31}\right] = 0.870 \text{ psi/ft}.$$

This technique should not be used for any other than its intended purpose. The poroelasticity constant is an elusive parameter, and using Eq. 3.82 to estimate its value for use in a linear-elastic equation is a mistake. This is another empirical tool having some successful application in the gulf coast.

Daines'[60] Method. The method proposed by Daines has been used successfully in some areas. It offers a procedure for estimating fracture gradients in variable lithology and in those regions where little information is available. He started by dividing the minimum in-situ horizontal stress into two components: the elastic response to overburden and a superposed tectonic stress.

TABLE 3.8—POISSON'S RATIO VALUES SUGGESTED BY DAINES[61]

Lithology	Poisson's Ratio
Wet Clay	0.50
Dry Clay	0.17
Conglomerate	0.20
Dolomite	0.21
Graywacke	
Coarse	0.07
Medium	0.24
Fine	0.23
Limestone	
Medium, calcarenitic	0.31
Fine, micritic	0.28
Porous	0.20
Stylolitic	0.27
Fossiliferous	0.09
Bedded fossils	0.17
Shaly	0.17
Sandstone	
Coarse	0.05
Coarse, cemented	0.10
Medium	0.06
Fine	0.03
Very fine	0.04
Poorly sorted, clayey	0.24
Fossiliferous	0.01
Shale	
Calcerous (<50% lime)	0.14
Dolomitic	0.28
Siliceous	0.12
Silty (<70% silt)	0.17
Sandy (<70% sand)	0.12
Kerogenaceous	0.25
Siltstone	0.08
Slate	0.13
Tuff	0.34

$$(\sigma_{He})_{\min} \approx p_{fp} = \left(\frac{\mu}{1-\mu}\right)(\sigma_{ob} - p_p) + p_p + \sigma_{tec}, \quad \text{(3.85)}$$

where σ_{tec} = the tectonic stress component. Daines made no distinction between initiation and propagation criteria, implicitly assuming $p_{fp} = p_{fi}$. He further proposed that the tectonic stress component, if present, increases uniformly with depth so that

$$\sigma_{tec}/\sigma_{Ve} = \sigma_{tec}/(\sigma_{ob} - p_p) = \kappa, \quad \text{(3.86)}$$

where κ = a constant. This relation conflicts with a commonly held opinion that an imposed tectonic stress remains consistent with depth.

Rearranging Eq. 3.85 gives

$$\sigma_{tec} = p_{fi} - \left(\frac{\mu}{1-\mu}\right)(\sigma_{ob} - p_p) - p_p. \quad \text{(3.87)}$$

Thus, σ_{tec} at a given depth can be evaluated with leakoff-test pressure if other variables can be quantified. Overburden stress and pore pressure can be estimated readily in many cases; a Poisson's ratio number is required to apply the technique. Values specific to the area lithology could be derived from the later-generation sonic logs or the data suggested by Daines in **Table 3.8.**

In practice, an early leakoff test is used to determine σ_{tec} for a rank wildcat in an unknown area. An integrated-density log establishes the overburden gradient, which is extrapolated to greater depths, while the tectonic-stress profile is assumed to obey Eq. 3.86. Careful monitoring of drill cuttings and pore-pressure indicators yields the final two variables required to solve for the fracture gradient.

Example 3.18. The following data were obtained during or prior to a leak-off test in an offshore wildcat well[60]: D = 3,684 ft; σ_{ob} = 3,518 psig; p_p = 1,663 psig; p_{fi} = 2,906 psig; and rock type = medium graywacke.Drilling resumed, and kerogenaceous-shale drill cuttings were recovered at 8,327 ft. Estimate the fracture pressure in the shale if the overburden and pore pressure are 8,686 and 3,889 psig, respectively.

Solution. Obtain Poisson's ratio from Table 3.8 and use Eq. 3.87 to determine the tectonic stress at the test depth.

$$\sigma_{tec} = 2,906 - \left(\frac{0.24}{0.76}\right)(3,518 - 1,663) - 1,663$$

$$= 657 \text{ psig}.$$

The ratio κ at 3,684 ft is

$$\kappa = 657/(3,518 - 1,663) = 0.354.$$

Assuming the tectonic-stress profile with depth is parallel to the overburden, compute the σ_{tec} component acting on the shale as

$$\sigma_{tec} = (0.354)(8,686 - 3,889) = 1,698 \text{ psig}.$$

From Table 3.8, the estimated Poisson's ratio is 0.25, and the predicted fracture pressure is computed with Eq. 3.85.

$$p_{fi} = (0.25/0.75)(8,686 - 3,889) + 3,889 + 1,698$$

$$= 7,176 \text{ psig}.$$

The actual fracture pressure was determined to be 7,001 psi, which represents a 2.5% disparity.

Some technique assumptions are debatable. Setting the leakoff-test pressure equal to minimum horizontal stress is unjustified in most situations. The derived tectonic-stress component is invalid if the rock does not behave as a linear-elastic material and if this stress does not increase in proportion to the overburden. Obtaining the Poisson's ratio for stressed rock from an all-purpose table is questionable at best. However, good results have been reported from use of the method, at least in the North Sea,[62] and it is difficult to criticize any procedure that works in the field.

Aadnoy and Larsen's[62] Method. The Aadnoy and Larsen method can be applied in areas where fracture-pressure data are available, and offers a procedure for extending predictions to inclined wells. The first steps are to plot the overburden stress and pore-pressure gradients with depth. Then, leakoff-test or lost-circulation pressures are used to estimate the in-situ horizontal-stress gradient. Eq. 3.37 gives the theoretical fracture-initiation pressure for a vertical borehole in an isotropic horizontal-stress field. Recasting the relation in terms of gradient and dropping the tensile stress gives

$$g_H = \left[g_{fi(0)} + g_p\right]/2, \quad \text{(3.88)}$$

where g_H = the horizontal-stress gradient and $g_{fi(0)}$ = the fracture initiation gradient at zero inclination. Eq. 3.88 calculates g_H at depths where vertical-hole fracture pressures are

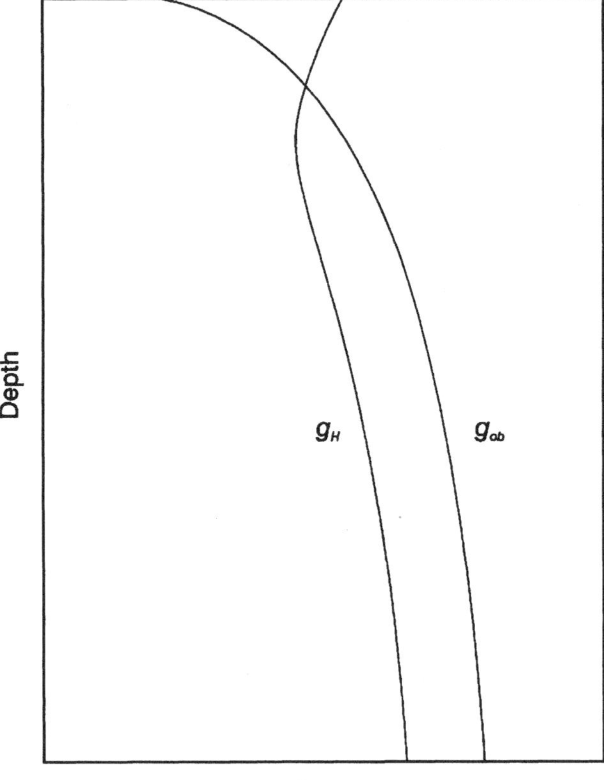

Fig. 3.50—Calculated horizontal-stress gradient in relation to overburden stress for gulf coast sediments.[24]

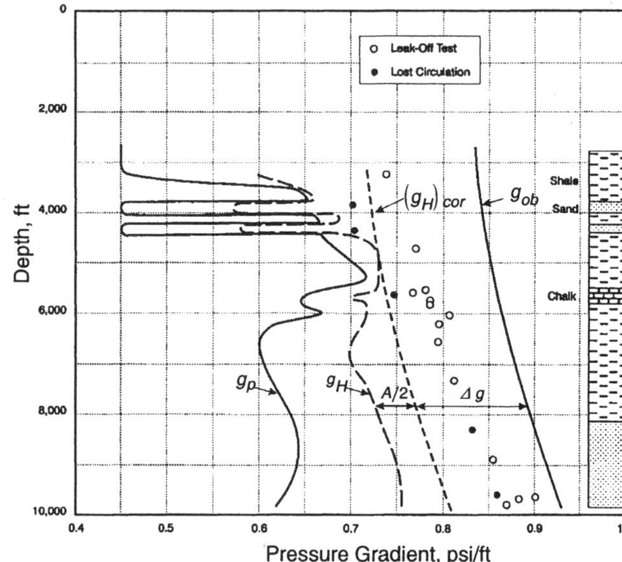

Fig. 3.51—Fracture-pressure data and gradient curves from Statfjord field, North Sea.[62]

available. The points are plotted and connected with a smooth curve.

Fig. 3.50 depicts the results of an earlier study by Aadnoy and Chenevert[24] calculating g_H with depth with Eaton's[47] overburden and fracture data in normally pressured gulf coast sediments. The authors noted, however, that using the calculated values to estimate fracture gradients in deviated holes as in Example 3.6 led to unrealistically conservative results and proposed modifying Eq. 3.88 to fit field data by including a correction factor.

$$g_H = \left[g_{fi(0)} + g_p + C\right], \quad \text{(3.89)}$$

where C = the correlation coefficient.

The overburden and horizontal-stress gradient curves in Fig. 3.50 are almost parallel in the deeper strata. On this basis, Aadnoy and Larsen's procedure requires constructing another horizontal-stress-gradient profile parallel to the overburden. This corrected horizontal-stress gradient is fixed by drawing the curve through the point representing the minimum distance between the g_H and g_{ob} curves.

The spacing between the overburden and corrected horizontal-stress gradients remains constant with depth, a relation expressed by

$$(g_H)_{\text{cor}} = g_{ob} - \Delta g, \quad \text{(3.90)}$$

where $(g_H)_{\text{cor}}$ = the corrected horizontal stress and Δg = the gradient spread. Setting $g_H = (g_H)_{\text{cor}}$ and combining Eqs. 3.89 and 3.90 yields the expression for C.

$$C = 2g_{ob} - 2\Delta g - g_{fi(0)} - g_p. \quad \text{(3.91)}$$

Fig. 3.51 shows the pore-pressure, overburden, and horizontal-stress gradient curves in the North Sea's Statfjord field. Compare the difference in appearance between the authors' overburden relation and Fig. 3.50. The distance between the two horizontal-gradient curves is given by

$$C/2 = (g_H)_{\text{cor}} - g_H, \quad \text{(3.92)}$$

where $C = 0$ at the depth used to establish $(g_H)_{\text{cor}}$.

The next step is to describe mathematically the relationship between the correlation coefficient and pore pressure for each lithology. In the Statfjord field case study, C values in sand and shale were determined with Eq. 3.92 and plotted vs. pore-pressure gradient as shown in **Fig. 3.52.** The correlation should be linear, and a least-squares fit results in a straight line for each rock type described by

$$C = b_C - m_C g_p, \quad \text{(3.93)}$$

where b_C and m_C = the line's intercept and slope, respectively.

Combining Eqs. 3.91 and 3.90 gives the fracture-gradient expression for vertical wellbores.

$$g_{fi(0)} = 2g_{ob} - 2\Delta g - b_C - (1 - m_C)g_p. \quad \text{(3.94)}$$

The fracture-initiation gradient, but not the propagation gradient, is expected to decline when a borehole is inclined with respect to the maximum principal earth stress. Matrix stress, however, governs failure, so pore pressure impacts how much the fracture gradient in a deviated well differs from that of a vertical well. Aadnoy and Chenevert[24] covered this phenomenon in detail and developed the empirical relation

$$g_{fi(\gamma)} = g_{fi(0)} + \frac{1}{3}\left(g_p - g_p^*\right)\sin^2\gamma, \quad \text{(3.95)}$$

where $g_{fi(\gamma)}$ = the fracture gradient at inclination γ and g_p^* = an empirically derived constant. If fracture-pressure data are available, the constant specific to the area can be obtained from deviated wells by computing $g_{fi(0)}$ with the described procedure and solving for the equation unknown, g_p^*. Lacking this information, obtain a conservative estimate using

$$g_p^* = 3g_{fi(0)} - 2g_p. \quad \text{(3.96)}$$

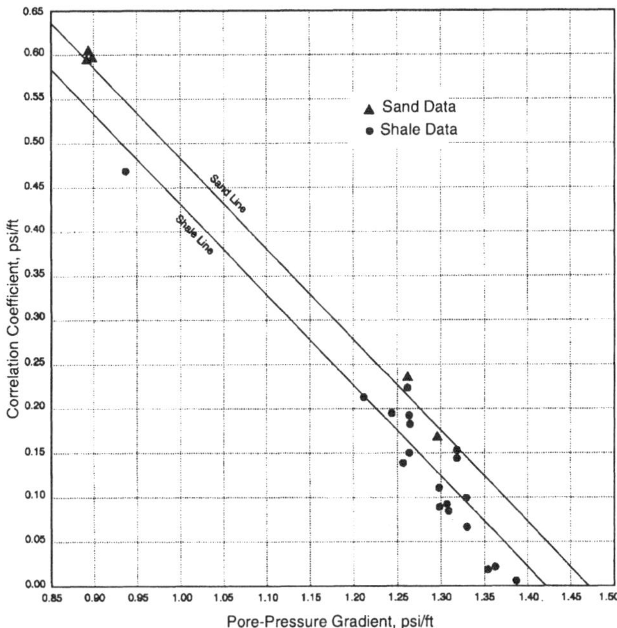

Fig. 3.52—Aadnoy and Larsen's[62] correlation coefficient for sand and shale as a function of pore-pressure gradient.

Combining Eqs. 3.95 and 3.96 yields the fracture-gradient prediction for deviated holes.

$$g_{fi(\gamma)} = 2(g_{ob} - \Delta g) - b_C - (1 - m_C)g_p$$

$$+ \frac{1}{3}\left(g_p - g_p^*\right)\sin^2\gamma. \quad \text{(3.97)}$$

Example 3.19 demonstrates the procedure for the North Sea application discussed in Aadnoy and Larsen's[62] paper.

Example 3.19. Refer to the diagram shown in Fig. 3.51. The Statfjord field overburden gradient was modeled using the following polynomial expressions:

$$g_{ob} = 0.8612 - \left(1.58 \times 10^{-5}\right)D + \left(3.52 \times 10^{-9}\right)D^2$$

$$- \left(12.59 \times 10^{-14}\right)D^3$$

The stress-gradient difference, Δg, was found to be 0.1205 psi/ft. The correlation coefficient in shale is described by

$$C_{sh} = 0.742 - 1.06g_p,$$

and the correlation coefficient in sand is described by

$$C_{sd} = 0.770 - 1.06g_p$$

for conventional units.

The pore-pressure constant, g_p^*, in the area has been determined to be 0.906 psi/ft.

1. Estimate the vertical hole fracture gradients for the shale and immediately underlying sand at 8,200 ft if the pore-pressure gradient is 0.641 psi/ft.

2. What is the estimated fracture gradient in the sand if the wellbore is horizontal?

Solution.

1. The estimated overburden at 8,200 ft is

$$g_{ob} = 0.8612 - \left(1.58 \times 10^{-5}\right)(8,200)$$

$$+ \left(3.52 \times 10^{-9}\right)(8,200)^2 - \left(12.59 \times 10^{-14}\right)(8,200)^3$$

$$= 0.899 \text{ psi/ft}.$$

The b_C and m_C terms in the shale are 0.742 psi/ft and 1.06, respectively, and the shale solution from Eq. 3.94 is

$$g_{fi(0)} = (2)(0.899) - (2)(0.1205) - 0.742$$

$$- (1 - 1.06)(0.641) = 0.853 \text{ psi/ft}.$$

The sandstone prediction will be 0.028 psi/ft lower than calculated for the shale (0.770 − 0.742). Thus, in a vertically drilled sand,

$$g_{fi(0)} = 0.853 - 0.028 = 0.825 \text{ psi/ft}.$$

2. Eq. 3.97 gives the solution for the horizontal wellbore.

$$g_{fi(90)} = (2)(0.899 - 0.1205) - 0.770$$

$$- (1 - 1.06)(0.641) + \frac{1}{3}(0.641 - 0.906)\sin^2(90)$$

$$= 0.737 \text{ psi/ft}.$$

This result approximately equals the in-situ stress.

Aadnoy and Larsen[62] do not recommended using this technique for depths shallower than approximately 3,300 ft. Most published correlations do not work well in shallow rock because of such factors as finite tensile strength, tectonic stresses, and pseudoplastic behavior in loosely consolidated rock. Anomalously high fracture gradients are usually the norm, often approaching and even exceeding the overburden. The difficulty in predicting shallow-fracture gradients poses some decision-making problems to both well planners and site supervisors. Chap. 8 covers this important topic in detail.

Holbrook* et al.*'s*[63] *Porosity Correlation. A glance at the stress-ratio curves shows that fracture gradients tend to increase with increasing sediment compaction. It is reasonable to expect that porosity could be correlated to fracture pressure in some way. Holbrook *et al.*,[63] used this concept to develop the empirical relation given by

$$\sigma_{He2} = \sigma_{Ve}(1 - \phi). \quad \text{(3.98)}$$

In effect, $(1 - \phi)$ is considered to be the matrix-stress ratio. Equating fracture-propagation pressure to $\sigma_{He} + p_p$ yields

$$p_{fp} = (1 - \phi)\left(\sigma_{ob} - p_p\right) + p_p. \quad \text{(3.99)}$$

Holbrook *et al.*,[63] found that results from Eq. 3.99 were in agreement with lower-bound fracture pressures evidenced in the North Sea and other areas. In practice, a continuous method for estimating pore-pressure and fracture gradients while advancing the hole was made available, by use of logging-while-drilling technology to measure the overburden and porosity.

This approach has attractive aspects. It is simple to apply if valid input parameters are available and the porosity function accommodates the theoretical upper fracture-pressure limit. At great depths, zero porosity implies zero pore pressure and the lithostatic state envisioned by Heim's[15] theory. Additionally, different fracture gradients observed in different lithologies often can be correlated to rock compactibility.

Example 3.20. Use the data and results of Example 2.27 to estimate the rock's fracture-propagation gradient.

Solution. The specified overburden stress and porosity were 3,875 psig and 24.1%, respectively, while the predicted pore-pressure result was 2,538 psig. Eq. 3.99 gives the fracture-pressure prediction.

$$p_{fp} = (1 - 0.241)(3,875 - 2,506) + 2,506$$
$$= 3,545 \text{ psig and}$$
$$g_{fp} = 3,545/4,500 = 0.788 \text{ psi/ft.}$$

3.5 Field Measurements

Fracture-pressure predictions, whether from one of the published relations or actual data from an offset well, are still only estimates. Some means of validating the predictions should be used in the drilling process, particularly in wells that have a pore-pressure transition. Engineering a hole is an inexact science, and modifications are required when field data are available. The reservoir engineer forecasting production performance may have to wait years before sufficient data are available for assessing the prediction basis. Drilling predictions are no different except that results show up more quickly and there are more opportunities to be proved right or wrong.

Pressure testing the shoe accomplishes at least three objectives. First, the cement job is tested and a squeeze job may be prescribed if the test information indicates a channeled cement sheath. Second, it may be necessary to modify the well plan and casing program if the predicted formation integrity does not match the field measurement. Finally, prudent decision making during a well-control problem requires knowing the rock strength.

3.5.1 Formation Leakoff and Integrity Tests. Leakoff or integrity tests are routine procedures on wells requiring a mud-weight increase before the next casing point. Not only are they a standard practice, but they also are decreed by law in many operating areas.[64] Leakoff testing involves pressuring the wellbore until the exposed formation fractures and begins to take whole mud. Integrity tests take the rock to a predetermined pressure dictated by the anticipated maximum mud weight. Both tests have their place, and the decision to fracture the rock depends on such factors as perceived risk, knowledge of the area, and certain aspects of the hole program.

The rock beneath the casing shoe and cement job normally is tested after drilling approximately 10 ft of new formation. Before conducting the test, the hole should be circulated clean and the mud should be of uniform density as verified by periodic measurement while circulating and a static mud column on both sides with the rig pumpoff. The bit is pulled back into casing and the well is shut in with the blowout preventer, leaving the wellbore as a closed container. Mud is pumped into the drillstring at a slow rate while frequent measurements of the pump pressure and volume are taken. **Fig. 3.53** shows these data plotted on a graph. Pumping continues until either the predetermined pressure is achieved or leakoff occurs.

In Fig. 3.53, data initially plot as a relatively flat curve, indicating that air bubbles in the drillstring and annulus are being compressed during this portion of the test. Thereafter, a straight-line relationship between pressure and volume develops. If the container is a truly closed system (no filtrate seepage), the slope of the line depends on the drilling fluid compressibility and wellbore elasticity. It is recommended that the supervisor or engineer do some calculations before the test and draft the anticipated straight line on the chart

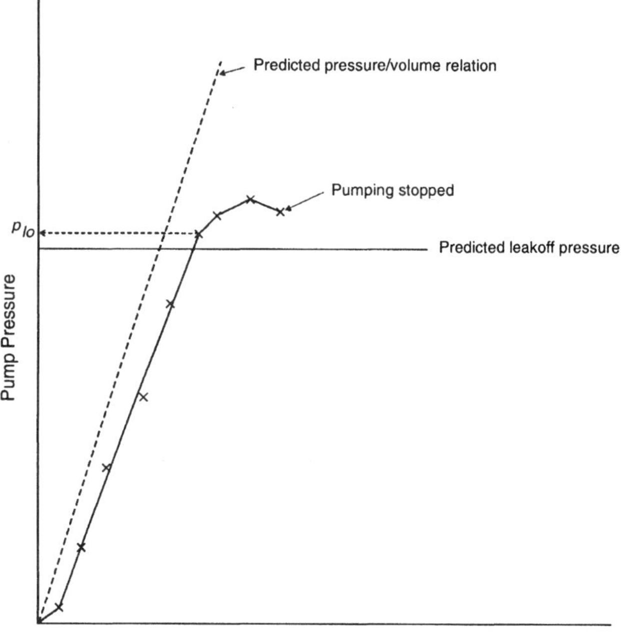

Fig. 3.53—Leakoff-test-data chart including pretest projection of the line slope and predicted leakoff pressure.

for comparative purposes and as a guide in selecting the pump rate.

Eq. 3.100 can be used to approximate the relation between volume added by the pump and pressure increase in the well, ΔV_m and Δp, respectively.

$$\Delta V_m = c_m \Delta p V_m, \qquad (3.100)$$

where c_m = mud compressibility and V_m = original mud volume in the well. Three components make up a drilling mud, and the combined compressibilities of each on a volume-fraction basis constitute the total compressibility.

$$c_m = c_w f_{Vw} + c_o f_{Vo} + c_s f_{Vs}, \qquad (3.101)$$

where $c_w f_{Vw}$, $c_o f_{Vo}$, and $c_s f_{Vs}$ = respective compressibility volume-fraction products of the water, oil, and solids components. Volume fractions can be obtained readily from the mud check sheet. **Table 3.9** lists compressibility values that should yield reasonable accuracy.

The open hole and casing strain in response to pressure increase, thereby enlarging the hole diameter. The volume change caused by wellbore elasticity is less significant than mud compressibility and is often ignored. This is reasonable because the equation variables and underlying assumptions are more suspect and computation precision is not needed.

The line slope for a unit pressure increase is the reciprocal of the volume change, given by

$$\frac{dp}{dV_m} = \frac{1}{\Delta V_m}. \qquad (3.102)$$

TABLE 3.9—TYPICAL DRILLING-FLUID-COMPONENT COMPRESSIBILITIES

Component	Compressibility, psi^{-1}
Water	3.0×10^{-6}
Oil	5.0×10^{-6}
Solids	0.2×10^{-6}

The line passes through the origin and is placed on the graph before the test.

Fig. 3.53 shows the formation is beginning to take whole mud when the measured data start to deviate from the straight line. The surface pressure at the breakover point is identified as the leakoff pressure and is used to compute the fracture-initiation gradient as

$$p_{fi} = p_{lo} + g_m D - \Delta p_f, \quad \ldots\ldots\ldots\ldots\ldots\ldots (3.103)$$

where p_{lo} = leakoff pressure, $g_m D$ = hydrostatic mud column pressure to the depth of interest, and Δp_f = pressure drop in the drillstring and bit. Given low pump rates, the pressure drop should be low and often is ignored. The effect of small jet nozzles or total flow area in the bit, however, should be considered when planning the test.

Shutting down the pump to record pressures eliminates friction pressure from consideration. However, the ability to obtain accurate data in this fashion depends on the filtrate bleed into the exposed rock, and the practice is not recommended unless the formation is known to be relatively impermeable. Chenevert and McClure[66] recommended using the pump pressure required to break the mud-gel strength in place of the friction pressure term. This value can be obtained by allowing the wellbore to remain static for 10 minutes and then noting the pump pressure when circulation initiates.

With a field approximation for the friction pressure, Eq. 3.103 can be rearranged to get the anticipated leakoff pressure on the basis of the predicted fracture gradient. A horizontal line corresponding to the anticipated pressure should be placed on the chart before the test begins. The rock immediately below the casing seat and the cement job are tested. Excessively low leakoff pressures may indicate mud loss into shallower rock through a deficient cement job and the need to squeeze the shoe.

The pump rate should be as low as is practically possible during the procedure, yet high enough to overcome any filtrate-loss rate. The rate should be such that the data fall on or slightly below the anticipated line after an appropriate shift for air-bubble compression is made. Pump rates when testing shale or tight rock typically range between 0.25 to 0.50 bbl/min, whereas porous-sandstone beds may require rates approaching 1.50 bbl/min. The effect of permeable rock and relatively-high-filtrate mud can be mitigated by spotting a low-filtrate pill across the openhole section before pulling up and starting the test.

The slow-rate pumping requirement and, at times, pressure limitations may preclude the use of the rig pumps. Normally, cementing units are used for leakoff tests because they have better low-rate capabilities and more accurate means for measuring volume and pressure. A continuous strip-chart recording of both drillpipe and annulus pressures can be provided as part of the service. Annulus pressure should closely match the drillpipe pressure when friction pressure is taken out, thus giving a quality-control check of the data.

Pumping should continue long enough to ensure that leakoff is achieved. Morita *et al.*[67] described the small increase in pressure following leakoff as the stable fracture-propagation period, suggesting that the fracture is extending through the altered near-wellbore stresses. This characteristic behavior may not be seen in every case, however. A sudden drop in pressure may indicate tensile strength in the rock or a rapid transition to an unstable fracture-extension mode.

Fig. 3.53 also depicts the stopping point where most operators shut down. Shut-in pressures are monitored for a short period and followed by opening the well and bleeding mud back. Return volume should be close to the volume pumped if filtrate losses are not too high. The advantage to having a cementing unit in place is that the bleed volume can be measured accurately in the unit's displacement tank. Sometimes, pumping is continued past the stable-propagation period until the fracture extends a considerable distance away from the wellbore.

There has been and still is some controversy over the maximum pressure to impose during a formation test. Some fear that taking a rock to leakoff pressure weakens the rock and that some shoe integrity is irretrievably lost. The tensile strength of the rock is exceeded when a fracture is initiated, and the resulting breach remains after the fracture has closed. However, this factor, if present, can be ignored in all but very shallow formations. Fracture principles discussed here have shown that the overriding component is not inherent rock strength, but rather its stress state. This stress does not change after fracturing, and fracture gradients should remain consistent in any followup testing.

Findings from DEA Project No. 13[40] cannot be dismissed. The laboratory fractures tended to heal with sealing-muds, whereas reopening pressures were lower for nonsealing OBMs. Planners must consider the type of mud in the hole at the time of the test. Testing with a WBM that has a significant solid fraction should be "safer" than testing with a low-colloid OBM, clear-water drilling fluid, or tight WBM. One thing in the operator's favor is evidence that fracture gradients tend to increase over time.

Pressure testing to some predetermined level below the expected leakoff pressure is a common and valid procedure. These pressure tests generally are based on anticipated pore pressure and the mud weights required in the hole section before another string of casing is set. The only problem introduced by a predetermined test pressure is in the situation where pore-pressure predictions while drilling are higher than those anticipated at the time of the test. In this case, the operator should either set casing early or conduct another test before increasing mud weight.

Pressure tests are valid only if the exposed formation has a lower fracture integrity than the upcoming hole section. This may or may not be the case with only a few feet or meters of open hole. Another pressure test should be run any time the suspicion arises that the fracture gradient is lower in the drilled formation than in the shoe. It is common in some areas to run two or more tests across the same hole section after drilling the first sand section or in a zone experiencing some pressure depletion.

Example 3.21. Surface casing is set on a well at 5,500 ft and the decision is made to test the rock again after drilling a sandstone at 6,900 ft. The bit will be tripped at 7,500 ft, at which time the test will be made. The following information is known or assumed. Measured fracture gradient at the shoe = 0.78 psi/ft; estimated fracture gradient in the sand = 0.79 psi/ft; mud type = water-based; mud density = 9.3 lbm/gal; solids content in the mud = 7%; total mud volume in the hole = 1,126 bbl; and pump pressure required to break gel strength = 50 psig.

1. Prepare the pretest graph.
2. Interpret the test results shown in **Fig. 3.54.**

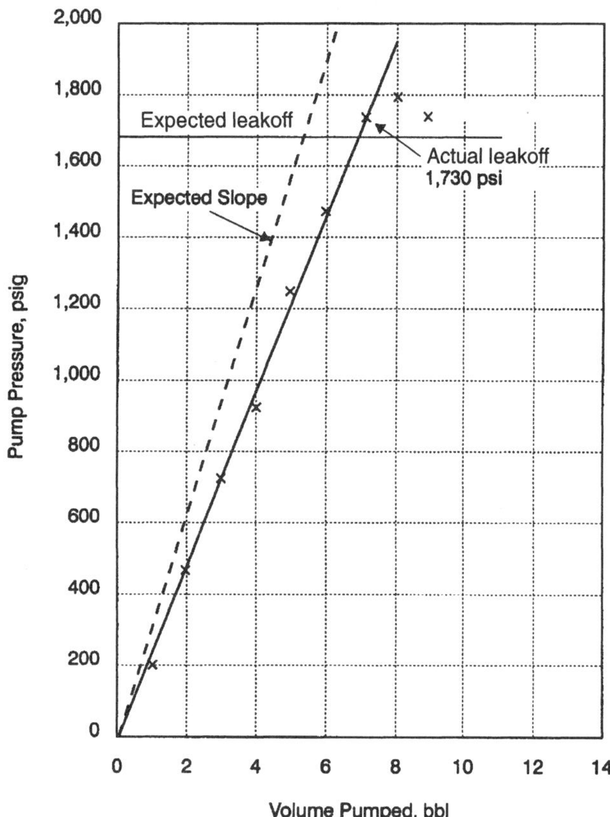

Fig. 3.54—Leakoff-test chart corresponding to Example 3.21.

Solution.

1. Start by computing the anticipated slope of the test data. With the tabulated values, Eq. 3.101 yields the estimated mud compressibility.

$$c_m = (0.93)(3.0 \times 10^{-6}) + (0.07)(0.2 \times 10^{-6})$$
$$= 2.8 \times 10^{-6} \text{ psi}^{-1}.$$

Eq. 3.100 gives the expected volume change for a 1.0-psi increase in pressure.

$$\Delta V_m = (2.8 \times 10^{-6})(1,126) = 0.00315 \text{ bbl/psi}.$$

The line slope is computed with Eq. 3.102 as

$$\frac{dp}{dV_m} = 1/0.00315 = 317 \text{ psi/bbl}$$

and drawn on the chart as shown in Fig. 3.54.

The shoe is still expected to be the weakest part of the hole because the predicted fracture gradient in the sandstone is slightly higher than from the shoe test. Accordingly, the estimated leakoff pressure is based on the shoe test and Eq. 3.103 yields the expected leakoff pressure.

$$p_{lo} = (0.78)(5,500) - (0.0519)(9.3)(5,500) + 50$$
$$= 1,685 \text{ psig}.$$

The prediction is placed on the chart before the test begins.

2. Leakoff pressure measured during the test is 1,730 psi. It is not certain, however, where the mud loss occurred. Fracture pressure at the shoe is likely to be higher than the original test, but the loss may also be in the sandstone or perhaps another formation. Taking the depth to be the shoe yields

$$p_{fi} = 1,730 + (0.483)(5,500) - 50 = 4,337 \text{ psig}$$

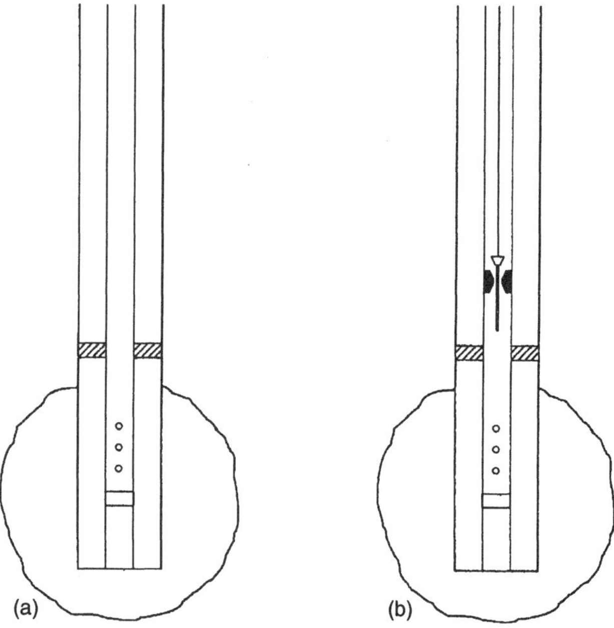

Fig. 3.55—Typical equipment arrangements for openhole microhydraulic fracturing procedures.

and $g_{fi} = 4,337/5,500 = 0.789$ psi/ft.

Assuming the losses are in sand gives

$$p_{fi} = 1,730 + (0.483)(6,900) - 50 = 5,013 \text{ psig}$$

and $g_{fi} = 5,013/6,900 = 0.727$ psi/ft.

The results are inconclusive. Surface pressure for the expected sandstone leakoff is 2,170 psig, so it is reasonable to conclude that the leakoff is at the shoe unless the sandstone prediction was greatly in error.

3.5.2 Microhydraulic Fracturing. Microhydraulic fracturing intentionally creates a small fracture by applying hydraulic pressure. Leakoff tests accomplish the same thing, but microhydraulic fracturing procedures are more data intensive and attempt to measure the parameters with more accuracy. The primary objective is to obtain minimum principal rock stresses, usually oriented to gathering data for designing large fracture-stimulation treatments in tight reservoirs. Microhydraulic fracturing entails some operational risks, such as stuck packers and lost circulation, and considerable expense. It is not a standard or routine procedure. Recently, however, drillers have taken an interest in the technique, applying the knowledge gained to understand and anticipate wellbore-stability problems in certain critical applications.

A microhydraulic fracturing job can be pumped into cased or open wellbores. **Fig. 3.55a** shows an openhole-equipment schematic described by Daneshy *et al.*[68] The formation is isolated from other rock with one or two inflatable packers, either in a straddle arrangement or similar to the anchor-pipe hookup shown in the diagram. A high-pressure, low-volume triplex pump supplies hydraulic energy at closely monitored pump rates while transducers on the drillpipe and annulus monitor surface pressure. Downhole Bourdon tube gauges record actual bottomhole pressures for retrieval and analysis.

Fig. 3.55b shows an alternative arrangement discussed by Warpinski *et al.*[69] During the pumping phase of the operation, bottomhole pressures are recorded and monitored in real

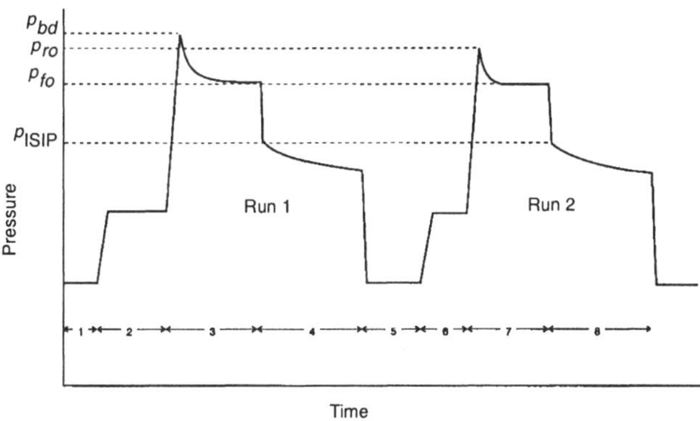

Fig. 3.56—Recorded pressures during a microhydraulic fracturing procedure.

time by wireline-conveyed quartz-crystal gauges. The data are stored with surface rate and pressure measurements in a minicomputer for retrieval and analysis. Downhole shutoff during the shut-in period is provided by a beveled mandrel and O-ring above the top gauge and matching seating nipple in the drillpipe. To close the device, wireline tools are lowered to the seat and kept there by differential pump pressure maintained from above.

Fig. 3.56 shows an idealized pressure-strip chart. Segment 1 represents the hydrostatic mud-column pressure; segment 2 depicts the packer test. Thereafter, pressure is applied at a low rate until a fracture is initiated. The first fracture-initiation pressure is designated as the breakdown pressure, p_{bd}, rather than the fracture initial pressure, p_{fi}. The fracture extends into the rock, and the pressure eventually stabilizes at propagation pressure, p_{fp}. Note that the initial fracture pressures shown in this chart do not have the small upward trend noted in the leakoff test example but break back to a flat extension pressure.

The pump is shut down at the beginning of Segment 4, and pressure rapidly declines to the ISIP pressure, p_{ISIP}, distinguished by a sharp break in the the falloff curve. In principle, ISIP represents the minimum principal earth stress plus any incremental pressure required to keep the fracture open. Assuming the created fracture is vertical, this incremental pressure is negligible in small-volume fractures and p_{ISIP} equates approximately to the minimum horizontal stress.

$$p_{ISIP} \approx \sigma_{H2}. \quad \text{(3.104)}$$

Note that the propagation pressure is somewhat higher than the minimum stress because of fracture toughness and friction pressure between the fracture walls.

Following ISIP, filtrate leaks off into the walls of the fracture and borehole until closure and into the borehole only after the fracture closes. The magnititude of the pressure decline is primarily a function of the rock permeability. The pressure is released after a time period, designated by Segment 5, and the process is repeated in several subsequent tests.

The fracture-initiation pressure in Run 2 is less than the initial breakdown pressure, p_{bd}. The difference may correspond to the tensile strength of the rock, a near-wellbore effective-stress reduction, or a combination of the two effects. If fluid does not penetrate the rock, it follows that

$$\sigma_{ts} = p_{bd} - p_{ro}, \quad \text{(3.105)}$$

where p_{ro} = fracture reopening pressure. The reopening pressure represents the minimum induced borehole stress if no fluid penetrates the rock during the buildup and initiation process. Adopting this limiting criterion and assuming linear-elastic behavior, p_{ro} can replace p_{fi} in Eq. 3.28 where σ_{ts} is now zero. This leads to an estimate for the maximum horizontal stress.

$$\sigma_{H1} = 3p_{ISIP} \approx p_{ro} - p_p. \quad \text{(3.106)}$$

Following the microhydraulic fracturing procedure, an oriented core may yield information pertaining to the minimum-stress direction. The created fracture should be visible if it extended through the bottom of the hole. If not, anaelastic strain recovery of the retrieved core should indicate the principal-stress orientation and, perhaps, a more accurate indication of the maximum principal stress.

Precise fracture-gradient data can be obtained with the microhydraulic fracturing technique, although such tests rarely are conducted solely for this purpose. A reasonably accurate minimum stress can be obtained if the ISIP break can be clearly identified. This point may be ambiguous if filtrate leakoff is excessive, and the best results are expected in rocks with permeabilities of less than 1 md. The maximum-horizontal-stress computation for vertical fractures, however, is based on some generally unsupported assumptions and should not be considered as more than a gross estimate.

Example 3.22. The microhydraulic fracturing data in **Fig. 3.57** is for Rollins sandstone in a well near Rifle, Colorado. The strip-chart diagram represents the first, second, and fifth runs. These are cased-hole tests in the interval between 7,530 and 7,532 ft, and the pore pressure is 6,000 psig. Estimate the sandstone's tensile strength and the two principal horizontal stresses.

Solution. From the chart, p_{bd} = 7,125 psig and p_{ro} = 7,020 psi. Assuming no fluid penetration, the estimated tensile strength is obtained from Eq. 3.105 as

$$\sigma_{ts} = 7,125 - 7,020 = 105 \text{ psi}.$$

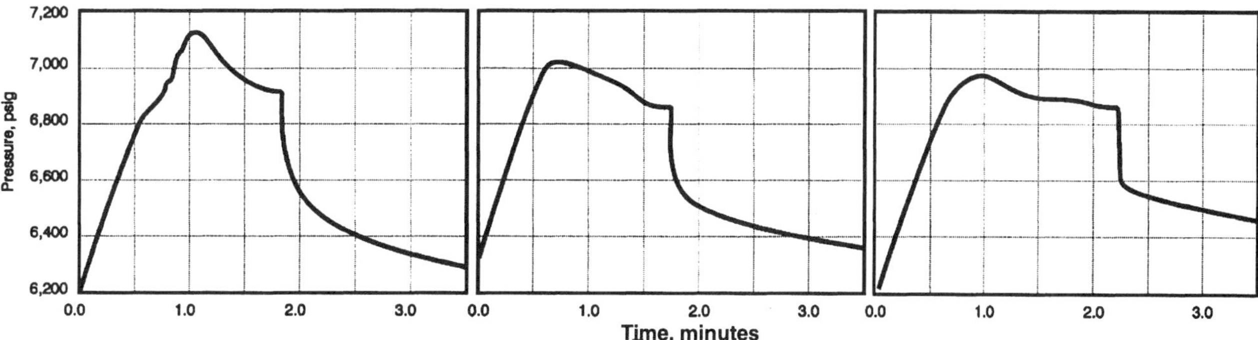

Fig. 3.57—Microhydraulic fracturing test data for the Rollins sandstone described in Example 3.22.[69]

TABLE 3.10—STRESS AND STRAIN MEASUREMENTS FOR A SANDSTONE CORE

Axial stress (psi)	Axial strain (in./in.)	Lateral strain (in./in.)
Zero Confining Stress		
1,000	0.00130	−0.00008
2,000	0.00222	−0.00018
3,000	0.00290	−0.00022
4,000	0.00342	−0.00034
5,000	0.00388	−0.00040
6,000	0.00422	−0.00048
7,000	0.00450	−0.00058
8,000	0.00474	−0.00068
9,000	0.00502	−0.00082
10,000	0.00530	−0.00100
11,000	0.00564	−0.00124
12,000	0.00602	−0.00158
13,000	0.00652	−0.00234
13,500	0.00680	−0.00304
1,450-psi Confining Stress		
1,000	0.00092	−0.00004
2,000	0.00154	−0.00008
3,000	0.00200	−0.00012
4,000	0.00236	−0.00018
5,000	0.00266	−0.00022
6,000	0.00296	−0.00026
7,000	0.00324	−0.00032
8,000	0.00352	−0.00036
9,000	0.00378	−0.00040
10,000	0.00404	−0.00044
11,000	0.00428	−0.00048
12,000	0.00456	−0.00056
13,000	0.00482	−0.00062
14,000	0.00504	−0.00070
15,000	0.00538	−0.00080
16,000	0.00562	−0.00092
17,000	0.00592	−0.00104
18,000	0.00624	−0.00124
19,000	0.00662	−0.00152
20,000	0.00704	−0.00186
21,000	0.00764	−0.00242
21,750	0.00880	−0.00330

The ISIP cannot be ascertained definitely until the Run 5 pressure decline, where its value is determined from the chart as 6,590 psig. Thus,

$$p_{ISIP} \approx \sigma_{H2} = 6,590 \text{ psig.}$$

Fracture-initiation-pressure decay from one run to the next tells us that rock weakening is occurring because of local pore-pressure buildup. The tensile-strength estimate is in doubt, and the σ_{H1} calculation will be suspect. Nonetheless, apply Eq. 3.106 for demonstration purposes and obtain

$$\sigma_{H1} = 3(6,590) - 7,020 - 6,000 = 6,750 \text{ psig.}$$

Problems

3.1 Refer to Fig. 3.2, the aluminum stress/strain diagram.

1. Determine Young's modulus of elasticity.
2. At what stress is the elastic limit achieved?
3. Determine the tensile strength.
4. What is the permanent deformation of the material at the breaking point?
5. Determine the transverse strain at 40,000-psi applied stress if Poisson's ratio is 0.33.

3.2 The stress and strain data shown in **Table 3.10** represent the curves shown in the text as Fig. 3.5. Prepare a plot similar to the diagram and estimate the following parameters.

1. Determine the tangent moduli, secant moduli, and Poisson's ratios for each confining stress at the 2,000-psi axial-stress condition.
2. Determine the tangent moduli, secant moduli, and Poisson's ratios for each confining stress at the 8,000-psi axial-stress condition.
3. Determine the elastic limit for each of the above confining stresses.

3.3 Compression and shear-wave travel-time measurements on a shale are 105 and 227 μsec/ft, respectively. The shale's bulk density is 2.20 g/cm^3.

1. Determine the dynamic Poisson's ratio.
2. Determine the dynamic Young's modulus.

3.4 Would you expect the dynamic Poisson's ratio to be higher or lower than the static value? Defend your answer.

3.5 Using geometry, derive the expression shown as Eq. 3.13.

3.6 Work Example 3.3, except assume uniaxial test conditions (σ_3 = zero).

3.7 Work Example 3.3, except use a 30° angle of internal friction and zero cohesion.

3.8 Construct the Mohr-Coulomb failure envelope for a rock having finite cohesion strength but zero angle of internal friction. In what type of rock would you expect to see this behavior?

3.9 Incorporate the poroelasticity constant in Eq. 3.16 and assume it is less than unity. Does this raise or lower the predicted horizontal stress?

3.10 Refer to the lithological sequence shown in **Table 3.11.** Use an overbuden gradient of 1.0 psi/ft and assume tectonic activity has created an incremental, consistent horizontal stress of 200 psi. Plot the pore pressures and horizontal stresses as functions of depth.

3.11 Work Example 3.4 except assume that both the pore-pressure and wellbore-pressure gradients are equivalent to a 15.0-lbm/gal mud.

TABLE 3.11—PORE PRESSURE AND POISSON'S RATIOS FOR PROBLEM 3.10

Interval (ft)	Poisson's Ratio (fraction)	Pore Pressure (psi/ft)
0 to 1,500	0.20	0.465
1,500 to 2,200	0.35	0.465
2,200 to 2,700	0.18	0.465
2,700 to 4,500	0.44	0.465
4,500 to 4,600	0.23	0.2
4,600 to 5,700	0.33	0.465
5,700 to 6,000	0.25	0.465
6,000 to 7,000	0.40	0.857
7,000 to 7,200	0.21	0.753
7,200 to 8,000	0.35	0.753

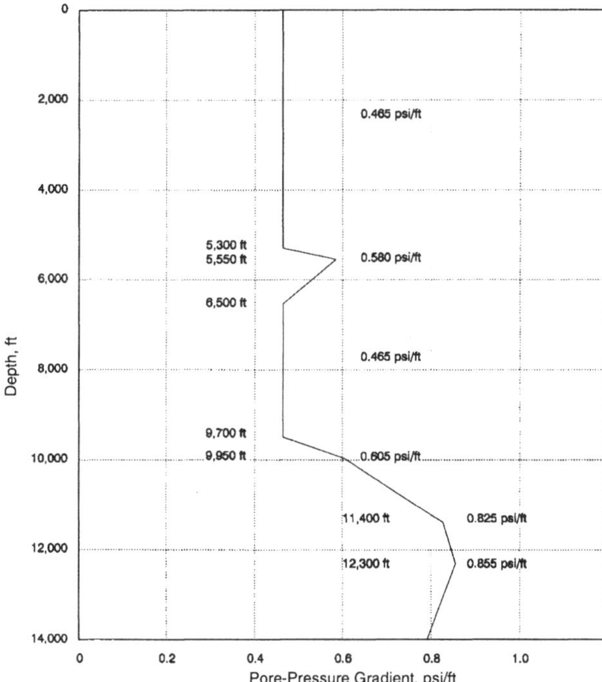

Fig. 3.58—Predicted pore-pressure gradients for the south Texas well described in Problem 3.25.[47]

3.12 Generate a dimensionless plot of effective tangential stress vs. radial distance if the in-situ horizontal stresses are equal. Do the same for wellbore positions of 0 and $\pi/2$ radians if the maximum effective horizontal stress is three times larger than the minimum.

3.13 Estimate the fracture-initiation and -propagation pressures for the conditions described in Part 2 of Example 3.4. Incorporate a 200-psi tensile strength.

3.14 Estimate the fracture-propagation gradient for the two depths described in Example 3.5.

3.15 Plot fracture-propagation gradient in equivalent-mud density vs. water depth for D_{sw} between 50 and 5,000 ft if $D = 7{,}0000$ ft; $D_a = 70$ ft; $g_{sw} = 0.44$ psi/ft; $g_s = 0.95$ psi/ft; $\mu = 0.25$; and$g_p = 0.465$ psi/ft.

3.16 Assume that in-situ stress σ_2 is equal to σ_3. Prove that the minor wellbore stress defined by Eq. 3.51 reaches a minimum when the borehole position angle θ is zero. Hint: substitute terms, differentiate with respect to θ, and set the result equal to zero.

3.17 Relative to the borehole axis, determine the theoretical fracture-plane-initiation angle for the well described in Example 3.6.

3.18 Reconsider the well described in Example 3.4 and plot fracture-initiation gradient vs. wellbore deviation from vertical. Assume zero tensile strength and isotropic stresses. Facilitate these calculations by using a programmable calculator or spreadsheet program.

3.19 Estimate the fracture gradient for a shallow, extended-reach well if $D = 2{,}300$ ft (true vertical); $\sigma_{ob} = 2{,}254$ psig; $\sigma_{H1} = 2{,}530$ psig; $\sigma_{H2} = 2{,}029$ psig; $p_p = 1{,}035$ psig; and $\mu = 0.19$. The inclination from vertical is 75°, the hole azimuth is N45E, and σ_{H1} acts in the N30W direction.

3.20 Would a high-filtrate or a low-filtrate mud be best suited for drilling in an area prone to compressive hole failures? Defend your answer.

3.21 Estimate the fracture-initiation gradient for Part 4 in Example 3.4 if the formation is permeable and the drilling fluid is clear water.

3.22 Of the following condition pairs, select the one for which you would expect lost circulation to be more of a problem.

1. Shallow or deep rock?
2. Active or relaxed tectonic region?
3. Vertical or inclined wellbore?
4. Large or small hole diameter?
5. Weighted or nonweighted mud?
6. Offshore or land operation?
7. Clear-water drilling fluid or an all-oil mud?
8. High-permeability or tight sand?
9. Young, plastic sediment or old, brittle rock?
10. Initial field development or future infill work?

3.23 Estimate the change in the fracture gradient for the well described in Example 3.4 if the sandface is cooled by circulating mud from 170 to 155°F. Assume E and b are 6.0×10^6 psi and 6.0×10^{-6} in./(in.-°F), respectively.

3.24 Derive an expression for fracture-initiation pressure based on the Hubbert and Willis[45] extension-pressure relation.

3.25 **Fig. 3.58** shows the predicted pore-pressure gradients for a prospect in the south Texas gulf coast. Plot the predicted fracture gradients as function of depth using the matrix-stress ratio correlations developed by

1. Matthews and Kelly.[46]
2. Pennebaker.[48]
3. Eaton.[47]
4. Pilkington.[51]

Use the 6,000-ft overburden curve in the Pennebaker computation.

3.26 The pore-pressure equivalent at 5,408 ft in the Green Canyon well described in Example 3.13 is 9.8 lbm/gal.

1. Estimate the fracture gradient using Constant and Bourgoyne's[52,53] technique. Use the coefficients that fit Eaton's Poisson's ratio curve.
2. What is the prediction if you use an a coefficient of 0.1 instead of 0.629?

3.27 Surface casing is to be set at 6,560 ft KB in a Gulf of Mexico well. The water depth is 2,000 ft, the rig air gap is 85 ft, and the pore pressure is normal. Estimate the RKB-fracture gradient at the surface-casing shoe using Simmons and Rau's[54] modification of the Eaton[47] technique (from Ref. 54).

3.28 The pore-pressure-gradient equivalent for a well in western Oklahoma is 15.9 lbm/gal at 10,208 ft. Estimate the fracture gradient using Zamora's[55] correlation.

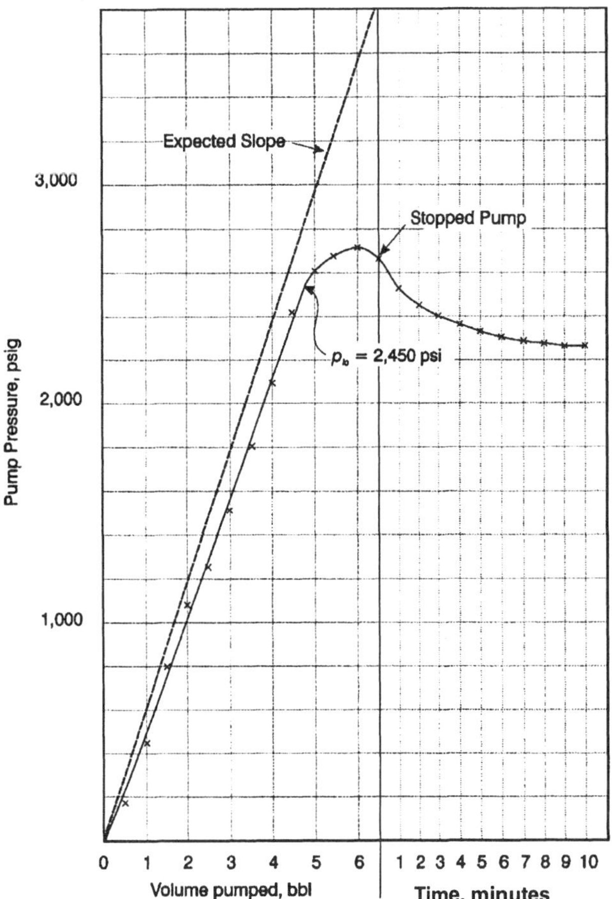

Fig. 3.59—Leakoff-test data for the well described in Problem 3.37.[66]

3.29 Estimate the fracture gradient for the conditions described in Problem 3.27 using the Brennan and Annis[56] method.

3.30 Use the appropriate equation in Table 3.7 to estimate the fracture-initiation and -propagation gradients for a 12.0-lbm/gal equivalent formation at 9,800 ft if the well is located in Brunei. Assume normal pore pressure is equivalent to an 8.5-lbm/gal mud.

3.31 The following data are obtained in a gulf coast well: $D = 11{,}522$ ft; $g_p = 0.685$ psi/ft; $g_{ob} = 0.950$ psi/ft; $\phi_D = 0.285$; and $I_{sh} = 0.087$. Estimate the fracture gradient using the Anderson *et al.*[59] correlation.

3.32 The following data correspond to a porous limestone in an offshore wildcat[60]: $D = 5{,}423$ ft; $\sigma_{ob} = 4{,}531$ psig; $p_p = 2{,}438$ psig; and $p_{fi} = 3{,}891$ psig. Use the Daines[60] approach to estimate the fracture pressure in a dry shale at 11,034 ft if the overburden and pore pressure are 10,590 and 5,629 psig, respectively.

3.33 Which is more sensitive to pore pressure, the Aadnoy and Larsen[62] technique or any one of the matrix-stress-ratio methods?

3.34 Use Aadnoy and Larsen's[62] technique and plot fracture gradient as a function of hole inclination (0 to 90°) for the sandstone described in Example 3.19.

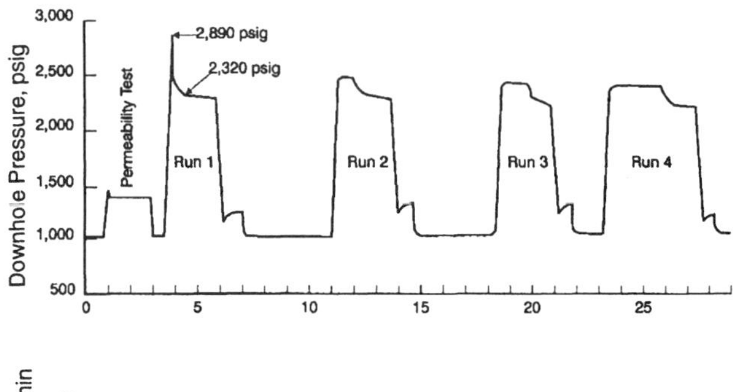

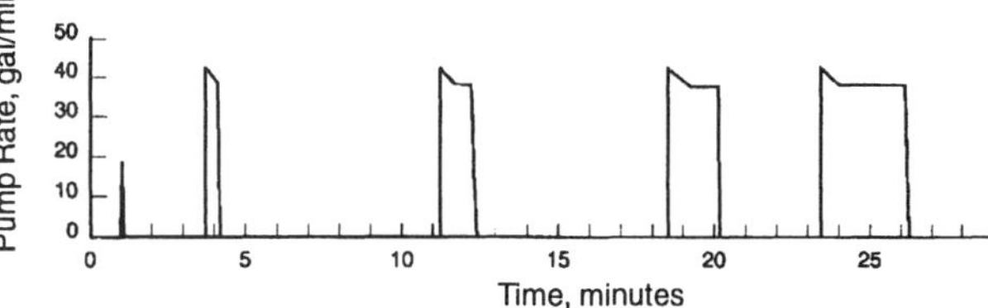

Fig. 3.60—Microhydraulic fracturing test data for Wilkins shale described in Problem 3.42.[70]

3.35 Estimate the porosity and fracture gradient for the conditions discussed in Example 3.20 if the rock is a clean sandstone. Use the Holbrook *et al.*[63] compaction table and fracture-pressure correlation.

3.36 Give two reasons why the actual leakoff-test slope in Fig. 3.54 is shallower than the computed slope.

3.37 Intermediate casing is set at 10,000 ft and the hole is subsequently deepened to 10,030 ft. **Fig. 3.59** shows the results of a leakoff test. Estimate the fracture gradient if the mud density at the time of the test is 13.0 lbm/gal.

3.38 For Problem 3.37, determine the test pressure if the objective is to go no higher than a 16.0-lbm/gal equivalent.

3.39 The mud density in the well discussed in Problem 3.37 is gradually increased to 15.0 lbm/gal as drilling progresses.

1. How much backpressure can be held safely on the annulus in the event of a well-control problem? Assume that the hole above the shoe is full of mud.
2. How much backpressure can be tolerated if a 1,000-ft gas column having a 0.1-psi/ft gradient is above the shoe?

3.40 Surface casing has been set at 200 ft in a land drilling operation. After drillout with an 8.8-lbm/gal mud, a shoe test is conducted with indicated leakoff at 100 psig. The RKB is 20 ft above the pump pressure gauge.

1. Determine the fracture gradient.
2. What maximum casing pressure can be tolerated with a 9.2-lbm/gal mud in the hole?

3.41 Casing is set 800 ft below the mudline in an offshore drilling operation. Water depth is 300 ft, and the air-gap distance is 100 ft.

1. Determine the fracture gradient at the casing seat if the mud density at the time of the test is 9.6 lbm/gal and the pump pressure at leakoff is 150 psig.
2. Recalculate the fracture gradient for the same parameters except reduce the air-gap distance to 50 ft.

3.42 **Fig. 3.60** represents a series of microhydraulic fracturing tests in a Wilkins shale section at 2,375 ft.

1. Estimate the shale's tensile strength and minimum principal stress.

2. What does your result lead you to suspect concerning fracture orientation?

Nomenclature

a = coefficient in the Constant and Bourgoyne stress-ratio equation, dimensionless
A = area, in.^{2}
b = coefficient of thermal expansion, T, in./(in.-°F)
b_C = Aadnoy and Larsen[62] C coefficient relation intercept, psi/ft
B = parameter in Zamora's[55] overburden stress equation, dimensionless
c = cohesion, psi
c_b = bulk compressibility, psi^{-1}
c_m = mud compressibility, psi^{-1}
c_{ma} = matrix compressibility, psi^{-1}
c_o = oil compressibility, psi^{-1}
c_s = solids compressibility, psi^{-1}
c_w = water compressibility, psi^{-1}
C = correlation coefficient, psi/ft
C_{sd} = correlation coefficient for sandstone, psi/ft
C_{sh} = correlation coefficient for shale, m/t^2, psi/ft
d = diameter, in.
D = depth, ft
D_a = air gap, ft
D_{eq} = equivalent depth, ft
D_s = sediment depth, ft
D_{se} = effective sediment depth, ft
$D_{s(eq)}$ = equivalent sediment depth of seawater, ft
D_{sw} = water depth, ft
E = Young's modulus of elasticity, psi
E_i = initial modulus, psi
E_s = secant modulus, psi
E_t = tangent modulus, psi
f_{Vo} = oil-volume fraction, dimensionless
f_{Vs} = solids-volume fraction, dimensionless
f_{Vw} = water-volume fraction, dimensionless
F_a = axial force, lbf
F_σ = matrix-stress ratio, dimensionless
g_{fi} = fracture-initiation gradient, psi/ft
$g_{fi(0)}$ = fracture-initiation gradient in a vertical wellbore, psi/ft
$g_{fi(\gamma)}$ = fracture-initiation gradient in an inclined wellbore, psi/ft
g_{fp} = fracture-propagation gradient, psi/ft
g_H = horizontal-stress gradient, psi/ft
$(g_H)_{cor}$ = corrected horizontal-stress gradient, psi/ft
g_m = mud hydrostatic gradient, psi/ft
g_n = normal-pore-pressure gradient, psi/ft
g_{ob} = overburden-stress gradient, psi/ft
g_p = pore-pressure gradient, psi/ft
g_p^* = pore-pressure-gradient constant in Aadnoy and Larsen's[62] relation, psi/ft
g_s = sediment gradient, psi/ft
g_{sw} = seawater hydrostatic gradient, psi/ft
I_{sh} = shale index, dimensionless
K_F = matrix-stress-ratio function constant, ft^{-1}
K_ϕ = porosity decline constant, ft^{-1}
L = length, in.
m_C = Aadnoy and Larsen[62] C coefficient relation slope, dimensionless
M = parameter in Zamora's[55] matrix-stress-ratio equation, dimensionless
p_{bd} = breakdown pressure, psi
Δp_f = pressure drop in the drillstring and bit, psi
p_{fi} = fracture-initiation pressure, psi
p_{fp} = fracture-propagation pressure, psi
p_{ISIP} = initial shut-in pressure, psi
p_{lo} = leakoff pressure, psi
p_n = normal pore pressure, psi
p_p = pore pressure, psi
p_{ro} = fracture reopening pressure, psi
p_{sw} = seawater hydrostatic pressure, ft
p_w = wellbore pressure, psi
r = radius, ft
r_w = wellbore radius, ft
s = poroelasticity constant, dimensionless
Δt_c = compressive transit time, μs/ft
Δt_s = shear transit time, μs/ft
T = temperature, °F
v_c = compressional acoustic velocity, ft/sec
v_s = shear acoustic velocity, ft/sec
V_m = mud volume, bbl
α = plane angle, degree
β = hole angle from the intermediate-principal-stress direction, degree
γ = hole angle from the maximum-principal-stress direction, degree
ϵ_a = normal strain in the axial direction, in./in.
ϵ_H = horizontal strain, in./in.
ϵ_p = plastic deformation, in./in.
ϵ_{tr} = normal strain in the transverse direction, in./in.
ϵ_x = strain in the x direction, in./in.
ϵ_y = strain in the y direction, in./in.
θ = angle from the center of a borehole, degree
κ = ratio of tectonic stress to effective vertical stress, dimensionless
μ = Poisson's ratio, dimensionless
μ_A = Anderson *et al.*'s correlated Poisson's ratio, dimensionless
μ_E = Eaton's correlated Poisson's ratio, dimensionless
ξ = angle between wellbore axis and fracture, degree
ρ_b = bulk density, g/cm^3
ρ_f = pore-fluid density, lbm/gal
ρ_{ma} = matrix density, lbm/gal
ρ_p = equivalent density of the pore-pressure gradient, lbm/gal
ρ_{sw} = seawater density, lbm/gal
σ = normal stress, psi
σ_1 = maximum principal stress, psi
σ_2 = intermediate principal stress, psi
σ_3 = minimum principal stress, psi
σ_1' = maximum principal stress in an inclined wellbore, psi
σ_2' = intermediate principal stress in an inclined wellbore, psi
σ_3' = least principal stress in an inclined wellbore, psi
σ_{e3}' = least effective principal stress at the wellbore, psi
σ_a = axial stress, psi
σ_e = effective normal stress, psi
σ_f = normal stress on incipient failure plane, psi

σ_H = horizontal stress, psi
σ_{H1} = maximum horizontal principal stress, psi
σ_{H2} = minimum horizontal principal stress, psi
σ_{He} = effective horizontal stress, psi
σ_{He1} = maximum effective horizontal stress, psi
σ_{He2} = minimum effective horizontal stress, psi
σ_{ob} = overburden stress, psi
σ_r = radial stress, psi
σ_{re} = effective radial stress, psi
σ_s = sediment overburden stress, psi
σ_t = tangential stress, psi
σ_{te} = effective tangential stress, psi
$\Delta\sigma_{te(q)}$ = change in effective tangential stress resulting from fluid flow, psi
$\Delta\sigma_{te(T)}$ = change in effective tangential stress resulting in a temperature change, psi
σ_{tec} = tectonic stress, psi
σ_{ts} = tensile strength, psi
σ_V = vertical stress, psi
σ_{Ve} = effective vertical stress, psi
σ_x = normal stress in the x orthogonal direction, psi
σ_y = normal stress in the y orthogonal direction, psi
σ_z = normal stress in the z orthogonal direction, psi
σ_{zz} = induced stress parallel to the borehole axis, psi
σ_α = normal stress on an arbitrary plane, psi
τ = shear stress, psi
τ_f = shear stress on incipient failure plane, psi
τ_{rt} = shear stress normal to the radial stress direction, psi
τ_{tzz} = induced shear stress on the t, zz surface, psi
τ_α = shear stress on an arbitrary plane, psi
τ_{xy} = shear stress on the x, y element surface, psi
τ_{xz} = shear stress on the x, z element surface, psi
τ_{yz} = shear stress on the y, z element surface, psi
ϕ = porosity, dimensionless
ϕ_D = porosity from a density log, dimensionless
ϕ_S = porosity from a sonic log, dimensionless
ϕ_{ml} = porosity at the mudline, dimensionless
ω = angle of internal friction, degree

References

1. Byars, E.F. and Snyder, R.D.: *Engineering Mechanics of Deformable Bodies*, Intext Educational Publishers, New York City (1975) 68.
2. Gidley, J.L. *et al*: *Recent Advances in Hydraulic Fracturing*, Monograph Series, SPE, Richardson, Texas (1989) **12,** 62–63.
3. Jumikis, A.R.: *Rock Mechanics*, second edition, Gulf Publishing Co., Houston, Texas (1983) 187.
4. *The Fracbook Design/Data Manual*, Halliburton Services, Duncan, Oklahoma (1971) 55–59.
5. Clark, S.P.: *Handbook of Physical Constants*, Geological Soc. of America, Inc., New York City (1966) 100.
6. Lama, R.D. and Vutukuri, V.S.: *Handbook on Mechanical Properties of Rocks*, Trans Tech Publications, Clausthal, Germany (1978) **2.**
7. Terzaghi, K.: *Theoretical Soil Mechanics*, John Wiley and Sons Inc., New York City (1943) 51.
8. Biot, M.A.: "General Theory of Three Dimensional Consolidation," *J. Applied Physics* (1941) **12,** 155.
9. Roegiers, J.: "Elements of Rock Mechanics," *Reservoir Stimulation*, second edition, M.J. Economides and K.G. Nolte (eds.), Prentice Hall Inc., Englewood Cliffs, New Jersey (1989) **2,** 5, 13–14.
10. Geertsma, J.: "The Effect of Fluid Pressure Decline on Volumetric Changes of Porous Rocks," *Trans.*, AIME (1957) **210,** 105.
11. Johnson, D.: "Evaluating SH with Biot's Poroelastic Theory," *The Technical Review*, Schlumberger Educational Services, (October 1986) 24–25.
12. Ward, C.D. and Holbrook, P.W.: "Author's Reply to Discussion of Brief: Pore- and Fracture-Pressure Determinations: Effective-Stress Approach," *JPT* (October 1995) 914.
13. Mohr, O.C.: *Abhandlungen aus dem Gebiete der technischen Mechanik*, second edition, W. Ernst & Sohn, Berlin, Germany (1914) 192–235.
14. Coulomb, C.A.: "Essai sur une application des régles de maximis et minimis á quelques problémes de statique, relatifs á l'architecture," *Mémoires de mathématique et de physique par divers savans*, Académie royale des sciences, Paris, (1776) **7,** 343-82.
15. Heim, A.: "Zur Frage der Gebris und Gesteinsfestigkeit," *Schweirzerische Bauzeitung* (February 1912).
16. Prats, M.: "Effect of Burial History on the Subsurface Horizontal Stresses of Formations Having Different Material Properties," *SPEJ* (December 1981) 658.
17. Kirsch, G.: "Die Theorie der Elastizität und die Bedürfnisse der Festigkeitslehre," *Zeitschrift des Vereines Deutscher Ingenieure* (1898) **42,** 797.
18. Timoshenko, S.P. and Goodier, J.N.: *Theory of Elasticity*, third edition, McGraw-Hill Book Co. Inc., New York City (1961).
19. Miles, A.J. and Topping, A.D.: "Stresses Around a Deep Well," *Trans.*, AIME (1949) **179,** 186.
20. Deily, F.H. and Owens, T.C.: "Stress Around a Wellbore," *Drilling*, Reprint Series, SPE, Richardson, Texas (1987) **22,** 126.
21. Cheatham, J.B. Jr.: "Wellbore Stability," *JPT* (June 1984) 889.
22. Gidley, J.L. *et al*: *Recent Advances in Hydraulic Fracturing*, Monograph Series, SPE, Richardson, Texas (1989) **12,** 66.
23. *International Petroleum Encyclopedia*, PennWell Publishing Co., Tulsa, Oklahoma (1998) 324–25.
24. Aadnoy, B.S. and Chenevert, M.E.: "Stability of Highly Inclined Boreholes," *SPEDE* (December 1987) 364.
25. McLennan, J.D., Roegiers, J., and Economides, M.J.: "Extended Reach and Horizontal Wells," *Reservoir Stimulation*, second edition, M.J. Economides and K.G. Nolte (eds.), Prentice Hall Inc., Englewood Cliffs, New Jersey (1989) Chap. 19, 7–10.
26. Daneshy, A.A.: "A Study of Inclined Hydraulic Fractures," *SPEJ* (April 1973) 61.
27. Aadnoy, B.S.: "Modeling of the Stability of Highly Inclined Boreholes in Anisotropic Rock Formations," paper SPE 16526 presented at the 1987 SPE Offshore Europe Conference, Aberdeen, 8–11 September.
28. Bradley, W.B.: "Failure of Inclined Boreholes," *J. Energy Res. Tech.* (December 1979) **101,** 232.
29. French, F.R. and McLean, M.R.: "Development Drilling Problems in High-Pressure Reservoirs," *JPT* (August 1993) 772.
30. Ewy, R.T., Myer, L.R., and Cook, N.G.W.: "Investigation of Stress-Induced Borehole Enlargement Mechanisms by a Liquid-Metal Saturation Technique," *SPEDC* (March 1994) 65.
31. Wong, S-W., Veeken, C.A.M., and Kenter, C.J.: "The Rock-Mechanical Aspects of Drilling a North Sea Horizontal Well," *SPEDC* (March 1994) 47.
32. Gnirk, P.F.: "The Mechanical Behavior of Uncased Wellbores in Elastic/Plastic Media Under Hydrostatic Stress," *SPEJ* (February 1972) 49.
33. Veeken, C.A.M. *et al.*: "Use of Plasticity Models for Predicting Borehole Stability," *Proc.*, Intl. Symposium on Rock at Great Depth, Pau (1989) **2,** 835.
34. Detournay, E. and Cheng, A.H-D.: "Poroelastic Response of a Borehole in a Non-Hydrostatic Stress Field," *Intl. J. Rock Mech. Min. Sci. and Geomech. Abstr.* (1988) **25,** 171.
35. Mouchet, J.P. and Mitchell, A.: *Abnormal Pressures While Drilling*, Elf Aquitane Manuels Techniques 2, Broussens, France (1989) 226.
36. Moore, P.L.: *Drilling Practices Manual*, second edition, PennWell Publishing Co., Tulsa, Oklahoma (1986) 115–16.

37. Hoberock, L.L., Thomas, D.C., and Nickens, H.V.: "Here's How Compressibility and Temperature Affect Bottom-Hole Mud Pressure," *Oil and Gas J.* (22 March 1982).
38. Peters, E.J., Chenevert, M.E., and Zhang, C.: "A Model for Predicting the Density of Oil-Based Muds at High Pressures and Temperatures," *SPEDE* (June 1990) 141.
39. Galate, J.W. and Mitchell, R.F.: "Behavior of Oil Muds During Drilling Operations," *SPEDE* (April 1986) 97.
40. Onyia, E.C.: "Experimental Data Analysis of Lost-Circulation Problems During Drilling With Oil-Based Mud," *SPEDC* (March 1994) 25.
41. Morita, N., Black, A.D., and Fuh, G-F.: "Theory of Lost Circulation Pressure," paper SPE 20409 presented at the 1990 SPE Annual Technical Conference and Exhibition, New Orleans, 23–26 September.
42. Haimson, B. and Fairhurst, C.: "Hydraulic Fracturing in Porous-Permeable Materials," *JPT* (July 1969) 811.
43. Simpson, J.P.: "A New Approach to Oil-Base Muds for Lower-Cost Drilling," *JPT* (May 1979) 643.
44. Griffith, A.A.: "The Phenomenon of Rupture and Flow in Solids," *Phil. Trans. Royal Soc. of London* (1920) **A, 221,** 163.
45. Hubbert, M.K. and Willis, D.G.: "Mechanics of Hydraulic Fracturing," *Trans.*, AIME (1957) **210,** 153.
46. Matthews, W.R. and Kelly, J.: "How to Predict Formation Pressure and Fracture Gradient from Electric and Sonic Logs," *Oil & Gas J.* (20 February 1967).
47. Eaton, B.A: "Fracture Gradient Prediction and Its Application in Oilfield Operation," *JPT* (October 1969) 1353.
48. Pennebaker, E.S.: "An Engineering Interpretation of Seismic Data," paper SPE 2165 presented at the 1968 SPE Annual Fall Meeting, Houston, 29 September–2 October.
49. Mitchell, B.J.: *Advanced Oilwell Drilling Engineering Handbook*, ninth edition, SPE, Richardson, Texas (September 1992) 180.
50. Christman, S.A.: "Offshore Fracture Gradients," *JPT* (August 1973) 910.
51. Pilkington, P.E.: "Fracture Gradient Estimates in Tertiary Basins," *Pet. Eng. Intl.* (May 1978) 138.
52. Constant, W.D. and Bourgoyne, A.T. Jr.: "Method Predicts Frac Gradient for Abnormally Pressured Formations," *Pet. Eng. Intl.* (January 1986) 38.
53. Constant, W.D. and Bourgoyne, A.T. Jr.: "Fracture-Gradient Prediction for Offshore Wells," *SPEDE* (June 1988) 136.
54. Simmons, E.L. and Rau, W.E.: "Predicting Deepwater Fracture Pressures: A Proposal," paper SPE 18025 presented at the1988 SPE Annual Technical Conference and Exhibition, Houston, 2–5 October.
55. Zamora, M.: "New Method Predicts Gradient Fracture," *Pet. Eng. Intl.* (September 1989) 38.
56. Brennan, R.M. and Annis, M.R.: "A New Fracture Gradient Prediction Technique That Shows Good Results in Gulf of Mexico Abnormal Pressure," paper SPE 13210 presented at the 1984 SPE Annual Technical Conference and Exhibition, Houston, 16–19 September.
57. Breckels, I.M. and van Eekelen, H.A.M.: "Relationship Between Horizontal Stress and Depth in Sedimentary Basins," *JPT* (September 1982) 2191.
58. Vuckovic, B.M.R.: "Prediction of Fracture Gradients Offshore Australia," paper SPE 19468 presented at the 1989 SPE Asia-Pacific Conference in Sydney, Australia, 13–15 September.
59. Anderson, R.A., Ingram, D.S., and Zanier, A.M.: "Determining Fracture Pressure Gradients From Well Logs," *JPT* (November 1973) 1259.
60. Daines, S.R.: "Prediction of Fracture Pressures for Wildcat Wells," *JPT* (April 1982) 863.
61. Weurker, R.G.: "Annotated Tables of Strength and Elastic Properties of Rocks," *Drilling*, Reprint Series, SPE, Dallas (1963) **6,** 23.
62. Aadnoy, B.S. and Larsen, K.: "Method for Fracture-Gradient Prediction for Vertical and Inclined Boreholes," *SPEDE* (June 1989) 99.
63. Holbrook, P.W., Maggiori, D.A., and Hensley, R.: "Real-Time Pore Pressure and Fracture Gradient Evaluation in All Sedimentary Lithologies," paper SPE 26791 presented at the 1993 SPE Offshore European Conference, Aberdeen, 7–10 September.
64. "Regulation 250.54(a)(6) Governing Oil and Gas and Sulphur Operations in the Outer Continental Shelf," U.S. Dept. of Interior, Minerals Management Service, *Federal Register* (1 April 1988) **53, 63,** 10717.
65. Bourgoyne, A.T. Jr. *et al.*: *Applied Drilling Engineering*, second printing, Textbook Series, SPE, Richardson, Texas (1991) 293.
66. Chenevert, M.E. and McClure, L.J.: "How to Run Casing and Open-Hole Presure Tests," *Oil & Gas J.* (6 March 1978) 66.
67. Morita, N., Fuh, G-F., and Boyd, P.A.: "Safety of Casing Shoe Test and Casing Shoe Integrity After Testing," paper SPE 22557 presented at the 1991 SPE Annual Technical Conference and Exhibition, Dallas, 6–9 October.
68. Daneshy, A.A., *et al.*: "In-Situ Stress Measurements During Drilling," *JPT* (August 1986) 891.
69. Warpinski, N.R., Branagan, P., and Wilmer, R.: "In-Situ Stress Mesurements at U.S. DOE's Multiwell Experiment Site, Mesaverde Group, Rifle, Colorado," *JPT* (March 1985) 527.
70. *Reservoir Stimulation,* J.J. Economides and K.G. Nolte (ed.), Prentice Hall, Englewood Cliffs, New Jersey (1989) 2–18.

SI Metric Conversion Factors

bbl	× 1.589 873	E − 01	= m^3
deg	× 1.745 329	E − 02	= rad
ft	× 3.048*	E − 01	= m
ft^{-1}	× 3.280 840	E + 00	= m^{-1}
°F	(°F − 32)/1.8		= °C
gal	× 3.785 412	E + 00	= L
in.	× 2.54*	E + 01	= mm
sq in.	× 6.451 6*	E + 02	= mm^2
lbf	× 4.448 222	E + 00	= N
lbm	× 4.535 924	E − 01	= kg
psi	× 6.894 757	E − 03	= MPa
psi^{-1}	× 1.450 377	E − 01	= kPa^{-1}

* Conversion factor is exact.

Chapter 4
Kick Detection and Control Methods

4.1 Introduction

The intent of well-control theory and its practical application is to manage formation pressure. Keeping formation fluids out of the wellbore in a drilling operation is assured if the mud-column hydrostatic plus any annular-friction pressure exceeds pore pressure. Primary well control involves the efforts devoted to maintaining sufficient wellbore pressure to prevent a formation-fluid influx. Despite the best intentions, however, formation-fluid flows still occur and may even be anticipated in the well plan. This chapter focuses on secondary well control—i.e., kick detection, containment, and displacement from a well.

4.2 Kick Causes, Detection, and Containment

By convention, a formation-fluid influx is designated as a kick. Not all formation-fluid flows are kicks, however. Small quantities of gas, oil, or salt water may seep into a hole from tight formations while drilling or tripping. A continuous flow from more permeable horizons often can be safely managed with underbalanced drilling procedures and equipment. These flows do not constitute a kick, nor does a flow initiated to determine the productive nature of a potential reservoir in a drillstem test (DST). A kick is defined as any influx that constitutes a well-control emergency. Normally this means using the blowout preventers to shut the well in and subsequently removing the influx using a choke on the annulus to maintain enough backpressure to prevent further entry. Shallow gas kicks cannot be managed in this fashion, but these complexities will be discussed in Chap. 8.

There are three prerequisites for a kick. The borehole pressure must be less than the pore pressure, the formation must be sufficiently permeable such that fluids will flow at significant rate, and the pore fluid must have sufficiently low viscosity that it can flow. Given the last two criteria, those circumstances given in **Table 4.1** can lead to a negative pressure differential and the consequent potential for taking a kick.

Kicks may occur in the drilling phase of a well or in later cased-hole work. As a rule, well control during completion or workover is a much easier task because formation fracturing usually is eliminated from the picture. For this reason, most of the principles discussed in this text are concerned with drilling operations. On a drilling well, kicks may occur while drilling, tripping, or during any of the other procedures associated with the drilling process (logging, fishing, running casing, cementing operations, etc.). Secondary well control is greatly facilitated and fewer complications arise when the drillstring is below the kicking formation, thus it is appropriate to discuss first those kicks that occur while the bit is advancing the hole.

4.2.1 Kicks While Drilling. The classic well-control problem is detecting and displacing a kick that results from drilling with an insufficient mud weight. Published statistics, however, suggest that less than half of all kicks happen in the classical sense, and they may comprise as little as 20% of the total.[1,2] This training focus seems to fit the 80:20 rule—i.e., we as an industry concentrate 80% of our well-control-training effort on 20% of the problems. Still, kicks while drilling constitute a significant percentage of well-control events worldwide. It is important to understand and be able to execute the associated procedures.

Reasons for Taking a Kick. While drilling, the pressure at any point in the annulus is obtained by adding the hydrostatic pressure to the annular-friction losses above the depth of interest. Hydrostatics were discussed in Chap. 1. Appendix A gives the means for estimating wellbore-friction losses. In some special applications, surface backpressure is maintained beneath a rotating head or rotating blowout preventer (RBOP). The additive pressure can be converted to a convenient mud-weight equivalent called the equivalent circulating density (ECD).

Consider the case where a bit cuts into permeable rock having a pore-pressure equivalent density greater than the ECD. Formation fluid will flow and the well has taken a kick. Review the Darcy pseudosteady-state flow equation for gas and identify the controlling variables for kick size and intensity.

$$q_{gsc} = \frac{703 k_g h\left(p_e^2 - p_{wf}^2\right)}{\mu_g T z(\ln r_e/r_w - 0.75)}. \quad \text{(4.1)}$$

TABLE 4.1—PRIMARY REASONS FOR TAKING A KICK
Insufficient wellbore-fluid density
Low drilling- or completion-fluid density
Excessive gas cutting
Wellbore-fluid-column height reduction
Lost circulation because of excess static or dynamic wellbore pressure
Fluid removal because of hole swabbing
Tripping pipe without filling the hole
Excessive swab-friction pressures while moving pipe
Wellbore collision between a drilling and producing well
Cement hydration

TABLE 4.2—CONVENTIONAL KICK INDICATORS IN ORDER OF APPEARANCE AND THEIR RELATIVE IMPORTANCE

Indicator	Significance
Drilling break	Medium
Pump pressure decrease/rate increase	Low
Increase in mud return rate	High
Pit gain	High
Loss of drillstring weight	Low
Gas cutting or salinity change	Low
Flow with the pumps off	High

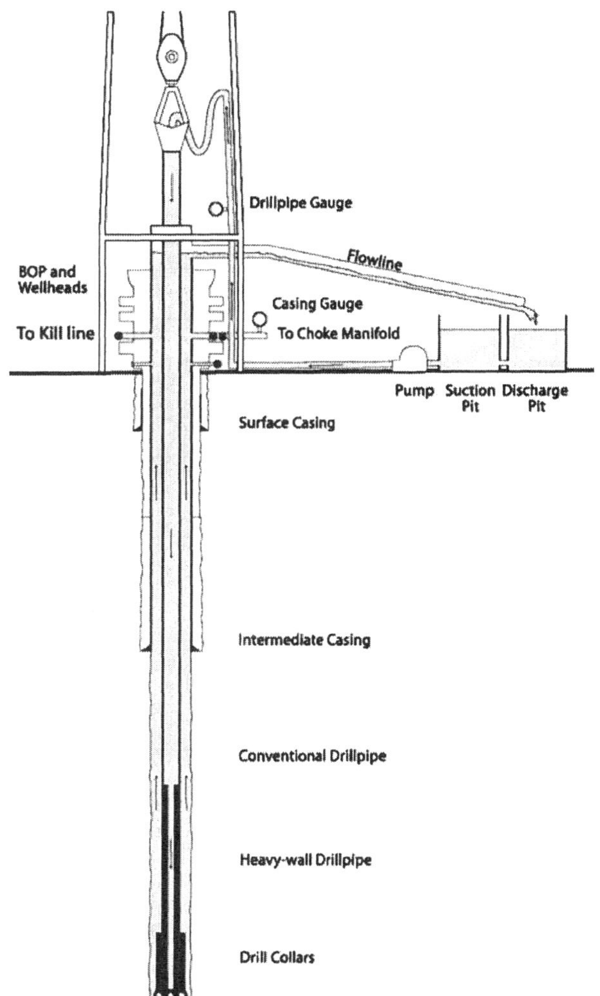

Fig. 4.1—Flow schematic for conventional circulation in a drilling well.

The pressure differential between the formation and wellbore—i.e., the degree of underbalance, is proportional to the influx flow rate and kick volume for a given flow period. The situation can only deteriorate with time because the (typically) less-dense formation-fluid volume further reduces p_{wf} and thereby serves to increase flow rate into the well. Permeability is another significant factor as are exposed thickness and fluid viscosity. We cannot control permeability or reservoir fluid properties, but the amount of exposed rock is governed by how long we continue to drill with kick entry.

Insufficient ECD is the principle cause of kicks while drilling. Fortunately, these events are relatively easy to manage when compared to other reasons given in Table 4.1. Severe lost circulation is one of the more difficult problems. Drilling into a depleted formation, pore-pressure reversal, or vugular rock can lead to a case where "the bottom falls out" and a subsequent loss of hydrostatic head sufficient to permit entry from a shallower formation. Alternatively, increasing the mud weight or annular friction may result in shallower losses and a kick from a deeper horizon. Serious well-control problems resulting from lost circulation generally can be avoided by adequate planning and supervision. Failure to prevent such kicks often leads to an underground blowout.

Finally, drilling into another wellbore has caused some major well-control problems. A surface or underground blowout is usually the outcome because wellbore collisions normally occur at shallow depths where shut-in or producing wellbore pressures are high and deliverability is virtually unrestricted. The risk is associated primarily with platform drilling or drilling in locations which involve a cluster of closely spaced wells. Accurate surveys on the offset wellbores, referenced from a common-depth datum, should prevent these occurrences if incorporated into the directional program.

Kick Detection. Detecting a kick early and limiting its volume by shutting in the well is critical to secondary control, and may make the difference between a manageable situation and one that leads to a loss of control. Thus it is important to have appropriate kick-detection equipment in working order and to have crews alert to the warning signs and trained in the shut-in procedures. **Table 4.2** lists conventional kick indicators in a general order of occurrence and classifies these as to their reliability.

Often, the first indication of a kick is the abrupt increase in penetration rate termed a drilling break. Porous and permeable sandstones (potential reservoir rock) generally drill faster than other strata. Also, flowing formation fluids provide an efficient bottomhole cleaning mechanism and tend to reduce pressure above the bit, thus causing the rock to drill faster. A change in drill rate may signify only a formation change and so drilling breaks are not considered to be a primary kick indicator. Given its timeliness, however, the driller should be especially alert to other, more positive, signals when a drilling break is encountered. In fact, company policy in some areas is to stop drilling and to conduct a flow check before proceeding very far into any drilling break, especially when drilling in a known transition interval.

Immediately following an influx, bottomhole pressure (BHP) is reduced to some extent by lightening of the annular-fluid column (if the mud is heavier than the kick fluids) and the added lift energy given by the formation-fluid flow. The net effect generally leads to an increase in pump rate and a decrease in pump pressure, though the change may be subtle and hard to detect until relatively late in the flow. The lower BHP also may be reflected as buoyancy loss at the weight indicator, but the change in string weight probably would not be detected unless the string is picked up off bottom and, even then, not until a large kick has entered the well.

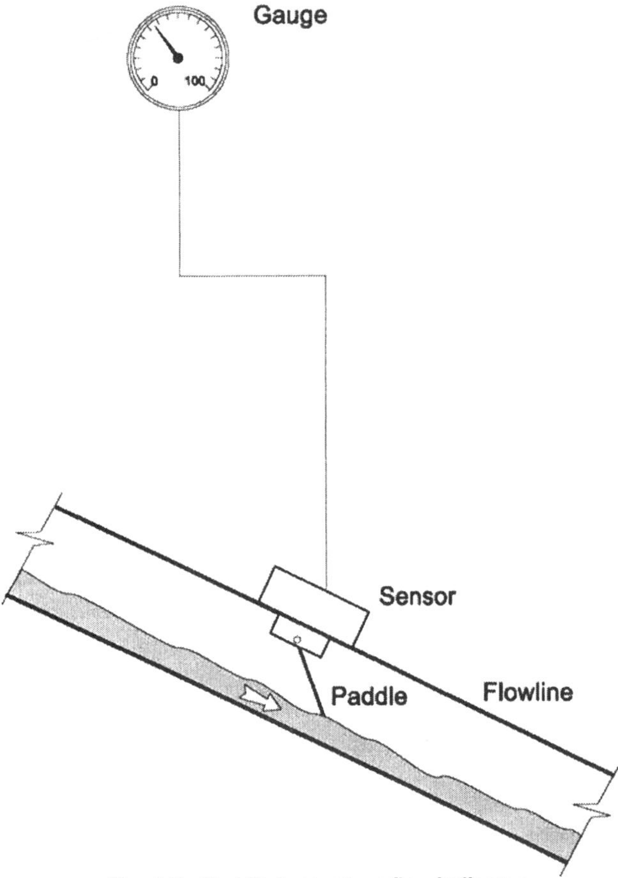

Fig. 4.2—Paddle-type return flow indicator.

Fig. 4.3—Example float and level sensor used in a PVT system.

Fig. 4.1 shows the circulation flow path for a rotary drilling operation. Drilling fluid is pumped from the suction pit and through the high-pressure surface piping, the swivel/kelly or top drive unit, and into the drillstring. The mud traverses the drillstring, passes through the bit, and returns up the annulus where it exits the well through the flowline for cleaning and treatment. While drilling, mud volume must be built continually to replace hole volume generated by the bit and the downhole and surface losses from filtrate seepage, cuttings absorption, and other processes. Over a short time period, however, we can consider normal circulation to be a steady-state process—i.e., what goes in comes out.

A kick violates this balance between well inflow and outflow, therefore an important early kick indicator is an increase in the return flow rate from the well. A flapper device, shown schematically in **Fig. 4.2,** is an old, reliable method for monitoring mud flow returns. Installed in the flowline, the paddle opens or closes in response to the flowline throughput. A sensor detects the paddle position and an electronic or pneumatic signal is relayed to the rig floor where the signal is converted to digital form. The gauge, LCD, or strip chart display gives the driller a continuous record of the return rate as percent of maximum flow (90° paddle position). Once the desired pump rate is achieved, high and low alarms can be set to give immediate warning of either a kick (high) or lost circulation (low).

Following the increase in return rate, mud displaced by the kick results in an increase in the surface mud volume. This volume increase, called a pit gain, usually is assumed to be equivalent to the formation-fluid volume at downhole pressure and temperature if solubility is negligible. An essential component on all but the lowest risk wells is a pit volume totalizer (PVT) system. Floats in each of the pits, similar to the one shown in **Fig. 4.3,** rise or fall in response to a change in the surface mud volume. A signal from each float sensor is passed to the rig floor, converted to digital form, and added to the other float responses. Calibration to the particular pit dimensions gives the display in barrels or cubic meters of surface volume. Thus continuous pit-volume monitoring is available, and alarms can be set to indicate a significant change in either direction. To avoid false alarms, the driller should be notified before transferring mud, jetting the pits, or changing surface volume for any other reason.

Fig. 4.4 shows a standard rental package combining the return flow and PVT instrumentation and alarms into one console. Pump speed in strokes per minute and cumulative strokes can be converted to flow rate and volume, and normally a stroke counter is incorporated into the package. One console is positioned near the driller while additional monitors can be placed at other locations such as the drilling supervisor's quarters.

Gas-cutting or salinity changes in the drilling fluid have been mentioned as warning signs in many well-control training manuals. Visual detection of these contaminants, of course, can be done only when formation fluids surface. For a true kick, this condition implies that most of the mud has been displaced from the annulus, the pits have run over, and a blowout is imminent if not occurring. Though useful in monitoring pore pressure, surface mud contamination is useless for proper kick detection.

Early kick detection and shut-in are important in all wells, and the methods discussed thus far should detect a 10-bbl kick if the equipment is in proper working order and the crews are alert to the indications. Ten-bbl kicks should pose no problem

Fig. 4.4 Example rig-floor console used to monitor return flow, pit volume, and pump strokes.

in most well-control events, though there are situations where detecting small kicks using conventional techniques is difficult, if not impossible. As discussed in Chap. 1, gas solubility in the drilling fluid is one case where pit gain does not reflect kick volume. The movement of floating drilling vessels also creates a condition where small-volume kick detection is beyond the capability of the standard methods.

A 10-bbl kick may be considered a severe kick in some cases. It is not the kick volume, but the height this volume occupies in the annulus that can lead to excessive pressures in a well during a control procedure. Slim holes have a narrow clearance in the annulus and a small kick may equate to an excessive kick height for maintaining pressure control. Small-kick containment is dictated in more conventional wellbores if pore pressure and fracture gradients are convergent.

The need to improve on traditional methods has led to the development and field application of more advanced systems over recent years. Several new-generation kick-detection methods rely on sonic measurements of the annular drilling fluid and the effect of gas on the acoustic-wave-transmission properties of the medium. The speed of sound in a water-based mud (WBM) especially is impacted when free gas is present. For example, a gas fraction of only 2% in a 10.0-lbm/gal WBM at 4,500 psi has been shown to reduce the mud's acoustic velocity from 4,600 to 3,180 ft/s.[3] Some acoustic techniques uses the damping effect of free gas. Gas diminishes the signal strength or amplitude of an oscillating wave in comparison to that of a clean mud.

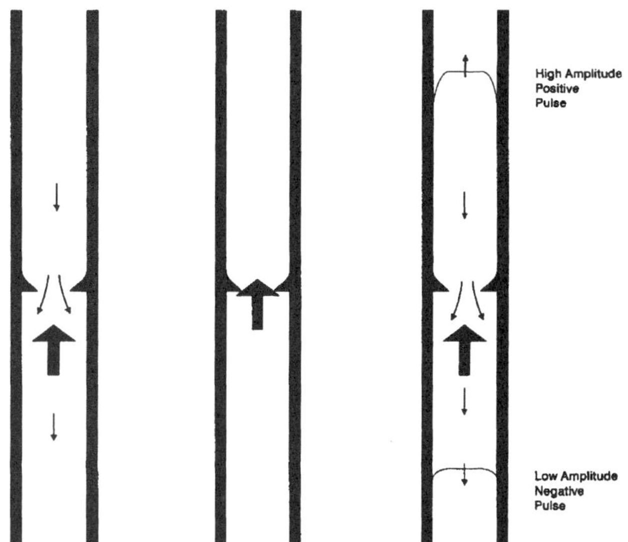

Fig. 4.5—Pressure waves created by an MWD positive pressure pulser.

Measurement-while-drilling (MWD) technology has been applied to the problem of detecting small gas concentrations in the annulus. Mud-pulse-telemetry tools take downhole measurements and convert the data into coded binary signals. Depending on the tool design, the information is transmitted through the mud column in the form of positive, negative, or continuous pressure-wave pulses. At the standpipe, a transducer feeds the raw data to a computer where extraneous noise is filtered and the information retrieved after the signals have been decoded.

Consider the action of the positive pulser shown in **Fig. 4.5.** Positive waveforms are created by the action of the poppet valve against the fixed seat. Pressure waves travel up the drillstring at a velocity which depends on the mud's compressibility and density. At the same time, a negative pulse of much lower intensity is produced and travels up the annulus where it can be detected a short distance below the flowline. Signal amplitude at the standpipe and flowline are reduced by viscous damping of the respective drillstring and annular systems.

The travel time and damping characteristics of both sides should correlate if the transmission media are the same. The kick-detection method described by Bryant *et al.*[4] uses both standpipe and annulus signals to generate a normalized amplitude and phase-angle response that should be characteristic of gas-free circulation. Free gas in the annulus, by reducing acoustic velocity and increasing the damping, results in a phase-angle increase (lag) and amplitude attenuation. Thus, gas kicks may be indicated very early in the flow period.

Other early-warning systems have been devised that monitor travel time through the entire circulation path using pressure pulses created at the surface. The technique discussed by Codazzi *et al.*[5] measures mud-pump pulsations at the standpipe and annulus and computes the change in acoustic travel time through the entire circulation path. As shown in **Fig. 4.6,** a kick is indicated when the travel-time increase reaches a preset alarm threshold. A similar method described by Stokka *et al.*[3] creates a negative pressure pulse at the standpipe with an automated variable choke branched from the standpipe. The generator is set to trigger at specific intervals, say once a minute, and the acoustic travel time for the signals is measured and compared to the known travel time of the gas-free mud.

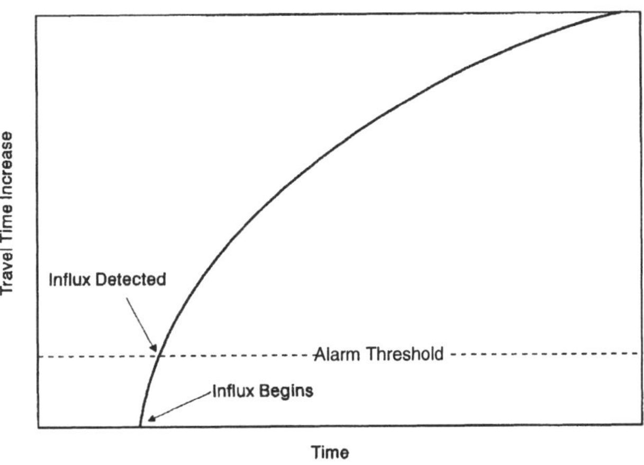

Fig. 4.6—Free-gas detection based on the change in acoustic travel time through a wellbore circulation path.[5]

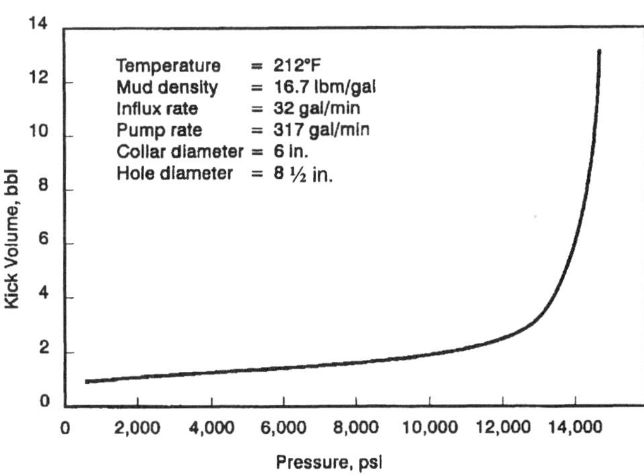

Fig. 4.7—Pit-gain detection limits associated with acoustic methods in a pressured WBM.[7]

The acoustic methods discussed thus far monitor the entire annulus for the presence of free gas. Orban *et al.*[6] discussed local gas kick detection using a conventional MWD data transmission path. In brief, an ultrasonic transmitter and receiver are incorporated in a short sub beneath an MWD tool. Acoustic pulses are transmitted into the annulus, reflected off the borehole wall, and picked up by the receiver where travel time and pulse amplitude are measured. A delay-line placed between the transmitter/receiver and annulus serves several functions, one being to echo a portion of the generated signal. Gas-cutting is evidenced by higher amplitude of the delay-line echo signal as well as by attenuation of the formation signal.

Under ideal circumstances, these surface and downhole acoustic methods can detect free gas concentrations on the order of 1 to 3% and reportedly have detected kicks as small as one barrel. However, the detection capability of these tools currently restricts their application to moderate well depths and WBMs because dissolved gas cannot be discerned readily by these systems.[7] Though solution gas changes a mud's acoustic properties, its effect is not striking and early acoustic-detection capabilities for oil-based muds (OBMs) are not currently available.

The competitive gap also narrows for WBM's in deep, high-pressure/high-temperature (HPHT) wells. At higher pressure, more gas is dissolved, the disparity in the density and compressibility of the two phases lessens, and the effect on the carrier's acoustic properties is thus reduced. **Fig. 4.7** depicts a calculated kick-detection sensitivity for a weighted WBM as function of pressure. Assuming a 10-bbl pit gain could be detected with a PVT, we see that the acoustic systems' advantage under the stated conditions is eliminated once the well achieves a vertical depth of approximately 16,700 ft.

One of the more reliable methods for rapid kick detection is similar in principle to the return-flow indicator. Delta-flow systems were proposed and developed in the late 1970s and have seen wide application and improved capabilities.[8-11] **Fig. 4.8** illustrates the concept and basic equipment requirements. While drilling, precise measurements of the pump rate and return-flow rate from a well are obtained and the difference between the two is computed and recorded automatically. The two rates should be approximately equal (delta-flow = zero) under normal circulation whereas a significant difference indicates that the well is either taking an influx or loosing circulation. The strip-chart recorder output for a kick event is similar in appearance to the diagram shown in **Fig. 4.9.** Field examples showing the kick detection and shut-in volume capabilities of the method are given in **Table 4.3.**

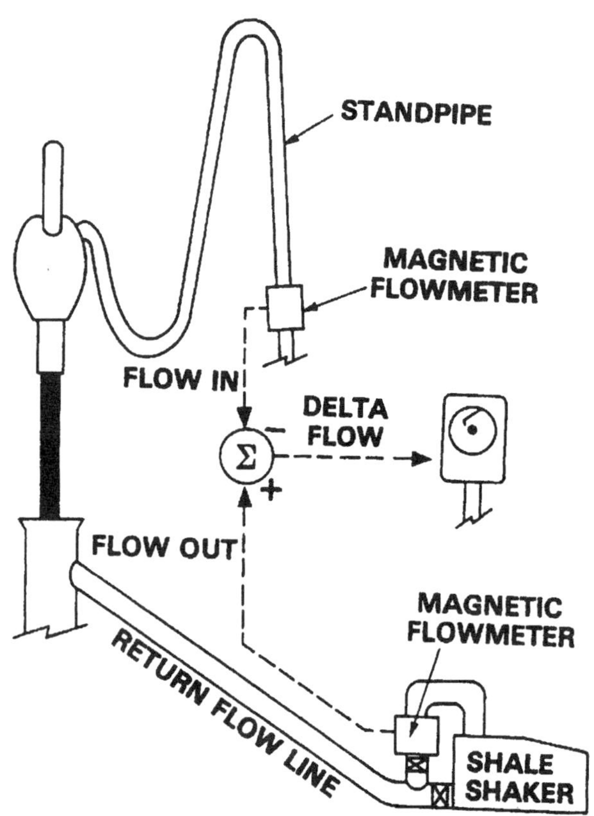

Fig. 4.8—Delta-flow method for early kick detection.[10]

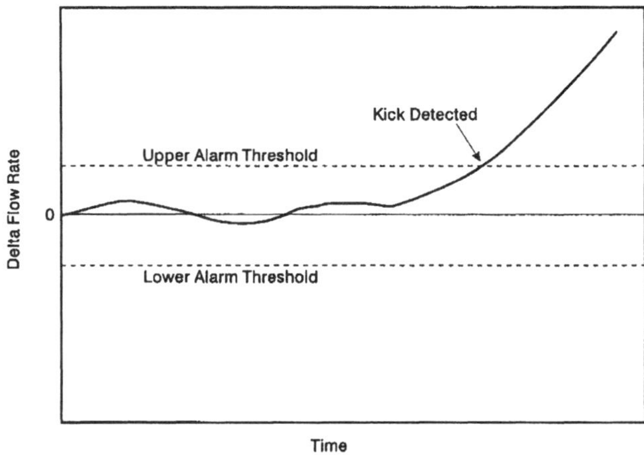

Fig. 4.9—Delta-flow strip-chart recording of a kick.

TABLE 4.3—FIELD EXAMPLES OF KICK DETECTION AND FINAL CONTAINMENT VOLUMES USING THE DELTA-FLOW METHOD[9]				
Hole size (in.)	Depth (ft)	Influx rate (gal/min)	Volume detected (bbl)	Volume contained (bbl)
5$^7/_8$	15,770	35	0.72	2.0
5$^7/_8$	14,005	7	0.70	1.5
5$^7/_8$	17,152	60	1.00	5.0

In most applications, the two rates are measured using magnetic flowmeters placed at the standpipe and flowline. As an alternative approach, pump rates computed from a stroke counter have been substituted for the standpipe meter with some success. A historic limitation to the technique has been the fact that magnetic flowmeters do not work with OBMs, but mass flowmeters will extend the kick-detection method to these muds as well.

High and low alarm thresholds must be set high enough to limit the number of false alarms, yet be sufficiently low to warrant its use as an early problem-detection tool. Delta-flow rate alarm settings on the order of ±25 gal/min fit within this window in most cases. However, positive and negative flow transients caused by pipe movement or changing the pump speed can trigger a false alarm unless the system is placed in the standby mode. The driller can deactivate the alarms simply by flipping a switch, but the more advanced systems anticipate these transients within the package software and shift to standby automatically whenever the pump throttle is engaged or a drawworks position sensor indicates rapid pipe movement. The periodic heave of floating drill vessels and the consequent variability to the flowline rate creates another set of problems for kick detection in general and, more specifically, to the delta-flow method. Means of compensating for vessel motion will be covered in Chap. 8.

Swanson *et al.*[12] recently described a similar kick-detection system designed specifically for slimhole application. The outflow from a well, however, is not compared directly with the pump rate but instead with the return rate predicted by a wellbore simulator. The pump rate is fed to the rig-site computer, but the annular-flow model also includes such effects as pipe-rotation speed and thermal expansion. A strip chart showing a comparison between the actual and predicted pump pressures is used as secondary means for identifying a gas kick. It is of interest that a typical slim hole differs from a conventional wellbore in that the increased annulus friction pressure resulting from the influx flow rate tends to override any hydrostatic-pressure reduction, which leads to an increase rather than decrease in standpipe pressure.

Kick Verification and Shut-in Procedures. When a kick has been verified, quick action by the driller and his crew is necessary to shut in and limit the volume of formation fluids in the well. There are situations—i.e., drilling below a shallow casing seat—where policy prohibits shutting a well in for fear of breaking down the shoe and having a surface blowout erupt around the rig. But the decision to divert will have been made long before the actual flow and this section presumes that the well can be shut in safely. Also, the considerations that follow apply to land and bottom-supported offshore rig operations (barges, platforms, and jackups). Floater drilling has unique equipment and operational aspects which will be addressed in Chap. 8. With these constraints in mind, the flow check and shut-in procedures given in **Table 4.4** should be referenced as we proceed.

A brief introduction to the primary equipment and their basic functions is appropriate. The large-bore valves used to close in and seal the annulus or open hole in the event of a kick are the blowout preventers (BOPs). The collective term for all the BOPs and other accessory equipment positioned directly over the wellhead is the stack. **Fig. 4.10** shows an example working stack on a land rig. A typical stack arrangement is illustrated in **Fig. 4.11.** There are no standards in the industry and Fig. 4.11 is one of many possible arrangements. The arrangement and number of components selected for a specific application depend on such factors as pressure rating, risk assessment, available space, contractor inventory, and operator preference (or prejudice).

TABLE 4.4—FLOW CHECK AND SHUT-IN PROCEDURES WHILE DRILLING: SURFACE STACKS

Hard Shut-in

1. Ensure beforehand that the choke manifold line is open to the preferred choke and that the choke is in the closed position.
2. After a kick is indicated, hoist the string and position the tool joint above the rotary table.
3. Shut off the pump.
4. Observe the flowline for flow.
5. If flow is verified, shut the well in using the annular preventer and open the remote-actuated valve to the choke manifold.
6. Notify supervisory personnel (company drilling supervisor and toolpusher or rig manager).
7. Read and record the shut-in drillpipe pressure (SIDPP).
8. Read and record the shut-in casing pressure (SICP).
9. Rotate the drillstring through the closed annular preventer if feasible.
10. Measure and record the pit gain.

Soft Shut-in

1. Ensure beforehand that the choke manifold line is open to the preferred choke and that the choke is in the open position.
2. After a kick is indicated, hoist the string and position the tool joint above the rotary table.
3. Shut off the pump.
4. Observe the flowline for flow.
5. If flow is verified, close the annular preventer and open the remote-actuated valve to the choke manifold.
6. Shut the well in by closing the choke.
7. Notify the supervisory personnel (company drilling supervisor and toolpusher or rig manager).
8. Read and record the SIDPP.
9. Read and record the SICP.
10. Rotate the drillstring through the closed annular preventer if feasible.
11. Measure and record the pit gain.

Fig. 4.10—Example BOP stack in operation on a land drilling rig.

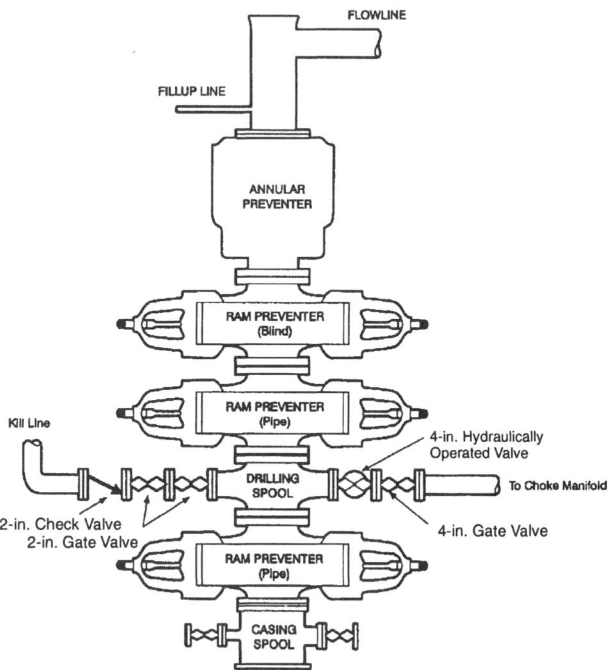

Fig. 4.11—A common surface stack arrangement.

The annular preventer is the topmost stack component and is designed to close and seal against any size or shape object, including an open bore. Ram preventers are equipped with ram blocks and sealing elements sized specifically for the drillpipe in use (pipe rams) or when pipe is out of the hole (blind rams). The closure and opening sequence for each BOP is achieved by stored hydraulic pressure activated from a control panel (**Fig. 4.12**) located near the driller or at another position away from the rig. Outlets for the kill and choke lines are provided at the drilling spool or mud cross and the entire assembly is flanged to the casing head or spool. On surface stacks, the kill line allows for pumping into the annulus if necessary whereas the choke line is the return conduit for the mud and formation fluids. One of the gate valves on the choke line side is remote-actuated by hydraulic pressure.

The choke line extends out to the choke manifold where circulating backpressure is held on the annulus by an adjustable choke. **Fig. 4.13** shows a choke manifold in operation. **Fig. 4.14** illustrates an overhead schematic of a common hookup. All choke manifolds provide one or more alternate flow paths in the event problems develop in the primary choke side, and include a large-bore vent line to release well pressure in an emergency. Desired redundancy and other factors dictate the overall design, but at least one of the chokes should be operated hydraulically. The adjustable choke controls are incorporated in a control panel (**Fig. 4.15**) located on the rig floor in view of the manifold and pits and in a place that facilitates communication with the driller. **Fig. 4.16** shows a control panel containing pressure gauges for the standpipe and choke manifold, pump stroke counter, choke position indicator, switches for selecting the pump and choke, and a supply-air-pressure gauge.

When a kick is suspected, the first step is to pull up and place the drillstring in the proper position for the shut-in and all subsequent control activities. In kelly-drilled holes, positioning a tool joint above the rig floor removes the kelly from the stack bore, usually ensures that a tool joint is not spaced opposite a pipe ram, and facilitates access to the lower kelly cock (a full-opening valve beneath the kelly). An accessible tool joint may be required for installing a wireline lubricator, cementing manifold, small-bore BOP, or other equipment packages.

Circulation should continue as the string is being hoisted. Primary reasons for keeping the pump on the hole are to disperse the kick volume, if present, in the drilling fluid and to stem the influx rate by maintaining circulating backpressure on the kick zone. In near-balanced holes, this difference between the ECD and static mud weight may provide the only overbalance to the pore pressure. A flow check in every drilling break, at every connection, and before starting a trip are recommended measures when drilling into a pressure transition.

An increase in the return flow and subsequent pit gain are considered normally to be solid indications of a kick. But the most definitive evidence is generally when annular flow is observed after the pumps have been shut down. Drill cuttings will densify the mud in the annulus to some extent and (absent a drillstring float) the natural tendency is for the mud level in the annulus to fall when circulation is stopped. Flow in the other direction is sufficient proof to warrant shutting the well in as soon as possible and thereby contain and limit further entry. A flow check is required whenever other signals indicate an influx or the suspicion arises for any reason.

A valid question is how long to observe the flowline before making the decision to shut the well in or return the bit to bottom. For example, severely gas-cut mud may flow for an extended period absent a kick. Also, the system is under pressure when circulating and it will take some time for the compressible mud and elastic wellbore to relax before the flow ceases entirely. The time requirement depends on mud type and hole volume, and probably will be no longer than two

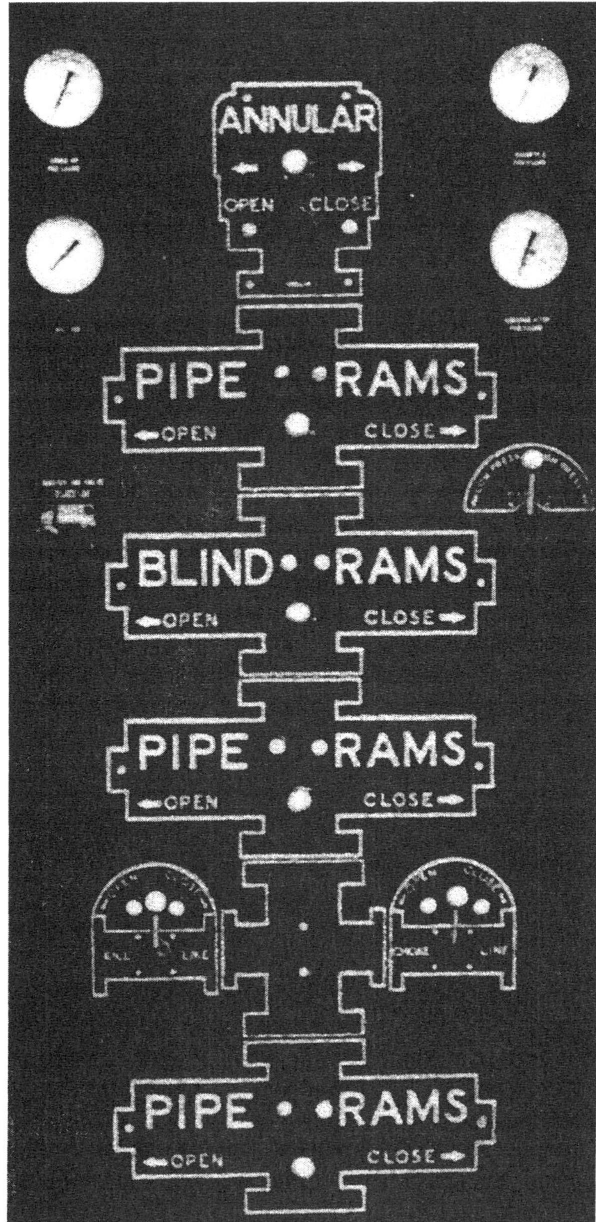

Fig. 4.12—Example BOP control panel. Courtesy of Koomey Inc.

or three minutes for WBMs at moderate well depths. At the high end, a stabilization time of 25 minutes or longer may be required for an OBM in a deep well.[13] Remember, however, that a kicking formation will continue to flow during the time this question is being resolved. If in doubt, prudence dictates shutting a well in and observing for a pressure buildup.

Two different shut-in procedures, the so-called "hard" and "soft" shut-in methods, are presented in Table 4.4. A hard shut-in simply means that the choke manifold line is closed when the well is shut-in at the stack whereas the soft technique uses the choke to stop the well flow after the preventer has been closed. **Fig. 4.17** shows the prearranged valve and choke settings for the two different methods.

The question of which technique is best has caused much controversy and confusion in this industry. Soft shut-in advocates argue that taking returns through the manifold before final shut-in minimizes any impact loading on the wellbore and surface equipment by slowly arresting the flow. Another argument for the procedure is that the choke operator can observe casing-pressure buildup as the choke is being closed and thereby is given the opportunity to take alternate steps if the pressure approaches the wellbore or equipment limitation. The disadvantage to the soft shut-in, however, is serious because the delayed shut-in leads to a larger influx volume.

Water hammer is an increase in pressure caused by a "rapid" change in a fluid's kinetic energy which, in a shut-in procedure, results from closing a preventer or choke on the flowing mud. The shock wave created by the closed valve propagates to the bottom of the hole at the speed of sound and then returns to surface at the same velocity. As shown in **Fig. 4.18,** the wellbore pressure increase is maximum at the closure point. The surface pressure surge in a rigid wellbore is given by

$$\Delta p_c = \rho_f v_f \Delta v / g_c. \qquad (4.2)$$

The relationship is valid only if the valve is fully closed before the shock wave has time to make the round trip from surface to total depth and back. If this condition is not met, closure is defined as "slow" as opposed to rapid and the resulting pressure surge will be lower. Regardless of the method, some pressure increase, however minor, cannot be avoided and the soft shut-in procedure in fact may be considered rapid in some cases.

Example 4.1. A kick is detected while drilling at 13,000 ft and the well is shut-in by the ram preventer in five seconds. Determine the water-hammer load at surface if the influx flow rate is 3.0 bbl/min, the mud's acoustic velocity is 4,800 ft/s, and the mud density is 10.5 lbm/gal. Compute the velocity assuming the annulus flow area corresponds to a 5.0-in. drillpipe inside an 8.921-in. inner diameter casing. Ignore the effect of the influx properties on the wave travel time and amplitude.

Solution. The time it takes for the pressure wave to return to surface is

$$\Delta t = (2)(13,000)/4,800 = 5.4 \text{ s}.$$

This would be characterized as a rapid shut-in; hence it is appropriate to use Eq. 4.2. The velocity change in the annulus is computed.

$$\Delta v = \frac{\Delta q}{A} = \frac{(3.0 \text{ bbl/min})(5.6146 \text{ ft}^3/\text{bbl})(144 \text{ in.}^2/\text{ft}^2)}{(60 \text{ s/min})[\pi/4(8.921^2 - 5.0^2)\text{in.}^2]}$$

$$= 0.94 \text{ ft/s}.$$

Eq. 4.2 gives the water hammer.

$$\Delta p_c = \frac{(10.5 \text{ lbm/gal})(7.48 \text{ gal/ft}^3)(4,800 \text{ ft/s})(0.94 \text{ ft/s})}{32.17 \text{ lbm–ft/lbf–s}^2}$$

$$\Delta p_c = 11,015 \text{ lbf/ft}^2 = 76 \text{ psi}.$$

The predicted 76-psi pressure increase would be significant only if a shallow casing seat were exposed to the imposed pressure and, even then, would not be a consideration if the annulus pressure later achieves a higher stabilized value. Thus rapid closure alone is not likely to cause lost returns except in those rare circumstances where flow velocity is extremely high. Also, the hydraulic closure time of a typical ram preventer is much faster than an annular preventer, which may take 20 seconds or longer to close completely. Hence the maximum pressure predicted by Eq. 4.2 rarely would be seen if a well is shut-in with the annular. Limiting kick volume should be the overriding consideration and, when feasible, the hard shut-in is the preferred technique.

Fig. 4.13—A choke manifold in operation on a land drilling rig.

Jardine *et al.*[14] developed a more rigorous model to study the effect of rapid shut-in on wellbore integrity and verified the results with experiments conducted on a test well. They concluded that water-hammer loads are usually negligible and any advantage to using the soft shut-in procedure probably is negated when considering the additional influx volume and its potential effect on the ultimate SICP.

The second reason for shutting in a well at the choke may have some validity in some cases. Many times, the maximum SICP is the initial SICP and numerous surface and subsurface blowouts have occurred on closing a well in. Monitoring the casing pressure gauge as the choke closes may seem like a good idea. But the only options other than closing in a kick are to allow the well to vent or to attempt a "low-choke" displacement procedure immediately. These actions are not to be approached lightly and may have consequences as serious as shutting the well in. As a general statement, it is better to have an underground blowout than risk a surface blowout and

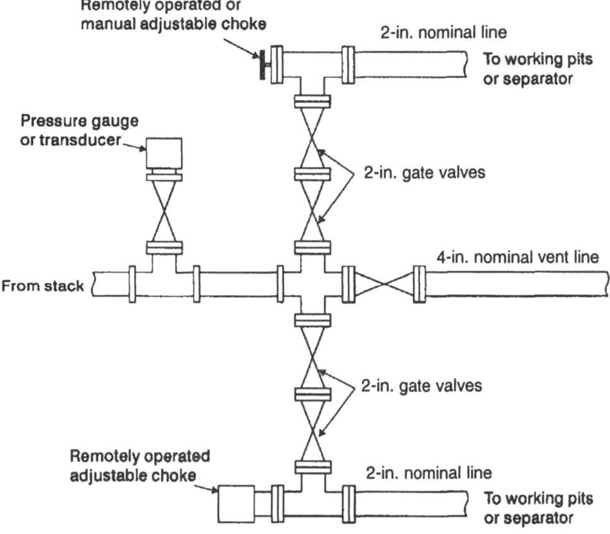

Fig. 4.14—A common choke manifold arrangement.

kicks should be contained as soon as possible unless there is a real possibility that doing so will result in ruptured casing, failure in the stack equipment, or an underground flow which is likely to broach back to the surface.

Using the annular preventer is recommended when initially shutting in a well because of the relatively slow closure time and all-around versatility of the device, but there are other reasons why the annular is preferred. When operating redundant valve systems, it is customary to assume that function cycles relate to increased wear and tear, and therefore to use the component most distant from the pressure source. This allows pressure to be isolated behind the secondary valve so that repairs or replacement can be effected in the primary valve. The same principle holds true in surface BOP stacks; the pipe ram or rams are considered as backup in case the topmost component fails or shut-in pressures are perceived to be excessive for the annular preventer.

Differential sticking is a concern whenever the drillstring is static, and the problem potential is aggravated by the backpressures imposed during a kill procedure. Moving the drillpipe through a closed preventer greatly minimizes this stuck-pipe risk. Pipe can be moved through a closed ram if the hydraulic closing pressure is reduced, but annular preventers are more suited for this type of work. In a kick situation, the drillstring is normally rotated at slow speed rather than reciprocated allowing the driller to accomplish other responsibilities.

After the kick is contained, the SIDPP and SICP can be read at the choke control panel or directly from the standpipe and choke manifold gauges. It is essential that the data be recorded at the time the well is first shut-in and at frequent intervals thereafter—i.e., once every minute. If not already alerted by the alarms, the senior supervisory personnel then can be notified of the problem by one of the rig crew.

The final step in this phase of the operation is to measure accurately the pit gain. Every school on well control emphasizes recognizing and shutting in a kick early, so there is a natural tendency to report a formation-fluid influx that is smaller than the actual feed-in. Precise pit-volume measurements are hard to come by, even under ideal conditions, and any inaccu-

Fig. 4.15—Example adjustable choke-control panel in use.

rate data make it that much more difficult to plan properly. Thus every effort should be made to ensure that the data are as accurate as allowed by the equipment limitations.

4.2.2 Off-Bottom Kicks. In Section 4.2.1, the observation was made that most kicks occur during tripping operations. In comparison to the few conditions that lead to a kick while drilling, several additional factors contribute to underbalancing a hole when pipe is removed from or placed into a well. Kick detection while tripping is less straightforward than drilling indications, the shut-in procedures are more complex, and regaining primary well control with pipe some distance up the hole requires procedures that fall outside the realm of the classical methods.

Reasons for Taking a Kick. As soon as the pumps are shut off, the wellbore backpressure furnished by the annular flow resistance is removed. Hence the beneficial aspects of the ECD will be lost during a trip and a flow check should be conducted on every well before starting out of the hole. It is common in most areas to slug the drillpipe while preparing for a

Speed
30
Total Strokes
2000
1 2 3
Reset
Pump
Drillpipe Pressure
Stroke Counter
Casing Pressure
Choke Position
Left
Right
Close
Open
Control Lever
Choke Selection
Air Pressure

Fig. 4.16—Basic components of a choke-control panel.

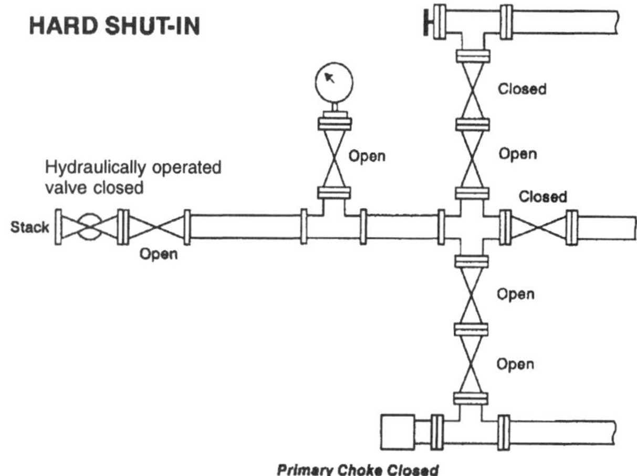

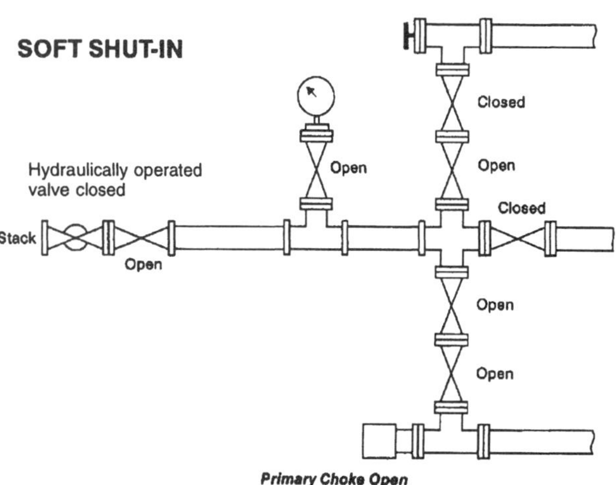

Fig. 4.17—Prearranged manifold settings for hard and soft shut-in procedures.

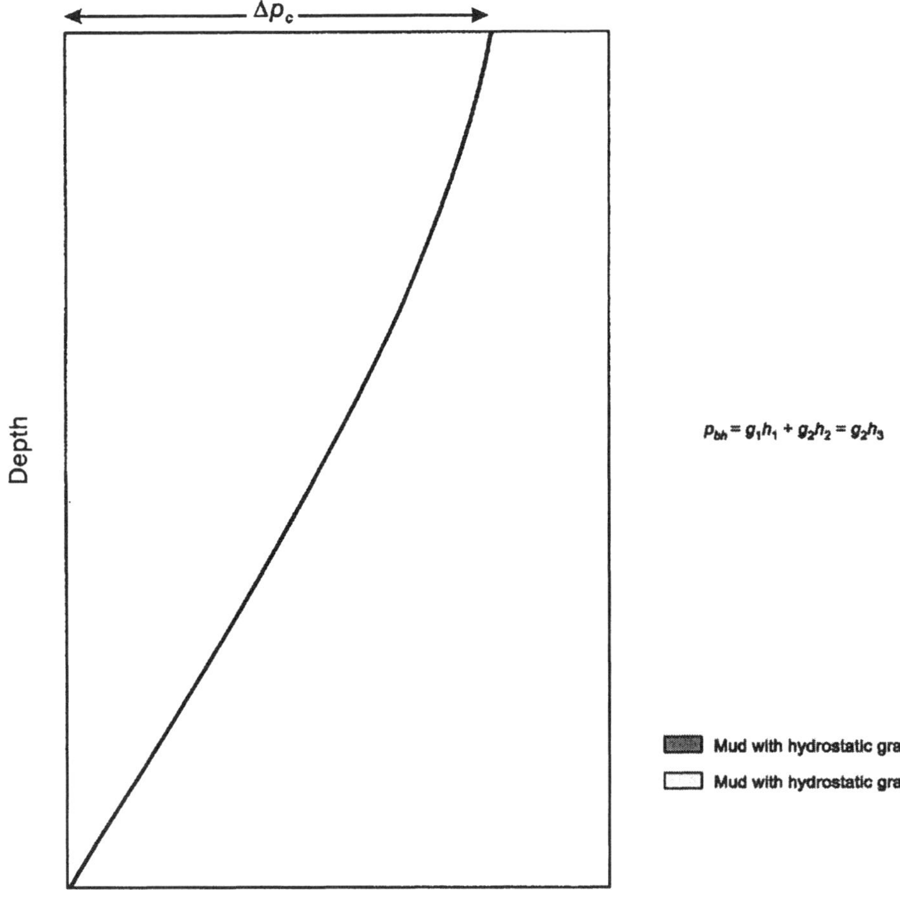

Fig. 4.18—Wellbore pressures induced by water hammer.

Fig. 4.19—Hydrostatic pressure balance after drillpipe has been slugged.

trip. Slugging refers to the procedure wherein a quantity of mud is densified relative to the mud in the hole and pumped down the drillpipe. As shown in **Fig. 4.19,** the mud in the drillpipe falls to a stable level which allows the string to be pulled in a dry condition. Kick detection will be more definitive and any subsequent well-control procedures will be less complicated if the pipe is slugged after rather than before the flow check.

Failing to keep a hole full of mud during a trip has caused more kicks and blowouts than any other single occurrence. **Fig. 4.20** demonstrates what happens to the hydrostatic pressure in a well as pipe is removed. With a full hole, a wellbore is occupied by the volume of the drillstring and drilling fluid. Removing a portion of the drillstring during a trip causes the mud level to fall to replace the volume evacuated by the steel, and the hydrostatic pressure is reduced by the same amount at every point below the final level.

To avoid an underbalanced condition, prudent tripping practices then dictate replacing the removed pipe volume on a continuous or periodic basis while pulling out of a well. The mud-volume requirement, of course, depends on dimensions of the string component and how much is pulled. Displacement factors, representing the pipe volume per unit length, are used to determine this mud volume according to the relation

$$V_d = C_d L_p, \quad \text{(4.3)}$$

where V_d = displacement volume, C_d = the pipe displacement factor, and L_p = the removed pipe length.

Displacement factors for drillpipe, casing, and tubing are obtained most readily by taking the mass per unit length, w_m, (including end finishing and connections) and dividing by the density of the pipe material. Steel has a density of 2,749 lbm/bbl and so the displacement factors for steel tubulars is given by

$$C_d \text{ bbl/ft} = (w_m \text{ lbm/ft})/(2,749 \text{ lbm/bbl}). \quad \text{(4.4)}$$

Tables 4.5 through 4.7 give the nominal dimensions and computed displacement factors for new American Petroleum Institute (API) and heavy-wall drillpipe. The average approximate weight given for Grade E and higher-strength drillpipe includes the internal and/or external upsets and tool joints attached to a 29.4-ft tube. These were obtained by taking the average of the approximate weights given by the API[15] for common drillpipe and upset/tool joint combinations, and so do not represent any specific connection type. Nonetheless, using these displacement factors will yield acceptable results given the inaccuracies inherent to the calculation assumptions and field measurements.

Drill collars usually are described by their calipered dimensions rather than weight. It may be more convenient to calculate the displacement factor directly from the tube geometry. For a cylinder of diameter d, the capacity per unit length C in bbl/ft is calculated as

$$C \text{ bbl/ft} = \left(\frac{\pi d^2}{4}\text{in.}^2\right)\left(\frac{12 \text{ in.}}{\text{ft}}\right)\left(\frac{\text{ft}^3}{1,728 \text{ in.}^3}\right)$$

$$\left(\frac{\text{bbl}}{5.6146 \text{ ft}^3}\right) = \frac{d^2}{1,029.4}. \quad \text{(4.5)}$$

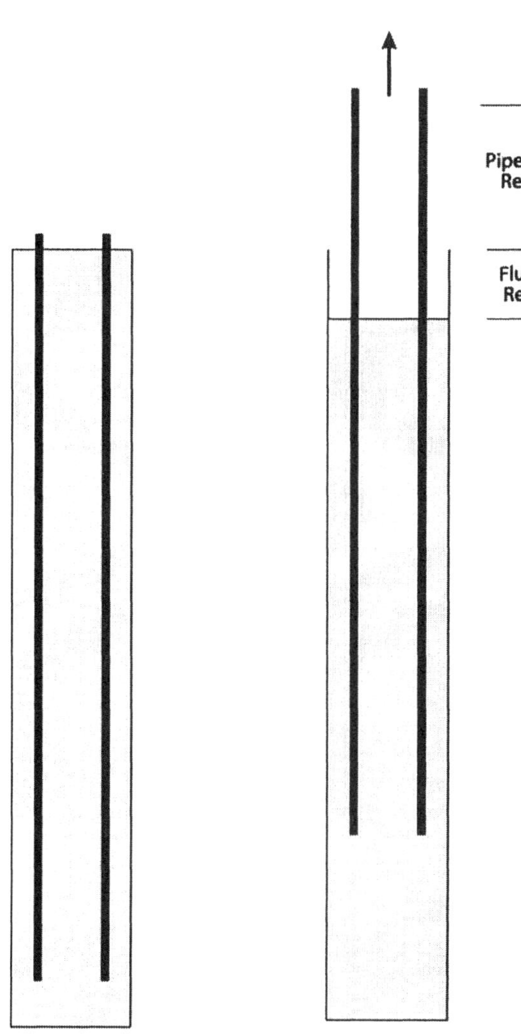

Fig. 4.20—Fluid-level reduction and hydrostatic-pressure loss after pipe volume is removed from a borehole.

The displacement factor in bbl/ft for a drill collar section is obtained as

$$C_d = \left(d_o^2 - d_i^2\right)/1,029.4, \quad \text{(4.6)}$$

where d_o and d_i are, respectively, the outer and inner diameters (OD and ID) of the tube. Apply Eq. 4.5 to give the expressions for the capacity factors for the inside of a tube, C_i, and an annulus, C_a.

$$C_i = d_i^2/1,029.4. \quad \text{(4.7)}$$

$$C_a = \left(d_h^2 - d_o^2\right)/1,029.4, \quad \text{(4.8)}$$

where d_h = the hole diameter.

Example 4.2. A well is drilled to a 9,500-ft vertical depth with a 10.0-lbm/gal mud. Casing with an 8.097-in. ID has been set at 1,500 ft. Determine the hydrostatic pressure loss at total depth if the following pipe sections are pulled without filling the hole with mud:

1. Ten 90-ft stands of 4½-in., 16.60-lbm/ft Grade E drillpipe.
2. Ten stands of drillpipe with a plugged bit.
3. One stand of 6¼ × 2½-in. drill collars.

Solution.

1. The displacement factor for the open drillpipe is obtained from Table 4.5 and the displacement volume is computed as

$$V_d = (0.00644)(10)(90) = 5.80 \text{ bbl}.$$

To determine the drop in fluid level, we must have capacity factors for the drillpipe and annulus. These can be obtained directly from a published table or by direct calculation

$$C_i = 3.826^2/1,029.4 = 0.01422 \text{ bbl/ft}.$$

$$C_a = \left(8.097^2 - 4.5^2\right)/1,029.4 = 0.04402 \text{ bbl/ft}.$$

TABLE 4.5—NOMINAL DIMENSIONS AND DISPLACEMENT FACTORS FOR API GRADE E DRILLPIPE

Outside Diameter (in.)	Nominal Inside Diameter (in.)	Nominal Weight (lbm/ft)	Approximate Weight (lbm/ft)	Displacement Factor (bbl/ft)
2³/₈	1.995	4.85	5.02	0.00182
	1.815	6.65	6.80	0.00247
2⁷/₈	2.441	6.85	7.09	0.00258
	2.151	10.40	10.53	0.00383
3¹/₂	2.992	9.50	10.15	0.00369
	2.764	13.30	13.86	0.00504
	2.602	15.50	16.39	0.00596
4	3.476	11.85	12.90	0.00469
	3.340	14.00	15.14	0.00551
	3.240	15.70	17.13	0.00623
4¹/₂	3.958	13.75	14.75	0.00537
	3.826	16.60	17.70	0.00644
	3.640	20.00	21.74	0.00791
	3.500	22.82	24.33	0.00885
5	4.276	19.50	21.58	0.00785
	4.000	25.60	27.58	0.01003
5¹/₂	4.778	21.90	23.77	0.00865
	4.670	24.70	26.33	0.00958
6⁵/₈	5.965	25.20	27.15	0.00988
	5.901	27.70	29.06	0.01057

TABLE 4.6—DISPLACEMENT FACTORS FOR HIGH STRENGTH DRILLPIPE			
Outside Diameter (in.)	Nominal Weight (lbm/ft)	Approximate Weight (lbm/ft)	Displacement Factor (bbl/ft)
2 3/8	6.65	6.95	0.00253
2 7/8	10.40	11.01	0.00400
3 1/2	13.30	14.51	0.00528
	15.50	17.02	0.00619
4	14.00	15.85	0.00577
	15.70	17.50	0.00637
4 1/2	16.60	18.65	0.00678
	20.00	22.40	0.00815
	22.82	25.21	0.00917
5	19.50	22.34	0.00813
	25.60	28.60	0.01040
5 1/2	21.90	25.14	0.00914
	24.70	28.13	0.01023
6 5/8	25.20	28.33	0.01031
	27.70	30.58	0.01112

These values are approximate because the effects of the upsets and tool joints are not considered. The mud level will fall by

$$\Delta h_m = 5.80/(0.01422 + 0.04402) = 99.6 \text{ ft}.$$

The corresponding reduction in hydrostatic pressure is

$$\Delta p = 99.6(10.0/19.25) = 52 \text{ psi}.$$

2. Tripping a plugged bit implies that the string is pulled wet and, if no mud falls back in the hole, the drillstring inner capacity is being evacuated along with the steel. The volume removed after pulling 10 wet stands is

$$V = V_i + V_d = (0.00644 + 0.01422)(10)(90)$$
$$= 18.59 \text{ bbl}.$$

Calculate the mud-level drop in the annulus,

$$\Delta h_m = 18.59/0.04402 = 422.3 \text{ ft},$$

and pressure loss,

$$\Delta p = (422.3)(0.519) = 219 \text{ psi}.$$

3. The drill collar displacement factor and volume are computed.

$$C_d = (6.25^2 - 2.5^2)/1,029.4 = 0.03188 \text{ bbl/ft},$$

and $V_d = (0.03188)(1)(90) = 2.87$ bbl.

The pressure loss is determined in the same manner.

$$C_i = 2.5^2/1,029.4 = 0.00607 \text{ bbl/ft};$$

$$C_a = (8.097^2 - 6.25^2)/1,029.4 = 0.02574 \text{ bbl/ft};$$

$$\Delta h_m = 2.87/(0.00607 + 0.02574) = 90.2 \text{ ft};$$

and $\Delta p = (0.519)(90.2) = 47$ psi.

Tripping only a few stands of drillpipe without fill-up may allow formation-fluid entry if the formation is near the balance point and the effect is compounded if the drillstring is plugged. The need to fill the hole becomes acute once the collars or heavy-wall drillpipe are in the rotary table. For Ex. 4.2, removing one stand of the collars led to a hydrostatic pressure loss approximately equivalent to 10 stands of drillpipe. It follows that filling the hole, if done on a periodic basis with drillpipe, should be done more frequently or on a continuous basis when the bottomhole assembly (BHA) is retrieved. It should be apparent that a decided advantage of top drive systems is the ability to circulate continuously during a trip.

Other trip-induced phenomena are the transient pressures created in a well as result of the interaction between drilling fluid and a moving pipe string. Cannon[16] recognized that these pressures were a contributing cause for blowouts in normal-pressure wells, even when greatly overbalanced by the mud weight. Downward motion of pipe through a mud column causes a positive surge pressure whereas a negative surge termed the swab pressure is produced when the movement direction is reversed. Excessive surge or swab pressures are of major concern in tripping operations because surge pressure can fracture the rock while swab pressure can invite kick entry.

The actual mechanics are complex, but surge and swab pressures can be described sufficiently by considering three separate reactions: the pressure to initiate movement in a thixotropic mud, steady-flow viscous drag between moving pipe and a static borehole, and dynamic pressures resulting from mud acceleration or deceleration. Appendix B gives simplified expressions for these effects and an example problem. Top-drive circulation while tripping from a hole can mitigate or eliminate the problem potential.

Most drilling fluids develop some thixotropic character after being quiescent for a period; the mud's gel strength is the conventional measure of thixotropicity. Gel strength can be converted readily to the surface and downhole pressure required to break circulation for a given hole geometry and string depth. Swab or surge pressures are experienced whenever the gel strength is broken by pipe movement. Muds attain some gel strength almost immediately after shear has stopped, thus some influence can be expected on every joint or stand during a trip.

Downhole pressures are caused also by the displaced fluid's resistance to a constant-velocity pipe string. **Fig. 4.21** illustrates what happens in the annulus when pipe is run into a stationary liquid. In laminar flow, the mud moves at veloci-

TABLE 4.7—NOMINAL DIMENSIONS AND DISPLACEMENT FACTORS FOR HEAVY-WALL DRILLPIPE (HWDP)				
Outside Diameter (in.)	Nominal Inside Diameter (in.)	Connection	Approximate Weight (lbm/ft)	Displacement Factor (bbl/ft)
3 1/2	2.063	NC38	23.20	0.00844
	2.250	NC38	25.30	0.00920
4	2.563	NC40	29.70	0.01080
4 1/2	2.750	NC46	41.00	0.01491
5	3.000	NC50	49.30	0.01793

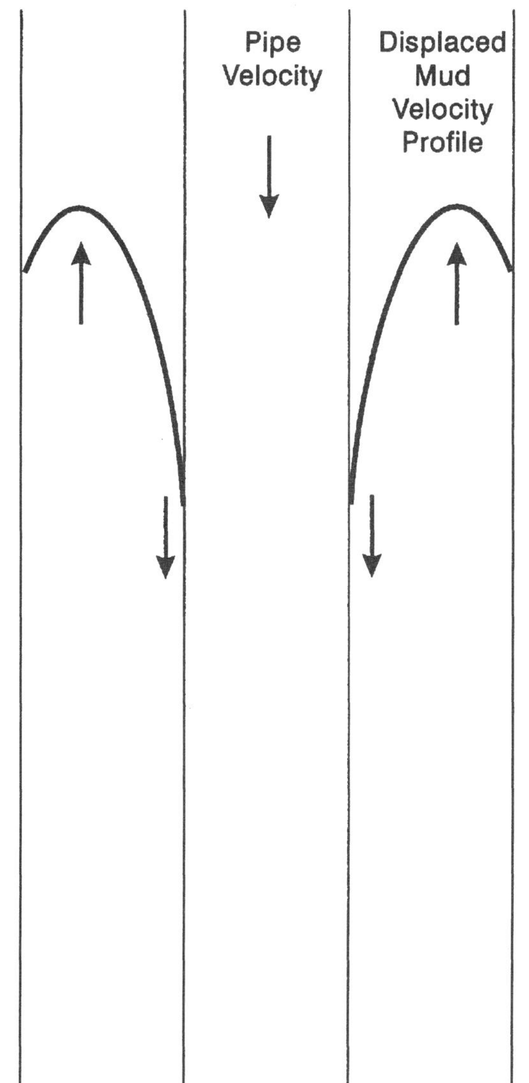

Fig. 4.21—Annular-flow profile when pipe is run into a mud at constant velocity.

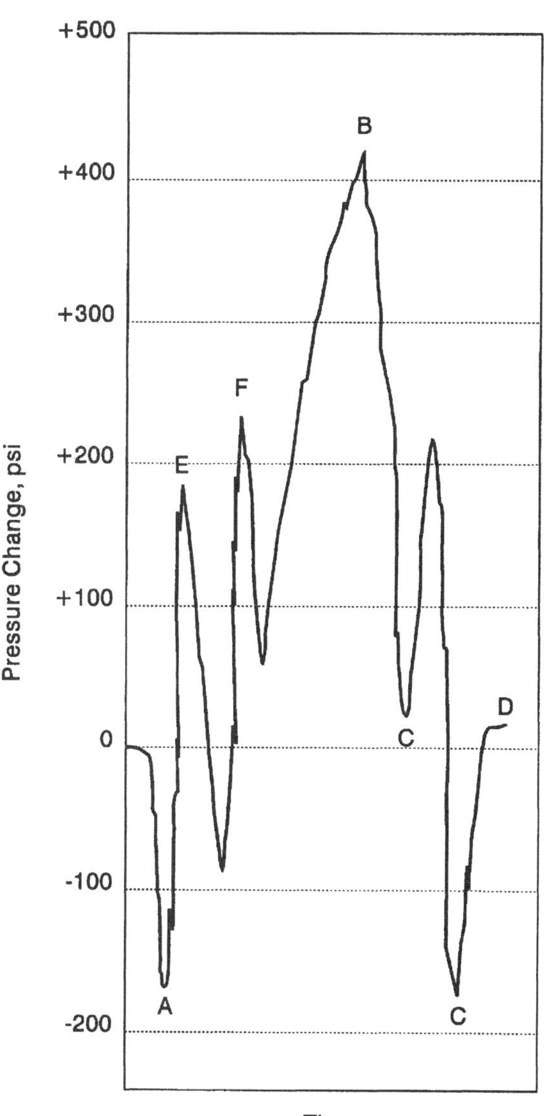

Fig. 4.22—Measured surge and swab pressures when running a joint of casing into a well.[17]

ties varying from zero at the walls of the hole to an upward displacement velocity in the annulus. At the pipe walls, the mud is carried along at downward velocity by the moving string. While tripping out, a similar velocity profile in the opposite direction is created by the mud as it falls to replace the vacated steel.

Burkhardt[17] developed a relationship expressing the net result as a single upward- or downward-effective velocity which is a function of the pipe speed, system geometry, flow behavior (laminar or turbulent), and whether the pipe is open or closed on bottom. The computed effective velocity or velocities above a specified depth in a well then are used in the appropriate annulus-friction pressure-loss equation to determine the surge pressure at that point. Swab pressures are calculated in the same manner except that the sign is reversed.

Inertial pressures are created by a change in kinetic energy and are present whenever a joint or stand is accelerated from its initial resting position, braked to a halt, and perhaps at other points in between. The displaced or falling drilling fluid undergoes a velocity change which, because the fluid has mass, must be balanced at the bottom of the string by a positive or negative pressure surge.

Fig. 4.22 is a chart from Burkhardt's paper[17] showing the downhole surge and swab pressures measured while running one joint of casing into a well. Each of the three reactions can be seen clearly. Starting at Point A, the joint was picked up off the slips and a large swab pressure was created by breaking the gel and the upward acceleration. The casing is accelerated then up to the maximum steady-flow velocity at Point B with swab deceleration effects caused by braking the string at Points E and F. Inertial effects, while the brakes were being applied to stop the string can be seen also at the two Points C.

As implied by Burkhardt's data, maximum surge and swab pressures are achieved often when the pipe reaches its maximum velocity. Pipe clearance and mud rheology are important considerations, but these parameters usually are dictated by the well plan or hole conditions. Thus the influence most under control of the wellsite supervisor lies in controlling the rate of pipe ascent or descent during a trip.

Inertial or dynamic pressure transients can be large enough to cause disequilibrium between the wellbore and formation. Pressure fluctuations approaching 300 psi in magnitude are evidenced on the acceleration side of the diagram while a large swab pressure is noted when the string is brought to a complete stop. This last effect coincides with the backflow often observed from the annulus when pipe is set in the slips. It is important to note and remember that swab pressures can occur while tripping in a hole.

Gel strength may become the overriding factor in deep wells. High temperatures usually aggravate gel strengths and some mud types exhibit a linear increase in gel strength over time. Steps should be taken to minimize the detrimental impact of the the progressive gel muds left at the bottom of a hot hole during a long trip operation. Chemical treatment and solids control will assist the effort while periodic circulation on the way back to the bottom will serve to reduce the average gel strength up the hole.

Surge and swab pressure predictions using the simplified equations in Appendix B are accomplished easily using only a calculator and graph paper. However, we had to assume that the pipe string was plugged on bottom, thus forcing all of displaced or falling mud to move through the annulus. The computed results therefore will be overly conservative if the assumption is unsupported. A dual-flow model should be used if an unacceptably high steady-flow pressure is calculated based on a closed bottom.

Limitations to the conventional steady-flow methods have been addressed in several subsequent computer-oriented models. Fontenot and Clark,[18] presented a steady-flow model which included variable mud rheology with depth and temperature. Wellbore and fluid elasticity (or compressibility) and their impact on downhole pressure transients were incorporated by Lubinski *et al.*[19] and extended by Lal.[20] Mitchell[21] later included some additional effects, most significantly the elastic strain of a dynamic pipe string. Chin[22] applied pressure-wave physics and non-Newtonian fluid behavior in eccentric annuli to surge/swab predictions.

Extremely close tolerances between a drillstring or casing component and the walls of a hole can be another source of major surge and swab pressures. Consider the extreme case where a balled bit or stabilizer isolates the annulus from the space beneath the bit. Severely restricting or completely stopping internal flow through the bit nozzles creates a system similar to that of a piston-cylinder configuration. For this condition, the wellbore beneath the bit will be pressured or depressured (depending on movement direction) as a function of the openhole volume beneath the bit, mud compressibility, and elastic properties of the rock. It should be apparent that close-tolerance conditions can lead to extremely high pressure surges, both positive and negative, if fluid cannot physically equalize across the restrictions commensurate with pipe speed.

Even with adequate equalization across the bit, tripping pipe from a well while the annular mud is being evacuated (also termed swabbing the hole) is a particularly hazardous situation. The annulus depicted in the schematic in **Fig. 4.23** is completely or partially packed off to the point that removing a portion of the drillstring also removes a mud volume corresponding to the dashed area. During the process, mud falls from the drillstring and replaces the space voided by the swabbed mud and drillpipe displacement. Removing a small mud volume from the annulus may create a large change in the drillstring-hydrostatic-column length and invite a kick before the swabbing is recognized. Example 4.3 demonstrates the problem nature.

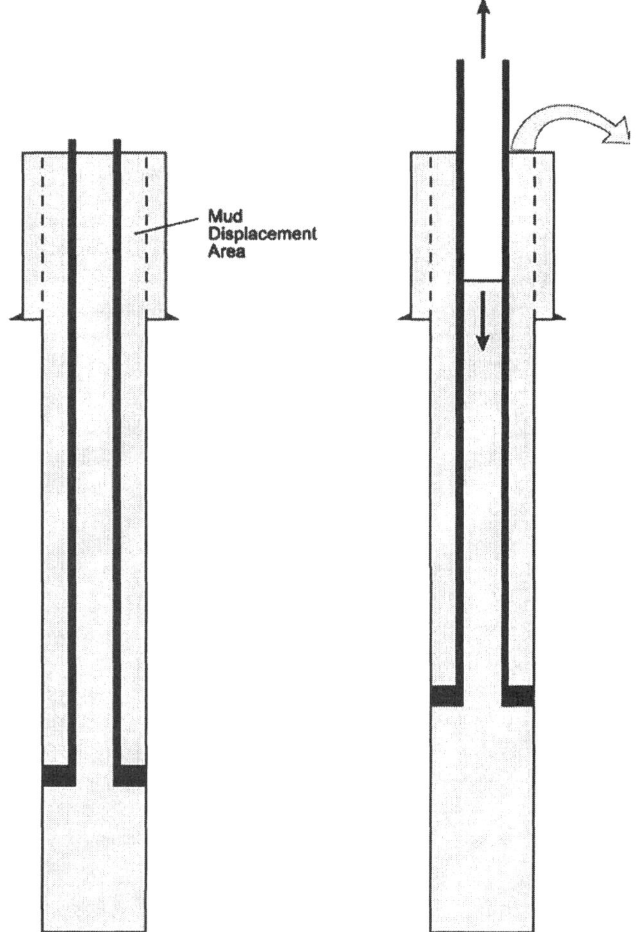

Fig. 4.23—Hydrostatic pressure imbalance created by swabbing the annulus.

Example 4.3. A trip operation commences at 5,010 ft and there is a gas sand at 5,000 ft that has a 0.45-psi/ft pore-pressure gradient. A 13³/₈-in., 54.5-lbm/ft surface casing is set at 2,000 ft and the hole diameter is assumed to be 12.25 in. The drillstring consists of 4½-in., 16.60-lbm/ft Grade E drillpipe and 600 ft of 7 × 3-in. drill collars. Excess hole drag is indicated some distance off bottom and the annulus soon becomes packed off. Determine the pressure gradient at the gas sand after pulling one more 90-ft stand of drillpipe if the mud density is 9.2 lbm/gal.

Solution. Along with steel volume, the mud in the space between the drillpipe and open hole is removed from the hole. The capacity factor for a 4.5 × 12.25-in. annulus is 0.12611 bbl/ft. The voided volume is calculated as

$$V = V_d + V_a;$$

$$V = 90(0.00644 + 0.12611) = 11.9 \text{ bbl}.$$

The mud-level change in the drillpipe is this volume divided by the internal capacity factor.

$$\Delta h_m = 11.9/0.01422 = 836.8 \text{ ft}.$$

The final wellbore-pressure gradient at the gas sand's depth is thus

$$g = \frac{(9.2)(5,000 - 836.8)}{(19.25)(5,000)} = 0.398 \text{ psi/ft}.$$

A swabbed kick appears to have been the initial event in the sequence that led to the famous 1969 blowout in the Santa Barbara channel. The example shows that a formation can become underbalanced by this mechanism very quickly. Recognizing that there was a problem while pulling one stand would have been difficult before flow started up the drillstring, a situation which can rapidly unload a well. A noticeable increase in drag usually precedes swabbing and should serve as a warning for the driller to increase his diligence in monitoring the flowline and detecting the condition before a packoff problem becomes a well-control problem, especially when the annular space is crowded with tools in an area prone to balling.

Drilling-fluid phase behavior during a trip can lead to an underbalanced condition, particularly when the geothermal temperature gradient is high. We normally consider the mud density as measured in the pits to represent what is in the hole. This, however, is not the case for any drilling fluid because all liquids compress under pressure and expand at higher temperature. Static temperature effects tend to predominate in hot wellbores, thus leading to an effective wellbore-fluid density lower than would be anticipated by a cool pit sample weighed at atmospheric pressure.

Babu[23] modeled the phenomena for both OBMs and WBMs in a hypothetical 25,000-ft well. Ambient surface and geothermal temperatures in the model were 70°F and 520°F, respectively. Raymond's[24] heat-transfer relationships for wellbore circulation were used to develop an effective fluid-density profile with depth while drilling with various muds. In every case studied, the net heating of the fluid column during a trip led to a predicted reduction in the hydrostatic pressure across the lower half of the wellbore. The maximum equivalent-density reduction occurred at total depth and ranged from 0.197 lbm/gal for an 11-lbm/gal diesel-based mud to 0.621 lbm/gal in an 18-lbm/gal WBM. Thermal expansion of the mud in a hot, near-balanced hole could easily induce a kick during a trip.

Kick Detection. During a trip, mud volume in the pits is not constant; it changes according to the amount of pipe in the hole. Detecting a kick with conventional PVT systems is impossible unless a means to account accurately for the string displacement is included. We can eliminate from consideration the delta flow and those acoustic techniques which require circulation below a free-gas influx. Kick-detection procedures during a trip must rely on less direct indications of extraneous wellbore volume.

Almost all methods used currently are based on an accounting between the mud volume required to fill a hole and what should be required for the string displacement. For example, assume 14 bbl of mud are needed to replace a twenty-stand pipe displacement. The mud fillup volume every twenty stands should equal 14 bbl if the hole is taking the correct amount of mud. A 10-bbl fillup would imply that a 4-bbl influx has entered the hole, whereas 20 bbl indicates a 6-bbl loss to the formation. Thus, kick and lost-circulation detection while tripping requires that the pipe displacement be known and that there be a way to measure accurately the fill and displaced mud volumes.

Historically, counting pump strokes has been the most common procedure for measuring fillup volumes during a trip. The positive-displacement duplex and triplex pumps used to circulate drilling fluid on a rig have a volumetric output which depends on the stroke length, cylinder bore diameter (liner size), and, in duplex pumps, the piston rod diameter. Eq. 4.9 equates the pump strokes to output volume.

$$N_p = V_p/F_pE_V, \quad \text{.......................... (4.9)}$$

where V_p = pump output volume, N_p = the number of complete stroke cycles, F_p = the rated output factor for the pump, and E_V = the pump's volumetric efficiency.

Example 4.4. A National 10-P-130 triplex pump has a rated 3.7-gal/stroke output when equipped with 6.0-in. liners. How many strokes should this pump take to fill the hole after pulling ten stands of 5-in., 19.50-lbm/ft high-strength drillpipe? Assume a 95% volumetric efficiency.

Solution. The displacement factor for the drillpipe is obtained from Table 4.6 and the volume corresponding to ten 90-ft stands is determined as

$$V_d = (0.00813)(900) = 7.3 \text{ bbl.}$$

Pump-stroke counters that come with a PVT rental package usually have a "trip" mode setting which causes the counter to stop the stroke count automatically when the flowline sensor detects return flow. From Eq. 4.9, when the hole fills, the stroke counter should read

$$N_p = (7.3)(42)/(3.7)(0.95) = 87 \text{ strokes.}$$

A trip tank is a more accurate way to measure fillup volumes. The basic equipment illustrated in **Fig. 4.24** includes a small measurement tank and low-pressure piping to a point in the annulus below the flowline. Fillup mud can be fed from the tank to the well by gravity, but the more common approach is to place the tank some distance below the fill connection and transfer the mud with a centrifugal pump. A means for diverting the return flow is needed to route the overfill and displaced mud back into the tank when tripping out or in of the hole.

Small tanks are desirable so that a minor volume change corresponds to a measurable change in the tank liquid level. A tank which is too small, however, will require more frequent filling from the active mud system. Trip-tank volumes

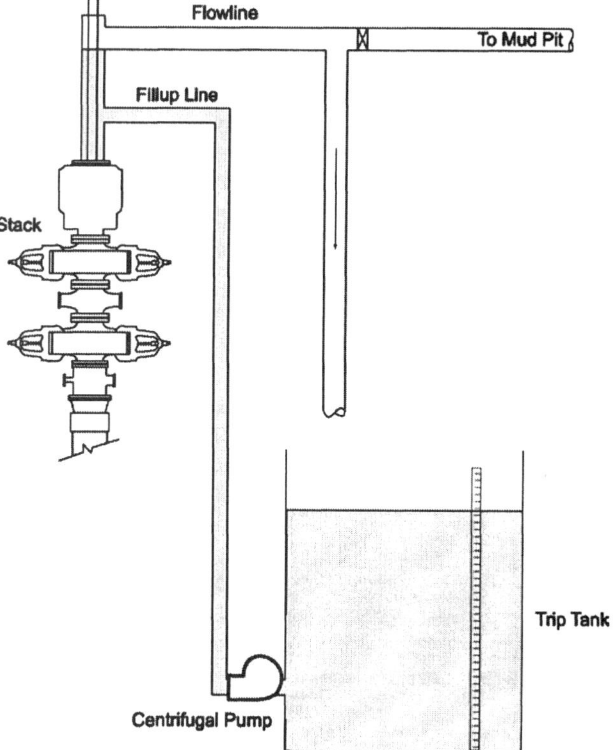

Fig. 4.24—Schematic of a typical trip tank.

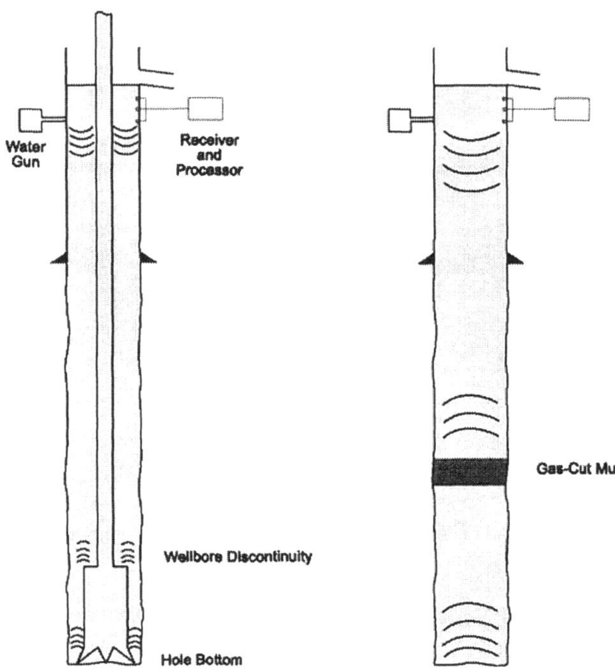

Fig. 4.25—Wellhead sonar kick-detection system.

between 10 and 50 bbls are typical. The dimensions are calibrated and scaled in bbl/in. (or fractions thereof) so that the driller can easily track the fillup or displacement volumes from a gauge or a PVT float reading at the console.

Counting strokes is a procedure used by many operators and drilling contractors today. However, Young[25] discussed several trip-tank advantages that should be considered in the overall risk assessment on a well. One weakness to counting strokes is that pump volumetric efficiency can be an elusive variable unless it is determined in the field by counting strokes for a known quantity of mud. Even so, exact calculated displacements are not absolutely essential because downhole filtration, drainage from the drillstring, and inexact displacement factors lead to some error despite surface volume measurement precision.

In the field, we must rely more on the actual volumes recorded during stable hole conditions and use calculated volumes only for comparison purposes. Standardized trip books are used by many companies to log the measured and calculated displacements during all of the trips on a well. Maintaining these records allows an operator to use prior data as the means for detecting any problems in the current trip operation. Increasing seepage losses on a well and other long-term trends may be noted by tracking the previous measurements.

A more serious problem with counting pump strokes is that the procedure only applies to half of a trip—i.e., when the string is pulled. Recall that surge pressures probably will be highest when running pipe into a well; thus, some way of detecting any consequent mud losses is desirable. A trip tank may not be needed to detect a major lost-circulation problem because flowback would cease altogether; but smaller losses, say those induced only at maximum surge, can be noted from the volume comparison. Furthermore, significant swab pressures and, hence the potential for taking an influx, can occur while running pipe. This is sufficient justification for using a trip tank.

Trip tanks offer some versatility in that they can be used in other applications requiring accurate liquid-volume measurements. For example, certain well-control-related procedures such as volumetric control, lubrication, and snub/strip operations require precise meaurements of the liquid going into or returning from a well. Other uses described include drillstem test liquid measurements and measuring the water volume required to fill the annulus when determining the tolerable equivalent density for a loss zone.

As for recent technology, an acoustic method has been developed that has the potential for direct gas-kick detection during a trip or when pipe is out of the hole. Bang *et al.*[26] presented test results for the wellhead sonar method, an idea similar in principle to the technique used to locate liquid levels in a pumping well. In **Fig. 4.25,** a water gun is used to generate a sound wave from a location at the kill line. The acoustic wave propagates down the hole and partially reflects off any wellbore discontinuities such as drill collars, stabilization tools, and the hole bottom. The return echos are caught by a series of hydrophones and the signals are amplified, filtered, and processed by computer.

In the experiments, free gas was intoduced into the mud near the bottom of a test well. Acoustic impedance caused by the gas-cut mud created another echo that, after processing the raw signals, could be detected not long after injection began. The wellhead sonar holds promise as a noncirculation kick-detection device, but also has the unique capability to indicate the location of gas. This information would be extremely valuable to the choke operator when circulating out a kick, particularly in a subsea-stack kill procedure. However, its success as an early-warning system will be subject to the same restrictions noted for the circulation-dependent acoustic methods—i.e., those conditions where the free-gas properties are greatly different from those of the carrier mud. It should be apparent also that the signal transmission medium must be continuous (the hole must be kept full of mud) before this technique will work.

Kick Verification and Shut-in Procedures. Kicks are verified by a flow check whether the pipe is on bottom or not, so this part of the procedure is no different than when in the drilling mode. How to go about shutting a well in, however, depends on where the string is when the flow is detected. The ease of regaining primary control is directly proportional to the amount of pipe in the well, and an operator may choose to delay shut-in and run pipe for a period with the well flowing. The decision will depend on the influx volume and flow rate, how much pipe is in the hole, how fast the string can be run safely, and perhaps other factors. For example, consider a trip situation where the drill collars are in the rotary when a small flow from the well is observed. Rather than closing in immediately, a reasoned assessment of the hole conditions may justify running drillpipe in the hole until some predetermined "safe" gain, say 10 to 20 bbl, is measured in the trip tank. However, it must be emphasized that failure to shut in a well immediately when flow is detected requires strong and calculated justification.

Shut-in procedures for three different phases in a trip are presented in **Tables 4.8 through 4.10.** Each table includes steps for both hard and soft shut-ins. Some variations to the described procedures are necessary if the rig is equipped with a top drive system. Other modifications may be dictated by company policy, so these procedures are offerred only as a guide.

Tables 4.8 and 4.9 assume that a float has not been previously placed in the string. A float can be considered as a string check valve that allows circulation and drainage from above, yet prevents backflow or kick entry from below. A float often is installed above the bit when drilling in areas known for

TABLE 4.8—SHUT-IN PROCEDURES WHILE TRIPPING DRILLPIPE: SURFACE STACKS

Hard Shut-in

1. Ensure beforehand that the choke-manifold line is open to the preferred choke and that the choke is in the closed position.
2. When a kick is verified, position upper tool joint above the floor and set slips.
3. Stab and makeup a full-opening safety valve in the open position.
4. Close the safety valve.
5. Shut the well in using the annular preventer and open the remote-actuated valve to the choke manifold.
6. Notify the supervisory personnel.
7. Install the kelly.
8. Open the safety valve. Read and record the SIDPP.
9. Read and record the SICP.
10. Rotate the drillstring through the closed annular preventer if feasible.
11. Measure and record the pit gain.

Soft Shut-in

1. Ensure beforehand that the choke-manifold line is open to the preferred choke and that the choke is in the open position.
2. When a kick is verified, position upper tool joint above the floor and set slips.
3. Stab and makeup a full-opening safety valve in the open position.
4. Close the safety valve.
5. Close the annular preventer and open the remote-actuated valve to the choke manifold.
6. Shut the well in by closing the choke.
7. Notify the supervisory personnel.
8. Install the kelly.
9. Open the safety valve. Read and record the SIDPP.
10. Read and record the SICP.
11. Rotate the drillstring through the closed annular preventer if feasible.
12. Measure and record the pit gain.

TABLE 4.9—SHUT-IN PROCEDURE WHILE TRIPPING DRILL COLLARS: SURFACE STACKS

More Than One Stand in the Hole

1. Assure beforehand that the choke-manifold line is open to the preferred choke and that the choke is in the open position.
2. When a kick is verified, position upper connection above the floor and set slips.
3. Pickup the last drillpipe or combination stand and makeup into collar.
4. Run stand into hole, position tool joint, and set slips.
5. Stab and makeup a full-opening safety valve in the open position.
6. Close the safety valve.
7. Close the pipe rams and open the remote-actuated valve to the choke manifold.
8. Shut the well in by closing the choke.
9. Notify the supervisory personnel.
10. Install the kelly.
11. Open the safety valve. Read and record the SIDPP.
12. Read and record the SICP.
13. Rotate the drillstring through the closed annular preventer if feasible.
14. Measure and record the pit gain.

One Stand in the Hole

1. When a kick is verified, pull the last stand from the well.
2. Proceed with a blind ram shut-in procedure given in Table 4.10.

shallow gas hazards, particularly offshore. However, its use entails several operational problems and the perceived benefits must be compared to the disadvantages.

Absent a float, we have two potential flow conduits when tripping: one up the drillstring and the other through the annulus. Both of these paths must be closed when an influx is detected, starting with the string bore. A full-opening safety valve (FOSV) is the preferred closure tool for the drillstring. An FOSV is a high-pressure ball valve with a bore large enough to not present a drillstring restriction when in the open position. The top and bottom threads are machined to mate with the tool joint connections in use, thus tapered strings (including multiple drillpipe sections, HWDP, and collars) require that an FOSV or appropriate crossovers be furnished for every connection in the hole. In the course of normal operations, the FOSV should occupy an assigned location on the rig floor so that the tool can be accessed immediately when needed.

The valve is left in the open position because stabbing a closed valve into a flowing stream is a difficult, if not impossible, task for the floor hands. Even with an open valve, stabbing and making up the tool may not be easy, particularly if the flow rate is high. Certain steps can be taken to expedite this portion of the procedure. One is to weld on a handwheel, perhaps fashioned from a ring gasket, so that two or three men can more easily carry, position, and turn the device. Another step worth considering is to install a spare and open FOSV on the bottom of a stand of drillpipe when the trip begins. When flow is detected, the dedicated stand can then be stabbed and made up in a conventional manner.

After achieving makeup torque, the FOSV is closed by a special wrench that should be located in a designated place known

TABLE 4.10—SHUT-IN PROCEDURE WHEN OUT OF THE HOLE: SURFACE STACKS

Choke Line Open with Blind Ram Closed

1. When a kick is verified, close the choke.
2. Close the manifold gate valve immediately upstream from the closed choke.
3. Notify the supervisory personnel.
4. Read and record the SICP.
5. Measure and record the pit gain.

Choke Line Closed with Blind Ram Open (Hard Shut-in)

1. When a kick is verified, close the blind ram.
2. Close the manifold gate valve immediately upstream from the closed choke.
3. Notify the supervisory personnel.
4. Read and record the SICP.
5. Measure and record the pit gain.

Choke Line Open with Blind Ram Open (Soft Shut-in)

1. When a kick is verified, close blind ram.
2. Close the choke.
3. Close the manifold gate valve immediately upstream from the closed choke.
4. Notify the supervisory personnel.
5. Read and record the SICP.
6. Measure and record the pit gain.

to all the crew. Following drillstring closure, the preventer (usually the annular) is closed. The kick is contained on both sides in a hard shut-in whereas flow is terminated at the choke in a soft procedure. The kelly can be safely installed now and the SIDPP read after opening the FOSV. Depending on subsequent control actions, an inside BOP may be installed above the FOSV prior to picking up the kelly. An inside BOP is similar to a float in that it serves as a check valve to prevent flow up the string bore after the FOSV is opened. It will be used if the string is to be stripped back to bottom, but reading the drillpipe pressure will be more difficult if it is installed beforehand.

Top drive rigs offer an advantage when it comes to shutting in drillpipe on a trip. The driller can set the slips midway through a stand if necessary, and automatically stab and make up the main shaft into the upper tool joint. Precious minutes are saved as a result, and the shut-in procedures do not depend on the floor crew having to manhandle an FOSV under the stress and potential hazards of working in the presence of a flowing stream. Unlike the kelly/rotary table configuration, reading the SIDPP is accomplished directly after the annulus is shut-in. A stabbing valve, no different than a lower kelly cock or FOSV, is run beneath the drive shaft in routine operations. The valve can be activated automatically by the driller in most units, and can be run into the hole with the drillstring in a subsequent stripping operation.

Detecting a kick when collars are being pulled probably means that warning signs were ignored or the hole was not kept full. An operator may be required to deal with this predicament and need to understand the associated problems. Possible shut-in procedures are shown in Table 4.9; it is impossible to write a standard procedure as the actions are more subject to judgement calls or company directives than when tripping drillpipe. The reason is that shut-in pressures now may result in forceful ejection of the remaining BHA from a well. Steps must be taken to avert this catastrophe; personal or company philosophy concerns govern the choice of preventative measures as opposed to any industry-recognized standards.

We can derive a force-balance relationship by drawing a free-body diagram for the drill collar section shown in **Fig. 4.26.** The wellbore pressure has been contained by sealing the annulus and collar bore with an annular preventer and a closed valve. System equilibrium requires that the force summation result in a net downward force if the section is to be supported by the rotary slips. Assuming consistent collar geometry and adopting a sign convention wherein upward forces are positive yields

$$\Sigma F + \uparrow = 0 = p_1 A_s + p_2 A_i - F_s - F_f - W_1, \quad \text{(4.10)}$$

where p_1 = pressure at the the bottom of the drill collars, p_2 = pressure at the the top of the drill collars, and W_1 = air weight of the collar section.

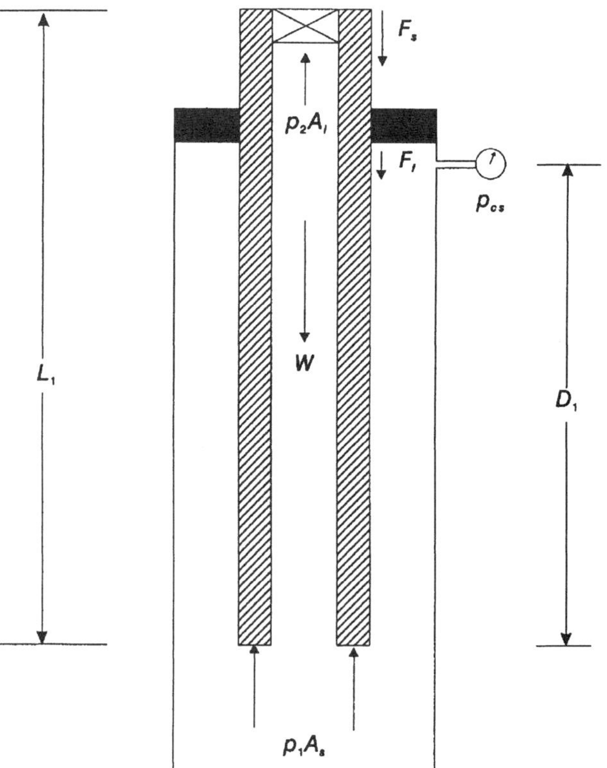

Fig. 4.26—Force balance for the case where a well is shut-in on drill collars during a trip.

We substitute terms and rearrange the equation to give

$$F_s = (p_{cs} + D_1 g_m)A_s + [p_{cs} - (L_1 - D_1)g_m]$$

$$A_i - F_f - w_1 L_1, \quad \ldots\ldots\ldots\ldots\ldots\ldots (4.11)$$

where D_1 = depth to the string bottom referenced from the casing pressure gauge, L_1 = collar length, and w_1 = unit air weight of the collars.

Equilibrium is ensured as long as F_s remains negative—i.e., the string is hanging in tension. However, a net upward force can be created if the shut-in pressure is too high, which means that the collars will be ejected forcefully through the closed preventer.

Example 4.5. Two 9 × 3-in. drill collar stands are left to be pulled when flow is detected. The bore is shut-in with an FOSV and the annular preventer is closed. At what shut-in pressure will the string be ejected from the hole if friction between the packing element and collar is 1,000 lbf and the mud density is 9.4 lbm/gal. Assume the casing pressure gauge is 20 ft below the closed valve.

Solution. Setting F_s equal to zero in Eq. 4.11 and solving for p_{cs} gives

$$p_{cs} = F_f + w_1 L_1 - D_1 g_m A_s - (L_1 - D_1)g_m A_i.$$

$$\ldots\ldots\ldots\ldots\ldots\ldots (4.12)$$

Note that the denominator term is equivalent to the cross-sectional area of the pipe OD, A_o. Solve for the other cross-sectional area terms.

$$A_o = \pi 9^2/4 = 63.617 \text{ in.}^2,$$

$$A_i = \pi 3^2/4 = 7.069 \text{ in.}^2, \text{ and}$$

$$A_s = A_o - A_i = 63.617 - 7.069 = 56.548 \text{ in.}^2$$

The unit air weight of the collar section can be determined by multiplying the steel volume over one foot by the steel density of 0.2833 lbm/in.^{3} and the conversion factor g/g_c.

$$w_1 = (g/g_c)(56.548 \text{ in.}^2)(12 \text{ in./ft})(0.2833 \text{ lbm/in.}^3)$$

$$= 192 \text{ lbf/ft.}$$

Assume a 180-ft stand length, substitute terms and solve for p_{cs} using Eq. 4.12.

$$p_{cs} = 1,000 + (192)(180) - (160)(0.488)(56.548)$$

$$- (20)(0.488)(7.069)/63.617$$

$$= 488 \text{ psig.}$$

The force balance is much more favorable if some drillpipe is beneath the closed preventer. For example, the allowable shut-in pressure for the previous problem increases to 2,166 psig if a stand of 4½-in., 16.60-lbm/ft Grade E drillpipe with NC46 connections had been above the collars. String weight is added, but more importantly, a large downward force is provided at the changeover between the drillpipe and collars. The calculation is left as a student exercise.

If flow conditions allow, placing a stand of drillpipe above the collars and closing the well in on drillpipe is probably the best approach. This step alone probably assures that the top joint will be hanging in tension, but our recommendation covers the worst-case scenario by closing the pipe ram rather than the annular preventer. A closed ram will not pass a moving tool joint, though the collision may damage the block and cause the elements to lose seal capability. Subsequent annular closure creates a seal if the ram is damaged as long as the wellhead and stack retain integrity after the shock load. If the collar and drillpipe connections are different, this procedure requires that the changeover or combination stand be set to one side so that it can be readily accessed when the need arises.

A soft shut-in is recommended. This may or may not reduce the incremental dynamic force, but it allows the operator opportunity to observe the string as the choke is being closed. Changing from a hard to a soft method during a trip will require that the choke be opened at some designated point in the operation.

A successful trip out of the hole does not imply that all will be well when the entire string is racked in the derrick. In fact, a well is most at risk having no pipe in the hole, at least from the standpoint of the available control options. This is no time to reduce surveillance diligence and someone should be in a position to monitor the wellbore when the string is on the bank for any reason. Standard policy is to close the blind rams after pulling the drillstring for a bit change, when changing stack components, or for other reasons. The flowline is isolated from what may be happening in the wellbore, so it is a good practice to open the choke line path back into the pits and to post a man at the line outlet.

Another step worth considering is to connect the trip tank to an outlet below the blind rams, perhaps at the casinghead flange. Flow can be detected in the trip tank, but this arrangement also provides the means for keeping the hole full when the blind rams are closed if seepage or filtration losses are excessive. However, it is important to remember to close the casinghead gate valve before resuming routine operations, and contingencies for shutting in a flow with this configuration must be considered in advance.

4.2.3 BOP Drills and Crew Responsibilities. Kick detection systems, from the most fundamental to the most advanced, are rendered useless if the crew is unable to respond quickly to warning indications. Any well-control problem is a stressful experience for the involved personnel, from the newly hired floor hand to the seasoned rig supervisor. Stress absent training leads to unwise decisions, serious mistakes, and panic. It is important to conduct surprise drills on a periodic basis to train each person in his/her respective control responsibility and to keep crews alert to the possibility of a kick at any time. Simulated BOP drills develop teamwork and coordination which, in turn, help to prevent kicks from developing into blowouts.

Drills conducted while drilling are called pit drills, as the process begins when the toolpusher or company supervisor activates the alarms by lifting one of the PVT floats. Trip drills are those drills conducted while pulling or running the drillstring. Drill frequency may be mandated by the government,[27] but every crew should be put through a drill at least once a week. More frequent intervals are advised if reaction times are unsatisfactory, when new personnel are hired or transferred in, or while drilling through a pore-pressure transition.

Choke drills are recommended before drilling out the casing and exposing new formation. One method for simulating a kill circulation is to close the preventers and apply a few hundred psi pressure with the rig pump or the cementing unit used to test the casing. In the drill, the toolpusher and company representative manipulate the choke and attempt to maintain a constant casing pressure as the driller brings the pump up to the kill-circulation rate. This gives the choke operator valuable practice in coordinating the choke-opening sequence with increasing pump rate while also getting a feel for the choke-handle response and lag time between the two pressure gauges.

Drilling contractors and their crews are often subject to different well-control policies, depending on the philosophies of the operator companies who contract their services. The preferred shut-in procedure and written assignments for each crew member should be posted on the rig floor. The supervisor should require all personnel to become familiar with the steps and their respective well-control responsibilities.

Starting when the alarm is given, the supervisor should time the drill up through the point when everyone is at their designated position and the well is closed in. The time required to react to the simulated kick and the time to shut the well in should be logged on the tour sheet. A crew accomplishing a shut-in in two minutes or less is considered to be doing a satisfactory job. Some operators do not carry the drill through shutting a well in, but stop once the driller is ready to activate the preventer. However, this is an opportune time to function test the equipment. Full closure is recommended if the well is not placed in jeopardy by doing so.

Drills should be conducted when the risk of endangering the operation is low. It would be unwise to carry out a pit drill immediately after penetrating a section known to be prone to differential sticking or when extremely fast penetration rates lead to the potential for packing off the BHA with cuttings. By the same reasoning, trip drills should be delayed until the drillstring is inside the casing.

All crew members should know how to operate the control equipment and should have specific task assignments in the event that the well takes a kick. Well-control problems rarely occur when the toolpusher or drilling foreman are on the floor; hence, the primary responsibility for detecting a kick and coordinating the shut-in lies with the driller and his crew. The driller's additional functions in the subsequent control include monitoring the shut-in pressures and mud volumes, operating the pump, moving the pipe string if so instructed, and overseeing and directing other crew members on whatever needs to be done.

The derrickman's primary charge during a well-control event is the mud and pit system. In general, the derrickman's responsibilities will be to inventory the barite on hand and obtain an accurate mud weight immediately after shut-in, verify that the pit level monitors are functional and that sufficient excess pit capacity is available, ensure that the mud/gas separator, degasser, and mixing equipment are lined out and in working order, and mix barite and other additives at the direction of the mud engineer.

One of the floor hands should verify that the choke manifold valves are in the intended open or closed position. The same person is the logical choice for the crew member whose main function is to observe the stack for leaks during the shut-in period and kill process. A fourth crew member should be stationed at the accumulator to look for leaks and to ensure that closing pressure is being maintained. Having a five-man crew, this leaves one man to help mix mud and carry out any other specific tasks as directed by the driller or toolpusher.

The toolpusher's responsibilities will depend on the nature of the drilling contract, but his primary function on a daywork job is to oversee the operation of the rig and work of the crew. He should ensure that backup tools and equipment (pump parts, ring gaskets, etc.) which may be required in the control procedure are available and order out more equipment or personnel if needed. There has been some debate on the question, but a capable toolpusher should be the designated choke operator. He may or may not have had more choke experience than the company representative, but he will be more accustomed to working with the driller and have more knowledge about the slow-speed characteristics of the rig pump. Also, the company representative needs to be mobile rather than fixed in one position during a control procedure.

After taking a kick, one of the first calls will be to the mud engineer. The mud engineer's function is essential to the control effort. He or she will be on hand throughout the kill procedure and thereafter until routine operations can be resumed. The job responsibilities include ensuring that enough barite is available for the required mud weight, scheduling the barite additions and other treatments when the mud weight is being increased, treating any contamination caused by the influx, and incorporating the trip margin (mud-density increase) after the well has been killed.

On a daywork job, the company representative is accountable for the safe conduct of the well operations. The crew's responsibilities are also those of the wellsite supervisor, and it is the supervisor's charge to see that the personnel are trained and capable of responding to an emergency. The supervisor's responsibilities include monitoring the well conditions and determining the proper control procedure, supervising the activities of the rig and service personnel during the preparation work and kill procedure, making or verifying the calculations and preparing the control schedule, keeping the home office informed, and relaying specific equipment or service needs to outside vendors.

4.2.4 A Blowout Case History. Many cases are documented where good drilling people became indifferent and neglected to pay attention to their business. This section concludes with a case history of a blowout from a shallow well. The purpose is to demonstrate the importance of maintaining adequate supervision on even the shallowest, low-pressure oil wells. It may be human nature to become complacent if the perceived risk is low, but this lapse of attention led to a major catastrophe involving numerous deaths in this instance.

Fig. 4.27 shows a schematic for the subject well. **Table 4.11** gives the description of the significant well events. Having only shallow conductor-casing set, the well was equipped to divert any flow rather than shut-in and risk broaching outside the conductor. The subsequent lost circulation was a serious problem, yet the hole would stand full and so the hydrostatic-pressure balance was apparently maintained until the trip operation. The influx may have been swabbed in, but it is more likely that poor tripping practices were used and the kick entered the well as a result of not filling the hole. Proper trip-tank use or even counting strokes in this case would have prevented the influx or at least detected the kick long before the hydrocarbons surfaced.

Inadequate equipment maintenance contributed to the later problems. Even had the flow been detected earlier, the diverter system would have been useless without the means for shutting in the drillstring bore and the work area likely would have been contaminated whether the diverter was activated or not. But the deaths from hydrogen sulfide poisoning resulted from inadequate crew training and lack of planning. Had a health and safety contingency plan been followed, all personnel on the rig would have been instructed and drilled in the proper use of breathing equipment. Audible alarms would

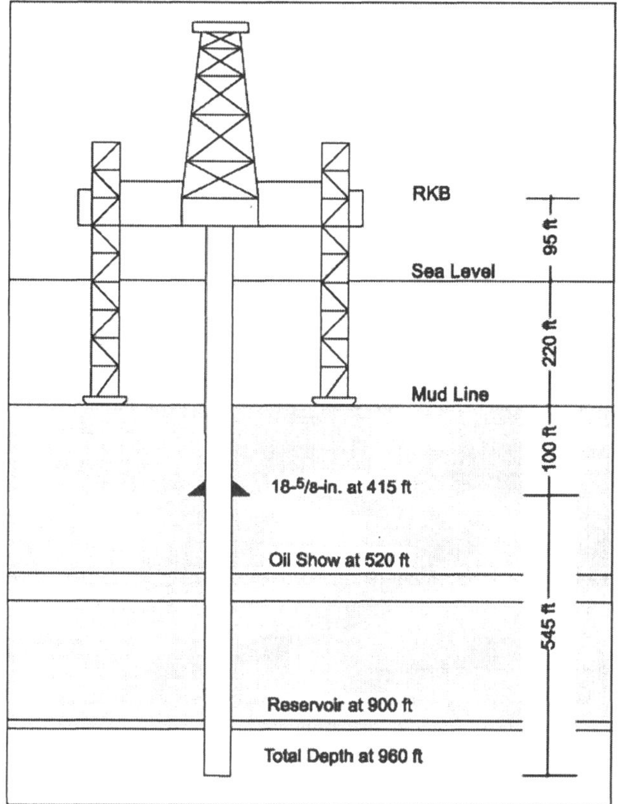

Fig. 4.27—Schematic of a shallow, offshore well at the time a blowout occurred during a trip.

have alerted the on- and off-duty personnel of the hazard. Finally, many lives would have been saved if the order to abandon the rig had been given when it was apparent that the situation was out of control.

The toolpusher and drilling foreman were competent and knowledgeable people. This event occurred because of complacent attitudes rather than a lack of knowledge. Paying attention to the work details, always remembering that there is potential for a well to flow, and remembering that downhole conditions can change will prevent such errors.

4.3 Shut-in Pressure Analysis

When a well is shut-in on a kick, the influx is contained and further entry will be prevented once the wellbore pressure adjacent to the kicking formation reaches equilibrium with the formation pressure. Shut-in surface pressures develop at the standpipe and casing. If the drillstring is below the kick, these surface measurements can be used to calculate the mud weight needed to balance the formation pressure and, perhaps, estimate the character of the influx.

The remainder of this chapter focuses on those well-control events that occur as a result of an underbalance condition with the drillstring near bottom. Assume that the control efforts can be achieved safely without undue complication and that no problems arise during the selected control technique. Many, if not most, well-control problems do not fit within this relatively narrow set of limitations, but a significant number do. Also, we must understand the basic principles outlined here before tackling more difficult cases.

A useful tool for analyzing hydraulics problems in both static and dynamic wellbores is to compare the pipe-annulus system to a U-tube. In the analogy, the pipe is considered to be one leg of the U-tube while the annulus is its counterpart. The pressure on bottom is equal to the sum of the surface and hydrostatic pressures taken from either the pipe or the annulus. Friction losses, if present, are added or subtracted depending on the flow direction. Also, it should be apparent that changing the surface or hydrostatic pressure on one of the legs will manifest a change in the surface pressure on the other side.

4.3.1 Kill-Mud Density Determination. Fig. 4.28 illustrates the U-tube analogy as applied to a well shut in on a kick. A well in the drill-ahead mode pumps clean drilling fluid of known density from the surface pits into the drillstring. When shut in on a kick, it should be apparent that the SIDPP directly measures the underbalance between the formation pressure and the hydrostatic pressure of the mud in the drillstring. Formation pressure can be calculated using the relation

$$p_p = p_{ds} + g_{om}D, \qquad \text{(4.13)}$$

where p_{ds} = the SIDPP, g_{om} = the hydrostatic gradient of the original mud within the drillstring, and D = true vertical depth (TVD) of the formation (and approximate depth of the drillstring). The mud gradient required to balance the pore pressure exactly is given by

$$g_{km} = (p_{ds} + g_{om}D)/D, \qquad \text{(4.14)}$$

where g_{km} = the kill-weight mud (KWM) gradient.

Example 4.6. A directional well takes a kick while drilling at 10,350 ft. After spacing the tool joint, the well is shut-in and the drillpipe pressure increases to 400 psig. Estimate the mud weight required to kill the well if the current mud density is 13.2 lbm/gal. The calculated TVD of the kick zone is 10,075 ft.

Solution. Vertical depth must be used instead of drilled depth when applying hydrostatics. Ignoring the spaceout distance (string depth ≈ total depth), Eq. 4.14 yields the required KWM gradient.

TABLE 4.11—SIGNIFICANT EVENTS PERTAINING TO A BLOWOUT FROM A SHALLOW OIL WELL

1. Set 18⅝-in. conductor casing at 415 ft (100-ft BML). Installed diverter equipment.
2. Had a good oil show in the samples from 520 ft.
3. Lost full returns at 650 ft. Attempts to regain circulation were futile. Resumed dry drilling with the hole standing full but not circulating.
4. Topped an oil reservoir at 900 ft and continued to drill to 960 ft. Began trip out of the hole.
5. Well began to flow at some point in the trip. The flow was not detected until oil began to impinge on the rotary table.
6. Poorly maintained FOSV could not be installed. No backup was available.
7. Panic ensued causing misuse of the diverter. Off-duty personnel were not alerted.
8. Crews failed to don breathing equipment in the presence of flowing hydrogen sulfide.
9. Failed to recognize that the rig equipment was inadequate to control the blowout and that abandonment was in order.

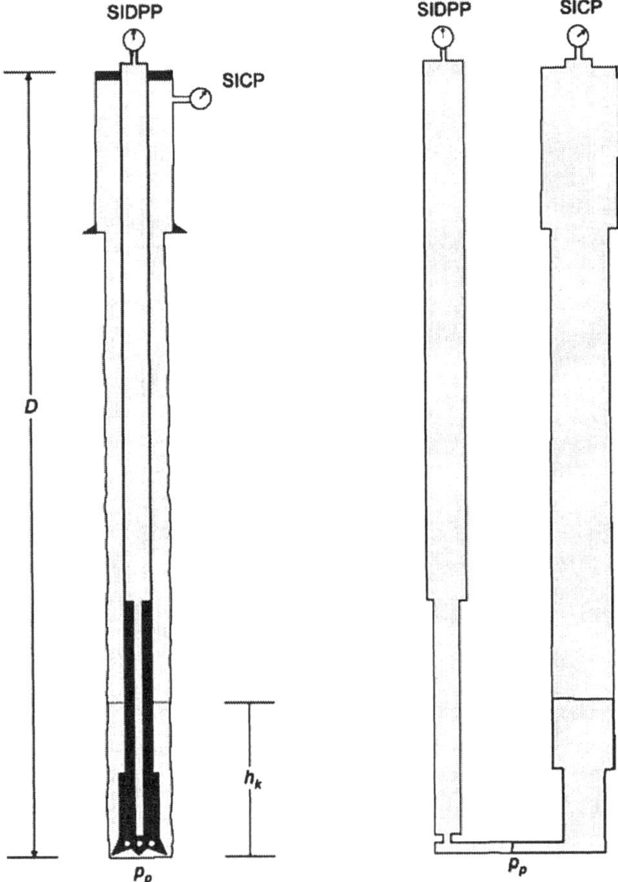

Fig. 4.28—Analogy between a shut-in well and a U-tube.

$$g_{km} = [400 + (0.0519)(13.2)(10,075)]/10,075$$
$$= 0.725 \text{ psi/ft.}$$

The gradient is converted to the required kill-mud density,

$$\rho_{km} = (19.25)(0.725) = 14.0 \text{ lbm/gal.}$$

A common way to characterize kick severity is in terms of the underbalance equivalent density or increase in mud weight required to effect a kill. The kick from the preceding example would be classified as a 0.8-lbm/gal kick. Defined in this manner, the same kick severity leads to higher underbalance and surface pressures with increasing depth.

4.3.2 Estimating the Nature of the Kick Fluid. According to the U-tube concept, the SICP must be the bottomhole pressure less the hydrostatic provided by the fluids in the annulus. Assume for the moment an influx enters a well as a discrete package and the mud density above the influx corresponds to drillstring-fluid density. It follows that

$$p_{ds} + g_{om}D = p_{cs} + g_{om}(D - h_k) + g_k h_k, \quad \ldots . (4.15)$$

where p_{cs} = the SICP while h_k and g_k denote the vertical height and pressure gradient of the formation fluid, respectively. The SICP will be recorded and we can determine the kick height if the influx volume and downhole dimensions are known. Rearranging Eq. 4.15 solves for the unknown parameter.

$$g_k = \frac{(p_{ds} - p_{cs}) + g_{om}h_k}{h_k}. \quad \ldots\ldots\ldots\ldots\ldots\ldots . (4.16)$$

This relation has been used to ascertain the type of formation fluid contained at the bottom of a shut-in well. Example 4.7 demonstrates the application.

Example 4.7. The initial SICP on the well described in Example 4.6 is 700 psig and the recorded pit gain is 25 bbl. The drillstring includes 690 ft of 7-in. drill collars and 5-in. drillpipe while the bit diameter is 9½ in. The hole inclination across the BHA is 20°. What is the expected formation fluid?

Solution. The influx volume and hole diameter must be known to determine the influx height. Assuming the hole is in gauge with the bit, we can compute the capacity factor in the drill collar annulus.

$$C_a = (9.5^2 - 7.0^2)/1,029.4 = 0.04007 \text{ bbl/ft.}$$

Assume also the increase in pit volume, G, is equal to the influx volume, V_k, so that V_k = 25 bbl. The influx length now can be determined as

$$L_k = 25/0.04007 = 624 \text{ ft.}$$

Its vertical height in the inclined hole is calculated as

$$h_k = 624\cos(20°) = 586 \text{ ft.}$$

Applying Eq. 4.16 yields

$$g_k = [400 - 700 + (0.686)(586)]/586 = 0.174 \text{ psi/ft.}$$

The gradient indicates this to be a gas kick.

It may be worthwhile to estimate the kick-fluid gradient using this procedure, but most of the underlying assumptions are unrealistic and the computed results are by no means conclusive. The disparity between the SICP and SIDPP is certainly a clue, and later signs may indicate a migrating gas bubble. But the definitive proof will not be realized until the formation fluids exit the choke. All kicks should be considered as gas kicks in control planning and selection of the circulation procedure.

Taking a more realistic approach, cuttings generated at the bit along with constituent formation fluids alter the mud's density to some degree; hence the mud densities in the annulus and drillstring are not the same. In a kick event, the formation fluids will be mixed with circulating mud rather than enter the well as an isolated bubble or slug. Influx flow will continue up through the time the kick is detected, after the pump has been shut down, and until equilibrium is achieved sometime after the well is shut-in. Also, a portion of any gas flow will be dissolved in the mud and so the pit gain and influx volume will not be equal if we account for solubility and post-shut-in flow along with other effects such as flowline storage and system elasticity. Finally, the openhole diameter is, at best, only a guess.

Example 4.8. Shale is being drilled when a thick sand is penetrated at 9,200 ft. Gas flow commences from the formation into the mud at a downhole rate of 2.0 bbl/min and 5% of the gas goes into solution. After flowing for eight minutes, the PVT alarm sounds and the driller starts picking up off bottom from a depth of 9,216 ft. It takes 30 seconds to space out the tool joint and shut down the the pump. The downhole-influx

TABLE 4.12—DATA FOR EXAMPLE 4.8	
Operational Parameters	
SIDPP	500 psig
Hole diameter	8½ in.
Drill collar OD	6¾ in.
BHA Length	380 ft
Drillpipe and HWDP OD	5 in.
Penetration rate	120 ft/hr
Mud density	10.2 lbm/gal
Circulation rate	350 gal/min
Shale-cutting density	20.8 lbm/gal
Formation Characteristics	
Sandstone porosity	26%
Sand water saturation	18%
Pore water density	9.1 lbm/gal
Sand gas saturation	82%
Gas specific gravity	0.60
Sand grain density	22.1 lbm/gal
Wellbore temperature	90°F + 0.009°F/ft
Flowline storage volume	2.25 bbl

flow rate then increases to 2.5 bbl/min and two more minutes elapse before the well is closed in. Another 0.5 bbl of gas entry occurs as the shut-in BHP builds up to the transient formation pressure at the wellbore. Using the operational parameters and formation characteristics in **Table 4.12,** estimate (1) pit gain, (2) influx volume, and (3) SICP.

Assume the dissolved gas does not affect appreciably the carrier mud's volume or density. Also assume the wellbore and mud are completely rigid and free gas does not expand or migrate relative to the mud.

Solution.

1. Assume that 5% of the gas dissolves in the circulating mud while the hole is being drilled and during the spaceout procedure and that all of the entry is in the free state during the flow check and closure. Following pump shut-down, the flowline drains another 2.25 bbl into the pit. The final pit gain is calculated.

$$G = (2.0)(8 + 0.5)(0.95) + (2.5)(2) + 2.25 = 23.40 \text{ bbl}.$$

Additional gain would be predicted had we considered the relaxation of the mud and wellbore when the annular friction pressure was eliminated.

2. The pit gain is not the same as the influx volume because the flowline storage is not part of the kick, a portion of the influx is dissolved, and flow from the formation did not stop with closure. Kick volume is determined approximately as

$$V_k = (16.15)/0.95 + 5.00 + 0.50 = 22.50 \text{ bbl}.$$

3. Applying Eq. 4.13 yields the formation pressure.

$$p_p = 500 + (0.0519)(10.2)(9,216) = 5,379 \text{ psig}.$$

The expected SICP is determined by subtracting the hydrostatic pressure of the segregated and mixed fluids in the annulus. The four segments to consider here are the gas, gas/mud mixture, gas/mud/sand mixture, and mud/shale mixture.

Assume the last 5.5 bbl of gas flow displaces the mud in a piston-like manner. The gas density on bottom is determined using the procedures discussed in Chap. 1. At bottomhole conditions, the pseudoreduced properties for a 0.6 specific gravity (SG) gas are determined.

$$T = 90 + (0.009)(9,216) = 173°\text{F}.$$

$$T_{pr} = T/T_{pc} = (173 + 460)/352 = 1.80.$$

$$p_{pr} = p_p/p_{pc} = (5,379 + 14.7)/677 = 7.97.$$

Fig. 1.6 gives a compressibility factor of 1.03. The universal gas constant is selected from Table 2.2 and Eq. 1.10 is used to obtain the gas density.

$$\rho_g = \frac{(29)(0.60)(5,394)}{(1.03)(80.275)(633)} = 1.79 \text{ lbm/gal}.$$

Simplify the problem by assuming the gas density is consistent across the kick region. The capacity factor in the drill collar annulus is 0.02593 bbl/ft and so the 5.5-bbl gas column occupies

$$h_1 = 5.5/0.02593 = 212 \text{ ft}.$$

The change in hydrostatic pressure across the gas is calculated as

$$p_1 = (0.0519)(1.79)(212) = 20 \text{ psi}.$$

Segment 2 contains mud and the gas that flowed into the well during the 30-second spaceout period. The mixture density of each annular fluid segment can be estimated using the mass balance equation for mixtures,

$$\rho_M = \frac{\rho_1 V_1 + \rho_2 V_2 + \ldots + \rho_n V_n}{V_1 + V_2 + \ldots + V_n}, \qquad (4.17)$$

where ρ_M = the mixture density and the other volumes and densities are those of the individual components. Assuming 5% of the influx is dissolved, the free gas volume during the 30-second circulation is

$$(2.0)(0.5)(0.95) = 0.95 \text{ bbl},$$

whereas the mud volume is

$$(350)(0.5)/(42) = 4.17 \text{ bbl}.$$

Applying Eq. 4.17 yields

$$\rho_{M2} = \frac{(10.2)(4.17) + (1.79)(0.95)}{4.17 + 0.95} = 8.64 \text{ lbm/gal}.$$

The 5.12-bbl mixture volume (4.17 + 0.95) is shared by the drill collar and drillpipe annuli. The mixture volume in the drill collar annulus is calculated.

$$(380 - 212)(0.02593) = (168)(0.02593) = 4.36 \text{ bbl}.$$

The capacity factor opposite the drillpipe is 0.04590 bbl/ft and so the total segment height and hydrostatic pressure are determined, respectively, as

$$h_2 = 168 + (5.12 - 4.36)/0.04590 = 185 \text{ ft}$$

and $p_2 = (0.0519)(8.64)(185) = 83$ psi.

Segment 3 contains mud, influx gas, and drilled rock with its constituent pore fluids. The rock volume removed during the eight-minute drill time is

$$\frac{\pi(8.5)^2(120)(8)}{4(144)(60)(5.6146)} = 1.12 \text{ bbl}.$$

Of this bulk volume, the respective sand grain, pore water, and undissolved pore-gas components are

$$(1.12)(1 - 0.26) = 0.83 \text{ bbl},$$

$$(1.12)(0.26)(0.18) = 0.05 \text{ bbl},$$

and $(1.12)(0.26)(1 - 0.18)(0.95) = 0.23$ bbl.

The effective influx is

$$(2.0)(8.0)(0.95) = 15.2 \text{ bbl},$$

and the mud volume is

$$(350)(8.0)/(42) = 66.67 \text{ bbl}.$$

Hence the density of the segment is

$$\rho_{M3} = (22.1)(0.83) + (9.1)(0.05) + (10.2)(66.67) + (1.79)(0.23 + 15.20)/$$
$$0.83 + 0.05 + 66.67 + 0.23 + 15.20$$

$$\rho_{M3} = 8.75 \text{ lbm/gal}.$$

The height is

$$h_3 = (1.12 + 15.2 + 66.67)/0.04590 = 1,808 \text{ ft}$$

and its hydrostatic pressure is

$$p_3 = (0.0519)(8.75)(1,808) = 821 \text{ psi}.$$

Above the influx, we have shale cuttings and 10.2-lbm/gal mud. Using a one-minute drill time as basis, the mud and cuttings volumes, respectively, are

$$350/42 = 8.33 \text{ bbl}$$

and $1.12/8 = 0.14$ bbl.

The mixture density above the influx is

$$\rho_{M4} = \frac{(10.2)(8.33) + (20.8)(0.14)}{8.33 + 0.14} = 10.38 \text{ lbm/gal}$$

and its hydrostatic pressure is

$$p_4 = (0.0519)(10.38)(7,011) = 3,777 \text{ psi}.$$

3. Finally, the expected SICP is

$$p_{cs} = 5,379 - (20 + 83 + 821 + 3,777) = 678 \text{ psig}.$$

The expansion of the gas as it moved up the wellbore was not considered in the pit gain and SICP calculation. Incorporating expansion would increase the pit gain by less than half of a barrel and reduce the annulus hydrostatic pressure by a small amount, but the computation errors in this particular case are not serious.

Ex. 4.8 highlights one of the principal reasons why casing pressures are not suitable for making accurate BHP predictions. Using Eq. 4.16 predicts a 0.256-psi/ft kick gradient using the calculated SICP, a 23.4-bbl measured pit gain, and a 10.2-lbm/gal mud weight in the annulus. One might conclude from this result that the influx was gas-cut oil or salt water rather than the actual gas kick.

4.3.3 Maintaining Well Control Before the Kill. Buoyancy of the free-gas portion of an influx and the consequent migration through circulating and static mud have not been considered as yet in this discussion. Recall from Chap. 1 that a migrating gas bubble in a shut-in well is evidenced by an increase in surface pressure. The occurrence can lead to extreme wellbore stresses if the gas retains its original volume during the migration process; thus we must keep gas migration in mind any time the wellbore is closed and take care of the well by letting any contained gas expand as it moves up the hole.

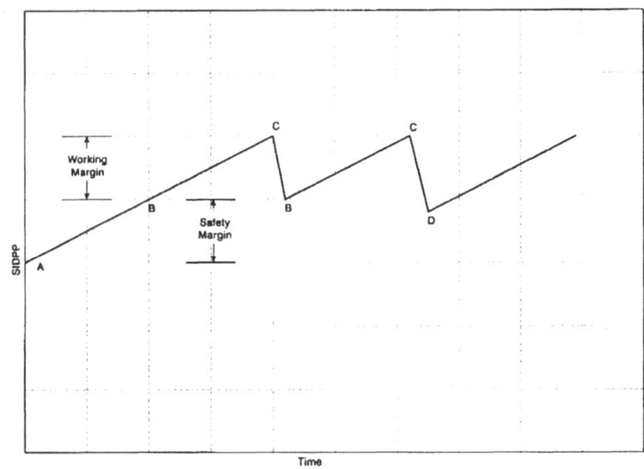

Fig. 4.29—Using drillpipe pressure to control gas migration and expansion prior to kick circulation.

Fig. 4.29 illustrates the means for maintaining a relatively stable BHP while allowing expansion to occur. Assume the bit is on bottom and the initial SIDPP at Point A reflects the underbalance pressure used to compute the KWM. The subsequent pressure increase indicates that gas has moved some distance up the hole. In the procedure, the driller is instructed to allow the SIDPP to rise by an amount defined as a safety margin, achieved on the diagram at Point B. An additional increase in the safety margin is permitted up to some maximum pressure at Point C. Fracture-gradient considerations normally control how much total buildup can be tolerated.

Once the "working margin" pressure is reached at Point C, the driller is instructed to open the choke partially and bleed mud back into the pits until the SIDPP falls back to the safety level. The process is repeated and pressures are maintained in the predetermined window until kick displacement can commence. During each bleed cycle, the gas expands by a volume equivalent to the released mud and BHP is reduced to the safety overbalance. The safety margin allows for some error in the procedure so that underbalancing the hole is avoided if excess mud is bled from the well at Point D. If need be, the procedure can be followed until gas reaches the stack. Migration has proceeded as far as it can at this point. No more fluids are bled from the well and the wellbore pressures will remain stable until the operator takes other action.

Example 4.9. Refer back to Examples 4.6 and 4.7. Assume casing is set above the kickoff point (hole is near-vertical) and the fracture gradient measured at the 8,700-ft setting depth was 0.81 psi/ft. What instructions should be left with the driller for taking care of the well until kick displacement can begin?

Solution. The fracture pressure at 8,700 ft is

$$p_{fi} = (0.81)(8,700) = 7,047 \text{ psig}.$$

The maximum SICP that can be tolerated is the fracture pressure minus the hydrostatic pressure above the critical depth.

$$(p_{cs})_{max} = 7,047 - (0.686)(8,700) = 1,079 \text{ psig}.$$

The initial SICP was 700 psig, so well safety dictates that no more than a 379-psi increase be permitted. An acceptable procedure in this case would be to tell the driller to monitor both pressure gauges and allow the SIDPP to increase from 400 psi to 600 psig. Thereafter he should bleed mud through the choke and maintain the SIDPP between 600 and 500 psig.

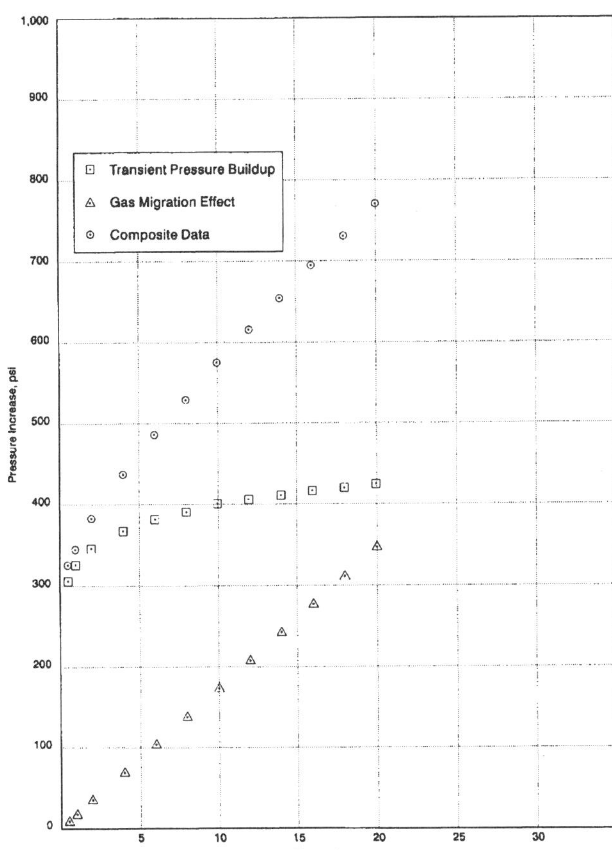

Fig. 4.30—Pressure buildup in a shut-in well caused by formation and gas migration effects.

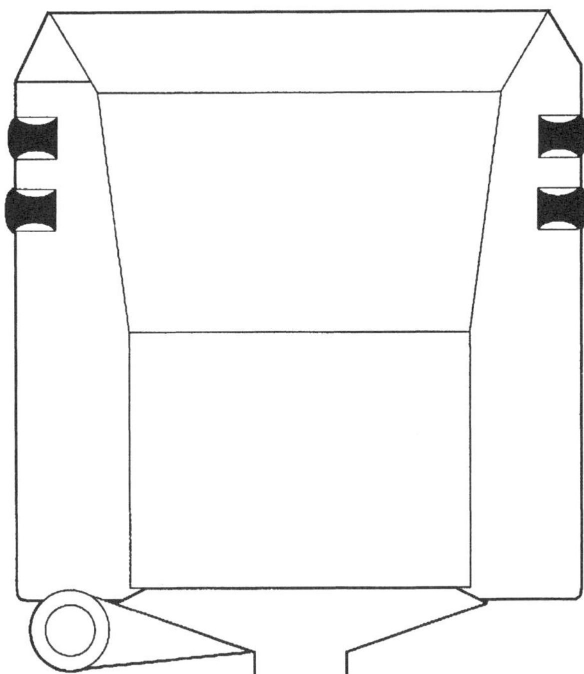

Fig. 4.31—Drillstring float design using a spring-loaded flapper.

4.3.4 Obtaining Valid SIDPP Measurements.

A step seen in many published well-control procedures is to record a "stabilized" SIDPP and SICP after closing in a well. However, we have seen that surface pressures will never stabilize if free gas is migrating through the mud and the effect may complicate the effort to calculate the proper kill-mud density. A suitable SIDPP commensurate with the accuracy of the gauges and other measurements can be obtained quickly in high-permeability formations but, as those drillers familiar with well-test analysis know, some time will be necessary before wellbore pressure approximates pore pressure in tighter rock.

The pressure increase vs. time plots in **Fig. 4.30** illustrate the problem for an 88-md formation. Square data points correspond to actual DST buildup measurements. Triangles represent the pressure increase resulting from gas migrating through a 10.0-lbm/gal mud at 2,000 ft/hr in a rigid, sealed wellbore. Circles reflect the sum of the two, a curve similar in appearance to the SIDPP response if this well were closed in on a kick. We can see that accurate determination of the pore pressure may be impossible in many cases. One suggested procedure is to record the SIDPP every minute and plot the data. Regardless of permeability, the pore pressure buildup in the early time period will dominate the migration buildup and a reasonable guess is to select a SIDPP somewhere near where the plot makes a major change in curvature. Fortunately, underpredicting the pressure in a relatively tight formation is not likely to present a serious control problem as any secondary kicks should be small and easy to manage.

A float presents a problem in obtaining the SIDPP because flow into the drillstring is obstructed. Various downhole check-valve designs are available, depending on the application, but a common float shown in **Fig. 4.31** uses a spring-loaded flapper to stop the flow. Some manufacturers bore small holes in the flapper or an operator may do the same with a conventional float. A vented float allows the SIDPP to be read directly, yet retains the primary function of preventing major flows from entering the string bore.

Other float designs require that the valve be opened before pressure beneath the string can be read. The most accurate way to obtain the SIDPP is to pump down the drillpipe at a low rate while monitoring the drillpipe and annulus pressures. As shown in **Fig. 4.32,** a subtle break in the drillpipe pressure should be detected when the valve opens. The SIDPP to use in the KWM calculation and displacement procedure is the pressure at which the valve opens. Continued pumping would verify downhole communication as evidenced by a rise in the casing pressure, but the procedure should be terminated soon to limit the wellbore pressure increase. Any excess, or trapped pressure thereafter should be bled from the annulus by reducing the SICP to its initial value.

Slow-rate pumping requirements restrict the suitability of most rig pumps for this application and a cementing unit is

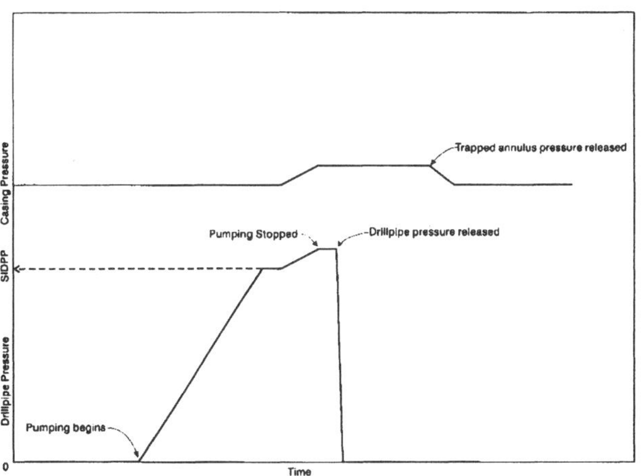

Fig. 4.32—Obtaining the SIDPP in a well equipped with a drillstring float.

recommended if at all possible. Using a rate/pressure strip chart is advisable. However, the proper equipment must be accessible and the pump-in procedure begun quickly or gas migration effects may lead to misinterpretion of the data. Another, less accurate means for estimating the SIDPP will be discussed in Sec. 4.5.1 for those cases where this condition cannot be met.

4.4 Increasing Mud Density

Pulverized barium sulfate—i.e., barite—is the most common substance used for increasing mud density. The pure material has a specific gravity used of 4.5, but other minerals also are included in the commercial grade used in drilling fluids. The API minimum SG specification for the product used in drilling fluids is 4.2.[28] Barite can be used to increase WBM densities up to about 19 lbm/gal before the high solids content leads to rheology control and other problems.

Hematite (SG range 4.9 to 5.3) typically is used if higher densities are demanded. Its abrasiveness, however, generally makes this an undesirable product unless the mud-weight requirement cannot be met with barite. Fluid densities in excess of 30 lbm/gal can be achieved with galena (SG range 7.4 to 7.7), but its use in well control normally is restricted to certain specialized procedures.

The mass balance equation as applied to densifying a mud system with barite is given by

$$\rho_{m1}V_{m1} + \rho_B V_B = \rho_{m2}V_{m2}, \quad \text{.................. (4.18)}$$

where the subscripts 1 and 2 denote the initial and final system density and volume, respectively, and B = barite. The final mud volume is calculated as

$$V_{m2} = V_{m1} + V_B, \quad \text{........................ (4.19)}$$

and the barite mass is given by

$$m_B = \rho_B V_B. \quad \text{............................ (4.20)}$$

Substituting Eqs. 4.19 and 4.20 into Eq. 4.18 and rearranging yields the barite mass needed to increase the density of a given mud volume.

$$m_B = \frac{V_{m1}\rho_B(\rho_{m2} - \rho_{m1})}{\rho_B - \rho_{m2}}. \quad \text{.................. (4.21)}$$

Barite additions can increase the mud system volume substantially and some economy may be realized by discarding or storing some of the original mud before weightup begins. Assume the intent is to keep the final system volume the same as it was before adding the barite. It should be apparent that the volume of mud to discard is the same as the barite requirement for the smaller system volume. Applying the mass balance equation to this situation yields

$$V_B = \frac{V_{m1}(\rho_{m2} - \rho_{m1})}{\rho_B - \rho_{m1}}. \quad \text{..................... (4.22)}$$

Example 4.10. An offshore well takes a kick while drilling at 6,785 ft vertical depth. The SIDPP and SICP are 300 and 550 psig, respectively. The current mud density 13.5 lbm/gal and the system volume is 1,593 bbl.

1. How much barite is needed to kill the well if no mud will be transferred out of the pits before weightup begins?

2. Determine the 13.5-lbm/gal mud volume to place in storage and the subsequent barite addition if the desired final volume is 1,593 bbl.

Solution.

1. The mud density needed to control the well is calculated as

$$\rho_{km} = \frac{300 + (0.0519)(13.5)(6,785)}{(0.0519)(6,785)} = 14.4 \text{ lbm/gal}.$$

Eq. 4.21 gives the barite requirement.

$$m_B = \frac{(1,593)(42)(4.2)(8.34)(14.4 - 13.5)}{(4.2)(8.34) - 14.4}$$

$$= 102,308 \text{ lbm}.$$

At 100 lbm/sack,

$$\frac{102,308 \text{ lbm}}{100 \text{ lbm/sack}} = 1,023 \text{ sacks}.$$

2. Eq. 4.22 gives both the transfer mud volume and the subsequent barite volume.

$$V_B = \frac{1,593(14.4 - 13.5)}{35.0 - 13.5} = 67 \text{ bbl}.$$

The corresponding mass and number of sacks of barite required are

$$m_B = (67)(42)(35.0) = 98,490 \text{ lbm} = 985 \text{ sacks}.$$

4.5 Kill Procedures

Secondary well control involves two separate tasks. First, the influx must be circulated from the hole while maintaining adequate backpressure at the choke to prevent further entry, not in a range that exceed the pressure integrity limits of the exposed formation, casing, and surface equipment. Then, before routine operations can be resumed, the mud weight in the wellbore and pits must be increased to the calculated requirement. These tasks may be accomplished separately or at the same time, depending on the selected kill procedure.

Numerous techniques for killing a well have been handed down over the years, but the fundamental principles used today have been understood and accepted in relatively recent time. The common basis for the approaches offered since about 1960 is that all endeavor to maintain a constant BHP while circulating the kick and increasing the mud weight.

It is of central importance to realize that keeping the BHP constant during a kill in no way means that the wellbore pressures at other depths will be constant. For one thing, the influx height is a function of the annular dimensions so a geometry change during displacement will result in pressure variation above the influx. But more importantly, maintaining a constant BHP means that we must allow a free gas influx to expand as it moves up the hole. Pressures above the gas kick must increase to make up for the loss in hydrostatic pressure.

BHP must be inferred from the gauge readings and so wellbore-pressure control is achieved by regulating the surface pressure.[29,30] Some earlier methods were based on controlling from the casing pressure.[30] In these procedures, Eq. 4.16 was used to ascertain the influx character. If a gas kick was indicated, volumetric expansion was predicted as a function of pump strokes and an increasing casing-pressure schedule was prepared for the kick circulation. A calculated incompressible fluid would be managed by little, if any, variance in the casing pressure. However, these historical procedures, being subject to the same limiting assumptions described in Sec. 4.4, did not work very well and the drillpipe-pressure control methods became predominant by the late-1960s.

Three kill procedures have been developed which use the drillpipe pressure as the means for controlling BHP. These are, in their most common designations, the Driller's Method, Wait and Weight Method, and Concurrent Method. The techniques are similar in many respects and certain procedural steps are identical, but the basic distinctions relate to when KWM is pumped with respect to influx displacement. The Driller's Method uses the mud currently in the pits to displace the influx from the annulus. Once the contaminants have cleared the wellbore, the original mud is displaced with KWM until the new mud is circulated around and the well is confirmed dead. The other two methods differ in that a heavier mud is pumped into the drillstring while the influx is still in the well. The Wait and Weight Method uses KWM whereas the Concurrent Method increases the mud densities by increments.

4.5.1 Circulation Principles. Kick circulation differs from the static condition in that the mud's flow resistance through the wellbore components must be considered. Calculating wellbore-friction pressures is generally unnecessary in most well-control applications, but there are occasions when annulus backpressures may be deemed excessive for the planned kill. In these cases, having a reasonable estimate of the annulus losses allows an operator to adjust the pump rate or, at times, the choke setting to maintain a safe margin below the fracture gradient. Friction-pressure relationships for a power law fluid are presented in Appendix A.

While drilling, the standpipe pressure gauge records the system friction pressure and any hydrostatic-pressure imbalance arising from spotty mud density or cuttings loading. Assuming a consistent wellbore fluid, the circulating drillpipe pressure (CDPP) is the sum of the pressure losses in the drillstring, bit, and annulus. When killing a well, the hydrostatic pressures are no longer balanced and additional backpressure is furnished at the choke. The circulating BHP as determined from the drillstring and annulus legs, respectively, of our U-tube are

$$p_{bh} = p_{dc} + p_{hd} - \Delta p_s - \Delta p_d - \Delta p_b \quad \ldots\ldots (4.23)$$

and $$p_{bh} = p_{ch} + p_{hcl} + p_{ha} + \Delta p_{cl} + \Delta p_a, \quad \ldots\ldots (4.24)$$

where p_{dc} = CDPP at the standpipe gauge, Δp_s = friction loss in the surface equipment downstream of the standpipe gauge, Δp_d = total friction loss in the drillstring, p_{ch} = pressure at the choke, p_{hcl} = total hydrostatic pressure in the choke lines, Δp_{cl} = total friction loss in the choke lines, and Δp_a = total friction loss in the annulus.

On surface stacks, the choke line provides no hydrostatic pressure and friction losses between the annulus and casing-pressure gauge are considered to be neglible. The circulating BHP is calculated as

$$p_{bh} = p_{ch} + p_{ha} + \Delta p_a. \quad \ldots\ldots (4.25)$$

Kill procedures in conventional wellbores usually are conducted at a pump rate between one-third and one-half the normal drill rate. Reasons for this practice include lower annulus-friction pressures, less pressure fluctuation in response to a change in choke setting, reduced risk of pump breakdown and other mechanical problems, ability to circulate with larger underbalance pressures for the given pump-liner rating, reduced instantaneous gas rates through the choke manifold and mud/gas separator, minimum cuttings throughput at the choke, and keeping up with the capabilities of the mud-mixing system. Equally important are the less tangible benefits which accrue from conducting a kill in a slow, methodical manner. The supervisor has more time to analyze the pressures and what they may be saying about the downhole conditions. Better judgement and wiser decisions invariably will follow.

The slow-circulation rate for each pump on the rig is selected before any well-control problems arise. Standard policy in drilling operations is to measure the pump pressure periodically at the designated kill rate for each pump. Knowing the kill-rate circulating pressure (KRCP) is a valuable and sometimes necessary adjunct to the information used to plan and execute a kill procedure. KRCP measurements are obtained and recorded by the driller at least once every tour (work shift). Rapid penetration rates may dictate more frequent measurements as will any changes in the mud density or thickness.

Table 4.13 contains information related to a kick on a hypothetical 12,000-ft well. The drillstring and annulus circulating pressure profiles while drilling at total depth are shown in **Fig. 4.33** and include the predicted dynamic-pressure losses and hydrostatic pressure of the 12.0-lbm/gal mud. The dashed line corresponds to the static wellbore condition and the difference between the static- and dynamic-annular profiles reflect the annulus-friction losses above a given depth.

In comparison, the effects of reducing pump speed to the kill-circulation rate are shown in **Fig. 4.34.** At the slower rate, predicted annular losses above total depth have been reduced

TABLE 4.13—KICK DATA FOR A HYPOTHETICAL WELL

Wellbore Configuration	
Vertical Depth	12,000 ft
Surface Casing Information	
Description	9⅝-in., 40.0-lbm/ft K-55 ST&C
Setting Depth	3,500 ft
Open Hole Diameter	8½ in.
Drillstring Information	
Drill Collar Size	6¼ × 2¾-in.
Drill Collar Length	600 ft
Drillpipe Description	4½-in. 16.60-lbm/ft
Drillpipe Length	11,400 ft
Bit Nozzle Flow Area	0.2623 sq in.
Prekick Circulation Data and Mud Properties	
Drilling Circulation Rate	90 strokes/min
CDPP While Drilling	3,190 psig
Kill Circulation Rate	45 strokes/min
KRCP	980 psig
Pump Output	3.89 gal/stroke
Mud Type	Water Base
Mud Density	12.0 lbm/gal
Power Law Fluid Rheology	
Flow Behavior Index (n)	0.60
Consistency Index (K)	500 eq cp
Recorded Kick Data	
SIDPP	250 psig
SICP	620 psig
Pit Gain	25 bbl
Other Known or Assumed Information	
Fracture Gradient at Casing Seat	0.81 psi/ft
Assumed Wellbore Temperature at Kill Rate	70°F + 1°F/100 ft
Kick Fluid	0.60 SG Gas

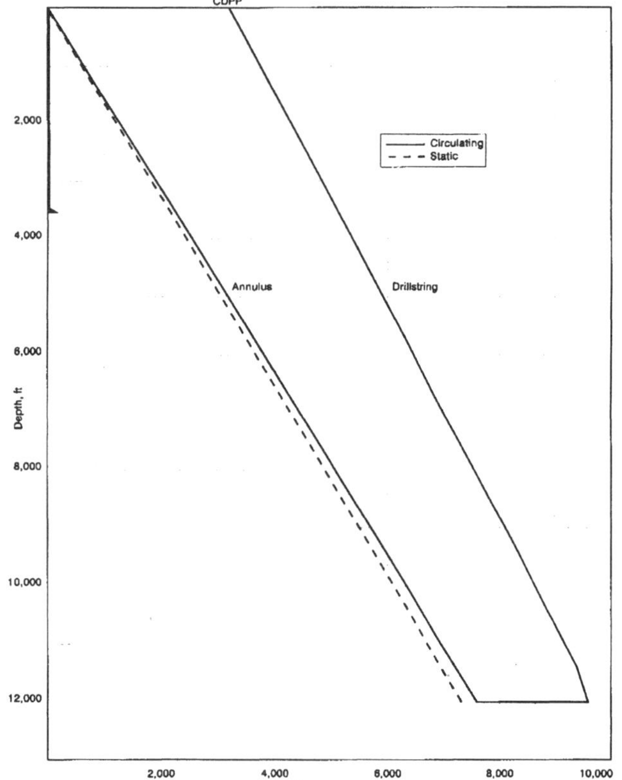

Fig. 4.33—Circulating-pressure profiles while drilling the well described in Table 4.13.

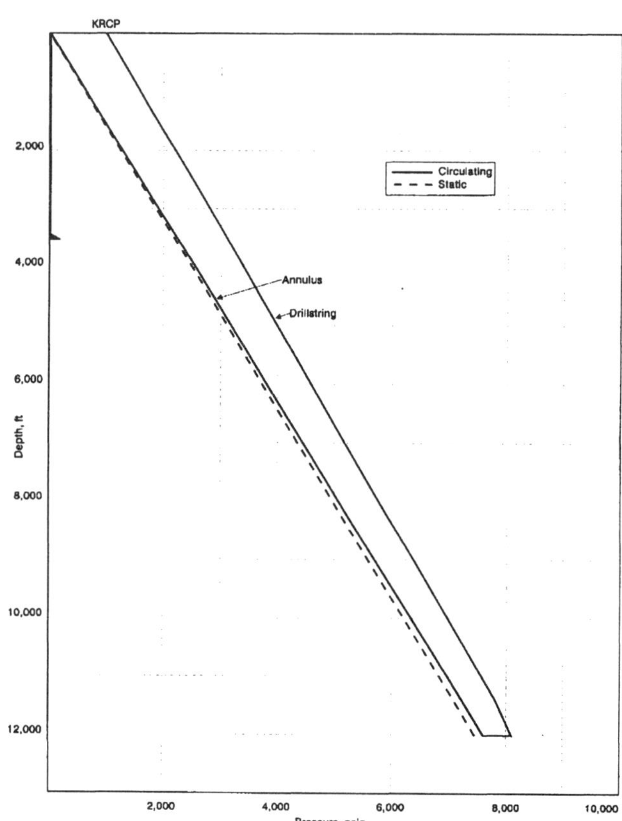

Fig. 4.34—Kill-rate circulating-pressure (KRCP) profile for the example well.

from 197 to 129 psi, thus reducing the ECD from 12.32 to 12.21 lbm/gal. The losses above the casing seat have been reduced from 45 to 29 psi, which equates to a change in ECD from 12.25 to 12.16 lbm/gal. Though not dramatic for these well conditions, the outcome could make the difference between losing and maintaining returns during a kill.

Conventional control methods require that the BHP be held constant by maintaining a drillpipe pressure which comprises the SIDPP and slow-rate resistance in the flow conduits and bit. Eq. 4.26 gives the initial circulating pressure (ICP) if weighted mud has not yet entered the drillstring.

$$p_{dci} = p_{ds} + p_{kr}, \qquad (4.26)$$

where p_{dci} = the ICP and p_{kr} = the KRCP. The entire system-pressure loss is incorporated in the p_{kr} term, meaning that the circulating BHP under this criterion will exceed the formation pressure by an amount equivalent to the annulus friction. Subsequent drillpipe-pressure-reduction schedules are based on the KRCP and so all control methods have a built-in overbalance "safety factor."

Consider again the well information in Table 4.13. The shut-in pressure profile for the drillstring and annulus (assuming a single gas bubble on bottom) are shown by the dashed lines in **Fig. 4.35.** The target ICP calculated from Eq. 4.26, 250 + 980 = 1,230 psig, was used to compute the drillpipe-pressure profile shown by the solid line in the diagram. As indicated, the dynamic BHP exceeds the static pressure by the 129-psi annulus friction at the kill rate. The density equivalent for this overbalance is 0.21-lbm/gal.

Normally, after a well is killed the mud density is increased by some margin over the KWM to counter the effect of wellbore-pressure reduction while making a connection or during a trip. Some operators may be tempted to go ahead and incorporate the trip margin before the kill or to add another increment to the ICP backpressure during the control procedure. As a rule, this practice is discouraged.[31,32]

Say that the operator of the well in Table 4.13 opted to increase the circulating backpressure by 200 psi and initially

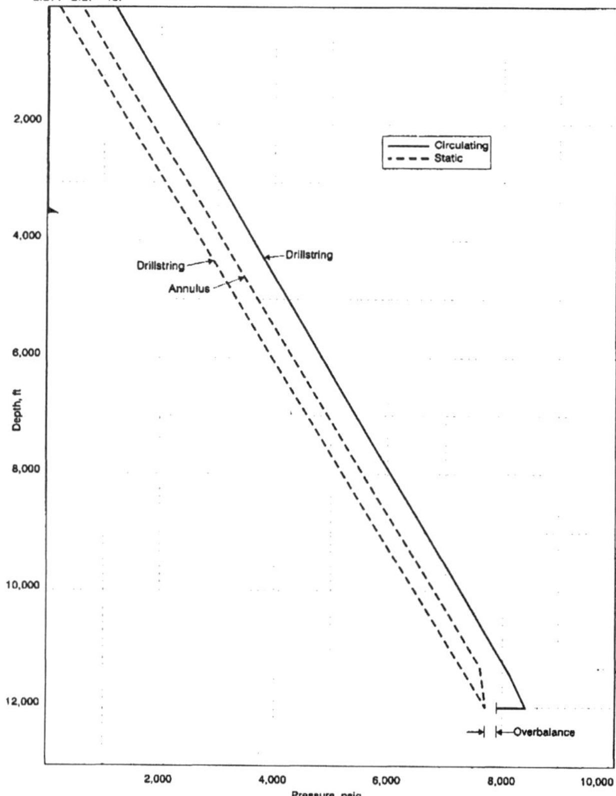

Fig. 4.35—Shut-in and dynamic BHPs for the example well.

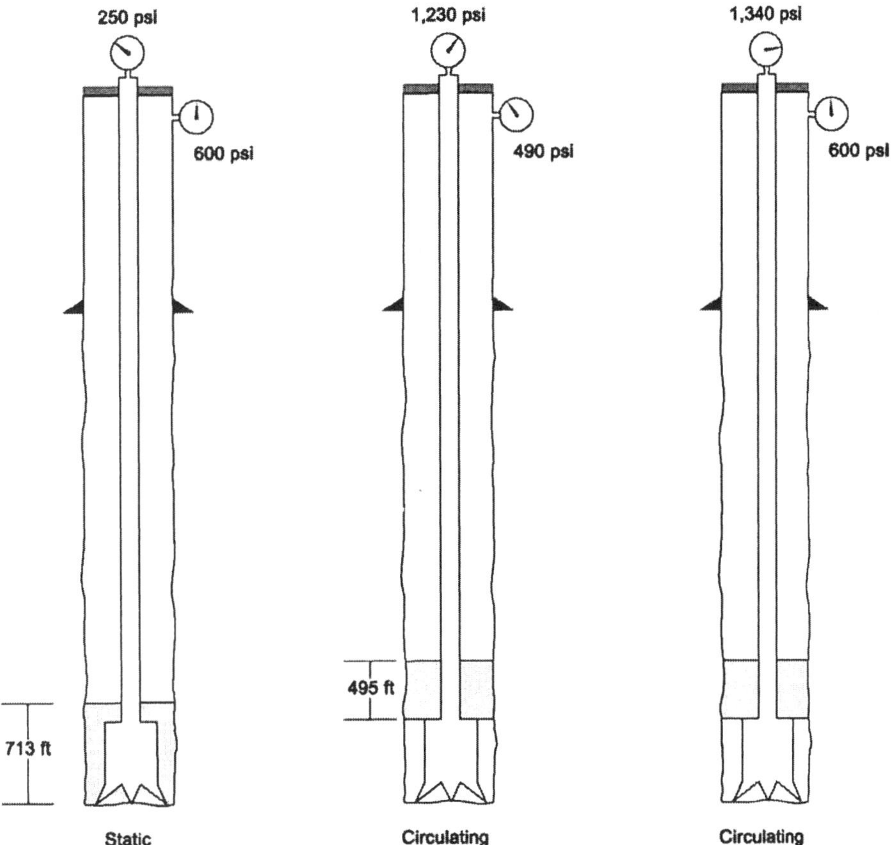

Fig. 4.36—The effect of maintaining a constant casing pressure when the kick described in Table 4.13 is circulated above the drill collars.

pumped at 1,430 psig instead of the recommended 1,230 psig. The higher pressure would be reflected throughout the wellbore and increase the equivalent shoe pressure by another 1.10 lbm/gal. Fracture pressure would have been exceeded in this case and the arbitrary pressure increase would be called more appropriately a "hazard factor." Annulus friction is generally sufficient to maintain adequate overbalance during a well kill. Any mistakes leading to a temporary underbalance may lead to a secondary kick, but these are normally small and easily managed.

Adjusting circulation rate will cause the CDPP to vary, thus it is desirable to pump at a constant rate throughout a kill operation. At minimum, however, a speed change will be necessary whenever the pump is brought on or off line and sometimes additional modifications to the circulation rate are recommended or needed. Keeping BHP constant when a speed change transpires requires that the casing-pressure gauge be monitored during the short time interval. Controlling from the casing gauge is valid as long as the hydrostatic pressure in the annulus remains fairly constant while the pump-speed adjustment is carried out, a condition generally met if the influx does not pass through different annular dimensions and before significant gas expansion occurs or weighted mud enters the annulus.

Normally we let the well dictate what the ICP should be rather than forcing the drillpipe pressure to obey the ICP calculated from Eq. 4.26. The operator will observe the drillpipe pressure immediately upon achieving kill rate and consider this to be the true ICP. The two values should be approximately the same and the calculated number is used only for comparison, but a significant disparity between the calculated and observed ICP should serve warning that something is amiss. The operator always has the option of interrupting a kill and shutting a well in if things do not seem right. The SIDPP should go back to its initial value if no procedural mistakes were made and the influx height did not change during pump startup. If it is different, bleeding the excess pressure through the choke and starting over would be in order.

The wellbore schematic on the left side of **Fig. 4.36** shows the precirculation condition for the problem well if we consider the influx as an isolated gas bubble. The drillpipe pressure would rise automatically to the target 1,230-psig ICP if we could bring the pump up to kill speed instantaneously—i.e., the bubble stays on bottom. Holding to this ICP and keeping the input mud density constant until the gas bubble clears the drill collars would result in the gauge pressures shown in the middle diagram.

Now take the same bubble position but hold the casing pressure constant at 600 psig while circulating the influx above the collars. As shown in the right-hand schematic of Fig. 4.36, the operator, upon looking at the drillpipe gauge, would not see the desired ICP but rather a circulating pressure of 1,340 psig. Holding this drillpipe pressure constant for the duration of a Driller's Method kick displacement would add 110-psi pressure to the wellbore, thus creating a hazard potential. This is an extreme situation which would not fit an actual well-control problem. Gas mixing and migration would serve to minimize the annular-pressure change deriving from bubble collapse; kick displacement would probably be no more than a couple of barrels during the pump speed change, and so the additional pressure caused by a slight shortening of the influx likely will be of no consequence.

In Sec. 4.3.4, we discussed the recommended procedure for obtaining the static BHP reading if the drillstring had a back-pressure valve. Rearranging Eq. 4.26 yields another way,

$$p_{ds} = p_{dci} - p_{kr}. \quad \text{(4.27)}$$

To use the relation, an operator opens the choke and initiates circulation immediately after shutting in a kick, records the ICP, and closes the well in again. The SIDPP then may be estimated by subtracting the KRCP obtained previously from the measured ICP. The procedure is less accurate, but may be necessary if the rig lacks the low-volume pump needed to pressure the drillstring slowly.

Fluid and wellbore elasticity cause a delay between the time a choke adjustment is made and when the result is observed at the drillpipe pressure gauge. An oft-quoted rule of thumb is to expect a two-second lag for every 1,000 ft of string depth, but the time can be much greater with OBMs or if a significant volume of free gas is present. The recommended technique for obtaining a change in the CDPP is to make the choke adjustment while monitoring the casing side until the desired incremental response is seen on the annulus pressure gauge. The same pressure change will be seen ultimately at the standpipe. Say, for example, that a 100-psi reduction in the CDPP is desired. The choke is opened until this pressure drop is seen at the casing gauge. The choke operator then turns his attention back to the drillpipe gauge and simply waits until the effect is registered.

Certain well-control actions depend on how far weighted mud has travelled down the drillstring. All methods require that drillstring capacities be determined in advance and converted into pump strokes. A pump stroke counter should be deemed an essential piece of equipment and it is a good idea to have a redundant counter should the one on the choke panel fail. Lacking a stroke counter, the choke operator is forced to rely on the less accurate technique of pump time to predict volume. Estimated annular volumes are not essential to well control, but are worthwhile in that the influx position and leading edge of the weighted mud can be predicted at critical points in the procedure.

Standardized calculation forms, or kill sheets, are of great assistance to the drilling supervisor in planning a kill. Relevant prerecorded data and information pertaining to the kick are entered and used to determine the KWM density and generate the desired CDPP schedule as a function of pump strokes and, in the Concurrent Method, mud weight. Kill sheets provide a methodical way to calculate the needed parameters and assure that no steps are forgotten—important considerations in the stressful environment associated with a well-control event.

Annulus pressures are allowed to change during a drillpipe-controlled kill procedure, but with some limitations. Free gas displacement results in a continually increasing casing pressure as the influx expands in moving up the hole and casing pressure reaches a maximum when the majority of the gas slug first hits the choke. By the same token, the maximum pressure applied to the casing shoe or other weak horizon occurs when the gas column below the critical depth reaches a maximum height. Depending on the casing depth and wellbore geometry, fracture-gradient concerns will be most pronounced either at initial shut-in or when the influx first reaches the weakest formation.

A standard prekick calculation determines the so-called maximum allowable annulus surface pressure (MAASP). The limit is based on the minimum pressure which potentially could exceed the structural integrity of the BOPE, casing, or open hole. Full working pressure, if tested previously to this value, is used for the stack and manifold while a pressure less than the API rating is applied to the casing. The MAASP dictated by the open hole is the leak-off test pressure less the hydrostatic of the wellbore fluids above the shoe (assuming the shoe has the lowest fracture gradient).

Example 4.11. Determine the MAASP for the well described in Table 4.12 if a 5,000-psi stack is flanged to a wellhead of the same pressure rating. Assume the casing was tested to 70% of its minimum internal yield pressure (MIYP) before drillout.

Solution. We can apply 5,000 psi safely to the stack and choke manifold equipment. The MIYP is 3,950 psi and so the MAASP from the standpoint of equipment failure is calculated as

$$(p_c)_{max} = (0.70)(3,950) = 2,765 \text{ psig}.$$

The MAASP limit to avoid exceeding the fracture pressure is readily determined as

$$(p_c)_{max} = (3,500)[0.81 - (0.0519)(12.0)] = 655 \text{ psig}.$$

The MAASP is the least of the three, or 655 psig.

Most kill sheets include a line that specifies the MAASP, but a word of warning needs to be inserted about its use in well control. Of course, rupturing the casing or exceeding the surface equipment pressure rating could lead to a catastrophe. The formation-fracture gradient, however, should not control the MAASP in a kick displacement procedure unless exceeding this pressure poses a real threat of fluids broaching to the surface outside the casing. This is a difficult question to address, particularly if working under government directive,[33] but it is generally better to lose circulation than to open the choke and allow additional formation fluid entry. Increasing the influx volume will cause the surface pressure to increase even more and magnify the problem.

Also, the MAASP depends not only on the last-measured fracture gradient but also on the annular fluid density between the weak point and casing pressure gauge. Hence the posted MAASP has little meaning when gas is above the shoe and higher casing pressures can be tolerated than would be implied by the kill sheet calculation. Cases where "low choke" methods may be acceptable and the issues concerning the risks of broaching will be covered in Chap. 5.

4.5.2 Driller's Method. Refer to the Driller's Method procedure in **Table 4.14**. The immediate hazard has been checked when a kick is contained in a shut-in wellbore and time is available for making the necessary phone calls, filling out and double-checking the kill sheet information and calculations, verifying the position of the pump line and manifold valves, meeting with the crew and all concerned personnel, and otherwise getting ready to pump. Step 1 is simply a reminder to be prudent and avoid excessive wellbore pressures during this period. Moving the pipe through the closed annular is recommended if possible.

The pump-stroke counters should be reset and the designated pump is brought up to the kill rate while holding casing pressure constant at the stabilized SICP reading. It is a good

TABLE 4.14—DRILLER'S METHOD PROCEDURE FOR KILLING A WELL

1. Take care of the well until the kill procedure can be started by maintaining a constant drillpipe pressure and allowing migrating gas to expand.
2. Open the choke and slowly start the pump. Coordinate choke setting with pump speed so that the original SICP is maintained until the pump is brought up to the kill rate.
3. Read the drillpipe pressure and compare to the computed ICP. Hold ICP constant by choke manipulation until the kick fluids are circulated from the hole and original mud density is measured at the choke outlet.
4. Reduce pump speed while closing the choke so that a constant casing pressure is maintained. When the pump is barely running, shut off the pump and finish closing the choke.
5. Observe pressure gauges and verify that both record the original SIDPP. If not, check for trapped pressure.
6. Recalculate the kill mud weight and increase the density of the mud in the pits.
7. Open the choke and slowly start the pump. Increase the pump speed to kill rate while maintaining constant casing pressure.
8. Continue to hold casing pressure constant until KWM enters the annulus.
9. Read the drillpipe pressure and hold constant until kill mud density is measured at the choke outlet.
10. Shut off the pump and close the well in.
11. Open the choke and check for flow.
12. Resume operations. Incorporate trip margin density in the mud.

Note: An alternate procedure involves replacing Step 8 so that a predetermined drillpipe-pressure-reduction schedule is obeyed while the string is filled with KWM.

practice to indicate the desired pressure on the controlling gauge with a paint stick or marker pen so that the choke operator does not forget what pressure to maintain. Upon achieving kill rate, the choke operator's attention will turn immediately to the drillpipe gauge and the observed circulating drillpipe pressure will be marked and held constant until all the influx has been circulated from the well. Again, this circulating pressure should be about the same as the calculated ICP if no changes have been made to the circulating system and if influx height did not vary significantly while bringing up the pump.

Casing pressures will begin to rise in the case of a gas influx; slowly at first but rapidly later on as the expanding influx approaches surface. Maximum casing pressure will occur when the annulus hydrostatic pressure is minimum—i.e., when the bulk of a gas kick reaches the choke. Gas offers less resistance through the choke and manifold equipment than drilling mud and so the choke operator should be prepared to pinch back the choke when gas surfaces. The overall pit gain will increase also with gas expansion as more mud is displaced from the annulus than is picked up at the pump suction. It is wise to estimate the maximum surface volume increase if overflowing the pits is a concern.

Gas voidage through the choke combined with increasing annular mud height from bottom will lead to a steady reduction in casing pressure until all of the gas has been cleared from the wellbore. A laminar-flow profile in the annulus may cause some gas to lag, but the stroke counter should correspond to approximately one annulus volume when the influx has been evacuated. Casing pressure now should be the original SIDPP plus any small friction losses downstream of the choke.

Usual practice is to close the well back in again after completing the first circulation. The pump rate is stepped down by controlling from the casing gauge and the choke is closed with hopefully noncontaminated mud in the annulus. Both pressure gauges should be the same if the procedure has been successful thus far. A secondary kick, identified by a disparity between the casing and drillpipe pressure, may be displaced by repeating the kick-displacement procedure. However, a small kick may not be discerned at this time and diligence must be maintained by monitoring the hole until KWM is ready to be pumped.

The SIDPP after circulating a kick should be close to the original measurement, but inconsistent mud densities throughout the system, trapped pressure, or transient pressure buildup in the formation can lead to a different reading. Frequent density measurements during kick displacement will quantify better the downhole mud weight and we can check to see if pressure has been trapped in the well, perhaps as result of closing the choke before the pump completely stopped rolling. The way to confirm and eliminate any trapped pressure is to open the choke momentarily and bleed a small volume of mud from the annulus. The drillpipe pressure should fall and stabilize at a lower value if the hole is overbalanced. The open-and-bleed process is repeated until the drillpipe pressure remains constant. Release of mud past this point will lead to an underbalanced condition and allow more entry.

In any event, the kill-mud density should be based on the second stabilized SIDPP reading because many of the error sources will have been eliminated at this time, most importantly those resulting from gas migration. After verifying or recalculating the kill mud weight, the density in the pits will be increased while continuing to monitor the shut-in pressures. Any increase beyond that needed to kill the well is not recommended until the hole has been circulated with KWM and observed to be dead when open at the annulus.

The stroke counter is reset to zero and the second circulation of the Driller's Method begins when KWM is pumped into the drillstring at the desired pump rate. If no mistakes were made, we should have a full column of original mud in the annulus. The easiest way to maintain a constant BHP is by holding casing pressure constant while the drillstring is filled with KWM. Once heavier mud reaches the bit, the drillstring side is balanced and continued pumping at constant casing pressure will start to increase BHP. Hence we can no longer control off the casing pressure and thus we turn our attention to the drillpipe gauge when the stroke counter reflects the calculated drillstring volume. The observed CDPP is held constant until the mud engineer verifies kill-mud returns. This CDPP should be a little higher than the KRCP because the heavier mud leads to a higher pressure drop through the drillpipe, BHA, and bit.

Some authorities contend that a wellbore free from secondary kicks cannot be guaranteed, which means that an alternate means for filling the drillstring with KWM must be used. Holding casing pressure constant during this portion of the procedure would not maintain the desired BHP if a gas influx was in fact present. The hole likely would become underbalanced under this scenario and so the pipe-filling procedure taught in many well-control schools is to control from the drillpipe gauge and manipulate the choke so that drillpipe pressure is reduced along a predetermined schedule. The technique is exactly the same as the one used in the Wait and Weight Method and the procedural discussion is presented in Sec. 4.5.3.

The argument may have some validity, but we lose two of the more compelling reasons for using the Driller's Method—its flexibility and simplicity, if a step-down pressure schedule has to be determined and followed during the job execution. It may be a judgment call, but there should be indications of significant incidental entry after the first circulation has been completed and the operator should have some understanding by this time of how well the kick-displacement operation was conducted. A secondary kick in most wellbores, if relatively deep, will not expand much during the relatively short time required to displace the string. Absent significant gas expansion, the casing gauge is suitable for control.

Regardless of how the drillstring is filled, the casing pressure will decline as the KWM column height increases in the annulus. A continuing trend of larger choke openings must follow as constant CDPP is maintained until the original mud has been displaced completely from the hole. The choke likely will be wide open when KWM surfaces and the casing gauge should read zero. The well has been killed if the calculations were valid and the procedure successfully carried out. The best way to demonstrate the application is to work through an example problem.

Example 4.12. Write a procedure to kill the well described in Table 4.13 using the Driller's Method.

Solution. The drilling supervisor probably has a standardized kill sheet showing the relevant prekick information, the shut-in pressures, and pit gain. At minimum, the prekill calculations will include KWM determination, the ICP, and the pump stokes to fill the drillstring. The KWM based on the initial SIDPP is

$$\rho_{km} = \frac{250 + (0.0519)(12.0)(12,000)}{(0.0519)(12,000)} = 12.4 \text{ lbm/gal}.$$

This parameter will be confirmed after displacing the kick. The calculated ICP for this case was determined by applying Eq. 4.26 as 1,230 psig (250 + 980). The drillstring volume is the sum of the length—i.e., capacity factor products of each string component, or

$$(11,400)(0.01422) + (600)(0.00735) = 166.5 \text{ bbl}.$$

Converting the drillstring volume to pump strokes gives

$$(166.5)(42)/3.89 = 1,798 \text{ strokes}.$$

Similarly, the annulus volume and equivalent pump strokes are computed as

$$(3,500)(0.05616) + (7,900)(0.05051) + (600)(0.03224)$$
$$= 614.9 \text{ bbl}$$

and $(614.9)(42)/3.89 = 6,639$ strokes.

One complete circulation will require

$$1,798 + 6,639 = 8,437 \text{ strokes}.$$

The first circulation will be complete after pumping one annulus volume,

$$6,639/45 = 148 \text{ min},$$

whereas a full circulation will be necessary for the second phase of the procedure.

$$8,437/45 = 187 \text{ min}.$$

At least 5½ hours of circulation plus the weightup time will be devoted to killing this well. The barite quantity needed to increase the density of the 1,500-bbl system is given by a simplified form of Eq. 4.21,

$$\frac{(1,470)(1,500)(12.4 - 12.0)}{35.0 - 12.4} = 39,017 \text{ lbm}$$
$$= 391 \text{ sacks}.$$

After weighting up the surface pits, the barite addition rate during the second circulation can be determined by dividing the barite required for the 781-bbl hole volume by the corresponding pump time.

$$\frac{(781 \text{ bbl})(391 \text{ sacks})}{(1,500 \text{ bbl})(188 \text{ min})} = 1.1 \text{ sacks/min}.$$

The procedure follows.

1. Reset the cumulative stroke counter. Maintain 620 psig casing pressure until the pump speed is increased to 45 st/min.
2. Read the drillpipe pressure and verify that the ICP is about 1,230 psig. Hold this ICP constant until the kick fluids are circulated (at least 6,639 strokes).
3. Maintain constant casing pressure (should be 250 psig) while reducing pump speed to zero. Close the choke.
4. Observe the SIDPP and SICP. Release trapped pressure if necessary.
5. Verify the 12.4-lbm/gal kill-mud weight using the current SIDPP. Increase the density of the mud in the pits.
6. Reset the cumulative stroke counter. Maintain the SICP constant and increase pump speed to 45 st/min. Start increasing the density of the return mud.
7. Continue holding constant casing pressure until the stroke counter reads 1,798 strokes.
8. Read the drillpipe pressure and maintain constant until the KWM circulates (at least 8,437 strokes).
9. Shut off the pump and close the well in. Check for flow.

Fig. 4.37 shows the drillpipe pressures corresponding to the kill; the predicted casing-pressure response is shown in **Fig. 4.38.**

Annulus pressures were predicted using an iterative technique discussed in Sec. 4.6 of this chapter. The simple algebraic equations used to calculate these casing pressures are based on some generally unsupported assumptions summarized in **Table 4.15.** Thus the kick displacement curve represents a set of circumstances that would not be seen in an actual well-control problem. Nonetheless, the response characteristics should be similar to a genuine kill.

The initial casing-pressure decline between zero and 209 strokes reflects the collapse of the gas bubble when displaced from the bottom of the drillstring up to the larger annular space opposite the drillpipe. The subsequent pressure increase, signifying gas expansion, occurs until a maximum 1,187-psig casing pressure is reached when the influx hits the choke at 5,679 strokes. Note that the break in the curve at 4,096 strokes indicates bubble-height reduction when moving from the gauge openhole section into larger-diameter casing. Annular mud starts to replace the gas when influx voidage begins and a linear pressure decline follows until kick displacement is complete at 6,639 strokes. From this point,

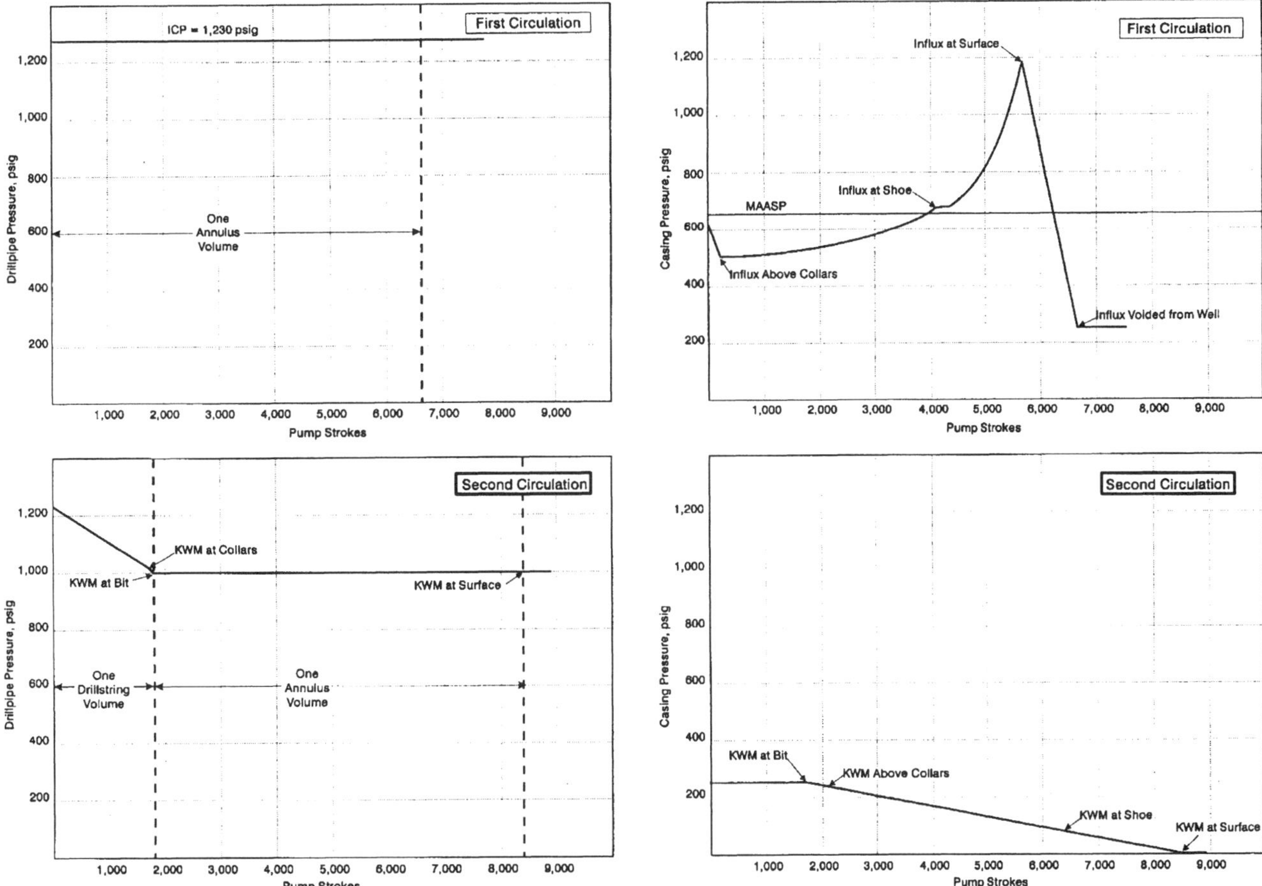

Fig. 4.37—Drillpipe pressures corresponding to a Driller's Method kill of the well described in Table 4.13.

Fig. 4.38—Ideal casing pressures corresponding to a Driller's Method kill of the example well in Table 4.13.

the annulus backpressure should agree with the original 250-psig SIDPP.

We should always be concerned with pressures imposed at the casing seat and their impact on wellbore integrity during a kill operation. **Fig. 4.39** shows the calculated shoe pressures for the example well during the first and second circulation. During the first circulation, maximum pressure is achieved when the top of the gas bubble has reached this point in the well. The effective gas-column length below the shoe shortens with continued circulation until the bubble completely enters the casing. The pressure is stable throughout the remaining kick displacement because the mud density below casing depth is constant. The same pressure is realized in the second circulation until KWM starts to replace the original mud in the annulus and a pressure decline follows until the heavier mud enters the casing.

We see that the leak-off test (LOT) pressure at the shoe and the MAASP for a 12.0-lbm/gal were exceeded by some 20 psi at maximum shoe pressure. Thus the calculated data suggests that returns would be lost during a Driller's Method procedure and another means for killing the well should be employed. The single-bubble assumptions, however, generally introduce a conservative element to calculated annulus pressures and so the actual imposed pressure at the shoe probably would be less than the predicted value. Also recall that fracture gradients, especially when measured at relatively

TABLE 4.15—ASSUMPTIONS USED IN ANNULUS PRESSURE CALCULATIONS

1. The influx enters the well and resides in the annulus as a continuous slug or bubble.
2. The influx remains in slug form throughout the displacement process—i.e., no mixing occurs with the mud.
3. The influx is a consistent fluid in one phase.
4. The influx does not change phases during the displacement process.
5. The influx is at the bottom of the drillstring when circulation begins.
6. Free gas does not slip or migrate through a circulating or static mud column.
7. No free gas dissolves in the mud.
8. Free gas behaves according to the real gas law.
9. Free-gas density does not vary over a gas column height.
10. The wellbore and mud are rigid.
11. Annulus-friction losses are negligible.
12. Inertial effects during free-gas expansion are negligible.
13. The annulus temperature profile during circulation can be quantified.
14. Openhole diameters can be quantified and are consistent.
15. The choke operator maintains constant BHP.

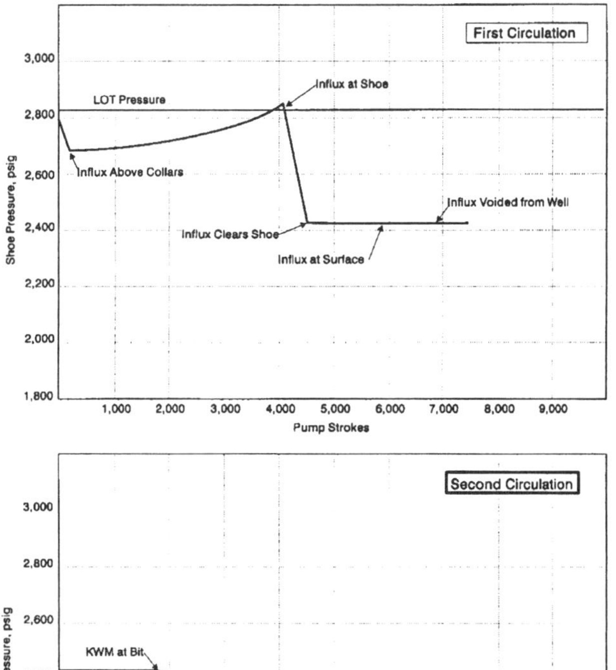

Fig. 4.39—Ideal casing-seat pressures corresponding to a Driller's Method kill of the example well in Table 4.13.

shallow depths, tend to increase with circulation time. These factors may come into consideration when deciding the proper kill approach.

Considering gas migration and mixing into the mud would yield more realistic casing pressures similar to the dashed curve shown in **Fig. 4.40.** The initial SICP would not be as high because most of the influx was commingled with the circulating mud before the flow check, plus migration tends to further disperse the gas into the carrier drilling fluid. Gas continues to slip faster than the annular mud velocity during the kick circulation and a laminar-flow-velocity profile will increase gas dispersion, or "string the gas out" through the mud column. Thus peak casing pressures will arrive sooner and be lower than predicted for the ideal case. The dashed line in Fig. 4.40 shows gas passage through the choke after pumping enough mud to fill the annulus, an expected outcome from laminar flow in the annulus. Though realism can be better approximated, an accurate kick model is not suited for hand calculations and we suggest basing well-control decisions on the conservative approach unless a kick simulator is available.

Again based on the single-bubble model, the computed pit-volume increase while circulating the gas is shown in **Fig. 4.41.** The initial 25-bbl pit gain when circulation began rises to a maximum of 88.9 bbl when the gas has reached maximum expansion in the wellbore. A linear decline reflecting the kill pump rate follows as gas vents through the choke and the prekick pit volume should be noted at the PVT monitor when influx displacement is complete.

Figs. 4.42 through 4.45 illustrate the effect of some of the key variables on calculated ideal casing pressures during kick

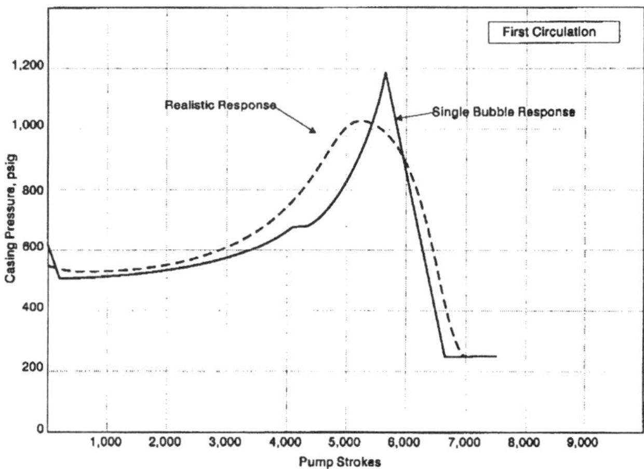

Fig. 4.40—Comparison between a single-bubble kick displacement with the Driller's Method and the case where gas migrates and mixes in the mud.

displacement. These are, in order of appearance, initial kick volume, kick intensity, annular dimensions, and formation fluid type. Keep in mind that all the gas-influx curves are based on the single-bubble concept.

The benefits of early kick detection and containment are illustrated clearly in the comparison shown in Fig. 4.42 where, keeping all other parameters the same, the 25-bbl influx is compared to initial pit gains of 15 and 35 bbl. What these curves suggest is that returns probably would be lost on initial shut-in or later during the circulation had the initial pit gain been much higher than 25 bbl.

A greater differential pressure between the rock and wellbore will lead to higher casing (and drillpipe) pressures while controlling a well. Fig. 4.43 compares the 0.4-lbm/gal kick intensity from Table 4.13 to the calculated casing pressures resulting from a 0.9-lbm/gal underbalance. As in the preceding comparison, exceeding fracture pressure is possible at a 0.4-lbm/gal underbalance, but probably an ensured outcome at the higher value. Two curve features having to do with gas volume increase in the well are of interest here; the 0.9-lbm/gal kick arrives at the choke later and has a smaller difference between maximum and minimum casing pressure. Less gas expansion will occur if higher wellbore pressures must be maintained to keep BHP constant.

The effect of different wellbore geometries is illustrated in Fig. 4.44 where the example drillstring is placed in a larger and smaller hole size; namely a $7^7/_8$-in. open hole below

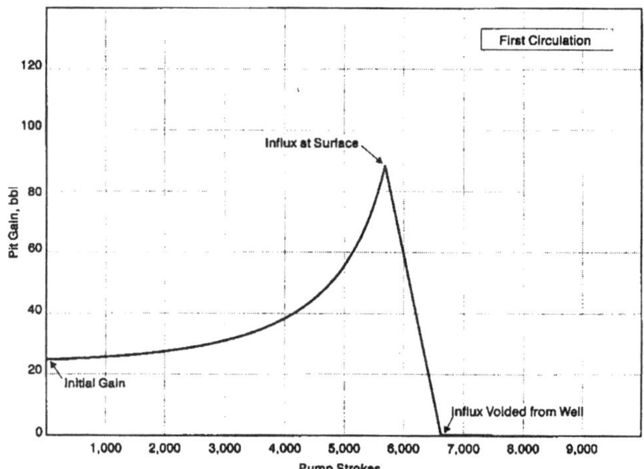

Fig. 4.41—Ideal pit-volume gains corresponding to a Driller's Method kill of the example well in Table 4.13.

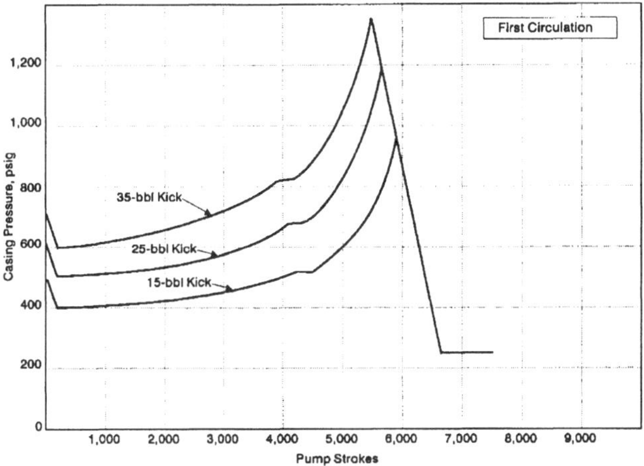

Fig. 4.42—The effect of influx volume on casing pressures while circulating a gas kick using the Driller's Method.

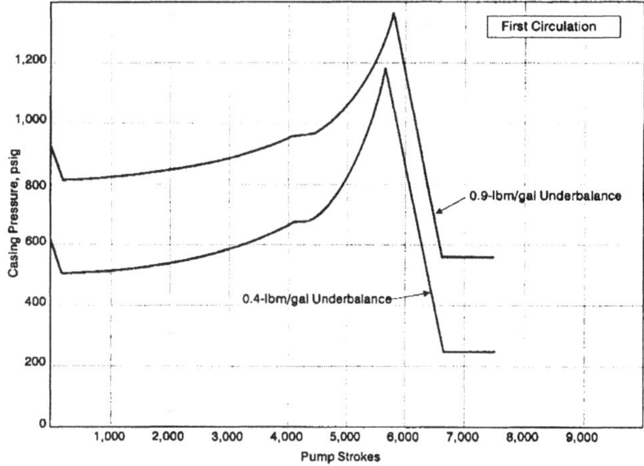

Fig. 4.43—The effect of kick intensity on casing pressures while circulating a gas kick using the Driller's Method.

$8^5/_8$-in. casing and a $9^1/_2$-in. open hole below $10^3/_4$-in. casing. As indicated, influx length in the smaller annular clearance leads to a significant increase in casing pressures, whereas kick circulation would be deemed safe in the $9^1/_2$-in. hole. By extension, the importance of limiting kick volumes in slimhole wells becomes apparent.

Finally, Fig. 4.45 compares the gas kick with a 25-bbl saltwater influx weighing 9.0 lbm/gal. Because the saltwater influx is relatively incompressible, very little volume change results from reducing pressure on the salt water during the displacement and the only casing-pressure fluctuations before the influx surfaces occur when the kick fluid passes through a changed annular dimension.

4.5.3 Wait and Weight Method. The Wait and Weight Method, sometimes called the Engineer's Method, realizes the same basic goal as the Driller's Method in that the procedure is designed to maintain a constant BHP somewhat higher than the formation pressure. The task, however, is completed in one wellbore circulation because KWM is pumped down the drillstring at the same time as influx displacement. The simultaneous handling of both objectives leads to certain distinct advantages in well control. The Wait and Weight Method is the preferred technique among many operators and drilling contractors, but the Driller's Method does retain some favorable attributes.

Fig. 4.46 illustrates the four consecutive stages during a Wait and Weight kill of a gas formation. Time 1 is the wellbore condition for an isolated gas influx immediately after KWM starts down the hole at the designated kill rate. The drillpipe gauge, measuring both flow resistance and underbalance, indicates the ICP. At Time 2 KWM has filled the drillstring and the CDPP must be at some lower final circulating pressure (FCP) if the constant BHP requirement is to be met. It is important to note and remember that the wellbore pressures in the annulus from Time 1 until Time 2 are the same as would be expected when displacing a kick with the Driller's Method.

Time 3 reflects the point when the gas first reaches the presumably weakest point in the hole at the last casing seat. KWM in this particular wellbore configuration that now occupies some height in the annulus serves to increase the hydrostatic pressure on the kick formation. Thus the casing pressure and imposed shoe pressure would be less than experienced in the Driller's Method if in each case the bubble started at the same depth. Continuing to Time 4, we have maximum casing pressure as the gas bubble first reaches the choke. Because of the KWM height in the annulus, the maximum surface pressure using the Wait and Weight Method generally will be less than for the Driller's Method. All that remains for this well to be killed is to finish displacing the influx and lower-density mud that was in the drillstring when the process began.

Table 4.16 gives a detailed kill procedure. The first step is identical to the Driller's Method. On most rigs, weighting up a portion of the surface mud system will add significantly to the time before circulation can begin. Monitoring the hole and

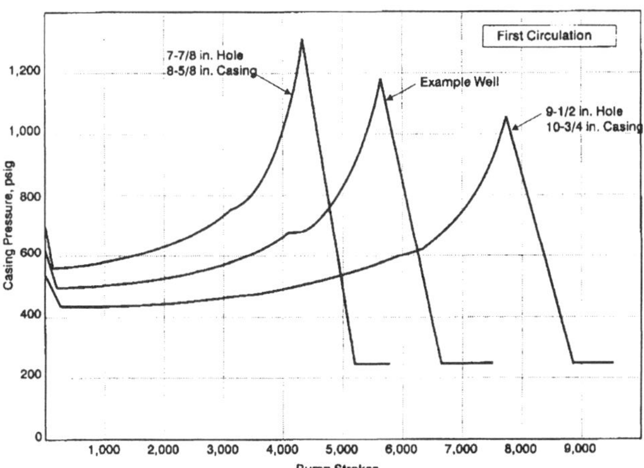

Fig. 4.44—The effect of hole geometry on casing pressures while circulating a gas kick using the Driller's Method.

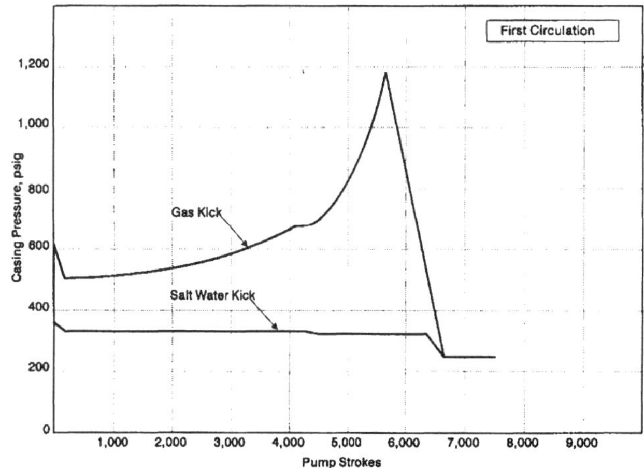

Fig. 4.45—The effect of influx type on casing pressures while circulating a kick using the Driller's Method.

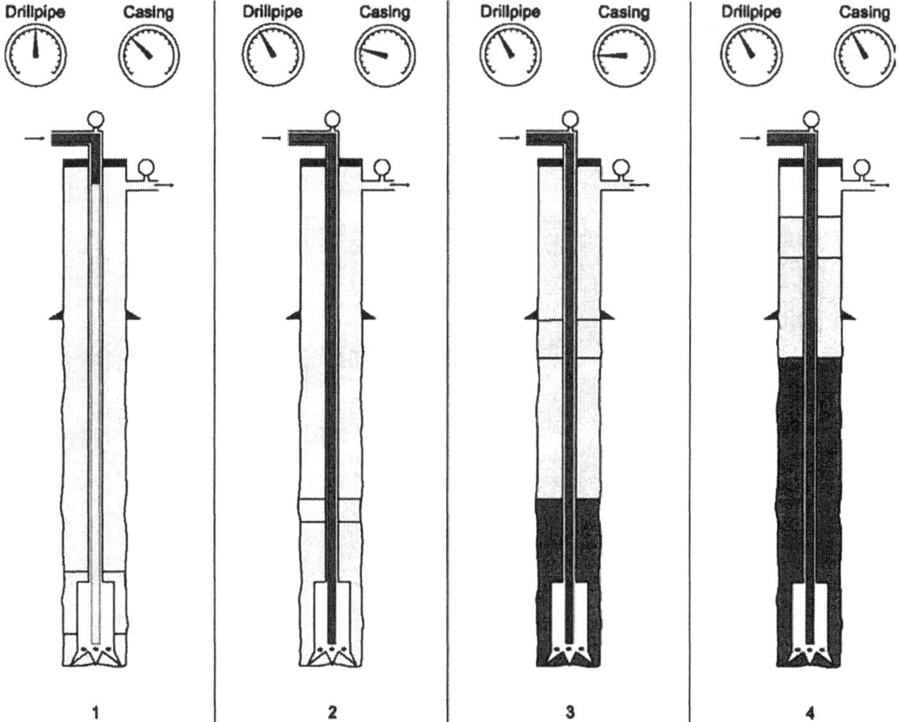

Fig. 4.46—Kick displacement and KWM circulation at four consecutive stages in the Wait and Weight Method.

letting any migrating gas expand gains importance in the Wait and Weight Method.

Filling the drillstring with KWM is more complicated than the equivalent operation in the Driller's Method procedure because the influx in the annulus must be allowed to expand during displacement. Maintaining a constant BHP cannot be accomplished with any accuracy by controlling from the annulus and the drillpipe gauge is used throughout the procedure (excepting pump rate changes). The drillpipe pressure cannot remain constant while different fluids are in the drillstring because the hydrostatic pressure is changing continually and so the operator must force the CDPP to decline as a function of pump strokes for one drillstring volume.

The small increase in friction and, possibly, hydrostatic pressure resulting from the negligible weighted mud volume pumped during the startup period is ignored. Accordingly, the ICP should reflect closely the SIDPP and KRCP components applicable to the original mud weight. When heavier mud exits the bit, the SIDPP component reduces to zero and all that remains is the flow resistance. However, the FCP value will not be the same as the original KRCP measurement because mud properties have changed, and so an adjustment must be made to arrive at the final targeted drillpipe pressure. The customary practice is to modify the KRCP by a density-multiplying factor according to the relation

$$p_{dcf} = p_{kr}(\rho_{km}/\rho_{om})^{a}, \quad \text{(4.28)}$$

where p_{dcf} = the final circulating pressure and ρ_{om} = the original mud density. The equation seems straightforward enough but, unfortunately, the exponent a is not an easily obtained parameter.

Eq. 4.28 implicitly assumes turbulent flow in the pipe bore at the kill rate for both original and final mud weights and no change in the other flow parameters that affect friction losses. The friction-pressure drop, Δp_f, for a Bingham plastic fluid at condition 2 when mud weight has changed from condition 1 is given by

$$\Delta p_{t2} = \Delta p_{t1}(\rho_2/\rho_1)^{0.75}$$

TABLE 4.16—WAIT AND WEIGHT METHOD PROCEDURE FOR KILLING A WELL

1.	Take care of the well until the kill procedure can be started by maintaining a constant drillpipe pressure and allowing migrating gas to expand.
2.	Calculate the kill mud weight and increase the density of the mud in the suction pit. Determine the weight material addition rate needed for the pumping operation and generate a drillpipe pressure-reduction chart or table.
3.	Open the choke and slowly start the pump. Coordinate choke setting with pump speed so that the original SICP is maintained until the pump is brought up to the kill rate.
4.	Read the drillpipe pressure and compare to the computed ICP. Manipulate the choke so that the CDDP follows the pressure reduction schedule.
5.	Maintain drillpipe pressure constant at the FCP value until kill mud density is measured at the choke outlet.
6.	Reduce pump speed while closing the choke so that a constant casing pressure is maintained. When the pump is barely running, shut off the pump and finish closing the choke.
7.	Observe pressure gauges and verify that both read zero. If not, check for trapped pressure.
8.	Open the choke and check for flow.
9.	Resume operations. Incorporate trip margin density in the mud.

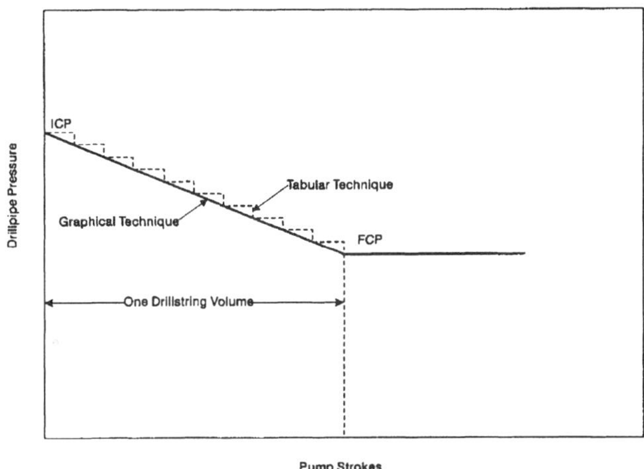

Fig. 4.47—Drillpipe-pressure-reduction schedule for the Wait and Weight Method.

However, the pressure drop through the bit yields the ratio function

$$\Delta p_{b2} = \Delta p_{b1}(\rho_2/\rho_1)^{1.00}.$$

Thus a should be approximately 0.75 before KWM reaches the bit and somewhere between 0.75 and 1.00 thereafter. Most people conclude that bit losses predominate and use an exponent of unity. In general, this is appropriate because the calculation is simplified and any associated errors lean toward conservatism.

Fig. 4.47 demonstrates two ways of accomplishing the drillpipe-pressure-reduction schedule, both customarily using a straight-line relationship. One is to construct a graph with the ICP at pump startup and the FCP at the pump stroke equivalent of the drillstring volume. The choke operator then attempts to comply with the straight line connecting the two endpoints. Another approach is to step the CDPP down at periodic intervals by determining the incremental pressure decline over a specified volume, and a table is constructed for the choke operator to follow.

As with the Driller's Method, we normally consider the observed drillpipe pressure as being the ICP needed to maintain the desired BHP and a significant disparity between actual and calculated values should be cause for investigation. The predetermined pressure-reduction schedule will be invalid if the ICP is modified, and so the well must be shut-in and another graph or table prepared before circulation can continue.

Let us examine the conventional schedule and its effect on circulating BHP during a kill. **Fig. 4.48a** depicts a straight-line fit between the ICP and FCP for respective a exponent values of 1.00 and 0.75. The chart shown in **Fig. 4.48b** is on an expanded scale and shows the circulating BHPs during the schedule for each selected exponent. Note that consistent pipe dimensions and a Bingham plastic fluid in turbulent flow are assumed in the BHP curves.

Initially, the BHP exceeds formation pressure by an amount equivalent to the annulus friction losses at the KRCP. The line fit for both schedules incorporates the bit pressure drop and so we would expect to see a linearly increasing BHP as KWM moves down the string. A discontinuity occurs when the bit-pressure drop suddenly increases resulting from the heavier mud and both curves drop to a lower overbalance for the remainder of the kill.

The final BHP for the larger exponent is higher than experienced at the ICP, whereas there is an overbalance reduction with the smaller exponent. It is reasonable to desire a final BHP at least as high as initially, thus setting $a = 1.00$ is a logical decision. The added backpressure at the casing seat while reducing the drillpipe pressure will be negligible in most cases and certainly within the tolerance afforded by the choke operator's capabilities.

A modified Wait and Weight procedure that did away with the pressure-reduction schedule gained some popularity for a period. The argument, justified by calculated data in many cases, held that gas expansion in most wells is small enough during the relatively short fillup time that little sacrifice in BHP reduction would result by holding casing pressure constant until weighted mud reached the bit. This reduced the preparation work and was easier to execute, but secondary

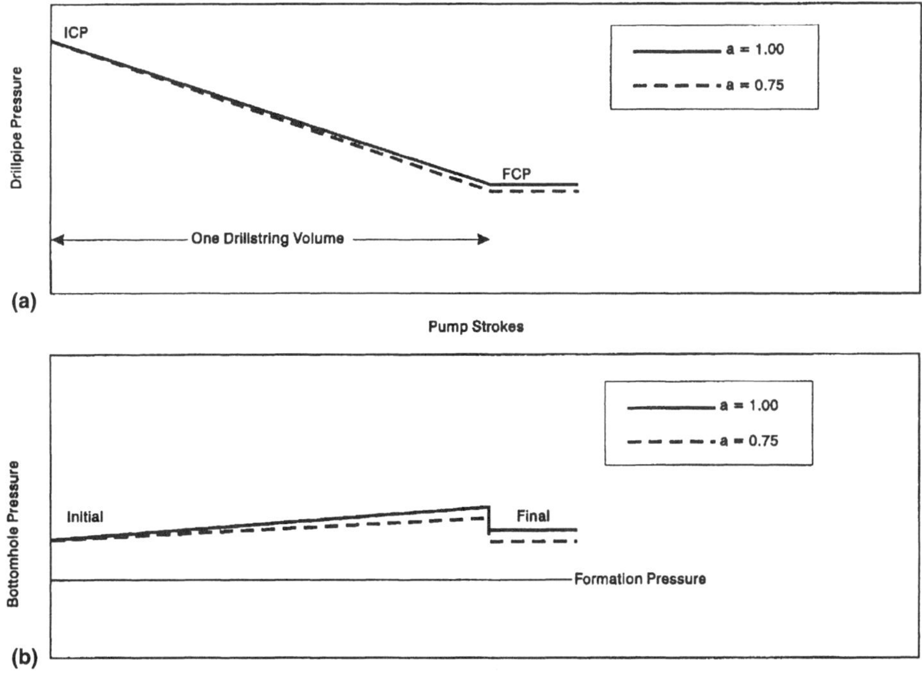

Fig. 4.48—The effect of a straight-line drillpipe pressure reduction on circulating BHP with consistent string ID.

TABLE 4.17—DRILLPIPE PRESSURE REDUCTION SCHEDULE FOR EXAMPLE 4.13	
Drillpipe Volume (pump strokes)	Drillpipe Pressure (psig)
0 to 180	1,230
180 to 360	1,208
360 to 540	1,186
540 to 720	1,164
720 to 900	1,142
900 to 1,080	1,120
1,080 to 1,260	1,098
1,260 to 1,440	1,076
1,440 to 1,620	1,054
1,620 to 1,800	1,032
1,800 +	1,013

kicks seemed to occur more often when the modified technique was employed. Gas migration during the preceding shut-in period is the likely culprit, as the pressure reduction will be lower if gas is assumed to be on bottom than if it is assumed that the gas has moved some distance up the hole.

Example 4.13. Write a procedure to kill the well described in Table 4.13 using the Wait and Weight Method.

Solution. Most of the basic data acquistion and calculation requirements are no different from the Driller's Method. However, we must develop a drillpipe-pressure schedule to follow when circulation begins. The calculated ICP is still 1,230 psig, whereas Eq. 4.28 yields the FCP,

$$p_{dcf} = 980(12.4/12.0) = 1,013 \text{ psig}.$$

We choose in this problem to build a table for the choke operator's use. A stair-step pressure reduction over ten drillstring-volume increments makes sense and so the drillpipe pressure will change every

$$1,798/10 = 180 \text{ strokes}.$$

The pressure reduction at each pump stroke increment will be

$$(1,230 - 1,013)/10 = 22 \text{ psi}.$$

Table 4.17 gives the schedule for this kill. The rest of the procedure follows.

1. Increase the density of the pit mud to 12.4 lbm/gal.
2. Reset the cumulative stroke counter. Maintain 620 psig casing pressure until the pump speed is increased to 45 st/min. Start increasing the density of the return mud.
3. Read the drillpipe pressure and verify that the observed ICP is approximately 1,230 psig.
4. Manipulate the choke so that the drillpipe pressure obeys the schedule shown in Table 4.17.
5. At 1,798 strokes, hold the 1,013-psig FCP constant until the KWM circulates (at least 8,437 strokes).
9. Shut off the pump and close the well in. Check for flow.

Fig. 4.49 shows the drillpipe pressures; **Fig. 4.50** illustrates the predicted casing pressures.

Fig. 4.51 compares the predicted casing-shoe pressures for the first circulation of the Driller's Method with those of the Wait and Weight Method. As shown, the pressures imposed in the latter procedure never exceed the LOT pressure and, for this reason, the Wait and Weight is the preferred way to kill this well. An operator should abandon the notion of using the Driller's Method and consider other options whenever the potential for lost circulation is indicated. Granted, the calculations used to generate these particular casing-pressure curves are conservative and the fracture pressure measured after drilling the shoe have likely increased. But why gamble when it is not necessary?

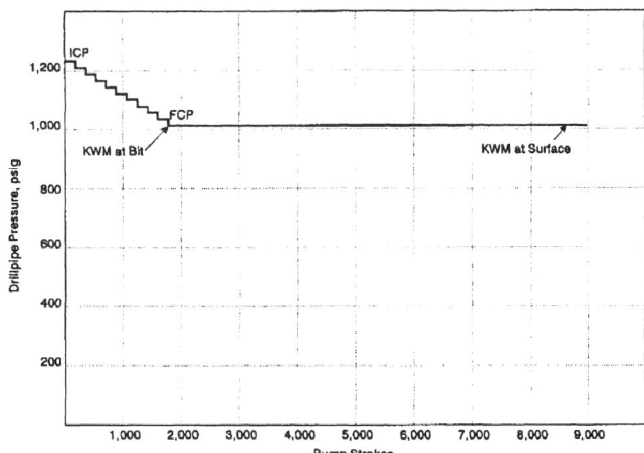

Fig. 4.49—Drillpipe pressures corresponding to the Wait and Weight Method kill of the well described in Table 4.13.

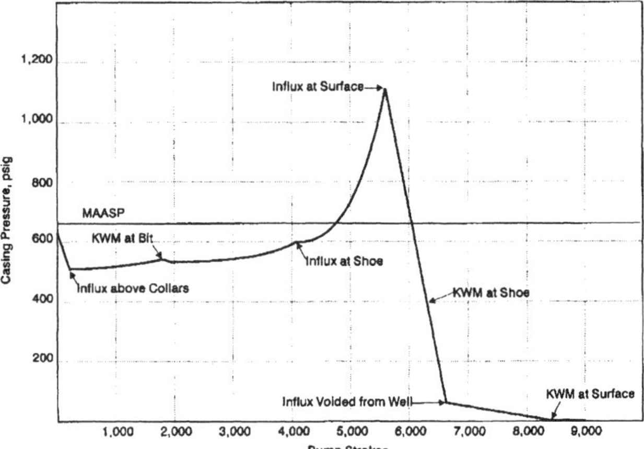

Fig. 4.50—Ideal casing pressures corresponding to a Wait and Weight Method kill of the example well in Table 4.13.

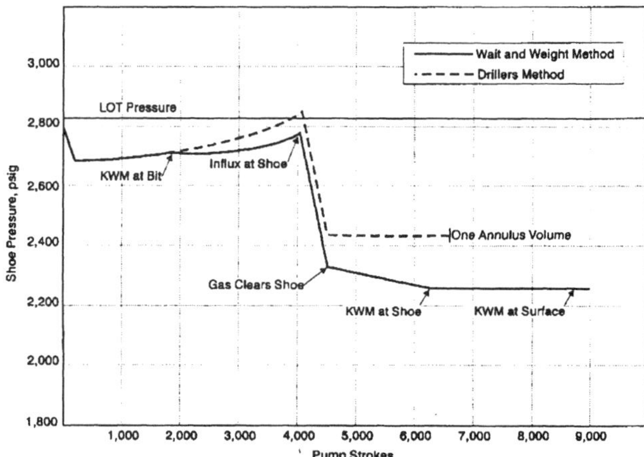

Fig. 4.51—Comparison between predicted casing-shoe pressures for a Driller's Method and Wait and Weight Method kill.

Say that the operator of this well wanted to assure that no secondary kicks would occur and added 0.3 lbm/gal of excess mud weight before pumping the kill. Accordingly, a drillpipe-pressure reduction is prepared so that the CDPP declines from 1,230 psig to

$$980(12.7/12.0) = 1,037 \text{ psig}.$$

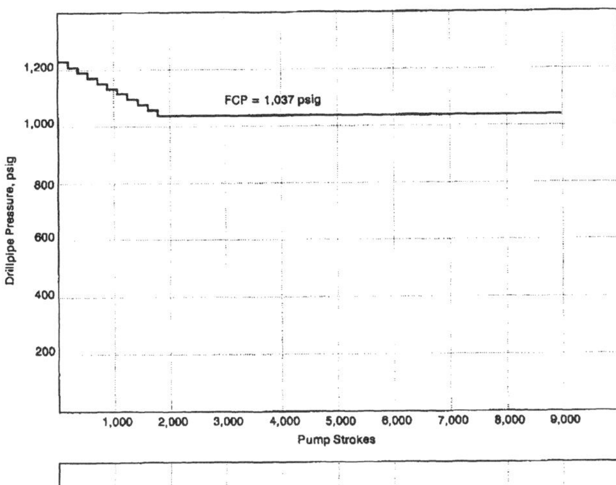

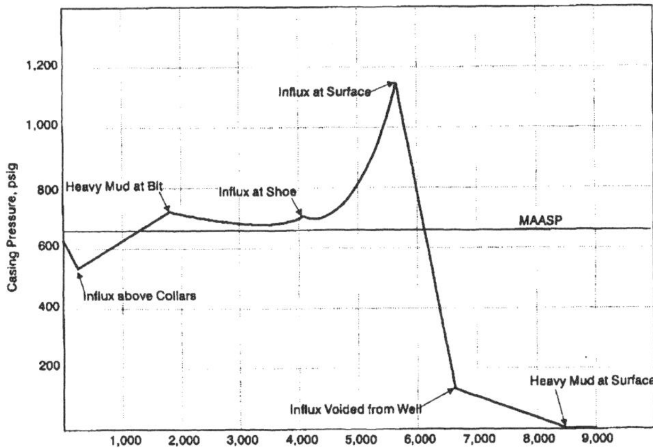

Fig. 4.52—The effect of adding excess mud density on surface pressures while conducting a Wait and Weight kill on the example well in Table 4.13.

Fig. 4.52 illustrates the drillpipe-pressure schedule and the consequent casing pressures. As shown, a precarious situation was created by taking this action because the excess BHP is reflected throughout the annulus.

One way to avoid extreme shoe pressures while pumping an overkill mud is to prepare a CDPP schedule that reduces the SIDPP component of the FCP to a negative value. For example, our well with the 0.3-lbm/gal overkill would need a target FCP of

$$1{,}037 - (0.0519)(0.3)(12{,}000) = 850 \text{ psig.}$$

However, the perceived advantages of pumping the heavier mud would be lost because the circulating BHP is back to the same pressure as with using KWM. Also, the well could not be shut-in without risking lost returns if a problem arose in the middle of the kill. Assume that this reduced pressure schedule was adopted, but the pump was lost at sometime during the operation. After closure, the static BHP relates to whatever hydrostatic pressure is in the drillstring that serves to place the shut-in casing pressure back on the curve shown in Fig. 4.52. In general, adding excess mud weight should be done only when circulating without the need for a choke—i.e., after the well is killed.

The drillpipe-pressure schedule must be modified if the Wait and Weight Method is employed in directional holes or when using a tapered drillstring. In these wells the problem is that a barrel or cubic meter of mud does not equate to the same vertical height at every point in the string. Thus hydrostatic pressure reduction is not a linear function of the volume pumped and we have to select different step-down rates if the BHP is to remain constant. Example 4.14 demonstrates the procedure for a tapered drillstring in a deep well.

Example 4.14. A 23,000-ft well takes a kick while drilling with a 9.2-lbm/gal mud and the resulting SIDPP is 1,300 psig. The prerecorded KRCP was 1,100 psig and the calibrated pump output is 2.85 gal/stroke. The following drillstring was necessary because of a 7-in. drilling liner: 570 ft of 4¾ × 2-in. drill collars; 5,952 ft of of 3½-in., 15.50-lbm/ft drillpipe; 13,423 ft of 5-in., 19.50-lbm/ft drillpipe; and 3,055 ft of 5-in., 25.60-lbm/ft drillpipe.

Determine the drillpipe-pressure schedule needed to maintain a constant BHP if the Wait and Weight Method is the selected procedure.

Solution. Starting at the top, the segment string volumes are determined.

$$(3{,}055)(0.01554) = 47.5 \text{ bbl} = 700 \text{ strokes;}$$

$$(13{,}423)(0.01776) = 238.4 \text{ bbl} = 3{,}513 \text{ strokes;}$$

$$(5{,}952)(0.00658) = 39.2 \text{ bbl} = 577 \text{ strokes; and}$$

$$(570)(0.00389) = 2.2 \text{ bbl} = 32 \text{ strokes.}$$

The calculated formation pressure at 23,000 ft = 12,295 psig, yielding a 10.3-lbm/gal equivalent density at total depth. The hydrostatic underbalance (SIDPP component of the circulating pressure) after filling each string segment with KWM will be

$$\Delta p_{ds}(700 \text{ strokes}) = 12{,}295 - [(3{,}055)(0.535) + (19{,}945)(0.477)] = 1{,}139 \text{ psi,}$$

$$\Delta p_{ds}(4{,}213 \text{ strokes}) = 12{,}295 - [(16{,}478)(0.535) + (6{,}522)(0.477)] = 372 \text{ psi,}$$

$$\Delta p_{ds}(4{,}790 \text{ strokes}) = 12{,}295 - [(22{,}430)(0.535) + (570)(0.477)] = 32 \text{ psi, and}$$

$$\Delta p_{ds}(4{,}822 \text{ strokes}) = 0 \text{ psi.}$$

The scheduled ICP and FCP, respectively, are

$$p_{dci} = 1{,}300 + 1{,}100 = 2{,}400 \text{ psig, and}$$

$$p_{dcf} = 1{,}100(10.3/9.2) = 1{,}232 \text{ psig.}$$

The increase in the KRCP component across the drillstring is

$$1{,}232 - 1{,}100 = 132 \text{ psi.}$$

Use depth ratios to approximate the KRCP increase after filling each segment. Though by no means rigorous, the calculation is accurate enough for our purpose. Thus,

$$\Delta p_{kr}(700 \text{ strokes}) = 132(3{,}055/23{,}000) = 18 \text{ psi;}$$

$$\Delta p_{kr}(4{,}213 \text{ strokes}) = 132(16{,}478/23{,}000) = 95 \text{ psi;}$$

$$\Delta p_{kr}(4{,}790 \text{ strokes}) = 132(22{,}430/23{,}000) = 129 \text{ psi;}$$

and $\Delta p_{kr}(4{,}822 \text{ strokes}) = 132(23{,}000/23{,}000) = 132 \text{ psi.}$

The desired circulating drillpipe pressures at the changeover points in the string are obtained.

$$p_{dc}(700 \text{ strokes}) = 1{,}100 + 1{,}139 + 18 = 2{,}257 \text{ psig;}$$

$$p_{dc}(4{,}213 \text{ strokes}) = 1{,}100 + 372 + 95 = 1{,}567 \text{ psig;}$$

$$p_{dc}(4{,}790 \text{ strokes}) = 1{,}100 + 32 + 129 = 1{,}261 \text{ psig;}$$

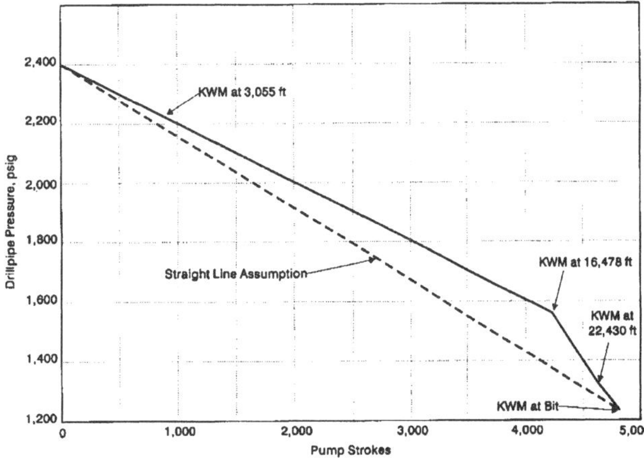

Fig. 4.53—Drillpipe pressure schedule for the well described in Ex. 4.14 with comparison to a straight-line reduction.

and $p_{dc}(4,822 \text{ strokes}) = p_{dcf} = 1,100 + 0 + 132$

$= 2,232 \text{ psig}.$

A relatively constant BHP will be maintained in Ex. 4.14 if drillpipe pressures correspond to the graph shown in **Fig. 4.53.** A straight-line assumption for a well with a tapered drillstring could underbalance the hole and invite a secondary kick. The same situation exists if small-diameter drill collars are beneath otherwise consistent drillpipe, but the effect is small enough in most cases to allow a conventional schedule. Even so, there is a trend in the drilling industry towards furnishing most of the bit weight requirements with HWDP and relying on drill collars solely for the BHA stiffness needed for trajectory control. This trend has obvious implications to well-control procedures that need to be addressed in well-control training.

The reverse condition holds true when the Wait and Weight Method is applied to a directional well because the height (not length) of a specific mud volume shortens as function of the inclination angle cosine. Thus a straight-line pressure schedule in a directional well will create overbalance pressures at lower hole angles; consequently, wellbore integrity may be unduly placed at risk. The pressure corrections across angle-hold sections are determined easily, but having KWM within a build or drop section is problematic because strictly adhering to a constant BHP would dictate a continually changing CDPP. Software is available for generating this type of schedule, but an approximation for hand calculations is to break the build or drop sections into suitable vertical-height increments of 500 or 1,000 ft and to use the average angle between the current and preceding depth selections. More accuracy will be obtained by shortening the height increment when the hole attitude increases with respect to vertical.

4.5.4 Concurrent Method. What later became known as the Concurrent Method was first described in a 1960 paper by O'Brien and Goins.[34] This was the earliest published technique to discuss using drillpipe pressures for maintaining a constant BHP and thus represented an important advance in well-control technology. Similar to the Wait and Weight Method, the technique involves pumping heavier mud at the same time an influx is in the annulus. The difference is that mud weight is increased to kill density by intermediate stages while circulating rather than during the shut-in period. **Table 4.18** describes the procedure generally used.

Mud density is increased "on the fly" once circulation begins and, as in the Wait and Weight Method, drillpipe pressures are stepped down until the pipe is filled with KWM. The ICP and FCP are determined in the same manner, but we may have several different mud densities of varying height in the string at any one time during the intermediate range. It is impossible to predict when a specific mud weight will be available at the suction pit; the schedule must be developed during the kill—a requirement greatly complicating the job execution.

Starting downhole with a new mud density would demand, in an ideal sense, that drillpipe pressures fall on a continuous basis because the heavier mud hydrostatic is increasing with time. The Concurrent Method procedure, however, makes some compromises and maintains a constant drillpipe pressure for a specified number of pump strokes dictated by the point at which an incrementally higher mud weight reaches the bit or, according to some procedures, another point in the string.

In practice, the mud engineer notifies the choke operator when the suction-pit mud density has increased by 0.1 or 0.2 lbm/gal. The choke operator records the cumulative pump strokes when the new mud weight has started, adds the string

TABLE 4.18—CONCURRENT METHOD PROCEDURE FOR KILLING A WELL	
1.	Take care of the well until the kill procedure can be started by maintaining a constant drillpipe pressure and allowing migrating gas to expand.
2.	Calculate the kill-mud weight and prepare a graph or table showing desired CDPP as function of mud weight at the bit.
3.	Open the choke and slowly start the pump. Coordinate choke setting with pump speed so that the original SICP is maintained until the pump is brought up to the kill rate.
4.	Read the drillpipe pressure and compare to the computed ICP. Start increasing the mud density in the suction pit.
5.	Note the pump strokes whenever a 0.1- or 0.2-lbm/gal density increase starts downhole and add the string volume to reflect the cumulative strokes at which the incremental mud density will reach the bit. Repeat the process for each mud density increment and keep track of this data with a table.
6.	From the CDPP graph or table, adjust the drillpipe pressure when a 0.1- or 0.2-lbm/gal density increase reaches the bit. Repeat the process for each mud density increment until KWM fills the drillstring.
7.	Maintain drillpipe pressure constant at the FCP value until kill mud density is measured at the choke outlet.
8.	Reduce pump speed while closing the choke so that a constant casing pressure is maintained. When the pump is barely running, shut off the pump and finish closing the choke.
9.	Observe pressure gauges and verify that both read zero. If not, check for trapped pressure.
10.	Open the choke and check for flow.
11.	Resume operations. Incorporate trip margin density in the mud.

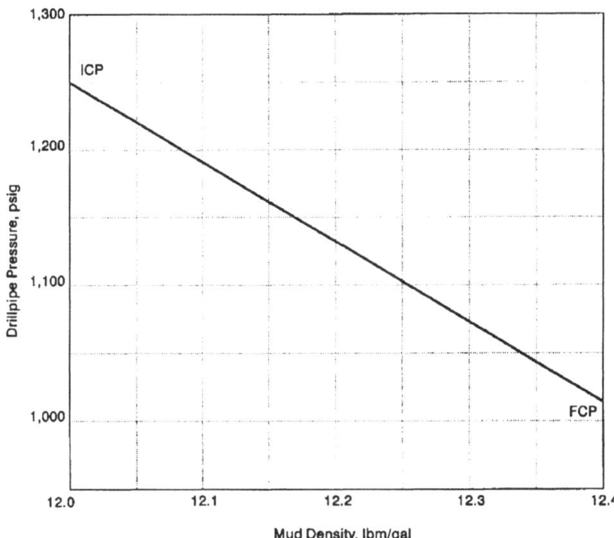

Fig. 4.54—Circulating drillpipe pressure as function of at-bit mud weight for a Concurrent Method kill of the example well in Table 4.13.

volume, and adjusts the CDPP once the new mud has reached bit depth. This procedure is repeated for each density increment until the KWM completely fills the drillstring. The FCP is held constant then until the well has been killed.

Desired pressures are determined before the job starts and in the same manner as the FCP except we have a positive SIDPP component for all densities less than kill-weight. The customary way to obtain the target drillpipe pressure is to plot the ICP and FCP at original and final mud weights and connect the two with a straight line. The CDPP at the intermediate mud weights then can be obtained directly from the chart. Alternatively, the values can be calculated eliminating the need for the graph. It is important to realize that BHP will not remain constant if heavier fluid is pumped behind the mud that was the basis for the CDPP, but will increase until the next choke adjustment is made.

Example 4.15. The operator of the well described in Table 4.13 decides to conduct the kill using the Concurrent Method. Determine the desired drillpipe pressures if changes are to be made when 0.1-lbm/gal density increments reach the bit. Prepare a table showing the final schedule if the mud engineer gave the following information during the circulation: 12.1-lbm/gal mud started downhole at 675 strokes, 12.2-lbm/gal mud at 1,550 strokes, 12.3-lbm/gal mud at 2,210 strokes, and 12.4-lbm/gal mud at 3,060 strokes.

Solution. The ICP and FCP determined in the preceding examples are plotted as shown in **Fig. 4.54** and the drillpipe pressures are obtained from the graph. p_{dci} = 1,230 psig; p_{dc} = 1,175 psig with 12.1-lbm/gal mud at the bit; p_{dc} = 1,121 psig with 12.2-lbm/gal mud; p_{dc} = 1,067 psig with 12.3-lbm/gal mud; and p_{dcf} = 1,013 psig with 12.4-lbm/gal mud.

TABLE 4.19—DRILLPIPE PRESSURE REDUCTION SCHEDULE FOR EXAMPLE 4.15

Pump Strokes	Drillpipe Pressure (psig)
0 to 2,473 (675 + 1,798)	1,230
2,473 to 3,348 (1,550 + 1,798)	1,175
3,348 to 4,008 (2,210 + 1,798)	1,121
4,008 to 4,858 (3,060 + 1,798)	1,067
4,858 +	1,013

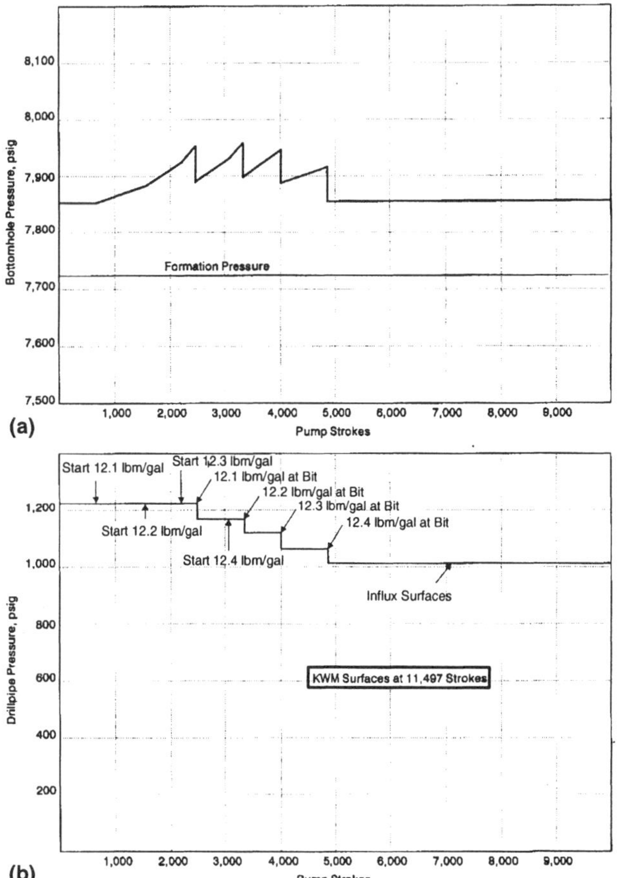

Fig. 4.55—Drillpipe and BHPs for the Concurrent Method kill described in Example 4.15.

Table 4.19 gives the drillpipe-pressure schedule observed during the kill. Adding the 1,798-stroke string volume to the stroke-counter reading when a given mud density was measured at the pump suction yields the period when constant drillpipe pressure is observed.

Fig. 4.55b presents the graphical representation of Table 4.19. **Fig. 4.55a** shows the computed BHPs. When pumping original mud, the circulating BHP exceeds the formation pressure by the 130-psi overbalance provided by annulus friction at the kill rate. The ICP is maintained as per the procedure, but we have as many as four different densities in the string before the next scheduled reduction occurs and BHP has increased as result. A maximum overbalance of 240 psi occuring in the interval may be tolerable in a properly designed hole but increases the risk of lost returns in this particular well. The impact of the Concurrent Method on shallow wellbore integrity should always be a consideration. The effect will be most pronounced in deep wells with fast mixing rates.

Cited advantages for the Concurrent Method include reduced kill time in comparison with the other two procedures, the ability to begin circulation quicker than the Wait and Weight Method, and lower wellbore pressures than the Driller's Method. The latter may or may not be true, depending on the well geometry, mud mixing rate, and influx characteristics. It should be apparent that the Concurrent Method is the least successful in maintaining a stable BHP and, due to the bookkeeping requirement, the most difficult to perform. For these reasons, this kill technique has fallen out of favor with most operators and drilling contractors.

4.5.5 Kill Technique Comparisons. No two well-control problems are the same and it is impossible to say that one control method has the advantage over the other under all conceivable circumstances. Procedure selection may not be within the control of the wellsite supervisor or engineer as many companies dictate, or at least influence, the choice as a matter of policy. Ideally, actual surface and subsurface conditions should be the governing considerations. Some key factors to contemplate and weigh for the situation at hand include the maximum anticipated wellbore pressure as it impacts fracture gradient, maximum anticipated surface pressure, time needed to begin and conclude the kill process, site or rig-specific logistics, and ease of execution. We focus on the Driller's and Wait and Weight methods and describe the primary advantages and disadvantages of each.

The Wait and Weight Method, in many cases, will result in a lower pressure at the last casing seat. However, this is not always true and is not a determining factor in those well configurations where the drillstring volume is larger than the annulus volume below casing. This advantage also may be lost if gas migration during the weightup time results in gas displacement to the shoe before weighted mud can enter the annulus. An exception occurs if the influx is a liquid and remains a liquid during the circulation, but we cannot verify fluid type until after the decision has been made. An adequately engineered hole design should nullify this argument, but the Wait and Weight Method is the best choice if there is real concern about exceeding the fracture pressure with the Driller's Method.

As a rule, maximum surface pressures with a gas kick will be less when using the Wait and Weight Method. Even so, a rapid migration rate combined with an exessively long mixing time may narrow or eliminate the comparative pressure margin. The incremental surface-pressure increase when pumping a Driller's Method kill should not be of major concern if the casing design and equipment rating are sufficient.

At minimum, the Driller's Method involves circulating one annulus volume followed by a full wellbore circulation whereas the Wait and Weight procedure can be accomplished in one circulation. Hence the Wait and Weight Method requires less time to complete, thereby resulting in cost savings and a reduced risk of mechanical breakdown or leaks in the control system. However, the increased shut-in time required by the Wait and Weight Method may present some hazards. Drill cuttings may settle out and packoff the annulus or plug the bit during the weightup period; mud contamination caused by the formation fluids may induce a gellation or thickening problem that deteriorates only with time. More attention to tracking gas migration is needed when shut-in time is extended. Hence the ability to begin circulation almost immediately is one attractive feature of the Driller's Method.

Adequate weight material must be on hand and the rig's mixing equipment should be capable before considering a Wait and Weight kill. Well control becomes a much easier task once an influx has been removed from a hole. Common sense dictates circulating out a kick as soon as possible if barite is in short supply on a remote or inaccessible rig. The same is true if equipment limitations present a situation where gas has migrated more than halfway up the hole before KWM can be pumped.

Well-control problems create stress and anxiety, from those making the decisions and supervising the operation to the personnel whose job is to follow orders. Procedural complexity induces greater risk of judgment errors and execution mistakes, so there is much to be said for simplifying well control as much as possible. Herein lies the main and, perhaps, overriding benefit to the Driller's Method: computations are minimal and the choke operator is not forced to follow a drillpipe-pressure-reduction schedule. This can get complicated if the well is deviated significantly from vertical or has a tapered drillstring. The well geometry is irrelevant to the Driller's Method because annulus backpressures are held constant when filling the string with KWM, regardless of the string dimensions or directional aspects of the hole.

4.5.6 Combined Method. A new approach to well control is the Circulate and Weight Method,[35] but the technique is essentially a hybrid procedure combining the Driller's and Wait and Weight methods. (The Concurrent Method has gone by the same name, but the technique described here is different.) Circulation starts off using the Driller's Method until KWM has been mixed in a dedicated tank or is otherwise ready to be pumped downhole. An immediate switch is made to KWM and the choke operator follows the drillpipe pressure-reduction schedule and subsequent steps of the Wait and Weight Method until the well has been killed.

Some operators recommend this technique because it combines many of the favorable attributes inherent to the separate methods. As in the Driller's Method, kick displacement can begin before anything is done to the mud, thereby avoiding the risks associated with the Wait and Weight shut-in period. Pump startup is simplified in comparison to the Wait and Weight Method because KWM will not affect initial circulating pressures. Also, the pressure reduction schedule, if generated after circulation begins, uses actual ICP data rather than the predicted pressure and so eliminates the need to shut the well in and recalculate parameters if a disparity is observed.

Wait and Weight advantages can be approximated, though not equaled, by the modified approach. Wellbore pressures may be lower than experienced in the Driller's Method if KWM can be mixed and pumped soon enough to place heavier mud in the annulus before free gas has reached a specified depth. This benefit, of course, depends on factors such as the bit and casing point depths, annulus volume, pump rate, and mixing rate. The operation can be completed more quickly than the Driller's Method if KWM is started while the influx is in the hole.

Some disadvantages are apparent, however, and the combined procedure would be difficult to execute on many rigs without modifying the existing pit system. For one thing, the simplicity of the Driller's Method's second circulation is lost because drillpipe pressures must be reduced according to a schedule. Also, most land rigs are not equipped to circulate one mud density while another system is being weighted up. Thus the kill technique has some physical constraints to its use if the tanks cannot be isolated and manifolded directly to the pump suction.

4.6 Annulus Pressure Prediction

The gas law can be used to estimate annulus pressures and expansion characteristics while circulating free gas in a well-control procedure. Some simplifying assumptions that must be adopted if the equations are to be tractable for hand calculations are listed in Table 4.15, and the relations generally

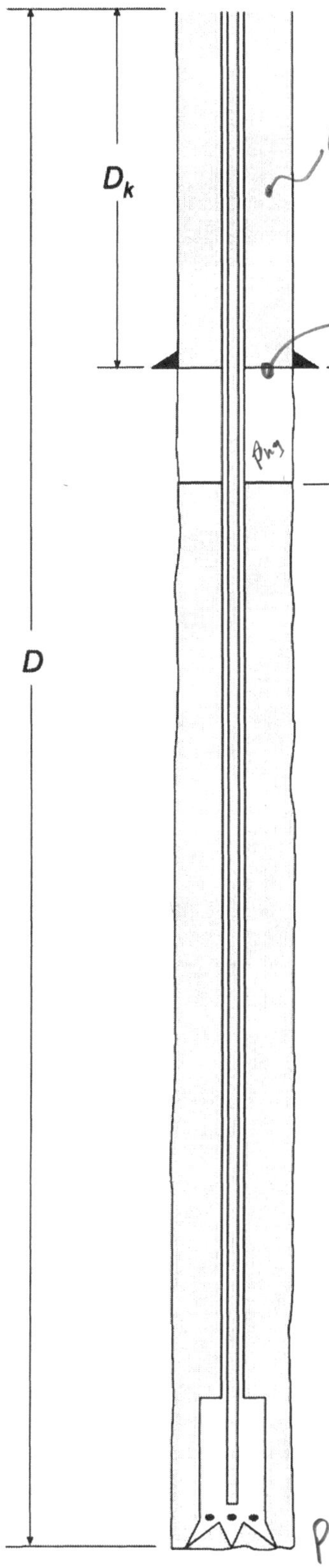

Fig. 4.56—Dimensional variables used in the annular-pressure-prediction equation when annulus mud density is constant.

yield conservative (high) results. Nonetheless, going through the exercise after a well kicks may be worthwhile for identifying the potential impact on surface and wellbore integrity or to quantify how much excess pit capacity will be needed at maximum gas expansion.

Thus we have a reasoned basis for deciding which conventional control method to use in an actual well-control problem or the need to consider one of the alternative, nonclassical techniques discussed in Chap. 5. Predictions can be used on a look-ahead basis as drilling progresses into a pore-pressure transition for purposes of assessing the effect of the variables on the ability to circulate out a kick safely and, therefore, better defining where the next string of casing should be set.

The wellbore schematic shown in **Fig. 4.56** depicts a gas influx after displacement or migration to shallower depth. As indicated in the diagram, the pressure at the top of the gas bubble is the BHP less the combined hydrostatic pressures of the gas influx and the underlying mud.

$$p = p_{bh} - g_{om}(D - h_k - D_k) - p_{hg}, \quad \text{........ (4.29)}$$

where p = pressure at the top of a gas kick, D = total depth, h_k = gas kick height, and p_{hg} = gas column hydrostatic pressure. This equation would be applicable to a Driller's Method kill or to the Wait and Weight Method before KWM enters the annulus. The term p_{bh} is the BHP when the gas kick reaches D_k and thus comprises the formation pressure plus any excess backpressure carried during the circulation or static migration procedure.

The hydrostatic pressure of the gas column is determined by multiplying its hydrostatic gradient by height. In customary units, we substitute the density expression from Eq. 1.22 and obtain

$$p_{hg} = 0.0519\left(\frac{\gamma_g p}{2.77zT}\right)h_k = \frac{\gamma_g p h_k}{53.29zT}. \quad \text{........ (4.30)}$$

The influx height is the gas volume at the depth of interest divided by the annulus capacity factor,

$$h_k = V_k / C_a. \quad \text{.......................... (4.31)}$$

Substitution into Eq. 4.30 yields a relationship which tells us that p_{hg} does not depend on the pressure/volume/temperature (PVT) conditions at the depth of interest. Because the number of moles in an influx does not change, its weight must be constant. The gas hydrostatic gradient does depend on the geometry and its value at any depth is

$$p_{hg} = p_{hgi}(C_{ai}/C_a), \quad \text{...................... (4.32)}$$

where p_{hgi} = the original hydrostatic pressure and C_{ai} = the annulus capacity factor where the influx initially resides.

The gas volume is obtained from the gas law,

$$V_k = \frac{p_{bh} V_{ki} z T}{p z_{bh} T_{bh}}. \quad \text{.......................... (4.33)}$$

The equivalent expression in terms of gas height is

$$h_k = \frac{p_{bh} z T h_{ki} C_{ai}}{p z_{bh} T_{bh} C_a}. \quad \text{...................... (4.34)}$$

The relation is not exact because we have defined the initial PVT properties at the bottom of the influx whereas the properties at the second condition apply to the top of the gas bubble. Even so, we accept the generally minor discrepancy. Substituting terms into Eq. 4.29 yields the quadratic expression,

$$p^2 - p\left[p_{bh} - g_{om}(D - D_k) - p_{hg}\right]$$

$$- \frac{g_{om} p_{bh} z T h_{ki} C_{ai}}{z_{bh} T_{bh} C_a} = 0. \quad \text{.................... (4.35)}$$

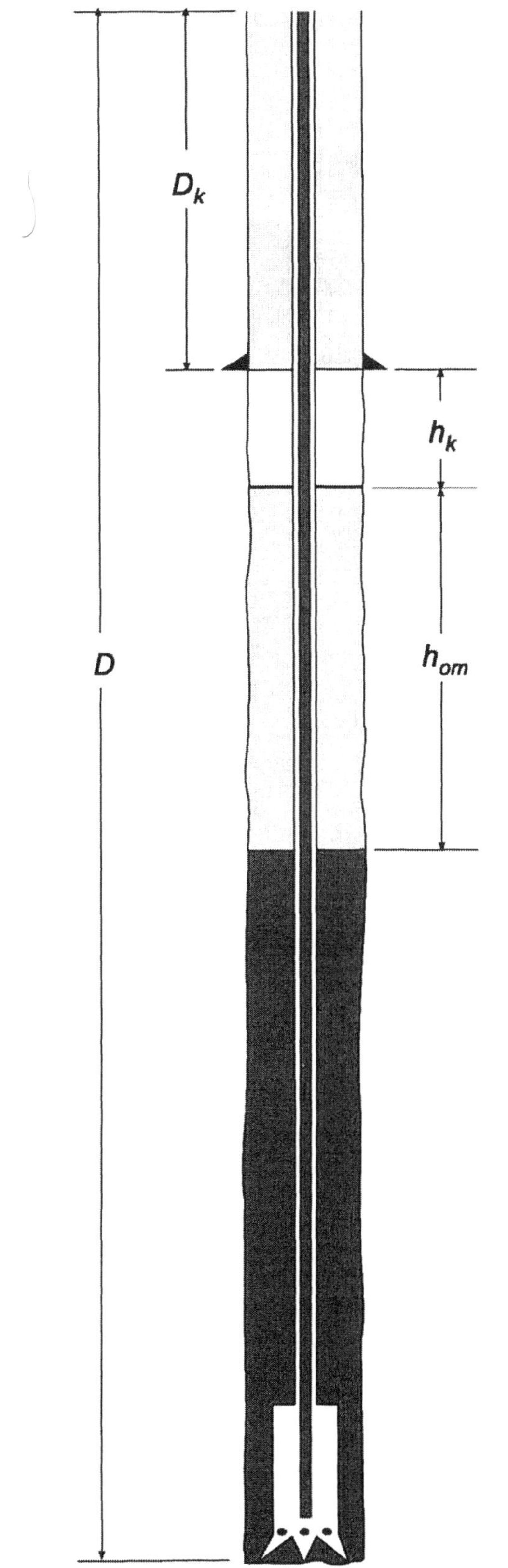

Fig. 4.57—Dimensional variables used in the annular-pressure-prediction equation when heavier mud is in the annulus.

Solving for the equation root yields

$$p = \frac{X}{2} \pm \sqrt{\frac{X^2}{4} + \frac{g_{om} p_{bh} h_{ki} z T C_{ai}}{z_{bh} T_{bh} C_a}}. \qquad (4.36)$$

The negative sign is ignored because it results in a negative pressure. The intermediate variable X is defined by

$$X = p_{bh} - g_{om}(D - D_k) - p_{hg}. \qquad (4.37)$$

Fig. 4.57 illustrates the case where a different mud density is pumped into the annulus behind the influx. In a Wait and Weight kill, the mud volume originally in the drillstring is displaced ahead of the heavier mud and thus occupies some vertical height in the annulus described by the term h_{om}. The counterpart to Eq. 4.29 is then

$$p = p_{bh} - g_{km}(D - h_{om} - h_k - D_k) - g_{om} h_{om} - p_{hg}, \qquad (4.38)$$

where g_{km} = the gradient of the KWM. Using the same reasoning, we obtain the expression for the pressure at the top of the influx when KWM has been placed in the annulus.

$$p = \frac{Y}{2} \pm \sqrt{\frac{Y^2}{4} + \frac{g_{km} p_{bh} h_{ki} z T C_{ai}}{z_{bh} T_{bh} C_a}}, \qquad (4.39)$$

where Y is given by

$$Y = p_{bh} - g_{km}(D - D_k) + h_{om}(g_{kwm} - g_{om}) - p_{hg}. \qquad (4.40)$$

These relationships require that something is known about the gas temperature in the well. A circulating wellbore is a complex heat-transfer system and an accurate estimate of the annular temperature at any depth depends on too many variables to make any generalizations. A nonlinear circulating-temperature profile similar to the curve shown in Fig. 3.32 would be expected immediately after wellbore closure. The static fluid will start to approach, though not achieve, the geothermal gradient during the precirculation period. The profile in the subsequent kick displacement would be expected to approach, but not achieve, the circulating temperatures seen earlier at the higher pump rate. Things become even more complicated if we consider the thermodynamics of a rapidly expanding gas bubble near surface.

Most kick simulators assume a constant temperature profile, either taking the characteristic parabola shape predicted by circulation models[24,36-41] or a straight-line approximation for static wellbores. Our general approach is to base annular pressure predictions on a constant temperature gradient, starting at the surface with a temperature somewhat higher than the mud inlet temperature and finishing with something slightly cooler than undisturbed earth temperature at total depth. Fortunately, any associated errors do not seriously affect wellbore pressure estimates because the gas law uses absolute temperature.

The gas SG must be assumed unless the operator has knowledge of the gas properties in the source formation. Estimated annulus pressures will increase with lower SGs and a conservative approach is to use a 0.6-gravity gas corresponding to a predominantly methane kick. Note also that the equations contain the compressibility factor at the depth of interest so we must iterate to the desired pressure. The results usually converge to a solution after one or two iterations because z factors are not very sensitive to small pressure changes.

Example 4.16. Refer to the kick event described in **Table 4.20.**

TABLE 4.20—KICK DATA FOR EX. 4.16	
Wellbore Configuration	
Vertical Depth	19,400 ft
Protective Casing Information	
Description	9⁵/₈-in. 53.5-lbm/ft P-110
API MIYP	10,900 psig
Setting Depth	15,200 ft
Open Hole Diameter	7⁷/₈ in.
Drillstring Information	
Drill Collar Size	6¼ × 2¾ in.
Drill Collar Length	660 ft
Drillpipe Description (Bottom Section)	4½-in. 16.60-lbm/ft
Drillpipe Length (Bottom Section)	17,900 ft
Drillpipe Description (Top Section)	4½-in., 20.00-lbm/ft
Drillpipe Length (Top Section)	840 ft
Mud Properties and Recorded Kick Data	
Mud Type	Water Base
Mud Density	15.5 lbm/gal
SIDPP	500 psig
SICP	2,300 psig
Pit Gain	100 bbl
Other Known or Assumed Information	
Fracture Gradient at Casing Seat	0.98 psi/ft
Assumed Wellbore Temperature at Kill Rate	100°F + 1.5°F/100 ft
Kick Fluid	0.60 SG Gas

1. Estimate the surface and shoe presures for a Driller's and Wait and Weight kill procedure when the gas bubble reaches the casing seat.

2. Estimate the surface pressure, shoe pressure, and pit gain for both methods when the gas bubble surfaces.

Solution.

1. The mud's hydrostatic gradient is 0.8052 psi/ft and so the shut-in BHP is

$$p_{bh} = 500 + (0.8052)(19,400) = 16,121 \text{ psig}$$

$$= 16,136 \text{ psia}.$$

The bottomhole temperature and pseudoreduced properties at bottomhole conditions are

$$T_{bh} = 100 + (0.015)(19,400) = 391°\text{F} = 851°\text{R};$$

$$T_{pr} = 851/352 = 2.42; \text{ and}$$

$$p_{pr} = 16,136/677 = 23.83.$$

The z factor is found to be 1.697 from Fig. 1.7. Capacity factors are now determined.

$$C_i(20.00 \text{ lbm/ft}) = 3.640^2/1,029.4 = 0.01287 \text{ bbl/ft};$$

$$C_i(16.60 \text{ lbm/ft}) = 3.826^2/1,029.4 = 0.01422 \text{ bbl/ft};$$

$$C_i(\text{DC}) = 2.25^2/1,029.4 = 0.00492 \text{ bbl/ft};$$

$$C_a(\text{DC} \times \text{hole}) = (7.875^2 - 6.25^2)/1,029.4$$

$$= 0.02230 \text{ bbl/ft};$$

$$C_a(\text{DP} \times \text{hole}) = (7.875^2 - 4.5^2)/1,029.4$$

$$= 0.04057 \text{ bbl/ft; and}$$

$$C_a(\text{DP} \times \text{csg}) = (8.535^2 - 4.5^2)/1,029.4$$

$$= 0.05109 \text{ bbl/ft}.$$

The volumes in each section are therefore,

$$V(20.00 \text{ lbm/ft}) = (840)(0.01287) = 10.8 \text{ bbl};$$

$$V(16.60 \text{ lbm/ft}) = (17,900)(0.01422) = 254.5 \text{ bbl};$$

$$V(\text{DC}) = (660)(0.00492) = 3.2 \text{ bbl};$$

$$V(\text{DC} \times \text{hole}) = (660)(0.02230) = 14.7 \text{ bbl};$$

$$V(\text{DP} \times \text{hole}) = (3,540)(0.04057) = 143.6 \text{ bbl; and}$$

$$V(\text{DP} \times \text{csg}) = (15,200)(0.05109) = 776.6 \text{ bbl}.$$

The pit gain exceeds the capacity outside the drill collars. The initial bubble height is

$$h_{ki} = 660 + (100 - 14.7)/0.04057 = 2,763 \text{ ft}.$$

The effective annulus-capacity factor is thus

$$C_{ai} = 100/2,763 = 0.03619 \text{ bbl/ft}.$$

Eq. 4.30 yields the initial hydrostatic pressure of the gas bubble,

$$p_{hgi} = \frac{(0.60)(16,136)(2,763)}{(53.29)(1.697)(851)} = 348 \text{ psi}.$$

All bottomhole parameters have been described.

The gas bubble will be within the drillpipe/openhole annulus when displaced to the shoe and Eq. 4.32 gives the gas hydrostatic pressure at this point in the kill.

$$p_{hg} = (348)(0.03619)/0.04057 = 310 \text{ psi}.$$

Solve for X using Eq. 4.37.

$$X = 16,136 - (0.8052)(19,400 - 15,200) - 310$$

$$= 12,444 \text{ psia}.$$

At 15,200 ft,

$$T_{15,200} = 100 + (0.015)(15,200) = 288°\text{F} = 788°\text{R},$$

and $T_{pr} = 788/352 = 2.24$.

Assume a z factor of 1.500 at 15,200 ft and predict the pressure for the Driller's Method kill using Eq. 4.36.

$$p_{15,200} = \frac{12,444}{2} + \left[\frac{12,444^2}{4} + \frac{(0.8052)(16,136)(2,763)(1.500)(788)(0.03619)}{(1.697)(851)(0.04057)}\right]^{0.5}$$

$$= 14,280 \text{ psia}.$$

Accordingly,

$$p_{pr} = 14,280/677 = 21.09.$$

Thus $z_{15,200} = 1.625$ from Fig. 1.7. Iterate again and solve for the pressure.

$$p_{15,200} = \frac{12,444}{2} + \left[\frac{12,444^2}{4}\right.$$

$$+\frac{(0.8052)(16,136)(2,763)(1.625)(788)(0.03619)}{(1.697)(851)(0.04057)}\Bigg]^{0.5}$$

$$= 14,414 \text{ psia} = 14,400 \text{ psig.}$$

The new pseudoreduced pressure yields the same z factor and so this is our answer. Ignoring annulus friction, the casing pressure is

$$p_c = 14,400 - (0.8052)(15,200) = 2,161 \text{ psig.}$$

For the Wait and Weight kill, the drillstring volume is 268.5 bbl while the annulus volume below the casing is 158.3 bbl. Hence KWM is still in the drillpipe when the gas bubble enters the casing and the maximum shoe pressure will be the same as in the Driller's Method.

2. The gas hydrostatic pressure when the kick reaches surface is

$$p_{hg} = (348)(0.03619)/0.05109 = 247 \text{ psi.}$$

Determine X.

$$X = 16,136 - (0.8052)(19,400 - 0) - 247$$

$$= 268 \text{ psia.}$$

The pseudoreduced temperature at surface is

$$T_{pr} = (100 + 460)/352 = 1.59.$$

Assume a z factor of 1.000 and apply Eq. 4.36.

$$p_0 = \frac{268}{2} + \Bigg[\frac{268^2}{4}$$

$$+\frac{(0.8052)(16,136)(2,763)(1.000)(560)(0.03619)}{(1.697)(851)(0.05109)}\Bigg]^{0.5}$$

$$= 3,277 \text{ psia.}$$

$$p_{pr} = 3,277/677 = 4.84.$$

Fig. 1.6 gives $z_0 = 0.835$. Iterate and solve for the pressure.

$$p_0 = \frac{268}{2} + \Bigg[\frac{268^2}{4}$$

$$+\frac{(0.8052)(16,136)(2,763)(0.835)(560)(0.03619)}{(1.697)(851)(0.05109)}\Bigg]^{0.5}$$

$$= 3,007 \text{ psia.}$$

$$p_{pr} = 3,007/677 = 4.44.$$

Once again using $z_0 = 0.820$,

$$p_0 = \frac{268}{2} + \Bigg[\frac{268^2}{4}$$

$$+\frac{(0.8052)(16,136)(2,763)(0.820)(560)(0.03619)}{(1.697)(851)(0.05109)}\Bigg]^{0.5}$$

$$= 2,981 \text{ psia} = 2,967 \text{ psig.}$$

One way to obtain the maximum pit gain is to solve for the unknown volume by direct application of the gas law. However, we are interested also in the bubble height and so take an indirect route using Eq. 4.34.

$$h_0 = \frac{(16,136)(0.820)(560)(2,763)(0.03619)}{(2,981)(1.697)(851)(0.05109)}$$

$$= 3,369 \text{ ft.}$$

Hence the maximum pit gain is

$$G_0 = (3,369)(0.05109) = 172 \text{ bbl.}$$

The pressure at the shoe can be determined by adding the casing pressure to the hydrostatic pressure of the two fluid columns.

$$p_{15,200} = 2,967 + 247 + (0.8052)(15,200 - 3,369)$$

$$= 12,740 \text{ psig.}$$

An alternative solution (and result verification) is obtained by working from the bottom up.

$$p_{15,200} = 16,136 - (0.8052)(19,400 - 15,200)$$

$$= 12,754 \text{ psia} = 12,740 \text{ psig.}$$

For the Wait and Weight Method, it should be apparent that the 268.5 bbl of 15.5-lbm/gal mud originally in the drillstring is contained totally in the casing when the gas bubble reaches surface. Thus

$$h_{om} = 268.5/0.05109 = 5,255 \text{ ft.}$$

The gradient of the 16.0-lbm/gal KWM is 0.8312 psi/ft. Eq. 4.40 yields

$$Y = 16,136 - (0.8312)(19,400 - 0) + (5,255)$$

$$(0.8312 - 0.8052) - 247 = -100 \text{ psia.}$$

The $z_0 = 0.820$ and apply Eq. 4.39.

$$p_0 = \frac{-100}{2} + \Bigg[\frac{-100^2}{4}$$

$$+\frac{(0.8312)(16,136)(2,763)(0.820)(560)(0.03619)}{(1.697)(851)(0.05109)}\Bigg]^{0.5}$$

$$= 2,840 \text{ psia.}$$

$$p_{pr} = 2,840/677 = 4.19.$$

The z factor changes to 0.815 and the final iteration yields $p_0 = 2,831$ psia $= 2,817$ psig. The maximum bubble height and pit gain are computed, respectively,

$$h_0 = \frac{(16,136)(0.815)(560)(2,763)(0.03619)}{(2,831)(1.697)(851)(0.05109)} = 3,526 \text{ ft}$$

and $G_0 = (3,526)(0.05109) = 180$ bbl.

The shoe pressure is computed from both directions.

$$p_{15,200} = 2,817 + 247 + (0.8052)(5,255)$$

$$+ (0.8312)(15,200 - 8,781) = 12,631 \text{ psig}$$

and $p_{15,200} = 16,136 - (0.8312)(19,400 - 15,200)$

$$= 12,645 \text{ psia} = 12,631 \text{ psig.}$$

Some observations regarding the results from the preceding example are in order. The maximum surface pressures may seem surprisingly low given the size of the influx. A 100-bbl kick is an extremely large volume of gas to handle. Without making the prediction, the operator of this well may conclude falsely that the kick could not be circulated safely. An unwise decision such as pumping into the annulus may have followed.

One reason for the relatively low magnitude of the computed surface pressures is that the limited gas expansion while circulating a kick from a deep, high-pressure well tends to suppress the difference between maximum and minimum casing pressures. Another major influence is the effect of the high compressibility factor at the bottomhole conditions in an HPHT well. Some published prediction methods choose to treat gas behavior as ideal. In the example, assuming a z factor of 1.00 in the Driller's Method equation yields a 4,130-psig maximum casing pressure as compared to the real gas prediction of 2,967 psig. The assumptions in Table 4.15 ensure that results will be on the high side and any further conservatism starts to place the pressure predictions outside the realm of reason.

We see also that wellbore and surface pressures alone would not have dictated the control procedure in this well. Shoe pressures were the same, regardless of whether the Driller's or Wait and Weight Method was selected, and the difference in maximum casing pressure was only 150 psi. Maximum pressure considerations should not influence the choice of control method if the well design is adequate, particularly in HPHT wells.

4.7 Well Control and OBMs

The factors that govern natural gas solubility in a drilling fluid were discussed in Chap. 1. Gas will dissolve to some extent in all drilling-fluid liquids, but we generally ignore the effect in WBM's unless H_2S or CO_2 comprise a significant fraction of the gas composition. Most of the studies to date have focused on the solubility characteristics of natural gas in No. 2 diesel or mineral oils. The new synthetic-oil systems exhibit similar behavior when combined with gas.

It is worthwhile to review some of these principles and to address how solution gas impacts kick detection and the subsequent well-control efforts. Recall that dissolved gas "hides" in the mud as the volume increase in the base oil does not reflect the gas volume in a free state. Thus a given pit gain may be only a small fraction of the influx volume and early kick detection becomes more critical to well safety. However, the unique characteristics of a solution gas mixture can lead to some definite benefits when displacing a kick.

4.7.1 Effect on Kick Detection. How much gas can be dissolved into a quantity of oil depends primarily on the chemical makeup of the gas and oil, pressure, and temperature. In general, gas solubility increases with increasing pressure, decreasing temperature, and when the molecular weights of the two components become closer. The bubblepoint pressure for a given temperature defines the pressure at which the first bubble of free gas breaks out of solution. By definition, oil is at its saturation limit when at the bubblepoint pressure and undersaturated—i.e., capable of dissolving more gas—at pressures below bubblepoint. Increasing the dissolved gas concentration will reduce the bubblepoint pressure of the mixture. Laboratory tests or sophisticated equations of state are necessary to describe gas/oil solubility behavior.

The gas/liquid ratio (GLR) conveniently describes the gas content in a fluid mixture. In conventional usage, GLRs are expressed in standard gas units and surface or stock tank liquid volumes. Consider the situation where a gas kick enters a mud while drilling. The gas/mud ratio at the formation face is given by

$$r_m = q_{gsc}/q_m, \quad \ldots\ldots (4.41)$$

where r_m is the gas/mud ratio, q_{gsc} the gas flow rate at standard conditions, and q_m the mud circulation rate. The gas/oil ratio (GOR) for an OBM is given by

$$r_o = r_m/f_o, \quad \ldots\ldots (4.42)$$

where r_o is the GOR and f_o is the mud's oil volume fraction.

The dissolved gas content in a liquid is described by the solution GLR and any gas concentration which exceeds this ratio will be in the free state. In drilling fluids, solids are incapable of dissolving any gas and we can generally neglect the water phase solubility if acid gases are not present. All of the gas entry then will be dissolved if conditions are such that $r_o \le r_{so}$ at the bubblepoint (r_{so} is the solution GOR). We have neglected gas solubility in the surfactants, but can include a composite term for the oil + emulsifiers or accept the small error.

Formation volume factors (FVFs) describe the ratio of a specific quantity of fluid in the well or reservoir to the same quantity as measured at standard conditions or the stock tank. Oil compressibility reduces the oil FVF with increasing depth and pressure whereas thermal expansion has the opposite effect. The FVF in the wellbore therefore may be greater or smaller than unity, depending on which factor has the greatest influence. Mud solids are essentially incompressible and we can generally neglect any water contribution because the volume fraction is smaller and brine is less subject to volume change with depth. Accordingly, we can estimate the volume factor of an OBM using

$$B_m = f_o B_o + (1 - f_o), \quad \ldots\ldots (4.43)$$

where B_m and B_o are the FVF of the whole mud and oil phase.

Dissolved gas increases B_o and the difference between the solution and gas-free factors describes the net volume increase associated with the solution-gas addition. The influx volume (stock tank units) of the gas in solution can be estimated from the pit gain using Eq. 4.44:

$$V_k = \frac{G r'_{so}}{B_{og} - B_{ong}}, \quad \ldots\ldots (4.44)$$

where B_{og} = the FVF of the oil containing dissolved gas and B_{ong} = the gas-free volume factor. The GOR term r'_{so} differs from r_{so} in that the gas volume in the former is determined at downhole conditions as opposed to standard pressure and temperature.

Circulation across a flowing gas zone provides a continuous supply of uncontaminated oil for taking gas into solution, and so PVT monitors and return flow indicators become less reliable kick-detection tools while drilling with an OBM. Delta flow measurements better detect the smaller flows and thereby limit the ultimate gas volume which must be handled, but the incremental flowline returns are still less than the formation flow rate when all or a portion of the influx is being dissolved. A stagnant mud, however, ultimately will be saturated and a free-gas flow will start to displace an equivalent mud volume to the pits. It follows that more frequent flow checks with extended observation time should be instituted when drilling a transition zone.

Under certain conditions, a gas kick into a static oil mud may result in the same pit gain as would be experienced in a WBM.[42] The reason goes back to the critical pressure and

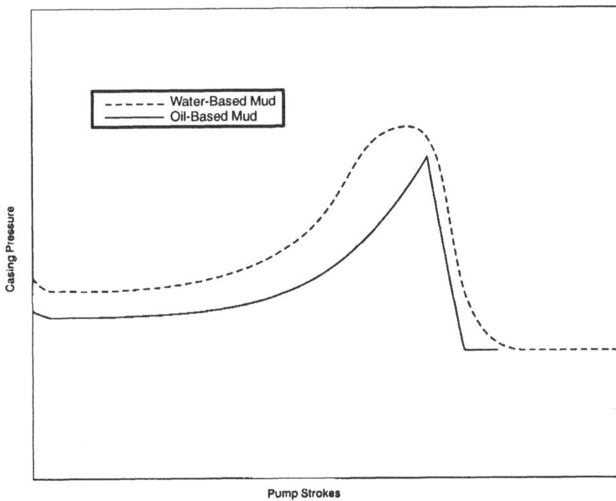

Fig. 4.58—Casing pressures when displacing a gas kick with two mud types when free gas is initially present.

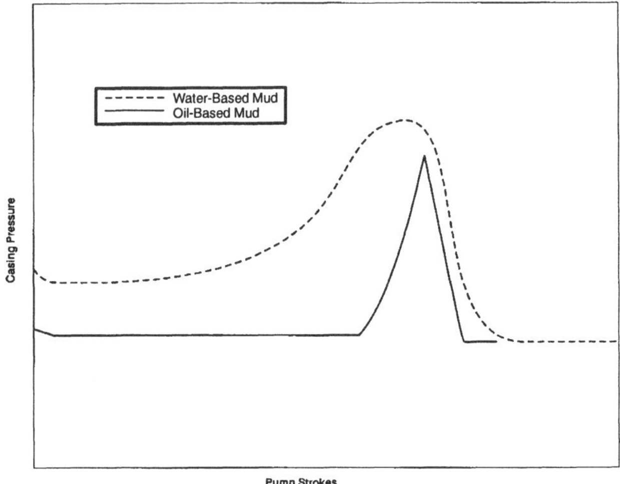

Fig. 4.59—Casing pressures when displacing a gas kick with two mud types when bubblepoint pressure is achieved in the annulus.

miscibility concepts discussed in Chap. 1. Refer to Fig. 1.18 and note that methane solubility in No. 2 diesel becomes infinite at higher pressures. Higher pressures imply complete miscibility of the two fluid phases and a region where the properties of the gas are indistinguishable from those of the oil. Say that a methane kick entered a deep well on a trip where a diesel-based mud was in use. Furthermore, the bottomhole temperature was 300°F and the hydrostatic pressure was in excess of 7,000 psia, a condition that falls to the right of the infinite solubility pressure. We would expect little natural mixing or dissolution of the gas to occur; consequently, the influx would enter the well as a free-gas slug and displace the overlying drilling fluid.

4.7.2 Effect on Conventional Control Procedures. The basic U-tube principles do not change if part of a gas influx has gone into solution. The SIDPP still reflects how much the hydrostatic pressure in the drillstring is underbalanced to the formation while the SICP is the formation pressure less the combined hydrostatic of the fluids in the annulus.

The density of a gas-free mud at any depth can be determined from

$$\rho_m = \rho_{msc}/B_m, \quad \text{.......................... (4.45)}$$

where ρ_{msc} is the density at ambient surface conditions. A numerical technique may be used to estimate the hydrostatic pressure with variable mud-column density. We can estimate the density of a mud containing solution gas using

$$\rho_m = \frac{\rho_{msc} + m_g/V_{msc}}{B_m}, \quad \text{.................... (4.46)}$$

where m_g/V_{msc} = the dissolved-gas mass in a unit volume of the surface mud. Some density reduction will occur when a gas kick is in the oil and emulsifiers, but the overall hydrostatic pressure reduction usually will be less than if all the gas has been in the free state. Hence the casing shoe limitation is more likely to be exceeded when a well is closed in with a WBM.

Any free gas at the bottom of a shut-in well will tend to rise through an OBM until contact is made with an undersaturated oil. Over time, all the gas may be dissolved and further migration will cease. It can be argued that nil gas migration is a well-control advantage because maintaining wellbore pressures in the desired window is an easier task and less uncertainty should arise regarding the pore pressure component of the

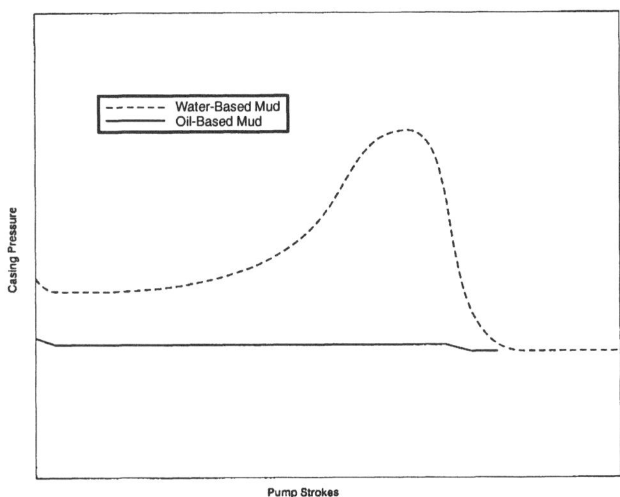

Fig. 4.60—Casing pressures when displacing a gas kick with two mud types when bubblepoint pressure is achieved in the choke manifold.

SIDPP. However, the SIDPP may not stabilize for a long time when using an OBM. This is because of the higher system compressibility plus the fact that the wellbore volume does not increase as much in response to a gas flow.

Dissolved gas tends to remain more in a "package" rather than migrate and disperse into smaller bubbles. This is not to say that a dissolved-gas influx will not string out in the annulus to some extent. Laminar flow implies that the fluid velocity near the center of the flow profile will exceed the average annular velocity whereas fluid near the walls of the hole or drillstring will lag behind. Also, density inversion resulting from having solution gas on bottom may lead to some gravity segregation of the annular fluids. These two factors result in a longer kick-contaminated region during the displacement process, but with much less dispersion than a gas/WBM mixture. Thus instantaneous gas rates at surface are expected to be higher when using an OBM. This means that the design criteria for the mud/gas separator may differ according to mud type. Even so, the choke manifold and other equipment will experience less slugging over a shorter time period—a decided advantage.

Figs. 4.58 through 4.60 show Driller's Method casing-pressure (or pit gain) curves for three conceivable

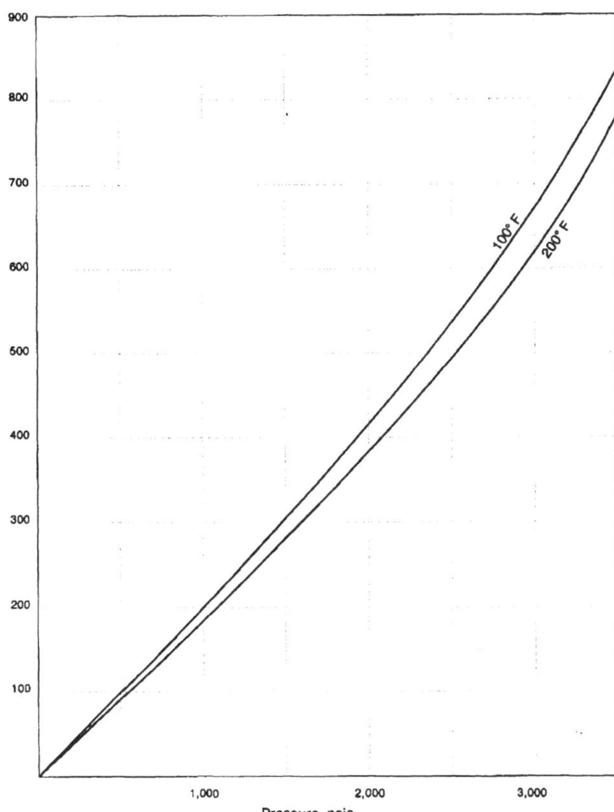

Fig. 4.61—Solubility of a CO_2-rich natural gas in a mineral oil.[43]

well-control situations with an OBM. Response curves for a WBM having the same kick volume and SIDPP are shown for comparison. Fig. 4.58 represents a case in which a free-gas column is present initially in the annulus. This situation could transpire when a large gas kick saturates the mud and circulation commences before any migrating gas has the opportunity to contact any fresh oil. A characteristic increase in casing pressure will occur as the first gas bubble expands and as new gas starts to break out of solution. But the applied backpressure assures that some of the gas will remain in solution until downstream of the choke. Thus maximum casing pressure may be less than seen with a WBM, even considering the more pronounced dispersion in the latter fluid.

Fig. 4.59 shows the case in which all the gas is in solution at the start of the procedure and bubblepoint pressure is achieved later at some point in the annulus. The gas-contaminated mud will expand to some extent as B_{og} is reduced during displacement, but the effect on casing pressure is minor when compared to free-gas expansion. Annulus pressures are relatively stable during the displacement process until the bubblepoint pressure has been reached, evidenced by the rapidly increasing casing pressure. In many cases, the bubblepoint depth will be fairly shallow in the well, a distinct advantage if the casing has been set deeper. The maximum pressure seen by the open hole then will be the pressure imposed at initial circulation.

Note that we could replace the casing-pressure curve in Fig. 4.59 with that of an oil kick in WBM in which the oil reaches bubble-point pressure in the annulus. Phase changes can complicate the annulus pressure and return flow response during a kick circulation. Consider a wet-gas kick that does not go into solution at bottomhole conditions. Reducing the pressure and temperature during circulation may result in crossing the dewpoint line on the composition-phase diagram at some point up the hole. We would expect to see a flattening or at least a less rapid buildup in the casing-pressure curve as condensate drops out of the gas.

Fig. 4.60 corresponds to a gas kick in an OBM where a relatively low solution GOR is combined with a large kick intensity (high SIDPP). Bubblepoint pressure for this solution mixture is not achieved until the contaminated mud has passed through the choke, thereby leading to a relatively stable casing pressure throughout the displacement. Maximum casing pressure in this case is in fact the initial SICP.

TABLE 4.21—DATA FOR EXAMPLE 4.17

Operational Parameters	
SIDPP	990 psig
Top of Gas Sand	9,158 ft
Total depth	9,241 ft
Gas SG	0.832
Hole diameter	12¼ in.
Drill collar OD	9 in. and 7 in.
Collar section lengths each diameter	90-ft
Drillpipe/HWDP OD	5 in.
Circulation rate	450 gal/min
Surface temperature at KRCP	120°F
Bottomhole Parameters While Circulating	
Pressure	4,800 psia
Temperature	200°F
Gas compressibility factor	0.949
Gas-free oil volume factor	1.014 bbl/STB
Oil volume factor at r_{so}	1.042 bbl/STB
Bottomhole Parameters While Shut In	
Pressure	5,700 psia
Temperature	200°F
Gas compressibility factor	1.015
Gas-free oil volume factor	1.009 bbl/STB
Oil volume factor at r_{so}	1.037 bbl/STB

Example 4.17. A 15-ft gas sand is penetrated while drilling with a 10.0-lbm/gal mineral OBM and the formation flows into the mud at a rate of 1.3 MMcf/D. The formation fluid is a CO_2-rich natural gas with solubility curves in the base oil as shown in **Fig. 4.61.** The kick is detected and the well is shut-in with a recorded 10-bbl pit gain. Estimate the (1.) influx volume, (2.) initial SICP, and (3.) maximum casing pressure during kick displacement using the data in **Table 4.21**.

Solution.

1. Eqs. 4.41 and 4.42 give a mixture GOR of

$$r_o = \frac{(1,300,000)(42)}{(0.71)(1,440)(450)} = 118.7 \text{ scf/STB}.$$

From Fig. 4.61, the bubblepoint pressure for this GOR at 200°F is 675 psia. Thus all the gas is being dissolved upon contact with the circulating mud and $r_o = r_{so}$. The downhole GOR corresponding to 118.7 scf/STB concentration is obtained by the gas law.

$$r'_{so} = \frac{(14.65)(118.7)(660)(0.949)}{(4,800)(520)} = 0.436 \text{ ft}^3\text{/STB}$$

$$= 0.0777 \text{ bbl/STB}.$$

Eq. 4.44 yields the kick volume associated with the 10-STB surface manifestation.

$$V_k = \frac{(10 \text{ STB})(0.0777 \text{ bbl/STB})}{(1.042 - 1.014) \text{ bbl/STB}} = 27.8 \text{ STB}.$$

2. A rigorous prediction of the SICP would require that we calculate the height of the influx using variable volume factors for the gas/mud solution, determine the pressure at the top

of the influx, and numerically work our way back to the casing pressure gauge. Instead, we will approximate the answer by assuming the composite FVF across the influx region is the same as the bottomhole factor at shut-in. We also take the FVF in the drillpipe opposite the influx region to be a constant gas-free factor at bottomhole conditions and assume the volume factors in the drillpipe and annulus above this region are the same. After shut-in, the FVF of the kick-contaminated mud is obtained from Eq. 4.43 as

$$B_m = (0.71)(1.037) + (1 - 0.71) = 1.026 \text{ bbl/STB}.$$

The gas density at bottomhole conditions is

$$\rho_g = \frac{(0.832)(5,700)}{(2.77)(1.015)(660)} = 2.56 \text{ lbm/gal}.$$

The 118.7-scf/STB solution GOR now is equivalent to

$$r'_{so} = \frac{(14.65)(118.7)(660)(1.015)}{(5,700)(520)} = 0.393 \text{ cu ft/STB}$$

$$= 0.0700 \text{ bbl/STB}.$$

Hence the mass of the gas dissolved in one STB of mud is

$$m_g/V_{msc} = (2.56)(42)(0.0700)(0.71) = 5.34 \text{ lbm/STB}.$$

Eq. 4.46 yields the density of the kick-contaminated mud at the bottom of the hole,

$$\rho_m = [(10.0)(42) + 5.34]/1.026 = 414.6 \text{ lbm/bbl}$$

$$= 9.87 \text{ lbm/gal}.$$

In the drillpipe, the mud-volume factor and density at the bottom of the hole are determined, respectively, as

$$B_m = (0.71)(1.009) + (1 - 0.71) = 1.006 \text{ bbl/STB}$$

and $\rho_m = 10.0/1.006 = 9.94$ lbm/gal.

The 27.8-STB gas volume is converted to standard cubic feet using the gas law.

$$V_{gsc} = \frac{(4,800)(27.8)(5.6146)(520)}{(14.65)(0.949)(660)} = 42,458 \text{ scf}.$$

Hence the kick transpired over a period of

$$(42,458)(1,440)/1,300,000 = 47.0 \text{ min}.$$

The mud volume pumped over this time is

$$(450)(47.0) = 21,150 \text{ gal} = 503.6 \text{ STB}.$$

The volume of the influx-contaminated mud after closing the well in is

$$(503.6)(1.037) = 522.2 \text{ bbl},$$

and we determine its annular height as

$$180 + \frac{522.2 - (90)(0.06709) - (90)(0.09818)}{0.12149}$$

$$= 4,356 \text{ ft}.$$

The estimated SICP is, therefore,

$$p_{cs} = 990 + (0.0519)(9.94 - 9.87)(4,356)$$

$$= 1,006 \text{ psig}.$$

Actually, the expected SICP would be somewhat higher because the swelling of the mud/gas solution increases up the hole and the temperature should be higher in the annulus than in the drillstring.

3. From Fig. 4.61, the bubblepoint pressure for the solution GOR at 120°F is about 640 psia. Thus sufficient backpressure would be maintained during the kill to keep all wellbore gas in solution and the maximum casing pressure is the initial SICP.

The large volume of the influx region may seem surprising, but this example illustrates the importance of early kick detection in OBMs. Bubblepoint pressure would have been achieved at the top of the contaminated mud had the influx volume been much higher or the hole size much smaller. A rapid unloading of the hole would have commenced and led to a much more difficult well-control problem.

Problems

4.1 Approximate the effect on influx rate if the permeability is doubled. Holding the remaining variables constant, what effect is seen if you double the formation thickness, pressure differential, viscosity, and wellbore diameter?

4.2 You are drilling at 15,000 ft with a 12.0-lbm/gal mud when returns are lost into a vugular limestone. What fluid-level reduction will invite a kick from an exposed 9.0-lbm/gal (pore-pressure gradient) sandstone at 3,000 ft? Repeat the calculation, assuming the permeable sandstone is at 6,000 ft.

4.3 A drilling well intersects the production tubulars in an offset wellbore. Determine the theoretical mud density needed to prevent an influx if the collision occurs at 900 ft where the flowing tubing pressure is 3,500 psig.

4.4 A well equipped with a delta flow kick-detection system takes an influx while drilling ahead. The inflow rate increases linearly from zero to 45 gal/min influx over three minutes.

1. What is the pit gain after this flow period?
2. Determine the contained kick volume if the inflow rate remains at 45 gal/min and it takes another two minutes to close in the well.

4.5 What kick detection equipment would be most suitable for drilling an 18,000-ft slimhole section with a 16.0-lbm/gal OBM?

4.6 Flow velocity at a given flow rate depends on the conduit geometry. In calculating water-hammer loads, where should the velocity be determined (drillpipe × casing annulus, drillpipe × BOP bore, choke line, etc.)?

4.7 For the water-hammer pressure calculated in Example 4.1, calculate the resulting equivalent mud density seen at the casing shoe if pipe is set at 100 ft. What if the casing seat is at 1,000 ft?

4.8 Calculate the weight or mass per unit length for an 11 × 3.0-in. drill collar. What is the displacement factor?

4.9 Determine the change in equivalent density at total depth and at the last casing seat for each of the pressure losses determined in Example 4.2.

4.10 A 5-in., 19.50-lbm/ft high-strength drillpipe is being tripped through 9⅝-in., 36.0-lbm/ft casing.

TABLE 4.22—DATA FOR PROBLEM 4.11	
Last casing setting depth	14,200 ft
Last casing inner diameter	6.184 in.
Openhole diameter	6.0 in.
Drillpipe and HWDP OD	3.5 in.
Drillpipe/HWDP length	14,470 ft
Drill collar OD	4.5 in.
Drill collar length	330 ft

TABLE 4.23—FANN VISCOMETER RHEOLOGIES	
θ_{600} =	46 lbf/100 ft^2
θ_{300} =	29 lbf/100 ft^2
θ_{200} =	18 lbf/100 ft^2
θ_{100} =	9 lbf/100 ft^2
θ_6 =	4 lbf/100 ft^2
θ_3 =	3 lbf/100 ft^2

1. Assume the hole is not being filled and prepare a table showing hydrostatic-pressure losses after pulling twenty stands for mud densities of 10.0, 12.0, 14.0, and 16.0 lbm/gal.
2. Generate another table showing the effects if the stands are pulled wet.

4.11 You are planning to trip for a bit change at 14,800 ft. The current mud density is 14.7 lbm/gal and there is a permeable formation at 14,650 ft with a known pore-pressure equivalent of 14.0 lbm/gal. The drillstring and hole dimensions used are listed in **Table 4.22.**

Assume the downhole mud has an average gel strength of 20 lbf/100 ft^2 and the Fann viscometer rheologies listed in **Table 4.23.**

Can the bit be tripped safely at the present mud weight? If so, what is the maximum safe trip velocity for a plugged drillstring? An iterative procedure is required and the problem is solved most easily by writing a spreadsheet program.

4.12 Without resorting to Appendix B and assuming you were supervising the job, how might you estimate the separate gel strength and inertial contributions to the total swab pressure indicated at Point A in Fig. 4.22?

4.13 The pressure data represented in Fig. 4.22 was obtained in 9$^5/_8$-in. casing at a depth of 2,100 ft while 7-in. casing was being lowered from 1,812 to 1,856 ft. Determine the maximum and minimum equivalent mud weights experienced at the measurement depth if the mud density was 10.8 lbm/gal.

4.14 In general, does ignoring fluid compressibility and borehole elasticity lead to underpredicting or overpredicting surge and swab pressures? Is the effect more significant with the string near top or near bottom?

4.15 In general, does ignoring pipe string elasticity lead to underpredicting or overpredicting surge and swab pressures? Is the effect more significant in deep wells or in shallow wells?

4.16 A 7$^5/_8$-in. flush-joint liner is being run in an 8.50-in. gauge hole when the bottom of the string completely packs off with drill cuttings at 8,000 ft. Estimate the pressure increase below the liner if another 44-ft joint is run at this condition. Total depth is 10,200 ft and the mud compressibility is 2.78×10^{-6} psi^{-1}. Ignore borehole expansion.

4.17 You are tripping off bottom from a depth of 6,000 ft. The string is pulling extremely tight and it is apparent that the stablizers are balled. After working your way 1,000 ft out of the hole, the annulus starts swabbing. How far can you continue to trip before risking an influx into the drillstring if the conditions in **Table 4.24** apply.

TABLE 4.24—DATA FOR PROBLEM 4.17	
Gas zone depth	5,500 ft
Pore pressure	7.0-lbm/gal equivalent
Mud density	9.0 lbm/gal
Average hole diameter	8.50 in.
Drillpipe	4½-in.,16.60-lbm/ft Grade E
Drill collar	6×2 in.
Drill collar length	500 ft

4.18 Compute the volumetric output in gal/stroke for a triplex pump equipped with 6-in. liners if the stroke length is 9 in.

4.19 Taking the pump from Problem 4.18, how many strokes are required to fill a hole after pulling five stands of 3½-in., 13.30-lbm/ft Grade E drillpipe? Assume 95% volumetric efficiency.

4.20 Consider a trip operation from 15,000 ft where the string is composed of a 540-ft drill collar section and conventional drillpipe. The hole is to be filled every ten stands while pulling drillpipe and every stand while pulling the collars. Assuming it takes three minutes to fill the hole, what cost is associated with the downtime if the daily operating cost for the operation is $20,000 a day?

4.21 You plan to shop-fabricate a cylindrical trip tank using a 6-ft diameter shell.

1. Determine the scaling factor (bbl/in.) for the tank.
2. How tall must the tank be if you need 20 bbl of capacity?

4.22 Derive an alternate expression for Eq. 4.12 by assuming the string depth and depth below the casing pressure gauge are equivalent.

4.23 For a 10.0-lbm/gal mud, plot the slip load as a function of SICP for a single stand of drill collars with the following dimensions.

1. 4¾ × 1¾ in.
2. 6¼ × 2¼ in.
3. 11 × 3¼ in.

4.24 Determine the allowable shut-in pressure for Example 4.5 if one 90-ft stand of 4½-in., 16.60-lbm/ft Grade E NC46 drillpipe is placed above the 9-in. collars and run in the hole. The approximate actual weight of the drillpipe is 18.37 lbm/ft.

4.25 A well takes a gas kick at 10,000 feet resulting in an initial influx length of 1,000 ft. The SICP is 400 psig and the mud density is 9.0 lbm/gal. Plot the equivalent density for the shut-in conditions in 500-ft increments between 1,000 ft and total depth. Assume a gas hydrostatic gradient of 0.1 psi/ft.

4.26 For a kick severity of 1.0 lbm/gal, plot the expected SIDPP for well depths ranging between 1,000 ft and 20,000 ft.

4.27 Refer to the well information shown in **Table 4.25.**

1. Determine the kill-mud density.
2. Estimate the nature of the influx.
3. With the influx on bottom, how much casing pressure can this well tolerate before the formation fractures?
4. What instructions should be left with the driller before the supervisor leaves the rig floor?

TABLE 4.25—KICK DATA FOR A WELL IN PROBLEMS 4.27, 4.35, 4.36, 4.38, AND 4.42	
Wellbore Configuration	
Vertical Depth	14,577 ft
Intermediate Casing Information	
Description	10¾-in. 55.50-lbm/ft
Setting Depth	10,370 ft
Open Hole Diameter	9½ in.
Drillstring Information	
Drill Collar Size	7×2½ in.
Drill Collar Length	540 ft
Drillpipe Description	5-in. 19.50 lbm/ft
Drillpipe Length	14,037 ft
Mud Properties and Recorded Kick Data	
Mud Type	Water Base
Mud Density	11.3 lbm/gal
SIDPP	1,200 psig
SICP	1,400 psig
Pit Gain	18 bbl
Other Known or Assumed Information	
Fracture Gradient at Casing Seat	0.88 psi/ft
Kill Circulation Rate	35 strokes/min
KRCP	700 psig
Pump Output	5.83 gal/stroke
Assumed Annulus Temperature at Kill Rate	110°F + 1.2°F/100 ft

4.28 In Example 4.8, how would the computed pit gain and SICP change (up or down) if we had considered the following:

1. The change in the mud's properties with dissolved gas.
2. Elasticity in the wellbore.
3. Compressibility of the wellbore fluids.
4. Filtrate loss to a shallower formation.
5. Gas slippage through the circulating mud.
6. Gas expansion? Defend your reasoning.

4.29 Your well kicks and the following shut-in drillpipe pressures are recorded by the driller in **Table 4.26.** What SIDPP value would you use to calculate the BHP?

4.30 Derive Eq. 4.22.

4.31 You are drilling at 9,500 ft with a 9.4-lbm/gal mud. A 13³/₈-in. surface casing with a 12.515-in. ID has been set at 3,800 ft. Assume a 650-bbl mud volume in the pits with pipe in the derrick and an average openhole diameter of 15 in. How much barite should be on hand if company policy requires enough inventoried weighting material to increase the mud density by 1.0 lbm/gal?

4.32 List at least eight reasons why the casing-pressure gauge is not suitable for maintaining constant BHP in a kill operation.

4.33 Use the data given in Table 4.13 and verify the circulating pressure curves shown in Figs. 4.33 and 4.34.

4.34 A kick occurs while drilling a well with a float positioned in the drillstring. The initial SIDPP and SICP are 0 and 500 psig, respectively, while the KRCP recorded on the previous tour sheet was 700 psig. The CDPP reading is 1,100 psig after the pump is brought up to kill rate while maintaining 500 psi on the casing-pressure gauge. What SIDPP value should be used to calculate the formation pressure?

TABLE 4.26—DATA FOR PROBLEM 4.29	
Time (min)	SIDPP (psig)
1	640
2	810
3	880
4	940
5	500
6	550
7	590
8	630
9	670
10	700

4.35 The well described in Table 4.25 is to be killed using the Driller's Method. Outline the control procedure.

4.36 Outline the procedure for a Wait and Weight kill on the well in Table 4.25. Include a drillpipe-pressure-reduction graph or table.

4.37 A gas kick is displaced while holding constant choke pressure as the drillstring is being filled with kill mud. What effect does this procedure have on the BHP?

4.38 Refer to the data shown in Table 4.25 and assume the well is to be killed with the Wait and Weight Method. Estimate the effect on annulus pressures if an incremental 1.0 lbm/gal is added to the kill-weight density before circulation begins.

4.39 A horizontal wellbore crosses an unexpected fault plane while drilling with a 10.0-lbm/gal mud. TVD of the lateral is 4,578 ft. The shut-in drillpipe and casing gauges both read 100 psig. The drillstring component lengths and wellbore configuration are shown below (all depths are measured).

4½-in., 16.60-lbm/ft drillpipe from surface to 3,980 ft, 6¼×2¼-in. collars from 3,980 ft to 4,280 ft, 4½-in. HWDP with a 2⁷/₈-in. ID from 4,280 ft to total depth, vertical section from surface to 4,100 ft, 12°/100 ft build to 4,850 ft, and horizontal section to 6,200 ft.

1. Determine the kill-fluid density.
2. Can you ascertain the influx type from the information provided?

4.40 The horizontal well described in Problem 4.39 is to be killed using the Wait and Weight Method. Prepare a drillpipe-pressure-reduction schedule if the pump output is 3.1 gal/stroke.

4.41 Discuss advantages and disadvantages of each of the circulation-kill techniques discussed in this chapter: Driller's, Wait and Weight, and Concurrent Methods.

4.42 The Driller's Method will be used on the well described in Table 4.25.

1. Estimate the maximum casing-seat pressure.
2. Estimate the maximum surface pressure on the annulus.
3. What is the estimated maximum pit gain?

4.43 Answer the questions posed in Problem 4.42 if the well is killed using the Wait and Weight Method.

4.44 Refer to the data shown in Table 4.25 and plot anticipated casing pressures vs. pump strokes for both kill techniques.

4.45 O'Bryan and Bourgoyne's[42] FVF data for a 0.62-SG gas in No. 2 diesel at 300°F are presented in **Fig. 4.62.** Assume the curves apply to a well-control event while using an OBM in

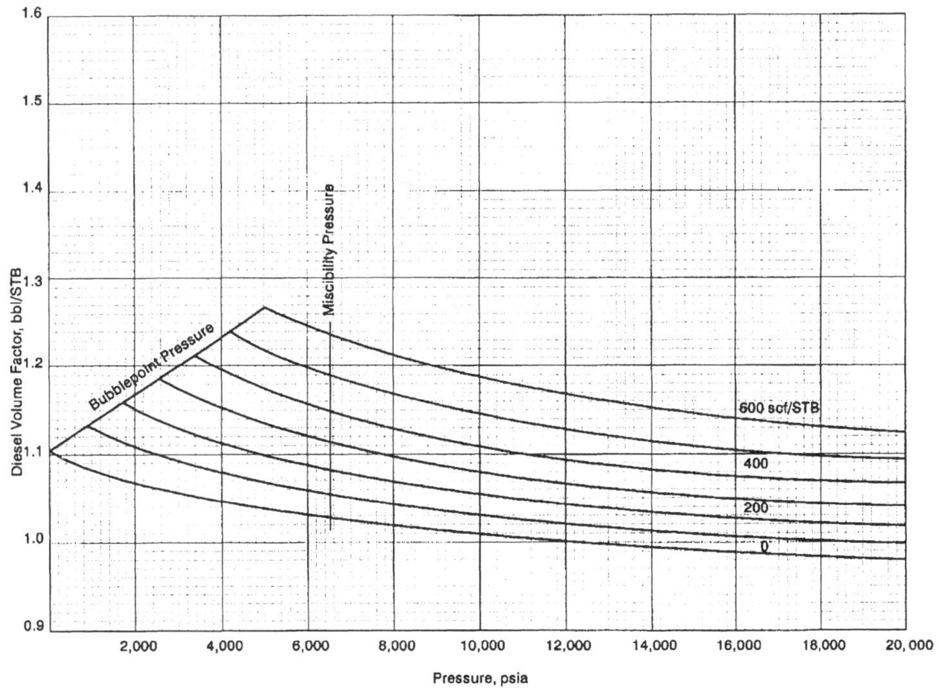

Fig. 4.62—Volume factors for a 0.62-SG gas dissolved in No. 2 diesel at 300°F.[42]

a hot hole. Estimate the influx volume and change in the bottomhole mud density if pit gain = 14 bbl, circulation rate = 350 gal/min, gas influx rate = 3,200,000 scf/D, BHP = 7,500 psia, surface mud density = 14.4 lbm/gal, and oil volume = 0.59.

Nomenclature

a = pressure drop exponent, dimensionless
A = area, sq in.
A_i = ID cross-sectional area, sq in.
A_o = OD cross-sectional area, sq in.
A_s = wall cross-sectional area, sq in.
B_m = mud volume factor, bbl/STB
B_o = oil volume factor, bbl/STB
B_{og} = oil volume factor including dissolved gas, bbl/STB
B_{ong} = oil volume factor excluding dissolved gas, bbl/STB
C = capacity per unit length (capacity factor), bbl/ft
C_a = annulus capacity factor, bbl/ft
C_{ai} = annulus capacity factor at initial conditions, bbl/ft
C_d = displacement factor, bbl/ft
C_i = internal pipe capacity factor, bbl/ft
d = diameter, in.
d_h = hole diameter, in.
d_i = inner diameter, in.
d_o = outer diameter, in.
D = depth, ft
D_k = kick depth, ft
E_V = volumetric efficiency, dimensionless
f_o = oil volume fraction, dimensionless
F = force, lbf
F_f = friction force, lbf
F_p = pump output factor, bbl/stroke or gal/stroke
F_s = slip load, lbf
g = acceleration of gravity, 32.17 ft/s^2
g_c = gravitational system conversion constant, 32.17 (lbm-ft)/(lbf-s^2)
g_m = mud hydrostatic gradient, psi/ft
g_k = kick fluid hydrostatic gradient, psi/ft
g_{km} = kill mud hydrostatic gradient, psi/ft
g_{om} = original mud hydrostatic gradient, psi/ft
G = pit gain, bbl
h = permeable thickness, ft
h_k = kick fluid height, ft
h_{ki} = gas influx height at bottomhole conditions, ft
h_m = mud height, ft
h_{om} = original mud height in the annulus, ft
k_g = permeability to gas, darcy
L_k = kick fluid length, ft
L_p = pipe length, ft
m_B = barite mass, lbm
m_g = gas mass, lbm
N_p = number of pump strokes
p = pressure, psi
p_{bh} = BHP, psi
p_c = casing pressure, psi
p_{ch} = choke backpressure, psi
p_{cs} = shut-in casing pressure, psi
p_{dc} = circulating drillpipe pressure, psi
p_{dci} = initial circulating drillpipe pressure, psi
p_{dcf} = final circulating drillpipe pressure), psi
p_{ds} = shut-in drillpipe pressure, psi
p_e = pressure at the external radius, psi
p_{fi} = fracture initiation pressure, psi
p_{ha} = hydrostatic pressure in the annulus, psi
p_{hcl} = hydrostatic pressure in the choke lines, psi
p_{hd} = hydrostatic pressure in the drillstring, psi
p_{hg} = gas hydrostatic pressure, psi
p_{hgi} = gas hydrostatic pressure at initial conditions, psi
p_{kr} = kill-rate circulating pressure (KRCP), psi
p_p = pore pressure, psi
p_{pc} = pseudocritical pressure, psia

p_{pr} = pseudoreduced pressure, dimensionless
p_{wf} = flowing wellbore pressure, psi
Δp_a = annulus friction pressure
Δp_b = pressure drop across bit, psi
Δp_c = casing pressure increase at surface, psi
Δp_{cl} = choke line friction pressure, psi
Δp_d = drillstring friction pressure, psi
Δp_s = surface equipment friction pressure, psi
Δp_t = turbulent flow friction pressure, psi
q = volumetric flow rate, bbl/min
q_{gsc} = gas flow rate at standard conditions, scf/D or scf/min
q_m = mud flow rate, bbl/min
r_e = drainage radius, ft
r_m = gas/mud ratio, scf/STB
r_o = gas/oil ratio, scf/STB
r_{so} = solution gas/oil ratio, scf/STB
r'_{so} = downhole solution gas/oil ratio, bbl/STB
r_w = wellbore radius, ft
t = time, s
T = temperature, T, °F or °R
T_{bh} = bottomhole temperature, T, °R
T_{pc} = pseudocritical temperature, T, °R
T_{pr} = pseudoreduced temperature, dimensionless
v = velocity, ft/s
Δv = velocity change of the flow stream
v_f = acoustic velocity in fluid, ft/s
V = volume, bbl
V_a = annulus volume, bbl
V_B = barite volume, bbl
V_d = displacement volume, bbl
V_i = inner volume, bbl
V_k = kick volume, bbl
V_{ki} = initial kick volume, bbl
V_{gsc} = gas volume at standard conditions, scf
V_m = mud volume, bbl
V_{msc} = mud volume at ambient conditions, gal
V_p = pump output volume, bbl or gal
w = weight per unit length, lbf/ft
w_m = mass per unit length, lbm/ft
W = air weight, lbf
X = variable in the pressure-prediction relation where mud density is constant
Y = variable in the pressure-prediction relation where kill mud is in the annulus
z = compressibility factor, dimensionless
z_{bh} = compressibility factor at bottomhole conditions, dimensionless
γ_g = gas specific gravity, dimensionless
θ = Fann viscometer reading, lbf/100 ft^2
μ_g = gas viscosity, cp
ρ = density, lbm/gal
ρ_B = barite density, lbm/gal
ρ_f = fluid density, lbm/ft
ρ_g = gas density, lbm/gal
ρ_{km} = kill mud density, lbm/gal
ρ_m = mud density, lbm/gal
ρ_{msc} = mud density at ambient conditions, lbm/gal
ρ_{om} = original mud density, lbm/gal
ρ_M = mixture density, lbm/gal

References

1. Dahl, E. and Bern, T.J.: "Risk of Oil and Gas Blowout on the Norwegian Continental Shelf," NSFI/SINTEF Project No. 880354.14, Trondheim, Norway (15 February 1983).
2. Teel, M.E.: "Well Control: Plan for Uncommon Occurrences," *World Oil* (December 1993) 25.
3. Stokka, S. *et al.*: "Gas Kick Warner—An Early Gas Influx Detection Method," paper SPE 25713 presented at the 1993 SPE/IADC Drilling Conference, Amsterdam, 23–25 February.
4. Bryant, T.M., Grosso, D.S., and Wallace, S.N.: "Gas-Influx Detection With MWD Technology," *SPEDE* (December 1991) 273.
5. Codazzi, D. *et al.*: "Rapid and Reliable Gas Influx Detection," paper SPE 23936 presented at the 1992 IADC/SPE Drilling Conference, New Orleans, 18–21 February.
6. Orban, J.J. *et al.*: "New Ultrasonic Caliper for MWD Operations," paper SPE 21947 presented at the 1991 SPE/IADC Drilling Conference, Amsterdam, 28 February–2 March.
7. Harris, T.W.R., Hendriks, P., and Surewaard, J.H.G.: "Advanced Kick Detection Systems Improve HPHT Operations," *Pet. Eng. Intl.* (September 1995) 31.
8. Maus, L.D., Tannich, J.D., and Ilfrey, W.T.: "Instrumentation Requirements for Kick Detection in Deep Water," paper OTC 3240 presented at the 1978 Offshore Technology Conference, Houston, 8–11 May.
9. Maus, L.D., Peters, B.A., and Meador, D.J.: "Sensitive Delta-Flow Method Detects Kicks or Lost Returns," *Oil & Gas J.* (20 August 1979) 125.
10. Speers, J.M. and Gehrig, G.F.: "Delta Flow: An Accurate, Reliable System for Detecting Kicks and Loss of Circulation During Drilling," *JPT* (December 1987) 359.
11. Haeusler, D., Makohl, F., and Harris, T.W.R.: "Applications and Field Experience of an Advanced Delta Flow Kick Detection System," paper SPE 29344 presented at the 1995 SPE/IADC Drilling Conference, Amsterdam, 28 February–2 March.
12. Swanson, B.W. *et al.*: "Slimhole Early Kick Detection by Real-Time Drilliung Analysis," paper SPE 25708 presented at the 1993 SPE/IADC Drilling Conference, Amsterdam, 23–25 February.
13. Hornung, M.R.: "Kick Prevention, Detection, and Control: Planning and Training Guidelines for Drilling Deep, High-Pressure Gas Wells," paper SPE 19990 presented at the 1990 IADC/SPE Drilling Conference, Houston, 27 February–2 March.
14. Jardine, S.I. *et al.*: "Hard or Soft Shut-in: Which is the Best," paper SPE 25712 presented at the 1993 SPE/IADC Drilling Conference, Amsterdam, 23–25 February.
15. "Recommended Practice for Drill Stem Design and Operating Limits," *RP 7G*, 15th edition, API, Washington, DC (1 January 1995).
16. Cannon, G.E.: "Changes in Hydrostatic Pressure Due to Withdrawing Drill Pipe from the Hole," *Drill and Prod. Prac.*, API (1934) 42–47.
17. Burkhardt, J.A.: "Wellbore Pressure Surges Produced by Pipe Movement," *JPT* (June 1961) 595.
18. Fontenot, J.E. and Clark, R.K.: "An Improved Method for Calculating Swab and Surge Pressures and Circulating Pressures in a Drilling Well," *SPEJ* (October 1974) 451.
19. Lubinski, A., Hsu, F.H., and Nolte, K.G.: "Transient Pressure Surges Due to Pipe Movement in an Oil Well," *Revue de l'Inst. Fran. du Pét.* (May–June 1977) 307–47.
20. Lal, M.: "Surge and Swab Modeling for Dynamic Pressures and Safe Velocities," paper SPE 11412 presented at the 1983 SPE/IADC Drilling Conference, New Orleans, 20–23 February.
21. Mitchell, R.F.: "Dynamic Surge/Swab Pressure Predictions," *SPEDE* (September 1988) 325.

22. Chin, W.C.: *Wave Propagation in Petroleum Engineering*, Gulf Publishing Co., Houston (1994) 298–315.
23. Babu, D.R.: "Effects of p-ρ-T Behavior of Muds on Static Pressures During Deep Well Drilling," *SPED&C* (June 1996) 91.
24. Raymond, L.R.: "Temperature Distribution in a Circulating Drilling Fluid," *JPT* (March 1969) 333.
25. Young, G.A.: "Trip Tanks Provide Early Warning of Downhole Problems," *World Oil* (December 1981) 125.
26. Bang, J. *et al.*: "Acoustic Gas Detection With Wellhead Sonar," paper SPE 28317 presented at the 1994 SPE Annual Technical Conference and Exhibition, New Orleans, 25–28 September.
27. United States Department of Interior MMS Regulation 250.58 governing Oil and Gas and Sulphur Operations in the Outer Continental Shelf, *Federal Register*, **53, 63** (1 April 1988) 10720.
28. "Specifications for Drilling-Fluid Materials," *Spec 13A*, 15th edition, API, Washington, DC (1 May 1993).
29. Records, L.R. and Evert, R.E.: "New Well Control Unit Speeds Safer Handling of Blowouts," *Oil & Gas J.* (10 September 1962) 106.
30. Schurman, G.A. and Bell, D.L.: "An Improved Procedure for Handling a Threatened Blowout," *JPT* (April 1966) 437.
31. Kastor, R.L. and Ledbetter, S.C.: paper SPE 4973 presented at the 1974 SPE Annual Meeting, Houston, 6–9 October.
32. Grace, R.D.: "Why Well Control "Safety Factors" May Not Be Safe," *Pet. Eng. Intl.* (June 1990) 41.
33. United States Department of Interior MMS Regulation 250.58 governing Oil and Gas and Sulphur Operations in the Outer Continental Shelf, *Federal Register*, **53, 63** (1 April 1988) 10721.
34. O'Brien, T.B. and Goins, W.C. Jr.: "The Mechanics of Blowouts and How to Control Them," *Drill and Prod. Prac.*, API (1960) 41–55.
35. Sonnemann, P.: "Circulate-and-Weight Well Control Method Has Several Advantages," *Oil & Gas J.* (31 January 1994) 96.
36. Ramey, H.J. Jr.: "Wellbore Heat Transmission," *JPT* (April 1962) 427; *Trans.*, AIME, **225.**
37. Holmes, C.S. and Swift, S.C.: "Calculation of Circulating Mud Temperatures," *JPT* (June 1970) 670.
38. Keller, H.H., Couch, E.J., and Berry, P.M.: "Temperature Distribution in Circulating Mud Columns," *SPEJ* (February 1973) 23.
39. Wooley, G.R.: "Computing Downhole Temperatures in Circulation, Injection, and Production Wells," *JPT* (September 1980) 1509.
40. Beirute, R.M.: "A Circulating and Shut-in Well-Temperature-Profile Simulator," *JPT* (September 1991) 1140.
41. Kabir, C.S. *et al.*: "Determining Circulating Fluid Temperature in Drilling, Workover, and Well Control Operations," *SPED&C* (June 1996) 74.
42. O,Bryan, P.L. and Bourgoyne, A.T. Jr.: "Swelling of Oil-Based Drilling Fluids Resulting From Dissolved Gas," *SPEDE* (June 1990) 149.
43. Van Slyke, D.C. and Huang, E.T.S: "Predicting Gas Kick Behavior in Oil-Based Drilling Fluids Using a PC-Based Dynamic Wellbore Model," paper SPE 19972 presented at the 1990 IADC/SPE Drilling Conference, Houston, 27 February–3 March.

SI Metric Conversion Factors

bbl	× 1.589 873	E − 01	= m^3
cp	× 1.0*	E + 03	= Pa·s
ft	× 3.048*	E − 01	= m
ft/hr	× 8.466 667	E − 02	= mm/s
ft^3	× 2.831 685	E − 02	= m^3
°F	(°F − 32)/1.8		= °C
°F/100 ft	× 1.822 689	E + 01	= mK/m
gal	× 3.785 412	E + 00	= L
in.	× 2.54*	E + 01	= mm
$in.^3$	× 6.451 6*	E + 02	= mm^2
lbf	× 4.448 222	E + 00	= N
lbm	× 4.535 924	E − 01	= kg
psi	× 6.894 757	E − 03	= MPa
psi^{-1}	× 1.450 377	E − 01	= kPa^{-1}
°R	°R/1.8		= K

* Conversion factor is exact.

Chapter 5
Well-Control Complications

5.1 Introduction

The kick displacement procedures discussed in Chap. 4 are for those cases where the drillstring is below the influx and where nothing precludes shutting in the well or displacing the kick in a safe manner. However, conventional kick displacement methods do not always apply. An operator may be faced with a well-control situation where pipe is off bottom or out of the hole, problems develop during a conventional kill, or the annulus cannot withstand the backpressures imposed during kick circulation. This chapter focuses on the means for maintaining or regaining conventional control when the traditional concepts do not apply. The first method discussed can be used on a shut-in well where a migrating influx is indicated, but the ability to read bottomhole pressure (BHP) confidently has been lost for some reason.

5.2 Volumetric Control and Lubrication

Well control can be accomplished without circulation by a technique called the volumetric method. We have stressed the importance of allowing migrating gas to expand as it moves up the hole and offered a procedure for doing so when a well is closed in on a kick. Holding constant drillpipe pressure by bleeding mud from the annulus is in fact a volumetric technique in that a semiconstant BHP is maintained without circulation.

However, reading the BHP directly from the drillpipe gauge is not always possible. For instance, pressure communication between the drillstring and annulus would be eliminated if the bit becomes plugged upon shut-in or at some other point in the kick circulation. Other situations where the drillpipe pressure gauge is useless for control are when (1) a migrating influx is beneath the bit, (2) the drillpipe has separated or has a hole above the influx, (3) the well is closed in with the blind rams, (4) the pumps are inoperable and the drillstring is not full of mud, and (5) gas has entered the drillstring. For these problems, the volumetric method must be used to guide the operation and to maintain wellbore pressures within the desired window.

Lubrication is a procedure for replacing gas that has collected beneath a closed preventer or valve with mud or another liquid. Its distinguishing characteristic is that the job is done without circulating the hole from below the gas accumulation. Lubrication often follows a volumetric-control effort in which the migrating gas has reached the surface and it is commonly used in cased-hole work to kill a well. Done correctly, a lubrication procedure can be accomplished while maintaining wellbore pressures within a predetermined window.

5.2.1 The Volumetric Method. Table 5.1 outlines the volumetric-control procedure. Gas migration, if present, will start to drive up the casing pressure from its initial shut-in reading. This is allowed to occur until an incremental pressure called the safety margin is achieved on the pressure gauge. A safety factor is needed to avoid underbalancing the hole because the choke operator cannot precisely maintain the desired choke pressures in the procedure. Other inaccuracies inherent to the approach can lead to underbalance if some excess pressure is not supplied. Safety margins used in the field typically range from 100 to 200 psi. Continued migration under the closed preventer is permitted until the casing pressure achieves another arbitrary 50- to 200-psi increase called the working margin. Large safety and working margins are desirable, but the fracture gradient limits the maximum selected values.

Reaching the working-margin pressure signals the point in the procedure when we start to bleed mud from the well. The choke operator attempts to maintain a constant casing pressure while opening the choke periodically and releasing mud from the well. This process expands the gas and the intent is to flow enough volume so that the hydrostatic pressure of the released mud is equivalent to the working-margin buildup. The relation is expressed by

$$\Delta h_m = p_{wm}/g_m, \quad \text{(5.1)}$$

where Δh_m = incremental mud column height, p_{wm} = working-pressure margin, and g_m = mud hydrostatic gradient.

The desired change in hydrostatic pressure occurs at the depth where we want to replace the added casing pressure with a longer gas column. Thus Eq. 5.2 defines the incremental mud volume to release from the well,

$$\Delta V_m = \Delta h_m C, \quad \text{(5.2)}$$

TABLE 5.1—VOLUMETRIC-CONTROL PROCEDURE

1. Record the initial shut-in casing pressure (SICP).
2. Allow the casing pressure to increase by the predetermined safety margin.
3. Allow the casing pressure to further increase by the predetermined working margin.
4. Bleed mud from the choke manifold into a measuring tank while maintaining a relatively stable casing pressure.
5. Continue to bleed mud until the volume in the measuring tank is equivalent to the mud's hydrostatic pressure of the working-margin buildup. The hydrostatic pressure is based on the hole dimensions at the depth of the rising influx.
6. Repeat Steps 3 through 5 until choke pressures stabilize, conventional control can be regained, or the influx surfaces.
7. Stop the bleed process if gas exits the choke. Monitor annulus pressures for further buildup.

where ΔV_m = the mud volume increment and C = the capacity factor at the depth where the operator suspects the rising influx to be.

After bleeding this volume, the gas influx has expanded by an amount equal to ΔV_m and the BHP is reduced to the original shut-in value plus the safety margin. Subsequent steps repeat the build-and-bleed process, letting migration drive up the casing pressure by the working margin followed by flowing mud while holding constant casing pressure. Ultimately, gas will reach the top of the well or surface pressure will reach a stable value. Gas flow through the choke signals the end of the procedure if the gas has moved up the hole as a unit package. While the well is not dead, control efforts have been successful thus far and other measures can be undertaken now.

Volumetric control is analogous to the kick displacement portion of the Driller's Method. The fundamental difference between the two is that buoyancy is the primary drive mechanism to influx movement rather than the pump. The annular-pressure-prediction equations discussed in Chap. 4 can be used to evaluate the consequences a volumetric procedure may have on wellbore integrity and, at the same time, give a reasoned basis for determining how much safety and working margins can be tolerated when a gas kick approaches a critical depth.

A meticulous accounting of the released mud volume is imperative when using the volumetric-control method, particularly if hole dimensions are tight (small ΔV_m = large Δh_m). A measuring tank calibrated to read only a few gallons may be necessary in some cases, hence the process/volume/temperature (PVT) monitors in the active pit system will not be suitable and even the trip tank may be too large. On any well, thought and planning should go into the equipment requirements and how the tank will be rigged up to the well outlet should the need arise for controlling by volume.

Example 5.1 demonstrates the volumetric method in a fairly uniform wellbore where the gas kick remains in a slug while moving up the hole. In the next few pages, we will discuss some of the difficulties which are encountered often and how a theoretical treatment of the problem differs from what is seen in the real world.

Example 5.1. A well has been drilled to 10,000 ft with an 8.7-lbm/gal brine. A flow is detected with pipe out of the hole and the blind rams are closed after gaining 20 bbl in the pits. The initial casing pressure is 100 psig and a linear increase over time indicates a migrating gas influx. 9⅝-in., 36.0-lbm/ft casing is set at 2,500 ft where the fracture gradient is 0.70 psi/ft. Consider the openhole diameter to be 8¾ in. and assume the wellbore temperature at any depth is 70°F + 0.9°F/100 ft. Write a volumetric-control procedure for this well.

Solution. Select prudent safety- and working-margin pressures to use in the procedure. The maximum casing pressure with gas at or below the shoe is controlled by the fracture gradient and current mud weight. The fracture pressure at 2,500 ft is calculated

$$p_{fi} = (0.70)(2,500) = 1,750 \text{ psig}.$$

Hence the surface pressure limit with gas below the shoe is

$$(p_c)_{max} = 1,750 - (0.452)(2,500) = 620 \text{ psig}.$$

Arbitrary safety and working margins of 100 psi will be acceptable, at least initially. However, the maximum expected shoe pressure also should be estimated to assure that the wellbore pressure limit is not exceeded. The kick was not detected until all the pipe was pulled, so it is reasonable to conclude that the gas was swabbed in or the hole was not kept full during the trip. A gas entry at total depth where the hole was balanced originally gives the maximum pore pressure,

$$(p_p)_{max} = (0.452)(10,000) = 4,520 \text{ psig} = 4,534 \text{ psia}.$$

The BHP term in the Driller's Method equations is the wellbore pressure imposed during the displacement procedure and thus comprises the formation pressure plus any overbalance provided at surface. In a worst-case scenario, pressure at the shoe would be achieved in this particular problem if the gas bubble reaches 2,500 ft at the same instant the casing pressure has climbed to the working margin. Thus,

$$p_{bh} = 4,520 + 200 = 4,720 \text{ psig} = 4,734 \text{ psia}.$$

The initial influx height is estimated from the pit gain and 0.07438-bbl/ft openhole capacity factor as

$$20/0.07438 = 269 \text{ ft}.$$

Incorporating the 200-psi overbalance yields the initial influx height to use in the pressure-prediction relation.

$$h_{ki} = (269)(4,534/4,734) = 258 \text{ ft}.$$

The bottomhole temperature is determined as

$$T_{bh} = 70 + (0.009)(10,000) = 160°\text{F} = 620°\text{R}.$$

We assume the gas specific gravity (SG) is 0.60 and obtain z_{bh} = 0.972 using the technique discussed in Chap. 1. Eq. 4.30 gives the influx hydrostatic pressure at total depth,

$$p_{hgi} = \frac{(0.60)(4,734)(258)}{(53.29)(0.972)(620)} = 23 \text{ psi}.$$

The same hydrostatic pressure is expected after a controlled migration to 2,500 ft because the hole geometry does not change ($p_{hg} = p_{hgi}$). Solve for X using Eq. 4.37.

$$X = 4,734 - (0.452)(10,000 - 2,500) - 23$$
$$= 1,321 \text{ psia}.$$

At the casing seat,

$$T_{2,500} = 70 + (0.009)(2,500) = 93°\text{F} = 553°\text{R}.$$

Assume a z factor of 0.90 at the depth of interest, substitute into Eq. 4.36, and solve for the pressure.

$$p_{2,500} = \frac{1,321}{2} +$$

TABLE 5.2—WELLBORE PRESSURES WHILE VOLUMETRICALLY CONTROLLING THE WELL DESCRIBED IN EXAMPLE 5.1								
Shut-in Time (min)	Casing Pressure (psig)	Gain Increment (bbl)	Influx Length (ft)	Influx Top (ft)	Mud Hydrostatic (psi)	Gas Hydrostatic (psi)	BHP (psig)	Shoe Pressure (psig)
Initial Shut-in Condition								
0.0	100	20.0	269	9,731	4,398	23	4,521	1,230
Safety Margin Buildup								
4.4	200	0.0	269	9,510	4,398	23	4,621	1,330
First Build and Bleed Stage								
8.8	300	0.0	269	9,289	4,398	23	4,721	1,430
119.0	300	16.46	490	3,560	4,298	23	4,621	1,430
Second Build and Bleed Stage. Gas Enters Casing								
123.4	400	0.0	490	3,339	4,298	23	4,721	1,530
146.2	400	17.10	700	1,991	4,204	22	4,626	1,316
Third Build and Bleed Stage								
150.6	500	0.0	693	1,770	4,207	22	4,729	1,339
159.8	500	17.10	914	1,090	4,107	22	4,629	1,239
Fourth Build and Bleed Stage								
164.2	600	0.0	914	869	4,107	22	4,729	1,339
168.1	600	17.10	1,135	452	4,007	22	4,629	1,239
Fifth Build and Bleed Stage. Gas Surfaces								
172.5	700	0.0	1,135	231	4,007	22	4,729	1,339
172.9	700	16.43	1,347	0	3,911	22	4,633	1,243
	Total Gain	104.19						

$$\sqrt{\frac{1,321^2}{4} + \frac{(0.452)(4,734)(258)(0.900)(553)(0.07438)}{(0.972)(620)(0.07438)}}.$$

$$p_{2,500} = 1,605 \text{ psia.}$$

The compressibility factor, z, at this pressure and surface temperature = 0.815. This z factor is substituted into Eq. 4.36 and, after two more iterations, the relation gives a maximum pressure of 1,569 psig at 2,500 ft. This is well below the fracture pressure and so we conclude that the procedure can be carried out without risking lost returns.

The mud volume to release after achieving the first working margin pressure is obtained by combining Eqs. 5.1 and 5.2.

$$\Delta V_m = (100)(0.07438)/0.452 = 16.46 \text{ bbl.}$$

Note that the mud volume changes,

$$\Delta V_m = (100)(0.07731)/0.452 = 17.10 \text{ bbl},$$

after the gas bubble passes into the 9⅝-in. casing. A general volumetric-control procedure would read:

1. Allow the casing pressure to increase from 100 to 300 psig (200-psi safety and working margin).
2. Bleed 16.46 bbl of mud from the choke while holding constant casing pressure.
3. Close the choke and allow the casing pressure to increase by 100 psi.
4. Repeat Steps 2 and 3 until annulus pressures stabilize or gas is bled from the choke, except change the released mud volume in Step 3 to 17.10 bbl when gas enters the casing.
5. Monitor well pressure.

Table 5.2 assumes the procedure from Ex. 5.1 was executed perfectly and tracks the wellbore pressures and location of the gas bubble for a 3,000-ft/hr slip velocity. The respective casing and bottomhole pressures for the job are plotted against cumulative pit gain in **Figs. 5.1 and 5.2.** We see from the last diagram that, in fact, computed bottomhole pressures after

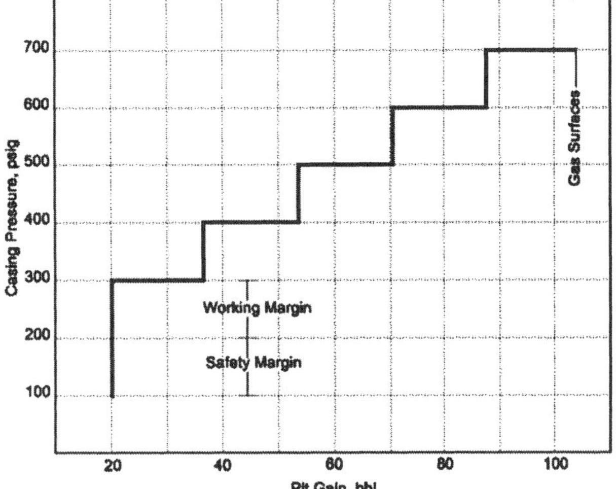

Fig. 5.1—Ideal casing pressure and pit gain schedule for the volumetric procedure used in Example 5.1.

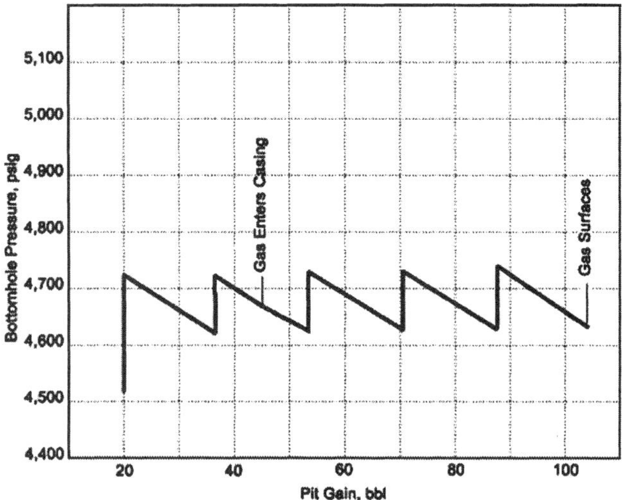

Fig. 5.2—BHPs for the pit gain schedule shown in Fig. 5.1.

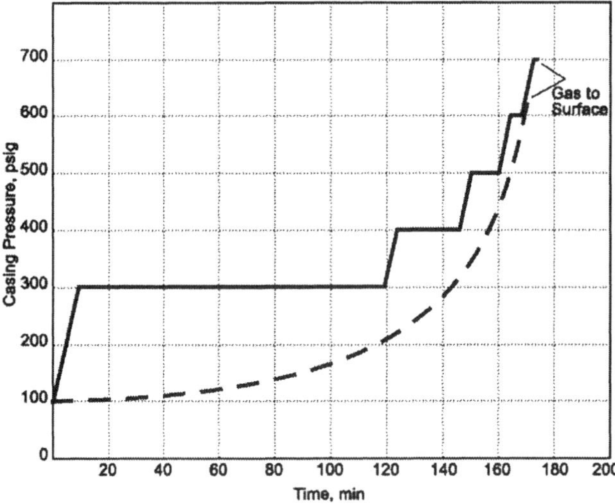

Fig. 5.3—Ideal casing pressures over time and comparison to the computed pressures for a Driller's Method displacement in Example 5.1.

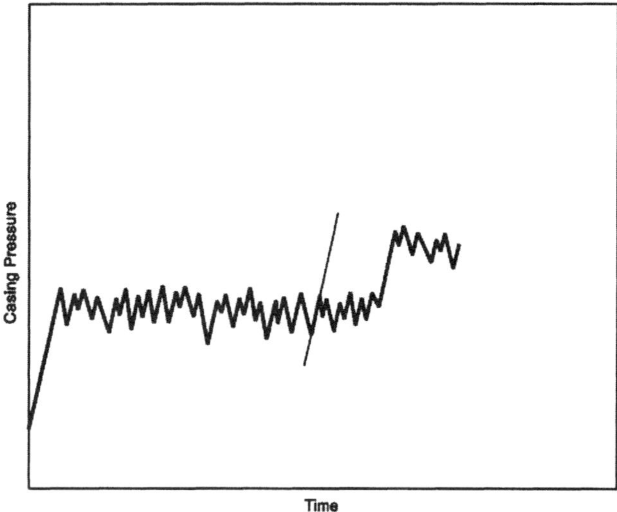

Fig. 5.4—Realistic casing pressures over time during a volumetric-control procedure.

reaching 200 psig do vary within the approximate working margin range and never dip below the safety overbalance.

The solid line in **Fig. 5.3** shows the casing pressures as a function of time. In comparison, a Driller's Method casing-pressure response for an ideal gas-bubble displacement is also shown as if the BHP could be read directly and held constant. The similarity between the two is clear. Both techniques allow an increasingly rapid bubble expansion to take place as the gas nears surface, but the volumetric-control method accomplishes the task in a step-wise fashion rather than by the smooth curve that results from guiding the operation using bottomhole pressure.

Some other items of interest are illustrated in Fig. 5.3. Note that almost two hours are devoted to the first bleed stage. The relatively minor gas expansion rate when the influx top moves from 9,289 to 3,560 ft leads to an average flowback rate of 0.15 bbl/min whereas the calculated release rate where most of the expansion occurs near surface is 41 bbl/min. Obviously, the latter result is unreasonable and would not be observed, much less achieved, in an actual operation. The outcome is a direct consequence of an unrealistic assumption that gas migrates up the hole as a slug.

Fig. 5.4 represents how a casing-pressure schedule would appear in a typical bleed stage. The choke operator does not attempt to find a choke setting that yields a flat surface pressure, but rather opens and closes the choke periodically to release a small quantity of mud. The BHP follows a similar pattern, thus one reason for the safety margin is to avoid an underbalance should casing pressure dip too far below the target value.

Another reason for carrying a substantial safety margin lies in the uncertainties as to where the bulk of the influx is at any time. In Example 5.1, the operator somehow knew when the gas entered the casing and automatically made the volume adjustment to keep incremental bubble lengthening the same. An operator probably does not know with any confidence the location of most of the gas at any point in the procedure. The consequences of not knowing may be serious or negligible, depending primarily on how much the annulus or hole dimensions change when gas passes through a discontinuity and how close the casing-seat pressure is to the fracture pressure.

Not accounting for a reduction in gas-column height will increase the pressure below the influx whereas a lengthening

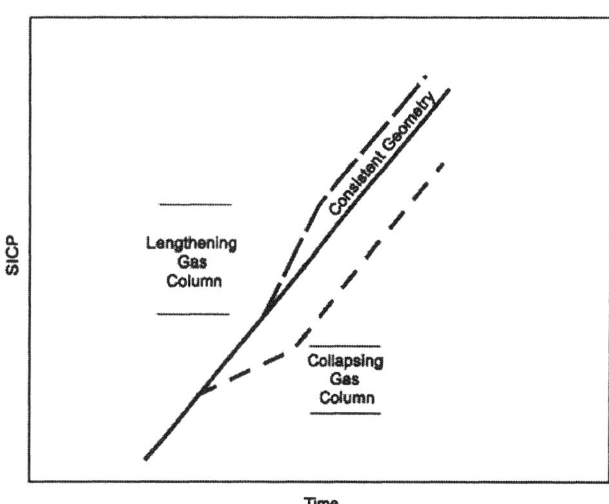

Fig. 5.5—Change in SICP build rate resulting from gas migration from one hole or annulus size into another.

will decrease the pressure. Even so, some errors can be tolerated if the selected pressure margins do not encroach on the fracture gradient and the safety margin is high enough to avoid an underbalance. In Example 5.1, keeping with the 16.46-bbl release volume after the influx cleared the casing increases BHP by only 4 psi at the start of each bleed stage and the cumulative effect is insignificant. But major differences in cross-sectional areas present a problem if the change is not anticipated and detected.

An option is to guess where the gas is at based on a calculated migration rate. Chap. 1 discussed how these estimates can be greatly in error and this exercise is not recommended for tracking influx position. A more reliable method is to monitor how fast the SICP increases and to look for a change. **Fig. 5.5** shows an increase in the build rate usually signifies that a rising gas kick is undergoing elongation while a decrease in the build rate represents a shortening of the gas column.

The written procedure is concluded when gas reaches the surface. This would be true if all the gas resides as a pocket beneath the closed preventer. However, this single-bubble model inadequately describes gas migration behavior in many actual events. Gas does not move up the hole as a slug and a void fraction less than 100% will be present because the overlying liquid must fall down the low side of the hole to replace the rising gas. In addition, we can expect gas bubbles to disperse and become distributed along the hole by the time free gas has migrated very far. A small gas kick may fragment into such small bubbles that migration stops and casing pressure stabilizes before any gas reaches the surface.

Dispersed gas can reach the surface while the lagging portion of the influx continues to rise from a significantly deeper portion of the hole. In this situation, casing pressure will not stabilize after gas is bled through the choke, but will be driven upward by continuing migration from below. Matthews and Bourgoyne[1] conducted a volumetric-control procedure in a 6,000-ft test well and demonstrated that commingled mud and gas had to be released from the well for a considerable time to maintain a constant BHP.

Unfortunately, there are no textbook answers. Migration is going on if casing pressure continues to increase after gas is observed at the choke; commingled fluids may need to be bled and routed through the separator if excessive wellbore pressures are of concern. One indication of a significant gas volume still migrating is if the predicted maximum surface pressure is much higher than the actual pressure at the time gas flow is observed. But how much to bleed and when to shut-in for an incremental build, if at all, are problematic. It may require a judgement call based on the current well conditions and whether it is better to risk overpressures or underpressures. Simulating migration characteristics for various kick volumes at the current or programmed hole configuration may help plan for these contingencies.[2]

5.2.2 Lubrication. Now consider the case of a free gas pocket at the surface, say from a large kick or when mud properties discourage dispersion. Lubrication is one way to replace the gas with liquid in a controlled manner. The procedure outlined in **Table 5.3** is accomplished without circulation and involves pumping a calculated mud volume directly into the gas column followed by a vent stage where gas is released to the atmosphere until the casing pressure falls to a calculated level. The pump and vent stages can be repeated until almost all the gas has been removed.

Pumping mud into the gas increases wellbore pressure by two mechanisms. The hydrostatic pressure rises as heavier fluid stacks up in the well and the surface pressure increases as a result of compressing the gas. We can solve for the maximum mud volume to pump in each stage by starting with the precept that the combined effect does not exceed the openhole

TABLE 5.3—PROCEDURE FOR LUBRICATING GAS FROM A WELL

1. Record the initial SICP.
2. Slowly pump into the annulus a mud volume calculated to be safe from the standpoint of fracturing the formation. Monitor the casing pressure buildup.
3. Give sufficient time for the mud to fall through the gas.
4. Vent gas until the casing pressure reduces to the pressure recorded in Step 1 less the hydrostatic addition of the mud. Verify that this surface pressure will not underbalance the hole.
5. Do not bleed any mud. Shut-in and allow more time if any mud is released before the pressure falls to the desired level.
6. Repeat Steps 1 through 5 until the surface pressure is reduced to the predicted value based on mud weight and pore pressure. The procedure is also complete if gas dispersion precludes venting any more free gas.

fracture pressure. Before adding to the system volume, the pressure at the shoe is given by

$$p_{sh1} = p_{c1} + g_{g1}h_{g1} + g_m(D_{sh} - h_{g1}), \quad \ldots\ldots\ldots (5.3)$$

where p_{sh1} = pressure at the casing shoe, p_{c1} = surface casing pressure, g_{g1} = gas hydrostatic gradient, h_{g1} = gas column height, and D_{sh} = casing shoe vertical depth.

The "1" subscripts in Eq. 5.3 denote the condition immediately before mud is pumped at any stage of the procedure. The maximum allowable pressure is defined by the fracture pressure,

$$p_{fi} = p_{c2} + g_{g2}h_{g2} + g_m(D_{sh} - h_{g2}), \quad \ldots\ldots\ldots (5.4)$$

where the "2" subscripts denote the condition after pumping the mud and before bleeding any gas. The gas-column heights before and after compression are

$$h_{g1} = V_{g1}/C, \quad \ldots\ldots\ldots (5.5)$$

and $h_{g2} = (V_{g1} - \Delta V_m)/C, \quad \ldots\ldots\ldots (5.6)$

where ΔV_m = the mud volume addition and g_g = a gas-hydrostatic-gradient constant $(\gamma_g p_c/Tz)$.

Substituting terms, subtracting Eq. 5.3 from Eq. 5.4, and rearranging yields a quadratic equation with the pumped mud volume as the desired root. The solution is expressed as

$$\Delta V_m = Z - \sqrt{Z^2 - \frac{(p_{fi} - p_{sh1})CV_{g1}}{g_m}}, \quad \ldots\ldots\ldots (5.7)$$

where the intermediate parameter Z is given by

$$Z = \frac{g_m V_{g1} + C(p_{c1} + p_{fi} - p_{sh1})}{2g_m}. \quad \ldots\ldots\ldots (5.8)$$

These relations assume the compressibility factors before and after compressing the gas are equal, but the accuracy is sufficient for our purpose

The mud volume to pump is arbitrary if wellbore-pressure limitations are observed. Fewer steps will be required and time will be saved if the selected mud volume is close to the calculated result. The compression component to the downhole-pressure increase escalates with reducing gas volume and so the allowable mud volume decreases in each subsequent stage. As in a volumetric control, accurate mud volume meaurements are essential.

Low viscosities are recommended and the mud should be thinned to the minimum properties needed to suspend any weighting material. Ideally, the mud should fall through the gas rather than compress the gas in a piston-like displacement. Slow pump rates on the order of ¼ bbl/min promote fluid separation in the well and also allow the operator to better monitor the surface pressures during the pumping cycle. Hence the rig pump probably is not suitable for lubrication purposes.

After mud placement, the compressed gas is given time to accumulate under the preventer and then is released through the choke until the surface pressure falls to its pre-pumping value less the hydrostatic pressure addition. The desired casing pressure after a vent cycle is given by

$$p_{c1}(i + 1) = p_{c1}(i) - g_m\Delta V_m/C. \quad \ldots\ldots\ldots (5.9)$$

It is important to halt the blowdown if mud returns are noted before the target pressure is achieved. Simply close the choke and give more time for the gas to replace the mud.

Voiding a portion of the gas bubble's mass lowers the column hydrostatic pressure and, accordingly, the BHP. The effect is not a factor if the relative amount of gas is low and an overbalance to the formation was present originally. However, some lubrication procedures start off in a near-balanced condition and the surface pressure obtained by Eq. 5.9 may need to be adjusted upward to account for the minor hydrostatic loss.

Example 5.2. Consider the final condition in Example 5.1 where all the gas has migrated to surface. Write a lubrication procedure for replacing the gas with mud.

Solution. The final pressure and gas-kick parameters shown in Table 5.2 define the starting point for the procedure. First, validate the computed values shown in Table 5.2. The pressure and temperature at the top of the gas influx when the well was first shut in, respectively, are

$$4,521 - 23 = 4,498 \text{ psig} = 4,512 \text{ psia, and}$$

$$70 + (0.009)(9,731) = 158°\text{F} = 618°\text{R}.$$

The z factor at these conditions is 0.960. The surface parameters after the volumetric procedure are

$$p_{c1} = 700 \text{ psig} = 714 \text{ psia},$$

$$T_1 = 70°\text{F} = 530°\text{R, and}$$

$$z_1 = 0.918.$$

The gas law yields the influx volume V_{g1}.

$$V_{g1} = \frac{(4,512)(20)(0.918)(530)}{(714)(0.955)(618)} = 104.19 \text{ bbl}.$$

Therefore the gas column height before pumping into the well is

$$h_{g1} = 104.19/0.07731 = 1,347 \text{ ft}.$$

The gas hydrostatic pressure and pressure at the shoe, respectively, are now calculated

$$p_{hg1} = \frac{(0.60)(714)(1,347)}{(53.29)(0.918)(530)} = 22 \text{ psi}$$

$$\text{and } p_{sh1} = 700 + 22 + (0.452)(2,500 - 1,347)$$

$$= 1,243 \text{ psig}.$$

Eq. 5.8 yields the intermediate variable Z.

$$Z = \frac{(0.452)(104.19) + (0.07731)(700 + 1,750 - 1,243)}{(2)(0.452)}$$

$$= 155.32 \text{ bbl}.$$

The largest mud volume that can be pumped in the first stage is given by Eq. 5.7.

$$\Delta(V_m)_{\max} = 155.32 - \sqrt{155.32^2 - \frac{(1,750 - 1,243)(0.07731)(014.19)}{0.452}}$$

$$= 32.48 \text{ bbl}.$$

Select an arbitrary lesser volume, say 31 bbl, and determine the shoe pressure at the end of the pumping stage. The new surface pressure is obtained by assuming a z factor and using the gas law.

$$p_{c2} = \frac{(714)(104.19)(0.918)}{(104.19 - 31.00)(0.918)} = 1,106 \text{ psia}$$

$$= 1,002 \text{ psig}.$$

The z factor is 0.852 at the higher pressure. We substitute and after two more iterations obtain $z = 0.865$ and $p_{c2} = 944$ psig. After pumping 31 bbl mud, the gas height reduction is calculated.

$$h_{g2} = (104.19 - 31.00)/0.07731 = 947 \text{ ft}.$$

The hydrostatic pressure of the gas has not changed and so the final shoe pressure is

$$p_{sh2} = 929 + 22 + (0.452)(2,500 - 947)$$

$$= 1,653 \text{ psig}.$$

Pumping the 31-bbl mud volume results in a shoe pressure almost 100 psi below the fracture pressure.

The compressed gas is allowed to work up through the pumped mud and gas is released from the well until the surface pressure falls to the value given by Eq. 5.9.

$$p_{c1}(\text{stage two}) = 700 - (0.452)(31)/0.07731 = 519 \text{ psig}.$$

We need to determine the BHP after closing the choke. At this point, the wellbore gas volume and height, respectively, are

$$V_{g1} = (104.19 - 31.00) = 73.19 \text{ bbl and}$$

$$h_{g1} = 73.19/0.07731 = 947 \text{ ft}.$$

The compressibility factor at the 533-psia shut-in pressure is found to be 0.830 and the hydrostatic pressure of the gas is determined as

$$p_{hg1} = \frac{(0.60)(533)(947)}{(53.29)(0.830)(530)} = 13 \text{ psi}.$$

Hence the pressure at 10,000 ft is

$$p_{bh1} = 519 + 13 + (0.452)(10,000 - 947)$$

$$= 4,624 \text{ psig}.$$

The reduced number of gas moles in the well leads to a BHP loss of 9 psi. Sufficient overbalance is still present and no adjustments to the casing pressure are necessary. The pressure at the casing seat is

$$p_{sh1} = 519 + 13 + (0.452)(2,500 - 947)$$

$$= 1,234 \text{ psig}.$$

The maximum mud volume for the second pumping cycle is determined now.

$$Z = \frac{(0.452)(73.19) + (0.07731)(519 + 1,750 - 1,234)}{(2)(0.452)}$$

$$= 125.11 \text{ bbl}.$$

$$\Delta V_m = 125.11 - \sqrt{125.11^2 - \frac{(1,750 - 1,234)(0.07731)(73.19)}{0.452}}$$

$$= 29.23 \text{ bbl}.$$

The calculations are repeated for this and subsequent stages with computed results shown in **Table 5.4.** Only 15 ft of gas remains in the well after the sixth vent cycle and, based on the conservative 4,520-psig pore-pressure estimate, bleeding the surface pressure to zero probably would not underbalance the

TABLE 5.4—MUD VOLUMES AND PRESSURES ASSOCIATED WITH THE LUBRICATION PROCEDURE DESCRIBED IN EXAMPLE 5.2

Calculated Mud Volume (bbl)	Pumped Mud Volume (bbl)	Casing Pressure (psig)	Gas Volume (bbl)	Gas Length (ft)	Gas Hydrostatic (psi)	Mud Hydrostatic (psi)	BHP (psig)	Shoe Pressure (psig)
			Initial Condition					
0.00	0.00	700	104.19	1,347	22	3,911	4,633	1,243
			First Pump and Vent Stage					
32.48	31.00	944	73.19	947	22	4,092	5,058	1,653
0.00	0.00	519	73.19	947	13	4,092	4,624	1,233
			Second Pump and Vent Stage					
29.23	26.00	854	47.19	610	13	4,244	5,111	1,721
0.00	0.00	367	47.19	610	5	4,244	4,616	1,226
			Third Pump and Vent Stage					
24.10	22.00	667	25.19	326	5	4,373	5,045	1,655
0.00	0.00	238	25.19	326	2	4,373	4,613	1,223
			Fourth Pump and Vent Stage					
16.24	14.00	531	11.19	145	2	4,454	4,987	1,597
0.00	0.00	156	11.19	145	1	4,454	4,611	1,221
			Fifth Pump and Vent Stage					
8.44	7.00	422	4.19	54	1	4,496	4,919	1,529
0.00	0.00	115	4.19	54	0	4,496	4,611	1,221
			Sixth Pump and Vent Stage					
3.42	3.00	419	1.19	15	0	4,513	4,932	1,542
0.00	0.00	97	1.19	15	0	4,513	4,610	1,220
			Seventh Pump and Vent Stage					
1.00	0.75	278	0.44	6	0	4,517	4,795	1,405
0.00	0.00	93	0.44	6	0	4,517	4,610	1,220

hole. From a practical standpoint, the lubrication job can be considered as complete at the conclusion of the sixth stage. A procedure corresponding to the schedule can be written.

1. Pump 31 bbl of 8.7-lbm/gal brine into the well while monitoring the casing pressure. Pump no faster than ¼ bbl/min and do not allow the casing pressure to exceed 950 psig.
2. Allow sufficient time for gas to accumulate at surface.
3. Vent gas until the casing pressure falls to 519 psig. Do not bleed any mud from the well.
4. Pump 26 bbl into the well. Do not allow the casing pressure to exceed 860 psig.
5. Allow sufficient time for gas to accumulate at surface.
6. Vent gas until the casing pressure falls to 367 psig. Do not bleed any mud.
7. Pump 22 bbl into the well. Do not allow the casing pressure to exceed 700 psig.
8. Allow sufficient time for gas to accumulate at surface.
9. Vent gas until the casing pressure falls to 238 psig. Do not bleed any mud.
10. Pump 14 bbl into the well. Do not allow the casing pressure to exceed 600 psig.
11. Allow sufficient time for gas to accumulate at surface.
12. Vent gas until the casing pressure falls to 156 psig. Do not bleed any mud.
13. Pump 7 bbl into the well. Do not allow the casing pressure to exceed 500 psig.
14. Allow sufficient time for gas to accumulate at surface.
15. Vent gas until the casing pressure falls to 115 psig. Do not bleed any mud.
16. Pump 3 bbl into the well. Do not allow the casing pressure to exceed 500 psig.
17. Allow sufficient time for gas to accumulate at surface.
18. Vent the remaining gas. Do not bleed any mud.
19. Observe for indications of flow.

5.3 Off-Bottom Well Control

Conventional control procedures do not apply if an influx is beneath the bottom of the drillstring. Adequate surveillance and no-flow indications up until the time of a trip suggest that the mud hydrostatic provided its primary control function. Thus we can often (but not always) assume that an off-bottom kick was induced by the trip and that restoring a balanced condition can be accomplished by replacing the kick fluids with original mud.

Killing a well in these situations may require that the string be run back to bottom before circulating. Moving a string in or out of a closed wellbore involves snubbing or stripping the pipe. Snubbing is the procedure wherein shut-in pressures acting on the pipe area are sufficient to eject the string from the well if surface restraints are not provided. Stripping can be performed when the string weight is high enough that gravity takes over and the pipe falls. Snubbing and stripping mechanics, equipment, and procedures are discussed in Chap. 9

Off-bottom volumetric control when followed by kick displacement or lubrication should kill a well if the original mud weight overbalances the formation. Volumetric-control techniques can be conducted with the string at rest or in conjunction with a snubbing/stripping operation. However, the combined procedure is a complex task and we delay this topic until snubbing/stripping groundwork is discussed in Chap. 9. As an alternative to stripping, a well with a nonmigrating influx or insufficient mud weight sometimes can be killed by circulating a mud weight high enough to overbalance the kick formation from the shallower depth.

5.3.1 Off-Bottom Volumetric Control. Example 5.1 demonstrates the effect of failing to make a volume adjustment when gas rises across a minor dimension change is probably insignificant to the overall control effort. However, the cross-sectional area difference between cased or open hole and a

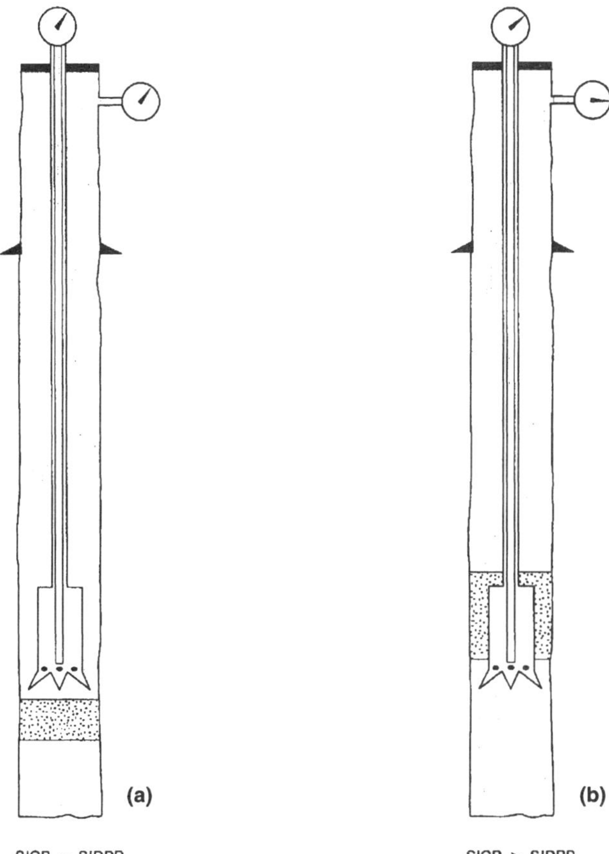

Fig. 5.6—Shut-in pressure comparisons when gas is below the bit and after gas has migrated into the drillstring annulus.

bottomhole assembly (BHA) annulus is important enough to warrant recognizing when gas has reached string depth.

Fig. 5.6a depicts a shut-in well with a gas kick beneath the bit. This particular string does not have a backpressure valve and pressure is transmitted from below the bit to the drillpipe gauge. The effective base of the U-tube is at the bit and the shut-in drillpipe pressure (SIDPP) and SICP readings will be equivalent so long as the fluid densities in the drillpipe and annulus are the same.

Fig. 5.6b illustrates the relative gauge response after a constant-volume gas bubble has migrated up alongside the drillstring. Pressure is retained in the bubble and the drillpipe pressure reading has increased by a small amount. In contrast the SICP has increased substantially because of the lengthened gas column. Comparing the two pressure gauges in an off-bottom kick situation may give an operator direct evidence of a migrating gas bubble's position if drillpipe pressures are available.

Other clues are provided at the casing gauge if the drillstring is equipped with a float, namely in the rate at which migration increases the SICP. The rate of pressure increase is expected to go up when a constant-velocity gas slug traverses into a smaller area. Also, reducing the annulus cross-sectional area in Rader *et al.*[3] test apparatus resulted in faster gas migration velocities. We can anticipate similar behavior in a wellbore when gas goes through a geometry transition at the bit, thus giving a further increase in the SICP build-rate.

Figs. 5.6a and 5.6b imply that gas will stay preferentially in the annulus rather than enter the bit, but this may or may not be the case. Absent a check valve on bottom, there is nothing to prevent drillstring entry other than the restriction offerred by the bit nozzles and BHA bore, and an operator may very well have to contend with gas migration up the drillstring. The occurrence should be easy to recognize by the drillpipe gauge response, but nothing fundamentally changes with respect to the procedure. Control is based on the casing pressure and the released mud volumes are based on the annular dimensions.

TABLE 5.5—OFF-BOTTOM KICK DATA FOR EXAMPLES 5.3 AND 5.4

Vertical Well Depth	9,600 ft
Estimated Kick Zone Depth	9,100 ft
Bit Depth	2,700 ft
Surface Casing Information:	
Description	13³/₈-in., 68.0-lbm/ft
Inner Diameter	12.415 in.
Setting Depth	3,900 ft
Fracture Gradient at Casing Seat	0.76 psi/ft
Assumed Openhole Diameter (Same as Casing)	12.415 in.
Drillstring Information:	
Bottom Section Collar Size	9×3 in.
Bottom Section Collar Length	270 ft
Top Section Collar Size	7×2½ in.
Top Section Collar Length	90 ft
Drillpipe Description	5-in., 19.50-lbm/ft Grade G-105 NC50
Mud Type	Water Base
Mud Density	10.2 lbm/gal
Assumed Static Wellbore Temperature	80°F+1.2°F/100 ft

Example 5.3. A flow is noted while tripping from the well described in **Table 5.5** and the shut-in procedure discussed in Table 4.8 is employed. A 45-bbl pit gain is estimated and both pressure gauges initially read 80 psig. Gas migration is evidenced and the operator decides to control the well using the volumetric-control method until gas surfaces. Write a control procedure.

Solution. The kick zone depth is based on the mud log or comparable data, but we cannot quantify the zone's pore pressure and we know little about the influx characteristics other than that at least some gas is present. The influx height is approximately

$$h_k = 45/0.14973 = 301 \text{ ft}.$$

Eq. 4.16 is used to estimate the hydrostatic gradient of the kick fluids. Assume the hole was balanced exactly by the 10.2-lbm/gal mud and obtain

$$p_p = (0.530)(9,100) = 4,823 \text{ psig and}$$

$$g_k = \frac{4,823 - (0.530)(9,100 - 301) - 80}{301}$$

$$= 0.264 \text{ psi/ft}.$$

We expect the influx to be primarily gas because smaller pore pressure estimates only serve to reduce the calculated kick gradient. Taking a conservative approach, we use the Driller's Method annular prediction equations and estimate the maximum expected shoe pressure by assuming this formation pressure and gas migration as a bubble. Arbitrary safety and working margins of 100 and 150 psi are deemed appropriate.

The incremental mud volume to bleed when gas is below the bit is determined.

$$\Delta V_m = (150)(0.14973)/0.530 = 42.38 \text{ bbl}.$$

TABLE 5.6—WELLBORE PRESSURES FOR THE VOLUMETRIC CONTROL PROCEDURE IN EXAMPLE 5.3

Time (min)	Casing Pressure (psig)	Gain Increment (bbl)	Influx Length (ft)	Influx Top (ft)	Mud Hydrostatic (psi)	Gas Hydrostatic (psi)	Pressure at 9,100 ft (psig)	Shoe Pressure (psig)
Initial Shut-in Condition								
0.0	80	45.0	301	8,799	4,663	25	4,768	2,147
Safety and Working Margin Buildup								
7.9	330	0.0	301	8,327	4,663	25	5,018	2,397
First Bleed Stage								
95.3	330	42.38	584	2,800	4,513	25	4,868	2,112
Second Build Stage with Gas at the Bit								
97.0	384	0.0	584	2,700	4,513	25	4,922	2,166
Second Build and Bleed Stage with Gas above the Bit								
98.3	480	0.0	679	2,519	4,463	29	4,972	2,216
104.8	480	20.10	970	1,836	4,309	34	4,823	2,067
Third Build and Bleed Stage								
110.4	630	0.0	891	1,579	4,351	31	5,012	2,256
118.1	630	35.50	1,140	868	4,219	30	4,879	2,123
Fourth Build and Bleed Stage								
122.8	780	0.0	1,140	585	4,219	30	5,029	2,273
124.8	780	35.50	1,423	180	4,069	30	4,879	2,123
Fifth Build Stage. Gas Surfaces								
127.8	876	0.0	1,423	0	4,069	30	4,975	2,219
	Total Gain	178.48						

We choose to make one bleed stage correspond to the largest drill collar. Thus

$$\Delta V_m = (150)(0.07104)/0.530 = 20.10 \text{ bbl}$$

with gas adjacent to the 9-in. collars and

$$\Delta V_m = (150)(0.12544)/0.530 = 35.50 \text{ bbl}$$

when the influx is in the drillpipe annulus. The volumetric-control procedure follows.

1. Allow the casing pressure to increase from 80 to 330 psig.
2. Bleed 42.38 bbl of mud from the choke while holding constant casing pressure. Monitor the casing pressure build rate and compare the drillpipe and casing pressures for indication of gas migration past the bit.
3. Close the choke and allow the casing pressure to increase by 150 psi. Monitor the casing pressure build rate and compare the drillpipe and casing pressures for indication of gas migration past the bit.
4. Repeat Steps 2 and 3 until gas enters the drillstring annulus. Adjust the incremental bleed volume to 20.10 bbl for one stage and 35.50 bbl for the succeeding stages. Continue the control until pressures stabilize or gas is bled from the choke.
5. Monitor well pressure.

Table 5.6 represents the calculated pressures for the procedure assuming the formation pressure equivalent at 9,100 ft is 10.1 lbm/gal, the influx is a 0.60-SG gas, and the gas slip velocity is a constant 3,600 ft/hr.

For illustration purposes, the problem was set up so that gas reached bit depth while the well was shut-in. We predicted both drillpipe and casing pressures as gas migrated into the annulus during the second buildup and **Fig. 5.7** presents a comparison between the two gauge readings. As discussed, the casing pressure build rate increases substantially when gas moves into the smaller space opposite the drill collars. The drillpipe pressure continues to increase as well, but at a much slower rate. Expanding the gas starting at 98.3 minutes leads to a reduction in drillpipe pressure because the bulk of the gas is above the bit.

Geometry transitions lead to a situation where predicted pressures below the gas do not fluctuate within the ideal limits. Indeed, the calculated overbalance to the kick zone dips below the safety margin to 55 psi at the end of the second bleed stage. But it is impractical (and unnecessary) to devise a schedule that maintains the precise working margin tolerance. The safety margin usually will handle the temporal condition. Also, it is important to remember that our constant-volume, single-bubble migration model does not match the true conditions in a well realistically.

Rather than allow gas to rise all the way to surface, an alternative recommended approach is to wait until the kick clears the bit and to displace the influx using the first circulation of the Driller's Method. The job is simplified, time is saved, and we can avoid the uncertainties associated with having gas at the choke when lagging gas continues to migrate up the hole.

5.3.2 Staging in the Hole. In most cases, off-bottom kicks are induced by the trip and restoring the pretrip hydrostatic pressure will kill the well. But there are situations where, for some reason (inadequate surveillance, tight reservoir), the original

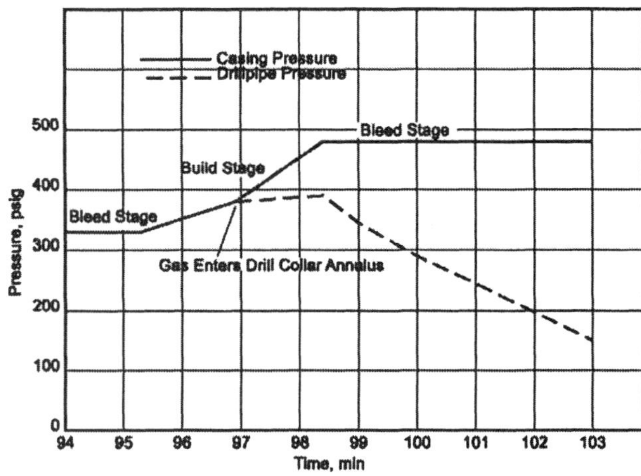

Fig. 5.7—Drillpipe and casing pressures for a portion of the volumetric-control procedure used in Example 5.3.

mud weight is too low for the pore pressure. Stripping back to bottom is always an option, but another solution to the problem may be to circulate with a mud weight heavy enough to balance pore pressure from the bit depth. The technique may work with a kick below the string, but should not be attempted if migration is evidenced.

Staging a kill entails placing a heavy mud on top of the original, lighter mud and the procedure requires that the dissimilar wellbore fluids retain their respective position in the well. One condition that must be satisfied is that the original mud have a rheology sufficient to prevent fluid swapping (heavy mud falling through lighter fluid).

The fluid density needed for an off-bottom kill is heavier than what would be needed if the string were below the kick formation. The mud weight requirement, however, involves more than the straightforward hydrostatics of the combined well fluids. When pipe is run into the hole, heavy mud in the annulus is displaced from the well and simultaneously replaced by lighter mud from below. To account for the hydrostatic-pressure reduction caused by string displacement, a stage margin pressure is incorporated in the density calculation.

The hydrostatic pressure required of the kill-weight mud (KWM) is the sum of the pore pressure and stage margin less the hydrostatic pressure of the mud below the string. The kill-weight hydrostatic gradient at a given string depth is given by

$$g_{km} = \left[p_p + p_{sm} - g_{om}(D - D_b)\right]/D_b. \quad \text{........ (5.10)}$$

D_b and D are vertical depths of the bit and kick formation, respectively, while p_{sm} is the stage margin pressure.

If the wellbore pressure imposed by the heavier mud exceeds the fracture gradient, the only option is to strip to bottom or to a depth where a staged kill can be executed safely. Large stage margins are desirable because the excess pressure controls how much pipe can be run before stopping to do another circulation. The choice, however, is not completely arbitrary but is governed by the fracture integrity. The maximum allowable stage margin is given by

$$(p_{sm})_{max} = p_{fi} - p_p + g_{om}(D - D_{sh}), \quad \text{......... (5.11)}$$

where D_{sh} = the casing seat depth or any other depth where the controlling fracture pressure has been determined. The relation does not consider the added stress imposed by circulation or surge effects and so the actual margin must be lower than the calculated value. Stage margins can be increased after the bit is below the last casing seat or other weak hole section.

The hydrostatic pressure reduction caused by string displacement may dictate more than one circulation before the string can be placed on bottom. The allowable height reduction of the KWM is given by

$$\Delta h_{km} = p_{sm}/(g_{km} - g_{om}). \quad \text{.................. (5.12)}$$

The equivalent volume for the case where the height change results solely from displacement is

$$\Delta V_{km} = \Delta h_{km} C_a, \quad \text{........................ (5.13)}$$

where C_a = the annulus capacity factor. Drillstring and hole discontinuities can result in a height change that must be considered in the volume calculation. The maximum length of pipe that can be run without introducing a kick is then calculated as

$$\left(L_p\right)_{max} = \Delta V_{km}/C_d, \quad \text{...................... (5.14)}$$

where C_d = the string displacement factor. For consistent dimensions, substitution yields the direct relation,

$$\left(L_p\right)_{max} = \frac{p_{sm}C_a}{C_d(g_{km} - g_{om})}. \quad \text{.................. (5.15)}$$

Example 5.4 demonstrates the fundamentals of a staged kill.

Example 5.4. Refer again to the data shown in Table 5.5. Volumetric procedures successfully removed the gas from this well, but the kick formation remains underbalanced by 380 psi (as reflected by the casing gauge). Determine if a staged kill can be conducted and, if so, the procedure for getting pipe on bottom.

Solution. Calculate the pore pressure at 9,100 ft.

$$p_p = 380 + (0.53)(9,100) = 5,203 \text{ psig}.$$

The kill-weight mud if the string was on bottom is therefore

$$\rho_{km} = (19.25)(5,203)/9,100 = 11.0 \text{ lbm/gal}.$$

Use Eq. 5.11 to determine if the well can be killed from the present string depth.

$$(p_{sm})_{max} = (0.76)(3,900) - 5,203 + (0.53)$$
$$(9,100 - 3,900) = 517 \text{ psi}.$$

A negative result indicates that fracture pressure would be exceeded regardless of the stage margin and, consequently, the need for a stripping job. We arbitrarily select a lower stage margin of 250 psi to provide a buffer for the friction and surge pressures and use Eq. 5.10 to determine the kill-mud density.

$$g_{km} = [5,203 + 250 - (0.53)(9,100 - 2,700)]/2,700$$
$$= 0.763 \text{ psi/ft}.$$

$$\rho_{km} = (19.25)(0.763) = 14.69 \text{ lbm/gal}.$$

A 14.7-lbm/gal (0.764-psi/ft) mud is selected for the first circulation. After the kill, the well is confirmed dead and a check valve is installed in the drillstring (if not already in place). We are now ready to determine the depth for the second kill. Eq. 5.12 gives the allowable height reduction of the 14.7-lbm/gal mud.

$$\Delta h_{km} = 250/(0.764 - 0.530) = 1,068 \text{ ft}.$$

The heavy mud volume originally opposite the drill collars will occupy less height when the string is run any deeper. The volume adjacent to the BHA annulus is computed as

$$\left[(90)(12.415^2 - 7^2) + (270)(12.415^2 - 9^2)\right]/1,029.4$$
$$= 28.4 \text{ bbl}.$$

The capacity factor in the drillpipe/openhole annulus is

$$C_a = (12.415^2 - 5^2)/1,029.4 = 0.12544 \text{ bbl/ft}.$$

The 14.7-lbm/gal height reduction after moving 28.4 bbl into the drillpipe annulus is determined as

$$360 - 28.4/0.12544 = 134 \text{ ft}.$$

The capacity factor between the drillpipe and casing is considered the same as the drillpipe/openhole annulus factor. Hence the allowable displacement volume is

$$\Delta V_{km} = (1,068 - 134)(0.12544) = 117.2 \text{ bbl}.$$

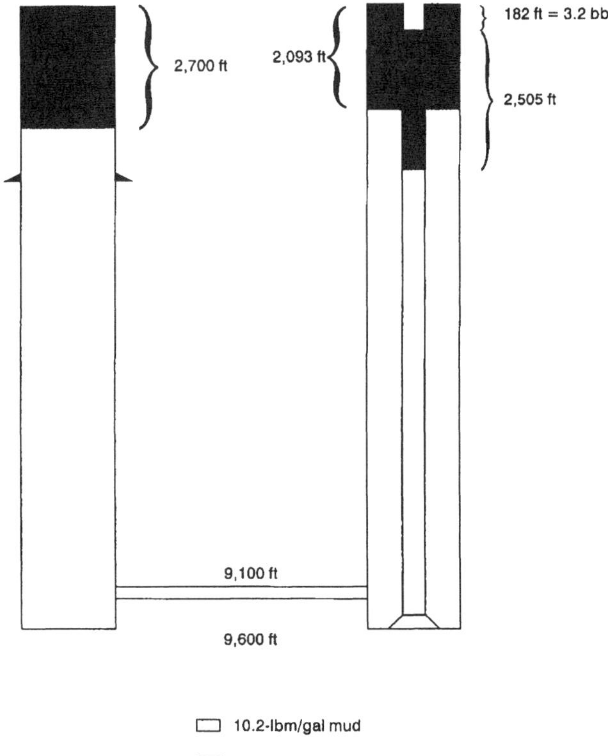

Fig. 5.8—Relative fluid positions after staging pipe to bottom without a check valve.

The displacement factor applies to plugged drillpipe because the string has a check valve which prohibits entry into the string. The factor for open drillpipe is obtained from Table 4.6. The plugged factor is determined by adding the contribution from the pipe bore:

$$C_d = 0.00813 + 4.276^2/1,029.4 = 0.02589 \text{ bbl/ft}.$$

Eq. 5.14 is used to calculate how much pipe can be run safely.

$$\left(L_p\right)_{\max} = 117.2/0.02589 = 4,526 \text{ ft}.$$

The drillpipe will be run in 90-ft stands so plan on running 50 stands or 4,500 ft. Repeat the mud-weight calculation for killing the well from the 7,200-ft bit depth.

$$g_{km} = [5,203 + 250 - (0.53)(9,100 - 7,200)]/7,200$$
$$= 0.618 \text{ psi/ft}.$$

$$\rho_{km} = (19.25)(0.618) = 11.90 \text{ lbm/gal}.$$

The allowable height reduction of the 11.9-lbm/gal mud is

$$\Delta h_{km} = 250/(0.618 - 0.530) = 2,841 \text{ ft}.$$

Repeat the remaining calculations.

$$\Delta V_{km} = (2,841 - 134)(0.12544) = 339.6 \text{ bbl}.$$

and $\left(L_p\right)_{\max} = 339.6/0.02589 = 13,117$ ft.

Pipe can be run to total depth where the final 11.0-lbm/gal stage is circulated.

Some engineers have suggested removing or not installing the check valve before running pipe any deeper. The idea has appeal because the string displacement factor is decreased, reducing the remaining number of stages. However, other factors must be considered before adopting this approach. The volume of heavy mud left in the string will remain constant as pipe is run deeper and its height may change as a result of moving into a larger or smaller diameter. In any event, string displacement causes a significant height reduction of the heavy-mud on the backside and the kill-weight hydrostatic pressure within the string probably will be much higher than in the annulus. A basic U-tube premise is for the hydrostatic pressures in the drillstring and annulus to be the same at bit depth. It follows that the drillstring will have some air space above the heavy mud and this voidage must be considered as part of the displacement.

Fig. 5.8 explains the concept as applied to the example well in Table 5.5 (the drill collars have been removed for illustration purposes only). Using an open-pipe displacement factor in Eq. 5.14 indicates that the string can be run to total depth after pumping the initial 14.7-lbm/gal kill. The 14.7-lbm/gal mud height within the drillstring will collapse to 2,505 ft after a portion moves from the drill collars into the drillpipe, but the same mud loses more height in the annulus because of string displacement. An iterative calculation gives 182 ft of air in the top of the drillstring and, effectively, a 3.2-bbl increase in displacement volume. The formation at 9,100 ft was not underbalanced as a result, but the effect would be a consideration in a well where conditions dictated a second off-bottom kill.

Another factor is the relative flow into the drillstring and annulus as the pipe is lowered. An effort should be made to remove all flow restrictions from the string bore and to lower the pipe at a slow velocity. Otherwise, more of the heavier mud will displace preferentially from the annulus and thereby limit the advantages to removing the valve. The trip tank should be used to determine if this is a problem and adjustments to the allowable run length made accordingly. After circulating heavy mud, the pipe may need to be tripped to remove a valve or flow restriction. If so, use the trip tank to ensure the hole is kept full of the heavy (not original) mud.

Maintaining wellbore pressures within tolerable limits while pumping a staged kill requires thought and planning. This is not a problem for the first circulation—we simply follow the procedure given for the second circulation of the Driller's Method. That is, control from the casing gauge until the string is filled with heavy mud and from the drillpipe gauge thereafter. Difficulties arise in the second and subsequent stages because the U-tube complexity has increased greatly.

The example well in Table 5.5 is depicted in **Fig. 5.9** (drill collars not shown) immediately after running the bit to 7,200 ft. We have 2,700 ft of 14.7-bm/gal (heavy) mud below the backpressure valve and 1,637 ft is left on top in the annulus. The desired outcome is 11.9-lbm/gal (intermediate) mud in and out of the string, but achieving this involves placing two different fluids in the drillstring and at least three in the annulus. The transition from the initial to final condition must be accomplished without inducing another kick or breaking down the shoe.

Every well is unique and no generalized procedure will apply to every situation. But we can analyze this particular case and offer some suggestions. **Fig. 5.10** presents the hydrostatic pressures at 9,100 ft for a complete circulation. Starting off, the drillpipe is dry down to 4,500 ft. While pumping intermediate mud down the drillpipe, the drillstring hydrostatic pressure exceeds that in the annulus before the pipe is filled, thus creating a situation where heavy mud starts moving out the annulus before any pressure is seen on the drillpipe gauge.

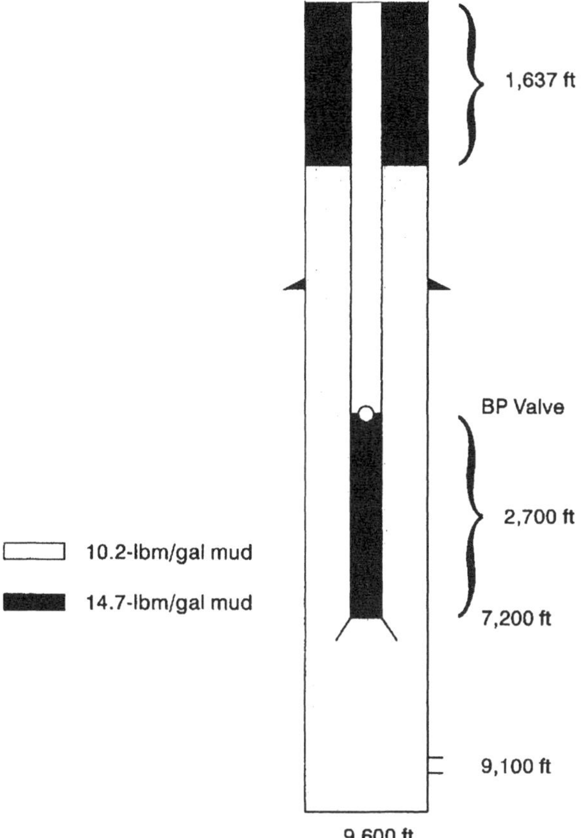

Fig. 5.9—Mud densities in the drillstring and annulus before circulating the second kill in Example 5.4.

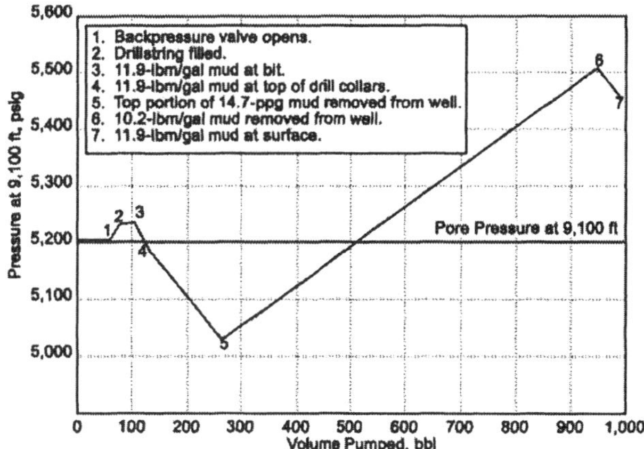

Fig. 5.10—Hydrostatic pressures at 9,100 ft during the second circulation of the staged kill in Example 5.4.

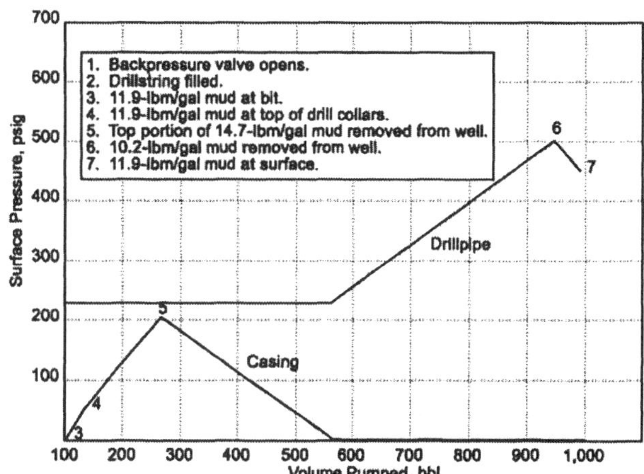

Fig. 5.11—A suggested method for controlling wellbore pressures while pumping the second kill in Example 5.4.

The BHP starts to climb when the first increment of heavy mud is displaced into the drill collar annulus and continues to increase until the intermediate mud exits the bit. Reducing the heavy-mud height by surface removal and (less importantly) changes in dimension results in a declining BHP until the top heavy-mud segment is voided. Thereafter, removing the 10.2-lbm/gal (light) mud leads to an increasing BHP. The final decline coincides with removing the heavy mud that was originally in the drillstring.

Annulus friction losses were not considered, but it is apparent we risk an underbalance condition for the time period when light-mud effects predominate unless some backpressure is provided by the choke. A casing-pressure schedule based on these predictions could be developed, but it is preferable to control the well from the drillpipe gauge because some gas may yet be in the hole and mud swapping may have occurred to some extent.

One way to pump this particular job safely would be to fill the string with intermediate mud, immediately observe the circulating drillpipe pressure (CDPP), and maintain a constant CDPP until the casing pressure declines to zero. Note that the U-tube fluids initially favor the drillstring (annulus hydrostatic < drillstring hydrostatic) and so the observed CDPP initially will be less than the combined pressure losses in the pipe and bit. The casing pressure will climb naturally until the top segment of heavy mud is voided from the well and thereafter decline as intermediate mud replaces the light mud in the annulus. The hydrostatic pressure is sufficient to prevent kick entry when the casing pressure falls to zero and the hydrostatic will increase until the heavy mud that was originally in the pipe reaches surface. **Fig. 5.11** show the schedule drillpipe and casing pressures assuming 450-psi circulating losses with intermediate mud.

The procedure would be facilitated by incorporating another step to the prior circulation at 2,700 ft. Say that the drillstring had been displaced with 11.9-lbm/gal mud after the 14.7-lbm/gal mud circulated to surface. The intermediate mud then would be retained during the trip to 7,200 ft and eliminate the procedural difficulties associated with having dissimilar fluids in the drillstring. Practical benefits would be realized by reducing the number of different muds to capture and process at surface.

5.4 Problems During a Conventional Kill

Operational and mechanical complications arising during a conventional kill can undermine well control if left unrecognized and unrectified. It is impossible to list every conceivable situation, but **Table 5.7** gives some likely occurrences and their surface indications. The diagnostics and corrective procedures may differ, but the initial step for almost any predicament is to shut down the kill operation and close the well in. Any consequent formation fluid entry is contained, the static U-tube can be analyzed, and time is gained to think and determine the best course of action. Planning for contingencies is the key for successful problem resolution and a wellsite supervisor should have some idea of what to do in advance.

TABLE 5.7—PROBLEMS THAT MAY BE ENCOUNTERED WHILE KILLING A WELL AND THEIR EFFECT ON SURFACE INDICATORS

Problem	Drillpipe Pressure	Casing Pressure	Pump Rate	Pit Level
Plugged Bit	Increase		Decrease	
Plugged Choke	Increase	Increase	Decrease	
Annulus Packed Off	Increase		Decrease	Decrease
Bit Nozzle Washout	Decrease		Increase	
Leak in Drillstring	Decrease		Increase	
Parted Drillstring	Decrease		Increase	
Loss of Pump	Decrease	Decrease	Decrease	
Cut Out Choke or a Stack/Manifold Leak	Decrease	Decrease	Increase	
Lost Circulation	Decrease	Decrease	Increase	Decrease

5.4.1 Blockage in the Circulating System. Plugging the bit or choke with formation solids or a clump of barite are two of the more common problems associated with pumping a kill. Both are evidenced by a rapid increase in the CDPP, but a plugged bit isolates the pressure increase to the drillstring whereas a plugged choke results in higher casing pressure as well. An unreasonably high initial circulating pressure (ICP) following a shut-in period may indicate that one or more of the bit nozzles became plugged before kick displacement began. Opening the choke and attempting to obey the predetermined drillpipe pressures can only lead to an underbalanced hole. It is imperative that the operator recognize the flow restriction and either shut the well in or, if the pump rating is not exceeded, continue circulation at a modified ICP and pressure reduction schedule.

A partial blockage at the bit may arise later in the displacement, signified by a drillpipe pressure increase with no change in choke pressure or drop in pit level. The next step depends on whether the well is being controlled from the casing or the drillpipe pressure gauge. Circulation may continue in the former case whereas recommended practice in the latter case is to shut the well in and redetermine the desired CDPP schedule based on the latest reading.

A completely plugged drillstring requires a wireline unit to regain communication with the annulus. Detonating a string shot at the bit is often effective in removing the nozzles. Otherwise, the string must be perforated, preferably deeper than the heavy mud if kill-weight fluid has entered the annulus. At minimum, the string should be perforated below the base of any kick fluids. The wireline setup logistics and job execution time dictate that a well be controlled volumetrically if migrating gas is indicated in the interim period.

A plugged choke is more critical because the potential for exceeding the fracture gradient is created by the excess annulus pressure. A rapid response on the part of the operator is essential and the kill procedure must be shut down immediately until the blockage is removed or the manifold valves rearranged.

Drilling fluid contamination by formation fluids can drastically alter rheological properties across the affected region. Muds may thicken with small salt water or free gas additions while adding a large fraction of the continuous phase can destroy a mud's suspension capabilities. Thus two possible side effects from a kick are for drilled and commercial solids to create a bridge or fall out of the mud. Both can lead to a restricted or completely packed-off annulus. The symptoms may mimic bit plugging, but the pressure increase resulting from an annulus restriction is usually more gradual and erratic. The problem can be identified conclusively by an increase in torque and/or drag if the drillstring is being moved during the kill procedure.

The pit level reduction mentioned in Table 5.7 implies that mud is being pumped into the formation below the packoff depth, an obvious outcome if the annulus is sealed completely. The pipe is probably stuck and now two problems must be dealt with rather than one. No generic solutions can be offered as every situation is unique, but normal considerations will involve first volumetrically controlling any gas above the packoff depth followed by regaining the ability to circulate from as deep as possible. The ultimate circulation may remove the packoff and allow pulling free, but releasing or fishing the stuck drillstring can wait until pressure control has been ensured.

5.4.2 Nozzle Washouts and String Leaks. The symptoms of a washed out bit nozzle and a leak in the drillstring are the same. Both are indicated by a drop in the CDPP without a corresponding change in the casing pressure or string weight. In any event, the situation demands shutting in the well and analyzing the U-tube. The shut-in pressures should indicate where the pressure loss occurred with respect to the position of the influx or kill-weight mud in the annulus. The assessment is based on the fact that the drillpipe gauge on a shut-in well reads the annulus pressure at the communication port less the overlying hydrostatic in the drillstring.

Refer to the two situations depicted in **Figs. 5.12a and 5.12b.** Both of these wells were being killed using the Wait and Weight procedure and had kill-weight mud in the annulus when the loss in CDPP was detected. In Fig. 5.12a, the SIDPP is zero whereas the SICP is much higher. The zero reading on top of the drillpipe leg indicates that the external pressure adjacent to the leak depth is the same as the internal pressure, which means that the leak is below the top of the kill mud in the annulus. Note that a washed out bit nozzle would give the same indication.

This situation can be handled without any major difficulties. The drillpipe pressure is adjusted downward based on the last observed reading and the kill is completed at a reduced CDPP. However, a string leak may washout further with continued pumping and lead to more pressure loss. An enlarging washout risks parting the string at the weakened section and it may be advisable to stop string movement for the duration of the kill.

In Fig. 5.12b the leak is between the base of the influx region and above the weighted mud. The hole will not be in equilibrium immediately after closure because the internal and external pressures at the leak depth are different. A flow circuit is created where the original mud enters the drillpipe and displaces KWM out the bit until the respective mud col-

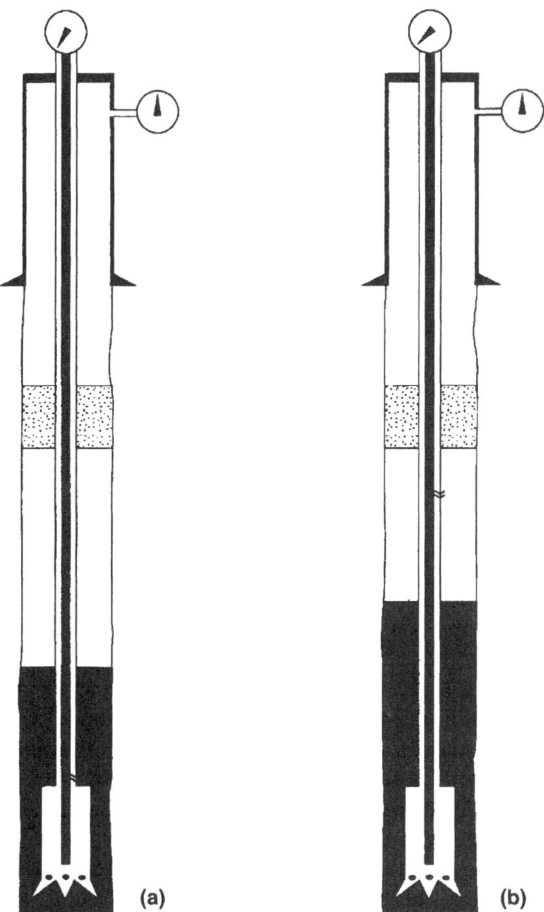

Fig. 5.12—Shut-in pressure relationships for a drillstring leak during a Wait and Weight procedure.

umn heights inside and outside the string are the same. Once this condition is achieved, the SIDPP is not zero because of the hydrostatic reduction in the drillstring. Assuming the gas does not migrate appreciably during this time, the SICP will be somewhat lower than if the well had been shut-in with an intact string integrity because there is more kill-weight mud in the annulus.

Example 5.5. Refer back to the kick event described in Table 4.12 and the Wait and Weight kill procedure kill discussed in Example 4.13. Assume that a drop in CDPP was experienced after pumping a total of 3,500 strokes and the well was closed in. The following fluids were in the annulus when the pump was shut down:

12.4-lbm/gal mud from 12,000 to 8,660 ft,

12.0-lbm/gal mud from 8,660 to 5,364 ft,

gas with a 0.079-psi/ft hydrostatic gradient from 5,364 to 4,677 ft, and

12.0-lbm/gal mud from 4,677 ft to surface.

Determine the SIDPP and SICP when the system reaches equilibrium if the leak is at 7,000 ft. Assume the gas does not migrate and that no kill-weight mud was pumped through the leak in the time it took to bring the pump down.

Solution. First determine the stabilized heights of the original and KWMs in the bottom 5,000 ft of the string. The following relationship must be satisfied.

$$g_{om}h_{om(p)} + g_{km}h_{km(p)} = g_{om}h_{om(a)} = g_{km}h_{km(a)}$$

$$0.623h_{om(p)} + 0.644h_{km(p)} = 0.623h_{om(a)} + 0.644h_{km(a)},$$

where the "(p)" and "(a)" subscript addendums designate the pipe and annulus legs of the U-tube, respectively. Also note that

$$h_{om(p)} = h_{om(a)}$$

and $h_{km(p)} = h_{km(a)}$.

For both sides,

$$h_{om} + h_{km} = 5,000 \text{ ft},$$

$$h_{om} = 5,000 - h_{km},$$

and $h_{km} = 5,000 - h_{om}$.

The stabilized height of the original mud inside the string is a function of the volume entry through the leak according to the relation

$$h_{om} = V_{om(p)}/0.01422 = 70.323V_{om(p)},$$

where 0.01422 is the capacity factor (bbl/ft) in the drillpipe. Initially, there was 3,340 ft of KWM in the annulus and each barrel of original mud that enters the leak will displace a barrel of 12.4-lbm/gal mud through the bit. Hence the final KWM height in the annulus is

$$h_{km} = 3,340 + V_{om(p)}/0.05051 = 3,340 + 19.80V_{om(p)},$$

where 0.05051 is the drillpipe/openhole annulus capacity factor. Substitution in the pressure balance equation gives the solution for the volume entry $V_{om(p)}$.

$$0.623\left[70.32V_{om(p)}\right] + 0.644\left[5,000 - 70.32V_{om(p)}\right]$$

$$= 0.623\left\{5,000 - \left[3,340 + 1980V_{om(p)}\right]\right\}$$

$$+ 0.644\left[3,340 + 19.80V_{om(p)}\right].$$

$$V_{om(p)} = 18.49 \text{ bbl}.$$

Therefore the final original and KWM heights below the leak are

$$h_{om} = 18.49/0.01422 = 1,300 \text{ ft}$$

and $h_{km} = 5,000 - 1,300 = 3,700$ ft.

The SIDPP is the formation pressure minus the hydrostatic pressure of the fluids in the drillstring.

$$p_{ds} = 7,723 - (0.623)(1,300) - (0.644)$$

$$(3,700 + 7,000)$$

$$p_{ds} = 22 \text{ psig}.$$

The SICP is obtained in similar fashion.

$$p_{cs} = 7,723 - (0.644)(3,700) - (0.623)$$

$$(1,300 + 1,636 + 4,677) + (0.079)(687)$$

$$p_{cs} = 543 \text{ psig}.$$

Eventually, circulation through the bit must be restored and several options may be considered for achieving this end. The first step is to allow gas to migrate and expand in a controlled manner using a volumetric-control procedure. Killing the well from the leak depth by pumping extra-heavy mud may be an option if the shoe can withstand the pressure and the leak depth is known. Doing so allows the drillpipe to be tripped

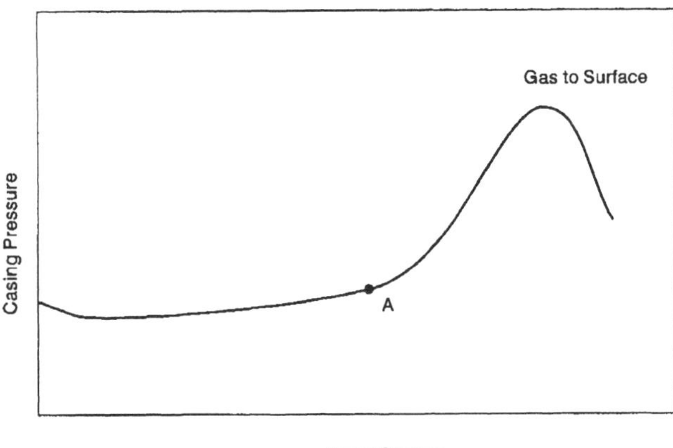

Fig. 5.13—Typical casing pressure response while displacing a gas kick with water-based mud (WBM).

and the bad joint to be replaced without having to strip or snub. Stripping/snubbing from the hole is an alternate approach if the leak depth is known and can be isolated with wireline-set plugs. Other measures mechanically isolate the leak from the pipe bore and thus allow the string to remain on bottom. One involves snubbing an isolation liner with top and bottom packoffs spaced out to straddle the leak. Finally, a packer and small-diameter tubing may be snubbed to a depth below the leak.

The situation becomes more complicated if the drillpipe leak is above the influx and the potential for problems escalates. A similar state of disequilibrium exists when the well is closed in, except gas will enter the drillstring through the leak until a pressure balance is achieved at the two communication ports. The final SIDPP in this case probably will be higher than the original SIDPP. Note also that the gas will retain a good part of its pressure if the well is shut-in while the gas is rising to the leak depth. The SICP should be monitored during the stabilization period and bleeding mud from the annulus may be necessary to avoid fracturing the rock. Volumetric control would keep wellbore pressures within the desired limits, but it may be difficult to execute the procedure if gas movement is rapid. The build rate should decline when gas starts to enter the drillpipe. However, the shut-in pressures will not stabilize if buoyancy drives further upward gas movement. As discussed in Sec. 5.3.1, the ensuing volumetric procedure does not change fundamentally if gas is migrating up both the drillpipe and the casing.

Fortunately, drillstring leaks are comparatively rare during a well-control event as those conditions which create leaks are expected to occur more often in a high-stress drilling environment rather than during a kill circulation. Even so, an operator should be prepared for this contingency. Knowing the leak depth is important to the successful outcome for both well control and repair efforts. But we have a dilemma if KWM has been placed in the well. Most diagnostic techniques require pumping into the drillpipe, yet we often need to know the leak depth to estimate how much backpressure to hold during the pump job. A wireline tracer survey is probably the most accurate method and may be the safest because low pump rates and volumes are usually successful. Temperature logs also have been used for this purpose.

Losing a portion of the drillstring is evidenced by a drop in pump pressure accompanied by a loss in string weight. The length of intact pipe in a near-vertical hole can be estimated from the change in the hook load, but we need to account for the friction through the closed preventer and hole drag. A way to measure the combined force is to pull up at constant speed and observe the weight indicator reading, W_{wi1}. Then lower the pipe, ensuring that the rate of descent is the same as the pulling speed, and obtain a second reading, W_{wi2}. Friction in the system is expressed by

$$F_f = (W_{wi1} - W_{wi2})/2. \qquad (5.16)$$

The effective or buoyant weight of the suspended string can be estimated by averaging the two readings and subtracting the weight of the traveling equipment.

$$W_e = (W_{wi1} + W_{wi2})/2 - W_t. \qquad (5.17)$$

The length of the suspended pipe is approximated using the relation

$$L_p = W_e/w_b, \qquad (5.18)$$

where w_b = the buoyed unit weight.

A severed drillstring in an underbalanced hole is handled basically in the same manner as an off-bottom kick: analyze the shut-in pressures to identify influx position and to manage gas migration using the volumetric method. Fishing tools must be snubbed in the well or the job conducted by pumping an off-bottom kill and staging in to the fish.

5.4.3 Loss of a Pump. Normally a pump breakdown is not a major problem. The well is shut-in and the standby pump is brought on line to complete the kill. However, the kill-rate circulating pressure (KRCP) characteristics may not be the same for the two pumps. The kill rate used before the breakdown may not be available from the standby. Hence the previous CDPP should not be used as the target pressure after the equipment is switched out.

The recommended procedure for starting up the standby pump is demonstrated by referring to the casing pressure curve shown in **Fig. 5.13.** For a WBM, the change in casing pressure is gradual until gas nears the surface at Point A. Prior to this somewhat nebulous point, only minor impact on BHP will be achieved by bringing the new pump up to speed while holding casing pressure constant. Upon reaching kill rate, the observed drillpipe pressure defines the schedule CDPP. A recent KRCP measurement for the standby pump should be available for calculating the pressure-reduction table (if applicable) and for comparing actual to calculated values.

The same procedure is followed if the operator wants to change the pump rate. For example, an increase in pump rate may be deemed necessary in a nonclassical procedure where annular friction losses are an integral part of the kill mechanics. Reducing the designated pump rate may be desirable before gas reaches the choke line in a subsea control operation. In either case, the casing pressure is held constant during the speed adjustment and the observed drillpipe pressure is the basis for the succeeding procedure.

5.4.4 Leaks in the Stack and Manifold Equipment. The blowout-preventer equipment (BOPE) is pressure-tested before drilling out the shoe and, perhaps, at periodic intervals thereafter to ensure performance reliability. A leak in the stack equipment should be a preventable complication. Nonetheless, nothing is guaranteed and one of the rig crew should have the designated responsibility for observing the stack and manifold for leak. A leak in the stack can be repaired if the component can be isolated from the well pressure by closing a lower ram preventer. If not, then a serious situation exists that could escalate into a blowout. Minor leaks can be tolerated for a period, but the small problem may intensify in severity. Pumping a sealing material into the kill line may plug the leak and allow safe conclusion of the kill, but always consider the wellbore pressure limits any time a pump is attached to the annulus.

The manifold lines and the choke in particular are subject to abrasion and possible cutting by solids in the mud and influx fluids. A cut-out choke can be isolated and the circulation path routed to a backup choke on even the most rudimentary of hookups. The backup choke then can be used while the damaged choke is replaced or for the remainder of the kill. The well must be shut-in at the cross before line repairs can be made upstream of the inboard manifold valves.

5.4.5 Lost Circulation. Lost circulation during a conventional kill threatens a loss of control on a well and severe losses generally escalate into an underground blowout if sufficient pressure cannot be held against the kick formation. The problem can be characterized as minor or major, depending on whether at least partial returns can be sustained while holding the required backpressure at the choke. Complete losses are discussed in the chapter which deals with underground blowouts. Minor losses do not necessarily constitute an eventual loss of secondary control.

Staying with the program is recommended if partial losses can be replaced by mixing or transporting in new mud. Recall that the maximum pressure at any point in a wellbore is experienced when the bulk of a gas reaches that depth. Thus full returns might be restored after the kick is displaced past the loss zone. If doing so is an option, reducing pump rate will lessen the pressure on the loss zone and improve the ability to circulate.

Completely removing the equivalent circulating density (ECD) backpressure by shutting down the pump and closing in the well may be beneficial. Given time, the loss zone may heal and allow circulation to resume with full or at least better returns. Monitoring the pressures is especially critical during the shut-in period. The procedure is no different from the initial shut-in: a constant BHP is achieved by bleeding mud from the choke to hold the SIDPP constant.

Lost-circulation material (LCM) may or may not be a viable option, depending on at what point in the kill the losses occurred. If LCM is considered, the material coarseness and concentration should be selected carefully to avoid plugging the bit. Losing the ability to pump and read BHP with losses above the bit will compound the problem.

5.5 Techniques Devised to Reduce Annulus Pressure

Ideally, engineering and design practices will ensure that the wellbore, casing, and BOPE will withstand the maximum stress imposed during a kill procedure. Unfortunately, not all well designs meet this criterion and some kicks cannot be displaced without exceeding the pressure limitations posed by the fracture gradient or equipment rating. This is not to say that the planning is always to blame. Perhaps a totally unexpected abnormally pressured horizon is encountered below a shallow casing shoe or an unusually large kick develops before the crew is able to close the well in. Whatever the circumstance, an operator may need to consider using one of the well-control techniques discussed in this section.

The methods are not foolproof and often carry the risk of allowing more gas to enter the hole. Some can cause the very failure the technique was designed to prevent. The use of any nonconventional procedure must be considered carefully and the risks weighed against the potential benefits. Most wells are equipped with adequate BOPE and most casings are designed to handle fairly large kicks. The open hole is generally the weak link and it is far better to chance lost returns than to create a situation that could result in a surface blowout. Some of the key issues to deliberate include deliverability of the kick formation, validity of the assumed fracture gradient (these tend to increase over time), casing wear, most likely influx character, how much casing is set, potential for an underground blowout to broach to surface, and validity of the pressure-prediction model (simple models yield conservative results).

5.5.1 Low-Choke Procedures. A "low-choke" method refers to any procedure where the operator intentionally opens the choke to keep the casing pressure below some predetermined value. The maximum choke pressure is based usually on the fracture gradient and the present mud density—the maximum allowable annulus surface pressure (MAASP) discussed in Chap. 4. We can characterize low-choke methods as those instances where opening the choke will or will not likely cause additional kick entry.

Annular friction losses during a well-control operation constitute a backpressure against the kick formation that is not considered in the Driller's or Wait and Weight Methods. These losses are usually minor for kill circulation rates in conventional hole geometries. But they are present and one application of the low-choke method is to estimate the annulus friction and to allow the choke pressure to be reduced by an equivalent amount. Another use of the method is where the choke operator reduces the choke pressure and fully expects a secondary kick to follow. The hope is that the additional entry will be small enough to manage safely and lead to lower pressures than experienced in the initial influx.

Slow pump speeds during conventional kills are advised for many good reasons. But low-choke methods are the exception where obvious benefits can be realized by circulating at some higher rate. Annulus friction losses increase as a function of pump rate and so higher rates serve to prevent or at least limit flow into the wellbore. In addition, a higher pump rate increases the proportion of mud to any gas entry. Gas is dispersed more and the mud hydrostatic pressure component is higher when the secondary kick expands.

Example 5.6. Refer to **Table 5.8.** Can the influx be circulated out of the well without exceeding the pressure limitations? If not, determine if the low choke method can be applied to prevent additional kick entry.

Solution. The estimated fracture pressure at 1,900 ft is

$$p_{fi} = (1,900)(0.69) = 1,311 \text{ psig.}$$

Equipment ratings are adequate and so the MAASP with kick fluid below the shoe is calculated as

$$(p_c)_{max} = 1,311 - (0.477)(1,900) = 404 \text{ psig.}$$

The SICP is only slightly less than the calculated maximum (our contrived intent). The next step is to estimate the maximum pressure at shoe depth during a Wait and Weight kick displacement. The formation pressure is calculated as

$$p_p = 230 + (0.477)(11,100) = 5,525 \text{ psig} = 5,539 \text{ psia}$$

and the KWM gradient is determined.

$$g_{km} = 5,525/11,100 = 0.498 \text{ psi/ft.}$$

TABLE 5.8—KICK DATA FOR A WELL WITH A WEAK SHOE

Wellbore and Stack Parameters	
Kick Formation Depth	11,100 ft
Surface Casing Information	
Description	$9^5/_8$-in., 36.0-lbm/ft K-55
Inner Diameter	8.921 in.
Setting Depth	1,900 ft
Bit Diameter	8¾ in.
Drillstring Information	
Collar Size	6¼×2¼ in.
Collar Length	540 ft
Drillpipe Description	4½-in., 16.60-lbm/ft, Grade E NC46
Wellhead, Stack, and Manifold Working Pressure	5,000 psi
Circulation Data and Mud Properties	
Kill Circulation Rate	4.0 bbl/min
KRCP	870 psig
Mud Type	Water Base
Mud Density	9.2 lbm/gal
Fann Multispeed Viscometer Data	
θ_{600}	60 lbf/100 ft^2
θ_{300}	40 lbf/100 ft^2
θ_{200}	28 lbf/100 ft^2
θ_{100}	17 lbf/100 ft^2
θ_6	3 lbf/100 ft^2
θ_3	2 lbf/100 ft^2
Recorded Kick Data	
Initial SIDPP	230 psig
Initial SICP	390 psig
Pit Gain	15 bbl
Other Known or Assumed Information	
Fracture Gradient at Casing Seat	0.69 psi/ft
Assumed Wellbore Temperature at Kill Rate	80°F+0.9°F/100 ft
Kick Fluid	0.60 SG Gas

The bottomhole temperature and z factor are determined as 640°R and 1.045. The initial influx height is calculated from the pit gain and capacity factor opposite the drill collars.

$$h_{ki} = 15/0.03643 = 412 \text{ ft}.$$

Its initial hydrostatic pressure is estimated as

$$p_{hgi} = \frac{(0.60)(5,539)(412)}{(53.29)(1.045)(640)} = 38 \text{ psi}.$$

The gas hydrostatic pressure after moving into the drillpipe annulus is therefore

$$p_{hg} = (38)(0.03643)/0.05470 = 25 \text{ psi}.$$

The original mud volume contained in the drillstring and its height in the annulus are determined.

$$V_{om} = (10,560)(0.01422) + (540)(0.00492) = 152.8 \text{ bbl}$$

and $h_{om} = 152.8/0.05470 = 2,793$ ft.

Eq. 4.40 yields the value for the intermediate variable Y.

$$Y = 5,539 - (0.498)(11,100 - 1,900) + (2,793)$$

$$(0.498 - 0.477) - 25 = 991 \text{ psia}.$$

The temperature at 1,900 ft is 557°R. Assume $z_{1,900} = 1.00$ and calculate the shoe pressure using Eq. 4.39.

$$p_{1,900} = \frac{991}{2} + \sqrt{\frac{991^2}{4} + \frac{(0.498)(5,539)(412)(1.000)(557)(0.03643)}{(1.045)(640)(0.05470)}}.$$

The z factor changes to 0.838 and another iteration is in order. Ultimately we reach the solution $p_{1,900} = 1,380$ psia $= 1,366$ psig. The calculated pressure at the shoe exceeds the fracture pressure by 55 psi.

The total annular friction loss is computed as 82 psi using the relations discussed in Appendix A. The predictions indicate that the choke pressure can be reduced by this amount and keep the rock from fracturing while preventing a secondary kick. Even so, the calculated failure margin is low and, given the conservative nature of the model, no adjustments to the Wait and Weight kill are probably necessary.

One problem with this application is that information is rarely available to determine annular friction losses accurately for even the simplest of cases. Openhole diameters are usually no better than an educated guess and the assumed downhole mud properties and flow patterns are often dubious. Sophisticated models are required when the situation is complicated by multiple fluids, two-phase flow, and acceleration effects.

A review of Darcy's law gives some insight as to when use of the method can be successful in limiting the size of a secondary influx. Four primary variables, the differential pressure, permeability, thickness, and time, govern the volume of a gas kick. Zone permeability and exposed height are fixed by well conditions and so a smaller kick can be ensured only if the formation flows into the well for less time, at reduced differential pressure, or at some ideal combination of the two.

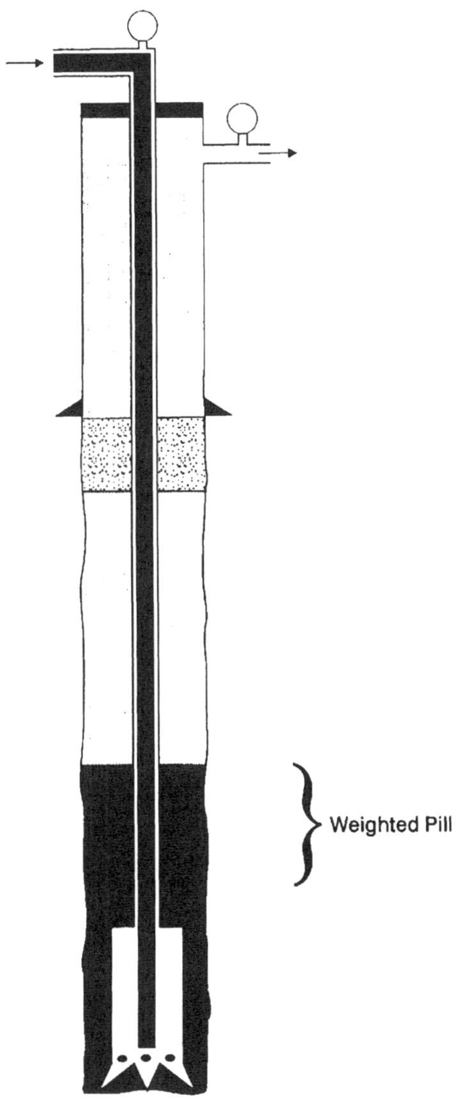

Fig. 5.14—The use of a weighted pill to reduce wellbore pressure during a gas kick displacement.

The flow period could easily exceed the initial detection and shut-in time, depending on when the backpressure is reduced and how long circulation continues with the well underbalanced. Opening the choke on an expanding gas bubble might in fact reduce the wellbore pressure to less than what it was when the zone first kicked. Thus low-choke procedures can lead to a much larger secondary kick, a problem that usually escalates to the point where the well is filled with gas and out of control.

The technique has been successful in west Texas and other areas where deliverability is low,[4] but letting any more gas into a well is a dangerous practice unless area knowledge ensures that the procedure will be met with success. The maximum pressure imposed on a potential loss zone occurs when the well is initially shut-in or when the leading edge of a gas influx is displaced to the critical depth. In the latter case, an MAASP based on the fracture gradient is meaningless if gas is expanding in the casing.

5.5.2 Pumping an Overkill Mud. Mixing and pumping an overkill mud weight can reduce the maximum surface pressure and, in many hole geometries, the maximum pressure at the shoe. The concept is a logical extension of the Wait and Weight Method except that the mud density pumped is higher than kill-weight. After the new mud is displaced into the annulus, the added hydrostatic serves to reduce the overlying pressures resulting from gas expansion. Mixing a volume sufficient to fill the hole completely is an option or, as shown in **Fig. 5.14,** a weighted pill calculated to give the desired increase can be pumped ahead of the kill-weight mud.

As in the Wait and Weight Method, the density used to replace that of the original mud does not start to affect wellbore pressures until the new mud enters the annulus. Thus the annular volume below the shoe must be greater than the drillstring volume if the intent is to protect the shoe. Also, recall from Chap. 4 that the drillpipe schedule must be revised if a mud heavier than kill-weight is pumped. Otherwise, the pressure held by the choke will reflect the overkill density and lead to higher pressures at the shoe and increase, not reduce, the risk of breaking down the shoe.

The target ICP is independent of mud weight and is still expressed by Eq. 4.26. However, the desired circulating pressure when the overkill mud reaches the bit must account for the excess hydrostatic in the drillstring and so Eq. 4.28 does not apply. The final circulating pressure (FCP) for an overkill mud is expressed by

$$p_{dcf} = p_{kr}(g_{okm}/g_{om})^{a} - D(g_{okm} - g_{om}), \quad \ldots\ldots. (5.19)$$

where g_{okm} = the hydrostatic gradient of the overkill mud. The practical maximum to the overkill mud density is achieved when the FCP is reduced to zero.

Example 5.7. Again consider the kick described in Table 5.8.

1. Estimate the maximum shoe pressure if a 10.0-lbm/gal mud was used to kill the well and prepare a drillpipe pressure schedule.

2. Calculate the pressure at the shoe if the well had to be shut-in at the instant the string was filled with the new mud.

Solution. 1. The objective is to circulate with the same BHP as would be imposed in a conventional kill. Hence all the parameters except the kill-mud gradient can be used in the Wait and Weight equation. Y is now

$$Y = 5,539 - (0.519)(11,100 - 1,900) + (2,793)$$
$$(0.519 - 0.477) - 25 = 856 \text{ psia.}$$

A z factor is assumed and Eq. 4.39 is used to compute the pressure. A couple of iterations yield $z_{1,900} = 0.858$. Thus,

$$p_{1,900} = \frac{856}{2} +$$

$$\sqrt{\frac{856^2}{4} + \frac{(0.519)(5,539)(412)(0.858)(557)(0.03643)}{(1.045)(640)(0.05470)}}.$$

$$p_{1,900} = 1,292 \text{ psia} = 1,278 \text{ psig.}$$

A safe kick displacement is indicated, but a revised drillpipe pressure schedule is needed to accomplish this result. Eqs. 4.26 and 5.19 yield

$$p_{dci} = 230 + 870 = 1,100 \text{ psig}$$

and $p_{dcf} = 870(0.519/0.477)^{1.0} - 11,100(0.519 - 0.477)$

$$= 481 \text{ psig.}$$

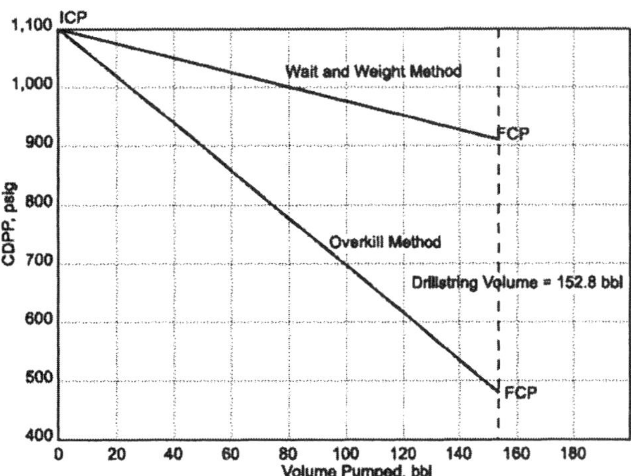

Fig. 5.15—Pressure reduction schedule for the overkill procedure in Example 5.7 and comparison to the schedule for a Wait and Weight kill.

Fig. 5.15 represents the schedule for this kill. For comparison, the schedule that would be employed in a Wait and Weight kill is shown also.

2. The BHP when the well is shut-in with overkill mud in the drillstring is

$$p_{bh} = (0.519)(11,100) = 5,761 \text{ psig.}$$

At this point in the kill we have 2,973 ft of original mud in the annulus. The base of the influx is at 8,127 ft and the pressure at this depth is

$$p_{8,127} = 5,761 - (0.477)(2,973) = 4,343 \text{ psig.}$$

We know that the pressure at the top of the influx is 25 psi less and obtain a 303-ft bubble height by iterating the gas law. Hence the pressure at shoe depth is

$$p_{1,900} = 4,323 - 25 - (0.477)(8,127 - 303 - 1,900)$$
$$= 1,492 \text{ psig.}$$

The shut-in results in a shoe pressure 181 psi higher than fracture pressure.

The primary weakness to this approach is demonstrated in Example 5.9. Shutting in a well with overkill mud in the drillstring could lead to excessive wellbore pressure and cause the failure the technique was designed to prevent.

5.5.3 Spotting a Heavy-Weight Pill. It may be feasible to spot a heavy mud column on bottom which, when combined with the original mud hydrostatic, is sufficient to overbalance the kick formation. After ensuring that the well is dead, the drillstring can be pulled above the balanced pill and the gas circulated from the hole. A reduced choke backpressure requirement in the second circulation thus minimizes wellbore pressures when the gas expands. In general, this nonconventional technique is applicable to relatively deep wells with a fairly long openhole section. Example 5.8 demonstrates why this is so and what needs to be considered in the plan.

Example 5.8. Can the well described in Table 5.8 be killed safely by spotting an 18.0-lbm/gal pill on bottom? If so, write a general procedure.

Solution. First, determine how much heavy-weight mud will balance the 5,525-psig formation pressure. The vertical height requirement for the mud is

$$0.935h_{km} + 0.477(11,100 - h_{km}) = 5,525.$$
$$h_{km} = 502 \text{ ft.}$$

Choose a larger height, say 700 ft, to account for possible hole enlargement and trip-induced pressure reduction. The pill volume for the 8¾-in. hole diameter is

$$(700)(0.07438) = 52.1 \text{ bbl.}$$

The balanced kill-fluid will be longer than 700 ft when the string is on bottom. The capacity of the drill collar bore and annulus is

$$(540)(0.00492 + 0.03643) = 22.4 \text{ bbl.}$$

The heavy-mud length above the drill collars will be

$$(52.1 - 22.4)/(0.01422 + 0.05470) = 431 \text{ ft.}$$

The total length is 971 ft and so the required volume of the 9.2-lbm/gal displacing fluid is

$$(11,100 - 971)(0.01422) = 144.0 \text{ bbl.}$$

The influx will be displaced up the annulus a distance that corresponds to the 196.1-bbl pumped volume. We assume the gas is on bottom when the job starts and does not slip relative to the mud during the procedure. Accordingly, the bottom of the influx at the end of the job is

$$11,100 - 540 - (196.1 - 19.7)/0.05470 = 7,335 \text{ ft.}$$

The influx is still fairly deep and we reasonably conclude that little expansion will occur during the displacement. A general kill procedure would read:

1. Mix and pump 52.1 bbl of 18.0-lbm/gal mud and follow with 144.0 bbl of the original 9.2-lbm/gal mud. Pump all fluids in a controlled manner.
2. Allow the plug to equalize and verify that the well is dead.
3. Slowly pull at least 700 ft of drillpipe. Ensure that the hole is kept full.
4. Circulate the influx from the well in a controlled manner and circulate the hole with 9.6-lbm/gal kill-weight mud.
5. Place the bit on bottom and circulate out the pill using kill-weight mud.

It would be desirable to maintain a constant BHP at all times when the pump is on the hole. But there is an operational problem in getting a heavy pill on spot because the hydrostatic pressure in the drillstring probably will exceed the combined formation pressure and circulating friction at some point. It follows that the drillpipe will be on a vacuum during part of the job and thus render the drillpipe gauge useless for control purposes.

Contrary to generally recommended practice, a workable plan for a relatively deep influx may be to control from the casing pressure. In doing so, we have to assume that the gas does not expand enough to overcome the overbalance provided by annulus friction and any "bubble collapse" phenomena. Otherwise the well will reach an underbalanced condition and allow a (probably) small secondary kick to enter the hole. The degree of gas expansion controls the reduction in BHP and gas expands faster at shallower depths. Hence the procedure should be instituted as soon as possible to limit the effect of gas migration.

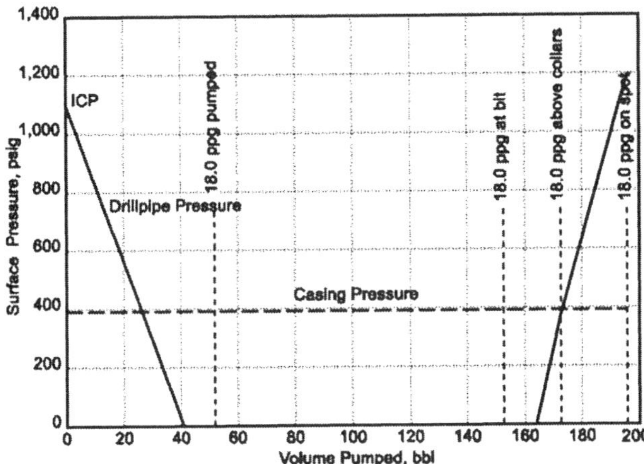

Fig. 5.16—Drillpipe and casing pressures while spotting the 18.0-lbm/gal pill on bottom in Example 5.8.

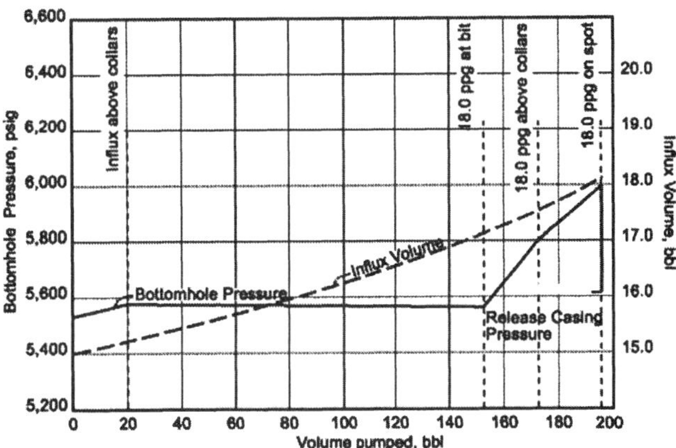

Fig. 5.17—BHP and influx volume resulting from holding constant casing pressure while spotting the weighted pill in Example 5.8.

Fig. 5.16 is a graph constructed to show both the constant-casing-pressure method as applied to the example well and the calculated drillpipe pressures. Note that the drillstring goes on a vacuum after pumping about 41 bbl of the heavy mud and does not catch pressure again until after 9 bbl enters the annulus.

Fig. 5.17 shows the BHP and influx volume over the same period. These predictions are based on a unit bubble model and do not consider gas migration. Holding constant casing pressure increases the BHP by 52 psi as the gas column shortens when moving into the drillpipe annulus. Thereafter, the gas volume expands to 17.1 bbl when heavy mud reaches the bit and, consequently, the BHP declines by 17 psi. Continuing with the same choke pressure leads to a steep increase in BHP as heavy mud enters the annulus.

Placing the pill should kill the well if the assumed hole diameter was not grossly in error. Even so, the well will continue to flow until the pill achieves equilibrium. Gas expansion may drive some mud out the annulus, but the flow should be minimal for a moderately deep influx. The string can be pulled above the pill after the hole has stabilized and the well is ensured dead. Stripping is advised if the operator is uncertain.

With the bit above the pill, the gas displacement can be completed and the hole filled with KWM using either the Driller's or Wait and Weight Method. In the former procedure, the gas is displaced by holding drillpipe pressure constant at the observed kill-rate pressure, which should be the same as or slightly less than the prerecorded KRCP. Similarly, the ICP for a Wait and Weight displacement schedule will be started at the KRCP, but will increase rather than decline to the FCP. Gas expansion will drive the casing pressure up in either method, but the choke backpressure will be much less than the case where pore pressure had to be maintained at the bit.

5.5.4 Reverse Circulation. Under certain conditions, it may be advisable to circulate out a kick by pumping into the annulus and taking returns from the drill string. **Table 5.9** lists perceived advantages and disadvantages to the so-called reverse circulation method. The main benefit is that a gas kick is kept within the string bore and the rock and hardware are protected from the expansion pressures. Reverse circulation is applied universally in cased-hole operations, but the technique definitely carries some major disadvantages when used for well control in a drilling operation.

TABLE 5.9—ADVANTAGES AND DISADVANTAGES OF REVERSE-CIRCULATING A GAS KICK

Advantages

1. Gas-expansion pressures are contained in the drillstring.
2. Drillstring pressure ratings generally yield higher safety factors.
3. Gas is removed from a well much faster.
4. Displacement efficiency is enhanced, resulting in less mixing of the kick fluids in the mud.

Disadvantages

1. Drillstring friction losses are imposed on the wellbore.
2. Plugging the bit or packing off the annulus is a possibility.
3. Reversing requires additional rigup time and equipment.
4. The technique promotes pipe sticking.
5. A leak in the drillstring could cause a catastrophic downhole failure.
6. The presence of a drillstring float precludes using the method unless the float can be removed.
7. Gas flow rates through the surface equipment are higher.

Fig. 5.18 depicts the well in Table 5.8 if the influx is reversed into the drillstring. The gas bubble length is much greater than it was in the annulus because of the small-diameter bore. The change in the SIDPP reflects the reduced hydrostatic in the drillstring, but the annulus is full of mud and so the SICP is equivalent to the original SIDPP. This illustrates the justification for the technique—the annulus is full of mud throughout the kick displacement and so the wellbore is isolated from the effects of the long gas column. A reverse displacement is still a constant BHP method, except the casing pressure is used for control rather than the drillpipe.

Fig. 5.19 compares the choke pressures while using the Drillers's Method to displace the gas in the normal direction and in reverse. Note that for the latter case, the volume is zeroed when the 15-bbl kick is placed in the pipe. A major difference between the two directions is that the kick can be cleared from the well in much less time when reversed.

The Driller's Method is used except the ICP as shown on the casing gauge is held constant until the hole is filled with original mud. As in Chap. 4, the ICP comprises the initial SICP (after gas enters the string) and the system friction losses. However, the bulk of the friction pressure now is imposed on the wellbore and therein lies a major weakness to the technique. The ECD at the kick formation includes the pressure drop across the bit and throughout the drillstring. The losses are additive farther up the hole and the ECD at the shoe also includes the pressure drop between the bit and the shoe. Hence the pressure applied at the weak point in the system

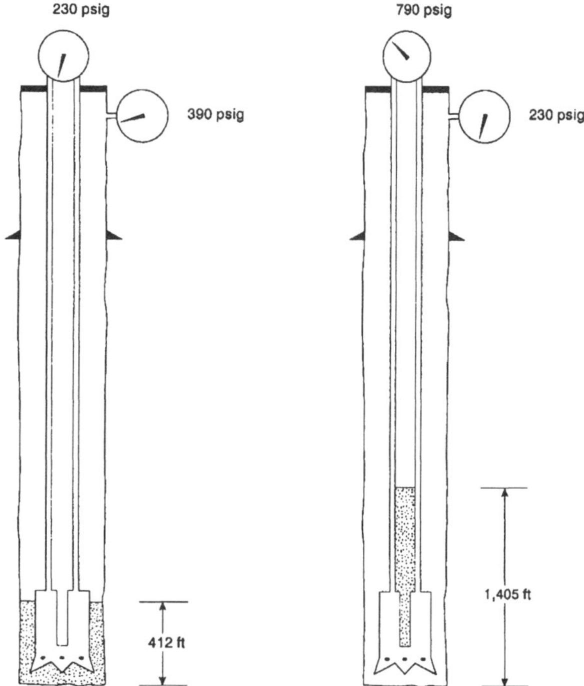

Fig. 5.18—Gas position and surface pressures before and after reversing the influx into the drillstring using Table 5.8 data.

must be less than the maximum pressure during a conventional displacement or nothing is gained.

The KRCP for the example well in Table 5.8 was 870 psig at the 4.0-bbl/min kill rate. We would expect the same pump pressure in reverse and it is obvious that reverse circulation is not an option for this well unless the job is accomplished at a reduced pump rate and/or something is done to remove the bit jets. A low-choke procedure would be appropriate, given the high overbalance across the kick zone. In any event, circulating slower than the normal kill rate is a probable requirement, which may dictate the use of a cementing unit or other low-volume, high-pressure pump.

As the initial step, it may be advisable to pump into the annulus while the well is shut-in. Bullheading the mud at a slow rate establishes a new leak-off criterion, thereby removing any uncertainties about the true fracture pressure and establishing a benchmark for the maximum allowed pressure during the procedure. As in all leak-off tests, the rig pump may not be capable of the rates needed to pump the job.

Drill cuttings and kick-contaminated mud create another set of potential problems and the possibilities of plugging the bit or forming a bridge in the annulus need to be addressed. Blasting out the bit jets would be ideal, not only to minimize the plugging risk but also to reduce annular backpressure. Time is of the essence if gas is migrating up the annulus and it may be difficult or impossible to remove the jets, drillpipe float, or other restriction before migration nullifies the benefit to reversing a kick. However, a plugged bit does not necessarily constitute a major difficulty if all or a major portion of the influx is displaced into the string before the obstruction occurs. The gas is isolated from the wellbore and the drillpipe pressures can be controlled using the volumetric method.

There is no advantage to killing a well in reverse using the Wait and Weight Method. Except for some slimhole geometries, the gas would surface before KWM enters the bit and so the maximum drillstring pressure would not be affected. In addition, pumping the heavier mud into the annulus increases the circulating friction and we have a practical

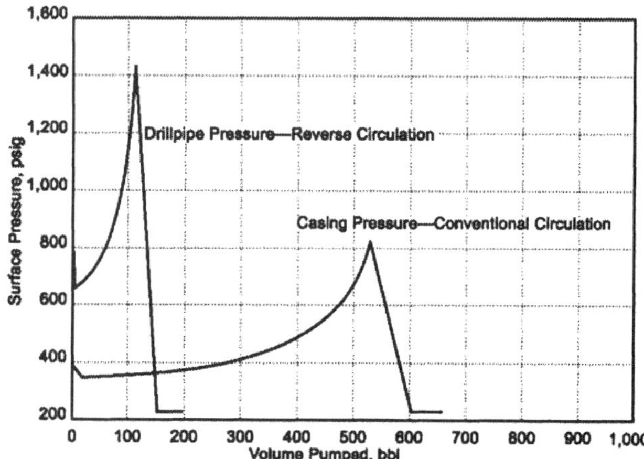

Fig. 5.19—Predicted choke pressures for conventional and reverse Driller's Method displacements using Table 5.8 data.

problem in developing a pressure schedule for the imprecise hole dimensions.

After voiding the gas, the original mud must be replaced. The reason for pumping in reverse has been eliminated and the recommended practice is to tie back onto the drillpipe and pump KWM in the normal manner. The risk of excessive annulus pressures greatly increases with the higher mud weights. Miska *et al.*[5] simulated the reverse kick displacement for a variety of well depths and kick conditions; in no case were they able to pump KWM into the annulus using either the Driller's or Wait and Weight Methods without breaking down the shoe.

Fig. 5.20 shows the recommended equipment configuration for a reverse kill operation. The drillpipe bore must be plumbed at the surface to both the choke manifold and the fluid pumps with piping rated for the maximum anticipated pressures. Extremely high flow rates should be expected when the kick surfaces. Structural support should be adequate, large diameter piping installed, and all turns and bends minimized.

A critical problem during a reverse-out procedure would be a shallow leak in the drillpipe. Should a leak develop, the pressure contained in the string would act on top of the fluid in the

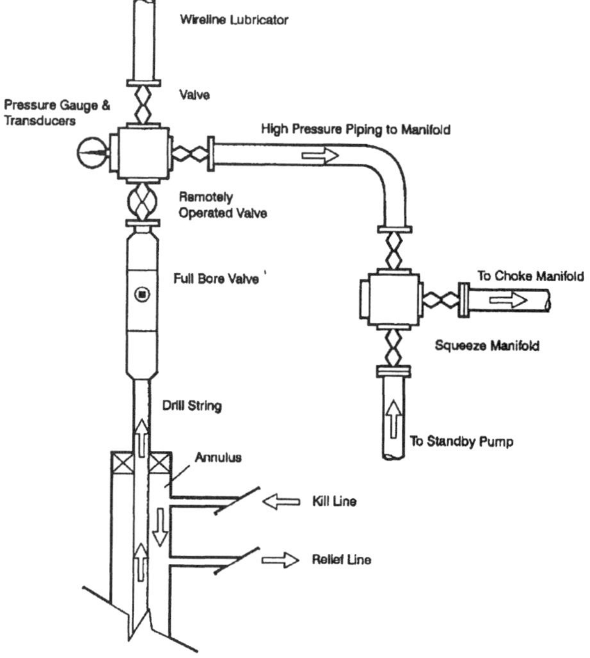

Fig. 5.20—Recommended surface equipment for the reverse circulation method.

annulus and potentially lead to a downhole failure. As shown in the diagram, one way to prevent this occurrence is to install a popoff valve so that excessive annular pressure can be relieved. However, venting a well usually means a complete loss of control and this provision should be considered only if a shallow leak will burst the casing or cause a broached shoe.

In summary, the reverse-circulation method has been applied successfully and is a feasible control option in some cases. The decision to displace a kick in this manner should be made only after weighing the potentially adverse consequences and developing contingency procedures for every conceivable problem. The well will not wait for plans to be written or allow for extensive rigup time after the fact. An operator should anticipate a reverse kill operation in the well-control planning process and have equipment on location and ready to be deployed before the need arises.

5.5.5 Bullheading. The intent of a bullhead-kill procedure is to pump into a shut-in well at sufficient rate and pressure to fracture the formation and place the kick fluids into the loss zone. The technique is fairly standard in cased hole work, but definitely a nonconventional method when applied to well control in a drilling operation. Adams[6] discussed several reasons why an operator might choose to bullhead rather than circulate out a kick including H_2S kicks, inability to circulate on bottom, a loss zone below the kick disallows adequate circulation rates for a kill, buying time, inability to withstand the maximum surface pressures during a conventional kill, and underground-blowout control.

It is unlikely that an influx/mud mixture can be pumped back into the source formation without exceeding its fracture pressure unless the zone has a vugular-type porosity. Placing a kick back where it came from requires that the zone fractures preferentially to the other exposed rock. Thus "pumping the bubble back into the formation" can be accomplished only if the kick transpired not long after drilling through a protective casing shoe. Only then will bullheading kill-weight mud balance the pore pressure.

This is not to say that a bullheading mud into a shallower interval is never an option, but successful control depends on such factors as the distance between the source and loss zones, how much gas is below the loss zone, available hydrostatic pressure, and the potential for inducing an underground blowout. The injected mud will enter the loss zone as well as any kick fluids which have moved above the exit point in the well. Gas migration may prescribe more than one job if gas remains below the loss zone after the pumps are shut down and the intent is to not allow any gas through the choke.

There is some minimum pump rate requirement. Obviously the pumping velocity down the wellbore flow conduit must exceed the upward gas migration rate before the influx can be flushed from the well. Also, sufficient rate is needed to displace a large free gas pocket, say underneath the preventers, or the mud will fall through the gas. But there is a tendency to pump too fast with possible detriment to the job objectives of maintaining control at minimum cost. High pump rates can waste mud if bullheading is done to buy time or in conjunction with underground-blowout control. Also, excessive friction pressures through the well tubulars needlessly consumes horsepower and increases the risk of rupturing the casing.

Wells have been lost during a bullhead procedure because the operator did not consider adequately the pressure ratings of the surface and downhole equipment. The surface breakdown pressure is the fracture-initiation pressure less the hydrostatic whereas we consider the fracture propagation pressure and incorporate friction during the injection phase of the job. The operator's plan must base the maximum allowable injection pressure not only on the surface equipment and casing ratings, but also on possible casing wear and the effect a higher mud weight has on the burst loads with depth.

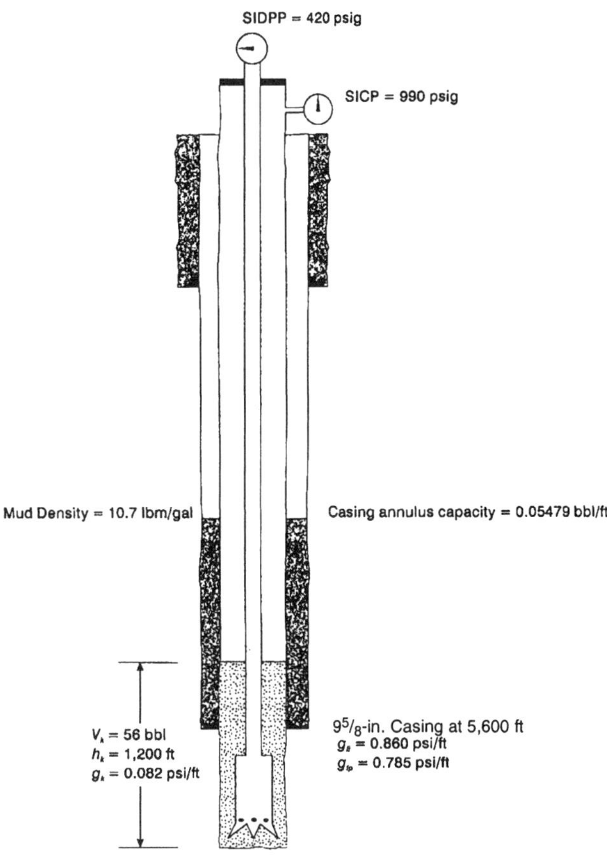

Fig. 5.21—Wellbore and relevant data for Example 5.9.

Example 5.9. Fig. 5.21 shows the problem conditions for a large H_2S kick.

1. Estimate the initial and final surface pressures required to bullhead a 13.3-lbm/gal KWM into the annulus.

2. Also determine the differential burst loads at 5,000 ft if the assumed backup density is 9.0 lbm/gal. Friction losses are 80 psi/1,000 ft at the 30-bbl/min final pump rate.

Solution. 1. The fracture-initiation pressure at the casing shoe,

$$p_{fi} = (0.860)(5,600) = 4,816 \text{ psig},$$

must be exceeded before the hole will start taking fluid. Determining the initial casing pressure would be a straightforward problem if incompressible fluids were above the shoe. But gas is present and so we calculate the gas bubble volume required to fracture the rock at 5,600 ft. Compressing the gas by 8.1 bbl from pumping an equivalent volume of mud into the annulus gets us close. At breakdown, the KWM height and gas height reduction are

$$8.1/0.05479 = 148 \text{ ft}.$$

The hydrostatic gradient of the compressed gas is 0.097 psi/ft. Ignoring the mud and wellbore elasticity, we obtain the surface pressure based on the hydrostatic pressures in the annulus.

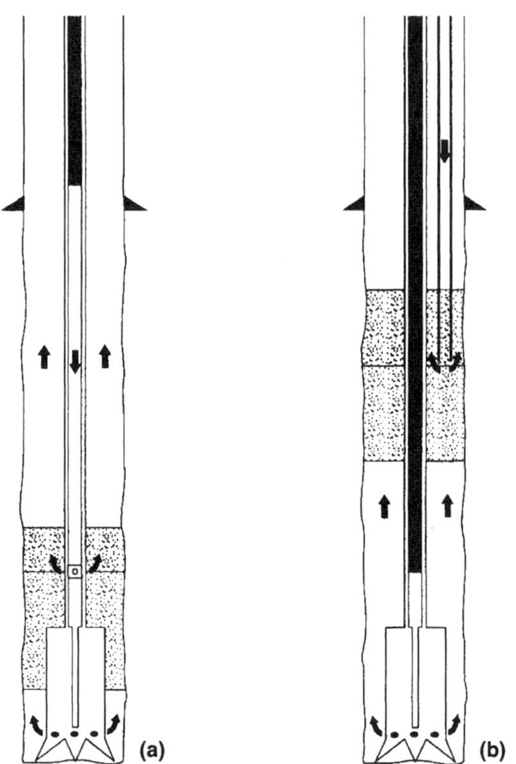

Fig. 5.22—Two concepts for injecting mud directly into a gas influx.

$$p_c = 4,816 - (0.092)(600 - 148) - (0.566)(5,000) - (0.691)(148) = 1,842 \text{ psig.}$$

At breakdown, the casing at 5,000 ft experiences an internal pressure differential of

$$\Delta p = [1,842 + (0.691)(148) + (0.566)(5,000 - 148)] - (0.468)(5,000);$$

$$\Delta p = (1,842 + 102 + 2,746) - 2,340 = 2,350 \text{ psi.}$$

The fracture propagation pressure is

$$p_{fp} = (0.785)(5,600) = 4,396 \text{ psig.}$$

The final surface pressure is obtained by subtracting the mud hydrostatic pressure and incorporating friction.

$$p_c = 4,396 - (0.691)(5,600) + (0.08)(5,600) = 974 \text{ psig.}$$

2. At this point in the kill, the burst load at 5,000 ft is

$$\Delta p = [974 + (0.691)(5,000) - (0.08)(5,000)] - (0.468)(5,000) = 1,689 \text{ psig.}$$

5.5.6 Dispersing or Segmenting a Gas Kick. Controlled circulation of a large, predominantly gas influx will lead to higher pressures in the annulus than would result if some way to distribute this same gas into a long mud column or remove the gas in smaller packages could be brought about. The premise has led to similar ideas for reducing the maximum shoe and surface pressures during a conventional kill. The techniques discussed in this section are mostly conceptual and we are not aware of any field applications at this time.

Bourgoyne[7] suggested injecting mud directly into the influx and two possible ways to accomplish this objective. Increasing the mud-to-gas ratio across the length of the kick would result in lower annulus pressures as the mixture is pumped up the well. **Fig. 5.22a** shows a proposed tool designed to divert a portion of the mud into the annulus. The device is positioned in the drillstring some distance above the bit, ideally above the top of some predetermined kick volume, and is opened by dropping an actuator. Tool design and procedural considerations would include a means to maintain the desired BHP while circulating through both communication ports (bit and diverter tool) and a flow regulator to keep a constant flow rate through the tool when the annulus pressure changes.

Fig. 5.22b depicts another proposal to pump mud through a small-diameter string in the annulus. Upon reaching the tube, the gas influx is dispersed into the mud flow and lower pressures above string depth follow. However, a means for injecting or snubbing the string into a shut-in wellbore is problematic with conventional BOP and wellhead equipment. One feasible and field-tested method would be to inject mud through a parasite string run in tandem with the surface or intermediate casing. But the pressures below the shoe would not be affected and the need for such a configuration to reduce the shallower pressures would have to be anticipated when the casing was run. Money would be better spent on the appropriate design of the casing and surface equipment.

Fleckenstein and Mitchell[8] proposed removing a large gas influx by partitioning the kick into smaller, more manageable segments. Three partitioning methods were suggested in the paper. The first was to perforate at the desired depth, set a plug above the bit, and circulate the upper portion of the kick in a conventional manner. Plugging the perforations and removing the obstruction at the bit would allow displacing the rest of the gas. However, the authors did not address the operational aspects for reliably isolating the perforations after the first circulation.

The second suggested procedure required a special tool invented for this purpose—i.e., one that would allow circulation ports to be opened in the drillstring so that all the mud is diverted into the annulus. After the first circulation, the ports would have to be closed and the mud directed back to the bit. Borrowing from completion technology, perhaps a slickline-actuated sliding sleeve with a plug profile below could serve this purpose or a pump-through ball device could meet the same objective. The suggested equipment, like Bourgoyne's tool, are not currently available for use in a drillstring.

The third partitioning method, stripping out of the hole to circulate the uppermost partition and stripping back for subsequent circulations, may allow separating the gas into two or more segments. However, stripping takes time (as does any wireline/slickline work needed for the other methods) and gas migration may not allow this option unless the pipe can be moved faster than the migration velocity. Stripping up into the influx seems to hold promise for cases where migration rates are slow or nonexistent, say for a large gas kick into an oil-based mud.

Problems

5.1 Refer to the information given in **Table 5.10.** A Driller's Method kill is instituted but the bit plugs when the top of the gas kick has been displaced to 11,000 ft.

1. Determine the expected surface casing pressure at the time the choke is closed and, starting from this point, write a volumetric-control procedure if the safety margin is 150 psi and the working margin is 100 psi.

TABLE 5.10—KICK DATA FOR PROBLEMS 5.1 AND 5.5	
Vertical Depth	15,000 ft
Mud density	14.5 lbm/gal
Openhole Diameter	6 in.
Last Casing Description	7-in., 29.00 lbm/ft
Casing Depth	14,200 ft
Fracture Gradient at the Shoe	0.86 psi/ft
Kill Pump Rate	2.0 bbl/min
KRCP	1,000 psig
Drill Collar Description	600 ft of 4¾×1½-in.
Drillpipe Description	3½-in., 13.30 lbm/ft
Wellbore Temperature	80°F+1.4°F/100 ft
SIDPP	300 psig
SICP	450 psig
Pit Gain	14 bbl
Kick Fluid	0.65 SG Gas

2. What is the maximum expected casing pressure if the gas migrates as a bubble?

5.2 Is breakdown of the shoe during a volumetric-control procedure more likely or less likely than with a conventional Driller's Method displacement? Explain your answer.

5.3 Verify the computed results shown in Table 5.2. The influx location at the end of any bleed stage is determined by trial and error: Assume a pressure at the top of the gas bubble; compute its wellbore location based on the mud gradient and casing pressure; determine the temperature and z factor at this depth and pressure; determine the pressure using the gas law; and iterate to a solution.

5.4 Derive Eq. 5.7.

5.5 The influx in Problem 5.1 migrates to surface. Starting from the computed final casing presure, write a procedure for lubricating the gas bubble. Prepare a schedule similar to the one shown in Table 5.4.

5.6 Confirm the drillpipe- and casing-pressure curves shown in Fig. 5.7.

5.7 Prepare a calculation sheet similar to Table 5.6 using a 35.50-bbl bleed volume for every stage after gas reaches bit depth. Would this have been a risky procedure?

5.8 The information in **Table 5.11** relates to a well-control problem on a vertical well.

A small flow was detected with drillpipe and collars completely out of the hole. The operator installed a backpressure valve and ran drillpipe to a depth of 3,000 ft. The net pit gain increased to 20 bbl by this time and the well was closed in. The volumetric method was successfully employed and the SICP reads 300 psig after all the gas has been cleared from the hole.

1. Can this well be killed with the bit at present depth?

2. If so, determine the staged kill procedure using a 150-psi stage margin. The backpressure valve will not be retrieved after the first kill is pumped.

5.9 Determine the staged-kill procedure for the well described Problem 5.8 if the backpressure valve is removed. Be sure to account for the effective displacement caused by a constant kill-mud volume in the drillstring.

5.10 For Problem 5.8, recommend a drillpipe or casing pressure schedule for the rig supervisor to follow when pumping the second circulation. The results must avoid a hydrostatic underbalance and excessive pressure at the shoe. Assume the backpressure valve is in place and that the prior kill-mud is retained below the valve.

5.11 Repeat Problem 5.10 except assume the kill mud required for the circulation is already in the string.

5.12 The well described in Table 4.25 is being killed using the Wait and Weight procedure and the pressure-reduction schedule prepared in Problem 4.36. Midway through the schedule the CDPP suddenly increases by 400 psi with no change in the annulus pressure.

1. What most likely caused the pressure increase?
2. Discuss the best way to complete the kill.

5.13 Refer to the shut-in well depicted in **Fig. 5.23.** A sudden drop in circulating pressure occurred while displacing a gas kick using the Driller's Method. The formation pressure at total depth is 4,000 psig, the mud density is 9.4 lbm/gal, and the hydrostatic pressure of the influx is 40 psi. Calculate the SIDPP and SICP for the case where

1. a bit nozzle has washed out,
2. the drillpipe has a hole at 6,000 ft, and
3. the drillpipe has a hole at 1,500 ft.

Discuss the options in each case for maintaining control and repairing the string.

5.14 The well shown in Fig. 5.23 suddenly lost string weight and the choke was closed. The neutral weight indicator reading before the kick was 167,000 lbf and the traveling equipment weighs 12,000 lbf. The driller reciprocated the pipe through the annular preventer and obtained weight indicator readings of 113,000 and 103,000 lbf on the up and down strokes.

1. Estimate the friction force between the annular preventer and the drillpipe.

2. Approximate the depth where the string parted.

TABLE 5.11—PARAMETERS FOR THE WELL IN PROBLEM 5.8	
Depth	12,000 ft
Mud density	10.0 lbm/gal
Openhole diameter	9½ in.
Last casing depth	5,000 ft
Last casing description	10¾-in., 51.00 lbm/ft
Fracture gradient at casing seat	0.75 psi/ft, and
Drillpipe description	5-in., 19.50 lbm/ft X-95

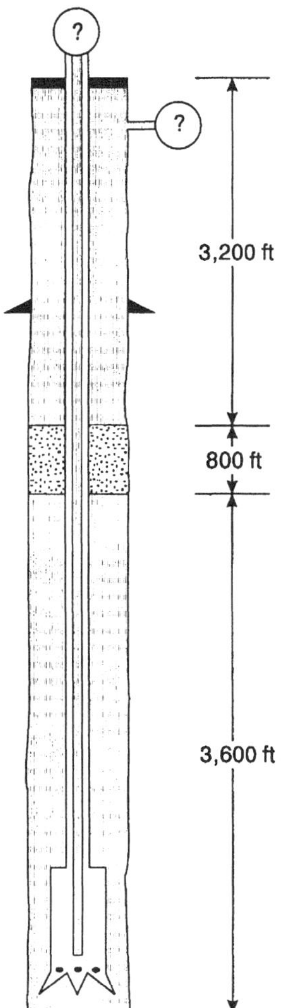

Fig. 5.23—Shut-in well described in Problems 5.13 and 5.14.

5.15 You were using the Driller's Method to displace a gas kick from a well at 3.0 bbl/min and a CDPP of 1,100 psig. The initial pressure at the choke was 700 psig. Gas was displaced halfway up the hole when the primary pump broke a rod. The well was shut-in with the same SIDPP and an SICP of 950 psig. The minimum rate of the backup pump is 4.0 bbl/min, but the driller neglected to measure its SICP on the last tour. How do you best complete the gas displacement using the backup pump?

5.16 Determine the allowable reduction in choke pressure if the kill rate in Example 5.6 is doubled.

5.17 What is the maximum mud weight one could pump in Example 5.7 and still be able to control BHP.

5.18 Refer to Table 5.8 and assume the intent is to use a 12.0-lbm/gal pill ahead of the KWM in a Wait and Weight procedure.

1. Determine the minimum pill volume requirement for protecting the shoe.
2. Prepare a drillpipe pressure schedule for placing KWM at the bit.

5.19 Consider where the kick is displaced in Step 4 of the procedure in Example 5.8. Estimate the maximum pressure at the shoe for a Driller's Method procedure and compare to the on-bottom case where the kick is circulated in a conventional manner.

TABLE 5.12—KICK DATA FOR PROBLEM 5.21

Kick Formation Depth	19,100 ft
Protective Casing Information	
Description	10¾ in., 60.70-lbm/ft Q-125
API Internal Yield Pressure	11,090 psi
Inner Diameter	9.660 in.
Setting Depth	12,000 ft
Drilling Liner Information	
Description	7⅝ in., 39.00-lbm/ft P-110
API Internal Yield Pressure	12,260 psi
Inner Diameter (ID)	6.625 in.
Setting Depth	15,600 ft
Liner Top Depth	11,600 ft
Bit Diameter	6½ in.
Drillstring Information	
Collar Size	4½×1½-in.
Collar Length	600 ft
Bottom Drillpipe Section	3,200 ft of 3½-in., 13.30-lbm/ft X-95
Top Drillpipe Section	15,300 ft of 5-in., 19.50-lbm/ft S-135
Wellhead, Stack, and Manifold Working Pressure	10,000 psi
Mud Type	Water Base
Mud Density	13.5 lbm/gal
SIDPP	3,400 psig
Initial SICP	5,200 psig
Pit Gain	90 bbl
Fracture Gradient at Casing Seat	1.10 psi/ft
Wellbore Temperature	100°F+1.1°F/100 ft
Kick Fluid	0.60 SG Gas

5.20 Recommend a way to reverse circulate a gas kick from the annulus into the drillstring while maintaining a constant BHP.

5.21 Refer to the wellbore and kick conditions shown in **Table 5.12.** Can the gas be displaced safely in a conventional manner or should the operator consider a reverse procedure?

5.22 Determine the minimum pump rate to bullhead a well if gas is migrating up a 6-in. openhole section at 4,000 ft/hr. What if the same migration rate is evidenced in a 12¼-in. hole?

5.23 Verify that an approximate 8.1-bbl volume reduction of the gas bubble in Example 5.9 will fracture the shoe.

5.24 A kick enters a well from a vugular limestone at 11,860 ft and 7⅝-in. protective casing with a 6.765-in. ID is set at 11,790 ft. The zone's estimated reservoir pressure is 6,000 psig. Pipe is essentially out of the hole and the operator elects to bullhead 18.0-lbm/gal mud into the well. Determine the maximum allowable pump rate if no more than 6,320 psi internal differential pressure can be tolerated on the casing. Assume the casing is backed up with a 9.0-lbm/gal fluid.

Nomenclature

a = pressure drop exponent, dimensionless
C = capacity per unit length (capacity factor), bbl/ft
C_a = annulus capacity factor, bbl/ft
C_d = displacement factor, bbl/ft
D = depth, L, t,
D_b = bit depth, ft
D_{sh} = casing shoe depth, ft
F_f = friction force, lbf
g_{fi} = fracture initiation gradient, psi/ft

g_{fp} = fracture initiation gradient, psi/ft
g_g = gas hydrostatic gradient, psi/ft
g_m = mud hydrostatic gradient, psi/ft
g_k = kick fluid hydrostatic gradient, psi/ft
g_{km} = kill mud hydrostatic gradient, psi/ft
g_{om} = original mud hydrostatic gradient, psi/ft
g_{okm} = overkill mud hydrostatic gradient, psi/ft
h_g = gas column height, ft
h_{ki} = gas influx height at bottomhole conditions, ft
h_{km} = kill-weight mud height, ft
$h_{km(a)}$ = kill-weight mud height in annulus, ft
$h_{km(p)}$ = kill-weight mud height in pipe, ft
h_m = mud height, ft
h_{om} = original mud height, ft
$h_{om(a)}$ = original mud height in annulus, ft
$h_{om(p)}$ = original mud height in pipe, ft
L_p = pipe length, ft
p = pressure, psi
p_{bh} = bottomhole pressure, psi
p_c = casing pressure, psi
p_{cs} = shut-in casing pressure, psi
p_{dci} = initial circulating drillpipe pressure (ICP), psi
p_{dcf} = final circulating drillpipe pressure (FCP), psi
p_{ds} = shut-in drillpipe pressure (SIDPP), psi
p_{fi} = fracture initiation pressure, psi
p_{fp} = fracture propagation pressure, psi
p_{hg} = gas hydrostatic pressure, psi
p_{hgi} = gas hydrostatic pressure at initial conditions, psi
p_{kr} = kill-rate circulating pressure (KRCP), psi
p_p = pore pressure, psi
p_{pr} = pseudoreduced pressure, dimensionless
p_{sh1} = pressure at the casing shoe, psi
p_{sm} = stage margin pressure, psi
p_{wm} = working margin pressure, psi
T = temperature, °F or °R
T_{bh} = bottomhole temperature, °R
T_{pr} = pseudoreduced temperature, dimensionless
V_g = gas volume, bbl
V_k = kick volume, bbl
V_{km} = kill-mud volume, bbl
V_m = mud volume, bbl
V_{om} = original mud volume, bbl
$V_{om(p)}$ = original mud volume entry into the pipe, bbl
w_b = buoyed weight per unit length, lbf/ft
W_e = effective (buoyant) weight, lbf
W_t = traveling equipment weight, lbf
W_{wi} = weight indicator reading, lbf
X = variable in the annulus pressure relation where mud density is constant, psia
Y = variable in the annulus pressure relation where kill-mud is in the annulus, psia
z = compressibility factor, dimensionless
z_{bh} = compressibility factor at bottomhole conditions, dimensionless
Z = variable in the lubricated mud volume equation, bbl
γ_g = gas specific gravity, dimensionless
θ_N = Fann viscometer reading, lbf/100 ft^2
ρ_{km} = kill mud density, lbm/gal

References

1. Matthews, J.L and Bourgoyne, A.T. Jr.: "How to Handle a Gas Kick Moving Up a Shut-in Well," *World Oil* (December 1983) 51–57.
2. Leach, C.P. and Quentin, K.M. "Gas Migration Modeling Improves Volumetric Method of Well Control," *Oil and Gas J.* (26 December 1994) 78–82.
3. Rader, D.W., Bourgoyne, A.T. Jr., and Ward, R.H.: "Factors Affecting Bubble-Rise Velocity of Gas Kicks," *JPT* (May 1975) 571–84.
4. Grace, R.D.: *Advanced Blowout & Well Control*, Gulf Publishing Company, Houston (1994) 178–80.
5. Miska, S., Beck, F.E., and Murugappan, B.S.: "Computer Simulation of the Reverse-Circulation Well-Control Procedure for Gas Kicks," *SPEDE* (December 1992) 247–53.
6. Adams, N.: "What to Remember About Bullheading," *World Oil* (March 1988) 46–48.
7. Bourgoyne, A.T. Jr.: "Bubble Chopping: A New Way to Control Large, Deep Gas Kicks," *World Oil* (December 1984) 75–79.
8. Fleckenstein, W.W. and Mitchell, B.J.: "Removal of a Kick With the Partition Method," paper SPE/IADC 21969 presented at the 1991 SPE/IADC Drilling Conference, Amsterdam, 11–14 March.

SI Metric Conversion Factors

bbl	× 1.589 873	E − 01	= m^3
bbl/ft	× 5.216 119	E − 01	= m/m
bbl/min	× 2.649 788	E + 00	= L/s
ft	× 3.048*	E − 01	= m
°F	(°F − 32)/1.8		= °C
°F/100 ft	× 1.822 689	E + 01	= mK/m
gal	× 3.785 412	E − 03	= m^3
in.	× 2.54*	E + 01	= mm
in.^{2}	× 6.451 6*	E + 02	= mm^2
lbf	× 4.448 222	E + 00	= N
lbm	× 4.535 924	E − 01	= kg
psi	× 6.894 757	E − 03	= MPa
°R	°R/1.8		= K

* Conversion factor is exact.

Chapter 6
Special Applications

6.1 Introduction

This chapter discusses those operations where the approach to well control differs from the fundamentals outlined in Chap. 4. Underbalanced drilling (UBD) fits in this category because the flow of formation fluids into the wellbore is tolerated or encouraged rather than avoided. Also, kick detection and control requirements may not be the same when the hole attitude approaches horizontal or when coiled tubing is the drillstring. Finally, pressure control is addressed for well completion, workover, and servicing operations.

6.2 Underbalanced Drilling

Wells drilled underbalanced are drilled intentionally with an equivalent circulating density (ECD) less than the pore pressure. Darcy's law states that an underbalanced rock will flow its pore fluids into a wellbore at a rate depending on permeability, pressure differential, and other parameters. UBD is not an exceptionally hazardous practice if the flow is manageable at the limiting surface backpressure and the well is equipped to process the produced fluids.

UBD is not a recently discovered method for drilling oil and gas wells. Many of the world's largest oil fields were developed using cable-tool drilling systems during the 50-year period preceding 1930. Tight, overpressured formations have been drilled underbalanced using rotary tools for decades.[1] The subject, however, has received much attention in the recent literature and the technique has been extended far beyond its original, somewhat limited, domain. The impetus for the new-found interest in UBD can be credited to factors such as concerns about formation damage and lost circulation in horizontal wells, better equipment for drilling with backpressure on the annulus, and improved systems for handling and separating the return fluids.

Some degree of formation damage is a given any time an oil or gas reservoir is drilled overbalanced and the effect is most pronounced in high-permeability, pressure-depleted rock. UBD prevents solids plugging and filtrate losses into a formation so long as an underbalanced condition can be maintained. Indeed, the current effort in many areas is to allow a continuous wellbore inflow from the time the bit penetrates the rock until product is sold down the pipeline.

The attempt to avoid skin damage is the primary reason operators consider UBD, but project economics may be improved also by faster penetration rates, longer bit life and fewer trips, eliminating one or more casing strings, reduced risk of lost circulation or differential sticking hazards, lower mud costs, and earlier product sales. Immediate and direct evaluation of formation fluid type and productivity is another incentive to drill underbalanced. The technique has been described as drilling with a continuous drill stem test (DST).

In general, daily operating costs are higher in a UBD operation; the success of a given project depends on a bottom-line economic comparison with offset wells drilled in the conventional manner. Not all wells benefit from the technique when all factors are considered[2] and the potential for well-control problems is too high on others. Nonetheless, trends suggest that UBD is an accepted practice and will be applied to an increasing number of wells.[3] It is incumbent upon the drilling engineer to understand and to assess the associated well-control risks.

6.2.1 Air and Natural Gas Drilling. Those drilling fluids in which gas is compressed at the surface and injected down the drillstring represent an extreme form of UBD. The first well in the petroleum industry to use straight air as a circulating drilling fluid was completed in 1953.[4] Air and natural gas drilling have become routine in many parts of the world where the drilled rock is strong enough to withstand the high collapse stress, formation pressures are normal to subnormal, and permeabilities are low. Generally, the technique is restricted to development wells in those areas where the risks are understood from years of local knowledge and experience. However, excessive flow from a zone with anomalously high permeability or pressure may constitute a hazard and planning should consider well-control contingencies.

Fig. 6.1 shows a diagram of a typical well-control-equipment arrangement on an air or gas drilling project. An essential addition to the stack is a rotating head positioned above the top preventer component. **Fig. 6.2** shows a conventional rotating head using a rotating packing element and bushing assembly

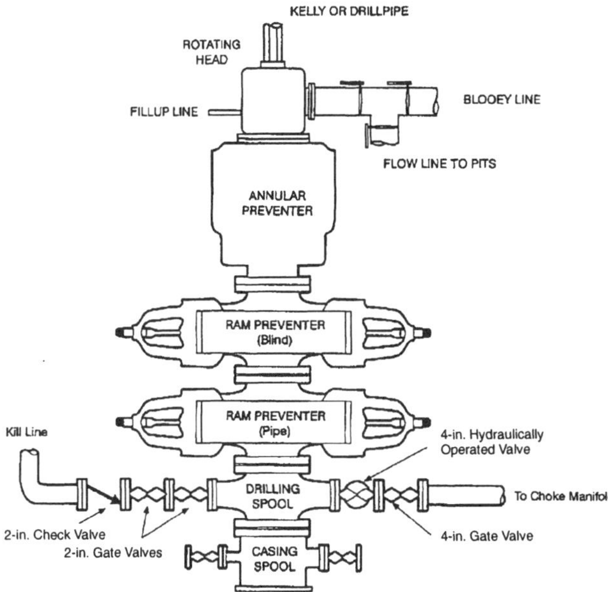

Fig. 6.1—Stack equipment for an air or natural gas drilling fluid.

to maintain a seal between the drillstem (kelly or drillpipe) and annulus. Gas returns and cuttings are diverted down a large-diameter blooey line to a flare pit. The compressors, boosters, and associated equipment may be furnished on a rental basis or come with the rig, but most air drilling hookups have pits and pumps tied into the standpipe to allow mud circulation. Downhole are two drillstring floats; one on the bottom prevents backflow through the bit and one near the top keeps compressed gas in the string when making a connection.

Some fluid flow into the wellbore is allowed or even encouraged on an air-drilling job; what presents a hazardous condition depends on several factors including the judgement of the wellsite supervisor. Flow rates and surface pressures

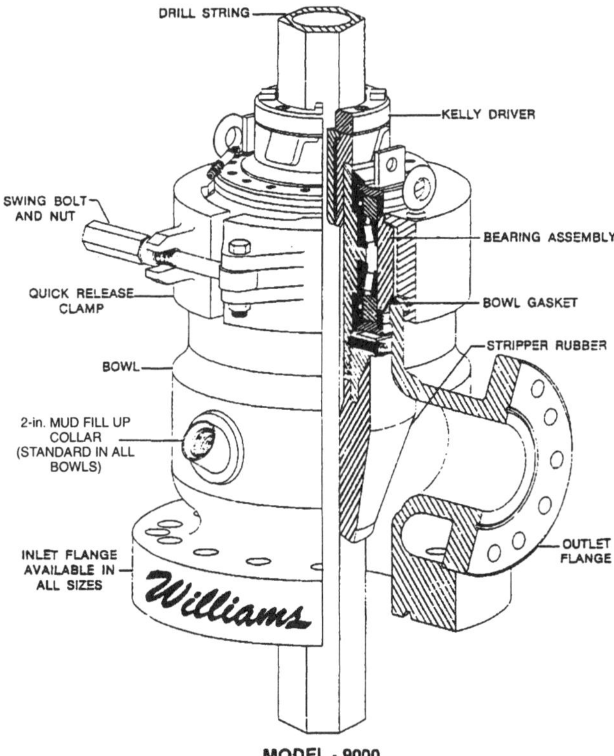

Fig. 6.2—Rotating head used in air drilling and moderate pressure applications. Courtesy of Williams Tool Co. Inc.

that exceed the capabilities of the surface equipment or prevent tripping the pipe will require closing the well in and putting mud in the hole. A prudent measure for any air drilling job is to store enough water and mud additives (or liquid mud) on location to fill the hole.

The required kill-mud density can be only approximated from the shut-in pressure readings. Unlike mud, wellbore fluids are highly compressible and the variable fluid density must be considered. Furthermore, the drillstring floats and lack of compressor throttle control precludes using the drillpipe gauge for reading bottomhole pressure (BHP). As shown in **Fig. 6.3,** the casing gauge must be used and the operator must guess at the nature of the annular fluids. The majority of wells drilled with a total or partial gas medium (including foam and aerated muds) are in a normal or subnormal pore-pressure environment and can be killed using low-density mud.

Eq. 6.1 gives the estimated pore pressure based on the shut-in casing pressure p_{cs}.

$$p_p = p_{cs}\exp\left(\frac{\gamma_g D}{53.3zT}\right). \qquad (6.1)$$

The relation assumes a consistent gas column with no liquids to the depth of interest, a condition which will not be met if

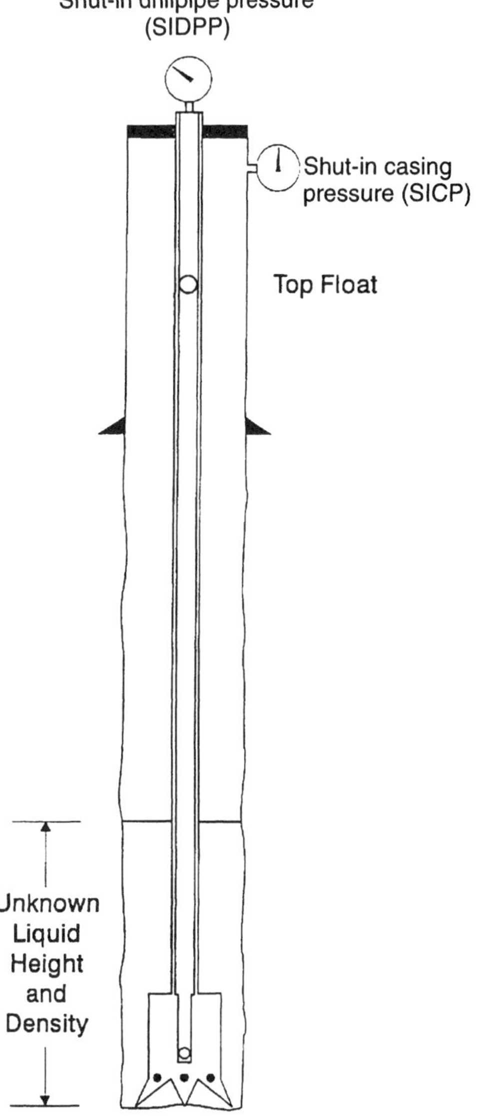

Fig. 6.3—Shut-in conditions on an air-drilled hole.

oil or water comprises part of the influx. Even so, accuracy is not of paramount concern at this point.

Classic well-control procedures do not apply to the situation where the drillstring is filled with gas and a modification of the Wait and Weight method is recommended. BHPs will be maintained as constant as the conditions allow if constant choke pressure is maintained until the drillstring is filled with mud. The choke operator's attention then turns to the standpipe gauge and the observed circulating drillpipe pressure (CDPP) is held constant until all the gas has been cleared from the well.

Example 6.1. Unexpected permeability is encountered in a 7,200-ft sandstone while drilling a well in the Arkoma basin with air. The well is shut-in; after wellbore temperatures reach equilibrium, the casing pressure gauge reads 1,250 psig. Assume the gas properties in the field are known and that the gas specific gravity (SG) = 0.69 (ignore the effect of the air mixed in with the hydrocarbons).

1. Estimate the kill-mud density requirement. Assume the average temperature in the annulus is 160°F.

2. Determine the maximum pressure at the 1,500-ft shoe depth before and during the kill procedure. Use the same average temperature as above.

Solution. 1. Iterate Eq. 6.1 by first assuming the z factor is unity.

$$p_p = (1,250 + 14)\exp\left[\frac{(0.69)(7,200)}{(53.3)(1.00)(620)}\right]$$

$$= 1,469 \text{ psia.}$$

Eq. 1.20 and 1.21 are applied to give $p_{pc} = 665$ psia and $T_{pc} = 375°\text{R}$. Take the pseudoreduced properties based on wellbore averages and determine the average z factor.

$$p_{pr} = (1,264 + 1,469)/(2)(665) = 2.05,$$

$$T_{pr} = 620/375 = 1.65, \text{ and}$$

$$z = 0.875.$$

Substitute this z factor into Eq. 6.1 and, after one more iteration, obtain $p_p = 1{,}501$ psia. The density equivalent at 7,200 ft is calculated as

$$\rho_{km} = (19.25)(1,501 - 14)/7,200 = 3.98 \text{ lbm/gal.}$$

The result is invalid if a substantial liquid column is above the kick zone, but a low-solids mud will be satisfactory.

2. The maximum shoe pressure will be the shut-in pressure if the casing side is held constant until the string is filled with mud. The preceding calculations are repeated for the shoe depth.

$$p_{sh} = (1,264)\exp\left[\frac{(0.69)(1,500)}{(53.3)(1.00)(620)}\right] = 1,304 \text{ psia.}$$

Also,

$$p_{pr} = (1,264 + 1,304)/(2)(665) = 1.93,$$

$$T_{pr} = 1.64, \text{ and}$$

$$z = 0.882.$$

Ultimately we obtain $p_{sh} = 1{,}296$ psig, which gives a pressure gradient of 0.864 psi/ft. The fracture integrity is higher or control would have been lost when the preventer was closed.

Shallow shoe depths present a problem with regard to how much casing pressure is acceptable. The recommended approach if shutting in the well poses undue risk is to vent gas through the choke manifold while holding a backpressure less than the maximum allowable annulus surface pressure (MAASP) and to maintain this casing pressure until mud exits the bit.

6.2.2 Mist and Foam Drilling. Air drilling success depends on keeping the hole dry. Small water flows can mix with the drill cuttings and cause mud rings to form on the drillpipe. Minor water seepages can be managed by converting to a mist system. The objective is to add enough water downstream of the compressors so that the injected air is nearly saturated with water vapor. A surfactant is injected also to assist in breaking up and preventing the rings from forming. Well-control procedures for a mist-drilling job should not be any different than those discussed for straight air.

Foam drilling fluids can tolerate higher formation-water loads and, compared to straight air or mist, offer other advantages such as reduced hole/pipe erosion and larger, more competent drill cuttings. Stiff foams are generated by shearing water and gas (air or nitrogen) together with a foaming surfactant and adding bentonite and/or polymers to improve cuttings transport. Correlations for achieving the desired foam properties are available and the recommended injection rates depend primarily on penetration rate, depth, hole size, and rotating head backpressure.[5,6] Typical air and water injection rates for a stiff foam range between 100 and 250 scf/min and 10 to 20 gal/min.[7]

Foams present well-control limitations similar to other pneumatic fluids. Gas density and foam quality (i.e., gas volume fraction) in the annulus change with pressure and a shut-in formation-pressure estimate could be obtained by starting at the casing gauge and numerically integrating foam density with depth. But this exercise is not necessary and would be worthwhile only if the appropriate software was available.

Foam drilling is rarely applied to abnormally-pressured sediments. Displacing the hole in a well-control procedure is best accomplished by using the mud on hand and by following the Wait and Weight modification discussed in Sec. 6.2.1. Any secondary kicks should be minor and an insufficient mud density will be discovered after the foam and kick fluids have been removed from the well. The second kill then can be more exacting in approach.

6.2.3 Underbalanced Drilling With Mud. Introducing a gas phase into the mud is another way to reduce equivalent mud weight in the annulus to less than the rock pressure. This approach to UBD has been known for a long time and remains one of the more popular methods. Air has been the most common choice because it is abundant and cheap, but nitrogen offers several advantages, including lower corrosion rates. The simplest and least expensive technique is to inject gas at the standpipe. However, obtaining reliable data from mud-pulse-telemetry measurement while drilling (MWD) tools is problematic when compressible fluids are in the drillstring. Electromagnetic telemetry or steering-tool data transmission may be options, but systems have been devised to place gas in the annulus without the need for standpipe injection. **Figs. 6.4a and 6.4b** illustrate parasite and concentric injection strings used by many operators.

In 1955, Poettmann and Begman[8] presented the chart shown in **Fig. 6.5** for estimating how much air is needed to reduce a base mud to the desired effective mud weight. Say that

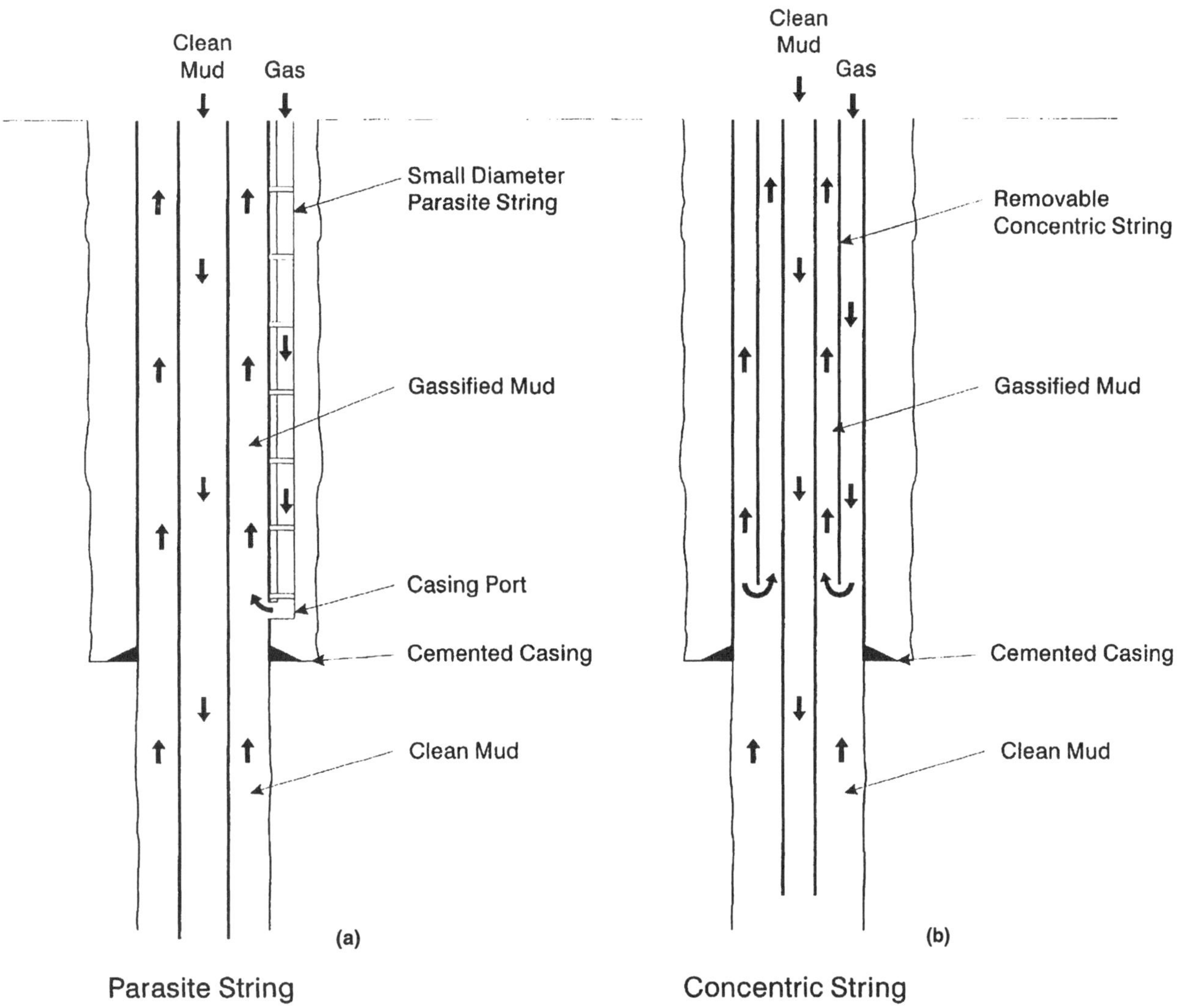

Clean Mud

Gas

Small Diameter Parasite String

Gassified Mud

Casing Port

Cemented Casing

Clean Mud

(a)

Parasite String

Clean Mud

Gas

Removable Concentric String

Gassified Mud

Cemented Casing

Clean Mud

(b)

Concentric String

Fig. 6.4—Injecting gas into the annulus using parasite and concentric injection strings.

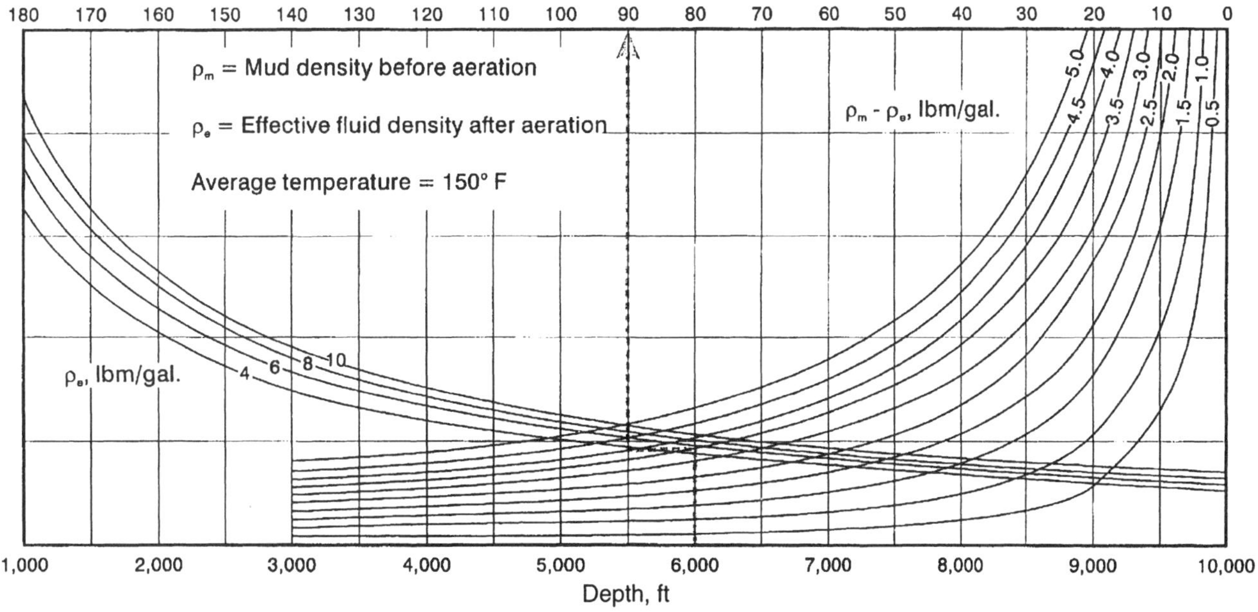

Fig. 6.5—Volumetric air requirements for reducing downhole effective mud density (after Poettmann and Begman[8]).

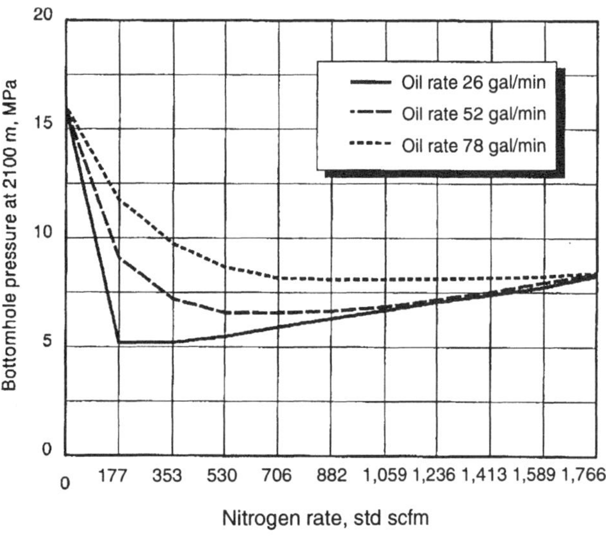

Fig. 6.6—Computer simulation showing the effect of increasing nitrogen rate on circulating bottomhole pressure.[9]

a drilling engineer is planning a 6,000-ft vertical well and intends to reduce a 9.0-lbm/gal mud to a 5.0-lbm/gal effective density at total depth. The chart indicates a target air/mud ratio of 90 scf/bbl. The air-injection rate for a 7.0-bbl/min pump rate is thus 630 scf/min.

Fig. 6.5 has several drawbacks that limit its use when reasonable ECD estimates are required. For instance, the correlation assumes the air at the top of the annulus is at atmospheric pressure. There may be substantial backpressure beneath the rotating head and so the chart loses some utility in UBD planning. Another limitation is that the simple model does not anticipate the effects of gas concentration and injection rates on two-phase friction losses.

Fig. 6.6 illustrates what actually happens to the BHP when gas injection is increased into a mud stream. The curves show the simulated BHP in a 6,900-ft horizontal well for an oil drilling fluid at three pump rates. Note that the model ignores inertial pressures resulting from near-surface gas expansion and assumes no gas slippage. Starting on the left, hydrostatic pressure reduction dominates as nitrogen rates increase. The curves start to flatten out as the increasing flow rate starts to increase the friction component to the ECD and a minimum BHP is achieved at a specific gas-injection rate. More gas increases BHP in the region dominated by friction.

Fine tuning a UBD program with a gassified mud requires BHP simulation in the planning and during execution. Multiphase flow prediction is a complex problem if all wellbore attitudes and possible flow patterns are considered.[10] However, the simplified computer models developed specifically for gassified muds should be accurate enough for most UBD applications.[11]

MWD tools are available for measuring annulus pressures above the bit and have been used in some UBD projects. Obviously, this capability could eliminate much of the guesswork and allow an operator to adjust mud- and gas-injection rates based on real-time measurements. It is also apparent that a UBD well-control problem would be simplified greatly if actual circulating bottomhole pressures were monitored. An operator could focus on the MWD measurements instead of the surface readings and adjust choke backpressure, injection rates, or mud densities as needed to maintain the desired ECD.

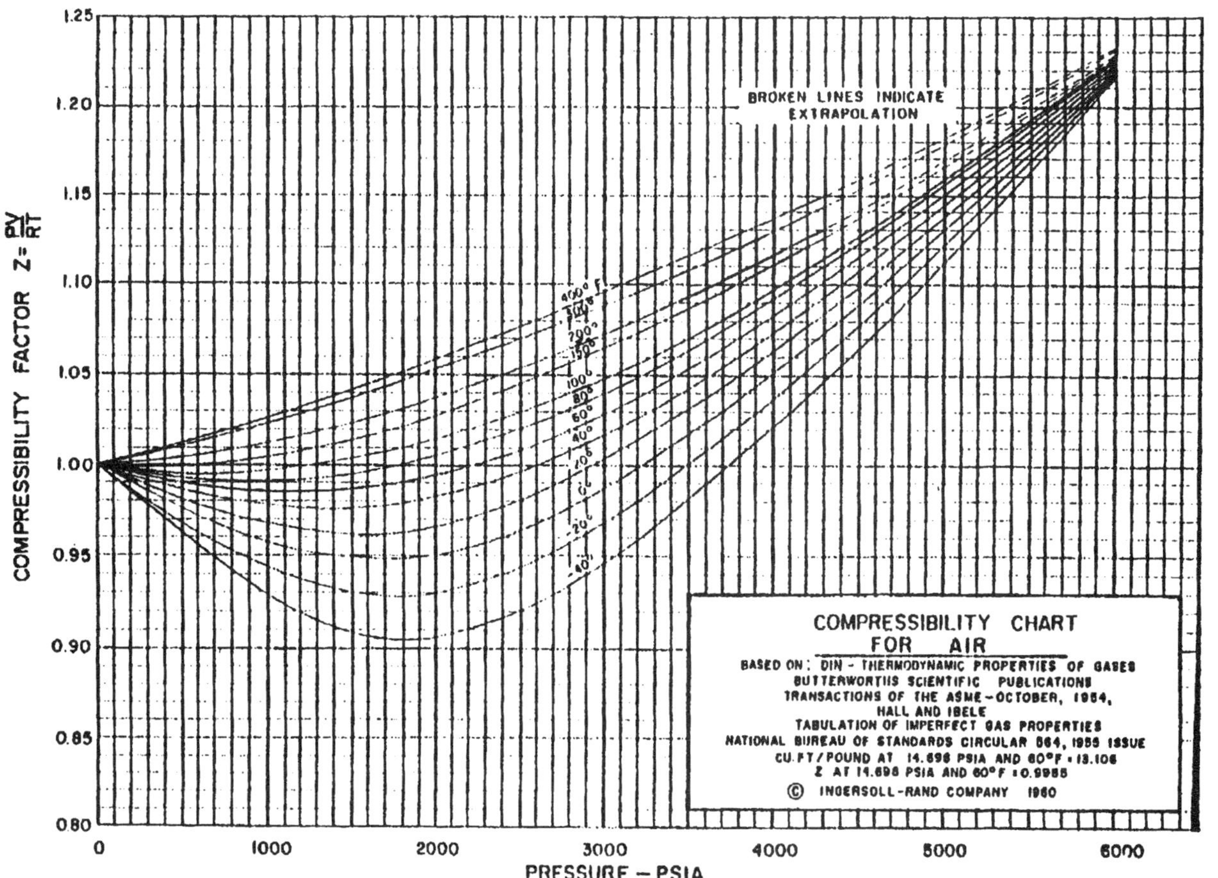

Fig. 6.7—Compressibility factors for air.[12]

TVD (ft)	p (psia)	z	V_g (gal)	f_g	ρ_g (lbm/gal)	ρ_{mg} (lbm/gal)
0	814	1.002	238	0.488	0.48	4.69
100	838	1.002	231	0.481	0.50	4.76
200	863	1.003	225	0.474	0.51	4.82
300	888	1.003	219	0.466	0.52	4.89
400	914	1.004	213	0.460	0.54	4.95
500	939	1.004	207	0.453	0.55	5.01
1,000	1,069	1.005	182	0.421	0.63	5.30
1,500	1,207	1.009	162	0.393	0.71	5.56
2,000	1,352	1.011	145	0.367	0.79	5.80
2,500	1,502	1.013	131	0.343	0.88	6.02
3,000	1,658	1.018	119	0.322	0.96	6.21
3,500	1,820	1.022	109	0.303	1.05	6.38
4,000	1,986	1.027	100	0.286	1.14	6.54
4,500	2,155	1.032	93	0.270	1.24	6.68
5,000	2,329	1.039	86	0.257	1.33	6.81
5,500	2,506	1.044	81	0.244	1.42	6.93
6,000	2,686	1.052	76	0.233	1.51	7.03
6,500	2,863	1.059	71	0.222	1.60	7.12
7,000	3,053	1.066	68	0.213	1.70	7.21
7,500	3,240	1.075	64	0.204	1.78	7.29
8,000	3,430	1.083	61	0.196	1.87	7.36
8.500	3,621	1.092	58	0.189	1.96	7.43
9,000	3,814	1.102	56	0.183	2.05	7.48
9,500	4,008					

TABLE 6.1—CALCULATED PRESSURES WITH DEPTH FOR THE SHUT-IN WELL DESCRIBED IN EXAMPLE 6.2

(By logical extension, it seems that the classical procedures discussed in Chap. 4 would become archaic technology if every well was equipped with one of these tools.)

Secondary well control using surface measurements presents some difficulties if gassified mud is in the drillstring. Even so, we can obtain a reasonable SIDPP by shutting down the compressors and slowly pumping clean mud into the drillstring until the floats open. Example 6.2 demonstrates that the static BHP can then be estimated by considering the variable density and volume fraction of the gas phase.

Example 6.2. A well has been drilled to 9,500 ft with aerated mud and the decision is made to shut-in and pump a kill. Air rate at the standpipe was 1,500 scf/min and an 8.7-lbm/gal mud was being pumped at 250 gal/min. Estimate the kill-mud density if the SIDPP is 800 psig. Assume the average temperature in the drillstring is 150°F and use the compressibility factors given in **Fig. 6.7.**

Solution. We choose to solve this problem by numerically integrating the mixture densities and pressures with depth. Eq. 1.22 gives the air density underneath the drillpipe gauge as

$$\rho_g = \frac{(1.0)(800 + 14)}{(2.77)(1.002)(150 + 460)} = 0.48 \text{ lbm/gal.}$$

The gas-volume fraction can be obtained from the respective mud- and air-injection rates. The gas law yields the surface gas volume over a one-minute period,

$$V_g = (1,500)(1.002)(610)(14.7)/(814)(520)$$
$$= 31.8 \text{ cu ft} = 238 \text{ gal.}$$

Hence the gas volume fraction is

$$f_g = 238/(238 + 250) = 0.488.$$

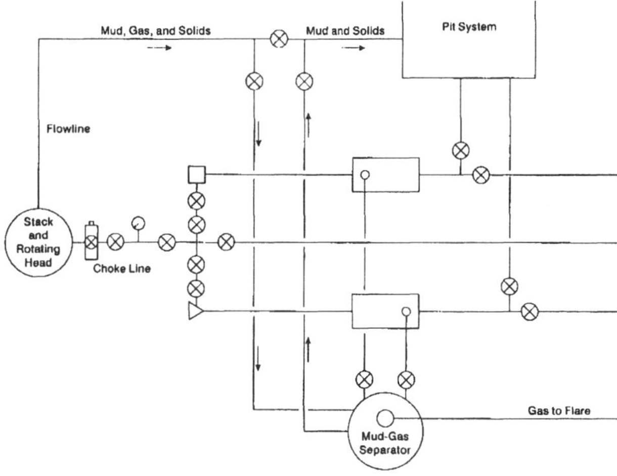

Fig. 6.8—Equipment traditionally used in low-pressure/low-risk UBD applications.

Eq. 2.26 yields a mixture density of

$$\rho_{mg} = (8.7)(1 - 0.488) + (0.48)(0.488) = 4.69 \text{ lbm/gal.}$$

Starting with a small depth increment, the pressure at 100 ft is calculated

$$p_{100} = 814 + (0.0519)(4.69)(100) = 838 \text{ psia.}$$

Repeating the procedure several times yields a bottomhole pressure of 4,008 psia (see **Table 6.1**) and the kill-mud density,

$$\rho_{km} = (4,008 - 14)/9,500 = 0.420 \text{ psi/ft}$$
$$= 8.09 \text{ lbm/gal.}$$

The numerical integration technique is best done with a computer program.

Most UBD applications discussed thus far have been used in subnormal or normal pore pressures; calculating a shut-in bottomhole pressure may be an academic exercise if the intent simply is to replace the gassified fluid with clean mud. Another class of UBD projects involves drilling with weighted or unweighted mud in abnormally-pressured rock. This has been a common practice for years in the Delaware and Anadarko basins and more recently has been adopted for the Austin Chalk and other formations.

Surface equipment and arrangements for drilling underbalanced can range from the simple to the complex. **Fig. 6.8** depicts the equipment typically used when backpressures are low and gas is the only produced fluid. Conventional rotating heads capable of holding well pressures up to 500 psi would be specified for this type of application. Depending on the backpressure, returns are taken either down the flowline or through the choke manifold and routed through the mud/gas separator (MGS) where the primary separation takes place. Gas is vented out the top of the MGS while mud and drilled solids are run back to the mud pits. Some entrained gas breaks out at the shale shaker and an atmospheric or vaccum-type degasser is used to completely degas the mud.

Fig. 6.9 illustrates a flow-drilling system used for higher backpressures with oil production.[9] The rotating blowout preventer (RBOP) is a recent well-control innovation devised in response to the increasing popularity of UBD in horizontal wells, especially in the Austin Chalk formation.[13] This tool is similar to an annular preventer in that the driller uses hydraulic pressure to open and close a bag-type seal element on the rotat-

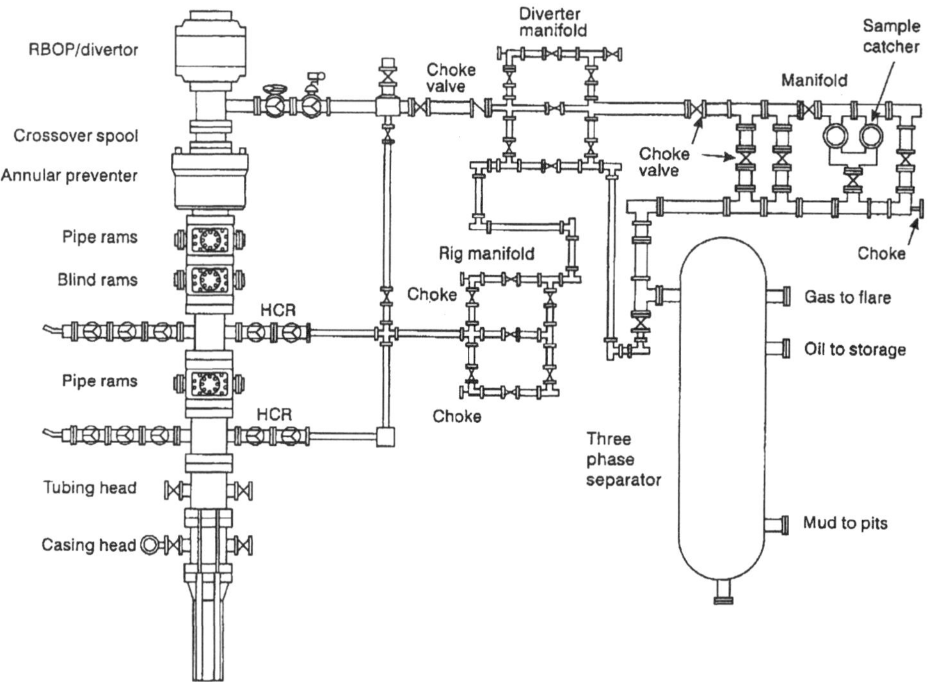

Fig. 6.9—Equipment used for UBD operations on a high-pressure well with oil production.[9]

ing or sliding drillstem (**Fig. 6.10**). An RBOP is reliable to approximately 1,500 psi and therefore can be used at higher backpressures than conventional rotating heads. High-pressure rotating heads using two seal rubbers have been developed and can achieve comparable performance in the field.

Tripping can be a problem, depending on the zone deliverability and the primary reason for drilling underbalanced in the first place. In tight formations, wellbore inflow may be small enough to allow tripping out and back to bottom without doing anything to change the BHP. Prudent trip practices and using a trip tank to account for pipe displacement gain importance on these wells. The operator should expect a large trip gas volume and plan bottoms-up circulation accordingly.

A trip with the mud density used in the drilling mode may be impossible if deliverability is too high. A high-density pill may be successful in killing the zone or at least arresting the flow in these cases. Another possibility is to displace the hole completely with kill-weight mud, but this requires maintaining two different mud systems—one for drilling and one for tripping.[14]

The last two options can be used on wells where formation damage is not an issue (UBD in nonpay formations) or is accepted (low-permeability rock that must be stimulated). Otherwise, other methods must be implemented. Bourgoyne[15] discussed a trip procedure where formation damage is addressed by using a clean, mimimum-solids drilling fluid. **Fig. 6.11** illustrates the sequence for killing an underbalanced well without placing weighted mud opposite the pay zone. The ultimate objective is to top-kill the well during the trip so that the formation is overbalanced when the string is out of the hole. First, the drillpipe is stripped up through the rotating head or RBOP to some point in the casing. Thereafter, weighted mud is used to replace the plugged string displacement volume as stripping continues. Casing pressure will decline as the weighted-mud height increases, eventually reaching zero.

Keeping a hole underbalanced while tripping is achievable. Basically, the well is never killed and a trip is conducted while

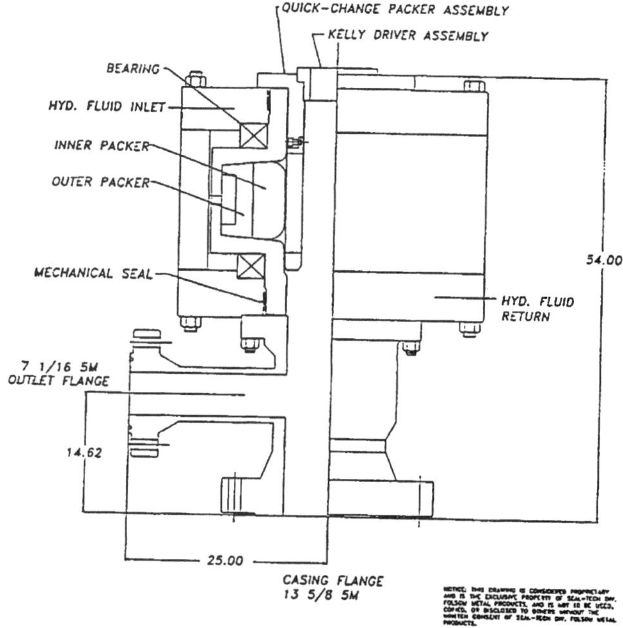

Fig. 6.10—Rotating blowout preventer. Courtesy of Seal-Tech.

the well flows out the annulus or is shut-in with surface pressure. It may be possible to trip conventionally if flow can be maintained at casing pressures low enough to keep the drillstring from ejecting. Otherwise, a snubbing unit must be used.[16] Another alternative is to drill with coiled tubing (CT) if the hole conditions are suitable.[17,18] The ability to trip with a live well is one of the major advantages of coiled-tubing drilling (CTD).

6.3 Unconventional Wellbores and Drilling Practices

Classical well-control principles are based on near-vertical wells having "conventional" hole sizes and drillstrings. Accordingly, the fundamentals invite scrutiny whenever well

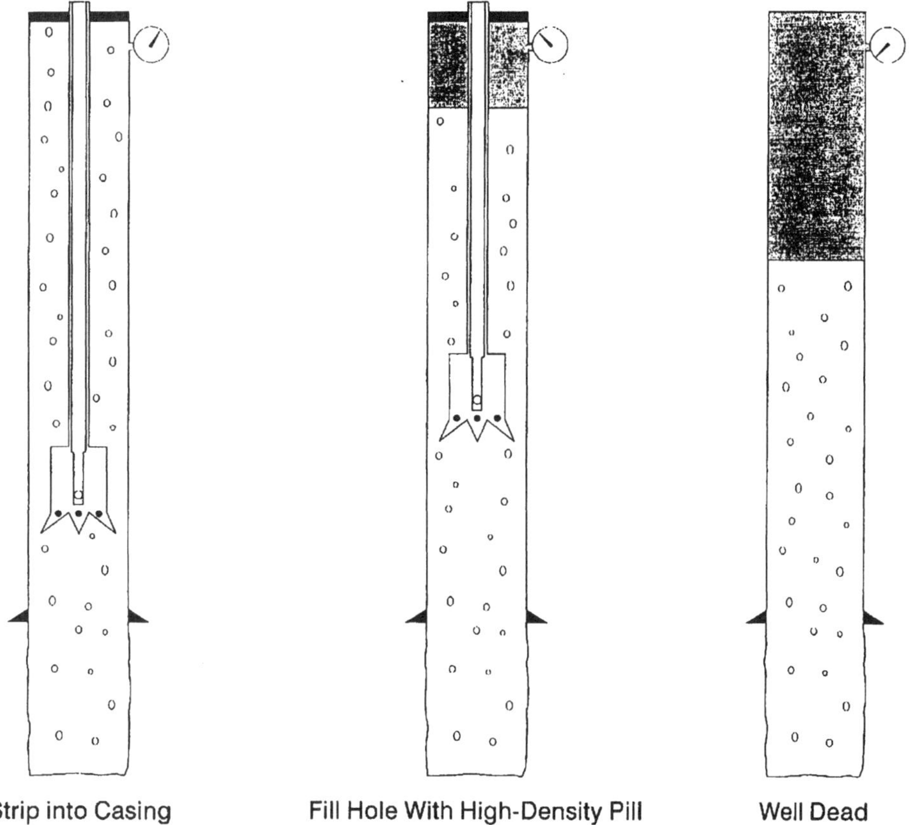

Fig. 6.11—Using high-density mud to top-kill a sensitive formation during a trip.

characteristics do not satisfy the underlying assumptions; modifications may be needed for well types that have gained recent prominance. This section focuses on unique well-control considerations and procedures for horizontal/extended-reach-drilled (ERD) wells and CT operations.

6.3.1 Horizontal and ERD Wells. The recent popularity of UBD has been driven in large part by the increasing number of horizontal wells and the attempts to avoid or minimize formation damage. Formations penetrated horizontally are particularly prone to damage, and adequately stimulating a long lateral section to remove drilling-induced damage is a major problem for completion designers.

Long exposure time to the mud is the most important damage-causing mechanism. Say that a 50-ft pay zone and 100 ft of rathole in a vertical well can be drilled with one bit at 50 ft/hr. The top of the pay zone is open to the mud for only three hours while drilling and maybe another day or two before casing is set and cemented. Consider the same penetration rate and mud conditions in an offset horizontal well drilled with a 3,000-ft lateral. We now have a total 60 hours rotating time in the horizontal, the potential for more trips, and additional time in logging, setting a liner, and so on. Thus more skin damage is expected, particularly in those productive intervals near the entry point, and log evaluation becomes more difficult because of the increased invasion-zone radius.

Fig. 6.12 illustrates the well paths of a vertical well and ERD well to the same objective. Consider how the two well paths relate with regard to the potential for rock failure. From Chap. 3, we normally expect fracture pressure to be reduced at higher hole angles, which means that the vertical well probably can withstand more wellbore pressure than the ERD well path. However, hole-collapse tendencies increase at higher hole angles and so the lower limit to the allowable wellbore pressure will be generally higher for the ERD program. Hence rock mechanics places the highly deviated hole at a disadvantage and an operator must consider the impact when planning and drilling such a well.

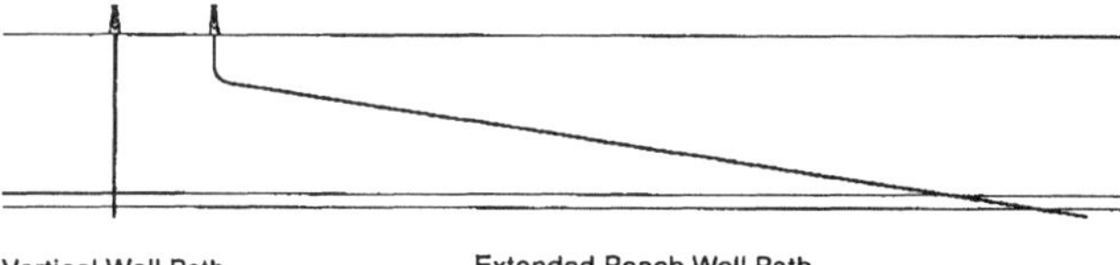

Fig. 6.12—Comparison between a vertical and extended-reach well.

The restricted mud-weight envelope is not the only problem. Every point on the ERD well from the build section through total depth has more drilled length than the vertical well and, all else being equal, the annular friction component losses will be higher. Thus we expect less margin for the fracture integrity and more filtrate invasion into the rock. Example 6.3 shows how in some cases the ECD could control how much lateral distance is achievable in a horizontal well.

Example 6.3. A 10,000-ft true vertical depth (TVD) horizontal well is constructed with a 9,642-ft kickoff point, a 563-ft arc length, and 3,500 ft of lateral section. Calculate the overbalance pressure at both the entry point and terminus if the mud weight is 9.4 lbm/gal, the pore pressure is a 9.0-lbm/gal equivalent, and annulus friction losses are 0.02 psi/ft.

Solution. The circulating BHP at the heel is determined as

$$p_{bh} = (9.4)(0.0519)(10,000) + (0.02)(9,642 + 563).$$

$$p_{bh} = 4,879 + 204 = 5,083 \text{ psi}.$$

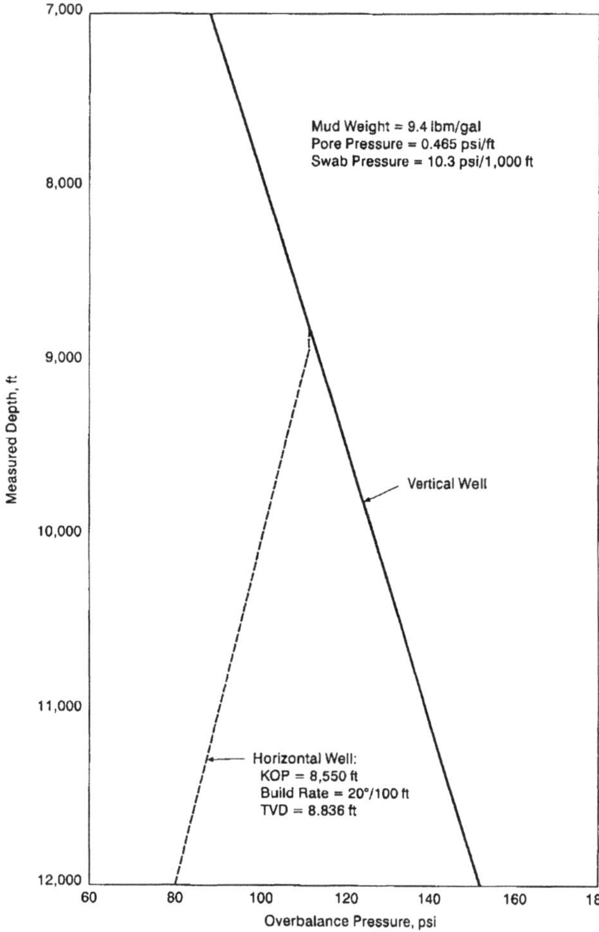

Fig. 6.13—The effect of swab pressure on overbalance pressure in a horizontal well. KOP = kickoff point.

This gives an overbalance of

$$5,083 - (9.0)(0.0519)(10,000).$$

$$5,083 - 4,671 = 412 \text{ psi}.$$

The pressure at total depth is

$$p_{bh} = 4,879 + (0.02)(9,642 + 563 + 3,500).$$

$$p_{bh} = 4,879 + 274 = 5,153 \text{ psi}.$$

The overbalance pressure increases by 70 psi across the horizontal section. Alternatively, the ECD opposite the pay increases from 9.8 to 9.9 lbm/gal.

Preventing or removing the cutting beds that deposit on the low side of a relatively flat wellbore is of paramount importance to ERD and horizontal project success and the measures taken to clean these holes impact the ECDs. Studies and field results[19-26] have shown that a high annular velocity is one of the most important variables in keeping these high-angle (inclinations greater than 60°) holes clean. High pump rates lead to higher friction losses in the annulus even though thin muds are usually used.

Increasing wellbore length leads to greater surge and swab pressures in these well types. Top-drive drilling systems offer the ability to rotate the pipe and circulate while tripping and have been essential to the success of many ERD and horizontal projects. The downside to lowering the drillstring if the pump is on is that surge pressures will be magnified by an amount equal to the circulating friction above the bit. An operator should evaluate any adverse effects on hole integrity.

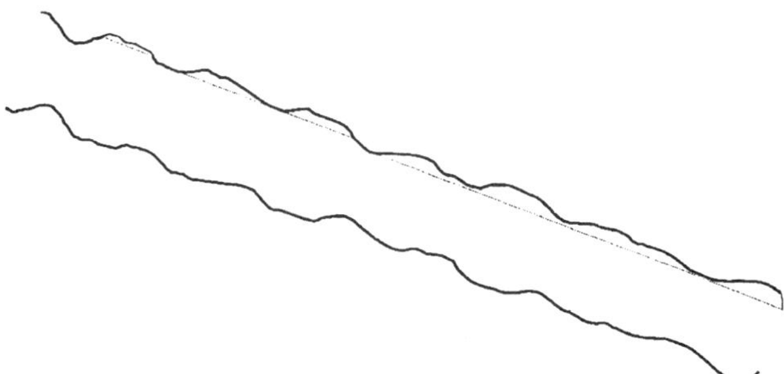

Fig. 6.14—Gas migration in a highly inclined and rugose borehole.

Swabbing is a greater concern in highly deviated wells because the pore pressure and mud-column hydrostatic depend on the TVD whereas steady-state swab pressures are a function of drillstring length. For a vertical well, the minimum overbalance in a swabbing condition should increase with depth if the hoist rate and mud properties remain constant. This is not necessarily true in highly deviated holes and, as shown in **Fig. 6.13,** the minimum overbalance actually decreases with drilled depth when the wellbore is horizontal. Backreaming off bottom helps to remove cutting beds and reduce drag, but another benefit realized from the practice is that the circulating friction serves to offset the swab effects.

ERD and horizontal holes are more prone to kick and lost-circulation problems, but their unique geometry creates some decided advantages in secondary control (i.e., after the well takes a kick). For instance, gas migration rates are usually much lower. Recall from Chap. 1 that gas slip velocities reach a maximum at approximately 45° and thereafter diminish with increasing hole inclination. The velocity rapidly drops to zero when the wellbore achieves a horizontal attitude because there is no uphill direction for the gas to move.

Migration experiments conducted in smooth pipes found significant migration rates at inclinations up to 80°.[27,28] Even so, the hole inclination that effectively stops gas migration in a drilled borehole is less than 90° and may be as low as 70°. **Fig. 6.14** depicts a wellbore inclined at 70°. Hole rugosity has been exaggerated for illustration purposes. Free gas will ride alongside the top of the hole and accumulate in "pockets" until its original volume is depleted. Two well-control advantages are apparent: gas migration ultimately stops and the phenomenon tends to string out the free gas over a considerable distance.

Maximum shoe pressures during a well-control procedure usually will be smaller for a horizontal well than for a comparable vertical well and choke pressures are lower for a longer period. **Fig. 6.15** depicts a horizontal well after shutting in on a gas kick. The volume represented by the kick occupies a length designated as L_k, but not any height. Thus we have the same hydrostatic pressure on both sides of the U-tube (if mud densities are equivalent) and the SICP should be approximately the same as the SIDPP. Furthermore, the pressure at the casing seat is lower than what would be seen in a vertical well at the same vertical depth.

Santos[29] discussed well control in horizontal wells and wrote a program to simulate wellbore pressures during a kill procedure. His model is more realistic than others because it considers two-phase flow (based on the Beggs-Brill[30] correlation) and inertial forces. Casing and shoe pressures during a Driller's Method kill were predicted for a shallow horizon-

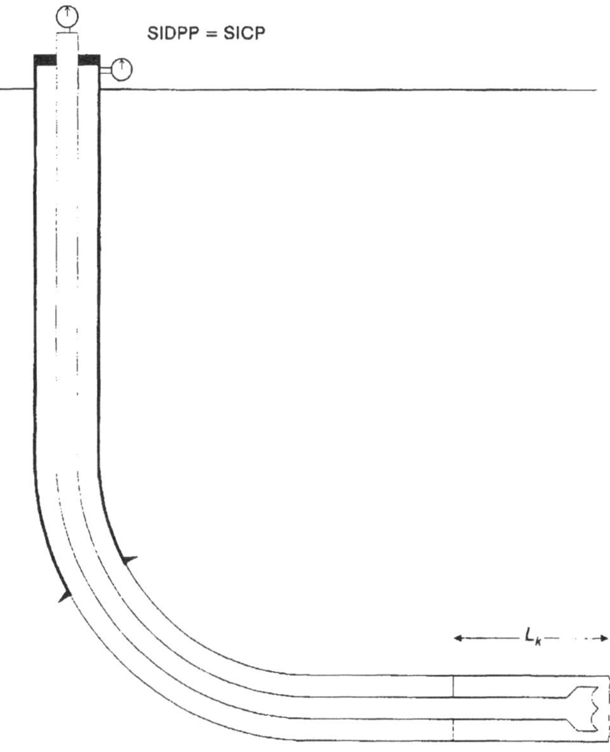

Fig. 6.15—Surface pressure relationship for a horizontal well closed in on a kick.

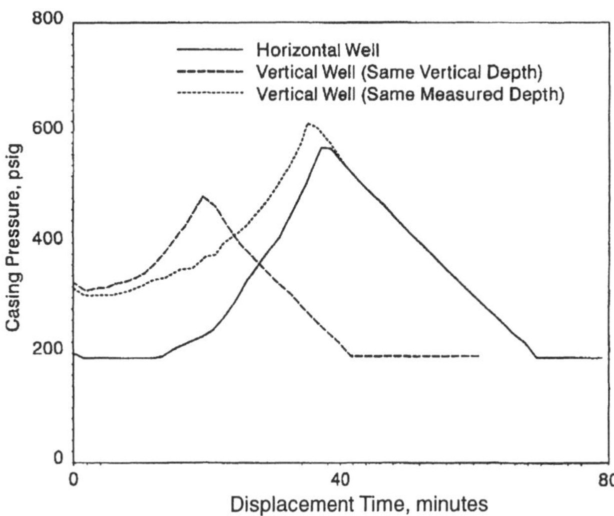

Fig. 6.16—Simulated choke pressures for the Driller's Method kill on a horizontal well and two vertical wells at the same depth (after Santos[29]).

tal well at respective measured and vertical depths of 5,400 and 3,246 ft. The same parameters were used to predict the pressures in vertical wells having the same vertical and measured depth.

Compare the three choke-pressure curves shown in **Fig. 6.16.** Our primary interest lies in the first ten minutes of the displacement. For the horizontal well, the lower density in the kick zone does not affect casing pressure until the gas starts gaining height and expanding in the build section. In comparison, the early-time pressures are higher in the two vertical wells because the kick zones have an initial height and gas expansion beginning immediately after the procedure is instituted.

The height and hydrostatic pressure of a specified weighted-mud volume in the drillstring are reduced at higher hole angles. Also, small-diameter heavy-wall drillpipe (HWDP) (or combined HWDP and drill collars) is normally used on horizontal wells from total depth into the vertical section. Accordingly, the unit mud volume's hydrostatic pressure will increase upon leaving the conventional drillpipe. For these reasons, we cannot keep BHP constant when filling the string with kill-weight mud by stepping the CDPP down in a linear fashion.

The initial circulating pressure and final circulating pressure are computed in the same manner as discussed in Chap. 4, but the pressures between the two end-points must account for the changes in hole inclination and string geometry. An operator could calculate the target CDPP at the string changeover points and at the measured and vertical depths corresponding to the end of the build section(s), tangent (if applicable) and lateral section. For medium- and short-radii designs, a series of straight lines connecting the calculated CDPPs will give acceptable results. Alternatively, a computer-generated schedule like the one shown in **Fig. 6.17** could be implemented.

Our general recommendation, however, is to kill a horizontal or ERD well using the Driller's Method and to control off the casing gauge until the pipe is displaced with kill-weight mud. Generally, casing is set deep enough that the maximum shoe pressure should not dictate the kill method. The operator may allow a small volume of gas into the well if a secondary kick is moving up the hole. The risks are minor compared to the potential problems that accompany a needlessly complicated well-control procedure.

6.3.2 CT Operations. This section applies to CT operations in general and not specifically to its use in drilling. CT is continuous, nonjointed pipe stored on a reel and transported to a wellsite to perform a specific operation. Downhole tools can be attached to the bottom of the string and specialized equipment is used to unspool and "inject" the tubing into a well. After the work is completed, the tubing is retrieved and spooled back onto the reel.

CT has been associated with through-tubing work in completions and workovers. Recently, the technology has seen widespread applications to drilling but, as with many other "new" methods, CTD is not entirely a recent concept.

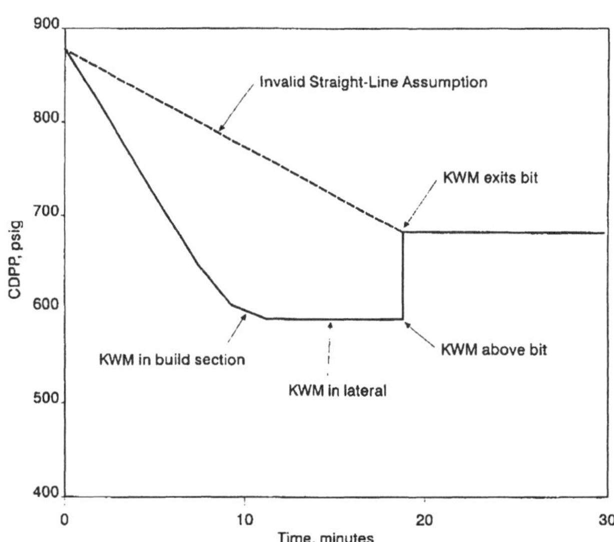

Fig. 6.17—Computer-generated CDPP schedule for filling the drillstring with kill-weight mud on a horizontal well (after Santos[29]).

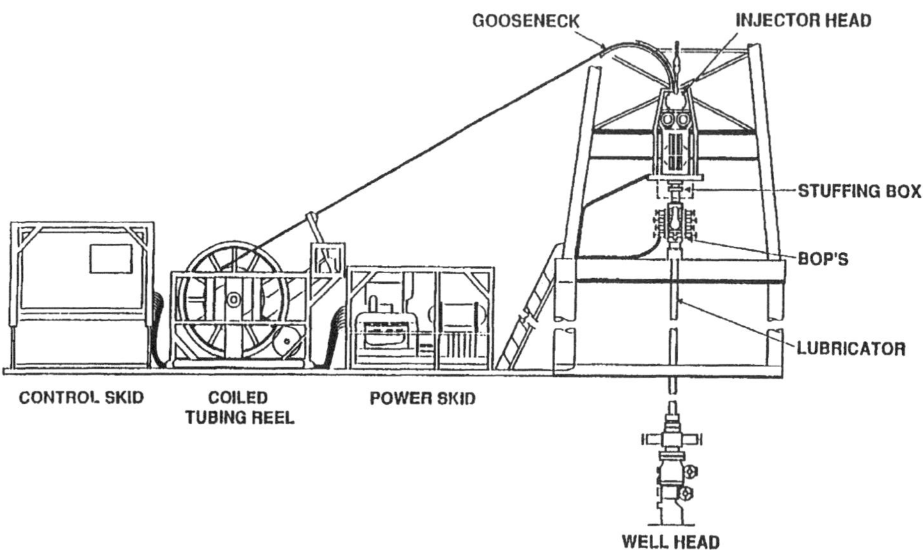

Fig. 6.18—Coiled-tubing equipment. Courtesy of BJ Services Co.

Continuous drillstring prototypes were developed in the 1960s and CT units were drilling shallow holes in Alberta during the late 1970s.[31-33] The technique lay dormant for more than a decade and did not reappear until 1991.[34]

Reawakened interest has been sparked primarily by its underbalance capabilities. CT's use as an effective drilling tool has been made possible by such advances as better CT manufacturing and weld technology, larger pipe diameters, better fixed-head bit designs, slimhole mud motors, slimhole MWD tools, and ways to orient the toolface without rotating the drillstring. The advantages, current limitations, and feasibility considerations for CTD have been cited extensively and we will not dwell on these here.[34-38] Suffice it to say that most of the CTD wells to date have been drilled as horizontal or deepening re-entries, in relatively soft rock, and with bit diameters smaller than 6¼ in.

Fig. 6.18 shows the basic components used to perform a CT operation. Reel diameters range from 90 to 180 in. and the maximum stored length depends on the pipe outside diameter (OD). For example, a 154-in. reel having a 92-in. core and a width of 65 in. can hold as much as 27,100 ft of 1½-in. tubing whereas the same reel can hold only 6,700 ft of 2⅞-in. tubing. Circulation is accomplished by directing fluid into a rotating connection mounted on the shaft and from there to the CT. Thus circulating friction in the string is essentially constant for a given pump rate and fluid rheology and does not depend on where the pipe is located in a well.

At the injector head, grooved blocks machined to fit the tubing are mounted on a contra-rotating chain-drive mechanism. Hydraulic pressure forces the blocks against the CT and the friction grip provides the thrust necessary to force the pipe into a pressurized well, controls the rate of descent when gravity overcomes the upward pressure-area forces, suspends the CT when movement stops, and retrieves the CT from the well. A tubing-guide assembly at the top of the injector head receives the tubing from the reel, bends it across the gooseneck, and guides the pipe into the gripping chains. A load cell at the bottom of the injector head measures the string compression or tension.

Diesel engines power the hydraulic pumps used to operate the unit components. The control and monitoring systems are located in the operator's console. The control console and power package are located typically on a truck for land jobs and are contained on separate skids for offshore applications.

The stuffing box (**Fig. 6.19**) is used for primary well control when drilling underbalanced or otherwise operating with pressure on a well. Hydraulic fluid forces the stripping rubber against the tubing and maintains a seal at rated working pressures of either 5,000 or 10,000 psi. The elastomer is subject to wear when stripping the CT and the packoff material can be replaced with pipe in the hole after the annulus pressure has been isolated by the blowout preventer (BOP). Annular preventers designed specifically for CT and wireline work (**Fig. 6.20**) serve the same purpose as a stuffing box and can be used as backup to the BOP or to close the well on different-sized or odd-shaped tools.

As in all drilling and cased-hole operations, BOPs provide secondary pressure control. **Fig. 6.21** depicts a standard quad BOP with four rams, each having a separate function. The topmost component is a blind ram which, in essence, acts like a gate valve in that it can be used to close in the well if CT or other tools are not in the way. Below the blind ram is a blind/shear ram designed to cut the tubing. High-pressure kill-line connections are made at the outlet located between the shear and slip rams. The slip rams, when activated, do not have any seal capability but are designed to suspend CT when in a pipe-heavy condition and can be reversed to keep well pressure from ejecting the string when the pipe is light—i.e., the pressure-area force is greater than the tube weight and elastomer friction. Finally, pipe rams sized and designed to seal across the tube OD are located on bottom.

The ram arrangement logic follows from the operating sequence followed in the event of an emergency. First, pipe movement is halted and the slip and pipe rams are activated thus isolating the annulus and restraining the tube. Pumping ceases and the blind/shear ram is closed to sever the pipe. Then the reel is spooled up to remove the cut-off section so that the blinds can be closed. The well is secured and the pipe rams are opened to allow control operations through the kill-line outlet.

Most quad BOPs are rated for 10,000-psi working pressure, but some 5,000-psi stacks are still in service. Hydraulic control fluid from the unit power supply is used to actuate the rams. Pressurized fluid stored in an accumulator bank can be used if CT unit power fails for some reason, or the rams can

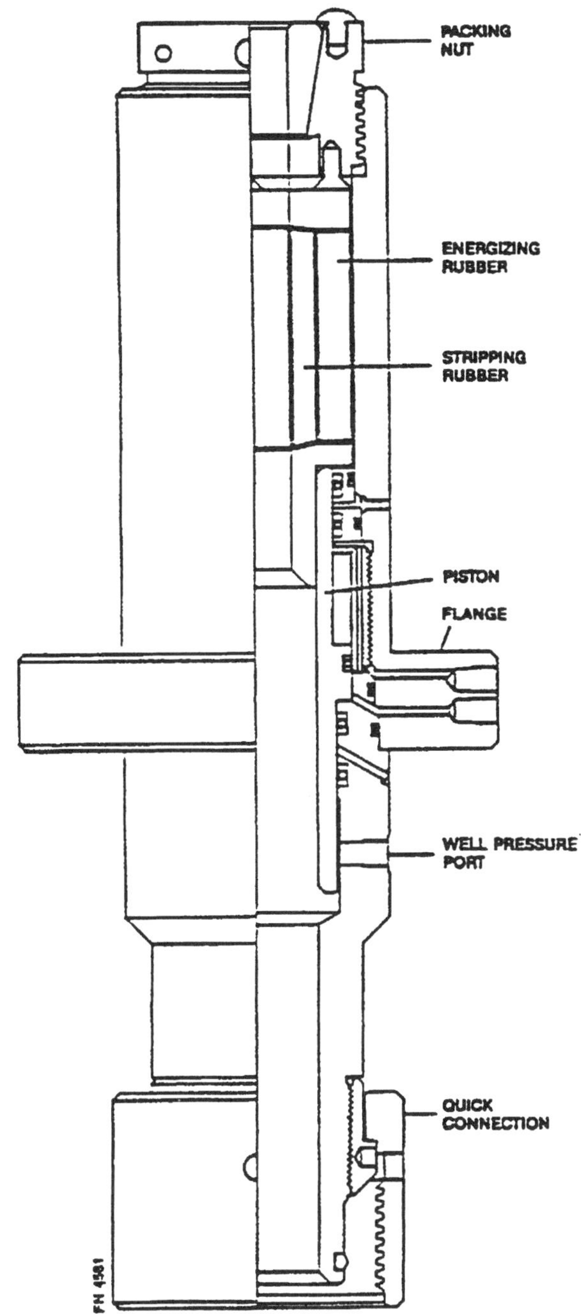

Fig. 6.19—CT stuffing box. Courtesy of Halliburton Services.

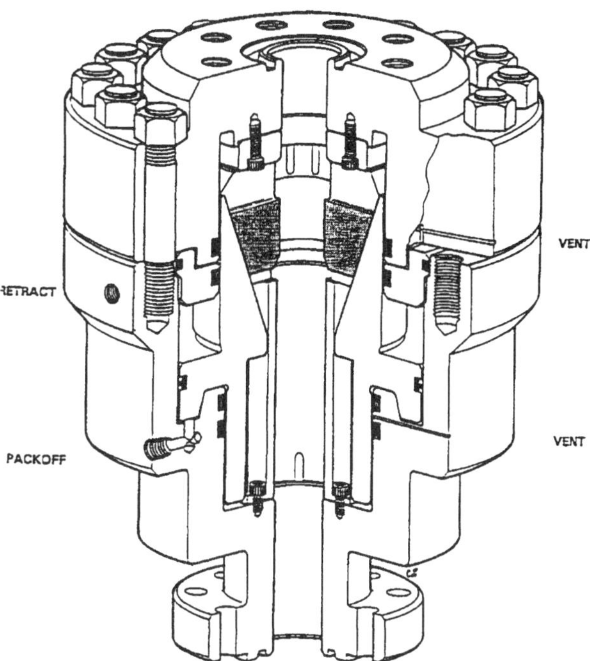

Fig. 6.20—Annular preventer used in CT and wireline operations. Courtesy of Texas Oil Tools.

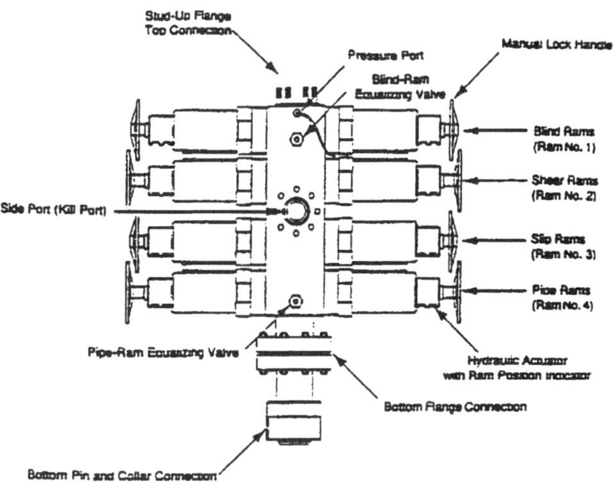

Fig. 6.21—Quad BOP.[39]

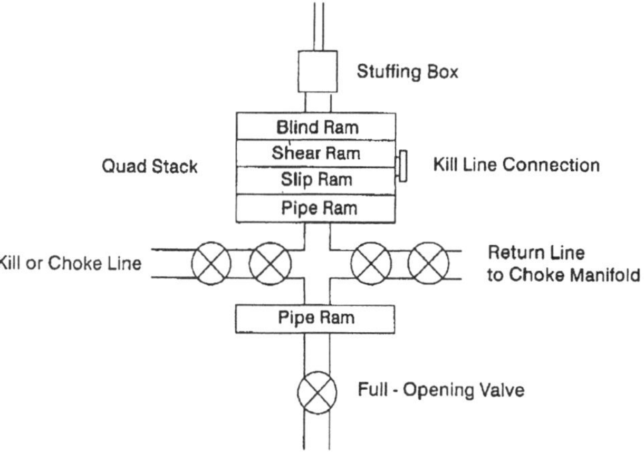

Fig. 6.22—Typical wellhead configuration for a high-pressure CT operation.

be closed and locked manually. Ports allow pressures to be equalized above and below a sealing ram so that the device can be opened.

A quad stack on top of a tree or wellhead would be considered the bare minimum for a routine cleanout job in a low-pressure well. High-pressures, H_2S, UBD operations, and other more risky undertakings require more redundancy and system flexibility. **Fig. 6.22** shows a typical configuration for high-pressure or UBD operations. The kill line outlet on the quad could be used as a circulation outlet (say while drilling underbalanced), but the recommended practice is to take returns through a flow tee installed beneath the quad. Alternatively, circulation may be routed through the wing connections on a tree if the work is done through-tubing.

The bottom pipe ram shown on the diagram in Fig. 6.22 allows the CT annulus to be isolated if replacement or repairs are needed in the quad or flow tee. A means to close in the well

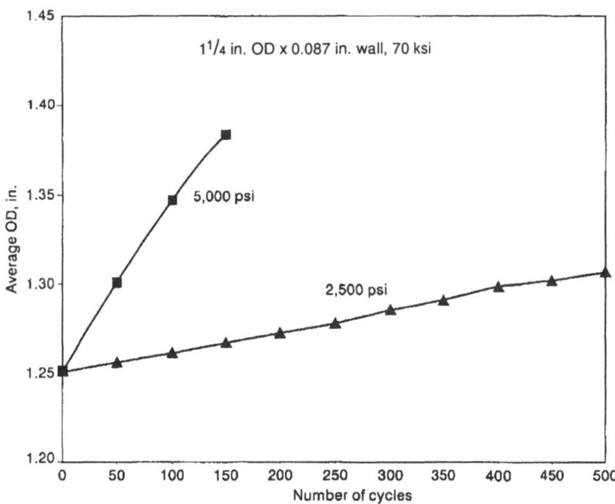

Fig. 6.23—CT OD as a function of internal pressure and number of bending cycles.[43]

fully must be provided somewhere below the return outlet. A full-opening valve or another set of blinds is suitable if the control equipment is attached directly to a wellhead whereas the tree gate valves can be used on a through-tubing job. A blind/shear ram immediately above the tree is recommended in offshore platform applications where risers or extensions are used to space the stack and flow equipment up to the servicing deck.

Bottomhole assemblies (BHAs) must be considered when planning for well-control contingencies. It may be necessary to install another larger-bore stack beneath the quad for possible closure on long BHAs (core barrels, drill collars, etc.) and a full-length lubricator should be provided for retrieving the shorter tools under pressure.

A rupture or split in the tubing above the stuffing box is a critical concern to well control. CT manufacturers do not use the same formulae in determining their published internal pressure ratings, p_{ip}. Even so, the relations are all based on Barlow's[40] equation,

$$p_{ip} = \frac{2\sigma_s t_w}{d_o}, \quad \text{(6.2)}$$

where σ_s = the material strength, t_w = the wall thickness of the tube, and d_o = its outer diameter. The American Petroleum Institute (API) uses a variation of Barlow's equation to calculate internal-yield pressures for jointed pipe[41] but, the internal pressure that will fail CT is not as straightforward or predictable.

The problem in predicting CT failure arises from the unique stresses placed on every CT string. The yield strength of the steel is exceeded the first time the pipe is spooled onto the reel and so permanent deformation is introduced before the string ever leaves the manufacturing facility. At the wellsite, the tubing experiences at least six bending-and-straightening cycles for every round trip in and out of a well. The pipe goes into tension and straightens as the injector head pulls the string off the reel. At the gooseneck, the tube is bent into approximately the same radius it had on the reel and is straightened again at the gripping chains. The repeated flexures eventually will cause a fatigue failure, limiting the service life of any CT string. Fatigue cracks tend to originate at the internal diameter (ID) surface on the compression side of the bend; failures generally occur in the vicinity of the gooseneck or injector head.[42]

TABLE 6.2—ACTIVITIES RELATED TO COMPLETION AND WORKOVER OPERATIONS

1. Perforating or re-perforating.
2. Stimulating the formation or near-wellbore region.
3. Adding new formations.
4. Isolating a zone (permanently or temporarily).
5. Cleaning out debris or fill.
5. Controlling unwanted fluid production (water or gas).
6. Controlling sand production.
7. Running logs to evaluate formation characteristics, cement sheath, pipe condition, or productivity.
8. Effecting mechanical repairs or replacement (casing, tubing, downhole or surface equipment).
9. Fishing jobs.
10. Improving or repairing the primary cement job.
11. Deepening, sidetracking, or drilling horizontal laterals.
12. Plugging and abandonment.

Internal pressure compounds the problem and reduces the cycles-to-failure. Internal pressure causes the pipe to balloon and reduces the wall thickness. It should not be surprising that the swelled diameter depends on the pressure but, as shown in **Fig. 6.23,** a tube actually grows with increasing number of bending cycles. The strains associated with the pressure combine with the bending deformation to create a situation that is difficult to model using mechanics principles, especially because factors other than pressure and cycle numbers also influence fatigue. These include corrosion, usage history in sour environments, and surface defects.

Based on these considerations, current industry practice limits surface pressures to no more than 5,000 psi for 70-ksi tubing. Fatigue-related phenomena are not as severe in higher strength materials and higher pressures are considered safe. The possibility of catastrophic failure can be minimized by placing at least one check valve in the string to prevent well fluids from entering and pressurizing the tubing. Any decision to inject flammable fluids should be scrutinized carefully.

6.4 Completion, Workover, and Well Servicing Operations

In the strictest sense, completion operations involve all activities from the time the decision is made to run production casing until the formation's initial potential is tested. Our focus, however, is on the completion work after the last string of casing has been cemented and pressure-tested. Workovers involve the subsequent efforts in a well's life to improve production (or injectivity) and include the final abandonment. In the United States, the Minerals Management Service (MMS) defines well servicing as coiled-tubing, snubbing, and small-tubing operations done without removing the tree.[44] We adopt their convention for this text. **Table 6.2** lists some of the specific activities related to completions, workovers, and well servicing.

The fundamental well-control concepts are similar to those for the well being drilled: to control wellbore pressures by either preventing, removing, or managing formation-fluid entries. An additional requirement is to minimize formation damage to the extent possible, by limiting completion-fluid losses to the productive formation and controlling its chemistry and solids content. Cased holes, however, are much more forgiving as the pressure limitations usually are dictated by the casing and surface equipment, not the fracture gradient.

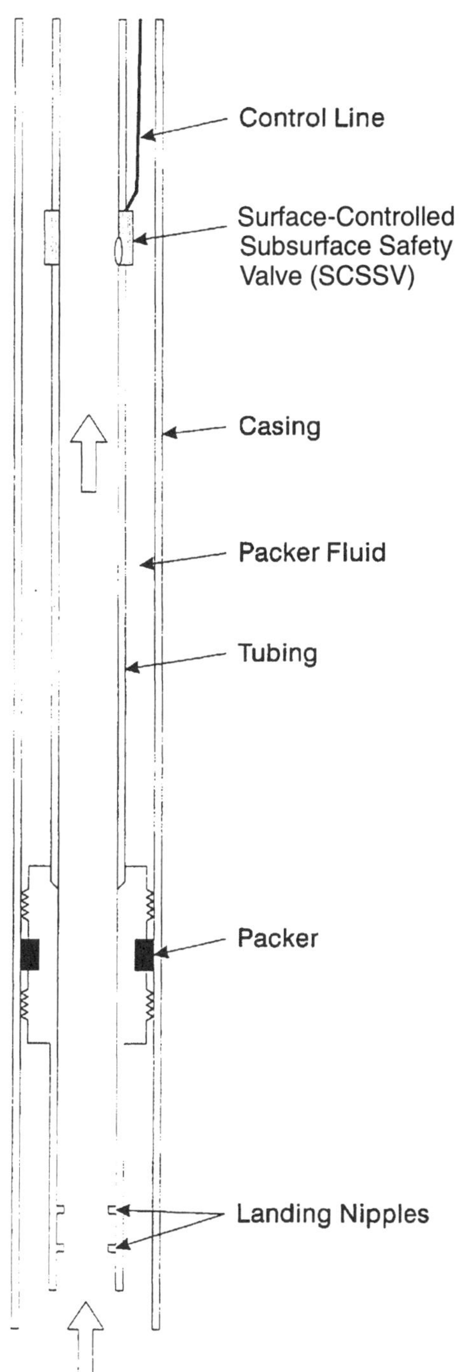

Fig. 6.24—Typical downhole production equipment used in a flowing oil or gas well.

6.4.1 Basic Pressure-Control Equipment. Typical subsurface pressure-containment equipment used in a flowing well producing from a single zone are illustrated in **Fig. 6.24.** The tubing and packer isolate well pressure from the casing and the casing gives secondary pressure containment if the tubing or packer leaks. Flow-control devices such as chokes, regulators, or check valves can be installed in the landing nipples during the normal course of operation or, in a permanent packer, used to seat a blanking plug before pulling the tubing. These plugs are designed to seal against high pressures and so provide a way to do shallower work without having to kill the producing formation.

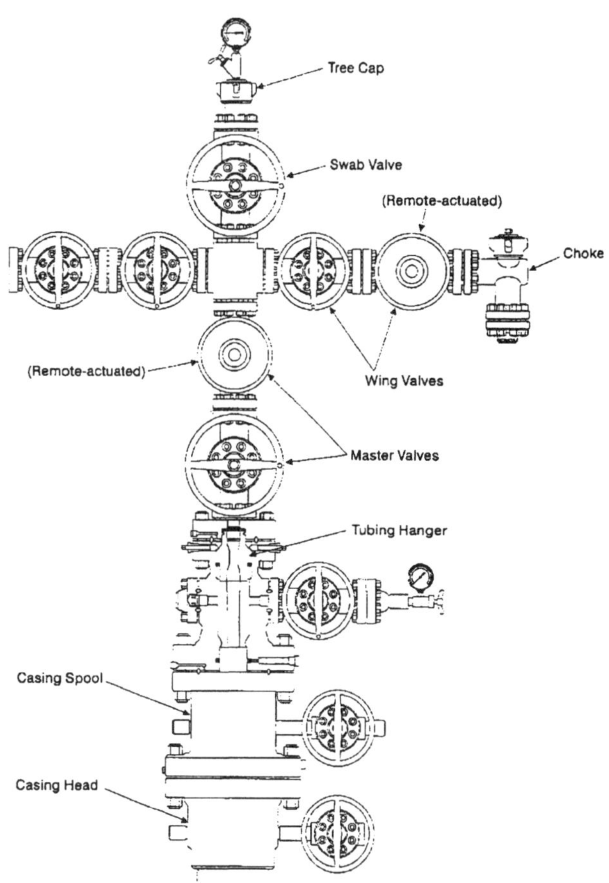

Fig. 6.25—Tree and tubing head arrangement for land and platform applications.

A surface-controlled subsurface safety valve (SCSSV) is a downhole closure-tool requirement on offshore platform wells. The valve itself is usually a flapper-type design and is kept open during production by hydraulic-fluid pressure transmitted through a control line. Releasing the pressure, say in the event of damage to the wellhead or platform, causes the SCSSV to close and prevents the well from blowing out. The control-fluid hydrostatic pressure limits the setting depth of an SCSSV; most are positioned within a few hundred feet below the mud line. Subsurface-controlled valves are available for deeper applications or subsea completions where closure is activated by excessive differential pressure across the valve, a condition that would arise if the surface flow rate was uncontrolled.

Fig. 6.25 illustrates a typical tree configuration used on a high-pressure land or platform well. The swab valve is used to shut the well in so that accessory equipment (wireline lubricator, CT riser, snubbing stack, etc.) can be attached. While producing, the well flows through the open master valves to the cross or tee and then out the wing valves to the choke. Remote-actuated valves remain open so long as air pressure is supplied to the valve actuator. Flowline sensors monitor flowline pressure and an abnormally high or low condition causes the emergency shut-down (ESD) system to block the control-line pressure and shut the well in. The hydraulic fluid supply and controls for the SCSSV are located at the ESD panel.

Fig. 6.26 illustrates the minimum BOP requirements for a high-pressure (5,000 psi or greater) completion/workover on a platform. Other pipe rams or a variable-bore ram will be

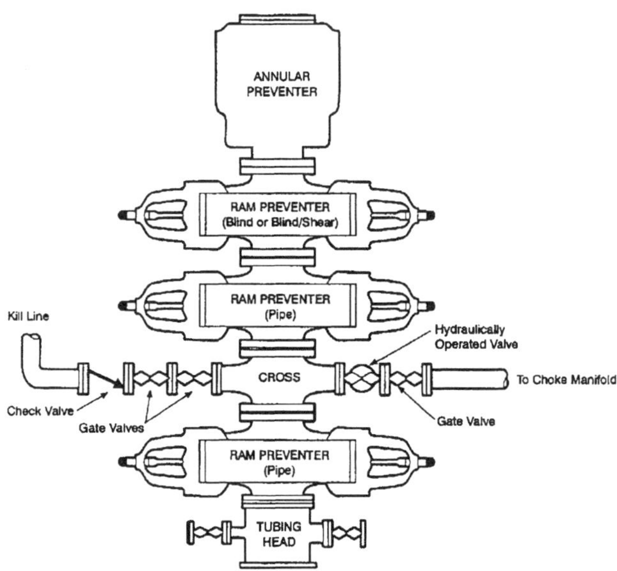

Fig. 6.26—Minimum BOP requirements for platform workover operations with 5,000 psi or greater stack equipment.

necessary when tapered strings are in use. The MMS allows an operator to eliminate the bottom pipe ram if the stack working pressure is less than 5,000 psi, but the decision to do so should be based on the assessed risk rather than on the potential shut-in pressure. Two ram preventers (pipe and blind) attached directly to the tubing head may constitute the stack in a low-pressure/low-risk application on land.

Wireline and slickline operations often are done under pressure using a lubricator assembly like the one shown in **Fig. 6.27.** The lubricator is flanged to the wellhead or tree and is of sufficient length to accommodate the wireline tools. A packing gland in the stuffing box is pumped up with grease and seals against the cable at working pressures as high as 15,000 psi in some systems. The wireline valve (normally called the wireline BOP) has a ram sized to close and seal against the cable should the element fail. As demonstrated in Example 6.4, solid rods called sinker bars must be added to counteract the well pressure and keep the tools from being ejected from the well.

Example 6.4. The stacked production logging tools on a well have an overall length of 12 ft and will be run on 5/16-in. cable.

1. Determine how many 1 11/16-in. sinker bars must be included in the assembly if the tubing pressure is 2,000 psig. The logging tools weigh 66 lbf in air; each bar is 5-ft long and weighs 38 lbf.

2. How many lubricator sections will be needed if each section is 8-ft long?

Solution. 1. **Fig. 6.28** illustrates a free-body diagram of the system. An upward pressure-area force across the tool diameter is trying to eject the tools from the hole, but the force is counteracted by the combined cable and tool weight *W*, the downward pressure-area force, and friction through the packing gland. Ignoring friction and the small hydrostatic pressure across the tool length, equilibrium requires that

$$W = p_t A_c, \quad \dots\dots (6.3)$$

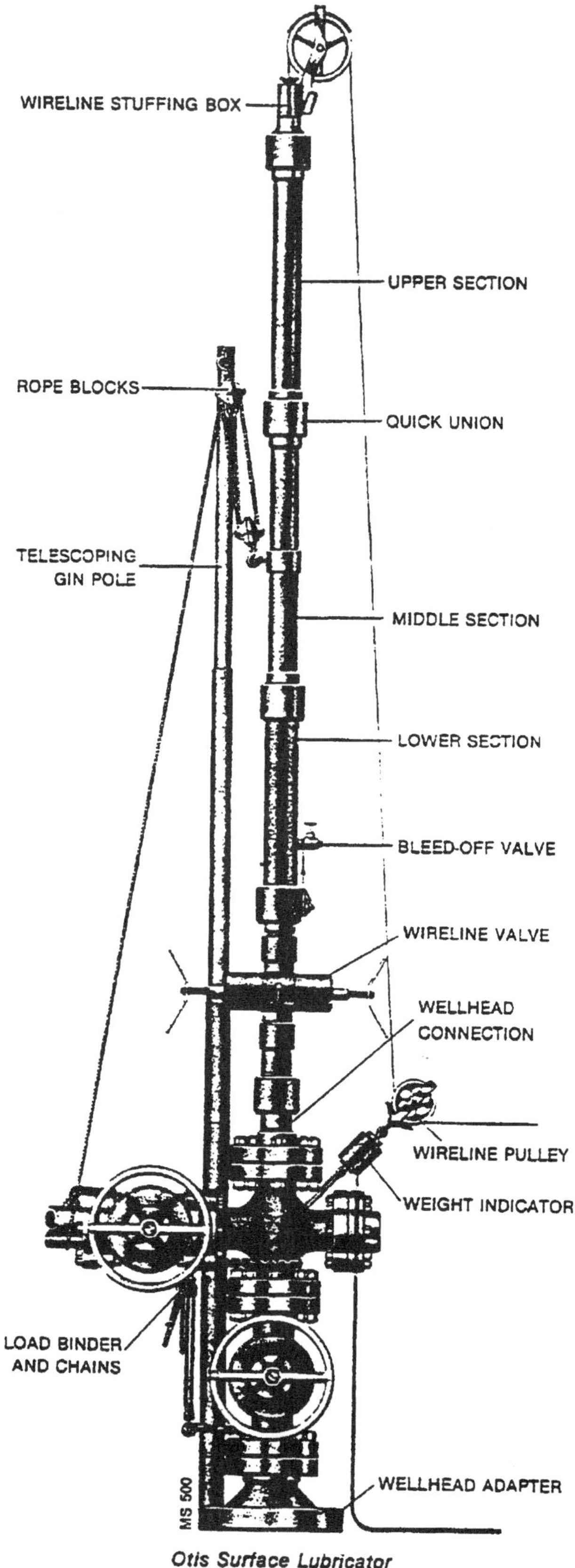

Fig. 6.27—Well-control equipment used in wireline work. Courtesy of Halliburton Services.

where p_t = the tubing pressure and A_c = the cable cross-sectional area. In the lubricator, cable weight is negligible and so

$$W = (2,000)(0.7854)(0.3125)^2 = 154 \text{ lbf}.$$

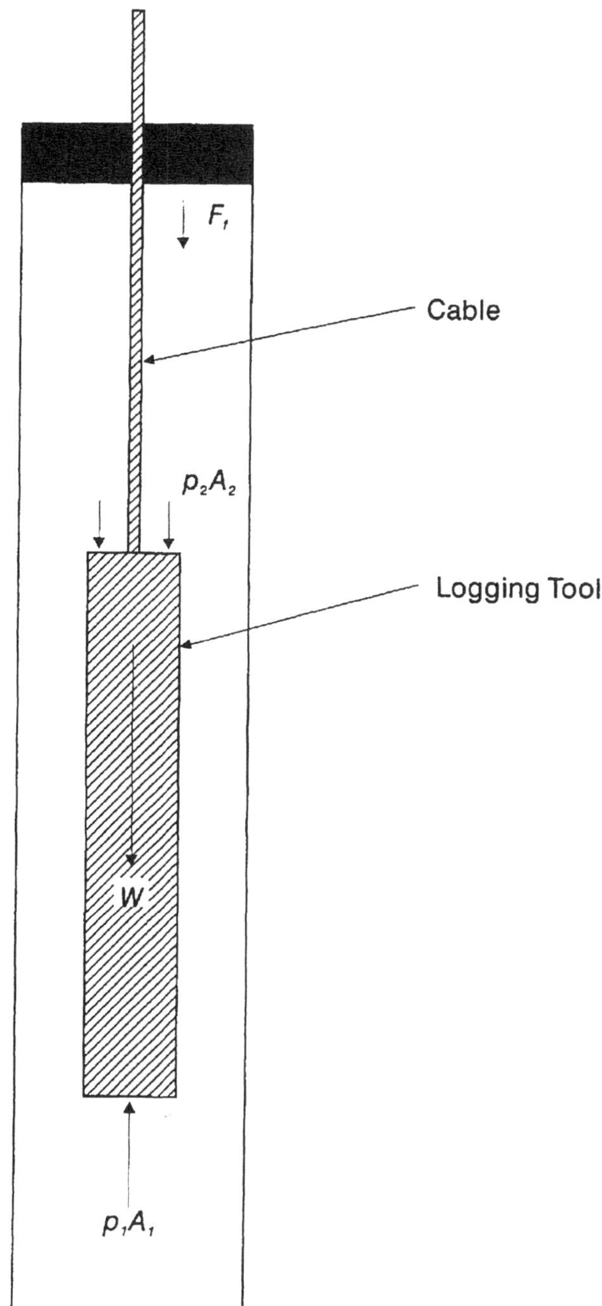

Fig. 6.28—Force balance for a wireline logging-tool assembly in a pressured wellbore.

The minimum number of sinker bars needed for this job are

$(154 - 66)/38 = 2.3$ (3 minimum).

2. The number of lubricator sections and lubricator length are, respectively,

$[12 + (3)(5)]/8 = 3.4$ (4 minimum)

and $(4)(8) = 32$ ft.

Tools conveyed on CT may be run during a completion, workover, or well servicing job. Equipment and operational considerations adequate for the scope of this text were discussed in Sec. 6.3.2. Snubbing and stripping, once a procedure applied to problem wells, is now considered a routine method for working on wells underbalanced. These topics are addressed in Chap. 9.

TABLE 6.3—QUALITIES OF AN IDEAL COMPLETION FLUID
1. Nondamaging to the reservoir.
2. Density sufficient to overbalance the exposed formations.
3. Stable with pressure, temperature, and time.
4. Noncorrosive.
5. Nontoxic to humans or the environment.
6. Nonhazardous.
7. Capable of transporting and suspending solids.
8. Inexpensive.

6.4.2 Completion, Workover, and Packer Fluids. Ideally, the fluids placed in a well during and after the initial completion will meet the qualities listed in **Table 6.3.** However, many of these goals may be incompatible and some tradeoffs must be made when selecting the fluid, keeping in mind the job objectives of safety and profitability. For example, the least-damaging fluid may not be heavy enough to control formation pressure and so the operator may need to accept some formation damage and try to remove this damage when kill-weight fluid is no longer needed.

Table 6.4 lists common completion/workover/packer fluids and their general characteristics. Detailed descriptions are beyond the scope of this text and we encourage the reader to consult other good treatises for guidance in selecting a fluid for a particular application.[45-47] Our primary concerns here deal with the density and filtration properties. Absent a way to work underbalanced, the fluid selected must have a density higher than the formation pressure equivalent. But density alone does not guarantee that a killed well will stay dead if losses to the formation are high. Unlike drilling mud, clean fluids do not build a filter cake and will flow into the rock at a rate proportional to the overbalance pressure and permeability. Example 6.5 demonstrates how the loss in head can create a situation where equilibrium is never realized and the well can take a kick.

Example 6.5. A 9.0-lbm/gal filtered brine is used to kill a high-permeability gas zone that has been depleted to an 8.0-lbm/gal equivalent. The top perforation is at 5,000 ft and the interval is 200 ft thick. Assume the gas has a hydrostatic gradient of 0.10 psi/ft. Is it possible for this well to kick later in the operation?

Solution. **Fig. 6.29** depicts the wellbore- and pore-pressure profiles, shown respectively by the solid and dashed lines. Initially, the top of the zone was overbalanced by 260 psi. Seepage losses over time, however, caused the overbalance pressure to diminish with head reduction. At some point, the zone top is exactly balanced with the kill fluid, yet losses still continue into the lower perforations. Hence a further drop in fluid level will occur and the zone will kick from the shallower perfs.

The only way to maintain the desired overbalance in Example 6.5 is to keep the hole continually filled with kill-weight fluid. This may be an undesirable situation for many reasons; namely that it requires close supervision by the rig personnel at all times, can cause a water block or other invasion-induced damage to the formation, increases load-recovery time and disposal costs, and can be very expensive with the high-cost fluids.

Filtration control may be able to limit these losses to something more tolerable. The viscosifying polymers used in brine waters are often effective fluid-loss agents, including hydrox-

TABLE 6.4—GENERAL COMPLETION/WORKOVER/PACKER FLUID CHARACTERISTICS

Fluid	Maximum Workable Density (lbm/gal)	Stability	Formation Damage Potential: Clay Inhibition	Solids Plugging	Emulsion Plugging	Wettability Alteration	Untreated Corrosivity	Solids Suspension	Filtration Losses	Environmental or Safety Concerns	Relative Cost
Air/Natural Gas	—	Long	—	Nil	Nil	Nil	Nil	Low	—	Low	Low
Mist	—	Nil	Variable	Low	Low	Low	Variable	Low	—	Low	Low
Foam	—	Short	Variable	Low	Low	Low	Variable	Variable	—	Low	Low
Crude Oil	8	Long	Excellent	Variable	Low	Low	Nil	Variable	Variable	High	Moderate
Diesel	7.03	Long	Excellent	Low	Low	Low	Nil	Low	High	High	Moderate
Invert Emulsion (Unweighted)	8.3	Long	Excellent	Excellent	Variable	High	Nil	High	Low	High	Moderate
Invert Emulsion (Weighted)	17	Long	Excellent	Moderate	Variable	High	Nil	High	Low	High	Moderate
Methanol	6.6	Long	Good	Low	Nil	Nil	Nil	Low	High	High	High
Fresh Water Mud*	22	Variable	Poor	High	Variable	Low	Variable	High	Low	Low	Low
Inhibitive Water-Based Mud*	22	Variable	Good	High	Variable	Low	Variable	High	Low	Low	Low
Fresh Water	8.3	Long	Poor	Low	Low	Low	Variable	Low	High	Low	Low
Seawater	8.5	Long	Variable	Low	Low	Low	Low	Low	High	Low	Low
KCl Water	9.7	Long	Good	Low	Low	Low	Low	Low	High	Low	Low
NaCl Water	10.0	Long	Variable	Low	Low	Low	Low	Low	High	Low	Low
$CaCl_2$ Water	11.7	Long	Good	Low	Low	Low	Low	Low	High	Low	Low
$CaBr_2$ Water	15.2	Long	Good	Low	Low	Low	Moderate	Low	High	Moderate	Moderate
$ZnBr_2$ Water	19.2	Long	Variable	Low	Low	Low	High	Low	High	High	High
Polymer-Viscosified Water	—	Variable	—	Variable	—	—	—	High	Low	—	Low
Weighted Brine—NaCl	13.5	Variable	—	Variable	—	—	—	High	Low	—	Low
Weighted Fluids—$CaCO_3$	14.5	Variable	—	Variable	—	—	—	High	Low	—	Low
Weighted Fluids—Fe_2CO_3	17	Variable	—	Variable	—	—	—	High	Low	—	Low
Weighted Fluids—Hematite	19	Variable	—	Variable	—	—	—	High	Low	—	Low

*Rarely used as workover/completion fluids but are a common packer fluid.

yethylcellulose (HEC), xanthan gum (XC), and guar. Polymers start losing stability at temperatures higher than 250 to 275°F and acid solubility is variable, depending on polymer type.

Extreme overbalance/permeability conditions may prescribe the use of a nondamaging solid to create a bridge in the perforations and block flow from the wellbore. Salt is removed easily and has been used for this purpose when the completion fluid is fully saturated with the salt. Calcium carbonate, on the other hand, is not dependent on fluid salinity and can be removed with acid. Also, there is a science to mixing different particle sizes to build an effective perforation bridge. An advantage to calcium carbonate is that it can be ground more easily to the specified diameter range. Finally, oil-soluble resins may be the best choice when trying to block an oil formation. Oil production dissolves the resin after the job.

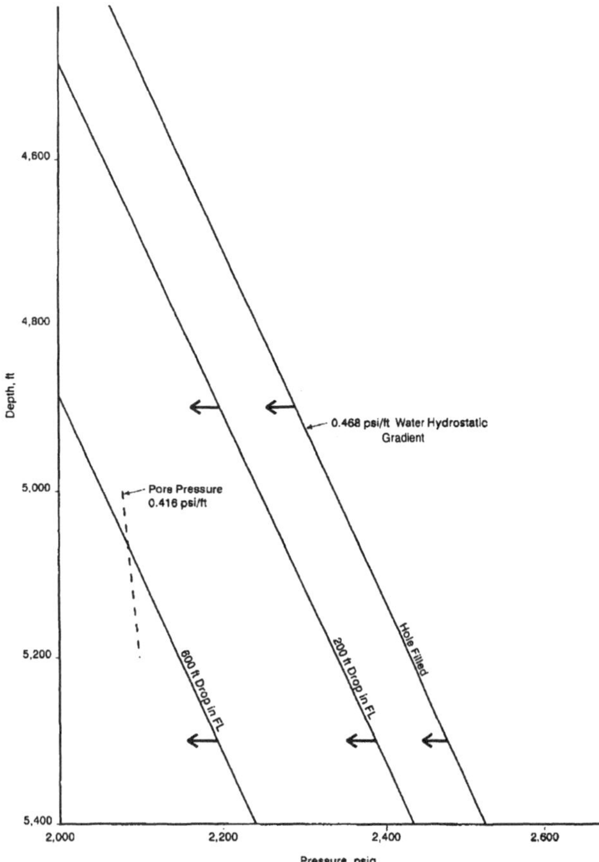

Fig. 6.29—Wellbore and formation pressure profiles for Example 6.5. FL = fluid level.

Barite is an undesirable weighting agent, but solids that are soluble in water or acid can be used to increase fluid density. Some of the more common ones for weighting both oil and water-based fluids are listed at the bottom of Table 6.4. XC polymers are used typically in fresh or saline water to give the fluid sufficient thixotropicity to suspend these solids.

Solids can be eliminated by using brine water as a completion or workover fluid. Single-salt systems can be run in many cases, but normally the brine will have two or three dissolved salts in the higher-density ranges. Freeze or crystallization temperature is one reason for running a blend. At low concentrations, dissolved salts reduce the freezing temperature from that of fresh water whereas the solubility dependence on temperature becomes more important at higher concentrations. A solidified mass can plug the flowline and set up in the tanks if the surface temperature falls below the crystallization temperature for the composition. **Table 6.5** lists the solidification temperatures for various salts and brine blends.

$CaBr_2$ and $ZnBr_2$ brines have several unattractive features. These are expensive fluids (particularly $ZnBr_2$), the untreated corrosivity is high, and special handling precautions are needed for crew protection. Thus it would be desirable to maximize the dissolved $CaCl_2$ content if a high-density fluid is required. However, Table 6.5 indicates that $CaCl_2$ increases the crystallization temperature, which leaves an operator working in a cool climate no choice but to increase the bromide salt concentrations. The fluid supplier should be consulted for as-

TABLE 6.5—FREEZING AND CRYSTALLIZATION TEMPERATURES (°F) OF VARIOUS BRINE SYSTEMS

Density (lbm/gal)	KCl	NaCL	Flake $CaCl_2$	$NaCl/CaCl_2$	$CaCl_2/CaBr_2$*	$CaCl_2/CaBr_2/ZnBr_2$**	$CaBr_2/ZnBr_2$†
8.5	29	29	30	—	—	—	—
9.0	21	19	21	—	—	—	—
9.5	32	5	9	—	—	—	—
9.7	50	−3	3	—	—	—	—
10.0	—	32	−8	−4	—	—	—
10.5	—	—	−38	−26	—	—	—
11.0	—	—	−15	−12	—	—	—
11.5	—	—	41	—	—	—	—
11.7	—	—	57	—	50	—	—
12.0	—	—	—	—	47	—	—
12.5	—	—	—	—	38	—	—
13.5	—	—	—	—	12	—	—
14.0	—	—	—	—	8	—	—
14.2	—	—	—	—	15	—	—
14.5	—	—	—	—	65	—	—
15.0	—	—	—	—	67	67	−22
15.5	—	—	—	—	—	59	−34
16.0	—	—	—	—	—	53	−33
16.5	—	—	—	—	—	47	−16
17.0	—	—	—	—	—	32	−4
17.5	—	—	—	—	—	41	5
18.0	—	—	—	—	—	43	9
18.5	—	—	—	—	—	32	17
19.0	—	—	—	—	—	18	21
19.2	—	—	—	—	—	16	16

*Crystallization temperatures after mixing 11.6-lbm/gal $CaCl_2$, 14.2-lbm/gal $CaBr_2$, and sacked $CaCl_2$. Different mixtures yield different results.

**Crystallization temperatures after mixing 15.0-lbm/gal $CaCl_2/CaBr_2$ and 19.2-lbm/gal $ZnBr_2$. Different mixtures will yield different results.

†Crystallization temperatures after mixing 14.2-lbm/gal $CaBr_2$ and 19.2-lbm/gal $ZnBr_2$. Different mixtures will yield different results.

sistance in designing the optimum mixture for the specific conditions and budgetary constraints.

Thermal expansion in the wellbore should be considered when determining what brine density to mix. The following relation can be used to determine the surface density required to achieve the correct downhole density.

$$\rho_s = \rho_{bh} + (T_{avg} - T_s)K, \quad \text{(6.4)}$$

where ρ_s and ρ_{bh} = the brine density (lbm/gal) at the surface and wellbore, T_s = surface temperature, and T_{avg} (°F) = average wellbore temperatures, respectively. K is a constant which depends on the dissolved solids concentration. Adequate results can be obtained using the K values listed in **Table 6.6.**

TABLE 6.6—BRINE DENSITY CORRECTION FACTORS FOR TEMPERATURE

Density Range lbm/gal	Density-Loss Constant lbm/(gal-°F)
8.4 to 9.0	0.0017
9.1 to 11.0	0.0025
11.1 to 14.5	0.0033
14.6 to 17.0	0.0040
17.1 to 19.2	0.0048

Example 6.6. A 10.2-lbm/gal density is needed to overbalance formation pressures in a well. Determine the surface density of a $CaCl_2$ brine if the mixing temperature is 80° F and the average temperature in the well is 170°F.

Solution. Obtain K from Table 6.6 as 0.0025. Setting the values of the terms in Eq. 6.4 yields

$$\rho_s = 10.2 + (170 - 80)(0.0025) = 10.4 \text{ lbm/gal}.$$

Maintaining the desired brine density requires accounting for the hygroscopic nature of the heavier brines; they absorb water from the air and will experience a loss in density if exposed to the atmosphere. In one case, open tanks caused a 15.1-lbm/gal $CaCl_2/CaBr_2$ brine to lose as much as 0.2-lbm/gal overnight.[48] This problem was corrected in part by covering the tanks.

Packer fluids are placed between the casing and tubing at the end of the job and do not necessarily contact the formation. One stated purpose of a packer fluid is to minimize the casing pressure should the tubing or packer leak. Many operators select a packer fluid density that is equivalent to or overbalances the pore pressure. Others opt for a density substantially lower than the pore pressure equivalent because lighter fluids mean less trouble and expense. Both policies are reasonable, but the potential ramifications of each should be explored.

Say that a packer leak causes the fluid to dump on the formation. The packer fluid falls and equalizes to a level depending on the original produced-liquid height and on the capacity factors in the casing, tubing, and annulus. Hence, the packer fluid will not keep the hole filled in a gas well or high gas/oil ratio oil well and a packer-fluid density that matches or even overbalances the formation pressure will not necessarily leave the well dead after the leak. Of course, an even higher pressure will be seen with a lighter packer fluid.

Casing designs should account for the effect of a heavyweight packer fluid deep in a well. Casings generally have a full column of drilling mud and cement in the annulus and the

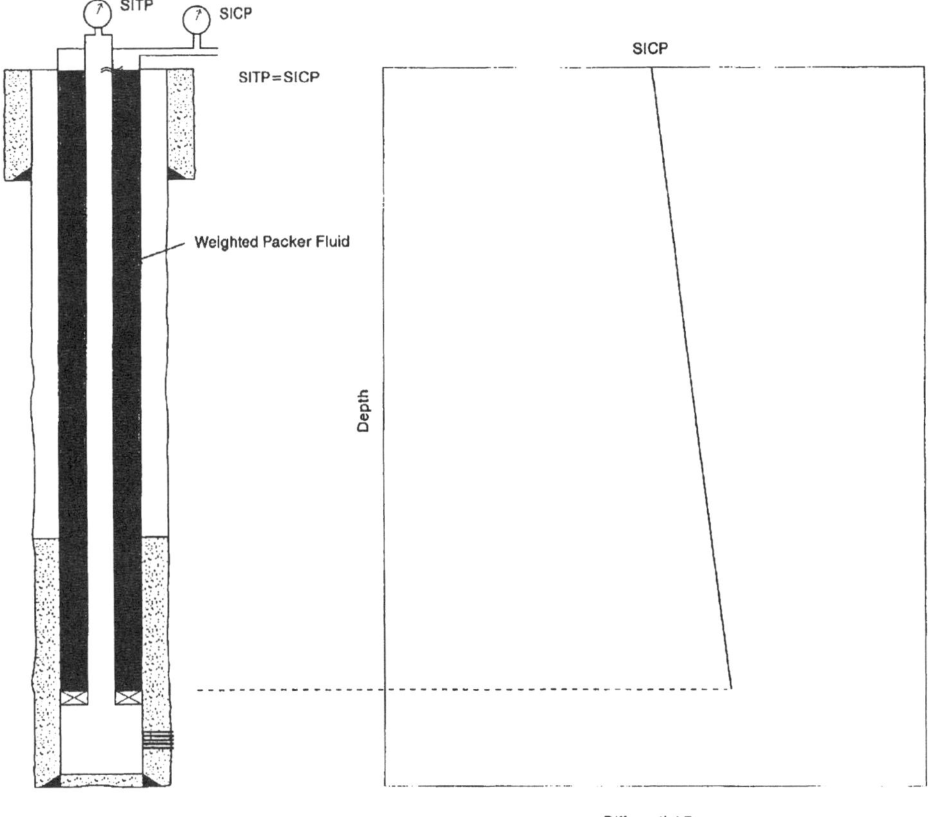

Fig. 6.30—Differential casing pressures when a shallow tubing leak occurs with a heavy-weight packer fluid.

external hydrostatic pressure gradient serves to backup the internal pressure. How much effective annulus pressure it provides is questionable because solids may settle out of a weighted mud and cement reinforcement is not a dependable parameter. Many operators therefore assume a relatively low backup density, say 9.0 lbm/gal, when calculating burst loads on casing.

The worst-case burst loads are on a shut-in well when a shallow leak occurs in the tubing string. **Fig. 6.30** demonstrates the differential internal-casing pressures for the situation where the tubing hanger loses integrity with a packer-fluid density higher than the effective backup-fluid density. As indicated, the shut-in tubing pressure (SITP) is applied now on top of the packer fluid and the differential is highest immediately above the packer. The well planner should anticipate this condition when designing the casing.

It is still fairly common for operators to treat the drilling mud with a biocide and corrosion inhibitor for use as a packer fluid. This is the least expensive approach, but the long-term stability in weighted muds may be questionable. Working the well over years later will require a fishing and washover job if the solids have settled to any extent. Also, we should expect any packer fluid to interact with the formation at some point in the well's life. Dumping mud on a formation may damage the formation permanently, especially later on when pore pressures are depleted substantially. A heavy-weight clear brine may be a better option if a high-density fluid is needed. These can be treated to yield acceptable corrosion performance in most cases.[49]

6.4.3 The Barrier Concept. Rike *et al.*[50] discussed some fundamental differences between drilling wells and cased holes work and presented a different way to approach blowout-prevention planning and execution called the Barrier Concept. A barrier is anything that prevents a well from flowing. Barriers considered "positive" are fairly reliable impediments to flow. "Conditional" barriers may not be as dependable or may require separate action (typically using another conditional barrier) before the well is shut-in. **Table 6.7** lists positive barriers to flow. **Table 6.8** gives conditional barriers to flow. Others come to mind.

The number and type of barriers recommended for a given job depend on the well's general risk classification. Minimum recommendations apply to sweet oil wells in nonsensitive

TABLE 6.7—POSITIVE BARRIERS TO FLOW[50]
1. Unobstructed master valve(s) on the tree.
2. Unobstructed swab valve, wing valve(s), or choke.
3. Wireline lubricator, if tested.
4. Insufficient formation pressure to cause the well to flow to surface.
5. Cemented and unperforated casing.
6. Cement-squeezed perforations if the squeeze has been tested differentially.
7. Cement plug above the top perforation if the plug has been tested differentially.
8. Mechanical bridge plug above the top perforation if the plug has been tested differentially.
9. Kill-weight fluid if losses are prevented by filtration control or perf-bridging materials.
10. Blind/shear rams if tested in the past week.
11. Blind rams if tested in the past week.
12. Annular preventer (openhole closure only) if tested in the past week.

TABLE 6.8—CONDITIONAL BARRIERS TO FLOW[50]
1. Cement-squeezed perforations if the squeeze has not been differentially tested.
2. Cement plug above the top perforation if the plug has not been differentially tested.
3. Mechanical bridge plug above the top perforation if the plug has not been tested differentially.
4. Kill-weight fluid if losses are not prevented by filtration control or perf-bridging materials.
5. With pipe in the hole, the pipe rams if tested in the past week. (Requires installing and closing a full-opening safety valve (FOSV) before the well is shut-in.)
6. Annular preventer (pipe closure only) if tested in the past week. (Requires installing and closing a FOSV before the well is shut-in.)
7. Blanking plug set in a landing nipple or a cast-iron plug set in tubing.
8. SCSSV if tested in the last six months.
9. Workstring FOSV or backpressure valve. (Requires setting a packer or closing the pipe ram/annular before the well is shut-in.)
10. Backpressure valve in a tubing hanger.
11. Stuffing box or other packoff device that allows well production or work under surface pressure.

locations where anticipated pressures are below 5,000 psi. For these conditions, a well should have in place one of the following minimum barrier combinations: two barriers with at least one being a positive barrier, or three conditional barriers, but only if the well has no surface pressure or the pressure is contained during a trip.

An additional barrier for each of these combinations was recommended for oil wells with anticipated pressures greater than 5,000 psi, gas wells with anticipated pressures greater than 3,000 psi, wells containing dangerous concentrations of H_2S, locations where the exposure liability is especially high, and offshore or other locations where mobilization logistics are difficult.

6.4.4 Operational Considerations and Potential Problems. This section addresses some of the more critical procedures from the standpoint of well control. Every well is different and it is impractical to cover every conceivable situation and how it relates to a specific well condition and equipment configuration. This discussion is general and focuses on single-zone as opposed to dual or triple completions, flowing wells as opposed to those on artificial lift, and land and platform operations as opposed to those conducted from a floating workover unit.

Testing the Pipe. Well control demands integrity in the work string and an operator needs to set some MAASP limitations on the casing. Older wells in particular are subject to the cumulative rigors of corrosion, cyclic stresses, wear from previous well work, and so on. It is imperative to test the casing and tubing to the maximum pressure anticipated for the upcoming procedure.

An opportune time to test the casing may be before the tree is removed if the applied pressure is within equipment limitations and does not cause the packer to unseat or the seals to pull free. Keep in mind, however, the effect of a weighted packer fluid on the burst differentials down the hole. Higher test pressures, if required, will have to wait until the weighted packer fluid has been replaced with something lighter.

Killing the Well, Removing the Tree, and Installing the BOPE. The well must be secured before the tree can be removed. Procedures up to the point where the stack is installed

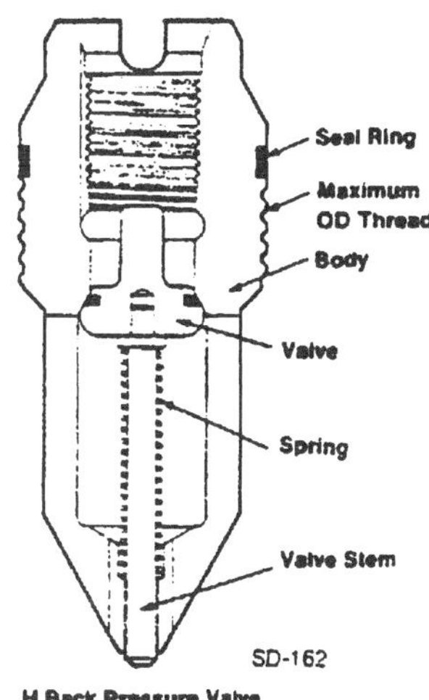

Fig. 6.31—Example backpressure valve. Courtesy of Cameron.

depend on several equipment and task-related factors. Specifically, the way the tubing is suspended makes some difference in how to conduct the job. Mandrel hangers are used in high-pressure completions and so our primary attention is on this type of suspension rather than slips, threaded adapter flanges, or other suspension methods. The tubing is screwed into the bottom of the mandrel; threads on top allow for installing a pickup sub or joint. The mandrel is prepped so that a threaded or expanding-lock backpressure valve (**Fig. 6.31**) can be installed to contain the well when the tree is off or to repair the bottom master valve. The well should be dead before removing the tree and the valve should be considered as a safety device, not a shut-in tool.

In a Type 1 configuration, the mandrel is suspended from the bowl of the tubing head. A tubing head adapter is flanged to the head and may have a seal bore sized to fit a slick or seal-neck extension on the mandrel, thereby containing well pressure and produced fluids to the tubing and tree. The tree and adapter are taken off or put on without affecting the suspended tubing string.

In the Type 2 style, the mandrel is located in the adapter itself. The hanger screws into heavy-duty threads cut into the adapter which, in effect, attaches the tubing to the tree. Hence the tubing moves while the tree is being stroked up or down. A secondary hanger called a wrap-around is positioned in the tubing-head bowl. Wrap-around seals isolate the tubing annulus when the tree and stack are not in place and the hanger can be used to suspend the tubing temporarily.

One of the first tasks in a workover is to take pressure off the tubing, remove the tree, and install the BOP equipment. **Table 6.9** gives two general procedures written with mandrel-type hangers in mind, but the well-control principles are not unique and can be adapted to other configurations. The first procedure is applicable in those wells equipped with a permanent packer and provision for seating a plug in the packer or tailpipe assembly. In addition, the workover does not require killing the producing zone as would be the case when replacing the tubing or recompleting a shallower zone. The second procedure applies to those jobs where the producing zone is killed initially with fluid.

The first step in both procedures determines tubing and packer integrity. A small tubing leak can be tolerated in the circulation step as long as the preferred flow path is at packer depth. Getting the packer fluid completely displaced from the well is a problem if the tubing has a large hole. Again, it may

TABLE 6.9—GENERAL KILL, TREE REMOVAL, AND BOP INSTALLATION PROCEDURES

Permanent Packer with Landing Nipples

1. Check to see if well has pressure on the tubing annulus. If so, bleed some pressure off and see if it comes back. If not, apply 200 to 300 psi and see if there is any leakoff. Release any pressure off the casing.
2. Close the swab valve. Install and test lubricator. Run blanking plug on slickline and seat in packer. Bleed all or some of the pressure off the tubing and verify plug isolation.
3. Fill tubing with kill-weight fluid. Allow time for any gas to break out.
4. Verify the well is dead. Place backpressure valve in tubing hanger. Close casing valve.
5. Remove the tree. Install and test the BOPE.
6. Install lubricator. Open backpresure valve and remove if the well is still dead. Verify casing pressure is zero.
7. Disengage tubing from packer. Open line to choke manifold and displace packer fluid from the annulus using the first circulation of the Driller's Method. Be prepared to handle any gas or oil originally in the tubing.
8. Pull the tubing.

Other Cases Where the Producing Zone Is Killed Initially

1. Check to see if well has pressure on the tubing annulus. If so, bleed some pressure off and see if it comes back. If not, apply 200 to 300 psi and see if there is any leakoff. Release any pressure off the casing.
2. Mix kill-weight fluid density, allowing for thermal expansion and incorporating the appropriate margin for trip and fluid-loss overbalance.
3. Bullhead the kill fluid at low rate with a volume sufficient to displace the tubing and casing to the bottom perforation.
4. Verify the well is dead. Place backpressure valve in tubing hanger. Close casing valve.
5. Remove the tree. Install and test the BOPE.
6. Open backpresure valve and remove if the well is still dead. Verify casing pressure is zero.
7. Establish communication with the annulus at or above the packer. Check casing pressure. Open line to choke manifold and displace packer fluid from the annulus using the first circulation of the Driller's Method.
8. Pull the tubing.

be advisable to pressure-test the casing before disengaging from or releasing the packer.

The planner needs to have a good basis for the kill-weight density prediction and then make adjustments for thermal expansion and the desired overbalance. The reservoir engineering group can probably be of assistance here. Otherwise the BHP must be calculated based on the known fluid properties and SITP.

If the zone is to be killed, pumping the produced fluids back into the formation—i.e., bullheading, is the most common method. The intent is to pump at a rate slow enough to avoid fracturing the formation, especially if the zone has a propped treatment in place. Depending on the near-wellbore permeability or fracture conductivity, it may not be possible to pump at a rate slow enough to avoid fracturing the rock while at the same time having sufficient kill-fluid velocity to keep from bypassing the gas and/or oil. A CT dispacement or a lubricate-bleed procedure may be necessary in these cases.

Once the well is dead, modifications of the procedures outlined in Table 6.9 are possible, depending on the surface and downhole configuration. The well can be circulated through a Type 1 hanger assembly at this point if a sliding sleeve is used to establish communication with the annulus or the tubing can be manipulated to access the annulus with a Type 2 hookup and the packer fluid displaced by circulating through the tree.

Whether circulation is accomplished beforehand or not, the backpressure valve is a necessary safety measure before removing the tree. It may be advisable to make up a pup joint into the tubing hanger and install a full opening safety valve, keeping in mind that the backpressure valve is the only flow barrier if seepage losses invite an influx during the procedure.

To review, the first circulation of the Driller's Method involves holding constant casing pressure until the pump is at designated kill speed followed by holding constant pressure on the tubing for the duration of the circulation. This allows us to circulate with a constant BHP if the fluid density in the tubing remains consistent. Seepage losses can be expected, however, and the tubing should be full before the packer is released and circulation started.

Keeping BHP constant does not necessarily mean that the pressure at the packer is equivalent to the kill-fluid density. Say that a weighted packer fluid had been in the annulus for a long time and that the formation pressure is now underbalanced to the fluid hydrostatic. There will be excess tubing pressure to start lifting the packer fluid which will be seen as additional overbalance across the formation. The planner should calculate the U-tube differential before the job to determine if the depleted fracture gradient will be exceeded during the displacement. If so, it may be necessary to set a plug in the tubing before exposing the formation to the hydrostatic pressure in the annulus.

Managing Kicks. With only slight modification, the shut-in procedures discussed in Chap. 4 are applicable to completions and workovers. There is no advantage, however, to the soft shut-in approach and all flow paths on the choke manifold should be closed during routine operations.

Only in special circumstances would the completion-fluid density be lower than the pore-pressure equivalent. Hence most kicks either are swabbed in or happen when the hydrostatic pressure is not maintained because of seepage losses or improper fillup practices. Normally, filling the hole with the selected completion fluid will restore well control. Bullheading is frequently a kill option and may be the best way to regain control if the work string is some distance off bottom. A Driller's Method kick displacement may also be instituted. If the string is above the entry, institute the procedure using the kill-weight fluid and wait until the gas or oil migrates up to pipe depth. Pumping an off-bottom kill or stripping in the hole could be done if the fluid properties discourage kick migration.

Completions and workovers conducted only in daylight present potential problems. Whether from underbalance flow, diffusion, or other processes, gas often migrates and collects underneath the closed preventers during the night. The shut-in workstring and casing pressures can be used to deduce how much gas has been trapped. If the shut-in pressures are low, an operator may choose to bleed a small volume through the choke to see if the pressure falls off to zero.

Bullheading is undesirable, considering the completion-fluid volume that would be lost to the formation. A Driller's Method circulation should be able to replace the formation fluids with a fluid of proper density. Top-killing the well using a lubricate/bleed procedure is another option and may be the only acceptable solution if pipe is out of the hole. Recommended practice is to leave the well with an appropriate amount of pipe in the hole when left unattended for a significant period.

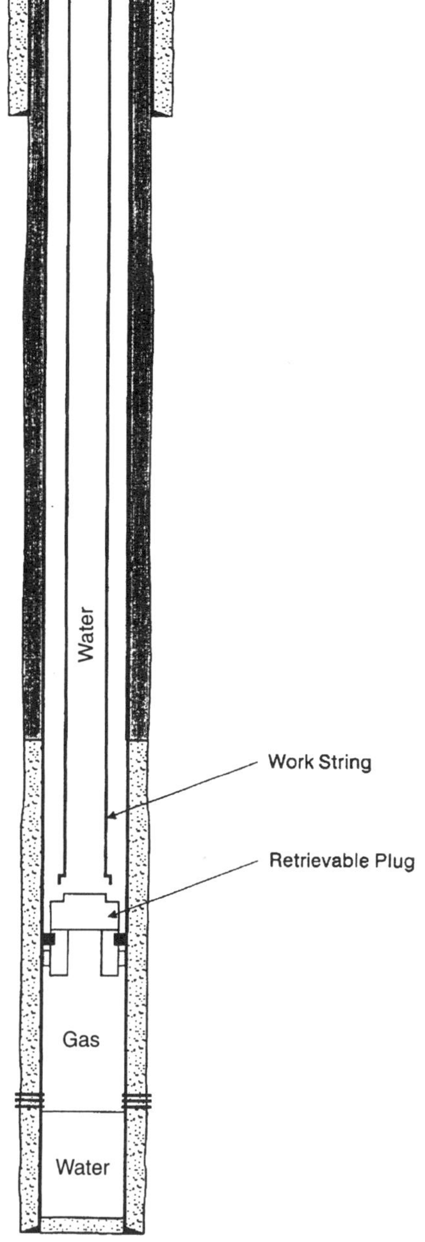

Fig. 6.32—The effect of an extensive gas column underneath a retrievable plugging device.

Reverse Circulation. Pumping down the casing and taking returns up the tubing is the most common circulation path in cased holes. The higher velocity up the relatively small-diameter work string overcomes many of the rheological limitations when using clear fluids to lift fill, cement cuttings, and other materials from a well. The ECD friction component, however, is the total loss through the work string including whatever device is attached to the bottom of the string. Hence the potential for an excessive overbalance is greater and should be considered in the job planning.

Removing Plugs in the Well. Drillable or retrievable bridge plugs sometimes are set above a zone to isolate that zone from shallower work or zone recompletion. Retainers, permanent packers, cement plugs, and other tools also can be used for this purpose. Regardless of how the well was plugged back, an operator should recognize the potential for trapped gas if the plug is to be drilled out or retrieved mechanically.

A kill-weight fluid corresponding to the isolated zone's pore pressure will closely balance the trapped pressure if the plug is set close to the perforations. **Fig. 6.32** illustrates a plug located a considerable distance above the perforations. In this case, seepage losses and gas entry have created a situation where the gas has completely filled the space between the bottom perforation and the plug. The pressure immediately below the plug is the pore pressure less the gas-column hydrostatic; the fluid density in use when the plug is removed may be greatly out of balance.

This does not imply necessarily that the completion fluid must be densified to account for the gas hydrostatic pressure, only that the operator should anticipate a significant flow and be ready to take returns through the choke. Ultimately, the fluids will swap and a fluid density designed with a customary overbalance will restore primary well control. A fast pump rate will help to restore primary well control.

Drilling openhole plugs, say while re-entering an abandoned well, is another story. It may be a good idea to calculate the potential pressure underneath the plug and, if feasible, to drill through the plug with a mud weighted up to balance this pressure. Gas may be encountered, but the risk of taking a large pit gain is reduced.

Plug and Abandonments. All commercial horizons may have been depleted or watered out when it comes time to plug and abandon a well permanently, but this is not the time to relax the well-control effort. Some of the more potentially hazardous operations are those involved in salvaging casing. Many casing pullers neglect to set BOPs, do not keep the hole full while pulling casing, or ensure that the well fluids are handled correctly at the surface while introducing cement plugs and plugging fluids. More than one blowout has occurred as a result of lax well-control efforts. Just keep the fundamentals in mind and, if contracting out the plugging work, specify the minimum standards in the contract.

6.5 Casing and Cementing Operations

Cementing the casing may be the most important factor in determining the economic success or failure of a well. This section concerns casing and cementing operations as they relate to well control. Common sense and engineering should guide the planning and job implementation, keeping in mind the well-control fundamentals of maintaining wellbore pressures within the limits set by the pore pressure and fracture gradient.

6.5.1 Running the Casing. Before beginning the job, an operator should consider replacing the drillpipe rams with pipe rams sized for the casing. Many operators, however, rationalize that the annular preventer can be used for closure on casing. This may be true, but shut-in pressures can result in a large upward force on the bottom of the string and, if the pipe is shallow, cause the casing to be ejected through a closed annular. Pipe rams should stop movement at a collar. Another practice sometimes used is to have a crossover swage back to the drillpipe on hand and ready to install if the well flows. Then a stand of drillpipe can then be made up and lowered into the well before shut-in.

Drilling a well "measures" the ability of the exposed formations to withstand pressure. However, the potential for exceeding the wellbore limitations are greater in a casing job than during a trip and the primary concern while running casing is to keep swab and surge pressures from introducing a well-control problem. Check valves in the float shoe and float

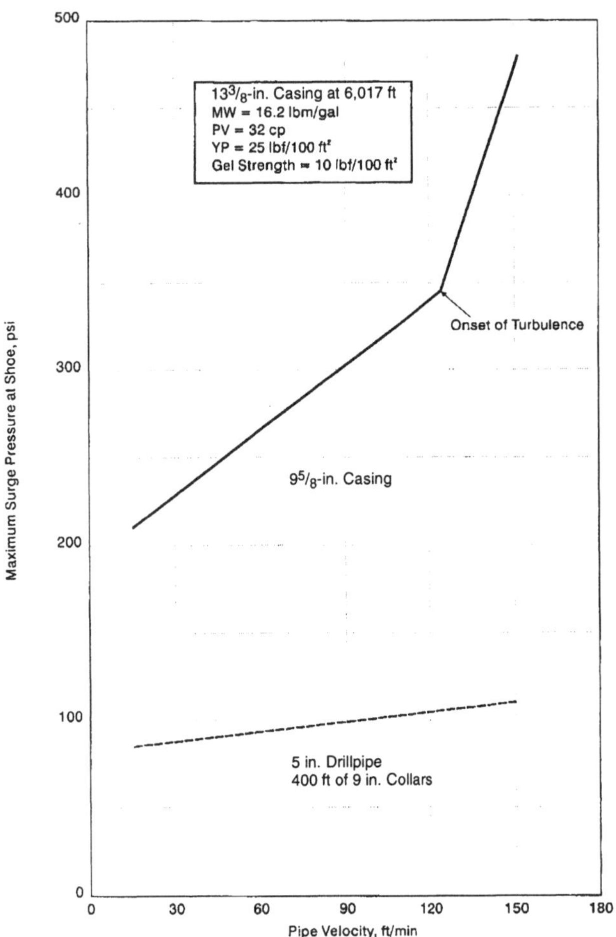

Fig. 6.33—Steady-state surge pressures as a function of pipe velocity for casing and a drillstring.

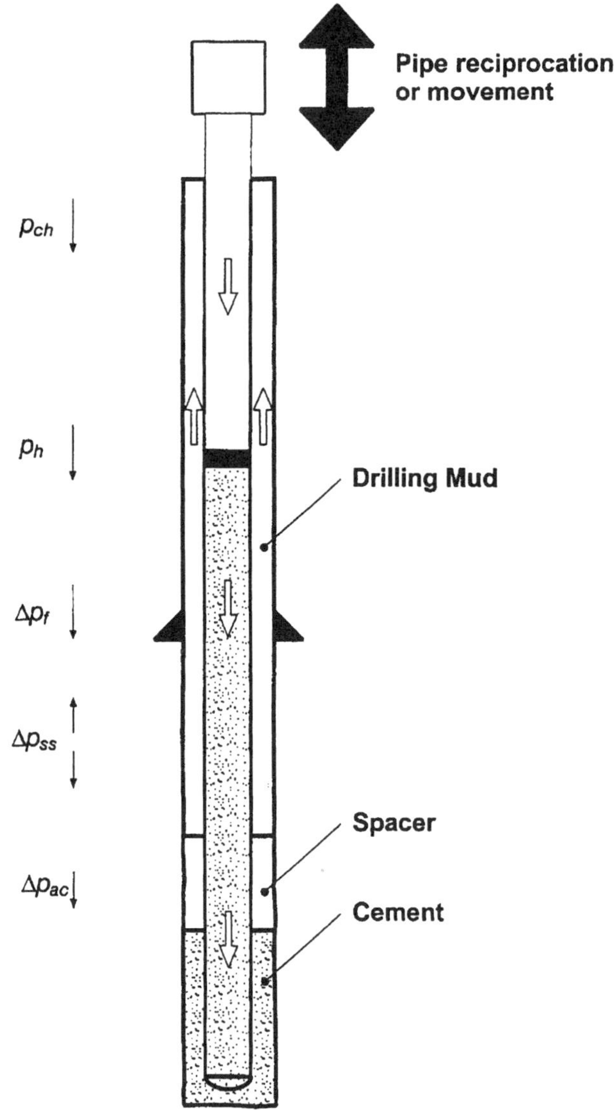

Fig. 6.34—Downhole pressures associated with a cement job.

collar force all the mud up the annulus (unless automatic or differential-fill equipment is used) and the reduced annular clearances lead to higher gel, steady-flow, and inertial pressures. As an example, the graph in **Fig. 6.33** compares the maximum predicted surge pressures at a 6,017-ft casing seat for open drillpipe and closed casing when both strings are lowered at constant velocity.

Keeping the mud as thin as feasible for the hole conditions is important to achieving a good cement job, but the benefits of proper mud conditioning also extend to surge-pressure control. Progressive-gel strengths are especially detrimental because the mud is static and cooking in the hole for a long period, particularly if the drillstring is laid down on the last trip out. The effect of excess thickening can be minimized by treating the mud and by periodically circulating the hole during the running procedure.

With pipe on bottom, the hole should be circulated at least long enough to displace the casing; in a gas well, bottoms up should be circulated to remove all gas from the annulus. Keep in mind the smaller annular dimensions and limit pump rates as needed to restrict ECDs opposite the shoe or other weak strata.

6.5.2 Cementing the Casing. Fig. 6.34 depicts the downhole pressures associated with a cement job. The pressure at any depth in the annulus is the sum of each component,

$$p_{bh} = p_{ch} + p_h + \Delta p_f \pm \Delta p_{ss} + \Delta p_{ac}, \quad \text{.......} (6.5)$$

where p_{bh} = bottomhole pressure, p_{ch} = choke backpressure, p_h = hydrostatic pressure, Δp_f = circulating friction pressure, Δp_{ss} = surge or swab pressure, and Δp_{ac} = pressure resulting from fluid acceleration.

The combined effect should be determined opposite the critical formations to ensure that well control is not compromised at any point in the operation. Service company simulators can assist in the job planning.

Cement densities are usually higher than the mud densities and so job design and execution concerns are focused more on the fracture gradient than the pore pressure. Well planners have got into serious trouble, however, by not considering the effect of the spacer density and volume on primary well control. A gassified liquid spacer is sometimes pumped ahead of cement to increase fillup across a weak zone, but the potential for allowing a shallower formation to flow must also be evaluated. Gassified preflushes are not the only potential problem. One offshore blowout occurred simply because a large fresh water spacer was pumped ahead of the cement. The hydrostatic pressure reduction in the annulus caused the well to go on production not long after cement displacement started.

Cement slurries are thick and generate significant friction pressures in the annulus. Putting cement in turbulent flow is recommended usually to enhance mud displacement, but the

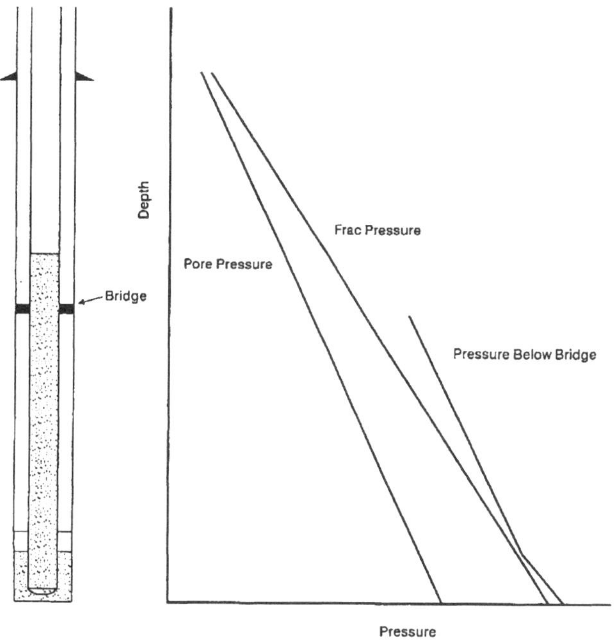

Fig. 6.35—Effect of a bridged annulus on wellbore pressures during a cement job.

allowable ECDs must be considered when deciding the pump rate schedule. The problem, however, is that annular velocities are independent from pump rate when cement is in the free-fall state. The cement density causes the slurry and mud in the casing to U-tube into the annulus so long as the annulus pressure at the shoe is less than the hydrostatic pressure in the casing. This causes higher annular velocities for a period and also introduces an acceleration-pressure component. Simulators are available for modelling these phenomena.[51]

Pipe movement is another prudent measure in achieving an adequate cement job. Reciprocation, however, introduces surge pressures, which may be important from a well-control standpoint. Finally, putting returns through the choke is done often when the well is underbalanced. In these cases, it is crucial for the design backpressure to account for the fracture integrity.

Fluid-loss control in the cement is important because a thick, dynamic filter cake may restrict the annulus and lead to higher ECDs. **Fig. 6.35** illustrates what can happen if the cement "flash-sets" because of sudden dehydration or contamination, or if the annulus completely bridges off. The diagram represents the condition when the annulus packs off with most of the cement still inside the casing. The hydrostatic pressure in the casing acts against the hole interval below the bridge and losses could result. Proper job designs and ensuring a clean hole should prevent problems like this.

Ineffective efforts to displace the mud completely with cement leaves mud channels in the cement that can cause well-control problems as well as problems in the completion phase. Under certain conditions, a slurry will bypass part of the mud and result in a channelled cement job. Lack of cement isolation can allow fluid migration in the annulus and thus the potential for a loss in subsurface or surface well control. As illustrated in **Fig. 6.36,** channeled cement yields a higher-than-calculated lift that could exceed the fracture pressure. The scope of this text does not permit full discussion of critical factors leading to incomplete mud displacement or poor bonding. **Table 6.10** lists some of the more important steps an

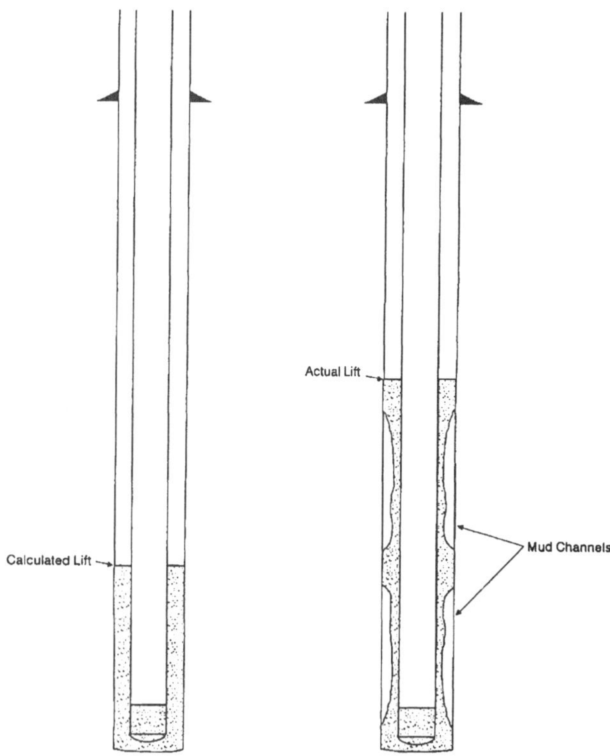

Fig. 6.36—Effect of channeled cement on hydrostatic pressure.

operator can take to address mud displacement and improve cement bonding.[52-54]

6.5.3 The Annular Flow Problem. Having a communicable pathway for formation fluids to migrate does not mean that entry will occur. Something must be done to effect a pressure

TABLE 6.10—WAYS TO ENHANCE MUD DISPLACEMENT AND IMPROVE CEMENT BONDING

1. Condition and treat the mud to limit viscosity to no more than is needed for the hole conditions.
2. Circulate before cementing with enough volume to break the gels completely and displace any enlarged hole sections.
3. Rotate or reciprocate the casing when cement exits the shoe. Rotation is especially desirable in highly deviated holes.
4. Incorporate mechanical aids (scratchers or wiper) to assist in mud-cake removal and agitate low-velocity mud in enlarged hole sections.
5. Centralize the pipe.
6. Place the cement in turbulent flow if conditions allow.
7. Pump a thin spacer with enough volume to achieve at least 10 minutes contact time and at a rate designed to achieve turbulent flow.
9. Ensure the slurry density is heavier than the mud.
8. Design the spacer density to be between that of the mud and cement.
9. Prepare the casing surface to remove mill scale and any oil film.
10. Employ a bottom wiper plug.
11. Eliminate free water, particularly in highly deviated holes.
12. Run an adequate number of shoe joints.
13. Pump excess slurry if conditions allow.
14. Release pump pressure before cement sets to avoid a micro-annulus.
15. Consider using an expanding cement to counteract the shrinkage associated with cement hydration.

TABLE 6.11—WAYS TO CONTROL ANNULAR FLOWS THROUGH CEMENT	
Method	Effect
Fluid-loss control	Limits cement water volume reduction. Reduces system rigidity.
Thixotropic slurries	Reduces transition time.
Delayed-gel-strength slurries	Allows for hydrostatic pressure transmission during volume loss. Reduces transition time.
Compressible slurries	Gas expansion compensates for water volume losses.
Low-permeability slurries	Creates a physio-chemical barrier to flow.
In situ foam generators	Immobilizes extraneous gas bubbles.
High slurry densities	Increases overbalance during transition period.
Limiting cement fillup	Reduces the detrimental effects of gellation and volume reduction.
External casing packers	Offers mechanical barrier to flow.

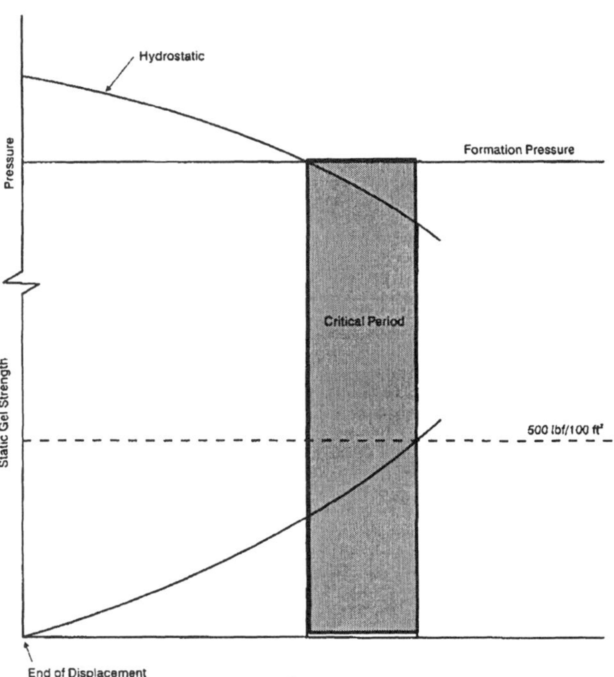

Fig. 6.37—Cement gel strength development and hydrostatic pressure reduction with time.

reduction in the wellbore. Cement undergoes some processes in changing from a liquid slurry to a hard solid that cause a hydrostatic pressure reduction in the cement column. During the transition period, gelation creates a self-supporting matrix that does not allow full transmission of the hydrostatic pressure from above. At the same time, water volume is reduced through hydration and filtrate losses which leads to a reduction in the pressure within the cement. A formation can thus become underbalanced and flow if the cement has not gained sufficient rigidity.

In an important study, Cooke *et al.*[55] measured cement-column pressures and temperatures in seven wells by clamping sensors on the casing and obtaining the data by means of wireline conductors. They found that pressures began falling soon after cement placement and declined steadily with time. Opposite permeable rock, a flow equilibrium tended to limit the lowest pressure to that of the formation. Across a dense bed, however, a very low value (2.5-lbm/gal equivalent) was achieved. By implication, we would expect the same behavior in a casing/casing annulus.

The critical period when flow can transpire may happen at an inopportune time. For example, the wellsite personnel may believe that the well is under control completely and proceed with the nipple-down procedure while cement sets. There is no way to shut in a well when the stack has been disconnected and one case history resulted in an offshore blowout while the rig crews were nippling down the BOP. The platform was lost along with all six wells on the platform.

The described blowout was an extreme example. Gas migration through cement may constitute nothing more than a nuisance if the channel conductivity is low. Even so, low flow rates can lead to excessive pressure buildup between casings and the need to vent gas on a continuous or periodic basis. While the upward migration may be arrested, a crossflow may charge a shallower zone and/or result in lost reserves. An operator should consider the likelihood of an annular flow based on local knowledge and attempt to mitigate the problem.

Table 6.11 lists common methods used to combat gas migration. The theory relating pressure loss to volume reduction has gained general acceptance in the industry. Preventative measures, however, are still the subject of some debate and source much competition in the service industry. Efforts to eliminate mud channels and improve bonding to the pipe and formation are requisite to problem prevention. But gas, being buoyant, can percolate upward and create flow channels even after all the mud has been removed.

Any water lost to the formation will lower cement pressure and so controlling fluid-loss has been recognized as an important measure.[56] Another benefit may be realized because a thick filter cake tends to increase the cement's self-supporting structure. Note, however, that filtrate losses were not a factor in the 2.5-lbm/gal equivalent measured in the Cooke *et al.*[55] well, and it is apparent that losses induced solely by hydration can be significant.

The bottom half of **Fig. 6.37** indicates that static gel strengths in a neat cement start developing immediately after movement ceases. Sabins *et al.*[57] showed that gas will percolate through cement if the gel strength is less than approximately 500 lbf/100 ft^2 and so we assume gas is immobilized at higher values. The top part of the chart describes a cement's hydrostatic-pressure reduction. The shaded area thus represents the critical period when the formation is underbalanced and gas can migrate. Many of the methods shown in Table 6.11 are directed towards eliminating or reducing this time.

Thixotropic cements retain fluid properties while being sheared but rapidly develop gel strenths if left static. The intent of thixotropic cements is to achieve a 500-lbf/100 ft^2 gel strength before the well becomes underbalanced or to at least reduce the critical period such that any gas entries are not able to coalesce and form a flow channel.

The so-called right-hand-set cements are designed to remain fluid and experience little gel-strength increase for a considerable time.[58] Hence the cement can transmit hydrostatic pressure while most of the volume reduction is taking place. Once they start developing, the gels increase rapidly and provide the same advantage as a thixotropic cement. A similar, though mechanical, approach is to rotate the casing at extremely slow speeds while cement is setting.[59] The agitation inhibits gel strengths until pipe movement stops at some predetermined maximum torque.

Cements are made compressible by introducing gas into the slurry under controlled conditions. The gas expands into the space voided by water losses and greatly reduces the impact on hydrostatic pressure. Foamed cements, prepared by shearing nitrogen with a surfactant, are compressible and experience less pressure reduction in response to a given water loss. Even so, foamed cements are used primarily to reduce downhole slurry density and would not be suitable for most applications where gas migration is a problem. The method described by Ott *et al.*[60] uses a solid or liquid additive that generates hydrogen gas bubbles after placement downhole. Intentionally putting gas in a cement to combat gas migration may seem a contradictory idea, but the gas represents only a 3% volume fraction; the bubbles are supposed to be small enough to remain discrete and not migrate.[61]

A fundamentally different class of cements uses an additive to form a flow barrier in the slurry. Some products used for this purpose include latex, specialty polymer/slag systems, and microsilica.[62,63] These additives tend to lower the filtrate and may offer other beneficial aspects (minimal shrinkage in the latex systems for instance). A similar method uses a foaming surfactant which, in theory, foams any gas entries and increases bubble viscosity to the point that the gas is immobilized.[64]

Simply increasing the cement density helps to control annular flows because the initial overbalance is higher and less mixing water means less overall effect from volume reduction. Holding backpressure on the annulus has been attempted, but the procedure is not totally effective because the gels do not transmit surface pressure unless the slurry can be moved. Another popular technique is to limit cement coverage over the problem intervals. If necessary, additional stages are pumped to get the required fillup.

Unfortunately, no universal solution to the problem of annular gas flows has been found. A combination of methods may work best in a given area. In severe cases, an operator may need to resort to mechanical means and position an external casing packer (ECP) somewhere above the offending zone. The ECP must be located in competent, near-gauge rock and, ideally, at a location below any shallower zone which would otherwise be a receptacle for crossflow.

6.5.4 Liner-Top Tests. Getting adequate cement isolation on a drilling or production liner can be difficult under the best of circumstances. Inadequate liner isolation has caused many well-control problems and it is imperative for the operator to test the liner top before proceeding with the drillout. Typically, the casing is cleaned out to the liner top with the drilling fluid used during the cement job and a packer is set above the lap. Pump pressure is applied to a surface pressure that equates to something higher than the fracture pressure in the liner annulus (at least 500 psi for wells drilled in United States Federal waters). However, the testing is only halfway complete when a positive test is verified. The running string then should be displaced fully or partially with a fluid lighter than the pore pressure to do a dry or differential test.

In Iran, a regressive pressure environment dictates the use of a drilling liner to case off and isolate a 14.5-lbm/gal zone before drilling the normally pressured reservoir. There is one case history from the area where the liner lap was not differentially tested. The mud weight was reduced to drill out, the well immediately kicked, and a blowout ensued. The liner top did pass a positive test, which points up the fact that differential testing is required to evaluate fully liner-top integrity.

It is common for leaky liner tops to hold pump pressure or experience only a slow bleed-off. Thus a squeeze job to shut off the gas may be precluded and the only options are to live with the problem or contain the flow beneath a liner-top-isolation (LTI) packer. Collapse and differential pressure ratings in the LTI packer components should anticipate the maximum potential pressure trapped beneath the elements. This will be either the wellbore hydrostatic-pressure column at the time the packer was set or the pore pressure of the leaking zone less a gas hydrostatic pressure. Thermal expansion effects during production should be determined if the liner top is a long way from the perforations.

Problems

6.1 Intermediate casing has been set at 10,800 ft and natural gas is used to drill below the shoe. A large gas flow enters the wellbore at 12,200 ft and the well is shut in with an initial SICP of 4,200 psig. Estimate the kill-mud density and write the kill procedure. Assume the average temperature in the annulus is 185°F and consider the annulus to be filled with a 0.65-SG gas.

6.2 Refer to the problem described in Example 6.1.

a. Determine the pore pressure in density equivalent if there is a 1,500-ft column of 9.0-lbm/gal water on bottom.

b. What will happen to the BHP if casing pressure is held constant as the water is displaced above the drill collars?

6.3 Determine the MAASP for the well described in Example 6.1 if the fracture gradient at the casing shoe is 0.75 psi/ft.

6.4 Use the Poettmann and Begman correlation and estimate the air volume required to reduce the effective mud weight from 8.8 lbm/gal to 6.0 lbm/gal in an 8,000-ft well. Verify the result by numerical integration. Assume ideal gas behavior and no backpressure under the rotating head.

6.5 A well is shut-in at 8,000 ft vertical depth with a nitrified oil column in the drillstring. The respective gas and oil injection rates before shut-in were 1,000 scf/min and 250 gal/min. Estimate the kill density if the SIDPP is 1,000 psig and the oil weighs 7.0 lbm/gal. Assume the average temperature in the annulus is 150°F.

6.6 Write a spreadsheet program to facilitate the calculations in Problem 6.5.

6.7 Write a kill procedure for the well described in Example 6.2.

6.8 The intermediate casing point on a deep well is to be at 17,200 ft. The mud weight at this depth will be 4.0-lbm/gal underbalanced to the 14.0-lbm/gal pore pressure.

1. Determine the height of a 20.0-lbm/gal pill which will regain a balanced condition and allow tripping the drillstring.
2. Discuss the primary considerations for determining the pill density and column height.

6.9 Intermediate casing is set on a horizontal well at 7,000 ft TVD and the pay zone is at 7,200 ft TVD. Fresh water has been used to maintain an underbalanced condition on the well while drilling and the well is allowed to flow as the string is stripped out to 6,500 ft. The SICP at this point is 300 psig. Determine the mimimum top-kill fluid density if the objective is to start filling the hole with weighted mud so that the casing pressure is zero when the bit reaches 1,000 ft. The plugged displacement factor of the drillpipe is 0.01246 bbl/ft. The annulus and casing capacity factors are 0.03401 bbl/ft and 0.04592 bbl/ft, respectively.

6.10 List five reasons why underbalanced drilling is not practicable for all projects.

6.11 A horizontal well is designed to kickoff from vertical at 11,834 ft and maintain a constant build rate to intersect the target at 12,120 ft TVD. Wellbore-collapse considerations dictate a minimum 11.0-lbm/gal mud weight whereas the formation will fracture at an 11.5-lbm/gal equivalent. How much lateral can be drilled before the operator loses returns if annulus friction losses are 0.02 psi/ft?

6.12 Refer to the well characteristics shown in Fig. 6.13.

1. Plot the pore pressure and mimimum wellbore pressure as functions of horizontal-section length.
2. Prepare another plot except use a 9.2-lbm/gal mud weight.
3. For the latter case, at what lateral distance would this well begin to produce formation fluids as the string is pulled off bottom?

6.13 Refer again to Fig. 6.13.

1. Prepare overbalance plots for the situations where maximum hole inclinations are 60° and 80° using the same kickoff point, build rate, and measured depth.
2. What effect does increasing the mud weight have on the curve shapes?

6.14 Use the information given in **Table 6.12** to answer the following:

1. Determine the kill-mud density.
2. Will this well experience any gas migration to surface? Why or why not?
3. Assume the drillpipe pressure will be used to control the well when filling the string with kill-weight mud and determine the target ICP and FCP.
4. Again assume drillpipe control when filling the drillstring. Recommend a drillpipe-pressure reduction schedule.
5. Which kill procedure would you recommend, the Driller's Method or the Wait and Weight Method? Why?
6. Estimate the maximum pressure at the casing seat for your recommended kill procedure.
7. Estimate the maximum surface pressure for your recommended kill procedure.

6.15 Consider CT with the following mechanical properties and dimensions: nominal OD = 2.00 in., maximum OD tolerance = +0.01 in., nominal wall thickness = 0.134 in., wall tolerance = ±0.005 in., and minimum yield strength = 70,000 psi.

1. Use nominal dimensions in Barlow's equation and determine the internal pressure rating if material yield is assumed to constitute failure.
2. Calculate the internal pressure rating for a maximum-OD, minimum-wall tube.
3. What maximum pressure do you specify if this string is going to be used to drill your well?

6.16 List at least three instances in which the fracture pressure could be a factor for well control in cased holes.

TABLE 6.12—KICK DATA FOR A HORIZONTAL WELL

Hole Architecture	
Kickoff point from vertical	7,219 ft
Upper build rate	16°/100 ft
Tangent inclination	55°
Measured depth at start of tangent	7,563 ft
Vertical depth at start of tangent	7,513 ft
Tangent length	250 ft
Measured depth at end of tangent	7,813 ft
Vertical depth at end of tangent	7,656 ft
Lower build rate	18°/100 ft
Lateral inclination	92°
Measured depth at start of lateral	8,019 ft
Vertical depth at start of lateral	7,714 ft
Lateral length	2,650 ft
Measured depth at total depth	10,669 ft
Vertical depth at total depth	7,621 ft
Casing and Drillstring	
Intermediate casing information:	
Description	7⅝-in., 29.70-lbm/ft, N-80
Measured setting depth	7,713 ft
Vertical setting depth	7,598 ft
Open hole diameter	6½ in.
Drillstring information:	
Heavy-wall and compressive-service Drillpipe description	3½ in. 25.30-lbm/ft
Heavy-wall and compressive-service Drillpipe length	4,150 ft
Drillpipe description	3½ in., 13.30-lbm/ft
Drillpipe length	6,519 ft
Prekick Circulation Data and Mud Properties	
Kill circulation rate	105 gal/min
Kill-rate circulating pressure	740 psig
Mud type	Water-based polymer
Mud density	8.7 lbm/gal
Recorded Kick Data	
SIDPP	320 psig
SICP	300 psig
Pit gain	14 bbl
Other Known or Assumed Information	
Fracture gradient at casing seat	0.69 psi/ft
Assumed wellbore temperature at kill rate	70°F + 1.0°F/100 ft
Kick fluid	0.60 SG gas

6.17 Work Example 6.4 again but recast the problem so that the tool is a 9-ft long chemical cutter conveyed on 7/16-in. cable. The tool weighs 95 lbf.

6.18 The pore pressure in your well is 5,243 psig at 7,352 ft TVD. The objective is to use a clear-brine completion fluid and to design its density to maintain a 400-psi overbalance.

1. Determine the mixture density if the surface water temperature is 70°F and the temperature at perf depth is 217°F.
2. What fluids or fluid blends may be considered?

6.19 The following conditions apply to a shut-in gas well.

Packer depth = 11,100 ft, perforation midpoint = 11,190 ft, shut-in pressure at midpoint = 9,190 psig, standing water level = 11,220 ft, packer fluid density = 16.2 lbm/gal, 3½-in. tubing ID = 2.992 in., casing ID = 6.276 in., gas SG = 0.72, and static wellbore temperature = 70°F + 1.0°F/100 ft.

1. Estimate the shut-in tubing pressure.
2. What is the pore pressure in density equivalent?
3. Assume the packer leaks and dumps the packer fluid on the formation. Will an overbalance be provided after the fluids equalize? If not, what is the expected shut-in pressure on the tubing annulus?
4. What is the expected shut-in pressure on the tubing annulus if the tubing hanger leaks?

5. Determine the differential burst load above the packer if the tubing hanger leaks and the casing annulus is backed up with a 9.0-lbm/gal fluid.

6. Assume the operator used 10.0-lbm/gal brine as the packer fluid. What is the expected shut-in pressure on the tubing annulus if the packer leaks?

6.20 Regulations require testing the intermediate casing in your well to 70% of its minimum internal yield pressure before drilling out the shoe. Someone neglected to do this at the opportune time and prematurely weighted up the mud from 9.0 to 12.0 lbm/gal.

1. Determine the applied surface pressure which will result in the specified load on bottom if the float collar is at 9,200 ft and the pipe is rated for 5,750 psi.
2. How do you best go about meeting the government's requirement?

6.21 Refer to the procedures given in Table 6.8. For each item, describe how many barriers are in place and the type of barrier (positive or conditional).

6.22 Write a bullhead procedure to kill the well described in Problem 6.19. Incorporate a 300-psi overbalance and adjust the fluid density for temperature effects.

6.23 A gas well was secured overnight during a routine cleanout job and the SICP the next morning was 200 psig. The perforation midpoint depth is at 6,100 ft and an 11.1-lbm/gal brine has been used to overbalance the 10.2-lbm/gal formation pressure.

1. What is the approximate gas column height?
2. Discuss options for removing the gas cap if tubing is in the hole.
3. Discuss options for removing the gas cap if all the tubing is standing in the mast.

6.24 The following conditions apply to the proposed work to drillout a bridge plug and commingle production from two abandoned gas zones with the existing horizon.

Plug depth = 7,670 ft, upper zone depth = 9,495 ft, upper zone pressure = 4,216 psig, lower zone depth = 10,477 ft, lower zone pressure = 4,656 psig, workover-fluid density = 9.0 lbm/gal, gas SG = 0.60, and static wellbore temperature = 70°F + 1.0°F/100 ft.

1. What fluid density will balance the pore pressures in the lower zones?
2. How much pressure do you expect underneath the plug?
3. How much pressure differential now exists across the plug?
4. Determine the drillout-fluid density if the objective is to balance the trapped pressure.

6.25 Refer to the casing surge pressures shown in Fig. 6.33.

1. Determine the lowering velocity that will break down the shoe if the fracture gradient is 0.88 psi/ft.
2. What type of float equipment would you recommend for this job?

6.26 You are planning a 7-in. casing job for a well with the following conditions.

Total depth = 7,200 ft [2194.6 m], surface casing = 9⅝-in. 36-lbm/ft, surface casing depth = 1,500 ft, loss zone depth = 4,100 ft, loss zone strength = 11.1 lbm/gal equivalent, and average hole diameter = 9 in.

Compute the maximum safe running speed at string depths of 1,000 ft and 7,000 ft if the following mud properties apply.

Density = 9.6 lbm/gal, gel strength = 10 lbf/100 ft², θ_{600} = 36 lbf/100 ft², θ_{300} = 23 lbf/100 ft², θ_{200} = 14 lbf/100 ft², θ_{100} = 10 lbf/100 ft², θ_6 = 5 lbf/100 ft², and θ_3 = 3 lbf/100 ft².

6.27 The following conditions pertain to a surface casing cement job on an offshore well.

Present depth = 2,265 ft, conductor pipe description = 20-in. 94-lbm/ft, conductor pipe depth = 766 ft, mud density = 9.4 lbm/gal, pore pressure gradient = 0.452 psi/ft, and hole diameter = 17½ in.

Fresh water will be used as the preflush fluid ahead of the 15.6-lbm/gal cement slurry. Determine the maximum spacer volume that can be pumped and still maintain well control.

6.28 Casing has been set at 5,600 ft on a well located in the Michigan basin and a 16.0-lbm/gal mud has been used to drill a series of tight carbonates containing 22.0-lbm/gal pore pressure. The cement design calls for circulating a 17.0-lbm/gal cement to 4,500 ft from the 8,100-ft shoe.

1. Determine the initial and final choke backpressures if the intent is to overbalance a zone at 6,200 ft during the cement job.
2. Can this be done if the fracture gradient at the last shoe is 1.3 psi/ft?

6.29 One cubic foot of a 15.6-lbm/gal neat cement slurry contains 59% water (by volume). Place this cement in the annulus at an initial hydrostatic pressure of 6,000 psi. Approximate the pressure drop in the slurry if gels prevent hydrostatic transmission and filtrate losses reduce the water content by 0.001 ft³. Water compressibility at downhole conditions is 2.6×10^{-6} psi^{-1}.

6.30 A production liner has been cemented at 10,061-ft vertical depth. A rental packer will be set within 40 ft of the liner top. Your job is to write the test procedure. The following conditions apply.

Liner top = 9,330 ft, intermediate-casing depth = 9,611 ft, mud density = 13.2 lbm/gal, fracture gradient at shoe = 0.831 psi/ft, and open hole diameter = 6½ in.

1. What minimum pressure do you recommend for the positive test?
2. After setting the packer and releasing drillpipe pressure, how much differential pressure is applied to the pay sand at 9,870 ft?
3. Assume gas flows at a prolific rate from the liner top during the differential test. Determine the minimum mud density requirement to trip the test packer safely if the pay sand was overbalanced originally by 0.3 lbm/gal.

Nomenclature

A = area, in.²
A_c = cable area, in.²
d_o = outer diameter, in.
D = depth, ft
f_g = gas volume fraction, dimensionless
F_f = friction force, lbf
K = brine density decline constant, lbm/(gal-°F)
L_k = kick fluid length, ft
p = pressure, psi
p_{bh} = bottomhole pressure, psi
p_{ch} = choke backpressure, psi

p_{cs} = shut-in casing pressure, psi
p_h = hydrostatic pressure, psi
p_{ip} = internal presure capacity, psi
p_p = pore pressure, psi
p_{pr} = pseudoreduced pressure, dimensionless
p_{sh} = casing shoe pressure, psi
p_t = tubing pressure, psi
Δp_{ac} = pressure change resulting from a change in a fluid's kinetic energy, psi
Δp_f = circulating friction pressure, psi
Δp_{ss} = surge/swab pressure, psi
t_w = wall thickness, in.
T = temperature, °F or °R
T_{avg} = average wellbore temperature, °F
T_{pr} = pseudoreduced temperature, dimensionless
T_s = surface temperature, °F
V_g = gas volume, ft^3 or gal
V_m = mud volume, gal
W = weight in air, lbf
z = compressibility factor, dimensionless
γ_g = gas specific gravity, dimensionless
θ_N = viscometer reading, lbf/100 ft^2
ρ_{bh} = density at wellbore conditions, lbm/gal
ρ_e = effective downhole density, lbm/gal
ρ_g = gas density, lbm/gal
ρ_{km} = kill mud density, lbm/gal
ρ_m = mud density, lbm/gal
ρ_{mg} = gas/mud mixture density, lbm/gal
ρ_s = density at surface conditions, lbm/gal
σ_s = material strength, psi

References

1. Grace, R.D.: "Pressure Control in Balanced and Underbalanced Drilling in the Anadarko Basin," paper SPE 5396 presented at the 51st Annual Technical Conference and Exhibition, New Orleans, 3–6 October 1976.
2. Bennion, D.B.: "Reservoir Screening Criteria for Underbalanced Drilling," *Pet. Eng. Intl.* (February 1977) 33.
3. Duda, J.R., Medley, G.H. Jr., and Deskins, W.G.: "Strong Growth Projected for Underbalanced Drilling," *Oil and Gas J.* (23 September 1996) 67.
4. Lyons, W.C.: *Air and Gas Drilling Manual*, Gulf Publishing Company, Houston (1984) 1-2, 115–123.
5. Krug, J.A. and Mitchell, B.J.: "Charts Help Find Volume, Pressure Needed for Foam Drilling," *Oil and Gas J.* (7 February 1972) 61.
6. Okpobiri, G.A. and Ikoku, C.U.: "Volumetric Requirements for Foam and Mist Drilling Operations," *SPEDE* (February 1986) 71.
7. Mitchell, B.J.: *Advanced Oilwell Drilling Engineering Handbook*, ninth edition, SPE, Richardson, TX (1992) 486.
8. Poettmann, F.H. and Begman, W.E.: "Density of Drilling Fluids Reduced By Air Injection," *World Oil* (August 1955).
9. Nessa, D.O., Tangedahl, M.J., and Saponja, J.: "Offshore Underbalanced Drilling System Could Revive Field Developments," *World Oil*, (July 1997) 61.
10. Brill, J.P. and Arirachakaran, S.J.: "State of the Art in Multiphase Flow," *JPT* (May 1992) 538.
11. Guo, B., Hareland, G., and Rajtar, J.: "Computer Simulation Predicts Unfavorable Mud Rate and Optimum Air Injection Rate for Aerated Mud Drilling," *SPEDC* (June 1996) 61.
12. *Compressed Air and Gas Data*, second edition, C.W. Gibbs (ed.), Ingersoll-Rand Co., Woodcliff Lake, NJ (1971) 34–12.
13. Tangedahl, M.J. and Stone, C.R.: "Rotating Preventers: Technology for Better Well Control," *World Oil* (October 1992) 63.
14. Koenig, R.L.: "An Extraordinary Drilling Challenge in the Anadarko Basin," paper SPE 22575 presented at the 1991 SPE Annual Technical Conference and Exhibition, Dallas, 6–9 October.
15. Bourgoyne, A.T. Jr.: "Rotating Control Head Applications Increasing," *Oil and Gas J.* (9 October 1995) 72.
16. Joseph, R.A.: "Planning Lessons, Problems Gets Benefits of Underbalance," *Oil and Gas J.* (20 March 1995) 86.
17. Ramos, A.B. Jr. *et al.*: "Horizontal Slim-Hole Drilling ith Coiled-Tubing: An Operator's Experience," *JPT* (October 1992) 1119.
18. Wodka, P. *et al.*: "Underbalanced Coiled-Tubing-Drilled Horizontal Well in the North Sea," *SPEDC* (May 1996) 406.
19. Tomren, P.H., Iyoho, A.W., and Azar, J.J.: "Experimental Study of Cuttings Transport in Directional Wells," *SPEDE* (February 1986) 43.
20. Gavignet, A.A. and Sobey, I.J.: "Model Aids Cuttings Transport Prediction," *JPT*, (Sept. 1989) 916.
21. Sifferman, T.R. and Becker, T.E.: "Hole Cleaning in Full-Scale Inclined Wellbores," *SPEDE* (June 1992) 115.
22. Hemphill, T. and Larsen, T.I., "Hole-Cleaning Capabilities of Water- and Oil-Based Drilling Fluids: A Comparative Experimental Study," *SPEDC* (December 1996) 201.
23. Mueller, M.D., Quintana, J.M., and Bunyak, M.J.: "Extended-Reach Drilling from Platform Irene," *SPEDE* (June 1991) 138.
24. Gao, E. and Young, A.C.: "Hole Cleaning in Extended Reach Wells: Field Experience and Theoretical Analysis Using a Pseudo-Oil (Acetal) Based Mud," paper SPE/IADC 29425 presented at the 1995 SPE/IADC Drilling Conference, Amsterdam, 28 February–2 March.
25. Payne, M.L., Cocking, D.A., and Hatch, A.J.: "Critical Technologies for Success in Extended-Reach Drilling," paper SPE 28293 presented at the 1994 SPE Annual Technical Conference and Exhibition, New Orleans, 25–28 September. Reprinted as SPE 30140 Brief, *JPT* (February 1995) 121.
26. Alfsen, T.E. *et al.*: "Pushing the Limits for Extended Reach Drilling: New World Record From Platform Statfjord C, Well C2," *SPEDC* (June 1995) 71.
27. Rader, D.W., Bourgoyne, A.T. Jr., and Ward, R.H.: "Factors Affecting Bubble-Rise Velocity of Gas Kicks," *JPT* (May 1975) 571.
28. Johnson, A.B. and Cooper, S.: "Gas Migration Velocities During Gas Kicks in Deviated Wells," paper SPE 26331 presented at the 1993 SPE Annual Technical Conference and Exhibition, Houston, 3–6 October.
29. Santos, O.L.A.: "Well-Control Operations in Horizontal Wells," *SPEDE* (June 1991) 111.
30. Beggs, H.D. and Brill, J.P.: "A Study of Two-Phase Flow in Inclined Pipes," *JPT* (May 1973) 607; *Trans.*, AIME, **255.**
31. Brawn, R.G.: "Unique Rig Uses Flexible Drillstem and Electric Motor," *World Oil* (May 1964).
32. Delacour, J.: "French Fexible Drilstem Tool and Technique Look Good," *World Oil* (July 1965).
33. Gates, G.: "A New Rig for Shallow Drilling," *Oilweek* (2 August 1976) 7.
34. Simmons, J. and Adam, B.: "Evolution of Coiled Tubing Drilling Technology Accelerates," *Pet. Eng. Intl.* (September 1993) 26.
35. Littleton, J.: "Horizontal Drilling with Coiled Tubing Gains Momentum," *Pet. Eng. Intl.* (July 1992) 22.
36. Gronseth, J.M.: "Coiled Tubing . . . Operations and Services, Part 14—Drilling," *World Oil* (April 1993) 43.
37. Leising, L.J. and Newman, K.R.: "Coiled-Tubing Drilling," *SPEDC* (December 1993) 227.
38. Blount, C.G.: "The Challenge for the Coiled-Tubing Industry," *JPT* (May 1994) 427.
39. Newman, K.R. and Allcorn, M.G.: "Coiled Tubing in High-Pressure Wells," paper SPE 24793 presented at the 1992 SPE

Annual Technical Conference and Exhibition, Washington, DC, 4–7 October.
40. Goodman, J.: *Mechanics Applied to Engineering*, eighth edition, Longman Greens, London (1914) 421–23.
41. *Bulletin on Formulas and Calculations for Casing, Tubing, Drill Pipe, and Line Pipe Properties*, *Bull. 5 C3*, fifth edition, API, Washington, DC (July 1989).
42. Tipton, S.M.: "Coiled-Tubing Surface Characteristics and Effects on Fatigue Behavior," *JPT* (June 1997) 612. Synopsis of paper SPE 38411 presented at the 1997 SPE/ICoTA North American Coiled Tubing Roundtable, Montgomery, Texas, 1–3 April.
43. Sas-Jaworsky, A. II: "Coiled Tubing . . . Operations and Services, Part 3—Tube Technology and Capabilities," *World Oil* (January 1992) 95.
44. United States Department of Interior MMS Regulation 250.210 governing Oil and Gas and Sulphur Operations in the Outer Continental Shelf, *Federal Register*, **56, 16** (1 April 1988) 2685.
45. Carney, L.L.: "Completion Fluids: Considerations for Proper Selection," *Pet. Eng. Intl.* (April 1977) 62.
46. Millhone, R.S.: "Completion Fluids for Maximizing Productivity—State of the Art," *JPT* (January 1983) 47.
47. Meyer, R.L. and Vargas, R.H.: "Process of Selecting Completion or Workover Fluids Requires Series of Tradeoffs," *Oil and Gas J.* (30 January 1984) 144.
48. Spies, R.J. *et al.*: "Field Experience Utilizing High-Density Brines as Completion Fluids," *JPT* (May 1983) 881.
49. Ezzat, A.M., Augsburger, J.J., and Tillis, W.J.: "Solids-Free, High-Density Brines for Packer-Fluid Application," *JPT* (April 1988) 491.
50. Rike, J.L. *et al.*: "Completion and Workover Well Control Needs Are Different!," paper CIM\SPE 90-22 presented at the CIM/SPE International Technical Meeting, Calgary, 10–13 June 1990.
51. Campos, W., Lage, A.C.V.M., and Poggio, A. Jr.: "Free-Fall-Effect Calculation Ensures Better Cement-Operation Design," *SPEDC* (September 1993) 175.
52. Sauer, C.W.: "Mud Displacement During Cementing: A State of the Art," *JPT* (September 1987) 1091.
53. Moore, P.L.: *Drilling Practices Manual*, second edition, PennWell Publishing Co., Tulsa, (1986) 432–36 (contribution by Dwight K. Smith).
54. Smith, D.K.: *Cementing*, Monograph Series, SPE, Richardson, Texas (1992) **4**, 94–97.
55. Cooke, C.E. Jr., Kluck, M.P., and Medrano, R.: "Field Measurements of Annular Pressure and Temperature During Primary Cementing," *JPT* (December 1984) 2181.
56. Christian, W.W., Chatterji, J., and Ostroot, G.W.: "Gas Leakage in Primary Cementing—A Field Study and Laboratory Investigation," *JPT* (November 1976) 1361.
57. Sabins, F.L., Tinsley, J.M., and Sutton, D.L.: "Transition Time of Cement Slurries Between the Fluid and Set State," *SPEJ* (December 1982) 875.
58. Sykes, R.L. and Logan, J.L.: "New Technology in Gas-Migration Control," paper SPE 16653 presented at the 1987 SPE Annual Technical Conference and Exhibition, Dallas, 27–30 September.
59. Sutton, D.L. and Ravi, K.M.: "Low-Rate Pipe Movement During Cement Gelation to Control Gas Migration and Improve Cement Bond," paper SPE 22776 presented at the 1991 SPE Annual Technical Conference and Exhibition, Dallas, 6–9 October.
60. Ott, W.K., Miller, G.W., and Sutton, D.L.: "Compressible Cement Slurry Helps Solve Annular Gas Flow," paper SPE 10481 presented at the 1982 Offshore Southeast Asia Conference, Singapore, 9–12 December.
61. Bour, D.L. and Wilkinson, J.G.: "Combating Gas Migration in the Michigan Basin," *SPEDE* (March 1992) 65.
62. Fery, J.J. and Romieu, J.: "Improved Gas Migration Control in a New Oil Well Cement," paper SPE 17926 presented at the 1989 Middle East Oil Technical Conference and Exhibition, Manama, Bahrain, 11–14 March.
63. Grinrod, M. and Dingsoyr, E.O.: "Development and Use of a Gas-Tight Cement," paper SPE\IADC 17258 presented at the 1988 SPE/IADC Drilling Conference, Dallas, 28 February–2 March.
64. Hibbeler, J.C. and Thay, M.: "Cost-Effective Gas Control: A Case Study of Surfactant Cement," paper SPE 25323 presented at the 1993 Asia Pacific Oil and Gas Conference and Exhibition, Singapore, 8–10 February.

SI Metric Conversion Factors

bbl	× 1.589 873	E − 01	= m^3
deg	× 1.745 329	E − 02	= rad
ft	× 3.048*	E − 01	= m
ft^3	× 2.831 685	E − 02	= m^3
°F	(°F − 32)/1.8		= °C
gal	× 3.785 412	E + 00	= L
in.	× 2.54*	E + 01	= mm
$in.^2$	× 6.451 6*	E + 02	= mm^2
lbf	× 4.448 222	E + 00	= N
lbm	× 4.535 924	E − 01	= kg
lbm/ft	× 1.488 164	E + 00	= kg/m
psi	× 6.894 757	E − 03	= MPa
psi^{-1}	× 1.450 377	E − 01	= kPa^{-1}
°R	°R/1.8		= K
scf	× 2.831 685	E − 02	= std m^3

* Conversion factor is exact.

Chapter 7
Well-Control Equipment

7.1 Introduction

High-pressure blowout prevention equipment (BOPE) is exposed to well pressure during a well-control event and comprises the circulating and control system between the pump and the choke. Low-pressure components operate at near-atmospheric pressure and include the circulation and processing equipment downstream of the choke. Closing units furnish the hydraulic energy to activate the BOPE while other accessory items support the overall control effort. Our objectives in this chapter are to address how this equipment operates and to offer some design considerations. This chapter is not intended to be an equipment catalogue and specific features available from a given make or model will receive less emphasis except when discussion serves the overall purpose. The primary focus is on surface stacks. Unique modifications for blowout preventers (BOPs) that sit on the seafloor will be covered in Chap. 8.

7.2 High-Pressure Equipment

The casing, last wellhead, BOP stack, choke and kill lines, and manifold must be designed to withstand the annulus pressures in a well-control application. Between the pump and the bottom of the string are other high-pressure components that must function reliably.

7.2.1 Casing. Perhaps the most important, yet most neglected, piece of equipment in pressure control is the last casing string set in the well. We can arrange our stacks and manifold with sufficient backup and redundancy to handle almost any contingency, test every component every week, and have the crews drilled to perfection. But all these efforts will be to no avail if the casing ruptures during a well-control procedure.

It all starts back in the office with the well planner. Casing designs need to anticipate well control in the base loading criteria. In surface and intermediate pipe, the worst-case approach is adopted by many engineers who base their burst designs on the assumption that procedural or equipment failure results in a situation where formation fluids completely evacuate the wellbore before the well is closed in. If gas unloads the hole, the shut-in casing pressure (SICP) is the pore pressure less the hydrostatic pressure of the gas column if this condition does not fracture the rock. Other designs are controlled by the fracture gradient, in which case the SICP is the fracture pressure less the gas hydrostatic pressure. Thus

$$p_{cs} = p_p - g_g D \quad \text{(7.1)}$$

in the former and

$$p_{cs} = p_{fi} - g_g D_{sh} \quad \text{(7.2)}$$

in the latter. The depth term in Eq. 7.1 is taken to be the vertical depth at the next casing point or perhaps a known gas interval. The shoe depth in Eq. 7.2 may apply to any other known weak point in the well. These relations do not account for a variable gas density but can be modified if so desired.

As we discussed in Chap. 6, design loads deeper in the well can assume some backup density in the annulus. The design load at the shoe is

$$p_{sh} = p_{cs} + D_{sh}(g_g - g_b), \quad \text{(7.3)}$$

where g_b = is the hydrostatic gradient of the selected backup fluid.

Example 7.1. Your hole program calls for setting 9⅝-in. intermediate casing at approximately 10,100 ft and drilling an 8½-in. hole to 12,900 ft. The expected fracture gradient at the shoe is 0.91 psi/ft and the maximum anticipated pore-pressure gradient is a 16.0-lbm/gal equivalent. Determine the minimum internal yield pressure (MIYP) requirements at the top and bottom of the string if a 1.2 design factor is applied to the discussed load criterion. Use a 9.0-lbm/gal backup density in the 9⅝-in. annulus.

Solution. A gas column at the shut-in pressures will have an average, conservative gradient of about 0.15 psi/ft. Check to see if the shoe will fracture and calculate the imposed pressure and gradient at 10,100 ft.

$$p_{sh} = (0.0519)(16.0)(12,900) - (0.15)(2,800),$$

$$p_{sh} = 10,712 - 420 = 10,292 \text{ psi},$$

$$g_{sh} = 10,292/10,100 = 1.02 \text{ psi/ft}.$$

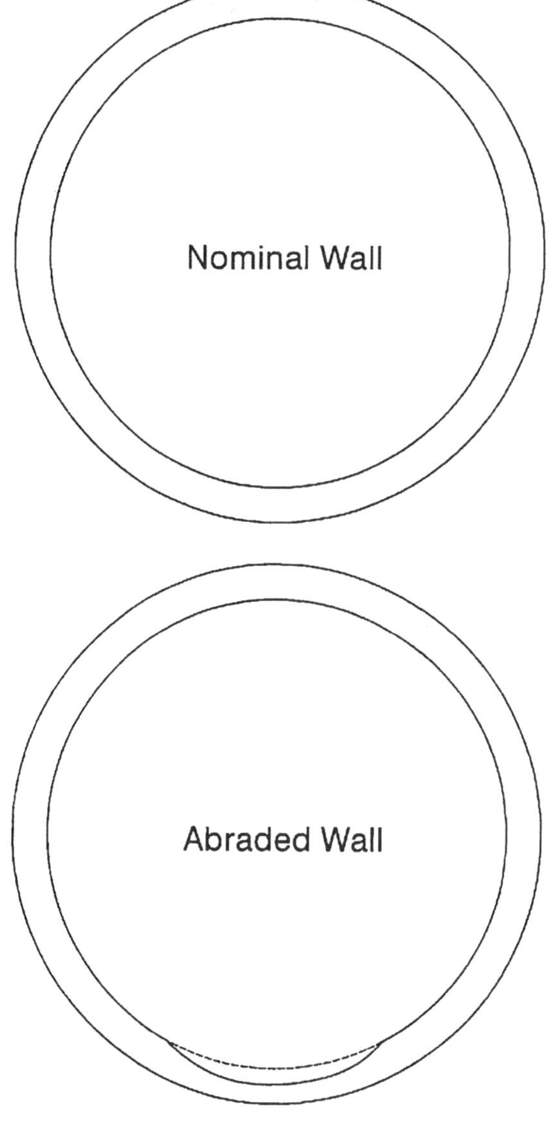

Fig. 7.1—Casing wear pattern resulting from uneven side force distribution.

Hence the fracture gradient controls this design and Eq. 7.2 yields

$$p_{cs} = (10,100)(0.91 - 0.15) = 7,676 \text{ psi}.$$

Eq. 7.3 gives the differential load on the bottom joint,

$$p_{sh} = 7,676 + (10,000)[0.15 - (0.0519)(9.0)]$$
$$= 4,473 \text{ psi}.$$

The designer multiplies both results by 1.2 and obtains top and bottom strength requirements of 9,211 and 5,368 psi, respectively.

Fully evacuating the mud may seem an unrealistic load assumption to some, but the situation can and has happened. Some companies adopt a less stringent policy and have their engineers design casings with at least some mud in the hole (say half-full) or base the design on predicted surface pressures with an extremely large gas kick.

In a gas well, the worst-case burst condition on the production casing may be encountered if the well is shut-in and the tubing has a shallow leak. The shut-in tubing pressure (SITP) is seen now on the casing gauge while, down the hole, this SITP is applied to the selected packer fluid. Hence the packer-fluid density is an important variable and the planner must know something about how the well is to be completed before submitting final pipe designs.

On critical wells, the casing body and end-areas should be inspected one last time before running in the hole. In some cases, hydrotesting the string or testing connections with gas during the running procedure is warranted. In any event, the string should be pressure-tested to a safe level before drilling out the shoe. For example, operators in the United States Outer Continental Shelf (OCS) must test casing to 70% of the MIYP.[1]

The initial test, however, does not guarantee that the casing will retain integrity through the next pipe job. Recall Barlow's equation in which the internal pressure capacity of a thin-walled tube is given by

$$p_{ip} = 2\sigma_s t_w / d_o, \quad \ldots\ldots\ldots\ldots (7.4)$$

where p_{ip} = internal pressure capacity, σ_s = material strength, t_w = tube wall thickness, and d_o = tube outer diameter.

We see from the relation that wall thickness is an important parameter in determining how much pressure the casing can withstand. The problem in intermediate and surface casings is that a rotating drillstring may abrade part of the wall and thereby degrade the string's MIYP rating over time.

This is a major concern if the drillstring is rotating in a dogleg. Resultant side forces from the hangdown weight can wear out a string quickly and the wall loss will not be uniform but, like a keyseat, concentrated in one area (see **Fig. 7.1**). Shallow doglegs where hangdown weight is high are especially harmful and preventative measures to keep shallow casing straight will go far towards eliminating the problem potential.

Similar wear patterns can be expected in the top joint or two if the drillstring is misaligned or drilling is conducted with a bent kelly. During a well-control problem, the wall loss is in a part of the casing where integrity is most needed—directly under the rig and personnel. It is important to keep the rig level, the pipe centered in the hole, the kelly straight, and to ensure the surface casing remains vertical by giving the cement sufficient wait-on-cement (WOC) time.

General wear can be mitigated by taking other steps such as using drillpipe protectors and smooth hardfacing on the tool joints.[2,3] Even so, some procedure should be instituted to monitor wear on casings exposed to hundreds or thousands of rotating hours. One way to track wear is to run a base casing-caliper log before drillout and on a periodic basis thereafter. Wall-thickness measurements based on the nominal outer diameter (OD) are obtained along with indications as to whether the loss is uniform or concentrated on one side. A derated MIYP can be estimated using the yield strength in Eq. 7.4 if the loss is uniform. Song *et al.*[4] presented a method to predict rupture (not internal yield) pressure for nonuniform wear. Setting a packer and retesting the casing is another approach that eliminates all doubt as to what pressure the string can withstand.

7.2.2 Casing Heads and Spools. After surface casing is cemented, the typical procedure is to cut off the pipe a predetermined distance below the floor and to screw, weld, or forge a casing head directly onto the casing. The first BOP stack on the well then is nippled up to this starting head or "bradenhead." The surface casing and its cement sheath are the structural foundation for the well and the system must be stout enough to support the weight of the BOP stack and suspend

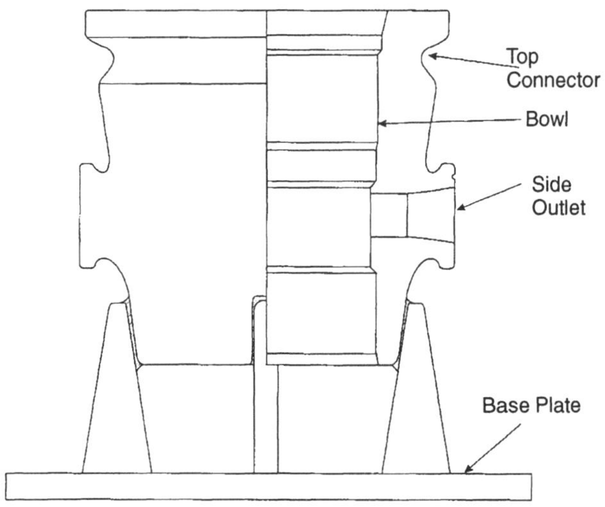

Fig. 7.2—Casing head with a base plate and clamp-type top connection. Courtesy of ABB Vetco Gray.

all subsequent tubulars. As shown in **Fig. 7.2,** the starting head may be equipped with a base plate to help distribute the compressive loads to the earth.

The casing-head bowl accommodates the slip or mandrel-type casing hanger for the next string. A reliable hanger seal demands that the bowl be in good condition and an operator can protect the bowl from the rigors of drilling by installing a temporary wear bushing. The top connection may be threaded for working-pressure ratings 2,000 psi and lower, but clamp or multibolt flange connections are necessary at higher working pressures. The pressure seal on a bolted flange is provided by a ring gasket that fits in a groove machined onto the flange face.

Threaded or flanged side outlets give access to the annulus of the suspended casing or can be used as as an emergency kill line connection if the stack equipment is nippled up to the head. The working-pressure rating of any valves, companion flanges, bull plugs, or plumbing coming off the side outlet must be commensurate with that of the head.

Each suspended string requires an additional head called a casing spool. The first intermediate casing string is hung from the starting head and is cut off a specified distance above the flange. The cut-off section then extends into the lower portion of the spool. The pressure ratings of the top and bottom flanges may stay the same if the maximum potential pressure does not increase. But pore pressures tend to increase with well depth and so the pressure rating of the top flange is usually higher. Secondary seals designed to pack-off against the casing and the spool may be used to isolate the well fluids and pressures from the bottom flange. As illustrated in **Fig. 7.3,** this crossover method gives the spool a working-pressure equivalent to the top flange.

The American Petroleum Institute (API) has established standards for wellhead manufacturers in Specification 6A.[5] API wellheads are rated for working pressures of 2,000, 3,000, 5,000, 10,000, 15,000, and 20,000 psi. The working-pressure rating is considered the maximum safe pressure to which a component should be used during service. The specified test pressure at the manufacturing facility is as much as double the working pressure so there is a built-in safety factor for a given application. Nonetheless, exceeding the rated working pressure should be avoided.

Table 7.1 lists the API specifications for some wellhead dimensions. The nominal size is the diameter through the flange

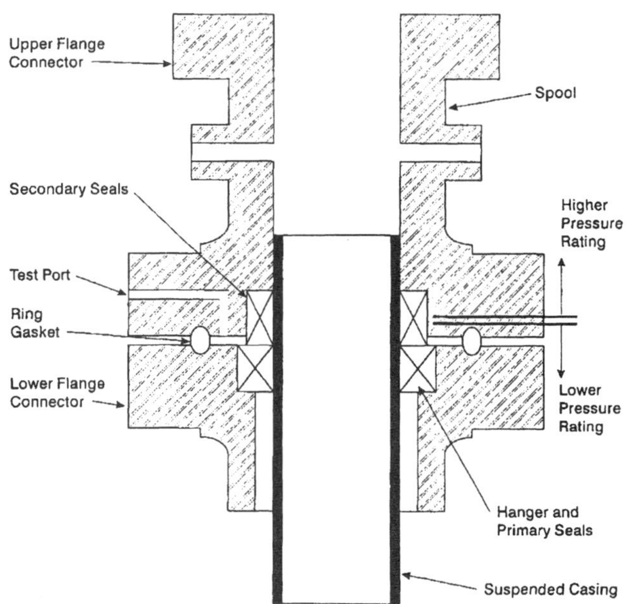

Fig. 7.3—Crossover casing spool with secondary packoff.

bore whereas the minimum vertical bore sets the maximum bit or tool diameter that can be run through the head. The casing sizes shown in Table 7.1 refer to the maximum casing OD that will fit into the lower flange prep. Exceeding the minimum weight for a given OD will result in a casing drift diameter less than the minimum vertical bore of the head.

7.2.3 Stack Equipment. The basic functions of a surface BOP stack are to seal the well against the drillstring or open hole and contain well pressure, provide a full-bore opening to allow passage of drilling and testing tools, permit unrestricted flow of fluids to the choke line while the preventers are closed, allow drillstring movement when the well is shut-in to prevent sticking or allow stripping, provide a way to allow fluids to be pumped into the well below a closed preventer, and convey drilling fluid to the bell nipple and flowline.

The working-pressure rating of the open wellhead and BOPE should be based on the same criteria discussed for the casing. Compromises may be necessary, however, in deep high-pressure wells where a dry-gas gradient yields an SICP higher than the available equipment can tolerate. Of course, one can argue that the chance of completely unloading a deep well is reduced as well.

Going back to Example 7.1, the 9⁵/₈-in. casing likely will be hung from a 13⁵/₈-in. head. The potential SICP indicates the need for a casing spool and BOPE rated for 10,000 psi. The top flange of the spool will have an 11-in. nominal size. Assuming the connecting flange on bottom was rated for 5,000 psi, conventional nomenclature would describe this as a 13⁵/₈-in. 5M × 11-in. 10M casing spool.

The BOP stack in Example 7.1 must have a minimum flange size equal to that of the casing spool. A 10M stack with a larger bore would be acceptable and may be the best choice, considering transportation costs, additional rig time, and risk of blowout to change out the stack completely. The stack used in the last hole section can be unbolted from the casing head and suspended while the casing is conditioned and the spool installed. A crossover flange is attached to the spool and the stack lowered and bolted on again.

Annular Preventers. The annular preventer is the topmost preventer and is usually the preferred shut-in device be-

TABLE 7.1—MINIMUM BORE DIMENSIONS THROUGH API WELLHEADS			
Nominal Size (in.)	Working Pressure (psi)	Maximum Casing OD and Minimum Weight/ft Below Head	Minimum Vertical Bore Through the Head (in.)
$7^1/_{16}$	2,000	7-in. 17-lbm/ft	6.45
$7^1/_{16}$	3,000	7-in. 20-lbm/ft	6.36
$7^1/_{16}$	5,000	7-in. 23-lbm/ft	6.28
$7^1/_{16}$	10,000	7-in. 29-lbm/ft	6.09
$7^1/_{16}$	15,000	7-in. 38-lbm/ft	5.83
$7^1/_{16}$	20,000	7-in. 38-lbm/ft	5.83
9	2,000	$8^5/_8$-in. 24-lbm/ft	8.00
9	3,000	$8^5/_8$-in. 32-lbm/ft	7.83
9	5,000	$8^5/_8$-in. 36-lbm/ft	7.73
9	10,000	$8^5/_8$-in. 40-lbm/ft	7.62
9	15,000	$8^5/_8$-in. 49-lbm/ft	7.41
11	2,000	$10^3/_4$-in. 40.5-lbm/ft	9.92
11	3,000	$10^3/_4$-in. 40.5-lbm/ft	9.92
11	5,000	$10^3/_4$-in. 51.0-lbm/ft	9.73
11	10,000	$9^5/_8$-in. 53.5-lbm/ft	8.41
11	15,000	$9^5/_8$-in. 53.5-lbm/ft	8.41
$13^5/_8$	2,000	$13^3/_8$-in. 54.5-lbm/ft	12.50
$13^5/_8$	3,000	$13^3/_8$-in. 61.0-lbm/ft	12.39
$13^5/_8$	5,000	$13^3/_8$-in. 72.0-lbm/ft	12.22
$13^5/_8$	10,000	$11^3/_4$-in. 60.0-lbm/ft	10.66
$16^3/_4$	2,000	16-in. 65-lbm/ft	15.09
$16^3/_4$	3,000	16-in. 84-lbm/ft	14.86
$16^3/_4$	5,000	16-in. 84-lbm/ft	14.86
$16^3/_4$	10,000	16-in. 84-lbm/ft	14.86
$18^3/_4$	5,000	$18^5/_8$-in. 87.5-lbm/ft	17.59
$18^3/_4$	10,000	$18^5/_8$-in. 87.5-lbm/ft	17.59
$20^3/_4$	3,000	20-in. 94-lbm/ft	18.97
$21^1/_4$	2,000	20-in. 94-lbm/ft	18.97
$21^1/_4$	5,000	20-in. 94-lbm/ft	18.97
$21^1/_4$	10,000	20-in. 94-lbm/ft	18.97

cause of its versatility and position in the stack. Annulars are designed to close and seal on multiple tool sizes and shapes and, in an emergency, will close and seal on wireline or an open bore. A closed annular is more suited for drillstring movement than a pipe ram; its slow closure time reduces the possibility of any excessive water-hammer loads.

Two basic annular preventer designs are the bag-type and the piston-driven preventers. Bag preventers are equipped with a cylindrical inner packing element that extrudes into the preventer bore when hydraulic fluid or nitrogen pressure is applied to an enclosing outer element. The elements relax back to their unstressed states whenever the closing pressure is released. The maximum working pressure available for bag preventers is 3,000 psi so their use is limited to moderate-pressure service. Large-bore bag preventers are popular in many diverter hookups.

Fig. 7.4 shows a cutaway view of a common piston-driven preventer. The piston on this style is the wedge-shaped device below the packing unit. Hydraulic pressure in the closing chamber creates a force on the bottom of the piston that drives the piston upward. The steel-reinforced rubber element is forced into the wellbore until closure is effected. The process is reversed when the preventer is opened. Fluid pressure is directed to the opening chamber and the pressure across the top side of the piston forces the wedge down again.

Other makes and models use hydraulic pressure to drive a piston, but the closure mechanisms are different. Whereas Hydril uses a wedge design, Shaffer preventers have a piston that pushes the element upward against a hemispherical bonnet and then out into the bore. Cameron annulars are designed so that movement of the piston compresses an outer donut element that extrudes against an inner element. A rotational movement of cam-shaped steel fingers is created until the inner element closes against whatever is in the bore (or against itself if closing on open hole).

Excessive closing pressures are harmful to the preventer element and may even collapse large-diameter pipe like casing. A regulator valve is used to adjust this pressure to the desired value, based on the preventer model and pipe diameter. Well shut-in pressure increases the closing force in some de-

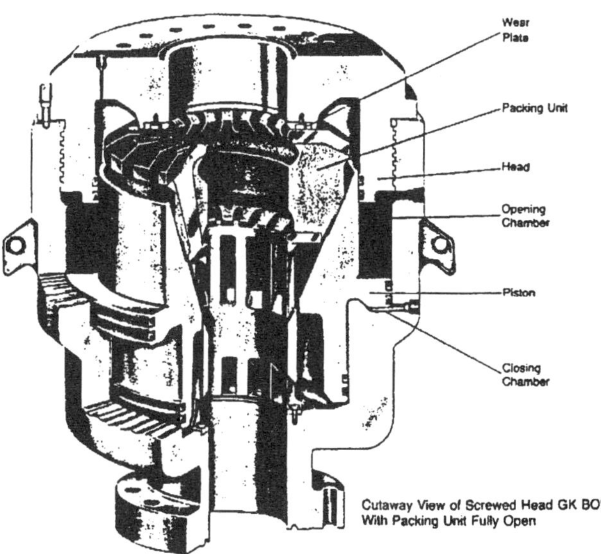

Fig. 7.4—GK annular preventer. Courtesy of Hydril Company.

Fig. 7.5—Type-U ram preventer. Courtesy of Cameron.

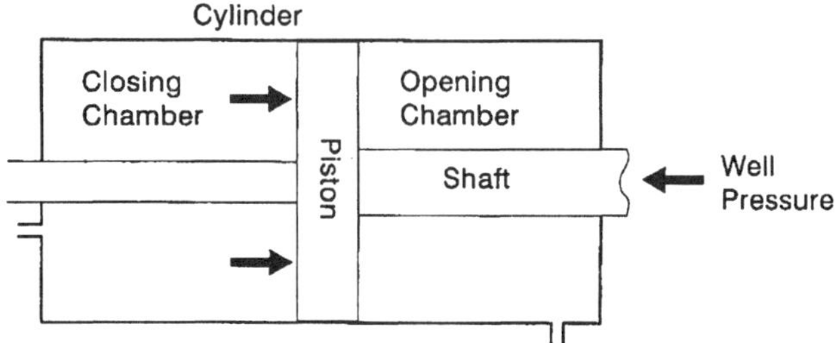

Fig. 7.6—Closing ratio parameters and effect on required closing pressure.

signs while in others it works in the opposite direction or has no effect. An operator or drilling contractor should become thoroughly familiar with the characteristics of his preventer and consult the manufacturer for the recommended regulator setting.

Moving the drillstring through a closed annular generates considerable friction and thus requires the minimum closing pressure that will achieve a seal. This can be found by adjusting the regulator until a small volume of fluid seeps past the closed element. The mud itself may be an effective lubricant if the string is being reciprocated. A bentonite/water mixture should be applied continuously from above if the preventer is closed on gas or pipe is being stripped in the hole.

Ram Preventers. Ram preventers are located below the annular in a conventional surface stack. **Fig. 7.5** illustrates a ram preventer equipped with pipe rams. Two opposing ram blocks are in a retracted position during routine well activities. When activated, hydraulic fluid pressure acts on the piston area and either closes or opens the preventer, depending on whether control fluid is routed to the closing or opening chamber. Manual actuation is an option on surface stacks. A prudent precaution is to have the operating shafts and handwheels installed and ready to use. Manual, hydraulic, or automatic mechanisms can be used to lock the preventer in a closed position.

The ram blocks can be removed or replaced with another type by opening the bonnet doors. Side outlets below the rams can be used to connect the choke and kill-line connections. This type of hookup may be desirable if space is limited and offers the advantage of having one less flange connection to be concerned about. A similar benefit is realized by "double" preventers that are equipped with two rams in the same housing.

An important parameter in some high-pressure applications is the preventer's closing ratio, r_{cl}, defined by

$$r_{cl} = A_{pe}/A_{sh}. \quad \text{(7.5)}$$

A_{pe} is the effective piston area (that area exposed to the closing fluid pressure) and A_{sh} is the cross-sectional area of the ram shaft. As shown in **Fig. 7.6,** the well pressure p_w creates a net opening force on the shaft area which must be overcome by pressure-area force acting on the piston before the preventer can close. An additional pressure p_{cp} is required to overcome the friction resistance between the cylinder and piston seals. Finally, the control fluid moves through relatively small-diameter piping and the accumulator must furnish the friction loss component Δp_{cf}. It follows that the minimum required closing pressure at the accumulator is

$$p_{cl} = p_w/r_{cl} + p_{cp} + \Delta p_{cf}. \quad \text{(7.6)}$$

Example 7.2. The manufacturer's catalogue indicates the closing ratio for the ram preventers on a snubbing stack is 7.3. High pressure ram-to-ram stripping operations are being conducted with a shut-in presure of 9,000 psig. Determine the minimum closing pressure if the friction resistance in the cylinder is 200 psi and the control-fluid friction losses are 300 psi.

Solution. Eq. 7.6 yields

$$p_{cl} = (9{,}000/7.3) + 200 + 300 = 1{,}733 \text{ psig.}$$

Initial closing-pressure requirements are much lower if the ram is used initially to close the well; Eq. 7.6 is applicable only if there is pressure in the BOP cavity. The master regulator on the accumulator typically is set to supply no more than 1,500 psi to the ram preventers and remote-actuated valves but, as demonstrated in the example, more closing pressure may be required in some cases.

Fig. 7.7 shows the different types of rams. Pipe rams seal against the drillpipe and serve as a backup to the annular preventer in the event of packing unit failure or if the shut-in pressure or temperature is deemed excessive for the annular. Pipe rams take much less fluid to close than an annular and may be the preferred shut-in device if a faster closure time is required. Shutting in with a pipe ram is needed when the drillstring is shallow because a closed pipe ram will not pass a tool joint, which keeps the string from ejecting if the shut-in pressure is high. Pipe rams are available that can suspend the drillstring weight. This capability is a necessity in subsea well control, but is used less often with surface stacks.

Variable-bore rams are a relatively recent innovation (since 1980) designed to close and seal on different pipe diameters or a kelly. Having one pipe ram where otherwise two would be required reduces the number of flanged connections and the overall stack height. Their primary value in surface stacks is when a tapered drillstring is being used to drill the well.

Annular preventers can close on open hole, but doing so severely stresses the packing unit. The preferred closure tool when the bore of the BOP is unobstructed is the blind ram, also called a "blank" or "master" ram. The standard practice of closing the blinds after a trip or when out of the hole for any reason gives an opportunity to function-test the ram. Also, closing the well prevents hand tools or other items from being accidentally dropped in the hole.

The blind/shear rams also completely seal off the wellbore. These serve the same function as a blind ram when pipe is out of the hole, but also can be closed on drillpipe to sever the string and retain a seal above the cut. Blind/shear rams are mandated in subsea stacks but are used less often in surface BOPs, especially on land.

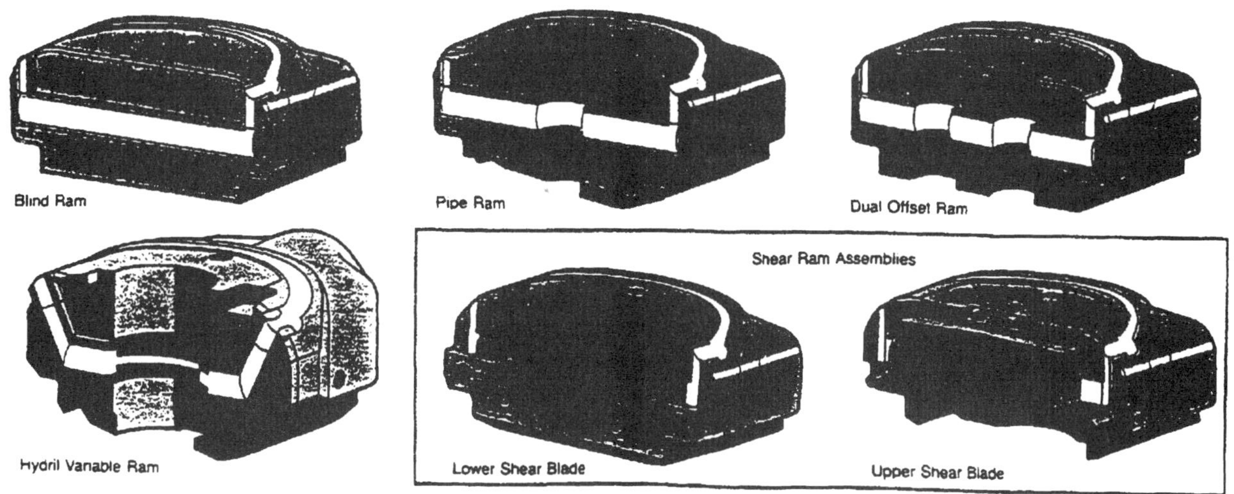

Fig. 7.7—Example ram styles. Courtesy of Hydril Company.

Fig. 7.8 illustrates the shearing process. An upper and lower blade close against each other, fail the pipe in shear, and mash the pipe together above the lower blade face in a recessed area of the lower block. Continued closure causes the rams to meet, compressing the semicircular seals against the BOP body and energizing the horizontal seal between the blocks. Shearing pipe requires more closing pressure than a conventional ram; how much depends on the tube's diameter, wall thickness, and strength (i.e., hardness). For instance, shearing-pressure tests using a Shaffer BOP with a 14-in. cylinder ranged from 1,311 psi for 3½-in., 15.50-lbm/ft Grade E drillpipe to 2,601 psi for 5-in., 25.60-lbm/ft Grade S-135.[6]

Drilling Spool. Some older ram preventers do not have the side outlets, leaving no choice but to connect the choke and kill lines to a separate drilling spool (or mud cross). Even with this capability, many operators and contractors still prefer to use a drilling spool because of the potential for wear and abrasion in the BOP body. Spools are part of the stack and must have the same working pressure and minimum bore as the other components.

7.2.4 Choke and Kill-Line Equipment. The functions of the choke line and manifold are to maintain backpressure on the annulus during a well-control procedure, to route the returning fluids to the desired location, and to provide a way to vent fluids away from the rig and personnel if control cannot be maintained by using the choke or shutting in the well. The kill line on surface stacks has one purpose only: to pump into the well in certain, nonroutine kill operations. All kill-line equipment and components inboard from the choke must have the same working-pressure rating as the stack.

Valves. Valves used in a choke manifold must be full-opening and capable of performing reliably under high-pressure service. Ball valves may have some application in a manifold hookup, but are notoriously difficult to open with differential pressure on one side of the ball. Gate valves (**Fig. 7.9**) are not as prone to this problem and offer other advantages such as

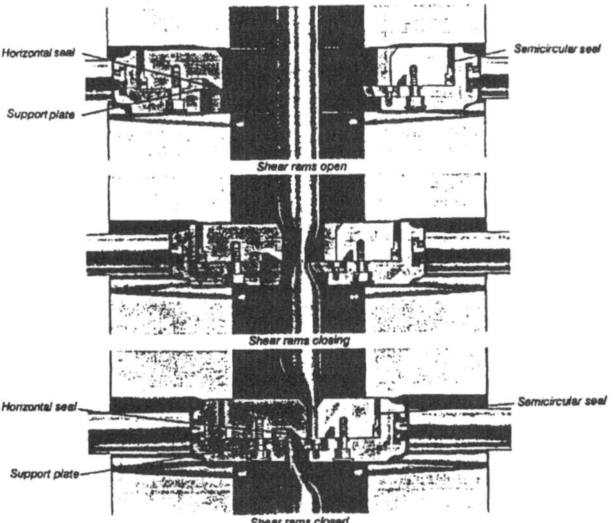

Fig. 7.8—Shear ram mechanism used in cutting pipe and sealing the wellbore. Courtesy of Shaffer.

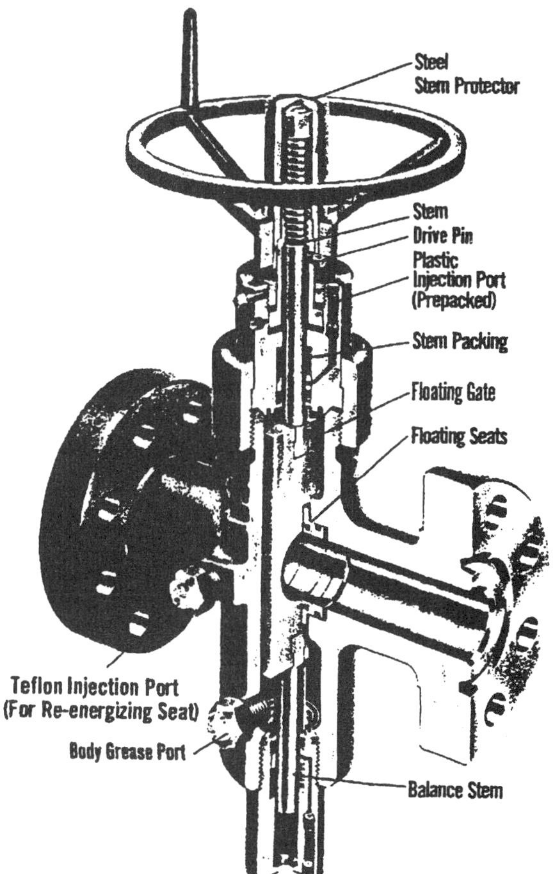

Fig. 7.9—Example gate valve. Courtesy of FMC Fluid Control Systems.

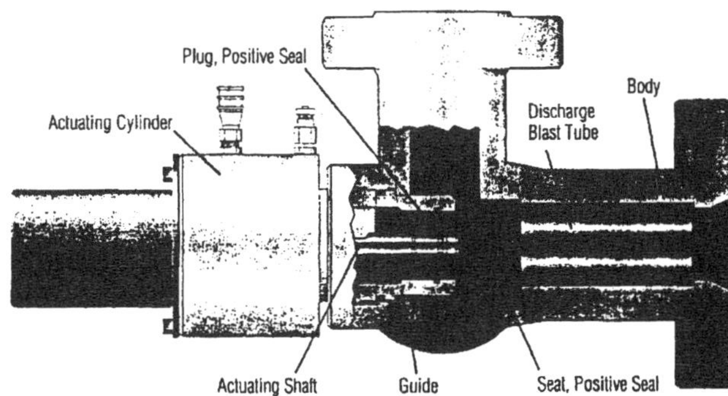

Fig. 7.10—Example adjustable choke. Courtesy of M/D Totco Instrumentation.

slower fluid release while opening a line to flow. The operation is simple; turning the handle moves the stem and gate up or down, thereby closing or opening the bore to flow. One of the gate valves adjacent to the spool is remote-actuated using power fluid from the accumulator. It is called the hydraulic-controlled-remote, or HCR, valve.

One or two check valves are normally placed in the kill line on the outboard side of the gate valves. Check valves allow flow in one direction only (i.e., into the well) and typically use a spring-loaded ball and seat design. An arrow on the valve indicates flow direction.

Chokes. Several different adjustable choke styles are available. All those suited for high pressure-service use an adjustable orifice made of tungsten carbide to create a flow restriction or positive shutoff. **Fig. 7.10** illustrates a type of choke in which a plug moves toward or away from a sealing donut seat. Another design uses a pair of rotating carbide plates with half-moon or circular openings to change the flow area. Others use a conical dart that is moved in or out of a carbide bean. Chokes may be operated manually or from a remote location using a hydraulic or pneumatic actuator.

Lines and Connections. The lines and components rated for 3M and higher working pressures are connected by multi-bolt API flanges, clamp-type connections, or welds. Threaded connections or hammer unions may be suitable in some low-pressure applications, but generally their use is discouraged. Some operators and drilling contractors use a high-pressure hose between the stack and the manifold. However, a fundamental precept in both high- and low-pressure equipment is that the lines be as rigid and straight as possible. Steel pipe is preferred over hoses, swivel joints, or other flexible piping.

7.2.5 Drillstem Control Equipment. Shutting in a well during a trip requires placing a full-opening safety valve (FOSV) in the drillstring while a subsequent stripping job means that an inside BOP must be installed above the FOSV. Well-control problems while drilling may involve shut-in pressures higher than the pump, kelly hose, or other equipment can withstand and so some way to close in the well at the drillstem must be provided.

Safety Valves. **Fig. 7.11** shows an example FOSV. An FOSV is a high-pressure ball valve with a flush OD and rotary shoulder connections matching the tool joint or BHA connections. A special wrench fits in the valve lug to open or close the valve. An FOSV is part of the drillstring and must have an ID and pressure rating commensurate with that of the drillpipe. An open FOSV and its wrench should be in a designated, accessible location at all times and one must be made available for every connection in the string.

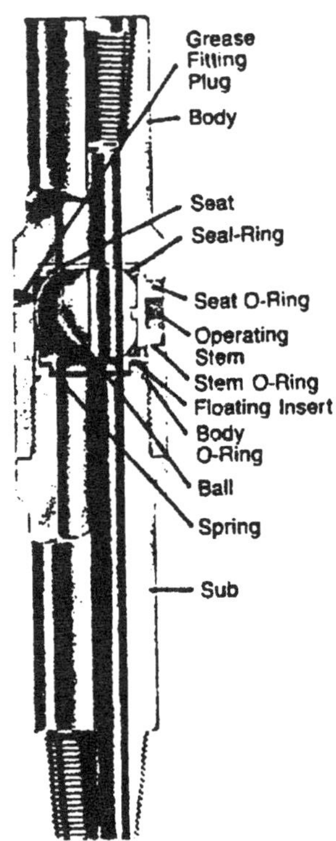

Fig. 7.11—Example full-opening safety valve. Courtesy of Texas Iron Works.

It is difficult for two or three men to stab and turn an FOSV quickly, especially when the drillstring is flowing, and some means should be considered for facilitating this task. Handles may be welded onto a removeable clamp or sub and attached to the FOSV. Another common practice is to make up a spare valve on the first stand of drillpipe and to set this stand aside. In the event of flow, the derrickman latches onto the stand and the valve is installed with the drillpipe.

Two other safety valves called kelly cocks are run as part of the kelly/swivel assembly and one of these valves is closed if the swivel or other upstream equipment fails. The lower kelly cock is another FOSV and is used to shut in a well if the valve is more accessible than the upper cock or if the kelly must be removed with pressure on the string. Lower kelly cocks are closed routinely during a connection to keep mud from draining out of the kelly. However, mud solids have been shown to interfere with the valve's ability to form a pressure seal when used for this purpose.[7]

In top-drive systems, a safety valve is run above the drillpipe as shown in **Fig. 7.12.** The unit can be attached quickly to the drillstring if the well flows while making a trip and the remote-actuated valve is closed at the driller's console. The ease by which a top-drive well can be shut-in during a trip is one of the system's advantages over the kelly-drive process.

Inside BOP. An inside BOP is no different in function than a conventional float. Both are check valves that prevent backflow into the drillstring and can be considered as interchangeable. Some are designed with a flapper valve whereas the BOP shown in **Fig. 7.13** has a spring-loaded dart which is pushed against a seat. When placed above an FOSV, an inside BOP allows the lower valve to be opened so that the hole can be circulated at the string's present depth or after stripping some distance. An added feature shown in the illustration is a releasing tool which holds the valve open, thus eliminating the need to install an FOSV first. After installation, disengaging the releasing tool closes the check valve.

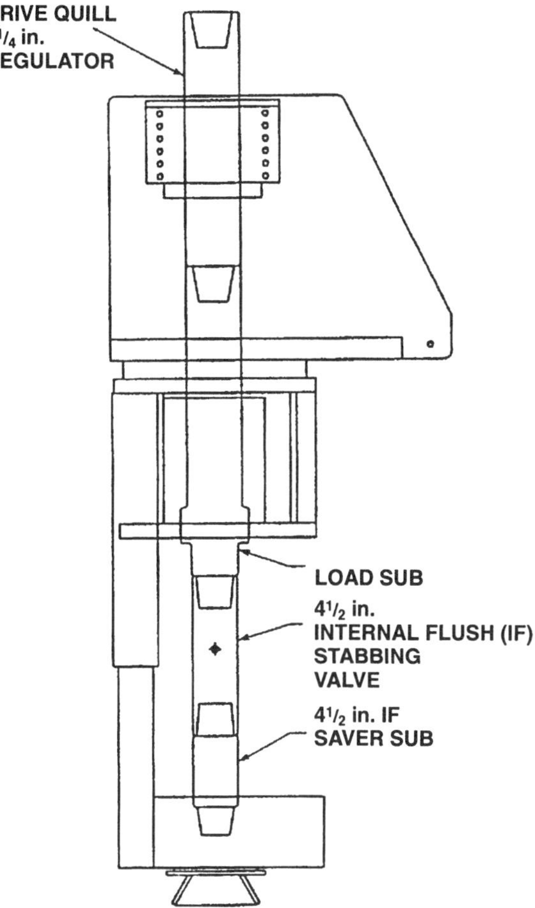

Fig. 7.12—Remote-operated safety valve position in a top-drive drilling system. Courtesy of Tesco Corporation.

7.3 Control System Equipment and Design

Fig. 7.14 is a photograph of a control unit. **Fig. 7.15** illustrates a typical equipment schematic. The control unit stores the hydraulic energy used to operate the BOPE in a series of pressurized bottles called the accumulator bank. Pumps transfer fluid from the reservoir to the accumulators when charging the system or transfer fluid to the stack components. Regulator valves adjust the closing pressure and a valve manifold directs power fluid to the intended stack function through high-pressure lines and fittings.

7.3.1 Accumulator Design Principles. The accumulator bank is the heart of the system. A properly designed accumulator will store ample control fluid to operate each component, perhaps several times, without any assistance from the charge pumps. Some units are equipped with large-capacity spherical accumulators whereas others have vertical cylinders like the ones shown in Fig. 7.14. The bottles are precharged with nitrogen and are filled with control fluid from a bottom port until the nitrogen is compressed to the accumulator's working pressure. Nitrogen and control fluid are physically separated by a rubber bladder, piston, or float mechanism.

Most accumulators in use today are precharged to 1,000 psig and have a 3,000-psig working pressure. The precharge supplies the minimum driving energy when the bottle is fully depleted; no control fluid is present when a bottle is at precharge pressure. Standard design procedures set a minimum

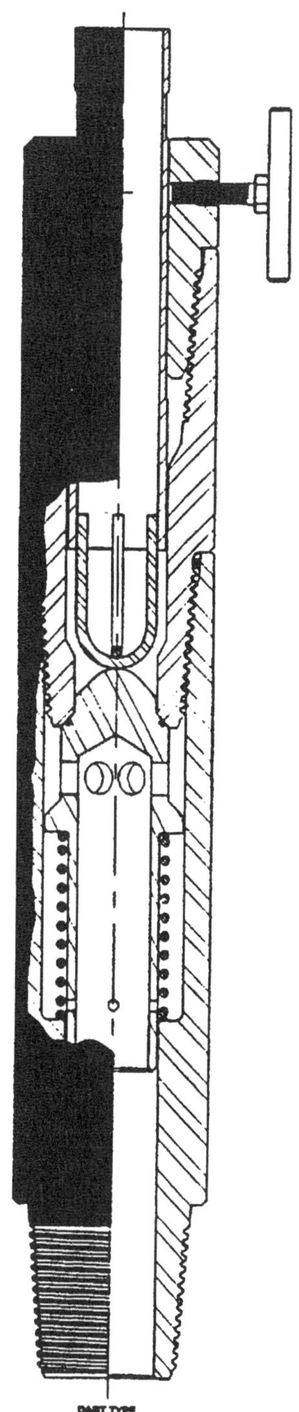

Fig. 7.13—Cut-away view of an inside BOP. Courtesy of Dreco Energy Services.

allowable accumulator pressure higher than the precharge and so the usable fluid is the control-fluid volume that can be recovered when pressure declines from the accumulator's fully charged state to the minimum design pressure.

The key to sizing an accumulator bank is to determine first how much usable fluid is required to manipulate the stack components adequately and to meet regulatory agency requirements. The illustration in **Fig. 7.16** suggests that the gas law can be used to calculate how many bottles will furnish the usable fluid volume. Assume for now that Boyle's law applies—i.e., nitrogen compressibility factors can be ignored and gas expansion is isothermal. The needed volumetric capacity (nitrogen + control fluid) is derived as

Fig. 7.14—Example hydraulic control unit. Courtesy of Shaffer.

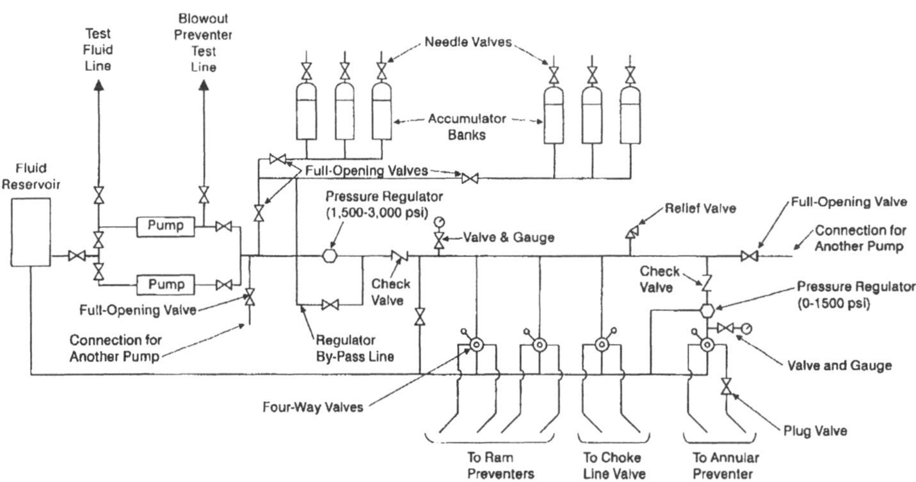

Fig. 7.15—Hydraulic-control-unit equipment schematic.[8]

$$V_{ab} = \frac{V_u}{p_{pch}/p_{\min} - p_{pch}/p_{fch}}, \quad \text{.................} (7.7)$$

where V_{ab} = total accumulator bank volume, V_u = usable fluid volume, p_{pch} = precharge pressure, $p_{\min}$ = minimum accumulator pressure, and p_{fch} = fully charged pressure.

The minimum required usable fluid volume depends on the stack equipment operating volumes and what the government regulations dictate. Absent statutory supervision, other procedures can be applied based on what minimum volume is deemed sufficient. For instance, the API's minimum recommendation is to have an accumulator capacity, with pumps out of service, that will close all of the preventers and still leave a 50% reserve at the minimum pressure.[9] Minimum pressure in this method is defined as the accumulator pressure that will close any ram preventer (excluding blind/shear rams) against the rated working pressure of the stack.

The Norwegian Petroleum Directorate (NPD) requirement for North Sea operators is more stringent. It requires an accumulator volume adequate to close, open, and close all preventers and thereafter leave a mimimum pressure reserve equal to 25% of the amount of pressure it takes to close all

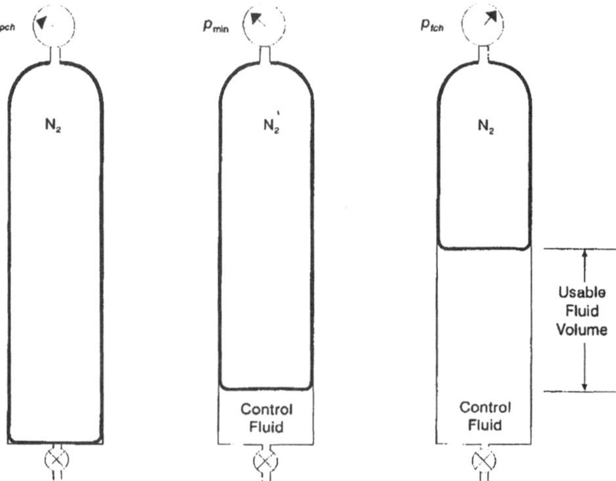

Fig. 7.16—Nitrogen and control-fluid volumes when an accumulator bottle is at precharge, minimum, and fully charged pressure.

TABLE 7.2—BOP EQUIPMENT AND OPERATING CHARACTERISTICS FOR EXAMPLE 7.3			
Component	Closing Volume (gal)	Opening Volume (gal)	Closing Ratio
$13^3/_8$-in. 10M Hydril GK annular preventer	37.18	12.59	—
$13^3/_8$-in. 10M Cameron Type U ram preventer (blind)	7.1	6.6	7.0
$13^5/_8$-in. 10M Cameron Type U ram preventer (pipe)	7.1	6.6	7.0
$13^5/_8$-in. 10M Cameron Type U ram preventer (pipe)	7.1	6.6	7.0

BOPs. This must be done with all pumps out of service and with a minimum pressure 200 psi over precharge.

The number of bottles required to meet the calculated accumulator volume capacity is obtained by dividing by the bottle capacity. A common nominal bottle size is 11 gal, which after taking out the bladder/float displacement yields an actual capacity of 10 gal.

Example 7.3. Assume the stack equipment listed in **Table 7.2** will be used on a well. Determine how many 11-gal accumulator bottles are recommended by the API and compare to the NPD requirement. Use fully charged and precharge bottle pressures of 3,000 and 1,000 psig, respectively, and assume the nitrogen behaves according to Boyle's law.

Solution. The control-fluid volume to close all four preventers is

$$37.2 + (3)(7.1) = 58.5 \text{ gal.}$$

The usable fluid volume recommended by the API includes a 50% reserve factor.

$$V_u = (58.5)(1.5) = 87.8 \text{ gal.}$$

The API's recommended minimum pressure is based on the ram preventer's closing ratio. We choose to ignore mechanical or fluid friction and obtain

$$p_{\min} = (10{,}000/7.0) = 1{,}429 \text{ psig} = 1{,}444 \text{ psia.}$$

Using absolute pressures, Eq. 7.7 gives the total accumulator volume.

$$V_{ab} = \frac{87.8}{(1{,}015/1{,}444) - (1{,}015/3{,}015)} = 239.7 \text{ gal.}$$

The number of 11-gal nominal bottles is,

$$239.7/10 = 23.97 \text{ or } 24 \text{ bottles.}$$

Now consider the NPD's directive. Determine the total usable fluid volume.

Close all preventers = 58.5 gal
Open all preventers = 32.4 gal
Close all preventers = 58.5 gal
Reserve = (0.25)(58.5) = 14.6 gal
Total = 164.0 gal

The minimum accumulator pressure is calculated as

$$p_{\min} = 200 + 1{,}015 = 1{,}215 \text{ psia.}$$

For the same stack, the NPD would require

$$V_{ab} = \frac{164.0}{(1{,}015/1{,}215) - (1{,}015/3{,}015)} = 330.4 \text{ gal.}$$

$$330.4/10 = 33.04 \text{ or } 34 \text{ bottles.}$$

Note that remote-actuated valves were not included in the calculation. It takes only a gallon or two of fluid to open one of these valves so ignoring this volume does not affect the outcome substantially.

3,000-psi accumulators have become the industry standard because their efficiency (usable fluid volume per bottle) and closing time is so much better than the 1,500- and 2,000-psi units. For example, a 3,000-psi bottle delivers about 50% more usable fluid volume than a 2,000-psi bottle at the same precharge pressure. Pressure ratings higher than 3,000 psi are available and may become the norm some day, especially on offshore locations where blind/shear rams are required.

The design procedures recommended by the API and required by certain governments may seem too conservative because they are predicated on several "extremely bad day" situations that are not likely to occur if redundant pumps and more than one source of operating power are made available. Recognize, however, that a larger accumulator volume provides other advantages such as a faster closure time. Furthermore, the reserve factor accounts for a couple of faulty assumptions in the conventional design calculation.

One of these is our assumption that the accumulator is at its rated pressure when called upon to shut in the well. This may not be the case. The API requires that the pumps kick in automatically after an accumulator loses 90% of its design working pressure.[10] Thus a 3,000-psi unit may be charged only to 2,700 psig and still meet specification.

Another potentially serious weakness is the use of Boyle's law to predict usable fluid volume. Nitrogen, as any other gas, does not behave ideally and a more rigorous approach would consider the compressibilty factors at the three pressure conditions. Furthermore, fast closure times make the gas expand under near-adiabatic conditions. Heat flow from the control fluid and container shell is limited and the expansion is accompanied by a drop in gas temperature. Hence the lower temperature leaves less available control fluid at the minimum pressure.

Brault and Rajabi[11] studied accumulator gas behavior and the effect of certain variables such as the gas compression ratio and expansion time. Their work indicates that adiabatic expansion in the example problem's 3,000-psi accumulator leaves about 24% less control fluid than the Boyle's law prediction. The 50% reserve factor adopted by the API and other regulatory agencies is sufficient for this case, but it may not be sufficient for higher accumulator pressures.

In a similar vein, Jones and LeMoine[12] assumed polytropic behavior (between isothermal and adiabatic) during expansion and included the combined effects of friction and a low switch-on pressure (2,700 psi for a 3,000-psi accumulator). A hypothetical design based on Boyle's law and a minimum pressure dictated by ram closure against a 15M working pressure was found to be inadequate, even with a 50% reserve factor.

In most applications, the design method discussed in this chapter should result in safe accumulator sizing for surface stacks if enough reserve is left at the minimum pressure. Consider using a more accurate procedure when using a higher accumulator pressure or if the minimum pressure criterion requires closing a ram against a 15M or 20M cavity pressure.

7.3.2 Other Components. Control fluid should be clean, noncorrosive, nonflammable, and relatively slick. The fluid is in a closed system when used to operate a surface stack, but the environmental impact may be a consideration in the event of a leak or spill. A light hydraulic oil is acceptable whereas most petroleum-based fluids can damage the seals and are not recommended. Fresh water may be used if mixed with a lubricant and antifreeze for cold-weather service. It is best to consult the equipment manufacturer for acceptable fluid types.

The fluid reservoir is a closed atmospheric chamber where control fluid is stored and returned after use. API Specification 16D[10] requires a minimum reservoir capacity equal to twice the usable fluid capacity of the accumulator system and requires air vents to prevent overpressuring the tank when fluid is transferred.

The accumulator is pressurized by air or electric pumps that transfer control fluid from the reservoir into the individual bottles. API Specification 16D[10] specifies having at least two pump systems operated by independent power sources. The pumps must be able to charge the system from precharge to the rated pressure in less than 15 minutes and open the hydraulic valve(s) on the choke line and close the annular on open hole in less than two minutes. However, we do not recommend closing an annular on open hole for the sake of a test and prefer the suggested two-minute closure on the smallest drillpipe as discussed in Recommended Practice 16E.[9]

The discharge pressure of the pumps must be no less than the rated working pressure of the closing unit while at least two pressure-relief devices are required to limit maximum discharge pressure. An automatic controller starts the pumps whenever the accumulator pressure falls below 90% of the accumulator operating pressure. Frequent charging without using the accumulator indicates a leak in the system; it needs to be found and repaired.

Refer to Fig. 7.15 and study the manifold flow schematic. In normal operation, the main line is open to the accumulators but, as indicated, it can be accessed directly by the pumps. Generally, ram preventers and HCR valves can operate with a 1,500-psi supply and the master regulator is set to this pressure. The by-pass line is opened if full accumulator pressure is needed to close a blind/shear ram. The annular preventer is operated at a lower pressure and so another downstream regulator is used to regulate the line pressure to that needed to effect a seal.

Each preventer and HCR valve is controlled by a four-way valve that directs the opening/closing fluid to the BOP and the return fluid to the reservoir. Note that the valve position on each control is standardized so that closing a BOP device is accomplished by moving the handle in the same direction. Protective covers or other measures that do not interfere with remote operation can be taken to prevent inadvertent operation of the blind or blind/shear rams. The preventer controls and an annular preventer regulator are provided at a remote station located for easy access by the driller. Fig. 4.12 illustrates one of these remote panels.

Goins[13] identified and discussed 17 key mistakes resulted in well-control problems. Four of these dealt with valves and connections in the accumulator manifold. One involved placement of a lock on the blind or blind/shear ram controls at the control unit. This was a fairly common practice at one time, but it does not allow the driller to close this ram from the remote panel.

Fig. 7.17—Control lines to a BOP stack in a protective "suitcase."

Other problems have developed when personnel forgot to open the valve to the accumulator bank after making repairs, did not cap a line coming off an unused control (which drained the accumulators when someone inadvertently threw the wrong handle), and reversed the lines on a control. Routine function tests and visual inspections would have prevented these three mistakes.

The piping and connections between the control unit and stack must have the same pressure rating as the accumulator and should be shielded from damage using an enclosure similar to the one shown in **Fig. 7.17.** API Specification 16D[10] requires a maximum closing time of 30 seconds for ram preventers and annulars smaller than 18¾-in. nominal size. The size and length of the control line are important variables in how fast a preventer will close. It follows that the accumulator should be placed as close to the stack as practical, yet far enough away to be accessed and protected in the event of a catastrophe at the well. A minimum line diameter of 1 in. is recommended, but a marked improvement in performance can be realized by using 1½-in. lines.

7.3.3 Test Procedures. With some modifications, the control-system test procedures outlined in **Table 7.3** were recommended by the API in Ref. 8. Before connecting the system to the stack, the unit should be visually inspected and the fluid reservoir checked to ensure that no foreign fluids, rocks, or

TABLE 7.3—RECOMMENDED CONTROL-SYSTEM TEST PROCEDURES

Precharge Pressure Verification
1. Open the bottom valve on each bottle and drain the operating fluid into the reservoir.
2. Measure the pressure on each bottle and adjust if necessary.
3. Charge the system to the unit's working pressure.

Accumulator Closure Test
1. Position a joint of drillpipe in the stack.
2. Shut off the power supply to the control-system pumps.
3. Verify that the accumulator pressure is within 100 psi of the rated working pressure.
4. Verify that the master and accumulator regulators are set correctly.
5. Simultaneously close the pipe ram and annular. Open the HCR valve(s).
6. Record the BOP closure times and the time to open the HCR.
7. Record the final accumulator pressure.
8. Restore the controls to their original position.
9. Restore power to the pumps and recharge the accumulator. Measure the time required to recharge the system to full operating pressure.

Pump Closure Test
1. Position a joint of drillpipe in the stack.
2. Isolate the accumulators from closing unit and pumps.
3. Isolate the rig air supply if it is being used to power any air pumps and connect it to a separate air or nitrogen source.
4. Simultaneously close the annular and open the HCR valves.
5. Record the annular closure time and the time to open the HCR.
6. Restore the controls to their original position.
7. Repeat the test using the other power source for the pumps.
8. Open the accumulators to the closing unit and pumps.

other debris are present. Precharge verification should be done at this time. Accumulator and pump closure tests should be conducted before pressure testing the BOPE.

7.4 BOPE Inspection and Test Considerations

BOPs are employed in an emergency and their performance during a problem depends on having quality equipment, the right tool for the application, proper installation, and routine followup maintenance. A thorough inspection before nippling up the stack is a prudent measure to detect flaws that may remain hidden in a pressure test. Also, it is much less expensive to find defects before everything is installed rather than during the test. Kandel and Streu[14] discussed the BOPE components that are often the weak links in the system and presented a thorough guideline for inspecting equipment in the field.

The only way to ascertain the capability of the BOPE to perform in a well-control event is to test the equipment immediately after installation. The rigors of drilling through the stack and using the equipment can cause wear, loosened bolts, etc., and so followup tests need to be conducted on a periodic basis. Testing involves isolating and pressuring up on each component to verify that no leaks are present and operating each device to verify that the equipment is functioning properly.

In general, drilling muds are a poor test fluid because solids can plug a small leak path and effect a seal at high pressure. Cleaning the moving parts and flushing all residue mud before testing will give more reliable results. Clear water is the recommended fluid in most cases. The equipment will be subjected to high pressures and the pump volumes required to get

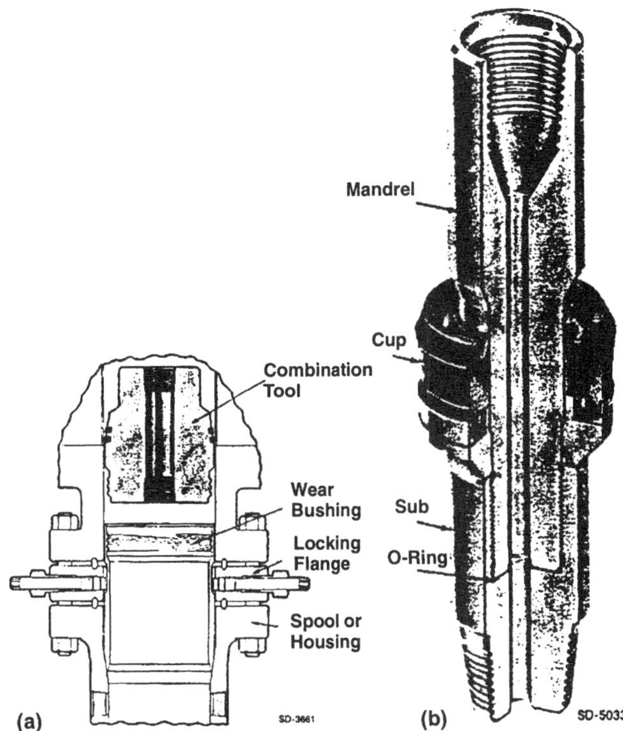

Fig. 7.18—Example BOP test tools. Courtesy of Cameron.

to these pressures are low. These demands are beyond the capability of the rig pumps and a cementing unit or reciprocating pump furnished by testing specialists must be used.

The means to isolate the test pressures from the wellbore and casing is provided by a test tool similar to one of the two different styles shown in **Figs. 7.18a and 7.18b.** The tool is lowered through the stack on a joint of drillpipe or specially-made test joint and is seated somewhere below the lowermost flange on the stack. The boll-weevil tester depicted in Fig. 7.18a lands in the wellhead bowl whereas the cup tool shown in Fig. 7.18b is positioned in the casing opposite the hanger. Each test tool has its advantages and disadvantages, but the boll-weevil tester is more common.

The specific procedure depends on the type of test tool and on certain variations available for a given type, but the general objective is to test each component in a logical sequence, which ensures that no part of the system is neglected. The sequence described in **Table 7.4** is suitable for boll-weevil test plugs and assumes the stack configuration depicted in Fig. 4.11. The methodology can be adapted to other stack arrangements.

Pressure is applied to the well-side of each component as this is the direction in which we need to seal against well pressure. This is best accomplished on surface stacks by pumping down the test joint and out a port somewhere above the seals. Using this method, the string bore must be plugged at or below the test tool. Boll-weevil testers are available with weep holes to access the test joint annulus or a perforated sub can be run above the tool.

The wellhead valves immediately underneath a boll-weevil tool should be left open during the test for two reasons. One is to avoid placing too much pressure on the casing or open hole if the tool malfunctions and the other is to verify that any leaks discovered in the test are not the fault of the tool or the seal seat. Opening the wellhead valve is not an option with a cup tool and the procedure should be modified so that the

TABLE 7.4—EXAMPLE BOPE PRESSURE-TEST PROCEDURE	
1.	Install test plug in the wellhead.
2.	Fill the BOP stack with water and flush the kill and choke lines.
3.	Open the wellhead valve below the test tool.
4.	Close the inboard choke and kill line valves. Open the outboard valves and chokes. Remove the check valve(s) from the kill line if present.
5.	Close the lower pipe rams. Test to the specified low pressure and to the working pressure.
6.	Close the upper pipe rams. Test to the specified low pressure and to the working pressure.
7.	With the upper rams closed, test each valve on the choke line/manifold and kill line to the specified low pressure and to the working pressure. Start with the inboard valves and proceed outwards.
8.	Install the check valve(s) on the kill line and test.
9.	Test the low-pressure side of the manifold.
10.	Close the annular preventer and test to the specified low and high pressure.
11.	Remove the test joint from the BOP stack, leaving the test plug in place.
12.	Close the inboard valve on the choke line. Pump down the kill line and test the blind ram to the specified low pressure and to the working pressure.
13.	Close the wellhead valves and retrieve the test tool.
14.	Pick up the kelly and flush the system with water. Install the FOSV and inside BOP on the kelly assembly connection.
15.	Open all lines back to the rig pump and open the bypass valve at the standpipe. Pump through the lower end of the kelly and test the inside BOP to the low and high pressures.
16.	Remove the inside BOP and close the FOSV. Test the FOSV to the low and high pressures.
17.	Remove the FOSV and close the lower kelly cock. Test the lower kelly cock to the low and high pressures.
18.	Close the upper kelly cock and test to the low and high pressures.
19.	Valve by valve, test each section of the standpipe and lines back to the mud pumps in the same manner.
20.	Drain the choke and kill lines and fill with freeze-protected fluid. Verify the open or closed position of all valves and the state of readiness of all accessory equipment. Configure equipment to the drill-ahead mode.
21.	Record test results.

string bore is open and pressure is applied from the kill line. Flow through the test joint will indicate if the tool is leaking.

One advantage to a cup tool is that it affords the opportunity to test the wellhead outlets to the full working pressure. This may not be of major concern, however, because the wellhead will see the pressure imposed during the casing test. A significant disadvantage is that the tool cannot be left in place to test the blinds, thus limiting blind-ram test pressures to what the casing can hold. Being free to move, test pressure acts on the cross-sectional area of the cup and increases tension in the test joint and a calculation is necessary to determine if its strength may be exceeded. Combination tools with a nonsealing boll-weevil above the cup are available and offer the advantages of both types.

A piece of well-control equipment can leak at a low pressure, yet test at a higher pressure. It is a good practice to first test each component to something much lower than the maximum potential pressure. The Minerals Management Service (MMS) requires a low-pressure test between 200 and 300 psi for BOPs in U.S. federal waters. Afterwards, a test to the full working pressure is required by the MMS and other jurisdictional authorities. This demonstrates that the equipment will hold together at a pressure higher than would be expected and, just as importantly, instills confidence in the rig crews.

An exception to this practice is the annular preventer, which is normally tested to no more than 70% of its rated pressure. This exception is justified because most operators will resort to a pipe ram long before shut-in pressures get this high. Depending on the make and model, closing pressure adjustments based on the manufacturer's recommendation may be necessary to prevent overstressing the element during the test.

At minimum, another pressure test should be conducted before drilling out subsequent casing strings or after a BOP component has been repaired or replaced. Depending on how much time has elapsed, another test may be considered before drilling into the transition zone. Additional tests may be specified by the local government agency. For instance, the MMS requires a full-scale test every 14 days for drilling operations in U.S. federal waters.

Operating the equipment tests the performance of the BOPE and a convenient time to function-test is while conducting a drill for the crews. The pipe rams may be closed every day while a weekly test of the annular is suggested. The blinds should be closed any time pipe is out of the hole.

7.5 Low-Pressure Equipment

The flow-control system downstream of the choke and the processing equipment are not subjected to high pressure, tens of psi at most, but this does not diminish their importance to well control. Improper design or failure in the low-pressure equipment can imperil the rig and personnel.

7.5.1 Manifold Lines. High-strength materials are not necessary downstream from the choke, but an operator should keep in mind the high velocities resulting from gas expansion through the choke and the associated potential for the entrained solids to cut or erode the piping. Selecting pipe with an adequate wall thickness will reduce the failure risks as will keeping the plumbing as straight as possible. Restrictions of any kind must be avoided and the downstream valves need to be full-opening.

It is common to see large-diameter expansion chambers (also called "boots" or "watermelons") placed in the low-pressure lines. An expansion chamber in many cases is nothing more than a common header for the manifold lines. Their use has been recommended to reduce effluent velocity before flow enters the separator, but this function is not present if line diameter is reduced again downstream from the chamber.

7.5.2 Mud/Gas Separators. The mud/gas separator (MGS), also called the gas buster and other descriptive terms, is the primary means to separate gas from the drilling fluid while controlling a kick, drilling underbalanced, or circulating large connection/trip gas volumes. **Fig. 7.19** shows an MGS in operation on a land rig. Many different separator designs and styles are seen in the field. Those used in drilling normally are modified two-phase production separators or shop-fabricated from scratch to an operator's or contractor's specifications. **Fig. 7.20** illustrates the features common to many closed-bottom separators.

Separators rely on gravity to segregate the gas from the liquid and the internal workings are uncomplicated. In Fig. 7.20, gas-cut mud is discharged into the chamber and strikes an impingement plate which protects the shell from erosion and assists the separation. The mud flows downward through a series of baffles, accumulates on the bottom, and flows out the

Fig. 7.19—Example mud-gas separator.

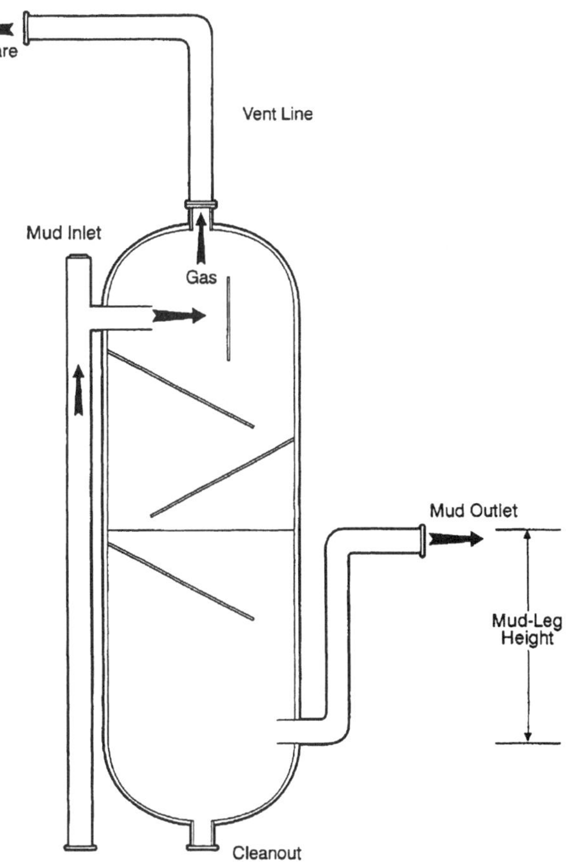

Fig. 7.20—Closed-bottom MGS operating principles.

side outlet to the flowline or directly to the pits. Some gas may remain entrained in the outlet mud, but the bulk exits the vent line on top and to the flare.

The liquid level is controlled by the height of the mud leg and pressure inside the separator. The separator's internal pressure is the friction drop of the gas flowing through the vent line; excessive pressure can suppress the mud level to the point that gas evacuates the separator and blows through the mud outlet. Significant gas volumes also can carryover with the mud when not given enough retention time to percolate free. Both of these undesirable situations can be avoided by designing the equipment properly.[15,16]

The allowable separator pressure is limited to the hydrostatic pressure of the fluid in the mud leg,

$$p_{ml} = g_m h_{ml}, \qquad (7.8)$$

where g_m = the mud's hydrostatic gradient and h_{ml} = the mud-leg height. The height of the mud leg is one factor we may be able to control to increase the allowable separator pressure. Even so, there is a practical limitation to what can be done here. A more important design consideration lies in maintaining a low separator pressure by engineering the vent-line diameter.

The Weymouth[17] equation for predicting gas-friction pressure is used in this chapter and we have modified and rearranged the relation to solve directly for the vent-line diameter.

$$d_{vl} = 0.3144\left[\frac{q_{vl}^2 p_a^2 \gamma_g L_e}{T_{vl}\left(p_{ml}^2 - p_a^2\right)}\right]^{0.1875}, \qquad (7.9)$$

where L_e = effective vent-line length and T_{vl} = average temperature in the vent line (°R).

The length term includes the effect of bends and diameter disruptions in the line. Most vent lines are consistent in diameter but three or more turns can be expected and it is appropriate to use an equivalent length of 70 ft for a sharp right-angle turn.[16]

Peak gas rates through an MGS can be predicted based on a hypothetical kick circulation, using a choke backpressure (p_{ch}) and upstream gas volume (V_{ch}) obtained from the single-bubble model discussed in Chap. 4 or a more exact solution. The pressure at the vent-line discharge is atmospheric and we determine the downstream kick volume using the gas law,

$$V_{vl} = p_{ch}V_{ch}T_{vl}/p_a T_{ch} z_{ch}. \qquad (7.10)$$

The subscripts denote conditions upstream of the choke and at the vent-line discharge. The time it takes to vent the gas is given by

$$t = V_{ch}/q_{kr}, \qquad (7.11)$$

where q_{kr} = the pump kill rate. It follows that the peak gas rate is

$$q_{vl} = V_{vl}/t. \qquad (7.12)$$

Sizing the separator diameter is an important design consideration based on giving the mud enough retention time for the gas to migrate into the upper part of the chamber. Sufficient time is given when the downward velocity of the drilling fluid in the separator is less than the rising gas velocity. Equating the two velocities yields the relation,

$$d_s = \sqrt{4q_m/\pi v_{sl}}, \quad \ldots\ldots\ldots\ldots\ldots\ldots\ldots\ldots\ (7.13)$$

where d_s = the minimum separator ID, q_m = the mud flow rate in the separator, and v_{sl} = the gas migration velocity.

As discussed in Chap. 1, mud rheology, conduit size, and other factors affect migration rates and add uncertainty to the sizing calculation. Furthermore, mud is accelerated as gas approaches the choke and the kill circulation rate cannot be used to determine the mud velocity. Absent detailed modeling, MacDougall[16] suggested using a mud return rate twice the pump rate ($q_m = 2q_{kr}$) and a conservative gas-migration velocity of 500 ft/hr.

Example 7.4. Size the vent line and separator to process a Driller's Method kick circulation on the well described in Table 4.20 and Example 4.16. Maximum mud-leg height is 7 ft and the 150-ft vent line will have three sharp bends. Assume the vent-line temperature is 70°F and the atmospheric pressure is 14 psia. The kill circulation rate is 175 gal/min.

Solution. The effective vent-line length including the turns is

$$L_e = 150 + (3)(70) = 360 \text{ ft}.$$

Now assume perfect separation efficiency (gas does not affect the fluid density) and calculate the allowable separator pressure using Eq. 7.8.

$$p_{ml} = (0.8052)(7) = 5.6 \text{ psig} = 19.6 \text{ psia}.$$

Annulus pressures and volumes were predicted using the conservative single-bubble model in the example from Chap. 4. When the kick was displaced to surface with the Driller's Method, we determined p_{ch} = 2,981 psia, V_{ch} = 172 bbl = 966 cu ft, z_{ch} = 0.820, and T_{ch} = 560°R.

The gas volume at the vent line is

$$V_{vl} = \frac{(2,981)(966)(70 + 460)}{(14)(560)(0.820)} = 237,400 \text{ cu ft}.$$

The peak gas rate is estimated by combining Eq. 7.11 with Eq. 7.12 and making a units correction.

$$q_{vl} = (237,400)(175)/(966)(7.48) = 5,750 \text{ ft}^3\text{/min}.$$

Eq. 7.9 gives the result for the first half of this problem.

$$d_{vl} = 0.3144\left[\frac{(5,750)^2(14)^2(0.60)(360)}{(530)(19.6^2 - 14^2)}\right]^{0.1875} = 6.9 \text{ in}.$$

Thus a separator vent line with a 7-in. ID will suffice.

Substitute the suggested gas-migration velocity and return flow rate into Eq. 7.13 to obtain the minimum separator ID.

$$d_s = \sqrt{\frac{(4)(2)(175)(60)}{(\pi)(500)(7.48)}} = 2.7 \text{ ft} = 32 \text{ in}.$$

A separator shell fabricated from 36-in. pipe will meet the design conditions.

Vent-line sizing criteria may suggest much smaller diameters than the example problem's result. Even so, a minimum 6-in. nominal line size is recommended for any installation. One other note concerning vent lines is in order. Separators have blown up in the field because a valve installed on the vent line somehow became closed. A vent-line valve has no purpose and under no circumstances should a valve be placed in a separator vent line.

7.5.3 Degassers. Some degree of gas cutting is expected downstream of the MGS and evidenced by a reduction in the drilling-fluid density. With the separator bypassed, drilled gas can contaminate a mud. For either case, a means to de-gas the mud is recommended to reduce the fire hazard around the pits and to prevent gas from being recirculated down the hole.

Fig. 7.21 illustrates a typical vacuum degasser. Gas-cut mud is picked up immediately downstream of the shale shaker and is drawn into the degasser where the jet pulls a vacuum on the mud and draws out the gas. Internal baffles spread out the mud into thin sheets and assist the breakout efficiency. The gas coming off the degasser is vented a safe distance away from the rig.

Atmospheric degassers work by the principle of gravity segregation. Gas-contaminated fluid is pumped into an atmospheric chamber and is forced through a valve disc that creates a high-velocity spray. The spray impacts the walls of the degasser, a hydrocyclone action is imparted, and centrifugal force separates the gas from the mud.

7.6 Equipment Arrangement: Design and Philosophy

There are no industry standards related to the arrangement of well-control equipment, how much redundancy is needed based on the perceived risk, and at what point the costs and system complexities begin negating the benefits of a design to handle remote (though possible) contingencies. These are not easy questions. Personal experience affects how people believe the equipment should be put together. Equipment-arrangement philosophies are determined also by the experience of others, company policy, governmental mandate, or

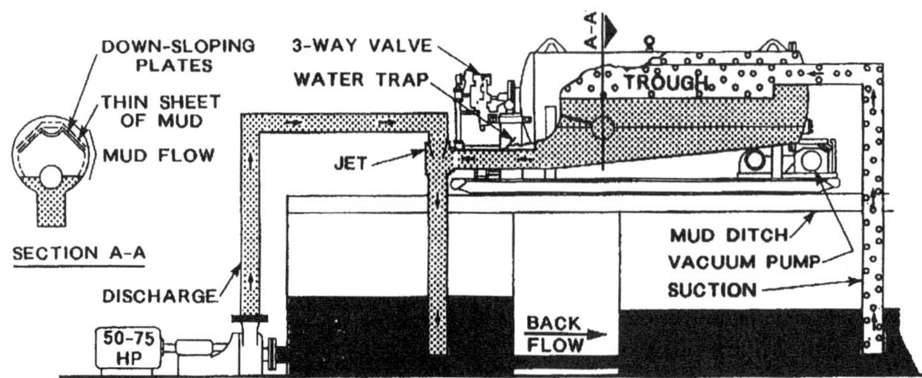

Fig. 7.21—Example vacuum degasser. Courtesy of Sweco Oilfield Services.

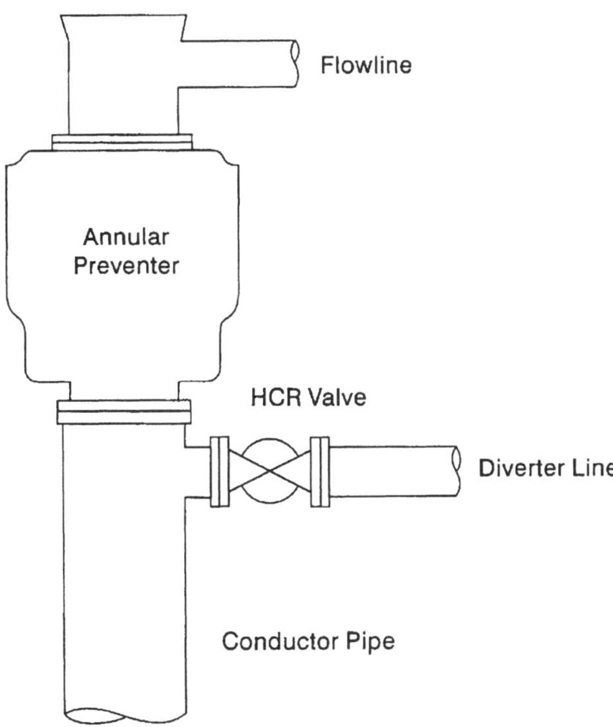

Fig. 7.22—Typical diverter equipment for a land operation.

perhaps because a hookup of unknown origin has been handed down through the ages.

Our intent is not to resolve these issues. We hope to point out why things are done certain ways, the advantages and disadvantages of selected arrangements, and suggest what we believe are certain minimum requirements.

7.6.1 Diverters. Managing a gas flow with only conductor pipe or shallow surface casing set in a well is one of the more perplexing problems facing the industry today. Rocks may be weak or unconsolidated and the effect of any SICP at all imposes a dramatic pressure gradient at the shoe. Thus shutting in the well may pose an unacceptable risk if an underground blowout is probable, particularly if an uncontrolled flow is likely to broach beside the casing or reach the surface by some other pathway.

Diverters are installed whenever the decision has been made not to shut a well in. Their sole purpose is to route flows away from the rig and personnel. **Fig. 7.22** illustrate a typical arrangement for a land operation. A bag-type preventer or other closure device is installed on the conductor pipe and a large-diameter diverter line is run out to the reserve pit below the preventer. Normally, the hydraulic or pneumatic controls are rigged so that an HCR valve on the diverter line opens when the preventer is closed. Diverters on bottom-supported offshore rigs are similar, but have two side outlets to allow routing the fluid overboard on the downwind side. Diverters used on floating rigs are discussed in Chap. 8.

Diverter lines should be sized as large as practical for two reasons: to keep two-phase friction losses down and to reduce the chance of plugging the line with rocks blown from the well. It has been suggested to size the line so that it has the same flow area as the drillpipe annulus, but this method yields impractical results for conductors larger than 13³/₈ in. A 10-in. nominal diameter should be considered the minimum. Larger sizes may be necessary based on the estimated maximum flow rate or plugging potential.

Diverters are not secondary control equipment. A well placed on a diverter is out of control and the objective is to restore equilibrium by pumping mud as fast as the pumps allow while increasing the mud weight. Fast pump rates are necessary to keep as much mud as possible in the well and to obtain what friction backpressure is available against the flowing formation. In many cases, the formation pressure depletes rapidly or a downhole bridge ultimately shuts off the flow.

Unfortunately, diverters fail about as often as they work. Sand, at the extreme velocities associated with wide-open flow, can rapidly cut the equipment and severe dynamic loads can shake the diverter or lines apart. Plugging the lines may be the biggest weakness when unconsolidated rock is exposed, even with 12-in. or larger diameter lines. Diverters and shallow gas hazards will be discussed in Chap. 8. Land operations are not immune to the problem, but at least the people have some place to run if things go wrong.

7.6.2 Stack Arrangement. Stack arrangements are not standardized and the different philosophies regarding equipment placement multiply as the potential pressures and job complexities increase. Inevitably, compromises must be made, especially when substructure height and equipment inventory limit the options. The reader should consider the described stack arrangements in the context of what basic features are required to furnish adequate redundancy and to manage the most likely well-control contingencies.

Low-pressure and low-risk wells on land where the likelihood of taking a kick is low or maximum potential surface pressures are less than 2,000 psi are usually equipped with minimum requirements. For example, a common low-pressure stack arrangement will have a drilling spool, a ram preventer equipped with blinds, and either a pipe-ram preventer or an annular. There is not enough room on most rigs used to drill these types of wells to do much more. Double ram preventers and using the BOP side outlets can reduce substructure height requirements.

A more elaborate system is needed when risks increase or when stack requirements increase to the 3M and 5M ratings. At least two ram preventers and an annular are recommended, but how the components are stacked together is subject to variation. Goins and Sheffield[18] discussed the four arrangements shown in **Figs. 7.23a through 7.23d** and the features offered by each. **Tables 7.5 and 7.6** summarizes their capabilities and limitations. Again, the spools shown in the schematics may be replaced by ram-preventer side outlets.

Overall, the configuration illustrated in Fig. 7.23a may offer the most versatility and it is the only hookup that allows shutting the well in with pipe in the hole if problems develop in the vicinity of the drilling spool. Most two-ram arrangements, however, are stacked together as shown in Figs. 7.23b and 7.23c. Flanges are the weak links in the system and these two configurations eliminate one connection if a double-ram preventer is used. Also, substructure height may preclude the options shown in Fig. 7.23a and 7.23d.

Three ram preventers and a pipe ram below the primary choke/kill outlets are recommended for 10M and 15M stacks. Four possible arrangements are shown in **Figs. 7.24a through 7.24d.** Note that configurations presented in Figs. 724c and 7.24d are the same as those presented in Figs. 7.23a and 7.23b except that secondary choke and kill lines have been attached to the bottom ram preventer.

One basic principle for any redundant system is to use the component farthest from the energy source and to keep the inboard component in reserve. As applied to the configuration

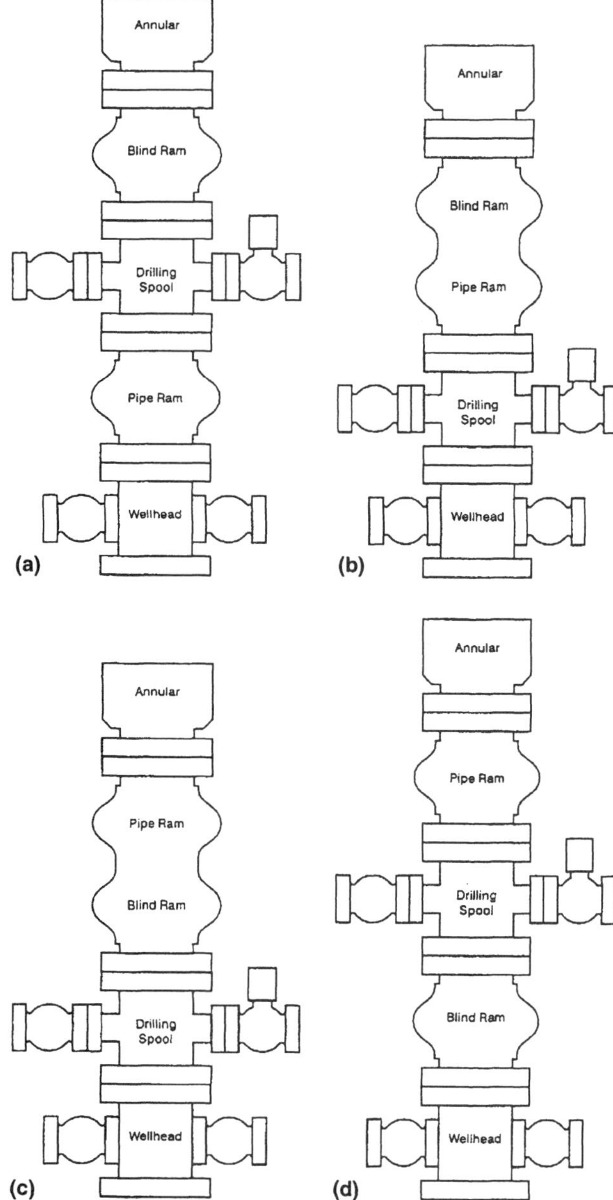

Fig. 7.23—Possible two-ram stack arrangements.

illustrated in Fig. 7.24a and 7.24b, the bottom pipe ram would be used only if problems developed in the overlying equipment. The same can be said for the bottom circulation outlets depicted in Fig. 7.24c and 7.24d. The wellhead outlets could be used in an emergency, but this is a last-resort measure. A better way is to have the bottom lines hooked up to the stack and tied into the manifold.

One advantage to the configurations illustrated in Figs. 724a and 7.24c is that annular-ram stripping can be accomplished while holding the lower ram as backup. One should anticipate the need to replace ram packing elements before any stripping job and these two arrangements allow closing in the well when the ram packing elements wear out.

Putting the pipe rams on top of the double probably does not give enough room to pass a tool joint between the annular and ram, nor does it offer a way to bleed well pressure between the two closing devices. On the other hand, having the blinds on bottom of the double preventer allows for closing in the well before a casing job so that the pipe rams can be replaced. Otherwise there would be no way to shut in if the well flowed while the changeout was taking place.

TABLE 7.5—CONSIDERATIONS FOR STACK CONFIGURATIONS PRESENTED IN FIGS. 7.23a AND 7.23b

Fig. 7.23a

Capabilities

1. If the drilling spool fails, the pipe rams can be closed to make repairs.
2. The pipe rams can be closed to replace the blinds with a pipe ram.
3. Annular-to-ram stripping is possible.*
4. If ram-to-ram stripping is necessary, the blinds can be replaced with pipe rams.*
5. If anything fails in the drillstem above the rotary, the drillpipe can be suspended from the pipe rams and the well circulated by pumping into the drilling spool.**
6. The drilling spool outlets can be used with the blinds closed.

Disadvantages or Limitations

1. Annular must be used to initially close on pipe.
2. With the blinds closed, there is no way to control the well if failure occurs near the drilling spool.
3. With the pipe rams closed, the wellhead outlets must be used to circulate the well.

Fig. 7.23b

Capabilities

1. Either the annular or pipe rams can be initially closed on the pipe.
2. The pipe rams can be closed to replace the blinds with a pipe ram.
3. Annular-to-ram stripping is possible.*
4. The drilling spool outlets can be used with either the blinds or the pipe rams closed.
5. Substructure height requirements and number of flanges are reduced if a double ram preventer is installed.
6. The drilling spool outlets can be used with the blinds closed.

Disadvantages or Limitations

1. There is no way to control the well if the drilling spool fails.
2. With the blinds closed, there is no way to control the well if failure occurs near the drilling spool.

*Ram-to-ram stripping is not recommended usually without having an extra set of pipe rams below the bottom working ram.

**Requires a drillstring float or pump-down check valve and connecting the wellhead outlet to the choke manifold.

Tapered drill strings require at least one ram preventer sized for each drillpipe diameter. Using a top and bottom variable-bore ram on top retains the flexibility of the arrangements shown in Figs. 7.24a through 7.24d without having to increase stack height and the number of flanges.

The configurations shown in Fig. 7.24a through 7.24d are common for high-pressure surface stacks, but are not the only options available to an operator. An additional ram may be placed in the stack, the double may be placed below the single, and other circulation outlets may be provided. Nor have we covered all the contingencies best handled by a given configuration (blind/shear ram use for instance). Altermann[19] discussed other arrangements and we encourage the reader to explore the topic further.

7.6.3 Kill-Line Considerations. As a minimum requirement, the kill line will be tied into the rig pumps with a manual valve next to drilling spool or ram outlet. A redundant gate valve is run normally in higher risk/pressure applications and placing a check valve on the outboard side allows the inboard valves to remain open so that pumping into the annulus does not require having somebody go up to the well to open a valve. Alternatively, a closed HCR valve next to the stack will keep the kill line between the spool and check from plugging with drill cuttings but still will allow remote operation in an emergency.

TABLE 7.6—CONSIDERATIONS FOR STACK CONFIGURATIONS PRESENTED IN FIGS. 7.23c AND 7.23d

Fig. 7.23c

Capabilities

1. Either the annular or pipe rams can be initially closed on the pipe.
2. The drilling spool outlets can be used with either the blinds or the pipe rams closed.
3. Substructure height requirements and number of flanges are reduced if a double ram preventer is installed.
4. By closing the blinds, drillpipe rams can be safely changed to casing rams with pipe out of the hole.
5. The drilling spool outlets can be used with the blinds closed.

Disadvantages or Limitations

1. There is no way to control the well if the drilling spool fails.
2. Annular-to-ram stripping is unavailable.*
3. With the blinds closed, there is no way to control the well if failure occurs near the drilling spool.

Fig. 7.23d

Capabilities

1. Either the annular or pipe rams can be initially closed on the pipe.
2. By closing the blinds, drillpipe rams can be safely changed to casing rams with pipe out of the hole.
3. Drillpipe can be dropped or set on bottom and the well closed in with the blinds if failures occur in the stack.
4. Exposed flanges below the blinds are minimized.
5. All overlying stack equipment can be repaired or replaced with the blinds closed.
6. The drilling spool outlets can be used with the pipe rams closed.

Disadvantages or Limitations

1. With pipe in the hole, there is no way to control the well if the drilling spool fails.
2. Annular-to-ram stripping is unavailable.*
3. The drilling spool outlets cannot be used with the blinds closed.

*Ram-to-ram stripping is usually not recommended without having an extra set of pipe rams below the bottom working ram.

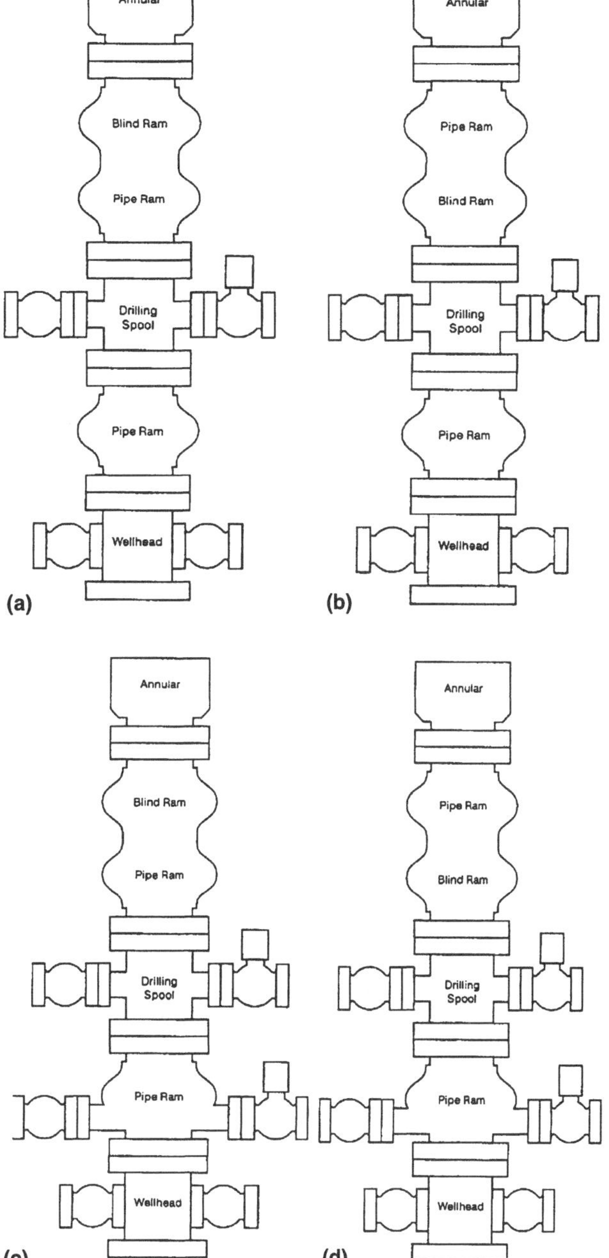

Fig. 7.24—Possible three-ram stack arrangements.

Though a convenient way to access the annulus, the kill line should not be used as a fillup line.

The means to pump into the well a safe distance away from the rig should be provided in high-pressure arrangements. **Fig. 7.25** illustrates a tee with two check valves that allow an operator to access the well with a high-pressure pump if the rig pumps are inoperable, inaccessible, or incapable of pumping against the well pressure. It is important to remember to pressure-test the remote line.

7.6.4 Choke Line and Manifold Design. The recommended layout of the choke line and manifold depends on the pressure rating, the expected service conditions (temperature, etc.), space limitations, and other factors. Three example arrangements described by the API[8] are reproduced here as **Figs. 7.26a and 7.26b and Fig. 7.27.** The manifolds depicted in Figs. 7.26a and 7.26b were suggested for low-pressure (2M and 3M) and 5M service while the layout illustrated in Fig. 7.27 is one way to put together a 10M manifold. We do not agree with all aspects of these drawings and offer discussion on the points of contention.

Two valves should be installed between the drilling spool and manifold with the ability to operate one from the control unit. Whether the HCR or the manual valve should be placed next to the stack is another debatable subject. One of these valves must be closed while circulating down the flowline and it is desirable to have an open line out to the manifold as soon as the HCR is opened. An open valve next to the stack is problematic in that the valve bore can be a repository for cuttings and barite accumulation. Hence many operators put the HCR inboard and leave the outboard valve open.

However, the HCR is the working valve and having it closest to the stack violates the principle of redundant systems. Some operators place the HCR on the outside, but doing so means that the manual valve is closed during routine operations. This negates one of the main reasons for having an HCR in the first place and requires one of the crew to go up next to the stack and open the valve before the casing pressures can be read and the manifold used. Alternatively, some operators put the HCR on the outside, leave the manual open, and flush the line periodically. This preventative measure is well advised, but it does not guarantee that the line will not be plugged when it is needed.

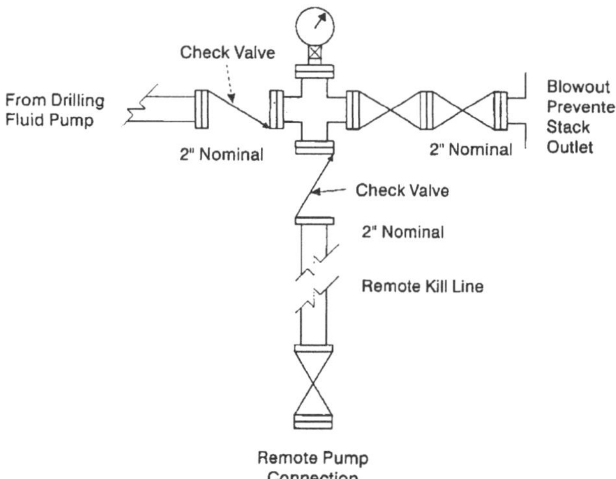

Fig. 7.25—Kill-line configuration for accessing a well from a remote location.[8]

The API illustrations in Figs. 7.26a and 7.26b show a 3-in. nominal choke line. A line this small may be acceptable for low-pressure service, but a minimum 4-in. diameter is recommended for most cases. An essential aspect of all choke manifolds is that they provide a vent line (also called the getaway or panic line) overboard or to the pit. Basically, the vent line is used as a diverter whenever conditions preclude shutting the well in and so the line should be laid out as straight as possible and have no restrictions or diameter changes. It is for this reason that a larger choke line is recommended. A remotely operated igniter should be installed at the end of the line.

Chokes are highly susceptible to plugging or cutting out and at least two adjustable chokes are needed to provide one backup in the event that the other fails. One side of the manifold then can be isolated and the other side used until the primary choke can be replaced or repaired (giving due consideration to safety) or the job completed.

Contrary to the illustrations in Fig. 7.26a and 7.26b, two valves normally are placed between the cross and choke on 5M manifolds. Also note that both chokes are manual on API's 2M and 3M manifolds. Keep in mind that the choke operator and driller must communicate with each other during a well-control procedure. This communication and thus the ability to control a well is curtailed or effectively eliminated if they cannot even see each other. The same can be said for using a manual backup choke. It may be advisable to shut the well in and remedy the problem with the remote choke before trying to control with a manual choke.

Additional redundancy is needed when manifold pressure ratings are 10M or higher. The flow schematic shown in Fig. 7.27 is one of many possible arrangements. Regardless of the hookup, at least two chokes should be remote-actuated. **Fig. 7.28** illustrates another high-pressure manifold discussed by Grace.[20] This manifold can handle almost any failure contingency. Either side can be used as the primary system and each side has its own redundant choke branch.

The manifold should be in a well-illuminated position and away from vehicle traffic. Drilling muds can solidify and so provisions must be available to drain and flush the lines with water. A slight elevation grade away from the well will facilitate drainage. Well-control problems have been intensified when freezing temperatures caused ice to form in the choke line and manifold and so antifreeze is needed in cold weather. Water-based mud can freeze very quickly in extremely cold climates unless fluids are kept moving. A better system is to enclose and heat the manifold. The area beneath the substructure probably is weatherproofed and it is a simple task to enclose the choke line between the sub and manifold house in a heated duct.

The low-pressure lines are arranged for the flexibility to route return fluids to any number of desired areas. At minimum the available destination points will be the separator, mud pits, and reserve pit. Gas and associated materials will exit the choke at high velocities and so 3-in. lines are recommended over 2-in. lines to help reduce erosion and plugging potential from debris or hydrates. Turns must be protected from erosion by targets and, perhaps, hard metal trim. A target is a solid bull plug, thick blind flange, or other specially made device which is installed so that it faces the incoming flow stream and bears the brunt of the punishment. Whipping tendencies can be severe and the manifold, vent line, and low-pressure choke lines must be anchored to prevent movement.

It is common to see a surge chamber used as a header for two or more incoming low-pressure lines. The arrangement is convenient in that it reduces by half the number of lines to the pits, separator, etc. These are normally made from heavy-wall casing and have targets at the right places, but their use should be discouraged if employed as the sole header in the system. Otherwise, a cut-out chamber or failure in any one of the lines leaves no backup.

Problems

7.1 Calculate the top and bottom MIYP requirements for the casing described in Example 7.1 except assume the well is half-full of 16.5-lbm/gal mud.

7.2 The API relation for MIYP ratings substitutes the specified minimum yield strength for the σ_s term in Barlow's equa-

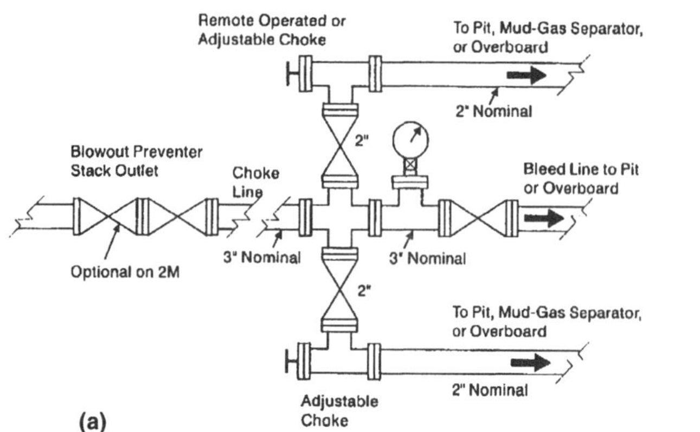

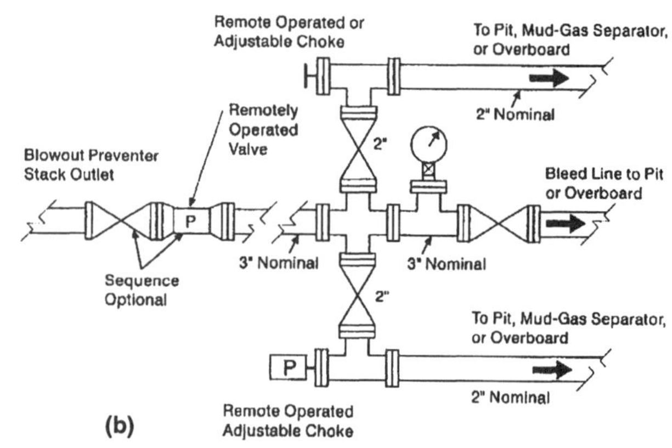

Fig. 7.26—Example choke manifolds for 2M, 3M, and 5M service.[8]

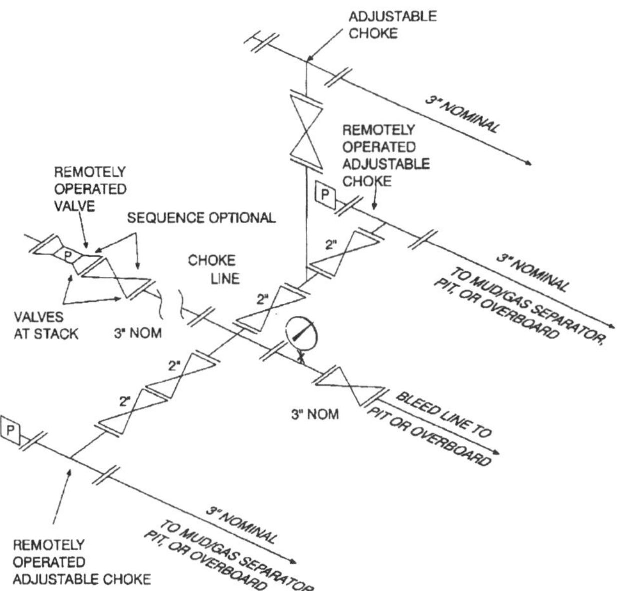

Fig. 7.27—Example choke manifold for 10M and 15M service.[8]

tion and multiplies the result by 0.875 to account for a 12.5% tolerance on wall thickness.

1. Determine the API's MIYP rating for 7-in. 32.0-lbm/ft N-80 casing (nominal wall = 0.453 in.).

2. Determine the MIYP if a caliper log shows that the pipe wall is uniformly worn across an interval to 0.35 in. [8.9 mm].

7.3 You have designed the following hole program.

Set 13³/₈-in. surface casing at 3,700 ft.

Fracture gradient at 3,700 ft = 0.735 psi/ft.

Drill to 11,000 ft where the pore pressure equivalent = 13.3 lbm/gal.

Set 9⁵/₈-in. intermediate casing.

Fracture gradient at 11,000 ft = 0.895 psi/ft.

Drill to 13,800 ft where the pore pressure equivalent = 16.4 lbm/gal.

Set 7-in. intermediate casing.

Fracture gradient at 13,800 ft = 0.965 psi/ft.

Drill to 15,000 ft where the pore pressure equivalent = 17.5 lbm/gal.

Set 4½-in. production liner.

Specify the wellhead and BOP sizes and minimum working pressure ratings for each hole section.

7.4 Use the parameters discussed in Example 7.2 and plot minimum closing pressure as a function of shut-in BOP pressure.

7.5 Start with Boyle's law and derive Eq. 7.7.

7.6 The MMS requirement in United States federal waters is that an accumulator be sized to provide 1½ times the fluid volume necessary to close all BOP equipment units with a minimum pressure equal to 200 psi above precharge. Take the equipment described in Example 7.3 and determine how many 11-gal bottles will meet the MMS requirement.

7.7 The cross-sectional area of a cup-type test tool run on a 4½-in. test joint and set in 9½-in. casing is 44.1 sq in. Determine the maximum allowable test pressure if the limiting test-joint tension is 320,000 lbf. Ignore the weight of the test joint.

7.8 Derive Eq. 7.13.

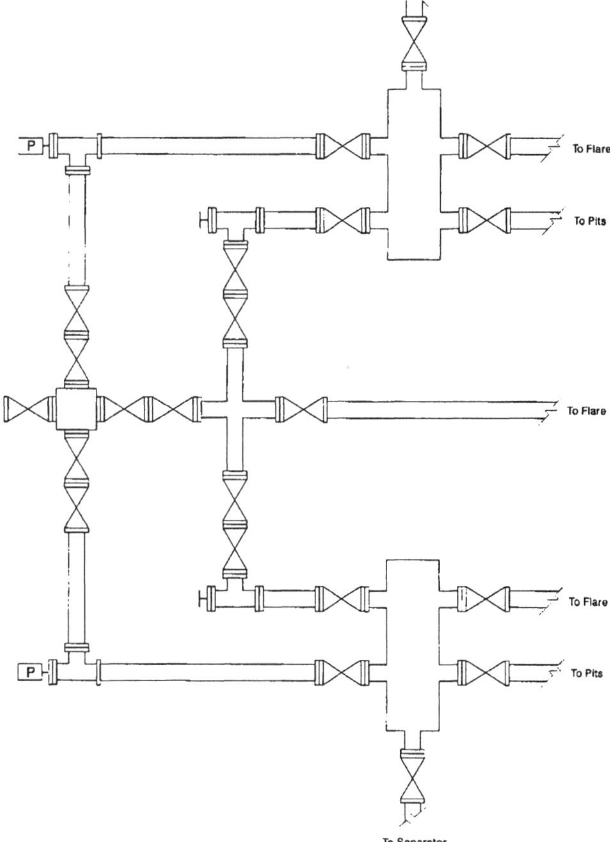

Fig. 7.28—Example high-pressure choke manifold.

7.9 Specify the minimum separator vent line and shell diameters for the following well-control conditions. Assume a Driller's Method displacement and that the vent line has three sharp bends.

Well depth = 10,500 ft;
drilling fluid density = 10.0 lbm/gal;
maximum anticipated kick size = 30 bbl;
pore pressure = 6,000 psig;
gas specific gravity = 0.70;
bottomhole annulus capacity factor = 0.03529 bbl/ft;
top-hole annulus capacity factor = 0.05017 bbl/ft;
bottomhole temperature = 225°F;
temperature upstream of the choke = 110°F;
temperature at the vent-line outlet = 85°F;
kill circulation rate = 3.0 bbl/min;
atmospheric pressure = 14.5 psia;
separator mud leg height = 5 ft; and
vent-line length = 100 ft.

7.10 Consider the kick conditions stated in the previous problem.

1. Calculate the initial SICP if gas entered the well as a package.

2. Start at 100 ft and plot the imposed pressure gradient with depth.

7.11 You are planning a well that will have the starting hole drilled below 16-in. conductor casing. Your supervisor asks you to design the diverter system and tells you to maintain equivalence between the flow area in the drillpipe annulus and the diverter line.

1. How large would a diverter line have to be if the conductor ID is 15.250 in. and the drillpipe OD is 5.0 in.?

2. What do you recommend?

7.12 Refer to the three-ram stack arrangements shown in Fig. 7.23 and discuss the relative merits and disadvantages of each.

7.13 A three-ram stack will be used on a well with a tapered drillstring. Assume variable-bore rams are not an option.

1. Which arrangement in Fig. 7.23 would you recommend?
2. Where should the larger pipe ram be placed?

Defend your answers.

7.14 Start from the stack and draw a choke manifold schematic for use in a high-pressure, high-risk application. Include all the valves, chokes, separator connections, and downstream destination points. Also, label each component as high pressure or low pressure.

Nomenclature

A_{pe} = effective piston area, in.^{2}
A_{sh} = ram shaft area, in.^{2}
d_o = outer diameter, in.
d_s = separator inner diameter, in.
d_{vl} = minimum vent-line inner diameter, in.
D = depth, ft
D_{sh} = casing shoe depth, ft
g_b = backup fluid hydrostatic gradient, psi/ft
g_g = gas hydrostatic gradient, psi/ft
g_m = mud hydrostatic gradient, psi/ft
h_{ml} = mud-leg height, ft
L_e = effective length, ft
p_a = atmospheric pressure, psi
p_{ch} = choke backpressure, psi
p_{cl} = closing pressure, psi
p_{cp} = pressure to overcome friction resistance between a cylinder and piston, psi
p_{cs} = shut-in casing pressure, psi
Δp_{cf} = control fluid friction pressure, psi
p_{fch} = fully charged accumulator pressure, psi
p_{fi} = fracture initiation pressure, psi
p_{ip} = internal pressure capacity, psi
p_{min} = accumulator minimum pressure, psi
p_{ml} = hydrostatic pressure in a mud leg, psi
p_p = pore pressure, psi
p_{pch} = accumulator precharge pressure, psi
p_w = well pressure, psi
q_{kr} = pump kill rate, gal/min
q_m = mud flow rate, gal/min
q_{vl} = gas flow rate in the vent line, ft^3/min
t = time, min
t_w = wall thickness, in.
r_{cl} = closing ratio, dimensionless
T_{ch} = temperature upstream of the choke, °R
T_{vl} = vent-line temperature, T, °R
v_{sl} = gas slip or migration velocity, ft/hr
V_{ab} = accumulator bank capacity, gal
V_{ch} = gas volume upstream of the choke, ft^3
V_{uf} = usable fluid volume, gal
V_{vl} = gas volume at the vent line discharge, ft^3
z_{ch} = compressibility factor upstream of the choke, dimensionless
γ_g = gas specific gravity, dimensionless
σ_s = material strength, psi

References

1. United States Department of Interior MMS Regulation 250.55 governing Oil and Gas and Sulphur Operations in the Outer Continental Shelf, *Federal Register*, **53, 63** (1 April 1988) 10717–718.
2. Lubinski, A. and Williamson, J.S.: "Usefulness of Steel or Rubber Drillpipe Protectors," *JPT* (April 1984) 628.
3. Best, B.: "Casing Wear Caused by Tooljoint Hardfacing," *SPEDE* (February 1986) 62.
4. Song, J.S., Bowen, J., and Klementich, F.: "The Internal Pressure Capacity of Crescent-Shaped Wear Casing," paper SPE\IADC 23902 presented at the 1992 SPE/IADC Drilling Conference, New Orleans, 18–21 February.
5. "Specification for Wellhead and Christmas Tree Equipment," *Specification 6A*, seventeenth edition, API, Washington, DC (1 February 1996).
6. Varcoe, B.E.: "Shear Ram Use Affected By Accumulator Size," *Oil and Gas J.* (5 August 1991) 34.
7. Tarr, B.A. *et al.*: "New Generation Drill String Safety Valves," paper presented at the 1996 IADC Well Control Conference for Europe, Aberdeen, 22–24 May.
8. "Recommended Practice for Blowout Prevention Equipment Systems for Drilling Wells," *RP 53*, second edition, API, Washington, DC (May 25, 1984); out of print.
9. "Recommended Practice for Design of Control Systems for Drilling Well Control Equipment," *Recommended Practice 16E*, first edition, API, Washington, DC (1 October 1990).
10. "Specification for Control Systems for Drilling Well Control Equipment," *Specification 16D*, first edition, API, Washington, DC (1 March 1993).
11. Brault, I.F. and Rajabi, I.B.: "How to Better Predict BOP Accumulator Performance," *World Oil* (May 1991) 61.
12. Jones, M.R. and LeMoine, J.: "What to Consider When Sizing BOP Control Units," *World Oil* (May 1990) 57.
13. Goins, W.C. Jr.: "Learning From Well Control Mistakes Can Help Prevent Future Blowouts," *World Oil* (October 1996) 45.
14. Kandel, W.J. and Streu, D.J.: "A Field Guide for Surface BOP Equipment Inspections," paper SPE\IADC 23900 presented at the 1992 SPE/IADC Drilling Conference, New Orleans, 18–21 February.
15. Butchco, D. *et al.*: "Design of Open-Bottom Mud/Gas Separators," paper SPE\IADC 13485 presented at the 1985 SPE/IADC Drilling Conference, New Orleans, 6–8 March.
16. MacDougall, G.R.: "Mud/Gas Separator Sizing and Evaluation," *SPEDE* (December 1991) 279.
17. Weymouth, T.R.: "Problems in Natural Gas Engineering," *Trans.*, ASME (1912) **34**.
18. Goins, W.C. Jr. and Sheffield, R.: *Blowout Prevention*, second edition, Gulf Publishing Co., Houston (1983) 121–23.
19. Altermann, J.A. III,: "Practical Considerations for Arranging, Testing BOP Stacks," *World Oil* (May 1980) 91.
20. Grace, R.D.: *Advanced Blowout & Well Control*, Gulf Publishing Co., Houston (1994) 17.

SI Metric Conversion Factors

bbl	× 1.589 873	E − 01	= m^3
ft	× 3.048*	E − 01	= m
ft^3	× 2.831 685	E − 02	= m^3
°F	(°F − 32)/1.8		= °C
gal	× 3.785 412	E + 00	= L
in.	× 2.54*	E + 01	= mm
in.^{2}	× 6.451 6*	E + 02	= mm^2
lbf	× 4.448 222	E + 00	= N
lbm	× 1.535 924	E + 00	= kg/m
psi	× 6.894 757	E − 03	= MPa
°R	°R/1.8		= K

* Conversion factor is exact.

Chapter 8
Offshore Operations

8.1 Introduction

Offshore drilling structures supported by the sea floor (platforms, jackups, and submersibles) use surface blowout preventer (BOP) stacks with equipment and operational considerations as discussed in Chap. 7. Floating drill vessels, on the other hand, normally require the use of subsea wellheads and BOP equipment. The distance between the rig and the stack, its relative inaccessibility, and the movement of the rig entail modifications to the equipment and well-control procedures.

In this chapter we address hazards associated with shallow gas. Blowouts from shallow-gas sands have led to more lost and damaged rigs than any other single well-control problem and opinions differ on how best to manage a kick below a shallow casing seat. Land operations in some areas are just as prone to the problem, but it is appropriate to discuss the shallow gas here as the safety, cost, and environmental issues are compounded whenever a rig is placed out in the water.

Finally, we will discuss briefly some special problems in deepwater well control. Conventional operations use a large-diameter flow conduit (the marine riser) to house the drillstring and convey mud from the subsea equipment to the drill vessel. New tools and procedures are attempting to address the difficulties associated with riser drilling at extreme water depths. Shallow water flows are also a major concern and some ideas for dealing with this problem are presented in the last section.

8.2 Equipment Used in Floater Drilling

Semisubmersible (**Fig. 8.1**) and ship-like (**Fig. 8.2**) mobile offshore drilling vessels each provide unique operating characteristics desirable in some applications and undesirable in others. For example, drillships have a large load capacity and can be transported more rapidly than a semisubmersible. Thus a drillship would probably be selected if a well is to be drilled in a location remote from the supply base. The storage capacity of a semisubmersible is lower and demands on the logistical support functions are greater. Semis, however, are more stable and will have less downtime than a drillship because of rough weather. Each type of vessel also offers specific advantages and disadvantages related to well control.

Before getting into the equipment specifics, we need to understand something about how a subsea well is drilled. A typical floater drilling sequence from the time equipment is first set on the sea floor until the BOPs are installed is represented in **Fig. 8.3** and a common hole program is shown in **Table 8.1.** Alternate programs and provisions for running additional casing strings are available.

First, a large temporary guide base (or template) is lowered on the drillstring and set on bottom. The guide base is levelled and the drillstring is released mechanically. Guidelines strung to the drill vessel are run with the temporary guide base to facilitate re-entry and for lowering accessory equipment such as TV cameras. Guidelineless systems are available for deepwater operations.

In Step 2, a pilot bit and hole opener or large-diameter bit is lowered to the temporary guide base. Slackoff force shears the drillstring from the break-away guide arms, the bit passes through a funnel in the guide cone, and the guide arms are pulled using retrieving lines. The first hole section is drilled with returns dumped at the sea floor and the hole is drilled deep enough to reach competent sediment needed to provide a good foundation for the structural casing. It is critical for this section to be drilled as straight as possible as the structural hole controls the verticality of the wellhead and BOPs. The bit is pulled and the structural casing and permanent guide base are run on the drillpipe as a unit. The assembly lands on the temporary guide base and the casing is cemented through the drillstring using shear and piggyback wiper plugs similar to those used to cement a liner. The running tool is released from the housing on top of the structural casing and the drillstring is retrieved.

Soft floor conditions allow an operator to modify the sequence depicted in the first three steps of Fig. 8.3. A common procedure in the Gulf of Mexico and other areas is to use an internal bit or jet head to jet the structural casing into place. For jetted casings, some operators elect to eliminate the temporary guide base and install a large mud mat some distance below the permanent guide base. The mat, upon contacting the sea floor, serves as a positive indicator that the pipe has been jetted to depth and provides some axial support to the structural casing during the subsequent operations. The annu-

Fig. 8.1—Semisubmersible drilling rig in operation.

lus silts in, which forms a fairly competent seal, and casing set in this manner is not cemented.

The conductor hole is usually drilled as shown in Step 4 with returns dumped at the sea floor. At the casing point, the conductor pipe is run, the wellhead is installed, and the assembly is lowered on drillpipe. The pipe is lined up with the hole using guide arms until the wellhead seats in the structural casing housing and the casing is cemented back to the mudline. Some operators prefer to have drilling fluid returns to surface when drilling the conductor hole and attach a riser to the structural casing housing using a hydraulic latch mechanism. Uncemented drive pipe must be set deep enough to prevent a pullout if the riser is connected at this point. The riser is released and pulled after the conductor hole is drilled.

The next step is to run the massive BOP stack and its accouterments on the riser. The stack rides down the guidewires which align the frame onto the permanent guide base posts. On guidelineless systems, the stack is guided to the wellhead by a combination of acoustic locating devices and direct observation by a subsea TV camera or acoustic imaging device. The bottom of the stack is hydraulically attached to the wellhead and, after testing the equipment, the surface hole is drilled with returns back to the vessel via the riser. In fact, routine operations and practices from this point forward are not greatly different from a surface drilling operation.

Floater operations can be conducted without excessive wear on the subsea equipment when vessel offset from the well is no more than about 5% of water depth. Station keeping—i.e., keeping the rig over the hole—is accom-

Fig. 8.2—An example drillship.

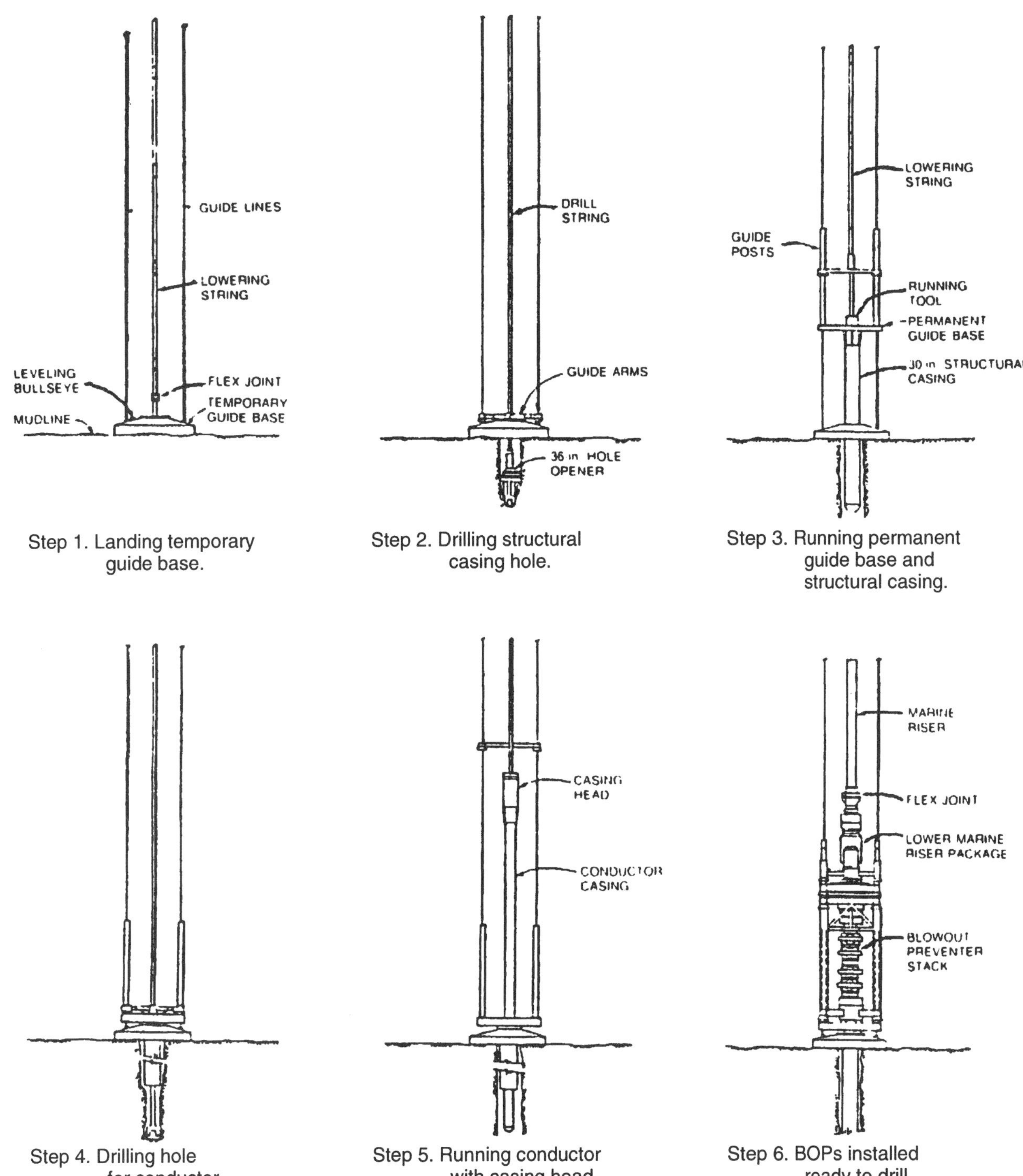

Fig. 8.3—Typical drilling sequence through setting the BOPs on a floating drilling operation.[1]

plished with anchors and mooring lines. Many deepwater vessels use dynamic positioning (DP) to maintain the desired position. In the latter method, acoustic beacons triangulate vessel location with respect to the well or other reference point and thrusters automatically keep the vessel within acceptable limits.

8.2.1 Subsea Wellheads. A drawing of an example subsea wellhead including the casings for a five-string hole program is shown in **Fig. 8.4.** The main difference between subsea and surface heads is that all the casing strings are suspended from the same housing. Note also that there is no way to access the

TABLE 8.1—COMMON OFFSHORE HOLE PROGRAM

Casing	Hole Diameter (in.)	Casing Outer Diameter (OD) (in.)	Below Mud Line Setting Depth (ft)
Structural	36	30	100 to 300
Conductor	26	20	500 to 3,000
Surface	$17^1/_2$	$13^3/_8$	2,000 to 6,000
Intermediate or Production	$12^1/_2$	$9^5/_8$	—
Production or Liner	$8^1/_2$	7	—

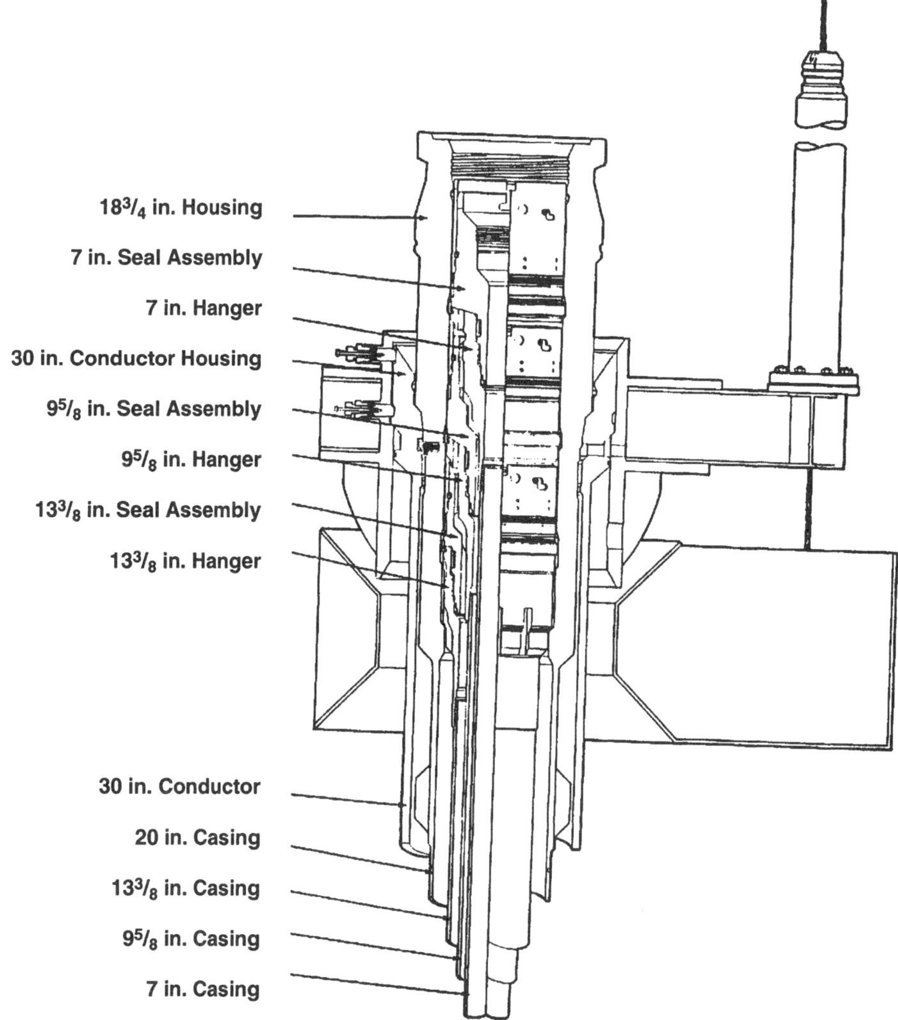

Fig. 8.4—Example subsea wellhead for a five-string hole program. Courtesy of Cameron.

annulus of a given string. This can present a problem during a drillstem test (DST) or after the well has been placed on production. Heat transfer from the produced fluids will increase the temperature at shallower depths, cause the mud and tubulars to expand, and thereby create potentially high burst and collapse pressures if the annulus fluids are confined.[2,3] The well planner should consider the effects and either design the pipe accordingly or provide a "relief valve" in the annulus by leaving some open hole above the cement top.

The features of a subsea wellhead depend on the make and manufacturer, but the general setting and sealing characteristics are similar in most respects. A hanger and running tool are installed on the top joint of casing and lowered into the wellhead with drillpipe. The surface casing hanger is landed on a shoulder in the housing while intermediate and production strings are usually seated on the last hanger. After cementing, the seals are compressed and energized by one of three ways: releasing the setting tool and slacking off; using the releasing torque to drive the seals against the hanger; or hydraulic actuation with a piston.

Modern equipment designs allow the seals to be tested without having to make a trip by using a test cup or O-ring extension below the running tool which seals against the hanger bore. The BOPs and riser are in place when surface casing is run and the standard procedure is to close a pipe ram and test the seals with water down the choke/kill lines. The bore of the drillpipe is open which allows a leak across the test tool to be detected. However, not having an outlet below the hanger seals presents a problem in that a leak could collapse the casing or rupture the outer string.

TABLE 8.2—SUGGESTED PROCEDURE FOR TESTING SEALS IN A SUBSEA WELLHEAD[4]

1. Set the test tool in the hanger bore.
2. Close a pipe ram above both choke and kill line outlets.
3. Pump down one of the lines with water and displace mud from the BOPs. Pump a large volume in attempt to purge the system of air.
4. Close the inboard valve on the lower side outlet.
5. Pressure the line up to the test pressure and record the volume pumped. Release the pressure.
6. Repeat Step 5 until volumes are repeatable.
7. Open the lower side outlet valve and close the valve on the upper side outlet.
8. Test the casing seal. Stop the test if the reference volume is exceeded before test pressure is achieved.

To monitor seal integrity, Goins and Sheffield[4] recommended the test procedure shown in **Table 8.2.** The objective is to measure the volume increase in the test lines at the designated test pressure resulting from fluid compressibility and other elasticity in the system. Taking the BOP cavity contribution to be negligible, the same volume should be observed when the seals are tested. These seals have proved highly reliable if the seals and seal surface are kept from damage.

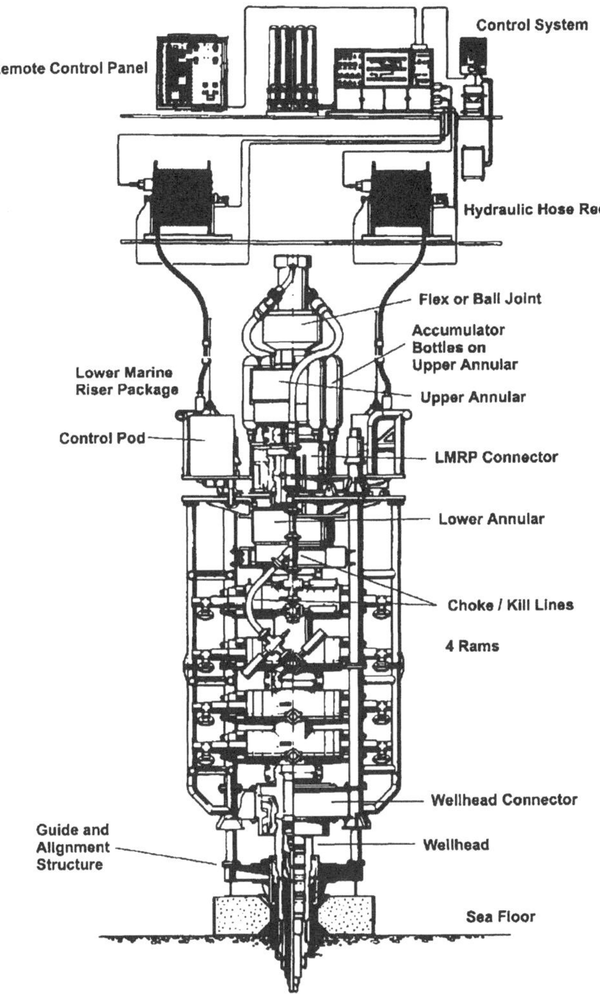

Fig. 8.5—Subsea well-control equipment. Courtesy of Cameron.

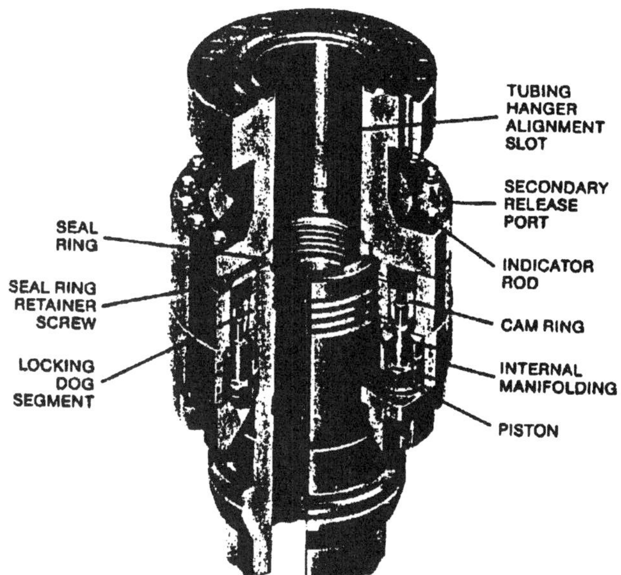

Fig. 8.6—Mandrel-type hydraulic connector. Courtesy of ABB Vetco Gray.

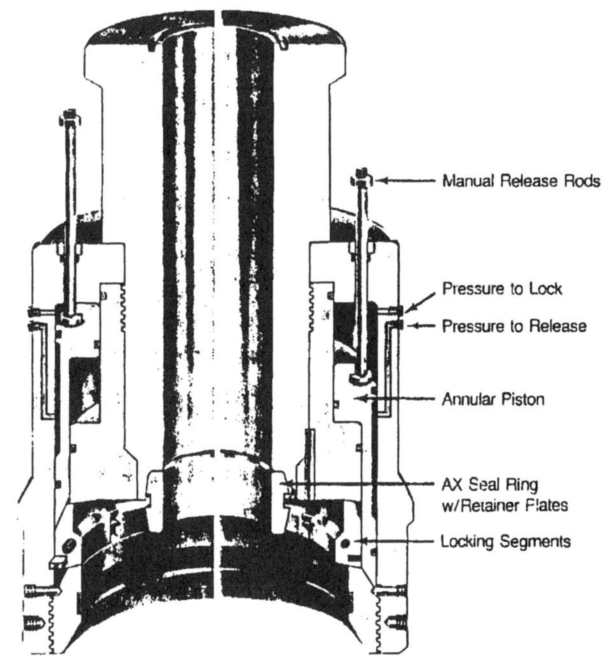

Fig. 8.7—Hub-type hydraulic connector. Courtesy of FMC.

It is essential to protect the housing bore from wear and abuse while drilling. A bore protector is run in place when the wellhead is first installed and is retrieved before running the surface casing. Wear bushings are run in the subsequent pipe jobs. Some designs allow the wear bushing to be run at the same time as the hanger while others require making an additional trip.

8.2.2 Stack Equipment. A schematic of the subsea well-control equipment used on a typical well is shown in **Fig. 8.5.** Several ram arrangements are possible, but placing a blind/shear ram above three pipe rams is the most common configuration. Changing out the stack or any of the components entails a great deal of time and expense and, once run, the equipment is usually left in place until drilling operations are completed. Accordingly, a single-stack operation requires that the equipment have a bore and working pressure suited for all hole sections, adequate redundancy, and provision for changes in drillpipe diameter (extra ram housings or variable-bore rams). Eighteen-and-three-quarter inch 10M and 15M stacks are used most often becuse they allow drilling a 17½-in. surface hole with a bit and have the pressure rating for the deeper hole sections.

The bottom of the stack connects to the wellhead by way of a hydraulic connector designed to mate with either a mandrel or hub profile on the wellhead. The mandrel type shown in **Fig. 8.6** has several dogs designed to accept matching grooves on the wellhead. The connector stabs over the wellhead and, by hydraulically moving a cam ring, the connector is pulled to the wellhead and compresses a metal seal ring on an inside taper at the top of the wellhead. A hub connector (**Fig. 8.7**) has finger-shaped locking segments that are actuated by hydraulic pressure to rotate under the hub and pull the connector to the wellhead.

Both connectors will remain attached if control pressure is lost and provisions can be made for diver or remote-operated-vehicle (ROV) actuation if need be. The two wellhead mating configurations are not interchangeable and, because the stack generally comes with the rig, the type of wellhead is dictated by the stack connector. Alternatively, the wellhead can be manufactured with connectors that will mate to the different type.

Subsea rams are essentially the same as rams in a surface stack but they have some added requirements. Hanging a tool joint on a ram is a rarely used option for surface stacks but the

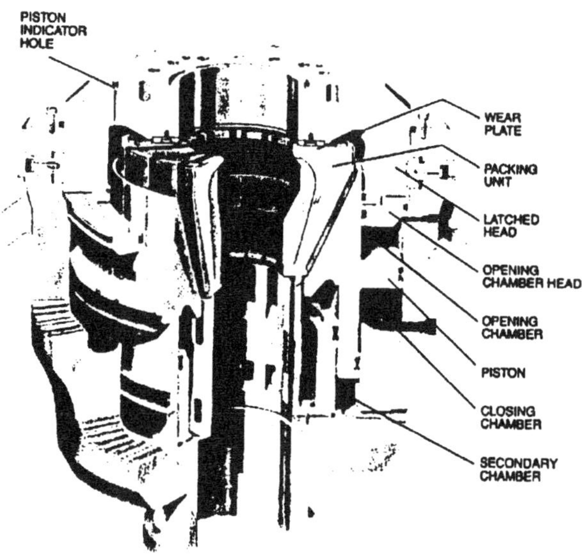

Fig. 8.8—Type GL annular preventer. Courtesy of Hydril Co.

ability to do so is a necessity in subsea applications. The drillstring must be kept stationary during a well-control procedure and suspending the string at the BOPs allows the drill vessel to be moved off location in a well-control or weather emergency. Subsea rams also must be equipped with a remote hydraulic locking mechanism to keep the ram in a closed position should closing pressure be lost. Finally, a blind/shear ram must be used instead of a blind ram.

Almost any piston-driven annular preventer can be used in a subsea stack, but water depth and mud weight may impact the closing-pressure requirements for a shut-in or stripping application. The pressure of the mud in the riser creates an opening force on most annular preventers while the control-fluid density comes into play if the opening and closing pistons have different cross-sectional areas.

Fig. 8.8 shows an annular preventer with design features that address these specific subsea considerations. The closing and opening sides have the same piston area, which balances forces across the piston regardless of water depth. A secondary closing chamber has a piston area roughly equal to the effective area exposed to the drilling mud.

For water depths up to 800 ft, the manufacturer recommends connecting the secondary chamber to the opening port. Thus the closing pressure is increased by an amount equal to the hydrostatic difference between the mud and control fluid. The surface closing pressure with this hookup can become excessive in deeper water and high mud weights; it may be desirable to reduce the closing pressure by connecting the secondary chamber to the closing port. Other annular-preventer makes and models are suitable and the manufacturer's literature should be consulted based on the mud weight in use and water depth.

The annular-preventer design must allow for an automatic closing-pressure adjustment when the sealing element passes a tool joint while stripping or when vessel heave pulls a tool joint into the annular before the drillstring can be hung off. Pressure regulators are too slow to react to the latter. The standard approach is to install a surge chamber (precharged accumulator bottle) on the closing side at the stack.

BOP tests are similar to the procedures used with surface stacks, but with some necessary modifications because of the economic consequences of a failed subsea test and the unique equipment. Before running, the stack is attached to a wellhead profile on the drill vessel and a stump test of every component is conducted. After placement, the process is repeated with an open-bore test tool located in the wellhead. Test pressure is applied through the choke and kill lines, hence any rams below the lowermost outlet cannot be tested any higher than the casing test pressure.

8.2.3 The Marine Riser and Associated Equipment. The primary functions of the marine riser are to set and retrieve the stack, return mud and cuttings back to the drill vessel, guide the drillstring into the wellbore, and support the choke and kill lines. **Fig. 8.9** depicts a schematic of the riser system. On the bottom, an assemblage of equipment called the lower marine riser package (LMRP) is the connecting link to the stack and provides for lateral movement of the riser with respect to the fixed stack. LMRP components consist of a hydraulic connector, an annular preventer (normally), and a ball or flex joint. On top there is a telescopic or slip joint and a flow diverter. Sometimes another ball or flex joint is installed below the slip joint.

The hydraulic connector used to attach the LMRP to the stack may be identical to the wellhead connector. However, the ability to release the riser under adverse conditions make certain designs unsuitable for use above the stack. High riser-deflection angles can cause conventional mandrel-type connectors to bind and fail to release. This occurrence has required the vessel to drive off and break the riser in two with consequent bending and damage to the wellhead and/or casing.[5] Hub-type connectors are less prone to this problem and specially designed mandrel connectors with minimum swallow of the pin are available.

Ball joints use a sealed ball and socket to accommodate lateral riser movement and allow for up to 10° vertical misalignment between the riser and stack. Ball joints, however, have experienced some reliability problems and are no longer used in some areas.[6] Flex joints similar to the one shown in **Fig. 8.10** bend in response to riser movement and are the preferred tool.

Riser joints are constructed of seamless pipe and typical outer diameters range from 16 to 24 in. A 21-in. riser will be used normally above an 18¾-in. stack. The joints are connected using dog-type, flanged, threaded union, or breech-block couplings. Fig. 8.9 shows the choke and kill lines are attached to the riser.

In most cases, risers do not need to be designed for high-pressure service. However, operating conditions in some areas have allowed floaters to be equipped with surface blowout-preventer equipment (BOPE), which dictates that the riser be designed as an extension of the casing string. All risers must have the strength and wall thickness to withstand the differential pressure between the drilling fluid and sea water. For example, a riser filled with 18-lbm/gal mud imposes a subsea differential pressure gradient of about 0.5 psi/ft. For this condition, it follows that a riser being used in 1,000 ft of water with a kelly bushing (KB) elevation 80 ft above sea level (ASL) must have an internal pressure rating of at least 600 psi.

Lost circulation or a gas flow can completely evacuate a riser. Large-diameter tubes are prone to collapse failure and deepwater applications should incorporate some way to flood the riser with sea water to prevent collapse. A fill valve using differential pressure to actuate a sliding sleeve is the most common type. Another design, which addresses conventional dump/fill valve limitations, is also available.[7]

Riser stability is a function of several variables including vessel offset, water depth, riser dimensions, mud density, and environmental forces. It is necessary to tension the riser to

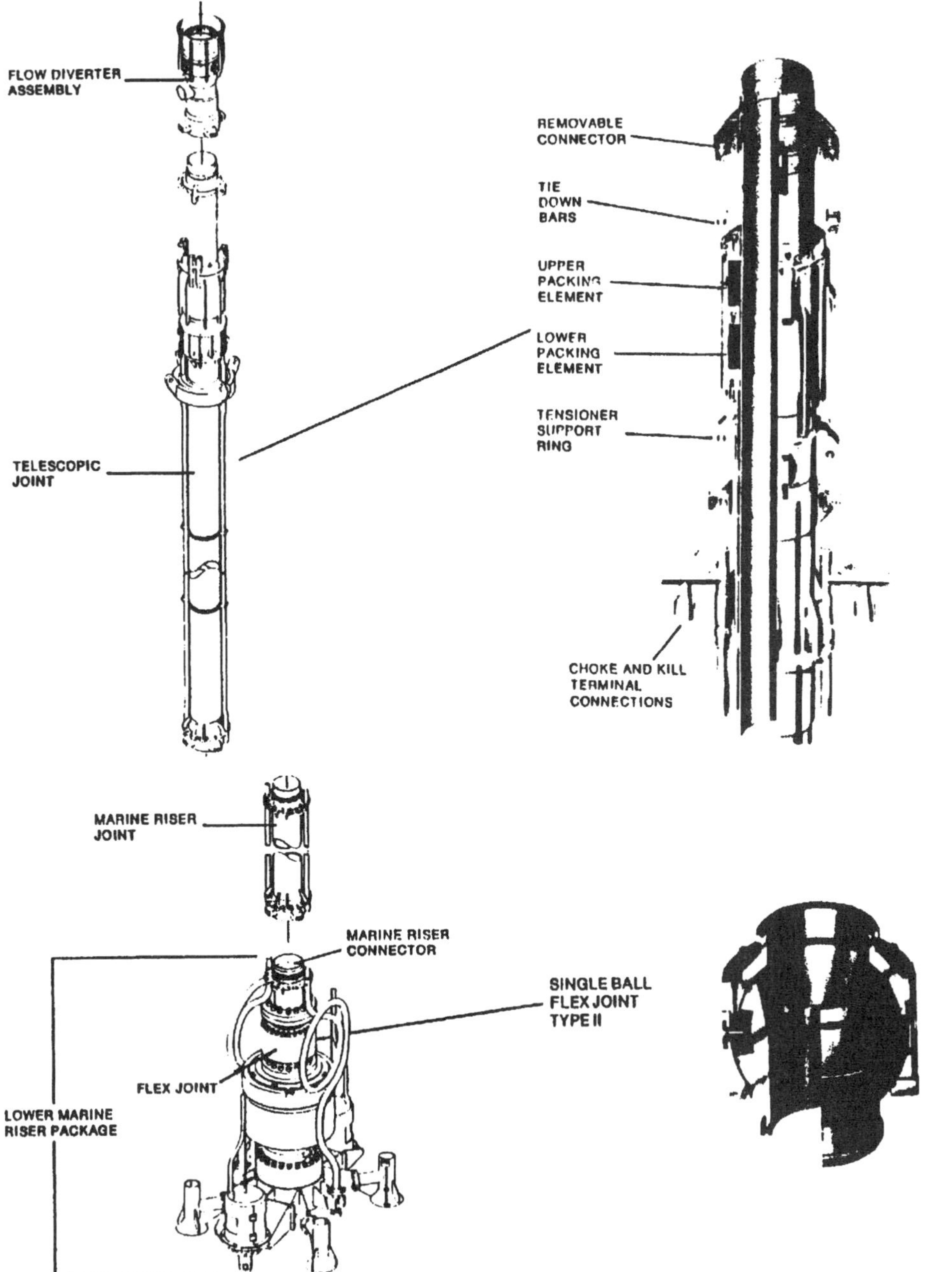

Fig. 8.9—Marine riser and associated equipment. Courtesy of ABB Vetco Gray.

limit riser sag on bottom and to maintain the top and bottom angles within acceptable limits. Pneumatic riser tensioners designed to maintain a relatively constant pull are used for this purpose. Tensioning requirements increase with water depth and buoyancy modules or air cans usually are added to the riser in waters deeper than 2,000 ft. This is a complex problem and computer programs are used to determine the optimum tension and maximum vessel offset for three conditions: when operations are not restricted (flex/ball joint angle $\leq 2°$), the riser is connected but operations are suspended, and the riser is disconnected.[8]

Flex joints are sometimes included above the riser to better accommodate lateral vessel movement. The telescopic joint accommodates the vertical movement and consists of two concentric cylinders with packer-type seals between the two barrels. The outer barrel is connected to the riser whereas the inner barrel strokes through the outer barrel. Riser tensioning lines are attached to the outer barrel. The diverter usually is installed above the inner barrel of the slip joint.

8.2.4 Choke and Kill Lines. The most common way to run the choke and kill (C and K) lines from the stack back to the surface manifold is to make the lines integral to the riser. When a riser joint is connected, a polished nipple on the bottom of each line is stabbed into an accommodating seal bore and the two ends are held together by the riser connector. The C and K lines and connections must be tested during the running procedure so that needed repairs can be made before the entire riser is installed.

C and K lines are constructed from rigid steel pipe except for the sections adjacent to the flex and slip joints. Contrary to our design criterion for surface stacks, the so-called jumper lines in these areas must be flexible to compensate for rig movement and bowing of the riser. Steel vertical or helical pipe loops normally are used near the LMRP. More flexibility is needed directly underneath the vessel because of heave and, as shown in **Fig. 8.11,** high-pressure hoses or steel piping with swivel connectors are required. The working-pressure rating of the C and K and jumper lines must match that of the stack

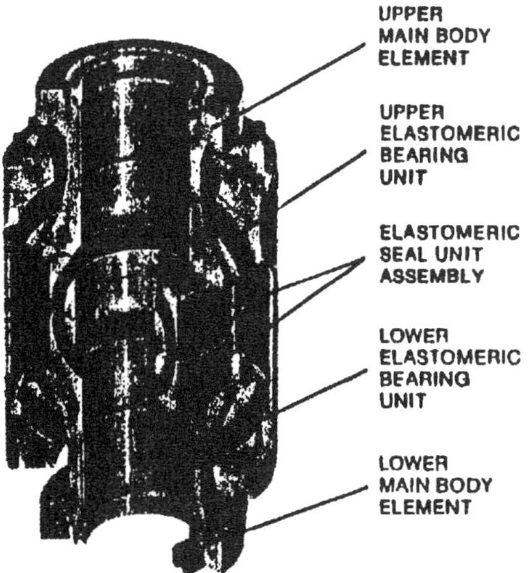

Fig. 8.10—Example of a flex joint. Courtesy of ABB Vetco Gray.

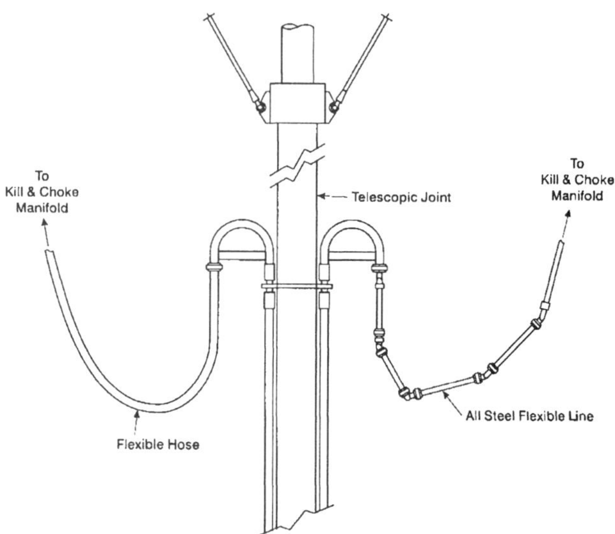

Fig. 8.11—C and K jumper lines above the marine riser.[9]

and the system design must be consistent with other equipment limitations. One blowout in the North Sea is suspected to have been caused by high temperatures which failed the end connections on a 15,000-psi hose.

The term "choke and kill lines" when applied to subsea use is called more accurately "choke *or* kill lines" as the surface manifold is arranged to permit a line to be used for either function. There are no drilling spools in a subsea stack and the ram outlets are used to connect the C and K lines. **Fig. 8.12** illustrates a common arrangement. For this system, two outlets are tied together in a single line with access to the stack immediately below the blind/shear ram. Access to the stack must be available to isolate and test an overlying ram. For this reason (and to give added redundancy), some operators install another C and K outlet beneath the bottom pipe ram.

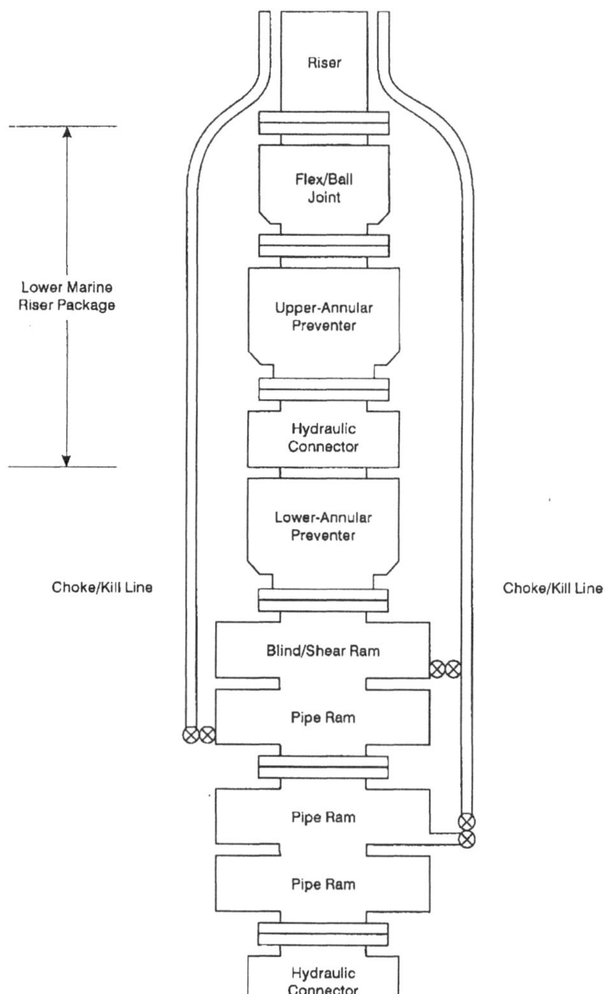

Fig. 8.12—Common C and K line arrangements on a subsea stack.

The lines are attached to the outlets on the stack using hydraulic connectors and two remote-actuated valves are provided at each stack outlet. **Fig. 8.13** shows a cutaway view of a subsea gate valve. The valve is designed to open or close when hydraulic control pressure is released or lost. For stack applications, the actuator is set so that hydraulic pressure holds the valve open. Releasing the pressure closes the gate making this a fail-close design.

As with surface stacks, limiting flow velocity and abrasion potential by sizing the line appropriately is important. Unique to a floater well-control operation are other advantages to having large C and K lines, namely the effects of line friction pressure and increased gas height when a kick enters the line. A 3-in. nominal diameter should be considered the minimum requirement. Four-inch lines can offer a substantial advantage in deep water.

8.2.5 Control Systems. The primary components in a conventional subsea control system were identified in Fig. 8.5. The basic functions of the system are the same as discussed for surface stacks, but the fact that the accumulator, pumps, and controls are on the drill vessel while the subsea stack components and connectors are hundreds or even thousands of feet away demand several modifications.

One major difference lies in how the power fluid is conveyed to the stack. Modern stacks may have as many as 100 remote-actuated functions; it would be impractical to have a dedicated line strung from the vessel to each one. This problem is solved by using a single line to carry the hydraulic fluid and a pilot-operated subsea control pod to direct the fluid to the desired function.

To operate a component, the driller pushes a button on the control panel and a separate, smaller line for the selected operation transmits hydraulic pressure to the pod. The signal then activates a three-way pilot valve that directs flow to the right place. The power-fluid line and pilot lines are carried together in a hydraulic hose bundle similar to the one illustrated

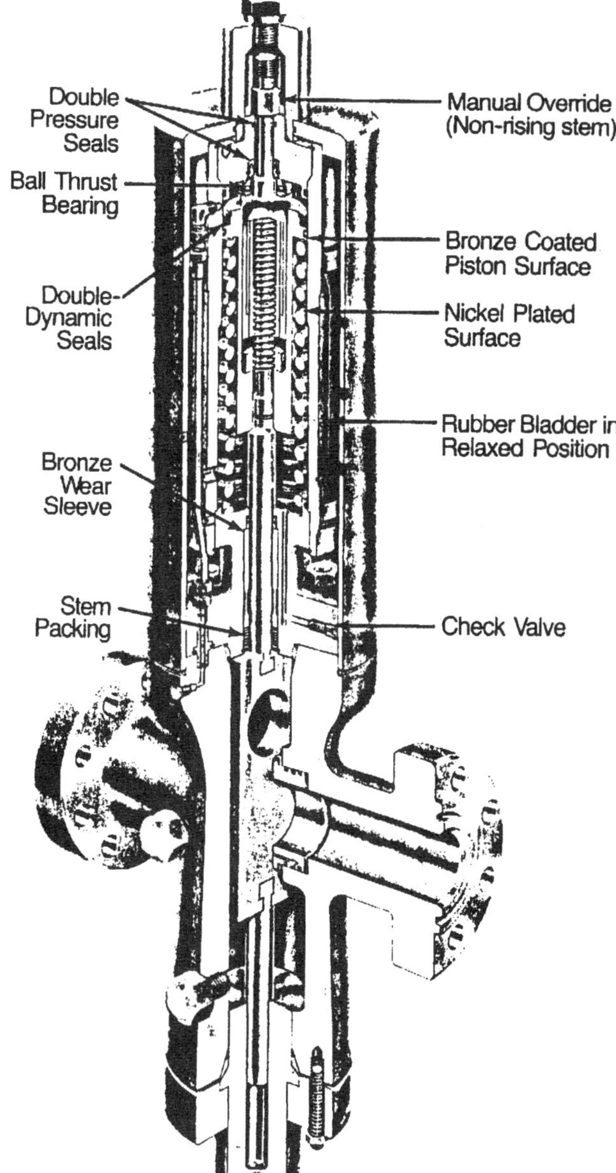

Fig. 8.13—Cutaway view of a fail-safe gate valve. Courtesy of FMC.

in **Fig. 8.14.** Fluid pressure is regulated to the component requirement at the pod. Dual pods in a parallel arrangement assure full backup for all the functions and allow one of the pods to be retrieved and serviced while the other is used to control the BOPE. Function lines from both pods connect at a shuttle valve that routes the fluid from the active pod and isolates the inactive pod.

In present systems, the power fluid is exhausted into the sea and so the fluid must be friendly to the environment. Fluid does not actually flow through a control line and it is important to filter the fluid at the accumulator to minimize the plugging potential. Corrosion, scale, and bacteria slime can also plug a line if not controlled and maintained. Typically, fresh water is mixed with a commercial blend containing a lubricant, corrosion inhibitor, and bacteriacide. Anti-freeze may be incorporated in cold-weather applications.

Subsea accumulators are recommended for two reasons. The control and supply lines in a conventional hose bundle typically are made of polyester and reinforced with a thermoplastic braid. The friction drop through the 1-in. conduit and line ballooning increase the time required to close a preventer,

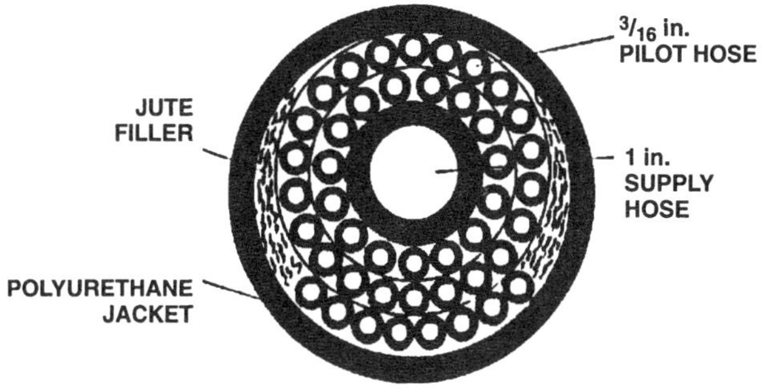

Fig. 8.14—Cross section of a hydraulic hose bundle. Courtesy of Shaffer.

a problem that increases with water depth. Having accumulators on the stack significantly improves this so-called fill-up time.

The other reason is to maintain a fluid reserve in a dedicated bank to operate the lower stack equipment if the vessel has to move off location or the control lines fail to do their job. An acoustic system backs up selected control-line functions and allows stack operation in such an emergency. The subsea electronics and solenoids are powered by a subsea battery pack.

Boyle's law is used often as the basis for sizing a stack-mounted accumulator bank. One manufacturer recommends providing 1½ times the usable fluid volume needed to close one annular preventer, the 50% factor being the safety reserve. Placing the accumulators underwater means considering the effect of subsea hydrostatics on the ambient pressure conditions. Ignoring the small density difference between the seawater and power fluid, Eq. 7.7 becomes

$$V_{ab} = \frac{V_u}{(p_{pch} + p_{sw})/(p_{\min} + p_{sw}) - (p_{pch} + p_{sw})/(p_{\min} + p_{sw})}, \quad \text{(8.1)}$$

where p_{sw} = the hydrostatic pressure of the seawater.

Example 8.1. The closing volume for an 18¾-in., 5M Shaffer spherical preventer is 48.16 gal. Assume this preventer is underneath 2,000 ft of water and use the discussed sizing criterion to determine the minimum number of 11-gal accumulator bottles to place on the stack. The precharge, minimum, and charged bottle pressures are 1,000, 1,200, and 3,000 psig, respectively. Assume a 0.445-psi/ft gradient for the seawater.

Solution. The usable fluid volume requirement is

$$V_u = (48.2)(1.5) = 72.3 \text{ gal.}$$

The hydrostatic pressure of the seawater is

$$p_{sw} = (0.445)(2,000) = 890 \text{ psi.}$$

Substitution into Eq. 8.1 gives the total accumulator volume,

$$V_{ab} = \frac{72.3}{(1,905/2,105) - (1,905/3,905)} = 173.3 \text{ gal.}$$

The number of 11-gal nominal (10-gal capacity) bottles is

$$173.3/10 = 17.3 \text{ or } 18 \text{ bottles.}$$

Water depth reduces subsea accumulator efficiency. For example, the design criterion used in the problem results in a 15-bottle requirement if the stack happened to be at atmo-

TABLE 8.3—SUGGESTED FLOW CHECK, SHUT-IN, AND HANG-OFF PROCEDURE

1. Place the drillstring in a predetermined position above the rotary table and shut off the pump.
2. Line up the flowline to the trip tank and observe for flow.
3. If the well is flowing, close the upper annular.
4. Open the necessary fail-safe valves on the stack and close the choke. (Closing the choke is applicable only if the manifold has been arranged for a soft shut-in).
5. Reduce the annular closing pressure to the minimum requirement.
6. Determine tool joint location and position the tool joint in the stack above the upper ram.
7. Close the ram, slowly lower the string, and hang it off on the ram.
8. Lock the ram in a closed position.
9. Reduce support pressure on the motion compensator and support the pipe weight above the rams with the compensator.
10. Measure shut-in pressures and pit gain.

spheric pressure. The efficiency can be increased greatly with higher accumulator pressures. Deepwater use presents one instance where 5,000-psig bottles could save considerable expense and space.

An alternative to using a dedicated subsea bank for improving reaction time is to convey the power fluid down a steel flow conduit clamped to the riser. Ballooning is not much of a problem in comparison to a flexible hose and, based on the friction drop and closing volume, the line can be sized to meet the desired closure time.[10]

Another time element is how long it takes for the signal pressure to be sensed at the pod. Acoustic velocity and line inflation are the two governing factors, both of which depend on water depth. Hydraulic signal time through a 3/16-in. pilot line is about one second in 400 ft of water and approximates a power law relationship up to 10 seconds at 3,000 ft. Obviously, signal time can become a critical problem in deep water and ways to reduce this factor are used when water depth exceeds 3,000 ft.

A biased hydraulic system is one way to reduce the signal time. Biasing involves precharging selected control lines to 200 or 300 psi below the pilot-activation pressure, thus eliminating most of the ballooning and compressibility effects in the precharge pressure. It has been used successfully in water depths up to 5,000 ft.

Signal time is practically eliminated with electrohydraulic systems. An electric signal is transmitted through a wire conductor to an electronic solenoid valve in the subsea control unit. The electronic solenoid valve converts the electric command to a hydraulic signal that travels a short distance to the pilot valve in the pod. Modern electrohydraulic systems use multiplex technology and eliminate the need for multiple conductors as coded commands are transmitted to the stack down a single pair of wires where the data is decoded and the selected solenoid is fired. Fiber-optic communication is the coming thing in the multiplex systems of the future.[11] Fiber optics allow both the command and status signals to be transmitted on the same channel and does not have the electronic "noise" problems associated with hard-wiring.

8.3 Well-Control Procedures From a Floater

Surveillance and control procedures from a floater demand consideration of vessel/drillstring movement and location of the stack. This section covers some of the operational concerns including kick detection, preventing pipe movement after shut-in, the effect of long choke and kill lines, potential for hydrate formation, measures to retain control after a well has been killed, and riser disconnect considerations.

8.3.1 Kick Detection. The kick-warning signals on a floater are the same as for any other drilling operation, but kick detection is more difficult than when the rig and pits are stationary. Vessel heave leads to a condition where mud is stored in the slip joint while the joint elongates and discharges in the trough. Because rate variations in the flowline can exceed 500 gal/min, paddle-type return indicators are useless as an early detection tool and visual flow checks with the pumps off are made more difficult. Also, pit volume totalizer (PVT) systems may have trouble keeping up with the constantly shifting mud level in the pits. PVT problems are partially solved by installing baffles and by adding more floats to each pit. Indicators should be located near the center of the tanks where the fluid level amplitude is smallest.

Delta-flow systems have been adapted to floaters. One concept uses low-pass filtering of the return-flow signals to average out the spikes and give a steady-state flow indication.[12] Another method compensates for heave-induced flow variations by using a computer to calculate a rate adjustment from the measured heave distance and riser diameter.[13,14] Other advanced kick-detection systems discussed in Chap. 4 also hold promise.

8.3.2 Shut-In and Hang-Off Procedure. A suggested procedure for shutting in a well and hanging off the drillstring is given in **Table 8.3.** Drillpipe suspended from the drill vessel is always moving in relation to the stack and the string is usually hung off on a ram soon after shutting in a well. Long-term movement through a closed annular induces unwarranted wear while movement through a closed pipe ram accelerates the wear problem and risks damaging the ram or parting the string if a tool joint is pulled into the ram. The disadvantage to having a motionless string is the risk of stuck pipe.

It is critical for the driller to know exactly where the tool joints in the stack are located before closing a ram. The distance to the top of the wellhead is confirmed and critical stack elevations are calculated and posted on the rig floor after the BOPs are set. Also posted are the number of stands and spaceout requirements for positioning a tool joint above each ram. Corrections for high or low tide conditions can be made at the time of the event. Tool-joint-position indicators are used to reduce any uncertainties associated with this part of the procedure.

8.3.3 Effect of the Choke and Kill Lines. On surface BOPE, the distance between the choke manifold and the stack is usually less than 75 ft and the friction pressure across this short distance is negligible. This is not so when controlling a well from a floater because the total chokeline is longer than the sea is deep. Pumping friction in a chokeline imposes unneeded and potentially hazardous backpressures on the annulus and the effect should be measured and considered in the well-control procedure.

Recall from Chap. 4 that the initial circulating pressure (ICP) for the selected kill procedure comprises the shut-in drillpipe pressure (SIDPP) and kill-rate circulating pressure (KRCP). Say that a well was closed in with a surface stack. Furthermore, assume the ideal case where the KRCP was measured by the driller immediately before the well kicked and bubble collapse is not a factor. The choke operator holds

TABLE 8.4—KICK DATA FOR A WELL DRILLED FROM A FLOATER	
Wellbore Configuration	
Water Depth	1,500 ft
Air Gap	85 ft
Vertical Depth from RKB Reference	8,200 ft
Surface Casing Information	
Description	13³/₈-in., 61.0-lbm/ft K-55 ST&C
RKB Setting Depth	4,585 ft
Openhole Diameter	12¼ in.
Drillstring Information	
Drill Collar Size	8×3 in.
Drill Collar Length	360 ft
Drillpipe Description	5-in., 19.50-lbm/ft
Bit Nozzle Flow Area	0.3313 sq in.
Choke and Kill Line ID	2.50 in.
Choke and Kill Line Length From Stack to Manifold	1,735 ft
Prekick Circulation Data and Mud Properties	
Drilling Circulation Rate	500 gal/min
CDPP While Drilling	2,750 psig
Kill Circulation Rate	250 gal/min
KRCP Through Riser	640 psig
Mud Type	Water Based
Mud Density	9.6 lbm/gal
Power Law Fluid Rheology	
Flow Behavior Index (n)	0.70
Consistency Index (K)	250 eq cp
Recorded Kick Data	
SIDPP	340 psig
SICP	580 psig
Pit Gain	30 bbl
Other Known or Assumed Information	
Fluid Density in the C and K Lines	8.55 lbm/gal
Fracture Gradient at Casing Seat	0.65 psi/ft
Kick Fluid	0.60 Specific Gravity (SG) Gas

constant casing pressure while the driller brings the pump up to speed and turns his attention to the drillpipe gauge upon reaching the kill rate. In this case, the observed ICP would exceed the calculated value by a small amount equal to the friction drop through the short choke line.

On a floater rig, KRCP measurements are taken while circulating through the large-diameter riser and the same choke-line effect occurs if the startup procedure discussed in Chap. 4 is followed. The problem, however, is that the line friction is significant and the observed ICP could exceed the target value by hundreds of psi. Also, most operators leave solids-free fluid in the C and K lines to minimize the risk of plugging. Holding casing pressure constant as heavier fluid is circulated into the line will increase the backpressure on the well. An example problem demonstrates what could happen.

Example 8.2. Well and kick information for a hypothetical offshore well are represented in **Table 8.4** and **Fig. 8.15.** Friction loss through one of the C and K lines at the designated kill rate is 240 psi and the loss in the casing/drillpipe annulus is 5 psi. Annular losses through the riser are negligible.

1. What pressure gradient will be seen at the casing shoe if casing pressure is held constant as the pump is brought up to kill rate?

2. If things proceed this far, what would be the observed ICP?

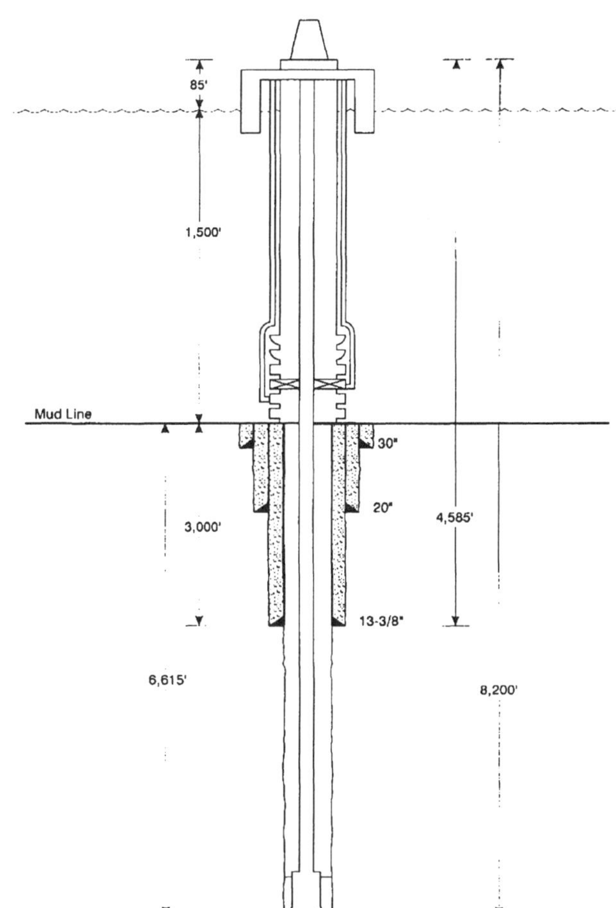

Fig. 8.15—Wellbore schematic for the well described in Table 8.4.

Solution. 1. The capacity of a choke line is only 10.8 bbl and, for convenience, we will assume that 9.6-lbm/gal mud displaces the line before kill rate is achieved. The pressure at the casing shoe then will be the sum of the casing-gauge reading, hydrostatic pressure of the mud, and friction losses above the shoe.

$$p_{sh} = 580 + (0.499)(4,585) + 240 + 5 = 3,113 \text{ psig}.$$

The pressure gradient referenced from the rotary kelly bushing (RKB) is calculated as

$$g_{sh} = 3,113/4,585 = 0.679 \text{ psi/ft}.$$

This exceeds the fracture gradient and lost returns would probably occur before the pump ever reached 250 gal/min.

2. Eq. 4.26 gives the target ICP,

$$p_{dci} = 340 + 640 = 980 \text{ psig}.$$

The observed ICP would exceed the calculated value by an amount equal to the sum of the choke-line friction loss and the hydrostatic pressure difference in the line:

$$p_{dci} = 980 + 240 + (0.0519)(9.6 - 8.55)(1,585)$$
$$= 1,306 \text{ psig}.$$

Having an accurate measurement of the circulating friction pressure in the C and K lines is essential information for some well-control operations. **Table 8.5** gives three ways to measure accurately these pressures. The first two procedures require that the well be shut-in and circulated through the bit. Choke-line friction is applied to the annulus, which does not

TABLE 8.5—PROCEDURES FOR DETERMINING CHOKE LINE FRICTION LOSS

Measurement Using the Drillpipe Pressure Gauge

1. Pump at the kill rate and measure the KRCP while circulating through the riser.
2. Close the BOP and open a choke line.
3. Pump at the kill rate and measure the drillpipe pressure while circulating through the choke line.
4. Determine the choke line friction as the difference between the Step 3 and Step 1 readings.
5. Open the other line and repeat Steps 3 and 4 with both lines open to the manifold.
6. Repeat Steps 1 through 5 using the alternate pump and, if necessary, a different kill rate.

Measurement Using the Kill Line as a Pressure Tap

1. Close the BOP and open both lines at the stack.
2. Displace the C and K lines with current drilling fluid by pumping down one side and out the other.
3. Open one of the lines through the manifold. Isolate the other line by closing a valve downstream of the pressure gauge.
4. Circulate the well down the drillpipe at the kill rate.
5. Directly measure the choke line friction at the pressure gauge on the static line.
6. Repeat Step 5 using the alternate pump and, if necessary, a different kill rate.

Measurement With the BOPs Open

1. Open one of the C and K lines at the stack.
2. Pump down the line at the kill rate, taking returns up the riser.
3. Directly measure the choke line friction at the pressure gauge on the active line.
4. Open both lines at the stack.
5. Pump down both lines at the kill rate, taking returns up the riser.
6. Repeat the procedure using the alternate pump and, if necessary, a different kill rate.

pose a difficulty if wellbore integrity is not jeopardized. A decided advantage to the third procedure is that mud is reversed down a C and K line and so the hole does not see the additional pressure.

Regardless of the selected method, the hydrostatics of the system must be considered. On the first and third procedures it is important to pump long enough to displace the lines with the current mud before any measurements are taken. Similarly, both lines must be filled with the same fluid before the second procedure will give an accurate measurement.

Implied in these procedures is that there may be more than one kill rate. Obviously, pump characteristics are different and, like recording a KRCP on a bottom-supported rig, a measurement should be made for each pump. But an aspect unique to floater well control is that it may be desirable to reduce the kill rate before the kick fluids reach the mud line. Accordingly, measurements should be made at more than one circulation rate. If two or, preferably, three rates are used the pressures can be plotted on logarithmic graph paper. The relationship between circulating pressure and pump rate should yield a straight line and the graph can be used to predict the C and K lines losses at any circulation rate.

It is suggested to measure the C and K line losses in conjunction with a routine KRCP measurement, though the former need not be done as often as the latter. Because line length is fixed, another test can be made only when the mud density or rheological properties change significantly. It is recommended to displace the lines after the measurements are made

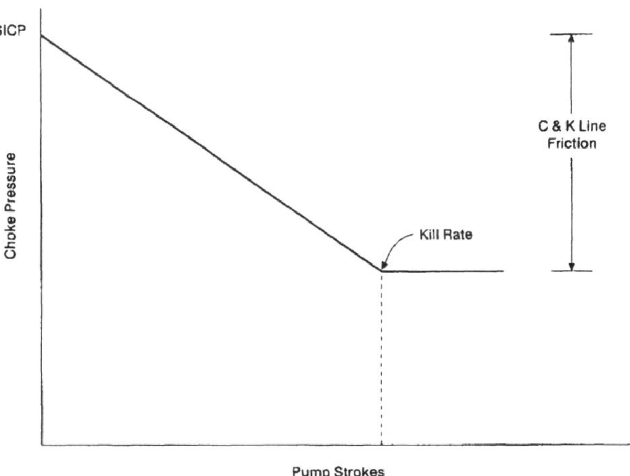

Fig. 8.16—Casing-pressure-reduction schedule to remove the effect of choke-line pressure during pump startup.

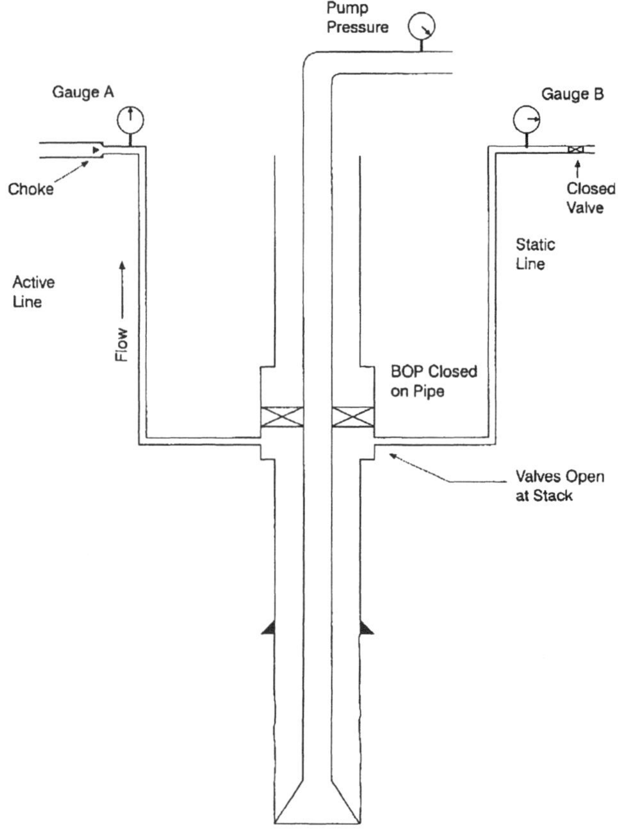

Fig. 8.17—Using the kill line as a pressure tap.

with a solids-free fluid, typically seawater with a corrosion inhibitor and antifreeze (if necessary).

Several different methods for removing the effect of choke-line friction on bottomhole pressure have been used with some success. One technique steps the casing pressure down during pump startup by an amount equal to the choke-line losses during pump startup. An ideal pressure-reduction schedule might look something like the chart shown in **Fig. 8.16.** A schedule should include any change in the choke-line hydrostatic pressure.

It is impractical to try to estimate precisely how many barrels will be pumped during the startup procedure and **Fig. 8.17** illustrates what may be a more workable procedure. The procedure uses one C and K line as a pressure tap while taking returns through the other line. Line B is static and Gauge B is

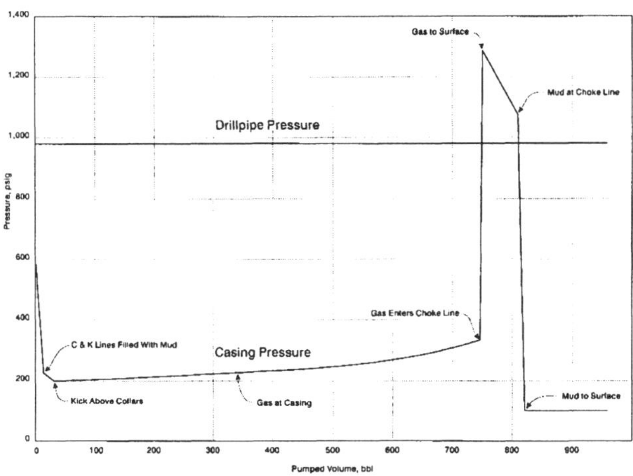

Fig. 8.18—Predicted pressures for a Driller's Method kill of the well described in Table 8.4.

held constant until the pump reaches kill rate. By doing so, Gauge A will fall automatically to the correct reading regardless of the fluid density in the two lines.

In some cases, an operator may have to accept some additional backpressure. Say that the shut-in circulating pressure (SICP) was less than the friction drop through one of the choke lines. At kill rate, the surface choke would be wide open, the manifold gauge would read zero, and the line, acting like a fixed choke upstream of the gauge, will add to pressure on the well.

Friction losses are reduced greatly if the rate through a flow conduit is halved and it may be recommended in these cases to take returns through both lines. The calculated pressure drop through the two lines in the example problem is approximately 70 psi, an amount the well could tolerate using a conventional startup procedure. Even so, one can still follow a pressure-reduction schedule based on the combined losses and include the effect of having seawater originally in the two lines. The latter will be the overriding concern with weighted muds.

Finally, reducing the circulation rate may be another option, although the time to kill the well will be increased and there is a lower limit to conventional rig pumps. The cementing unit could be lined out to kill the well except for the fact that communication between the pump operator and the man on the choke suffers, especially during the critical startup period.

After achieving kill rate, the casing/drillpipe pressure (CDPP) is held constant (Driller's Method) or reduced (Wait and Weight Method). The pressure response deriving from gas expansion and choke manipulation is similar to a surface-control operation. **Fig. 8.18** illustrates how a gas kick enters the choke line(s). Pressure predictions were made for a Driller's Method kill of the example well and are based on the single-bubble model and other simplifying assumptions adopted in Chap. 4 (i.e., no gas slip, constant bottomhole pressure (BHP), no friction pressure in the drillstring annulus, etc.). As indicated, the choke pressure must increase dramatically when gas enters the choke line to make up for the loss in hydrostatic pressure.

The choke operator's assignment to maintain a constant CDPP is made more difficult as a result of things that are going on in the choke line. Only minor adjustments are needed when all the gas is in the drillpipe annulus, but holding constant BHP demands a fast reponse in reducing the choke orifice when gas enters the choke line and

TABLE 8.6—PROCEDURE FOR OPENING UP ANOTHER C and K LINE IN THE MIDDLE OF A KILL

1. Coordinate between the choke operator and driller to shut off the pump and close the well in.
2. Measure the SICP.
3. Open second C and K line at the stack, keeping it closed at surface.
4. Switch the choke over to control returns from the new line and line up the pump to the old line.
5. Bring the pump up to kill rate and pump down the mud-filled line while displacing seawater from the other line. Hold the pump pressure equal to the SICP plus the total of the friction losses in both C and K lines.
6. Circulate until the new line is filled with mud.
7. Coordinate between the choke operator and driller to shut off the pump and close the well in.
8. Line up to take returns through both C and K lines and join the flow in front of the manifold choke.
9. Resume circulating out the kick controlling on the drillpipe pressure.

an even faster opening sequence when gas starts to vent at surface. Making the choke operator's job easier is the reason why the pump rate is sometimes reduced before gas reaches the mud line.

Opening the other C and K line will assist the operation, but the opportune time to do so is before the pump is engaged. Opening the second line is complicated if the active line contains mud or kick fluids while the other side is still filled with sea water. **Table 8.6** gives a procedure that could be followed. As written, the procedure assumes the wellbore is open below the bottom C and K line outlet. Assuming the stack is configured as shown in Fig. 8.12, closing the lower pipe ram will isolate the well and avert the risk of overbalancing or underbalancing the hole.

Bourgoyne and Holden[15] disussed the problems associated with subsea C and K lines and, to assess the procedural aspects, simulated well behavior in a 6,000-ft land well. They modeled a subsea well in 3,000 ft of water by placing a packer midway down the hole with tubing on bottom as the drillstring. The packer had a triple parallel flow tube and two other strings were run to the packer as the C and K lines. Simulated kicks were induced by injecting nitrogen down a string inside the middle tubing, a pressure transducer allowed BHPs to be monitored throughout a given experiment.

A choke manifold from the two C and K lines was installed and experienced choke operators were asked to bring kicks of varying size to surface through one or both lines at different pump rates. Based on the results, the authors concluded that choke control in an actual kill procedure is not as demanding as their simplified computer models indicated. For example, near-surface gas expansion increases the mud velocity which automatically increases the choke pressure and assists the choke operator who is pinching back the flow. Other factors such as gas slippage and dispersion also mitigated problem severity associated with the lead and trailing edges of the kick.

Benefits accruing from large-diameter C and K lines should be apparent by now. Many problems would be solved or reduced with nominal line diameters in excess of 4 in. Some practical limitations are involved here. Larger lines increase the overall weight of the marine riser, an important issue in deep and ultra-deep waters. Also, Barlow's equation states that large outer diameters need either more wall thickness or higher strength to get the same pressure rating as a smaller tube.

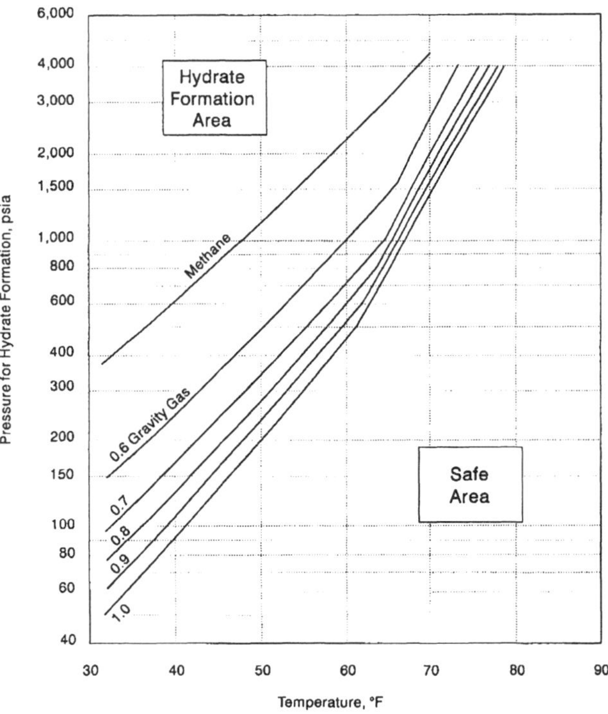

Fig. 8.19—Conditions favorable for the formation of hydrates.[17]

TABLE 8.7—SUGGESTED PROCEDURE TO CLEAR THE STACK AND KILL THE RISER	
1.	Assure that a tool joint is not in the way and close a set of pipe rams below the C and K line outlets.
2.	Pump sea water through the lower C and K line with returns via the upper line. Displace the mud and trapped fluids from stack and C and K lines.
3.	Close the lower line and allow time for gas to migrate to the choke manifold.
4.	Pick up the string weight. Unlock and open the upper pipe rams (vessel motion permitting).
5.	Allow the two C and K lines to equalize and become static.
6.	Open the surface diverter and open the annular preventer.
7.	Allow the riser fluid to U-tube into the C and K lines, taking returns into the waste tank.
8.	After the riser and C and K lines equalize, close the surface diverter and return flow path to the flowline.
9.	Circulate kill-weight mud down the C and K lines and fill the riser with kill-weight mud.
10.	Close the upper pipe rams.
11.	Displace the kill-weight mud from the C and K lines with the specified fluid (seawater + additives).
12.	Close the subsea valves at the stack outlets.
13.	Open the lower rams followed by the upper rams.
14.	Monitor the riser for flow.
15.	Resume operations. Incorporate the riser or trip margin to the mud density.

8.3.4 Hydrates. Hydrates are a solid crystalline material composed of water and natural gas molecules and resemble snow in appearance. Water + gas = hydrates if the temperature is sufficiently low and pressure is sufficiently high. **Fig. 8.19** shows what conditions favor hydrate formation for various natural gas compositions (read specific gravities). Hydrates are a problem in many gas production and transmission operations, but usually are not a consideration to drillers. Floater drilling in deep water is an exception. Hydrates in subsea equipment have caused some well-control problems and can plug C and K lines, keep BOPs from closing, stick the drillstring, or interfere with the ability to read wellbore pressures.[16] Those who drill in deep water should recognize the potential for hydrate formation and, if possible, take precautions to prevent its occurrence.

Ocean temperatures decrease rapidly with water depth. In the Gulf of Mexico, for instance, the temperature at 3,000 ft is about 41°F.[18] Consider a well in this area and at this water depth being drilled with a 9.2-lbm/gal fresh-water-based mud. The hydrostatic pressure is 1,400 psi. Fig. 8.19 indicates that with gas present, hydrates are possible.

The circulating drilling fluid may keep the stack and subsea equipment warm enough to be in the safe region. However, a choke line is an efficient heat exchanger with the sea and there will be some cooldown from gas expansion in a line. Thus, hydrate plugging of a chokeline can occur during a deepwater kill circulation. In any event, subsea temperatures approach ambient conditions soon after the pumps are shut down and there is potential for hydrate formation in the wellhead and stack equipment if gas is present and the well has been quiescent for a period.

Hydrates normally cannot be dislodged with pump pressure. Water solutes such as salts, glycols, and alcohols suppress or dissociate hydrates and special spotting fluids have been formulated specifically for use with muds.[19] Circulating hot mud has been successful also in dissolving a hydrate mass. But these are after-the-fact measures and depend on the ability to circulate and being able to contact the hydrate with the pill or hot mud. A hydrate plug in a C and K line would not be dissolved by these methods; pulling the stack or riser may be required to remove the blockage.

Preventative measures typically focus on the use of high-salinity drilling fluids to inhibit hydrate formation (i.e., reduce the hydrate-formation temperature at a given pressure).[20] Invert emulsion oil-based and synthetic oil muds do not prevent hydrates, but saturating the water phase with dissolved $CaCl_2$ has been found to eliminate the effects at the pressures most likely to be seen in a well-control problem.[21,22]

8.3.5 Post-Kill Considerations. Consider the conditions present immediately after a subsea well has been killed successfully. Kill-weight mud has been circulated to the surface and both gauges read zero, but opening the preventer could invite another kick because of the lower mud density in the riser. Also, gas and other liquids lighter than the mud have accumulated above the stack outlet and are trapped beneath the closed preventer. This trapped volume could be four or five barrels in an 18¾-in. stack. Allowing this much gas to enter the riser could evacuate much of the mud and thereby create a surface hazard, collapse the riser, or allow formation fluid entry.

Table 8.7 contains a logical procedure for getting the system back in order for routine operations. It may be helpful to refer to Fig. 8.12 while reading the procedure and following discussion. Step 1 isolates the well from what will transpire above the stack. Steps 2 and 3 should take care of removing most of the trapped fluids and, at the same time, place a low-density fluid in the C and K lines.

There may yet be a pocket of gas and/or oil between the upper line outlet and the closed preventer, but the volume will be small. Steps 4 through 6 provide the best method for managing any remaining formation fluids. Oil and perhaps some gas will be pushed into the C and K lines while the rest can migrate and safely vent through the diverter. The remaining steps of the procedure involve displacing the riser with kill-weight mud, placing a clean fluid in the C and K lines, and getting the well opened up. Be aware that during riser displacement some oil and gas may be left in the C and K lines from Step 5.

The riser margin is an incremental mud weight sometimes used to prevent an underbalance situation should the riser leak or become disconnected with the well open. With an open riser, the mud will U-tube out into the sea until hydrostatic pressures equalize at the depth of communication. The worst-case scenario is a leak in the bottom joint that gives a hydrostatic pressure above the stack equal to that of seawater. Eq. 8.2 gives the riser margin in terms of hydrostatic gradient for this condition.

$$g_{rm} = \frac{g_{km}D - g_{sw}D_{sw}}{D_s} - g_{km}, \quad \text{.......... (8.2)}$$

where g_{rm} = riser margin, g_{km} = kill-weight mud hydrostatic gradient, D = vertical well depth, g_{sw} = seawater hydrostatic gradient, D_{sw} = water depth, and D_s = sediment depth, below mud line (BML). The minor effect of the stack height has been ignored in the relation.

Example 8.3. Determine the riser margin for the well described in Table 8.4.

Solution: Eq. 4.14 yields a kill-weight mud gradient of

$$g_{km} = [340 + (0.0519)(9.6)(8,200)]/8,200$$

$$= 0.540 \text{ psi/ft}.$$

Eq. 8.2 gives the riser margin.

$$g_{rm} = \frac{(0.540)(8,200) - (0.445)(1,500)}{6,615} - 0.540$$

$$= 0.028 \text{ psi/ft}.$$

The incremental mud weight is thus

$$\rho_m = (0.028)(19.25) = 0.55 \text{ lbm/gal}.$$

Having a riser margin may be a good safety precaution, but may not be desirable or achievable. The incremental mud weight may be excessive in deep water or with shallow BML depths, needlessly hindering penetration rates or even exceeding the fracture gradient. If used, the operator should compare the riser margin to the trip margin and incorporate the higher of the two.

8.3.6 Disconnecting the Riser. Conditions may require that an operator disconnect the LMRP from the stack and move off location. These include station-keeping problems, severe weather, or a well-control emergency. Given adequate time (for instance, an approaching storm with plenty of advance warning), the usual procedure is to pull the bit up into the casing and run a square-shouldered tool to hang off the string in the stack. After releasing from the tool, the blind/shear rams can be closed and the riser disconnected in an orderly fashion.

There are some situations, for example a blowout or slipping anchors, where there is not enough time to use a hang-off tool. An emergency disconnect is required in these instances and the blind/shear rams must be used to cut the drillpipe and close in the well. Emergency disconnects in deep water with heavy mud can collapse the riser when the mud pressure equalizes with the seawater and it may be necessary to close the LMRP annular on the drillpipe to retain the mud. Later, the annular can be opened after seawater has been reversed into the riser above the riser booster line(s).

8.4 Shallow Gas Hazards

By definition, a shallow gas formation is penetrated before enough casing has been set to control a well using conventional procedures. Shallow gas sands may have several darcies of permeability. A kick from one of these zones may flow at prolific rates and unload a hole before kick-detection equipment and crews can respond. Indeed, the majority of shallow gas kicks rapidly escalate into full-scale blowouts.[23]

Most operators consider "enough" casing to be the surface pipe and will divert flows encountered in the conductor or surface hole. However, the track record of diverters is such that an operator should expect them to fail as part of the contingency planning. A dilemma exists here, which may never be completely resolved, but there are things an operator can do to avoid shallow gas or to mitigate the risk if these zones are unavoidable. First we must understand the nature of shallow gas before preventative or corrective measures can be instituted successfully. An appreciation of the problem will lead to more intelligent policy decisions, hole programs, drilling procedures, and equipment design.

8.4.1 Broaching Concerns. Shallow casing seats provide very little margin for shutting a well in on a kick because the effect of the SICP at the shoe gives such a high pressure gradient. After fracturing occurs, the risk of formation fluids broaching the shoe and reaching the surface is greater when the casing is set shallow. The dangers associated with diverter use are generally recognized today. Most operators would rather shut a well in on a shallow gas kick and take an underground blowout if they could ensure that the blowout would stay underground. Lacking this confidence, shutting in the well is not an option.

Vertical fracture propagatation is one possible way for a subsurface flow to broach to surface. However, a vertical fracture will likely encounter a high-stress confining layer, or a zone that will charge up as opposed to parting, or encounter a stress regime where induced fractures are horizontal. Fluids probably will stay confined to the rock when horizontal fracturing occurs, but vertical fracturing in the confining layer or permeable zone may ensue if pressures reach the critical level.

Walters[24] described some other rock-failure processes by which hydrocarbons can work their way to the surface. Faults and joints are one avenue for flow to migrate upward through the strata. Nonsealing faults with a permeable gouge zone can be a pathway and sealing faults can fail if the pressure exceeds the stress normal to the fault plane. Compressive failure is another possibility if we consider a permeable zone that has been overpressured by flow from below. The effective or grain-to-grain stresses may be reduced to the point where the bed's Mohr-Coulomb failure envelope is encountered (see Fig. 3.9) and the rock fails in shear. Formation collapse is then a possibility and fluids may flow to the surface through subsidence passages.

These are valid mechanisms, but another potential communication link has been the source for many cratering incidents. Flow through a channeled cement job or otherwise failed bond can travel all the way up the hole or exit into a shallower zone and thence to the surface by one of the rock-failure modes. It is therefore incumbent on an operator to recognize that doing the right things to get a good cement job is just as important in surface and conductor strings as it is in intermediate and production casings. Other avenues for blowouts to escape include casing leaks, fresh water wells, casing annuli in offset producers, platform pilings, and other conduits.

However a blowout gets to the surface, cratering underneath the rig or some distance away is the usual outcome. Rocha and Bourgoyne[25] described dilfferent ways in which craters are caused by shallow-gas blowouts. Flow up a casing annulus can erode a crater outside the pipe in the form of an inverted cone and grow large enough to topple a land or bottom-supported offshore structure. As an example, one blowout outside a 13³/₈-in. surface casing caused a crater at the mud line that was 125 ft in diameter and at least 765 ft deep.[23]

Caving failure is the primary reason why large boulders and individual sand grains dislodge and flow or erupt at the surface. The concept was introduced in Chap. 3 when we talked about borehole collapse deriving from insufficient mud weight. The same holds true in a blowout condition except that problem severity is increased because only a small backpressure is held against the rock by the flowing fluids. Compressive stresses induce shear failure of the material at the wall, rocks or sand spalls into the crater, and then are lifted by the flow.

Liquefaction is another process applicable to unconsolidated, near-surface soils. Friction drag between the flowing formation fluids and matrix grains can reduce the effective stress of the material to zero. Hence there is no overburden or cohesive force and a surface boil characteristic of quicksand develops into which a drilling structure can sink.

So-called "piping" craters derive their name from horizontal tunnels in the subsurface which are created by flow erosion. The blowout flow in these cases is horizontal, in a permeable stratum, and in the direction of the most favorable permeability. Cratering underneath the rig or platform can occur if the structure weight collapses the tunnel.

8.4.2 Shallow Fracture Gradients. Most of the matrix stress ratio correlations used to predict fracture gradients do not extend much shallower than 1,000 ft or so. The region-specific correlations discussed in Chap. 3 were based on leak-off test, lost circulation, and stimulation data. There may have been a lack of shallow-hole information, but another reason for stopping the correlations short is that they do not work very well. The question is subject to debate,[26] but conventional prediction methods may be off the mark down to 3,000 ft or deeper in some areas.[27,28]

In general, shallow frac gradients tend to be higher than the published methods would suggest and an operator may be overly conservative if these curves are extrapolated to ground surface or the mud line. Several possible reasons are evident including the singular or combined effects of rock with a finite tensile strength, the overburden being the least principal stress, water depth on sediment compaction, and plastic behavior in the younger clays and other rocks. The last is probably the predominant reason in the tectonically relaxed basins of the coastal and offshore regions.[29] The matrix stress ratio will approach unity if lithostatic conditions exist in the shallow sediments, which implies that the fracture gradient will be close to the overburden.

As part of an effort to secure better hole programs and operational procedures, a major oil company has compiled shallow leak-off test data from land and offshore wells in several different geographic regions.[30] **Fig. 8.20** shows the results of this investigation. The fracture gradients represented in the chart ranged from 0.80 to 3.73 psi/ft with respective mean and median values of 1.00 and 0.92 psi/ft. The data tend to support

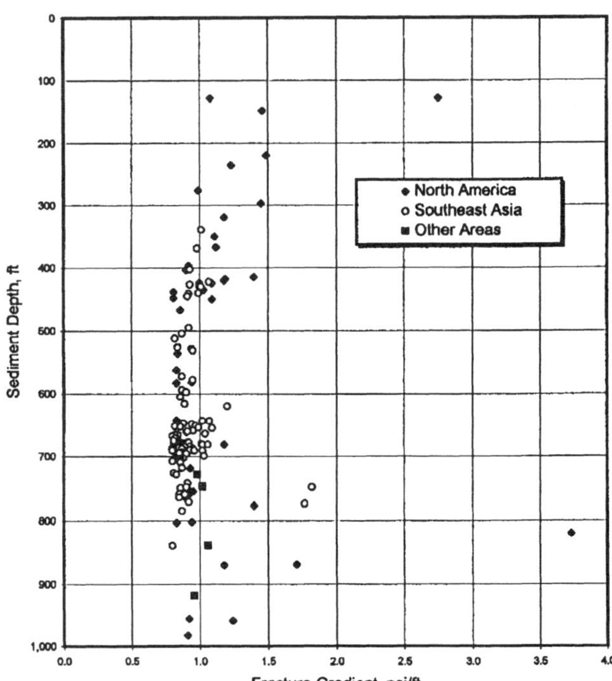

Fig. 8.20—Shallow leak-off test data compiled by a major oil company. (From Ref. 30).

the theoretical underpinnings as to why shallow frac gradients may be higher than one would suspect based on the older correlations. Other investigators have published similar findings for Alberta and the Gulf of Mexico.[31-33]

Many processes may be at work and the data scatter in any one area may be too large to ever develop a generalized correlation fitting every case. Even so, it will behoove the industry to spend the time and money to gather a large data base on shallow rock integrity. Empirical, yet conservative, prediction methods suitable for a specific region will lead to savings and increased safety in the long run. There will always be risk, but shallow casing-point selection (conductor and surface pipe) and diverter-use philosophy should be based on data applicable to the problem.

Fig. 8.21 illustrates the potential magnitude of the spread in predicted fracture gradients. The curve on the left used Eaton's[34] 1969 correlation for a land well on the U.S. Gulf Coast with a normal pore pressure for the area. The curve on the right has the same pore pressure and overburden relationships but assumes a matrix stress ratio equal to 1.00. Actual points may fall somewhere between the two curves. Even so, the implications to well design should be apparent.

Fig. 8.22 shows the effect of water depth on fracture gradients. The curves were generated using Constant and Bourgoynes's[28] offshore prediction method and their stress ratio coefficient that best fit the shallow sediments in the Green Canyon area of the Gulf of Mexico. This chart is presented for illustration purposes and is by no means universal.

The chart suggests some important considerations for offshore operators. One is that fracture gradients are impacted by the lower overburden and the effects are most significant in the shallowest part of the hole and in deep water. This situation leads to a narrower operating window for allowable mud weights across a given hole section and, consequently, a greater risk for losing returns on the high end and taking a kick on the low. After surface pipe is set in deep water, drilling into a pore-pressure transition often will require more casing strings for a given (BML) pressure profile.

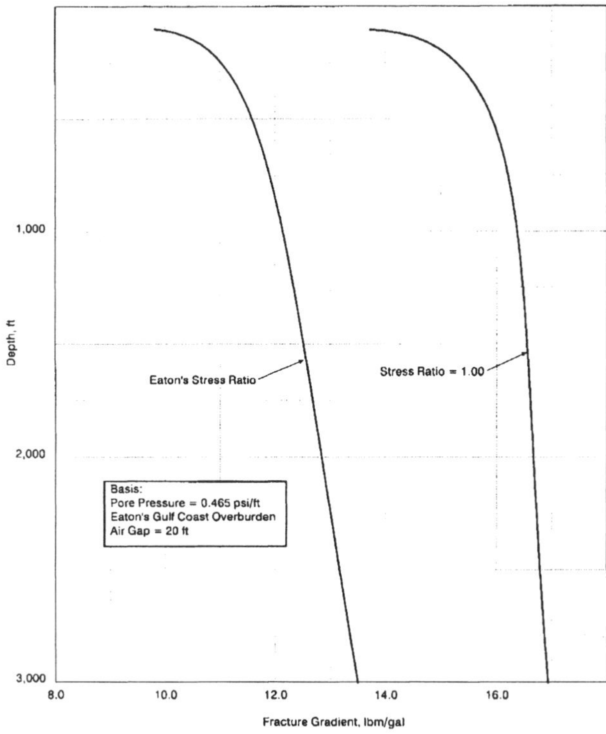

Fig. 8.21—Comparison between shallow frac-gradient predictions.

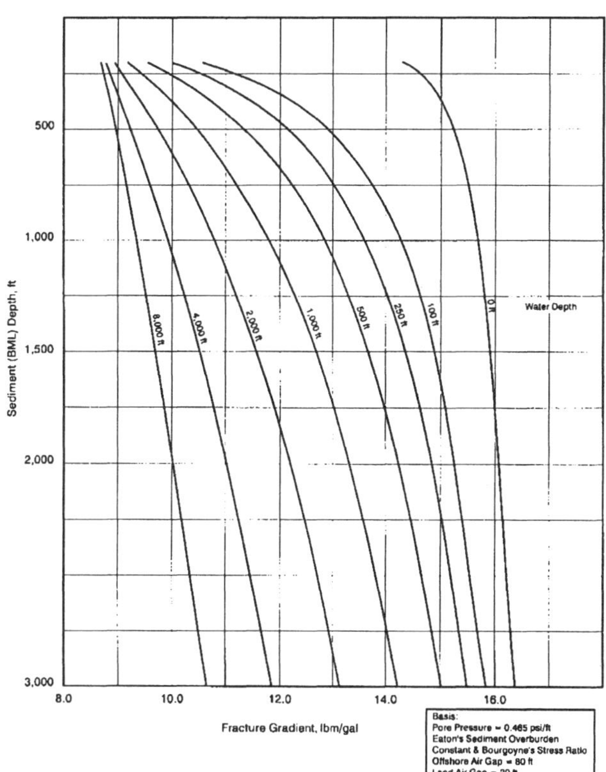

Fig. 8.22—Effect of water depth on fracture gradients.

It also may be impossible to drill the conductor hole with a riser in deep water. For the example conditions, the maximum allowable mud weight when drilling below structural casing at 300 ft BML is 9.2 lbm/gal in 2,000 ft of water and only 8.8 lbm/gal at 8,000-ft water depth. Add the effects of cuttings loading and annulus friction and the tolerable mud weights are reduced even further.

The air gap exerts considerable influence on allowable mud weights and the effect is most pronounced at shallow sediment penetrations in shallow water. Consider the 100-ft water-depth curve on the chart. The frac gradient referenced to an RKB 80 ft above the water is an 11.7-lbm/gal equivalent at 300 ft BML. The same rock at the same sediment penetration could be drilled with a 14.0-lbm/gal mud if we could somehow place the flowline at water level.

Comparing floater types, a drillship's deck is closer to the water than a semisubmersible's and the ability to drill with a higher mud weight is one of the few well-control advantages offered by a ship-like vessel. Assume the data shown in Fig. 8.22 are correct and that a floater in 500 ft of water was drilling below conductor pipe set at 1,600 ft BML. The maximum mud weight in the surface hole for a semisubmersible with an 80-ft air gap is 13.8 lbm/gal. The same hole program could be drilled with a 14.0-lbm/gal mud by a drillship with a 25-ft air gap. This may not seem a decided advantage, but the difference becomes significant in the conductor hole if the section is drilled with a riser.

Adding drill cuttings to the mud increases hydrostatic pressure in the annulus and penetration rates in the shallow part of an offshore well must be controlled because of the low frac gradients. **Fig. 8.23** illustrates the nature of the problem for

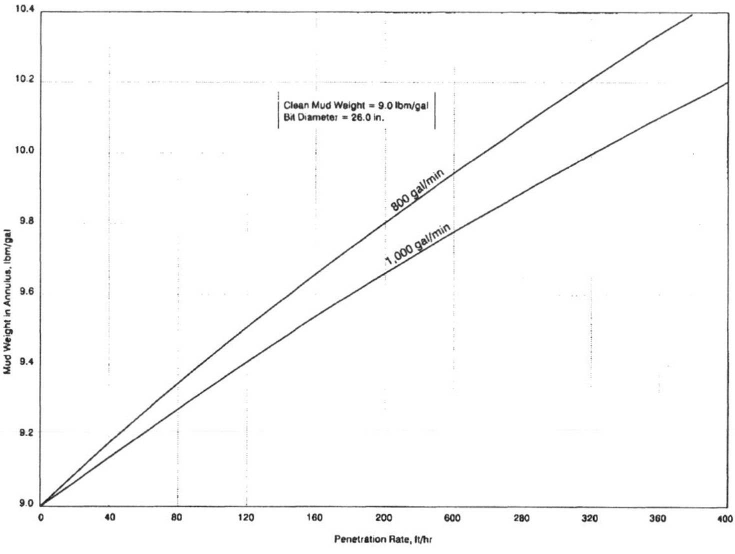

Fig. 8.23—The effect of penetration rate on annular mud density.

a 26.0-in. conductor hole drilled with a riser. For this and other reasons, offshore operators do not attempt to maximize drill rates in the structural or conductor holes and only begin to focus on optimum drilling practices after drilling out from under surface pipe.

The diagram in Fig. 8.23 indicates that increasing the circulation rate reduces the mixture density by diluting the cuttings concentration. However, the annular friction losses were not included in the diagram and the benefits of a higher pump rate are offset to some degree by the increased equivalent circulating density (ECD). Much lower effective mud weights can be realized by drilling the conductor hole without a riser because cuttings are dumped at the sea floor and the rig's air gap has no effect.

8.4.3 Shallow Gas Kicks: Causes and Prevention. Shallow gas sands are a rarity in some areas while they are expected in others. An operator can sometimes plan a well to minimize the risks associated with a problem area and, after drilling commences, conduct the operation so as to avoid underbalancing any sands that may be exposed.

Predrill Hazard Assessment. The best way to avoid taking a kick from a shallow gas zone is not to penetrate the zone at all. Stating the obvious is not meant to be facetious, but there are predrill methods for defining areas that pose greater risks. High-resolution seismic data across the shallow strata often can distinguish amplitude anomalies associated with gas-bearing strata. Seismic analysis, in conjunction with drilling knowledge of the area, can help identify the probability of encountering shallow gas. It may be possible to offset the surface location and develop a hole program to reach the objective while staying away from the high-risk area.

Pore-Pressure Environment. Troublesome shallow gas sands often are sandwiched between normally pressured shales and thus cannot be anticipated using the compaction methods discussed previously. Having normal pressures somewhere in the sand does not necessarily mean that abnormal pore pressures are precluded at the depth encountered by the drill. As discussed in Chap. 2, structural overpressuring at the top of a normally pressured rock can occur when gas overlies formation water. All else being equal, the structural overpressuring in terms of equivalent mud weight increases with shallower depth.

Example 8.4. An offshore well drilled from a jackup encounters a shallow gas sand at 670 ft RKB. The pore pressure at the 770-ft gas-water contact is 320 psig. Water depth at the location is 200 ft and the rig's air gap is 70 ft. Is this zone likely to kick if a 9.0-lbm/gal mud is in use?

Solution. A hydrostatic gradient of approximately 0.01 psi/ft is appropriate for the formation gas. Hence the pore pressure at the bit depth is

$$p_p = 320 - (0.01)(100) = 319 \text{ psig}.$$

The pore-pressure gradient (in mud-weight equivalent) referenced from the RKB is

$$g_p = (19.25)(319)/(670) = 9.17 \text{ lbm/gal}.$$

The sand will kick, either when encountered or after the pumps are shut down. In comparison, the mud weight needed to control the sand at the gas/water contact is

$$(19.25)(320)/(770) = 8.0 \text{ lbm/gal}.$$

TABLE 8.8—BASE OPERATING PARAMETERS AND GAS SAND CHARACTERISTICS WHILE DRILLING AN EXAMPLE CONDUCTOR HOLE

Operating Parameters	
Bit Diameter	26 in.
Circulation Rate	800 gal/min
Clean Mud Density	9.2 lbm/gal
Present Depth (RKB Datum)	2,000 ft
Penetration Rate	200 ft/hr
Gas and Sand Properties	
Porosity	0.30
Water Saturation	0.25
Formation Water Density	9.0 lbm/gal
Sand Matrix Density	21.65 lbm/gal
Pore Pressure	0.465 psi/ft
Gas SG	0.60
Wellbore Temperature	80°F
Gas Behavior	Ideal Gas Law

Gas-Cut Mud. Gas associated with drilled rock will enter the mud stream and lighten the fluid column in the annulus. Example 2.17 showed that severe gas-cutting from drilled gas is not normally an underbalance concern because of gas compressibility with depth. But Fig. 2.51 implies this may not be the case at shallow well depths. Goins and Ables[35] discussed how drilled-gas expansion can induce a kick when shallow gas sands are open and presented several charts to show the equivalent mud densities in a hypothetical well for a variety of drilling parameters.

We have done the same for the base set of conditions described in **Table 8.8.** It was assumed that the same gas mole concentration was consistent throughout the annulus. Starting at the top, the mixture densities with depth were computed numerically using small depth increments. The equivalent density at any given depth is not the true mixture density, but rather what the hole sees in terms of the hydrostatic pressure. **Fig. 8.24** shows the results along with three other curves to compare how a different penetration rate, pump rate, and hole diameter change the predictions.

Penetration rate is one of the more significant parameters and so we have another reason for control drilling when there is shallow gas potential. Cuttings are included in the hydrostatic calculations, but the gas contribution more than offsets that of the solids. Reducing the gas concentration by circulating the hole at 1,000 gal/min served to increase the simulated wellbore pressure. The effect may seem minor, but the annulus pressures would be higher if friction losses had been included. The drilled gas volume varies with the square of the hole diameter and hole size has a major influence. However, 800 gal/min is excessive for a 12¼-in. hole and the difference narrows considerably at a more reasonable 500-gal/min pump rate.

All four curves predict that an 8.95-lbm/gal gas sand being drilled at 2,000 ft will not kick. This usually will be the case if the zone is normally pressured because the increased solids content of the mud tends to predominate at bit depth. Up the hole is another story. The curve on the right side of **Fig. 8.25** simulates the wellbore pressures for a 1,000-ft penetration depth. The BHP is equivalent to a 9.2-lbm/gal mud, but the same sand becomes underbalanced when the 2,000-ft zone is drilled. Hence the primary gas-cutting hazard lies in shallower exposed sands.

The authors did not discuss riser vs. riserless conductor-hole drilling, but we will. Without a riser, gas expansion

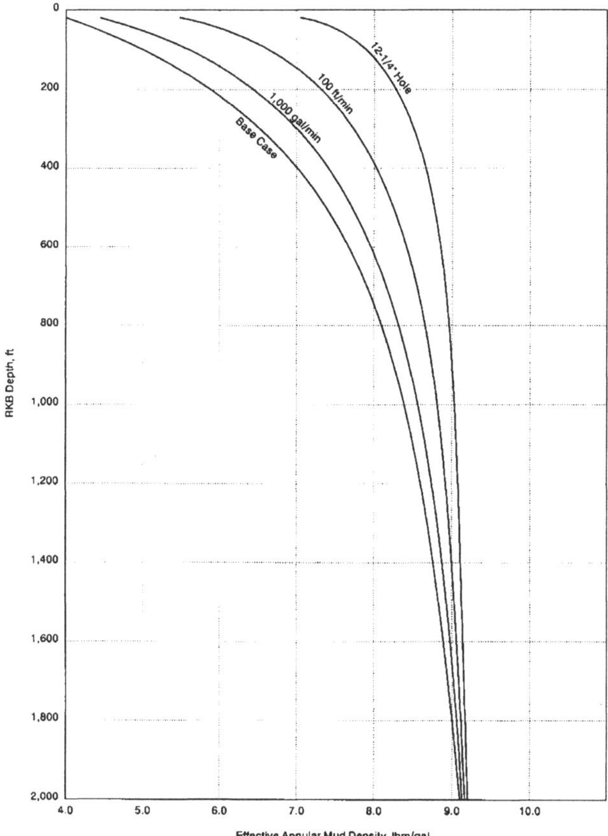

Fig. 8.24—Effect of drilled gas on effective mud density in the annulus.

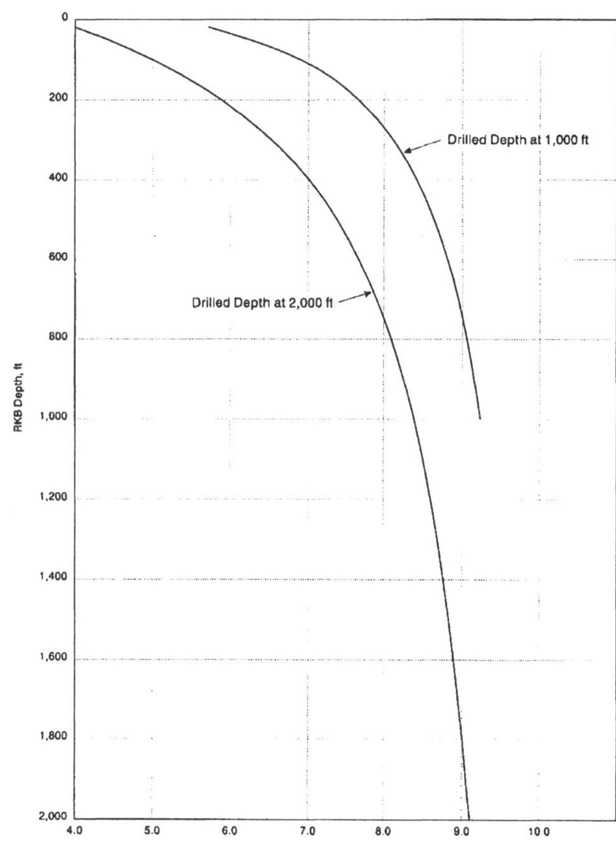

Fig. 8.25—Effective annular mud densities resulting from drilled gas at different total depths.

above the mud line does not affect wellbore pressures as the mixture density in the annulus is dictated by the hydrostatic pressure of the ocean. The effect on wellbore pressures depends on whether the seawater hydrostatic pressure is higher or lower than that provided by the gas-cut mud at the depth of the sea floor. **Fig. 8.26** suggests the seawater hydrostatic tends to predominate in shallower water *if mud is used in the riserless hole*.

Riserless Drilling in the Conductor Hole. Seawater is used commonly when the conductor hole is drilled without a riser. Seawater is the least expensive option because mud containing commercial solids and other additives will be pumped one time and then dumped at the sea floor. What is considered to be a normally-pressured gas sand for many areas will kick if the hole is filled with seawater. Mud can be used, but the effect of the sea pressure must be considered in the density calculation if the prospect is in an area known to contain shallow gas sands.

The drillstring will be tripped after reaching surface casing depth. Doing so means that seawater will fill the annulus by a volume equal to the string displacement and the mud-column height reduction also must be considered in the density calculation. The drop in mud level below the mud line is given by

$$\Delta h_m = V_d/C, \quad \text{(8.3)}$$

where V_d = the steel displacement volume and C = the capacity factor of the structural casing. The minimum mud gradient to balance the pore pressure after the drillstring is pulled above the mud line is given by

$$g_m = \frac{p_p - g_{sw}(D_{sw} + \Delta h_m)}{D_s - \Delta h_m}. \quad \text{(8.4)}$$

D_s in this case is the sediment (BML) depth of the shallowest zone which could flow. It is apparent that the density, and hence the cost per unit mud volume, can increase substantially with water depth.

Example 8.5. A conductor hole will be drilled without a riser in an area notorious for shallow gas problems. What minimum mud weight is recommended if the following conditions apply?

Conductor casing point = 2,625 ft, water depth = 600 ft, air gap = 75 ft, structural casing depth = 935 ft, structural casing capacity factor = 0.76161 bbl/ft, drillpipe displacement factor = 0.00813 bbl/ft, drill collar displacement factor = 0.05343 bbl/ft, drill collar length = 120 ft, and pore-pressure gradient = 0.452 psi/ft.

Solution. The worst-case scenario would be if a shallow gas sand were encountered directly below the structural casing. The pore pressure at 260 ft BML is

$$p_p = 0.452(600 + 260) = 389 \text{ psig},$$

with the gradient referenced from sea level. The steel volume that will be removed from the wellbore is

$$V_d = 0.00813(2,625 - 600 - 75 - 120) + (0.05343)$$
$$(120) = 21.3 \text{ bbl}.$$

This leads to a mud-height reduction of

$$\Delta h_m = 21.3/0.76161 = 28 \text{ ft}.$$

Eq. 8.4 yields the mud gradient needed to keep a sand at 260 ft BML from kicking.

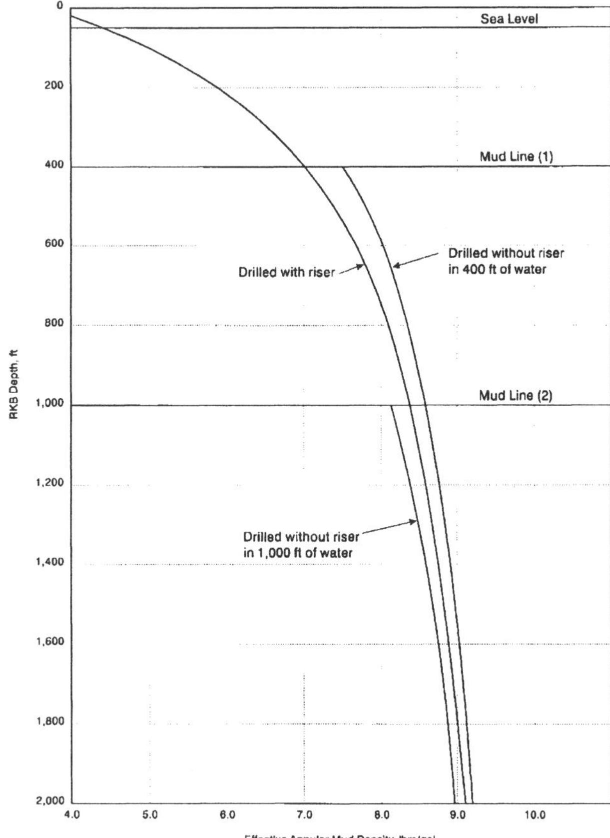

Fig. 8.26—Effective annular mud densities resulting from drilled gas with and without a riser.

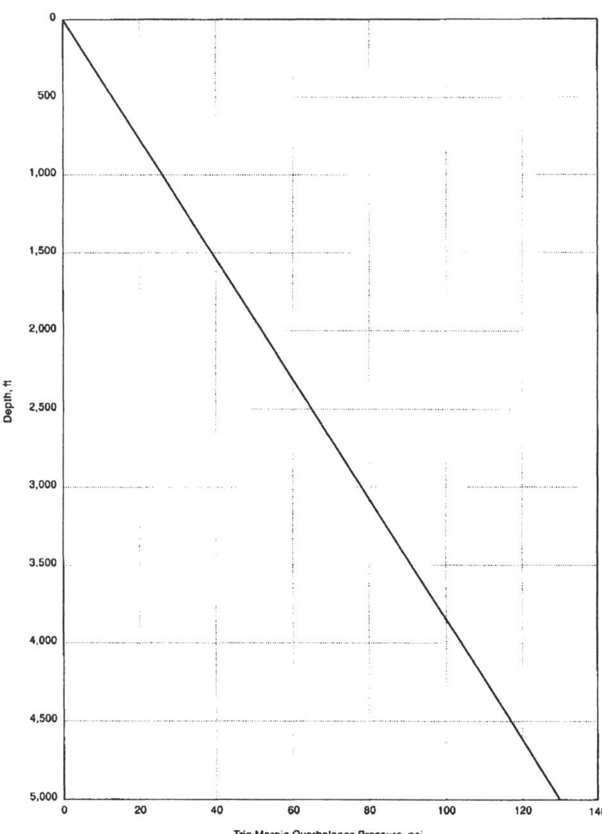

Fig. 8.27—Hydrostatic pressure furnished by a 0.5-lbm/gal trip margin.

$$g_m = \frac{389 - (0.445)(600 + 28)}{260 - 28} = 0.472 \text{ psi/ft}.$$

The minimum mud density is

$$\rho_m = (0.472)(19.25) = 9.1 \text{ lbm/gal}.$$

Lost Circulation and Trip-Induced Pressures. Lost circulation will depress the fluid level in a well and could invite a shallow-gas kick. Low fracture gradients are one area of concern as are losses of whole mud or seawater into permeable, loosely consolidated sediments. The limiting wellbore pressure for the latter event is the loss zone's pore pressure instead of its frac pressure.

Keeping the hole full and regulating surge and swab pressures become critical concerns when shallow gas sands are exposed. Trip margins are added to the mud density to compensate for the unavoidable pressure reduction when tripping, but the incremental pressure supplied by the margin diminishes linearly with depth and very little excess pressure is furnished at shallow well depths. **Fig. 8.27** depicts the hydrostatic pressure of a 0.5-lbm/gal trip margin. The overbalance at 5,000 ft is 130 psi whereas only 26 psi is realized at 1,000 ft. Higher trip margins would help but, again, the mud weight is limited by the frac gradient.

Hole geometries and mud properties in the shallow hole sections limit surge and swab pressures. Even so, an operator should pay special attention to the problem potential on trips. Slow trip speeds and slick bottomhole assemblies are recommended. Hole filling should be done more often if not on a continuous basis. Circulating during a trip with a top drive unit should be considered, both to keep the hole filled and to offset the swab-pressure losses with the ECD. The pump rate should be calculated to match the volumetric rate at which the BHA is removed.

Swabbing the annulus with a balled bit is of special concern and the trip should be stopped as soon as a hoisted string starts removing mud from the well. It is difficult to detect this condition when the conductor hole has been drilled without a riser. An indirect method is suggested in **Fig. 8.28.** The first step is to spot a densified mud in the hole and displace the mud to the drillstring float with seawater. Starting off bottom, there will be a differential pressure acting upward on the float and the drillpipe will stay full of water. The string will pull wet until the pressures equalize, at which time the pipe will start to drain.

Swabbing the annulus should not be a problem after the bit is pulled up into the structural casing and the mud weight needed to keep the drillpipe full can be calculated. Swabbing is indicated if the pipe pulls dry before the bit reaches the shoe. Note that the mud-height reduction from drillpipe removal must be considered, except in this case the dislacement factor relates to plugged drillpipe and the capacity factor is in the annulus.

Our criterion is that the drillpipe be full of seawater when the bit reaches the shoe. The hydrostatic pressure in the drillstring at the structural casing shoe depth is given by

$$p_{sw} = g_{sw} D_{sh}. \quad \text{(8.5)}$$

When the bit reaches the shoe, the pressure below the bit will be composed of the hydrostatic pressure of weighted mud and seawater.

$$g_m = (D_{sh} - D_{sw} - D_a - \Delta h_m) + g_{sw}(D_{sw} + \Delta h_m). \quad \text{(8.6)}$$

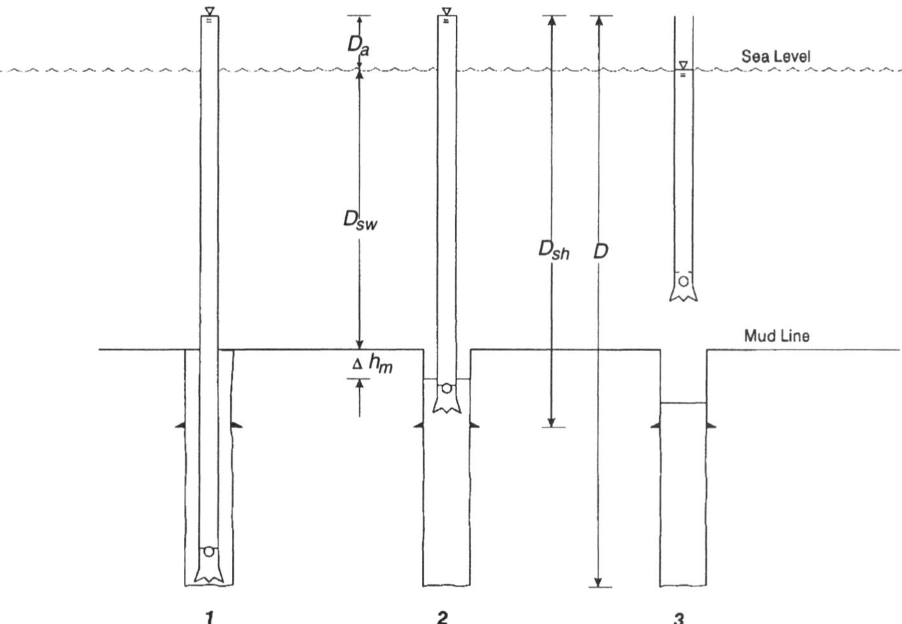

Fig. 8.28—Procedure for detecting swab conditions without a riser.

Equating pressures and substitution yields the required mud gradient,

$$g_m = \frac{g_{sw}(D_{sh} - D_{sw} - \Delta h_m)}{D_{sh} - D_{sw} - D_a - \Delta h_m}. \quad \text{.............. (8.7)}$$

Plenty of excess mud volume should be pumped if the hole volume below the casing shoe is unknown.

Example 8.6. Conductor casing point has been reached on a floater-drilled hole and the intent is to displace weighted mud to the drillstring float with seawater for the purpose of monitoring swab conditions during the trip. Determine the mud density if the following conditions apply.

Total depth = 4,515 ft, water depth = 2,320 ft, air gap = 85 ft, casing depth = 2,705 ft, casing-drillpipe capacity factor = 0.73732 bbl/ft, and plugged drillpipe displacement factor = 0.02429 bbl/ft.

Assume the BHA will be below the mud line when the bit reaches shoe depth and that the mud-level reduction in the annulus does not extend below the drillpipe.

Solution. We consider only the displacement factor of the plugged drillpipe because the collars will not leave the wellbore. Eq. 8.3 gives the mud-height reduction when the bit reaches the casing.

$$\Delta h_m = (4,515 - 2,705)(0.02429)/0.73732 = 60 \text{ ft}.$$

Substituting terms into Eq. 8.7 gives

$$g_m = \frac{(0.445)(2,705 - 2,320 - 60)}{2,705 - 2,320 - 85 - 60} = 0.603 \text{ psi/ft}.$$

Finally,

$$\rho_m = (0.603)(19.25) = 11.6 \text{ lbm/gal}.$$

Underbalancing a Riser-Drilled Conductor Hole. Drilling the conductor hole with a riser involves some considerations when it comes time to disconnect the riser to run casing. When seawater is exposed to the wellbore, the objective is to not allow wellbore pressures opposite any shallow gas horizons to be any lower than what was present when the riser was full of mud. Spotting a weighted pill in the well with a hydrostatic gradient determined from Eq. 8.8 will accomplish this result for any gas sands in the vicinity of the last casing shoe.

$$g_{wm} = \frac{g_m D_{sh} - g_{sw} D_{sw}}{D_{sh} - D_{sw} - D_a}, \quad \text{.................. (8.8)}$$

where g_{wm} and g_m refer to the hydrostatic gradients of the weighted pill and original mud, respectively.

The spotting procedure may risk fracturing the wellbore and the combined fluid hydrostatic must be compared to the fracture pressure. Another problem is that some excess volume must be pumped to account for overgauge hole and string displacement, yet too much excess will place undue stress at the shoe.

Example 8.7. Assume the well described in Example 8.6 was drilled to casing point with a riser. Determine the mud density to spot in the well to prevent underpressuring the shoe if the section was drilled with a 9.1-lbm/gal mud. Verify that the procedure will not break down the shoe if the fracture gradient is 0.494 psi/ft.

Solution. Substitute terms into Eq. 8.8 and obtain

$$g_{wm} = \frac{(0.473)(2,705) - (0.445)(2,320)}{2,705 - 2,320 - 85} = 0.824 \text{ psi/ft}.$$

The gradient is that of a 15.9-lbm/gal mud. At the shoe, the combined hydrostatic pressure of the two fluids is

$$p_{2,705} = (2,405)(0.473) + (300)(0.824) = 1,385 \text{ psig}.$$

This imposes a gradient of

$$g_{2,705} = 1,385/2,705 = 0.512 \text{ psi/ft}.$$

This job cannot be pumped without risking lost returns.

Example 8.7 emphasizes the difficulties associated with deepwater operations. Many times we cannot avoid reducing the shoe pressure after the riser is released without breaking the shoe down with the riser in place. If this is the case, openhole logs should be run to ascertain the depth of the shallowest sand and the mud-weight calculation based on this depth.

Another advantage to running logs is that an integrated caliper removes the uncertainties associated with how much mud to pump.

It is too late to do much about a flowing well after the riser has been disconnected. An alternate procedure is to reduce the wellbore pressure with the riser in place by opening a hydraulically operated dump valve.[36] The valve allows the mud to equalize with the sea while the bit is still in the hole and with the pumps ready to go if the well flows. The mud level in the riser is depressed, but flow can be detected by observing the open dump valve with a TV camera or, with the valve closed, using a transducer in the riser to monitor the hydrostatic pressure. A means to flood the riser has other functions including preventing riser collapse and maintaining some pressure on a well in the event of severe lost circulation or a shallow gas blowout.

Casing and Cementing Operations. Several offshore blowouts from shallow gas sands have occurred in conjunction with or sometime after cementing the conductor or surface casing. The fracture gradient may dictate an ultra-low slurry density and reduced pump rate to get cement back to the mud line. Recognizing the rate/density limitations, every other effort should be made to achieve adequate mud displacement (see Table 6.2), except that pipe movement usually is not an option with subsea wellheads. Finally, gas-migration control additives and procedures should be implemented any time the casing will be cemented across shallow gas sands.

8.4.4 Managing a Shallow Gas Flow. The standard way to deal with a shallow gas flow is to actuate the diverter and to start pumping seawater or mud as fast as the pumps allow. Planning considerations include designing the system to minimize failure risks and increase the likelihood of stopping the flow. There are no hard and fast rules; diverter-usage philosophies vary among different operators, drilling contractors, and well-control specialists. Complicating matters is that an advisable approach for one situation may be totally inappropriate for another. There is a school of thought that proposes shutting in and applying conventional well-control procedures. Such a method would have been unthinkable only a few years ago.

Diverters. Diverters have received much attention over the past few years and the performance of the well/diverter system during a shallow-gas event has been scrutinized because of numerous failures. Many insights have been gained from computer modeling and experimental work on the subject.[37-44] Diverter design and operational issues are also the topic of an API publication,[45] which we find to be an excellent reference in most respects.

Primary components of a diverter system include the annular sealing device, side outlets in the diverter housing or a separate spool, vent line(s), remote-actuated valve(s), and hydraulic or pneumatic controls. The equipment objectives are to close off flow from around the drillstem and route the well returns to a location away from the rig and personnel. Diverters are not a well-control device. A diverted well is out of control and the intent of a diverter operation should be to buy time until kill procedures can be instituted or the well kills itself by either depleting or bridging downhole.

It is imperative that the system precludes the possibility of an inadvertent shut-in. The vent outlets are below or at the same elevation as the flowline in most configurations, which may require that the vent line be closed during routine operations. If the vent line is closed, activating the diverter must include controls to assure that the well is open through at least one line before the packoff-element seals close on the drillstring. Various automatic and manual sequencing schemes have been used to meet this purpose. **Fig. 8.29** shows another method in which three actions are made simultaneously as the flowline is closed and vent line opened when the piston moves upward to close the preventer.

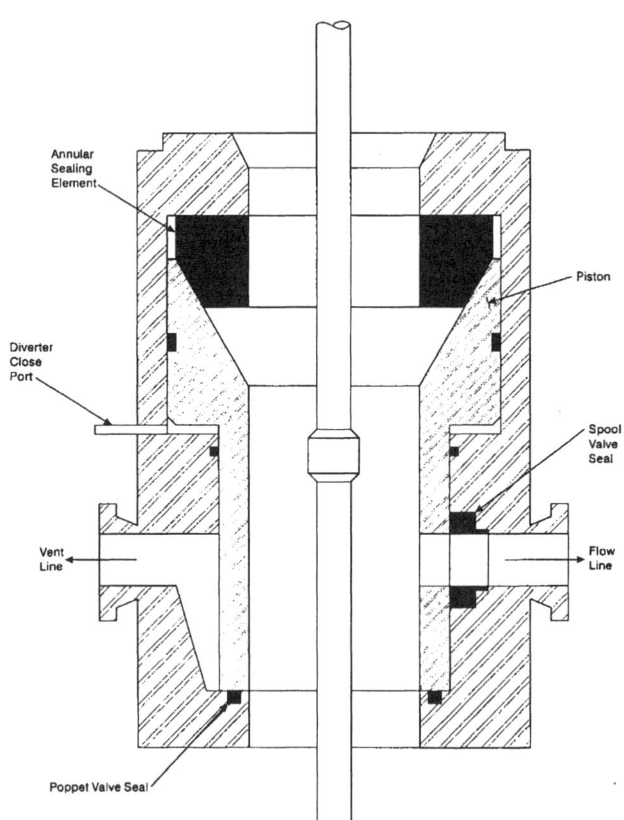

Fig. 8.29—Diverter that uses an integral sequencing mechanism.[45]

Valveless systems have been employed on some bottom-supported offshore rigs. One arrangement has vent lines coming off a spool at the drilling deck level and then making a long sweeping turn up to an elevation above the flowline. Another installs an open vent line a short distance above the fillup line and flowline.[46] Valves on these two lines are closed when the sequence starts.

The potential consequences of putting a well on the diverter are not to be taken lightly. One of the more fortuitous circumstances is if the hole geometry, flow rate, and available pump capacity allow for a quick, dynamic kill. A detailed discussion of dynamic-kill theory and execution is provided in a Chap. 10, but the procedure involves pumping water or mud at a rate sufficient to stem and ultimately stop the flow by the combined hydrostatic and annulus friction pressures.[47]

Many shallow-gas reservoirs are small and may rapidly deplete, but allowing a zone to kill itself by this manner is not a dependable outcome. Some wells have blown uncontrolled for years before finally depleting.[5] A downhole shut-in is another possibility when a well is diverted. Dislodged rocks may bridge and shut off the flow or the hole may simply collapse around the drillstem with the same result. A high percentage, maybe most, blowouts do bridge,[48,49] but experience has shown that a well will blow for an extended period if bridging does not occur within a day or two.[5]

Dynamic kills, zone depletion, and bridging far enough below the casing seat do not cause any serious problems except

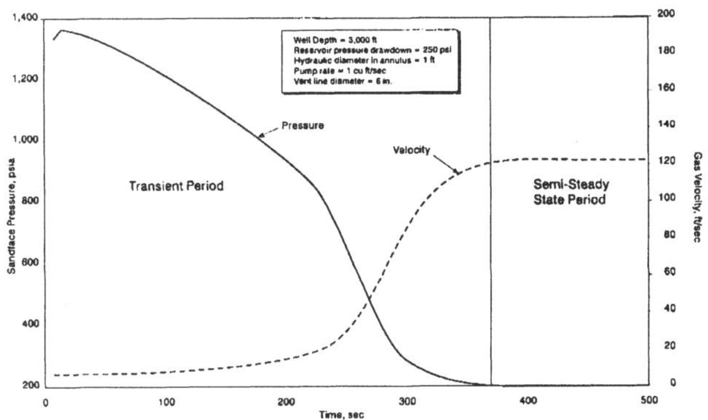

Fig. 8.30—Simulated BHPs and velocities for a shallow gas blowout through a diverter.[39]

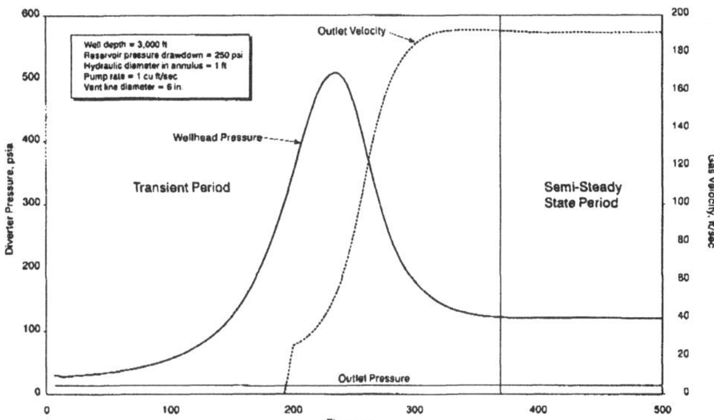

Fig. 8.31—Simulated surface pressures and flow velocities for a shallow gas blowout through a diverter.[39]

for, perhaps, the loss of the well. A failed diversion operation, on the other hand, can lead to casualties and the loss of the rig or an entire platform. Diverters fail to perform their function if any part of the system does not operate when activated, falls apart because of the dynamic loads, leaks, is cut by solids, becomes blocked, or imposes enough backpressure to cause formation fluids to broach the shoe. Human error is also a major contribution, a factor which gains importance with increasing system complexity.

Studies have shown that diverters fail about as often as they work while some failure-rate estimates run as high as 70%.[50,23] These statistics should be of considerable concern to anyone who rigs up one of these devices. Many improvements have been made, partly in response to some well-publicized disasters, and the performance track record is getting better. Even so, an operator should consider diverters as sacrificial equipment, expect a failure, and use the time allowed to evacuate the nonessential personnel.[51] Ultimately, the decision may be made to abandon the rig altogether and start planning a relief well or other means of intervening with the blowout.

Two separate and distinct flow periods have been identified for diverted wells. The flow is transient from the time the zone kicks until gas expels the original mud out the diverter line(s). During this time, a cycle develops in which flowing bottomhole pressure decreases as the liquid hydrostatic is removed, leading to an increased flow rate at the sand face, leading to an accelerated loss in bottomhole pressure, and so on until the mud is removed. Semisteady-state conditions prevail thereafter where the flow rates and pressures are essentially constant. The flow consisting of gas, formation water, circulated mud or seawater, and formation solids reaches a maximum during the latter period.

Fig. 8.30 shows the calculated BHPs and gas velocities for a shallow gas blowout simulated by Starrett *et al.*[39] Corresponding surface predictions are represented in **Fig. 8.31.** The charts, though somewhat idealized, are instructive and allow identification of typical responses during a diversion event and where problems are most likely to occur.

The influx rates into the well and BHPs are governed by the inflow performance relationship (IPR) of the reservoir and the flow is transient until the flowing BHP stabilizes. During the transient period, the diverter-spool pressures are driven up to a maximum by the mud acceleration and friction losses in the line. Santos and Bourgoyne[44] observed this behavior while conducting experiments on a test well and found that pressure peaks in the unloading period can exceed the semisteady-state pressure by a factor of 18. After reaching a maximum, flowing pressure at the diverter inlet declines and effluent velocity increases.

The outlet pressure in Fig. 8.31 is atmospheric and the pressure at the diverter reflects the backpressure held by the vent line. Sonic velocity, however, is achieved in many diversion events that lead to a higher pressure at the discharge. The critical velocity for a single phase or mixture is the speed of sound through the flowing medium and defines the maximum achievable velocity for a given pipe diameter. Sonic conditions thus set a limit on the flow rate, regardless of how much additional upstream pressure can be furnished.

Considering the diverter inlet pressure in Fig. 8.31, we may have the greatest potential for breaking down the shoe during the unloading period. The semisteady-state pressures are much lower but are still substantial. The single most important thing an operator can do to reduce diverter pressure is to use a large vent line. For the example well, the predicted peak pressure goes from 510 psia to 160 psia when the line diameter is increased from 6 to 12 in. Similarly, the predicted semisteady-state inlet pressure is reduced with a 12-in. line.

The pressures seen while the well is unloading suggest that structural failures (blowing a diverter off a well, line separation or whipping, vibration-induced fatigue, etc.) may be more likely during the transient period. A larger line will reduce the transient forces, but liquid heading and ejected rocks during the later period can be more destructive. It is important to mount the diverter securely to the wellhead or casing, use flanged line connectors, and anchor the equipment to the earth or steel foundation. Structural support is most important where flow velocities change in direction or magnitude.

Gas and liquids can erode steel at the flow velocities of a wide-open well. Add sand to the flow stream and steel equipment can be severed in a matter of minutes. The areas most subject to erosion are at the 90° turn where the flow goes from vertical to horizontal, bends in the vent line, and any place where the cross-sectional area changes. Bends and diameter changes are detrimental not only from the standpoint of erosion, but also increase the friction backpressure.

Ceramic coatings and hardfacing metals have been applied inside the diverter spool and other problem locations to take the brunt of the erosion. These measures, in addition to increasing the steel-wall thickness or targeting the affected areas, are certainly worthwhile, but an operator should expect sand to erode any material eventually and plan accordingly. The ideal diverter line has a straight shot from the spool to the diverter outlet, a consistent diameter, and a wall thickness appropriate for its purpose. Flexible lines are discouraged and the number of bends should be minimized. Where unavoidable, turns should be gradual (bend radius equal to at least 20

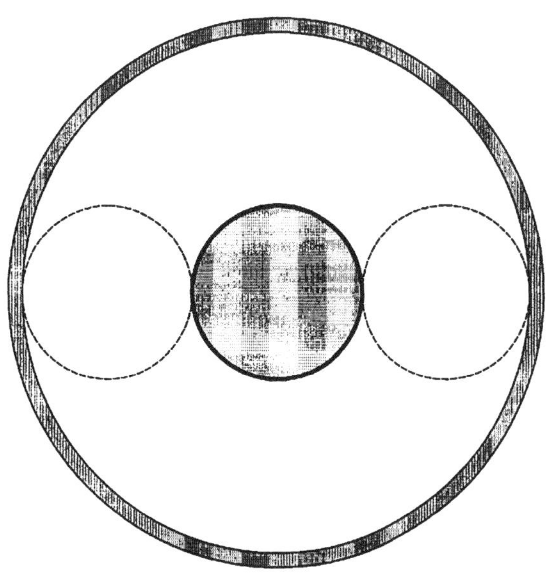

Fig. 8.32—Particle and conduit diameters conducive to bridging.

pipe diameters) or protected by harder material, thicker steel, or targets.

Erosion rates are exponentially proportional to sand velocity with an exponent value reportedly near 2.65.[52] Large vent lines are recommended for reducing backpressure, but the lower flow resistance can increase effluent velocity and cause more erosion than a smaller line. Placing a restriction at the discharge has been proposed and used for reducing the erosion potential.[50,52] It may seem bad advice to choke flow through a diverter, but the devices used for this purpose are relatively large and do not cause much additional well pressure: for instance, a 12-in. fixed choke at the end of a 16-in. line.

The normal practice on land rigs is to run a single diverter line out to a designated pit. DP drillships maintain a heading into the wind and may use a single line in the aft (i.e., rearward) direction. Bottom-supported offshore rigs and moored vessels usually have two lines to give the option of routing well returns with the wind direction. However, this arrangement has some drawbacks including added complexity and a system that may be more subject to erosion and valve failure. For these reasons, one North Sea operator equipped a platform diverter with a single large-diameter line.[46] The decision was justified by modeling the gas plume for a variety of flow and wind conditions and determining that the possible ignition sources presented a negligible fire hazard.

The pressure surge resulting from a plugged line can cause a structural failure or a broached shoe; the possibility of debris blocking the flow path may be the greatest weakness in a diverter system. From some old work by Coberly and Wagner[53] related to gravel-pack design, we can expect particles having a diameter d to form a bridge in a conduit with a diameter of approximately $3d$. This concept is illustrated in **Fig. 8.32.** Large-diameter lines will allow larger particles to pass through the diverter outlet, but vent lines simply cannot be sized large enough for the boulders ejected in some blowouts.

Pilot Holes. Small-diameter pilot holes are often drilled in areas known to have shallow gas sands. After drilling to the casing point, a hole opener or larger bit is used to drill the section again. The primary reason for drilling a pilot hole is to effect a dynamic kill more easily. Other advantages during a diversion operation include the choking effect of the wellbore and, according to Darcy's law, lower formation deliverability. While drilling, the smaller rock volume generated by the bit reduces the effects of drilled gas and cuttings on wellbore pressure.

A dynamic kill on a 17½-in. hole was executed successfully using 9.5-lbm/gal mud pumped at 45 bbl/min.[40] The flow rate from the well was estimated to be 44 MMcf/D. Rates through a diverter may reach hundreds of MMcf/D and flows of this magnitude cannot be dynamically killed in a hole this large. Computer modeling and field experience suggest a maximum pilot hole diameter of 9⅞ in. with 8½ in. being the preferred size.[5,36,38]

Designing a dynamic kill or planning a pilot hole without the help of a simulator is difficult, if not impossible. Numerous variables are involved, many of which are unknown to any degree of certainty. The vent-line diameter has a major impact on the kill rate (larger lines = faster pump rates) and the fracture gradient is always a consideration. As for unknowns, the gas flow rate is one of the more important design parameters. A range of reasonable $k_g h$ values are considered when planning for the zone IPR. In soft sediments, erosion can cause significant hole enlargement and the designer may need to consider a range of possible hole diameters.[41] In fact, erosion may cause a killable hole to expand large enough to preclude a dynamic kill, suggesting that an early effort will be most likely to succeed.

Pilot holes are recommended in problem areas, but smaller holes do carry some disadvantages.[5] For one thing, a shallow gas kick will unload a smaller hole much faster than a large one because of the lower mud volume and effect of influx height on transient zone deliverability. Also, the small clearances can greatly impact the surge and swab pressures and so tripping is a more critical concern.

Floater Considerations. Diverting a flow from a floating vessel can be more problematic than when the structure is in contact with ground surface or the sea floor. On the other hand, managing a shallow gas flow from a floater offers more opportunities for safety than bottom-supported rigs can achieve.

The slip joint is the weak link in a floater diverter system. Valve failure or line plugging has caused a force sufficient to extend the slip joint and drive the diverter out of its housing.[54] The slip joint also is prone to packing element leaks when subjected to the rigors of an uncontrolled flow. Improvements have been noted by replacing bladder-type packers with high-pressure annular-type elements while a dual packing design allows for backup capability.[51,36]

Problems associated with surface diverters tend to increase with water depth. Riser collapse is always a concern when formation fluids replace the drilling fluid. Opening a riser dump valve to flood the riser with seawater addresses the collapse issue and reduces the surface pressure peaks. However, severe slugging forces may be a consequence of flooding the riser. The riser may also lead to an annulus configuration that affects the ability to pump a dynamic kill. Finally, failure to disconnect the riser could be catastrophic in the event of a failed diverter.

Because of these hazards, a trend over the past several years has been to eliminate using surface diverters as a means to manage shallow gas flows from a floater. Drilling the conductor hole without a riser is one method that achieves this objective. Subsea diverters were developed in the mid-1980s and are an option for holes equipped with a riser.[55,56] A subsea diverter system offers many of the well-control advantages of

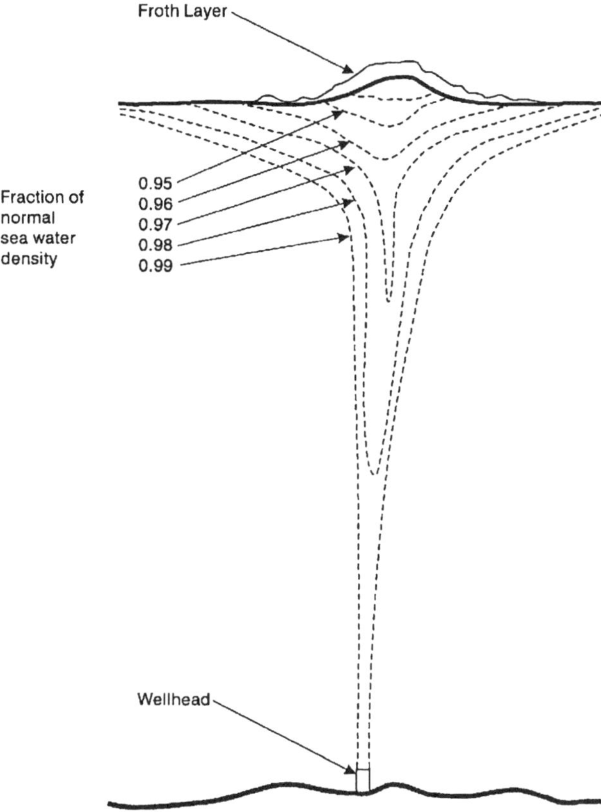

Fig. 8.33—Seawater density reduction associated with a blowout plume.[58]

drilling without a riser, yet retains the ability to maintain a higher wellbore pressure while drilling.

Subsea diverters operate in much the same way as a surface diverter. As in a riserless hole, fluid flow is suppressed by the hydrostatic pressure of the seawater. The pressure surges and velocities through the equipment are reduced greatly and venting at the sea floor eliminates the risks associated with surface diverters, as well as the noise and distraction. Subsea diverters may be held in reserve as backup to the surface diverter and activated when conditions are deemed unsafe.

Several ship-like vessels and a few semisubmersibles have sunk due to an underwater release of natural gas. For decades, conventional wisdom held that water density in the surface gas boil is reduced to the point where the vessel loses buoyancy and sinks like a rock. The concerns greatly limited the practice of riserless drilling, impeded the development of subsea diverters, and led some operators to avoid disconnecting a riser until it was too late. Theoretical models, experimental studies, and case history analyses have proved these fears to be groundless except in the most exceptional of circumstances.[57-59]

Fig. 8.33 shows the predicted seawater densities in a blowout plume for a well blowing 50 MMcf/D in 400 ft of water. Note that limited gas expansion yields only a slight density reduction until the gas approaches the atmosphere. Water density is reduced by only 5% in the center of the boil, not enough to cause a significant loss of draft. The effect becomes more important in shallower water and/or higher gas rates. For example, the same blowout in 100 ft of water reduces the predicted seawater density by 16%. Even so, floaters are not used very often in water depths where the loss in freeboard could cause a serious problem.

The gas concentration in an underwater plume and the size of a surface boil depend on water depth and gas flow rate. Thus we would expect a high-rate blowout in shallow water to pose more of a fire hazard than a small-rate blowout in deep water. Løes and Fanneløp[60] studied this question in a series of experiments where methane was released under 50 m of water at flow rates approaching 1.3 std m^3/s. Admittedly this gas rate is on the low side, but they found in all cases that the methane concentrations immediately above the water surface were small (less than 1% by volume) and were dissipated in the presence of a light wind.

Though retaining sufficient buoyancy to float, a single-hulled drillship is unstable when sitting directly on top of a plume. Invariably, the violent action of the boil will push the vessel to one side. The mooring line tension on the other side and small water-density reduction can cause the vessel to list into the plume to the point where water floods the deck from the side or through the moonpool. Any open hatches or manways then take on water and cause the vessel to sink.

Semisubmersibles, in comparison, are much more stable in the presence of a gas boil. The pontoons are below the zone of maximum aeration and the vessel/mooring characteristics lead to a self-righting behavior that tends to keep the semisubmersible positioned over the plume. In addition to the enhanced stability, the larger air gap results in a lower gas concentration where the potential ignition sources are located. These advantages make a semisubmersible the preferred rig for killing a well when it is possible to run a kill string or other tools directly into the blowout wellbore.[61]

Gas blowouts in deep water are easier to manage because the surface boil is expected to be less concentrated. Ocean currents usually will cause the boil to surface a considerable distance away from the rig and allow the vessel to stay over the hole. A blowout in water depths greater than 1,000 ft are favorable to hydrate formation. Thus much of the gas may be converted to a solid mound at the sea floor, one case where hydrates can work in favor of an operation.

Regardless of the water depth or vessel type, certain precautions should be taken when drilling shallow zones without a riser. **Table 8.9** gives a short list of the major considerations. Similar guidelines apply when subsea diversion is the riser-drilled option. Many of the recommendations are concerned with the ability to move off the hole at a moment's notice and a communication system must be established so that the watch person, driller, toolpusher, marine superintendent, and drilling supervisor are in immediate contact when this deci-

TABLE 8.9—SAFETY GUIDELINES FOR RISERLESS DRILLING A CONDUCTOR HOLE	
1.	Eliminate all discretionary ignition sources (smoking, cutting, welding, etc.).
2.	Run water spray on engine exhausts, including the crane engine.
3.	Post a watch in the moonpool area to observe for gas below the rig and downcurrent.
4.	If possible, observe the well area with a subsea camera.
5.	For moored vessels, man all anchor winches and prepare to use the stored energy in the mooring lines to pull the vessel to one side. Hold the anchors with the brakes and not the dogs.
6.	Keep all boats away from the rig and outside the anchor pattern.
7.	Keep the crew and supply boats on standby and ready for an emergency.
8.	Keep all water-tight doors and hatches closed.

sion needs to be made. The possibility of a sudden movement must be understood by all personnel and contingency plans rehearsed.

Containing the Gas. Several years ago an industry and government consortium sponsored research into the development of an inflatable packer designed to contain a shallow gas kick.[62] The packer, run in the drilling BHA, was inflated with mud when a slackoff action or a ball/seat mechanism closed the pipe bore and opened ports to the elements. The ability to circulate above the packer was provided by simply picking the string up again. The packoff range of the prototype tool allowed closure on a 20-in.-diameter hole, thus allowing for considerable enlargement of a 12¼-in. pilot hole.

The idea is sound but, to our knowledge, the concept did not go much beyond the initial tests. Inflatable packers have a proven track record in achieving or enhancing zone isolation on casing strings, but normally are inflated when the well is static. Circulation or flow past the elements can rapidly destroy the rubbers, which would tend to limit their application in a drillstring, especially given the velocities and entrained sand associated with many shallow gas flows.

Based on the problems associated with diverters, many operators have made a policy decision to shut wells in and use conventional control methods in hole sections previously deemed to have little or no integrity. Shut-in measures with a potentially low frac gradient initially became an accepted technique when drilling the surface hole (i.e., below conductor casing) from a floater.[63] The concept has extended to floater-drilled conductor holes, based on the reasoning that it is better to risk having the sea floor crater around the well, subsea shut-in equipment, and temporary guide base than to have a failed diverter.[5]

On bottom-supported rigs, there is an understandable reluctance to shut a well in with only conductor pipe set as the consequences of a cratered blowout are much more serious. Even so, the consequences of a failed diverter may be just as serious, if not more serious. At least one operator is in the process of adopting a shut-in policy for all rig types.[30] This somewhat radical (to some) approach to well control was driven by costly experiences with diverters and has been justified by a better understanding of shallow frac gradients and cratering mechanisms. Wells in selected operating areas are designed with control, rather than diversion, as the objective. Conductor casing shoes are tested to leak-off. Poor cement jobs are suspected whenever a shoe test is anomalously low and inadequate shoes are cement-squeezed until the desired integrity is obtained.

8.5 Trends in Deepwater Well Control

Two recent issues in deepwater well control will be addressed in this section, neither of which have been fully resolved. One has to do with the problems associated with using a marine riser to convey mud back to the drill vessel when drilling in deep water. Riser-drilling technology has been pushed to the limit and it appears that current practices must change when exploration efforts extend into water depths approaching 10,000 ft. The chapter concludes with a brief discussion of shallow water flows: their occurence, the hazards posed to the well structure, and possible means of preventing or otherwise managing these events.

8.5.1 Modifications to Conventional Riser Drilling. Low frac gradients in a deepwater operation result in using more casing strings to reach the objective depth than would be needed for an equivalent BML depth in shallow water. **Fig. 8.34** illustrates the planned casing points for a hypothetical subsea well in 6,000 ft of water. The hole program shown in the diagram uses the most common technique for designing a well (see Chap. 11 for a full discussion). The drilling engineer constructs pore-pressure and fracture-gradient curves using the best information available and then plots two other curves representing the minimum and maximum allowable mud weights. Typically a 0.5-lbm/gal margin is used to define the upper and lower pressure limits.

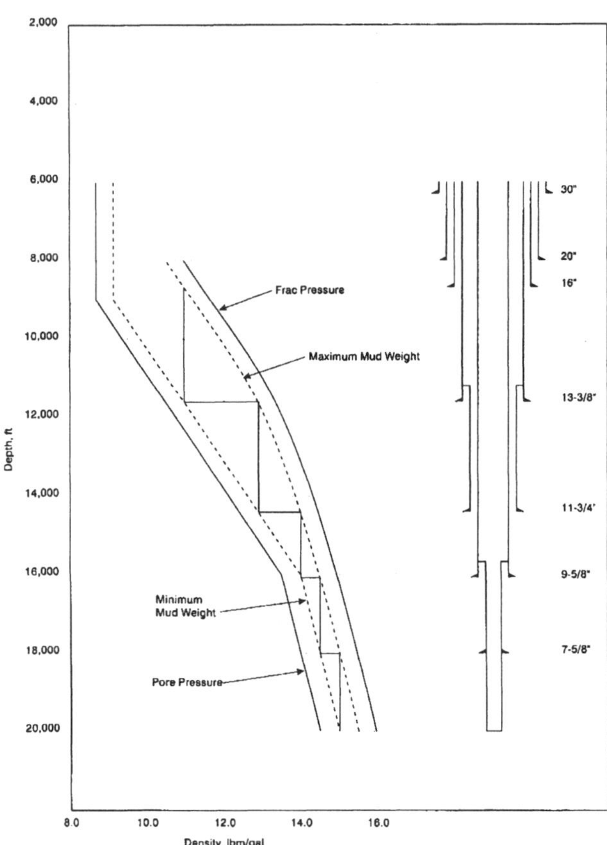

Fig. 8.34—Casing points for a hypothetical well drilled in 6,000 ft of water.

The mud-weight curves thus define an operating window for a given hole section and, starting at total depth, we can graphically identify where casing must be set to avoid exceeding the limits. In the example, the expected mud density at 20,000 ft is 15.0 lbm/gal. Moving up the hole, that 15.0-lbm/gal mud starts to exceed the maximum limit at 18,000 ft. Therefore, 18,000 ft defines a casing point. The process is repeated until the 2,000-ft conductor depth is reached.

It will require five full casing strings and two drilling liners to satisfy the design method. Seven casing points is a lot of pipe for any well and the costs and risks associated with this many casing jobs goes up accordingly. The program represents the most pipe a conventional subsea wellhead can suspend, involves some underreaming to get the drilling liners in the hole, and poses a limitation on the production tubular size—hence production rate. Being able to drill abnormally pressured sediments will require some novel solutions for water depths much greater than 7,000 ft.

One concept for eliminating some casing strings in a deepwater job is to reduce the wellbore pressure above the mud line while running a weighted system below. An example

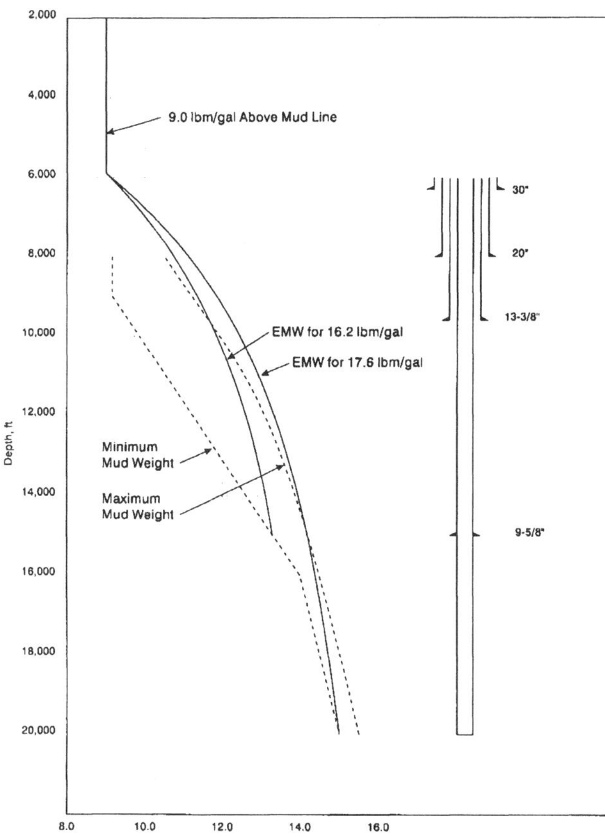

Fig. 8.35—Casing points for the deepwater well drilled with a lower mud density above the mud line.

problem demonstrates how this capability could impact a deepwater-hole program.

Example 8.8. Refer to the well described in Fig. 8.34 and assume there was a way to reduce the mud weight in the riser to a 9.0-lbm/gal equivalent at the stack.

1. Ignore the effect of the air gap and determine the mud density requirement in the wellbore to achieve a 15.0-lbm/gal equivalent at total depth.

2. How does the lower riser density affect the hole program if the same procedure is followed for every hole section?

Solution. 1. The target hydrostatic pressure for the combined mud column is

$$p_{20,000} = (0.0519)(15.0)(20,000) = 15,570 \text{ psig}$$

whereas the hydrostatic pressure in the riser is

$$p_{6,000} = (0.0519)(9.0)(6,000) = 2,803 \text{ psig}.$$

Calculate the mud gradient as

$$g_m = (15,570 - 2,803)/(20,000 - 6,000)$$

$$= 0.912 \text{ psi/ft}.$$

Equivalently,

$$\rho_m = (0.912)(19.25) = 17.6 \text{ lbm/gal}.$$

2. The next casing point for a conventional well is at 18,000 ft. The equivalent mud weight at 18,000 ft is less than 15.0 lbm/gal because of the lower hydrostatic pressure in the riser. For this problem, construct an equivalent density curve and determine the casing point where the curve intersects the maximum mud weight for the lowermost hole section. At 200 ft below the mud line,

$$\rho_{eq} = \frac{19.25[2,803 + (0.912)(200)]}{6,200} = 9.27 \text{ lbm/gal}.$$

Other depths are computed and the equivalent-density curve is prepared as shown in **Fig. 8.35.** The equivalent density equates to the maximum density at 15,000 ft and so we can safely drill the bottom 5,000 ft without having to set casing.

The process is repeated: determine the mud weight needed in the wellbore to achieve an equivalent density equal to the minimum limitation at 15,000 ft; construct an equivalent density curve corresponding to this mud weight; and obtain the next casing point from the graph. The curve and final hole program are represented on the diagram.

The potential savings in casing-related costs and the ability to drill smaller holes are enormous. Another benefit from placing a lower mud weight in the riser is the reduced riser-tensioning requirement, an important consideration in deep water. Also, the lower ECDs at shallower depths result in less overbalance and, therefore, lower filtrate losses and reduced risk for differential sticking.

Lopes and Bourgoyne[64] discussed one way to achieve the dual-mud concept. Their proposal was to inject nitrogen into the riser through gas-lift valves at a rate sufficient to lower the circulating density to the desired level. Such a configuration could be installed using off-the-shelf technology. Even so, some procedural difficulties must be addressed for operations other than routine drilling, including connections, trips, casing jobs, and well control.

Shutting off the nitrogen when making a connection will cause gas to break out of the mud and will create an overbalanced hole when gas injection resumes. The authors' suggestion was to continue gas injection at a much slower rate while making a connection. Another approach is to close the annular after making a connection and keep the riser isolated from the wellbore until the riser again reaches equilibrium pressure. Hollow glass spheres, long used as an additive for achieving ultralow cement densities, could solve this and other problems related to gassified muds.[65] However, the means to loop mud from the wellbore, add the beads in a controlled manner, and re-inject the low-density mud into the riser must be devised.

Pumping through the C and K lines to displace the gas-cut mud with seawater before making a trip was suggested. Doing so would mean that the mud left in the hole must have enough excess density to account for the reduced mud hydrostatic after seawater replaces the removed steel. Having seawater above stack has the opposite effect when running casing because mud displacement when pipe enters the wellbore will overbalance the hole. Floats held open by a release mechanism would reduce the displacement problem, but the system precludes the ability to circulate when going in the hole.

Kick detection and well-control procedures are other problems associated with the proposed method. Using a pressure transducer at the bottom of the riser was recommended for detecting kicks, a pressure increase indicating weighted mud displacement into the riser by formation fluids. It seems to us that the ability to read pressure at this point is also beneficial to drilling because it removes the guesswork about the effect of nitrogen rates on riser pressure.

Under the proposed system, heavy mud would be located in the drillstring and annulus up to the BOP stack with lighter

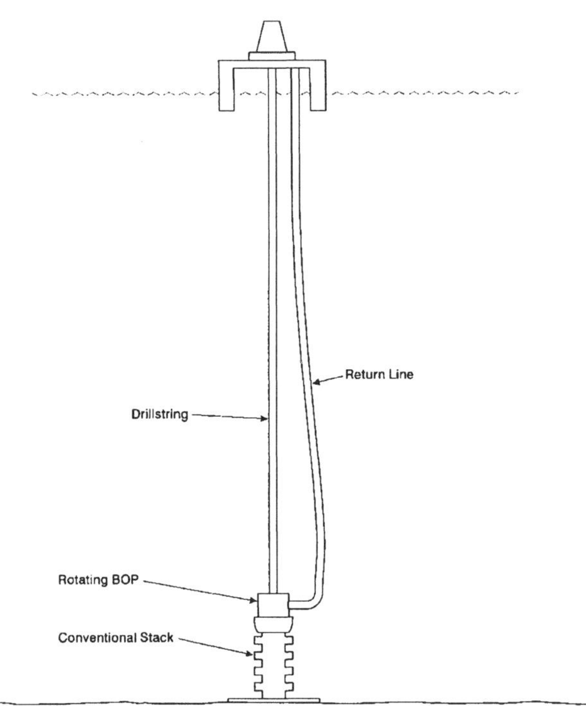

Fig. 8.36—Concept for drilling wells in deep water without a riser.

fluid above the stack. The unbalanced U-tube presents a situation where simply shutting the well in will result in an annulus gradient equivalent to the heavy mud. The last casing seat cannot withstand this much pressure. Incorporating additional pressure from an underbalanced formation and the kick-fluid hydrostatic only worsens the condition at the shoe. Thus conventional well-control procedures are not applicable and some other means of dealing with a kick must be addressed.

Other research efforts are focused on eliminating the riser.[66-68] The 21.0-in. risers currently used in floater drilling have nearly reached their depth limitation. The riser-design problems are formidable, and a drill vessel with enormous storage and weight capacity is required to transport and suspend a long riser. Thus smaller rigs will be able to drill in deeper water when a workable riserless system is available. Indeed, the advantages of a dual-mud system and riserless drilling extend to all water depths.

Fig. 8.36 illustrates a simple idea for a riserless drilling system. The wellbore is isolated from the sea by a rotating blowout preventer (RBOP) while the circulated returns are routed through the RBOP outlet to a return line and then back to the pits for processing. The return line could be used as a high-pressure conduit back to the choke if sized small enough and with sufficient strength/wall thickness. Alternatively, a larger line could be used with attached C and K lines, similar to how risers are now configured.

Another benefit offered is a relaxed stationkeeping requirement. The present guideline is to maintain vessel offset from the well within 5% of water depth. This restriction could be reduced to 15% or 20% of water depth with the configuration depicted in Fig. 8.36. (The drillpipe is pulled against the side of the moonpool at higher offsets.) It follows that the wait-on-weather time will be reduced greatly. Also, less expensive mooring systems will be needed whereas the dynamically positioned vessels will be less expensive to operate.

Consider also the savings in mud costs. The capacity factor of a 21.0 × 19.75-in. riser is 0.3789 bbl/ft. This represents almost 3,800 bbl of storage in a 10,000-ft riser, which is a volume not otherwise needed for drilling purposes. The excess volume at the same water depth would be reduced to less than 1,000 bbl if an 11-in. return line was used in a riserless drilling operation.

The dual-density concept and the advantages related to well design are applicable to a riserless drilling system. This could be accomplished by gas lifting the return line, hollow-glass-bead injection, or other means. However, the unbalanced U-tube presents the same problems in well control that were discussed for a reduced mudline density in a conventional riser. Choe[68] discussed these concerns and suggested some new procedures related to kick detection, preventing further entry, kill-weight mud determination, and controlled-kick circulation.

8.5.2 Shallow Water Flows. Overpressured water flows from shallow sands (relative to the mudline) are a common occurrence off the continental shelf in the Gulf of Mexico and other deepwater regions of the world. Typically, these sands are found between 300 and 2,000 ft BML which, in a conventional hole program, means they are most likely to be encountered while drilling the conductor hole below structural casing. Their pore-pressure gradient usually ranges between 9.3 and 9.4 lbm/gal, but can be higher.[69] The sands tend to be loosely consolidated and exhibit high permeability and, when underbalanced, can flow at prodigious rates with much entrained sand.

The primary hazard of a shallow water flow (SWF) is when the flow broaches the structural casing shoe and washes out the soft sediments outside the well. Failure of the structural support system can result in the consequent loss of well and seafloor equipment. These flows can be violent and erode an extensive trench on the seafloor. One deepwater operator had to relocate a number of subsea wells because of this problem and sustained losses of approximately \$150 million.[70]

Once started, a SWF is difficult to stop; every effort should be made to construct or drill a well so as to prevent an SWF event. Should a flow develop, it is important to act quickly to restore equilibrium. It may be possible to stop a moderately overpressured flow by weighting up and drilling with kill-weight mud returns to the seafloor. The ability to effect such a solution depends on water depth, the magnitude of the pore-pressure gradient, and the fracture integrity of the open hole. The following example demonstrates.

Example 8.9. A water flow from a shallow sand at 1,500 ft BML is detected while drilling the conductor hole below structural casing set 300 ft BML. The water depth is 3,200 ft and the sand's pore-pressure gradient referenced from sea level is a 9.4-lbm/gal equivalent.

1. Determine the mud weight needed to overbalance the flowing formation.

2. Can this mud weight be pumped safely if the fracture gradient at the casing seat is equivalent to a 9.3-lbm/gal density?

Solution. 1. Seawater will comprise most of the hydrostatic pressure because this hole section is being drilled without a riser. The mud-weight requirement is thus

$$(0.0519)(9.4)(4,700) = (0.445)(3,200)$$

$$+\ \rho_m(0.0519)(1,500)$$

$$\rho_m = 11.2 \text{ lbm/gal}.$$

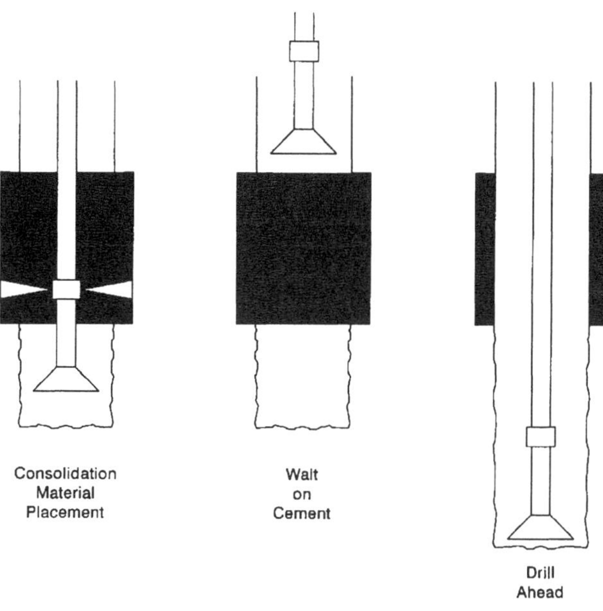

Fig. 8.37—Jet-stabilization process for SWF control.

2. The fracture pressure at the structural casing shoe is

$$p_{fi} = (0.0519)(9.3)(3,500) = 1,689 \text{ psig}.$$

Increasing the mud density yields the hydrostatic pressure at this depth.

$$p_{3,500} = (0.445)(3,200) + (0.0519)(11.2)(300)$$
$$= 1,598 \text{ psig}.$$

We conclude that the densified mud can be pumped without risking lost returns. The effect of cuttings loading should be considered, however, and the penetration rate controlled as necessary.

Increasing mud weight is an effective, though expensive, way to manage the flow in the well in Example 8.9. However, an overbalanced kill becomes more difficult in deeper water because of diminished fracture gradients and lessened contribution of the mud hydrostatic.

Based on seismic data or information from offset wells, an operator may know in advance where to expect an SWF zone and modify the hole program by setting an additional string of casing immediately above the problem sand. The procedure would be to drill/underream below the 30-in. structural casing to a depth just short of the SWF formation and then to run and cement 26-in. casing. The deeper casing mitigates the fracture-gradient problem and provides another flow barrier between the wellbore and structural casing. Having casing immediately above the sand also facilitates placement of the chemical treatments proposed for SWF control.[71]

Another suggested method for furnishing the required wellbore pressure is to install a SWF diverter on the last casing string.[70] Basically, the system consists of a subsea rotating head and adjustable choke allowing an operator to drill with backpressure against the formation. This arrangement has many of the same drawbacks as using weighted mud because the combined hydrostatic and choke pressures must be maintained at a level that will prevent a lost circulation and a broached shoe. It may be a practical option if the casing is not set too far above the overpressured interval.

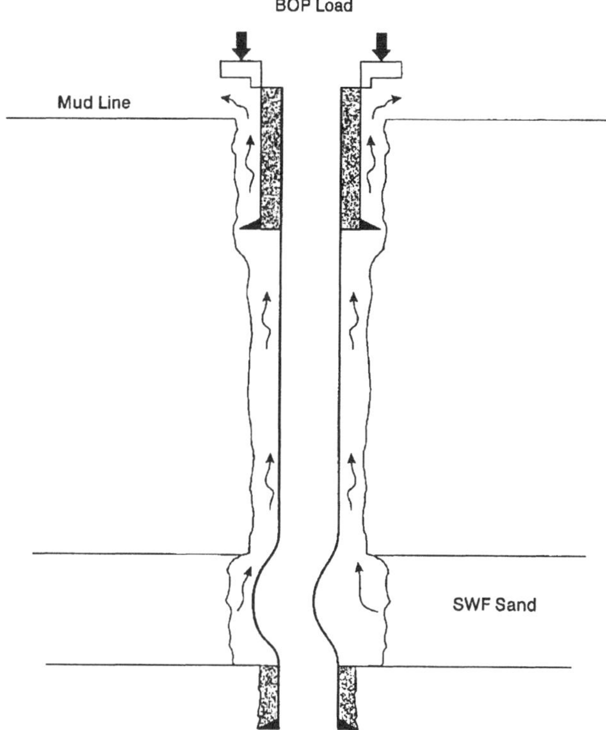

Fig. 8.38—Buckled conductor casing across an SWF sand.

Chemical agents such as stabilization polymers and resins are being considered as a way to block SWF zone permeability and consolidate the rock.[70-72] An acrylate monomer solution (AMS) is one of the chemical alternatives being studied. The suggested procedure is to place the AMS across an SWF sand after drilling through the zone. Ideally, the wellbore construction allows for enough overbalance that the fluid can be displaced a considerable distance into the formation through the rock matrix. After a period, a temperature-activitated initiator causes the AMS to change from a thin liquid to a rubber-like solid which blocks permeability and holds the sand together. Water-based epoxy resins have similar consolidation/sealing properties, yet offer another advantage in that resin-treated rock normally has increased fracture integrity.

Fig. 8.37 depicts a proposed jet-stabilization process using cement, resin, or other consolidating material. The objective is to pump the material at high velocity through side jets. Filter cake is removed and the technique ensures that the enlarged borehole is filled completely with the material. After placement, the drillstring is pulled up above the zone, the hole is circulated clean, and time is allowed for the cement or other material to achieve a set. The hole is then drilled out through the plug and on to the next casing point.

Getting a good cement job across the conductor casing is critical. **Fig. 8.38** demonstrates what can happen if conductor casing is cemented across an SWF zone that has not been fully contained. The cement above the zone has been displaced by the water or is at least channeled sufficiently to allow flow outside the structural casing. The surface equipment settles as a result of the sea floor erosion and the conductor pipe buckles under the weight of the BOPs into the enlarged hole across the sand.

Obviously, conductor casing should not be run into a flowing well. Even so, a dead SWF sand may not remain dead because of the loss in cement-sheath pressure during the hydration process. The cement properties and placement technique should be tailored using the same considerations discussed in

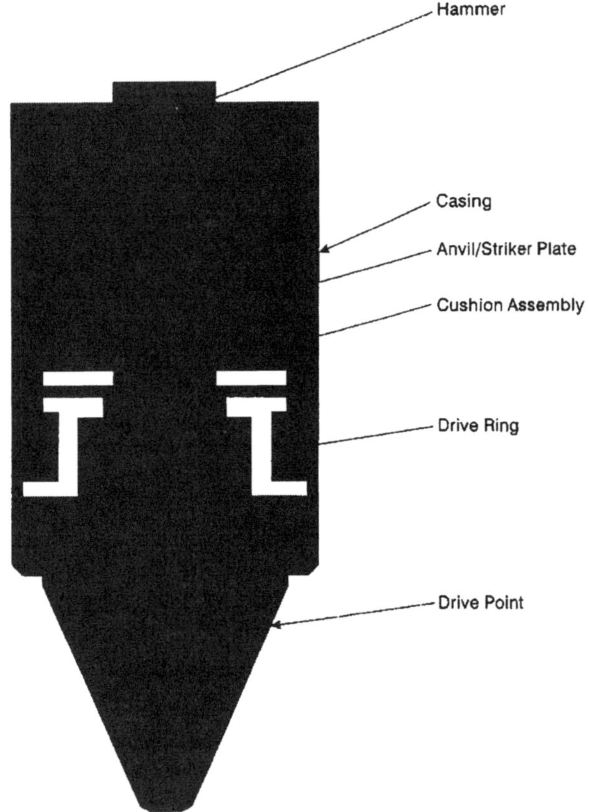

Fig. 8.39—Schematic of a proposed mechanism to bottom-drive casing through an SWF zone.[72]

Chap. 6 for combatting annular gas migration. Foamed cement has seen successful application in SWF control because the compressed nitrogen bubbles maintain hydrostatic pressure during the critical transition period. Other benefits of foamed cement are that it sets fast at low temperature and expands to fill an enlarged borehole.

Finally, researchers in academia and industry are working on a method that will eliminate an annulus through which an SWF zone can flow. This can be accomplished by driving structural pipe through the sand using a bottom-drive mechanism similar to the one depicted in **Fig. 8.39.**[72] As indicated, the basic equipment is simple. The hydraulic hammer, by repeatedly striking the casing shoe, can drive the casing to a depth where soil resistance matches the impact force. Drive depths up to 1,000 ft BML are feasible in soft sediments, which is deep enough to cover most of the known SWF horizons in the deepwater Gulf of Mexico.

By necessity, treatment of this subject has been rather cursory. We hope that the general overview given will encourage the reader to review the cited references (and their references) to gain more appreciation of current efforts to solve one of the many challenges facing deepwater operators today.

Problems

8.1 Intermediate casing has been hung from a subsea wellhead in 1,100 ft of water. It is time to test the casing seals using water as the test fluid. Estimate the volume required to pressure the system between the pump and wellhead if the volume in the lines is 25 bbl. Assume water compressibility is 3×10^{-6} psi^{-1} and ignore the effect of line ballooning.

8.2 Refer to the annular preventer shown in Fig. 8.8. Plot incremental closing pressure vs. water depth in 100-ft increments from zero to 800 ft for mud weights of 10 and 18 lbm/gal. Assume the control-fluid hydrostatic gradient is 0.445 psi/ft.

8.3 Plot usable fluid volume in an 11-gal accumulator bottle for water depths ranging from 0 to 10,000 ft. The precharge, minimum, and charged bottle pressures are 1,000, 1,200, and 3,000 psig, respectively. Do the same for a 5,000-psig bottle working pressure.

8.4 Assume the hydraulic signal time as function of depth can be described by a power law model (straight line on logarithmic graph paper) at water depths greater than 1,600 ft. Signal time at this depth is 4 seconds while the time in 3,000 ft of water is 10 seconds. Develop the corresponding equation and estimate the signal time if the water depth is 6,000 ft.

8.5 Bourgoyne and Holden[15] used the well data shown in **Table 8.10** to model pressure behavior during a kill procedure conducted from a floater. Answer the following related questions.

1. What is the circulating friction in the choke line at kill rate?
2. What is the target ICP?
3. What should the casing pressure gauge read when the target ICP is achieved? Assume there was seawater in the choke line before circulation commenced.

8.6 Draw a diagram to represent each one of the procedures discussed in Table 8.5.

8.7 The following friction-pressure data in 2,400 ft of a 3-in. choke line were recorded on the most recent measurement. 480 psig at 300 gal/min, 360 psig at 250 gal/min, and 260 psig at 200 gal/min.

TABLE 8.10—KICK DATA FOR A WELL DESCRIBED IN THE PROBLEM SET

Wellbore Configuration	
Water Depth	4,342 ft
Air Gap	50 ft
Vertical Depth from RKB Reference	11,540 ft
Surface Casing Information	
Description	13³/₈-in., 61.0-lbm/ft
RKB Setting Depth	7,410 ft
Openhole Diameter	12¼ in.
Drillstring Information	
Drill Collar Size	8×3 in.
Drill Collar Length	540 ft
Drillpipe Description	5-in., 19.50-lbm/ft
Choke and Kill Line ID	3.15 in.
Prekick Circulation Data and Mud Properties	
Drilling Circulation Rate	496 gal/min
CDPP While Drilling	2,400 psig
Kill Circulation Rate	226 gal/min
KRCP Through Riser	600 psig
KRCP Through Choke Line	880 psig
Mud Density	9.2 lbm/gal
Recorded Kick Data	
SIDPP	300 psig
SICP	440 psig
Pit Gain	30 bbl
Other Known or Assumed Information	
Fracture Gradient at Casing Seat	0.551 psi/ft
Kick Fluid	Gas

You are displacing a kick using the Driller's Method at 200-gal/min kill rate. The casing pressure has dropped steadily to zero and now the CDPP is starting to climb. The decision is made to slow the pump to 180 gal/min in an attempt to reduce friction losses in the choke line and stabilize the CDPP. Determine the expected choke-line friction at the reduced rate.

8.8 The well in Table 8.10 will be killed using the Wait and Weight Method. Use a 0.6-specific gravity gas, make your own assumptions with regard to wellbore temperature, and answer the following.

1. Determine the shut-in pressure gradient at the shoe.
2. Determine the required kill-mud density needed.
3. Estimate the pressure gradient at the shoe when gas first reaches the casing.
4. Estimate the pressure at the top of the influx when gas reaches the mud line.
5. Estimate the maximum choke pressure. Assume gas friction losses in the choke line are 50 psi at this point in the kill.

8.9 Consider the following conditional pairs. Which of the two conditions is most favorable for hydrate formation?

1. Water depth of 3,000 ft or 6,000 ft?
2. High-salinity polymer mud or fresh-water lignosulfonate mud?
3. Immediately after a kick or after reconnecting the riser after several days?
4. A dry gas or a gas-condensate kick?
5. Slow or fast circulation rate?

8.10 You have finished killing a well from a floating drilling rig and want to ensure that absolutely no gas is circulated down the flowline when the riser is displaced with kill-weight mud. How might you modify the procedure shown in Table 8.7 to accomplish this end?

8.11 Derive Eq. 8.2.

8.12 Calculate the riser margin for the well shown in Table 8.10. Is it advisable to incorporate a riser margin on this well? Why or why not?

8.13 A riser installed in 6,000 ft of water has a 1,310-psi collapse rating. The well is drilling ahead with 17.0-lbm/gal mud when the DP thrusters malfunction and it becomes necessary to disconnect the riser. What precautions should be taken before releasing the LMRP connector? Show any calculations that support your answer.

8.14 Refer to Fig. 8.22. Start with structural casing at 250-ft sediment depth and specify casing points down to 3,000 ft for a well drilled in the two water depths listed below. Base the design on safely shutting the well in on dry gas. Assume a 0.45-psi/ft pore-pressure gradient and use a gas hydrostatic gradient of 0.05 psi/ft.

1. Water depth = 100 ft and
2. water depth = 8,000 ft.

8.15 Work the preceding problem except assume the hole will be half-full of 9.0-lbm/gal mud when shut-in.

8.16 Assume the Brennan and Annis[73] equation discussed in Chap. 3 works for your area from ground surface all the way to 8,000-ft water depth. Generate a series of frac-gradient curves similar to the ones depicted in Fig. 8.22. Use the same pore-pressure gradient, sediment-overburden relation, and air gaps as described in the chart's text block. This problem is well-suited for a spreadsheet program.

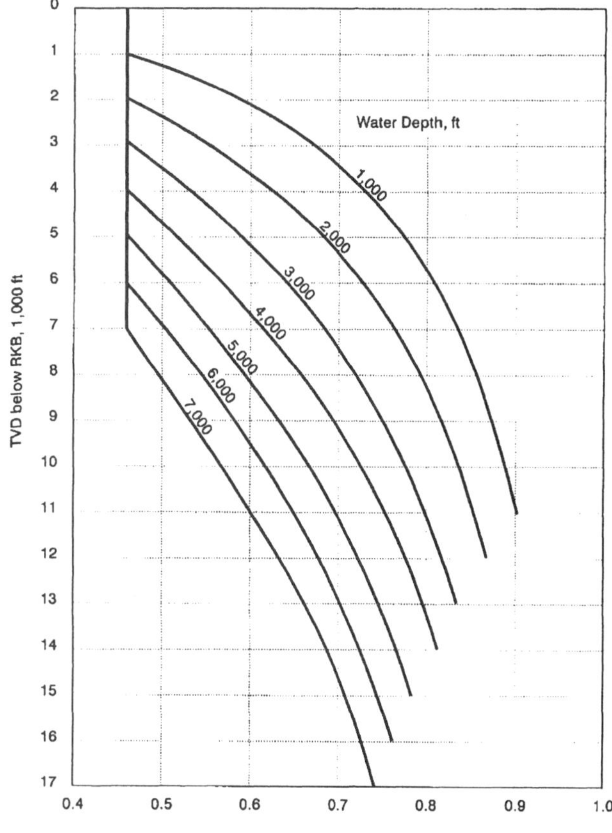

Fig. 8.40—Overburden-stress gradient correlation for deepwater Gulf of Mexico sediments.[33]

8.17 Modify the 500-ft water-depth curve in the preceding problem to reflect allowable mud weights with a 25-ft air gap.

8.18 Barker and Wood[33] used density logs in the deepwater Gulf of Mexico to generate the overburden gradient curves shown in **Fig. 8.40.**

1. How do the values compare to Eaton's[34] data (corrected for water depth)?
2. What do the curve shapes suggest about the effect of water depth on compaction below the mud line?

8.19 You are drilling a long shale sequence with a riser from a drillship. The present RKB depth is 3,250 ft, the air gap is 20 ft, and the water depth is 1,340 ft.

1. Determine the hydrostatic pressure at total depth if the clean mud density is 9.0 lbm/gal, the penetration rate is 120 ft/hr, and the circulation rate is 1,000 gal/min.
2. Determine the hydrostatic pressure and equivalent mud weight at total depth if returns were dumped at the ocean floor.

8.20 Reconsider Example 2.2 except place the top of the sand at 500 ft with the gas-water contact at 600 ft.

8.21 Seawater weighing 8.6-lbm/gal is being used to riserless drill the conductor hole of the well described in Table 8.8. The air gap and water depth are 50 and 400 ft, respectively. Assume the gas sand occupies the entirety of the annulus and estimate the equivalent mud weight at total depth by numerically integrating the mixture density. Use the technique given in Section 4.3.2 of *Applied Drilling Engineering*[74] to determine the effects of solids and formation water.

8.22 You are budgeting a well for which the conductor hole will be drilled from a floater without a riser. Drilling mud is to be used and discarded at the sea floor. The interval between 300 and 1,750 ft BML will be control-drilled at 100 ft/hr and the planned circulation rate is 1,000 gal/min. Estimate the effect this procedure will have on the mud bill if the mud costs $12.00/bbl.

8.23 Reconsider Example 8.5 except adjust the depths to reflect a water depth of 5,000 ft.

8.24 Determine the spot mud density needed to monitor swab conditions while pulling from a riserless conductor hole if the following conditions apply. Total depth (RKB) = 1,370 ft, water depth = 400 ft, air gap = 70 ft, casing depth (RKB) = 770 ft, casing-drillpipe capacity factor = 0.73732 bbl/ft, and plugged pipe displacement factor = 0.02429 bbl/ft.

8.25 Derive Eq. 8.8.

8.26 The following conditions pertain to a conductor hole drilled from a floater with a riser. Total depth (RKB) = 1,600 ft, water depth = 800 ft, air gap = 22 ft, casing depth (RKB) = 1,190 ft, and mud density = 9.3 lbm/gal.

Determine the mud density to spot in the wellbore if the intent is to keep from underpressuring the shoe. Can the procedure be implemented if the fracture gradient at the structural pipe shoe is 0.540 psi/ft?

8.27 Refer to the well discussed in Example 8.7. Assume that the hole was logged and that the shallowest potential gas sand was not encountered until 3,020 ft. Rework the problem except base the calculations on maintaining balance at the gas sand as opposed to the shoe.

8.28 Assume single-phase gas flow and subcritical velocity in a diverter line. Rearrange the form of the Weymouth formula presented in Chap. 7 to solve for the inlet pressure.

8.29 Using the Weymouth formula, plot the inlet pressure for 0.6-specific gravity gas flowing through a 6-in. nominal line at rates ranging from zero to 800 MMcf/D. Do the same for 10-, 12-, and 16-in. lines. Assume a 0.50-in. wall thickness in all cases.

8.30 According to Darcy's law, by what factor could flow rate from a shallow gas sand be reduced if a 9½-in. pilot hole was drilled instead of a 26-in. hole? Assume the reservoir drainage radius is 1,320 ft.

8.31 An 8½-in. pilot hole was drilled at a rate of 150 ft/hr and the hole will be enlarged using a 26-in. hole opener. What penetration rate will give the same mud density in the annulus as obtained in the pilot hole?

8.32 Consider a trip from a pilot hole drilled below structural casing. Does the smaller openhole diameter have an effect on how often the hole needs to be filled? Defend your answer.

8.33 Discuss the advantages and disadvantages of drillships and semisubmersibles as they relate to primary well control and managing a subsea blowout.

8.34 Make your own calculations and duplicate the 16.2-lbm/gal equivalent density curve shown in Fig. 8.35.

8.35 Say that the well shown in Fig. 8.34 kicked from a 15.5-lbm/gal formation at 20,000 ft while drilling with a 15.0-lbm/gal mud.

1. Assume returns are not lost and determine the pressure gradient at 18,000 ft when the well is shut-in if the respective kick height and hydrostatic gradient are 500 ft and 0.15 psi/ft, respectively.
2. Determine the SICP if the C and K lines are filled with seawater.

8.36 Refer to Example 8.9. Work the problem again except change the water depth to 7,000 ft and the fracture gradient to a 9.1-lbm/gal equivalent. Use the same BML depths for the casing shoe and SWF zone.

Nomenclature

C = capacity factor, bbl/ft
d = diameter, in.
D = depth, ft
D_s = sediment depth, ft
D_{sw} = water depth, ft
D_{sh} = casing shoe depth, ft
g_m = mud hydrostatic gradient, psi/ft
g_{km} = kill-weight mud hydrostatic gradient, psi/ft
g_p = pore pressure gradient, psi/ft
g_{rm} = riser margin, psi/ft
g_{sh} = pressure gradient at the casing shoe, psi/ft
g_{sw} = seawater hydrostatic gradient, psi/ft
g_{wm} = weighted-mud hydrostatic gradient, psi/ft
h = permeable thickness, ft
h_m = mud height, ft
k = permeability to gas, darcy
p = pressure, psi
p_{dci} = initial circulating drillpipe pressure, psi
p_{fch} = fully charged accumulator pressure, psi
p_{fi} = fracture initiation pressure, psi
p_p = pore pressure, psi
p_{pch} = accumulator precharge pressure, psi
p_{sh} = pressure at the casing shoe, psi
p_{sw} = sea water hydrostatic pressure, psi
V_{ab} = accumuator bank capacity, gal
V_d = displacement volume, bbl
V_u = usable fluid volume, gal
ρ_{eq} = equivalent mud density, lbm/gal
ρ_m = mud density, lbm/gal
ρ_{rm} = riser margin, lbm/gal

References

1. Silcox, S.H.: "Floating Drilling: The First 30 Years—Part 1," *JPT* (January 1983) 18.
2. Halal, A.S. and Mitchell, R.R.: "Casing Design for Trapped Annular Pressure Buildup," *SPEDE* (June 1994) 107.
3. Adams, A.J. and MacEachran, A.: "Impact on Casing Design of Thermal Expansion of Fluids in Confined Annuli," *SPEDE* (September 1994) 210.
4. Goins, W.C. Jr. and Sheffield, R.: *Blowout Prevention*, second edition, Gulf Publishing Co., Houston (1983) 150.
5. Adams, N.J. and Kuhlman, L.G.: "Case History Analysis of Shallow Gas Blowouts," paper SPE/IADC 19917 presented at the 1990 SPE/IADC Drilling Conference, Houston, 27 February–2 March.

6. Holand, P.: "Subsea Blowout-Preventer Systems: Reliability and Testing," *SPEDE* (December 1991) 293.
7. Roche, J.: "New Anticollapse Valve Protects Deepwater Risers," *Ocean Industry* (July 1988) 21.
8. Recommended Practice for Design, Selection, Operation, and Maintenance of Marine Drilling Riser Systems," *Recommended Practice 16Q*, first edition, API, Washington, DC (1 November 1993) 13–20.
9. "Recommended Practice for Blowout Prevention Equipment Systems for Drilling Wells," *RP 53*, second edition, API, Washington, DC (25 May 1984) 25 **(Out of print)**.
10. Bayless, J.H. and Bayless, J.S.: "Hand Calculations Define Subsea BOP Reaction Times," *Ocean Industry* (July 1987) 28.
11. Von Flatern, R.: "Subsea Drilling BOP Controls Ready for Ultra-Deepwater," *Offshore* (September 1997) 44.
12. Speers, J.M. and Gehrig, G.F.: "Delta Flow: An Accurate, Reliable System for Detecting Kicks and Loss of Circulation During Drilling," *SPEDE* (December 1987) 359.
13. Haeusler, D., Makohl, F., and Harris, T.W.R.: "Applications and Field Experience of an Advanced Delta Flow Kick Detection System," paper SPE/IADC 29344 presented at the 1995 SPE/IADC Drilling Conference, Amsterdam, 28 February–2 March.
14. Harris, T.W.R., Hendriks, P., and Surewaard, J.H.G.: "Advanced Kick Detection Systems Improve HPHT Operations," *Pet Eng. Intl.* (September 1995) 31.
15. Bourgoyne, A.T. Jr. and Holden, W.R.: "An Experimental Study of Well Control Procedures for Deepwater Drilling Operations," *JPT* (July 1985) 1239.
16. Barker, J.W. and Gomez, R.K.: "Formation of Hydrates During Deepwater Drilling Operations," *JPT* (March 1989) 297.
17. Katz, D.L.: "Prediction of Conditions for Hydrate Formation in Natural Gases," *Trans.*, AIME (1945) **160,** 141.
18. Churgin, J. and Halminski, S.J.: "Temperature, Salinity, Oxygen, and Phosphates in Waters Off the United States: Vol II Gulf of Mexico," *Key to Ocean Records Documentation No. 2*, Natl. Oceanographic Data Center, Washington, DC (1974) 117.
19. Hale, A.H. and Dewan, A.K.R.: "Inhibition of Gas Hydrates in Deepwater Drilling," *SPEDE* (June 1990) 109.
20. Kotkoskie, T.S. *et al.*: "Inhibition of Gas Hydrates in Water-Based Drilling Muds," *SPEDE* (June 1992) 130.
21. Grigg, R.B. and Lynes, G.L.: "Oil-Based Drilling Mud as a Gas-Hydrates Inhibitor," *SPEDE* (March 1992) 32.
22. Wood, T. and Billon, B.: "SBM Drilling Fluid System Sets Deepwater GOM Record," *World Oil* (October 1997) 71.
23. Adams, N.J. and Kuhlman, L.G.: "How to Prevent or Minimize Shallow Gas Blowouts—Part 1," *World Oil* (May 1991) 51.
24. Walters, J.V.: "Internal Blowouts, Cratering, Casing Setting Depths, and the Location of Subsurface Safety Valves," *SPEDE* (December 1991) 285.
25. Rocha, L.A. and Bourgoyne, A.T.: "Identifying Crater Potential Improves Shallow Gas Kick Control," *Oil and Gas J.* (27 December 1993) 93.
26. Eaton, B.A. and Eaton, T.L.: "Fracture Gradient Prediction for the New Generation," *World Oil* (October 1997) 93.
27. Aadnoy, B.S., Soteland, T., and Ellingson, B.: "Casing Point Selection at Shallow Depth," paper SPE/IADC 18718 presented at the 1989 SPE/IADC Drilling Conference, New Orleans, 28 February–3 March.
28. Constant, W.D. and Bourgoyne, A.T.: "Fracture-Gradient Prediction for Offshore Wells," *SPEDE* (June 1988) 136.
29. Rocha, L.A. and Bourgoyne, A.T.: "A New Simple Method to Estimate Fracture Pressure Gradient," paper SPE 28710 presented at the SPE International Petroleum Conference & Exhibition, Veracruz, Mexico, 10–13 October 1994.
30. Black, D. and Laurie, A.M.: "Control of Shallow Gas Kicks: A Case History," paper presented at the IADC 1995 Asia Pacific Well Control Conference, Jakarta, 29 November.
31. Baron, S. and Skarstol, S.: "New Method Determines Optimum Surface Casing Depth," *Oil and Gas J.* (7 February 1994) 51.
32. Bourgoyne, A.T., Kelly, O.A., and Sandoz, C.L.: "New Ideas for Shallow Gas Well Control," *World Oil* (June 1996) 50.
33. Barker, J.W. and Wood, T.D.: "Estimating Shallow Below Mudline Deepwater Gulf of Mexico Fracture Gradients," presented at the Houston AADE Chapter Annual Technical Forum, 2–3 April 1997.
34. Eaton, B.A.: "Fracture Gradient Prediction and Its Application in Oilfield Applications," *JPT* (October 1969) 1353; *Trans.*, AIME, **246.**
35. Goins, W.C. Jr. and Ables, G.L.: "The Cause of Shallow Gas Kicks," paper SPE/IADC 16128 presented at the 1987 SPE/IADC Drilling Conference, New Orleans, 15–18 March.
36. Sandlin, C.W.: "Drilling Safely Offshore in Shallow Gas Areas," paper SPE 15897 presented at the SPE European Petroleum Conference, London, 20–22 October 1986.
37. Beck, F.E., Langlinais, J.P., and Bourgoyne, A.T. Jr.: "An Analysis of the Design Loads Placed on a Well by a Diverter System," paper SPE/IADC 16129 presented at the 1987 SPE/IADC Drilling Conference, New Orleans, 15–18 March.
38. Koederitz, WL. *et al.*: "Method for Determining the Feasibility of Dynamic Kill of Shallow Gas Flows," paper SPE 16691 presented at the 1987 SPE Annual Technical Conference and Exhibition, Dallas, 27–30 September.
39. Starrett, M.P., Hill, A.D., and Sepehrnoori, K.: "A Shallow-Gas-Kick Simulator Including Diverter Performance," *SPEDE* (March 1990) 79.
40. Bourgoyne, A.T. Jr. and Abel, L.W.: "Two-Phase Modeling Improves Diverter Designs for Shallow Gas Hazards," *Oil and Gas J.* (24 July 1995) 29.
41. Murray, S.J. *et al.*: "Well Design for Shallow Gas," paper SPE/IADC 29343 presented at the 1995 SPE/IADC Drilling Conference, Amsterdam, 28 February–2 March.
42. Beck, F.E., Langlinais, J.P., and Bourgoyne, A.T. Jr.: "Experimental and Theoretical Considerations for Diverter Evaluation and Design," paper SPE 15111 presented at the 1986 SPE California Regional Meeting, Oakland, 2–4 April.
43. Bourgoyne, A.T. Jr.: "Experimental Study of Erosion in Diverter Systems Due to Sand Production," paper SPE/IADC 18716 presented at the 1989 SPE/IADC Drilling Conference, New Orleans, 28 February–3 March.
44. Santos, D.L. and Bourgoyne, A.T. Jr: "Estimation of Pressure Peaks Occurring When Diverting Shallow Gas," paper SPE 19559 presented at the 1989 SPE Annual Technical Conference and Exhibition, San Antonio, 8–11 October.
45. "Recommended Practices for Diverter Systems Equipment and Operations," *RP 64*, first edition, API, Washington, DC (1 July 1991).
46. Crawley, F.K. and Thorogood, J.L.: "Single Vent-Line Design Selected for Diverter System," *Oil and Gas J.* (14 September 1987) 43.
47. Blount, E.M. and Soeiinah, E.: "Dynamic Kill: Controlling Wild Wells a New Way," *World Oil* (October 1981) 109–26.
48. Westergaard, R.H.: *All About Blowout*, Norwegian Oil Review, Oslo (1987) 39.
49. Danenburger, E.P.: "Outer Continental Shelf Oil and Gas Blowouts," *Drilling-DCW* (August 1980) 48.
50. Nokleberg, L., Schulle, R.B. and Sontvedt, T.: "Shallow Gas Kicks, Safety Aspects Related to Diverter System," paper SPE 16545 presented at Offshore Europe 87, Aberdeen, 8–11 September 1987.
51. Roche, J.R.: "Better Diverter System Designs Stressed" *Oil and Gas J.* (7 March 1988) 37.
52. Mills, D. and Dyhr, E.: "Larger Diverters Safer for Shallow Gas Control," *Oil and Gas J.* (2 December 1991) 65.
53. Coberly, C.J. and Wagner, E.M.: "Some Considerations in the Selection and Installation of Gravel Pack for Oil Wells," *Well*

Completions, Reprint Series, SPE, Richardson, Texas (1970) **5**, 158.
54. Beall, J.E.: "Alternative Well Control Method Offerred," *Drilling Contractor* (January 1985) 45.
55. "Subsea Diverters Handle Shallow Gas Kicks," *Ocean Industry* (November 1986) 41.
56. Klavenes, P.T. and Leistad, G.H.: "Two Years Experience with Use of Seabed Diverter," paper SPE 16544 presented at the Offshore Europe 87 Conference, Aberdeen, 8–11 September 1987.
57. Milgram, J.H. and McLaren, W.G.: "The Response of Floating Platforms to Subsea Blowouts," MIT Dept. of Ocean Engineering Report No. 82-8 (July 1982).
58. Milgram, J.H. and Erb, P.R.: "How Floaters Respond to Subsea Blowouts," *Pet. Eng. Intl.* (June 1984) 64.
59. Hammett, D.S.: "Drill Vessels Float in Aerated Water," paper SPE 13994 presented at the Offshore Europe 85 Conference, Aberdeen, 10–13 September 1985.
60. Løes, M. and Fanneløp, T.K.: "Concentration Measurements Above an Underwater Release of Natural Gas," *SPEDE* (June 1989) 171.
61. Adams, N.J. and Kuhlman, L.G.: "Shallow Gas Blowout Kill Operations," paper SPE 21455 presented at the SPE Middle East Oil Show, Bahrain, 16–19 November 1991.
62. Askeland, R. and Nergaard, A.: "New Inflatable BOP Above Bit Contains Shallow Gas Kicks," *Ocean Industry* (April 1989) 31.
63. Erb, P.R. and Stockinger, M.P.: "Riser Collapse—A Unique Problem in Deep-Water Drilling," paper SPE/IADC 11394 presented at the 1983 SPE/IADC Drilling Conference, New Orleans, 20–23 February.
64. Lopes, C.A. and Bourgoyne, A.T. Jr.: "Feasibility Study of a Dual-Density Mud System for Deepwater Drilling Operations," paper SPE 39155 presented at the 1997 Offshore Technology Conference, Houston, 5–8 May. Synopsis published in *JPT* (November 1997) 1216–17.
65. Medley, G.H. Jr. *et al.*: "Field Application of Light-Weight Hollow-Glass-Sphere Drilling Fluid," paper SPE 38637 presented at the 1997 SPE Annual Technical Conference and Exhibition, San Antonio, 5–8 October Synopsis published in *JPT* (November 1997) 1209–11.
66. Gault, A.: "Riserless Drilling: Circumventing the Size/Cost Cycle in Deepwater," *Offshore* (May 1996) 49.
67. Choe, J. and Juvkam-Wold, H.C.: "Riserless Drilling Concepts, Applications, Advantages, Disadvantages, and Limitations," 1997 CADE/CAODC Drilling Conference, Calgary, Alberta, 8–10 April.
68. Choe, J.: "Analysis of Riserless Drilling and Well-Control Hydraulics," *SPEDC* (March 1999) 71.
69. *IADC Deepwater Well Control Guidelines*, International Association of Drilling Contractors, Houston (1998) 33.
70. Furlow, W.: "Shallow Water Flows: How They Develop; What to do About Them," *Offshore* (September 1998) 70.
71. Eoff, J. and Griffith, J.: "Acrylate Monomer Solution Stops Artesian Water, Geopressured Sand Flows," *Oil and Gas J.* (2 November 1998) 89.
72. Medley, G.H. Jr.: "Shallow Water Flow: A Technology Update," *Deepwater Technology* supplement to *World Oil* (August 1998) 37.
73. Brennan, R.M. and Annis, M.R.: "A New Fracture Gradient Prediction Technique That Shows Good Results in Gulf of Mexico Abnormal Pressure," paper SPE 13210 presented at the 1984 SPE Annual Technical Conference and Exhibition, Houston, 16–19 September.
74. Bourgoyne, A.T. Jr. *et al.*: *Applied Drilling Engineering*, second printing, Textbook Series, SPE, Richardson, Texas (1991) 117.

SI Metric Conversion Factors

bbl	× 1.589 873	E − 01	= m^3
cp	× 1.0*	E + 03	= Pa·s
darcy	× 9.869 233	E − 01	= μm^2
ft	× 3.048*	E − 01	= m
ft^3	× 2.831 685	E − 02	= m^3
°F	(°F − 32)/1.8		= °C
gal	× 3.785 412	E + 00	= L
in.	× 2.54*	E + 01	= mm
$in.^2$	× 6.451 6*	E + 02	= mm^2
lbf	× 4.448 222	E + 00	= N
lbm	× 4.535 924	E − 01	= kg
psi	× 6.894 757	E − 03	= MPa
psi^{-1}	× 1.450 377	E − 01	= kPa^{-1}

* Conversion factor is exact.

Chapter 9
Snubbing and Stripping

9.1 Introduction

Snubbing and stripping operations involve introducing or removing tubulars from a well under surface pressure. Snubbing refers to the operation where the pressure force acting on the exposed pipe area exceeds the string weight. An external force is applied to counteract this force and the workstring is pushed into the hole or restrained out of the hole. The state is designated commonly in the industry as a "pipe-light" condition. Pipe is stripped when the string enters the well because of its weight, hence stripping refers to a "pipe-heavy" condition. Either one or both of the techniques may be used in a given job.

Traditionally, snubbing and stripping indicate a well in trouble. Though blowouts and other less serious well-control problems are important applications, there are some advantages to using the techniques as an alternate method for drilling and completing wells. **Table 9.1** lists some common situations where working on a live well is necessary or desirable.

The ability to drill and trip from an underbalanced hole is one benefit that is receiving recent attention.[1,2] Snubbing economics cannot compete with conventional rotary rigs in most cases, but a higher cost per foot across the production hole may be justified by eliminating skin damage. On some offshore platforms, it may be more economical to drill with a snubbing unit than to bring in a jackup rig because hydraulic units are small in comparison to a drilling or workover rig and the weight of the equipment is carried by the wellhead.[3]

Snubbing and stripping have become routine completion practices in many areas because of the demonstrated effect on productivity.[4,5] Other savings from not killing a well include eliminating the use of expensive weighted brines and the associated cost to transport, store, and pump these fluids. The size and portability of these units have created another application on small platforms with limited load-bearing capacity or when it is difficult to mobilize a conventional rig into a jungle or otherwise inaccessible location.[6]

The first part of this chapter discusses the different types of snubbing units and workstring and tool considerations. Blowout prevention equipment (BOPE) systems and procedures for moving pipe through the surface sealing devices are covered next, followed by some important predictions related to planning a snubbing job. The chapter concludes with a discussion of the volumetric principles associated with moving pipe in or out of a pressurized well.

9.2 Equipment and Procedures

Specialized equipment and trained personnel are needed to carry out a snubbing job, but working under pressure need not be a hazardous undertaking if the details are thought out by the operator and service companies before the job begins. How the objectives are met by the equipment, major design considerations, operational procedures, and safety issues should be understood thoroughly by an operator who implements this well-intervention technique.

9.2.1 Snubbing Units. There are three basic types of snubbing units. Cable-type rig-assist units have been around since the late 1920s and are the oldest design. The downward thrust is achieved by mechanical action with power transmission through the rig drawworks, hence the origin of the term "rig-assist." A schematic showing how the unit works is given in **Fig. 9.1.** A platform is attached to the snubbing blowout preventer (BOP) stack and a set of sheaves are incorporated into the structure. Cables are strung from the traveling slips down through the platform sheaves and back up to the blocks.

To snub a joint in, the traveling slips grip the pipe, the blocks are pulled upward, and a reverse motion is translated to the pipe. The stationary slips are closed when the tool joint or collar is positioned; the traveling slips are opened, and the counterbalance weights pull the traveling slips up as the blocks are lowered. The unit is not required once a pipe-heavy condition is achieved and the rig traveling blocks and elevators are used for the stripping procedure.

Hydraulic snubbing units, also called hydraulic workover (HWO) units, came into being in the early 1960s. **Fig. 9.2** illustrates the equipment. **Table 9.2** gives a description of some of the available models. The components that form the HWO system are trailer-mounted on land jobs or packaged in modular skids for offshore applications.

Hydraulic pumps driven by a diesel engine boost power fluid to a pressure of 3,000 psi. Output pressure is adjusted to

TABLE 9.1—SNUBBING AND STRIPPING APPLICATIONS
Drilling Operations
1. Running the drillstring into a well following an off-bottom kick.
2. Retrieving a drillstring where a leak or blockage prevents pumping a kill.
3. Surface or underground blowout control.
4. Underbalanced drilling or washing operations.
5. Drilling applications where equipment portability is a dominant factor.
6. Drilling applications where a snubbing unit's load placement characteristics are an advantage.
Completion and Workover Operations
1. Live-well operations to keep from having to kill the well and damage the formation.
2. Through-tubing jobs where the depth or pressure exceeds coiled-tubing capabilities.
3. Well work where equipment portability is a dominant factor.
4. Well work where a snubbing unit's load placement characteristics are an advantage.
5. Blowout control.

the function needs by a series of regulators and dump valves. The main system pressure is directed to the jack assembly where hydraulic cylinders generate the snub or lift force. BOP control pressure operates the slips, preventers, and remote-actuated valves. Counterbalance pressure is used to move joints from the ground or deck and the work basket (or vice versa). Another power/control circuit may be used to operate auxiliary functions such as the power tongs.

As shown in **Fig. 9.3,** snubbing is accomplished by applying hydraulic pressure to the rod side of the piston whereas the weight of a pipe-heavy string is counteracted by pressure acting on the opposite side. The hydraulic pressure to produce a given snub force is described by

$$p_h = \frac{F_s}{(A_c - A_r)N_c}, \quad \text{(9.1)}$$

where F_s = snub force, A_c = cylinder cross-sectional area, A_r = piston rod cross-sectional area, and N_c = number of active cylinders.

Similarly, the hydraulic pressure to achieve the lift force F_l is given by

$$p_h = F_l/A_cN_c. \quad \text{(9.2)}$$

The operator may need to limit the maximum jack force to keep from exceeding the compressive (i.e., buckling) or tensile limitations of the workstring. It is common to set the lift pressure based on 80% of the pipe tension rating and the snub pressure for a maximum thrust equal to 70% of the calculated buckling load.

Jack speed is limited by the pump rate available from the power pack at the output pressure. When snubbing pipe into a well,

$$v_{js} = \frac{q_p}{(A_c - A_r)N_c}, \quad \text{(9.3)}$$

where v_{js} = the snubbing jack velocity and q_p = the flow rate from the power pack. The speed when lifting the work string, v_{jl}, is given by

$$v_{jl} = q_p/(A_cN). \quad \text{(9.4)}$$

Example 9.1. The respective cylinder and piston rod diameters for a hydraulic jack are 5.0 and 3.5 in.

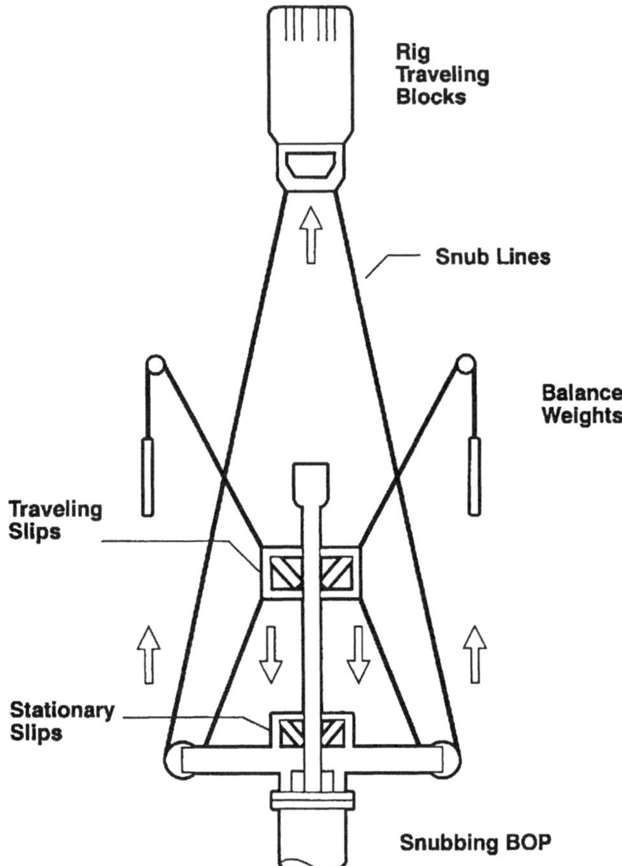

Fig. 9.1—Cable-type rig-assist snubbing equipment.

1. How much hydraulic pressure must be supplied to the jack to achieve a 20,000-lbf snubbing force if two cylinders are used?

2. Determine the snubbing jack speed if the pumps can deliver 125 gal/min at the power pack's output pressure.

Solution. 1. First calculate the area terms.

$$A_c = \pi(5.0)^2/4 = 19.63 \text{ in.}^2$$

and $A_r = \pi(3.5)^2/4 = 9.62 \text{ in.}^2$

Substitution into Eq. 9.1 yields

$$p_h = \frac{20,000}{(19.63 - 9.62)(2)} = 999 \text{ psig.}$$

2. The jack speed is obtained after inserting variables and conversion constants into Eq. 9.3.

$$v_{js} = \frac{(125)(144)}{(19.63 - 9.62)(2)(7.48)} = 120 \text{ ft/min.}$$

Most four-leg jack assemblies give the option of using two cylinders, which allows for a faster jack speed but reduces the load capacity by half. Another way to increase jack speed when pulling pipe from a well is to use a regenerative circuit. In this method, power fluid is circulated from the snub side of the piston to the lift side; the effect is to double the available flow rate and, hence, jack speed. However, the pressure now acts across the rod area reducing the lift capacity according to the relation,

$$F_l = p_hA_rN. \quad \text{(9.5)}$$

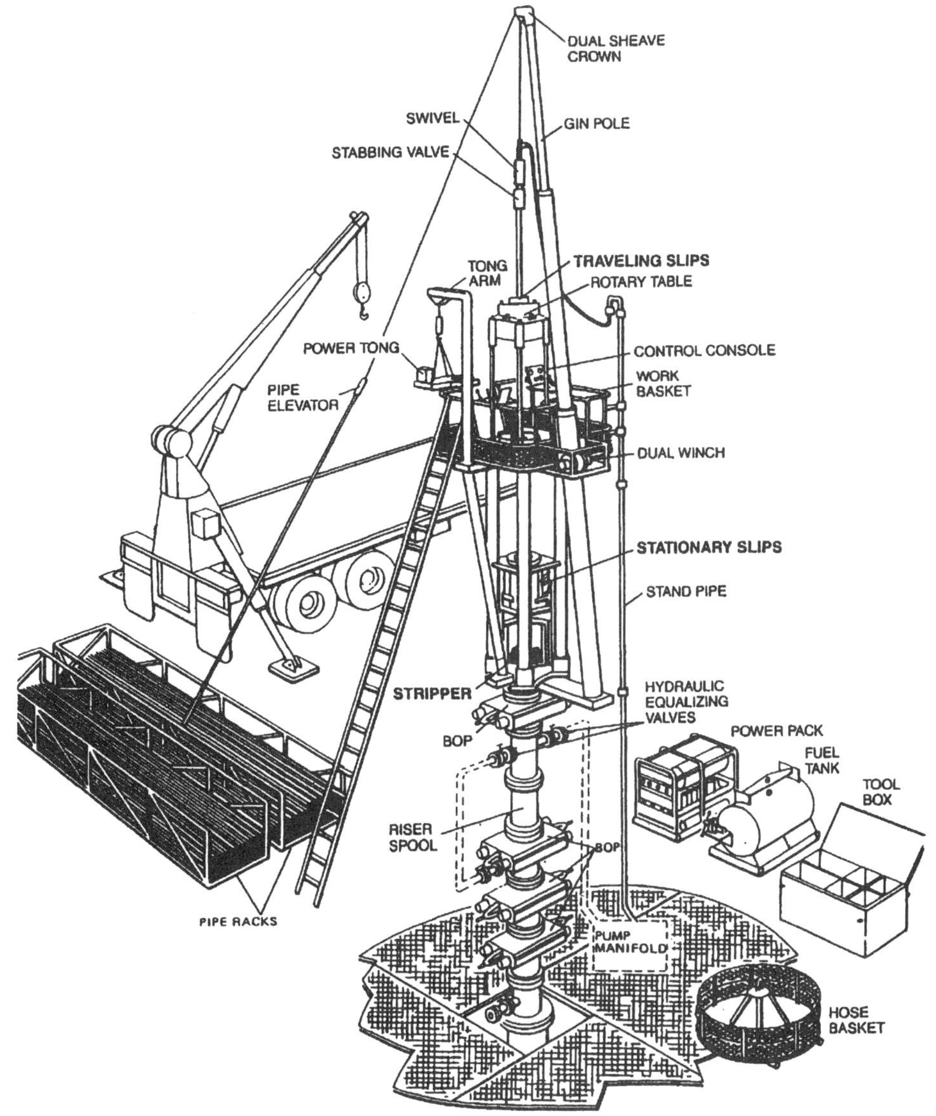

Fig. 9.2—Hydraulic snubbing unit and associated equipment. Courtesy of Cudd Pressure Control.

The stationary and traveling slips are considered part of the jack system. The stationaries are installed near the base of the jack frame and include two slip bowl assemblies—one for use in the pipe-light condition and the other for use when the pipe is heavy. Both bowls can be equipped to hold in the same direction if the pipe-heavy load risks damaging the pipe with only one set of slips.

An access window immediately below the stationaries is required if any downhole tools are larger than the jack bore. Other situations where a window might be necessary include strapping electrical submersible pump (ESP) cables or hydraulic control lines to a production tubing string. Buckling the work string during a snubbing job is a possibility if the pipe can bow or move laterally and the window is one area of concern. For this reason, removable pipe guides are installed in the window to provide lateral constraint.

The traveling-slip bowl assembly is attached to the piston rods and so the slips move up or down with the piston. Travelers are designed to be detached and inverted when pipe goes from the light to heavy condition. Two traveling assemblies

TABLE 9.2—PERFORMANCE DATA FOR REPRESENTATIVE HYDRAULIC SNUBBING UNITS

Model	120*	150**	200*	225**	340**	400*	600 (1)*	600 (2)**
Rated lift load (lbf)	117,810	150,720	199,110	235,560	340,000	381,690	580,770	600,000
Rated snub load (lbf)	57,727	65,940	95,430	120,000	188,400	199,080	294,510	260,000
Stroke length (in.)	120	116	120	116	116	120	120	168
Number of cylinders	2	4	4	4	4	2	4	4
Bore through jack (in.)	4.06	7.06	7.06	11.06	11.06	11.12	11.12	13.63
Operating pressure (psi)	3,000	3,000	3,000	3,000	3,000	3,000	3,000	3,000
Rotary torque (ft/lbf)	2,235	1,000	4,890	2,800	2,800	6,437	11,500	6,437
Jack weight (lbf)	4,725	5,800	7,900	8,500	16,000	13,750	34,000	17,750

*Data courtesy of Halliburton.

**Data courtesy of Cudd Pressure Control.

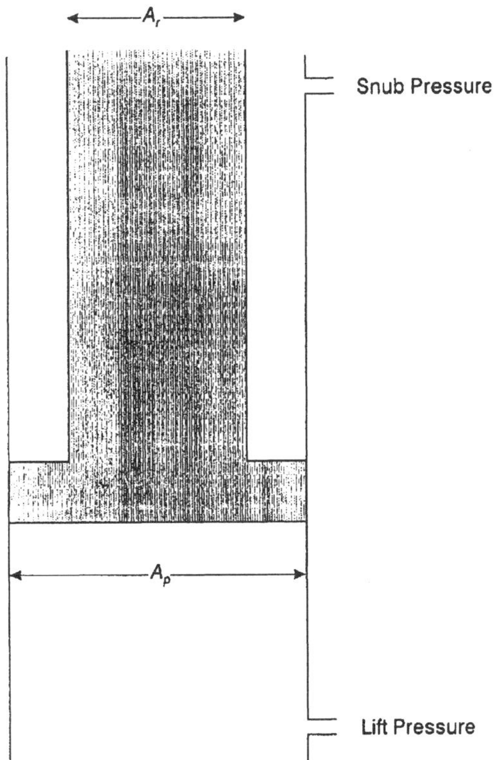

Fig. 9.3—Snub- and lift-force generation in a jack cylinder.

can be installed and it is fairly common to run dual opposed slips to keep from having to make the changeover at the balance point (i.e., the depth where the pipe is neutral). Dual opposed slips are necessary when working stuck pipe that would otherwise be light.

The maximum stroke length on most HWO units is approximately 10 ft, though some larger units may allow a 14-ft stroke and small units may have something shorter. A telescoping guide tube in the jack bore from the traveling assembly to the top of the stationary slips stabilizes the snubbed string across the stroke length and prevents buckling. The inner diameter (ID) of the guide tube should be as small as the coupling outer diameter (OD) allows.

String rotation can be accomplished with most HWO units by using a rotating head integral to the traveling assembly. The rotating head is driven by the hydraulic fluid with torque controlled by the hydraulic pressure and rotary speed by the volumetric flow rate. Standard practice limits maximum rotary torque to the lowest makeup torque in the work string.

The work basket is attached to the top of the jack frame and is the primary control center for the operation. A basket has two control panels, one for the jack functions and one for operating the BOPs and other components. The jack operator is in charge of throttling power to the jack and operating the snub/lift functions, slips, and rotary. The operator's console also contains gauges to monitor hydraulic pressure for each of the components and a weight indicator. Another snubbing specialist called the "helper" works at the other panel. The helper's duties include operating the preventers, equalizing loop, vent line, and pipe-handling winches.

Scaffolding, ladders, or a spiral staircase are the normal means to access the work basket. A snubbing job carries some inherent risks and, because the basket may be more than 100 ft above ground or deck level, a safe way to evacuate the personnel quickly should be provided. Various emergency escape devices have been used, including a nearby crane and personnel basket, Geronimo lines, and slides. Improvements to these methods, notably with regard to fire protection and the ability to remove injured personnel safely, have been incorporated by means of enclosed escape pods.[7]

TABLE 9.3—PERFORMANCE DATA* FOR A MODEL 170 HYDRAULIC RIG-ASSIST UNIT

Rated lift load	169,600 lbf
Rated snub load	94,200 lbf
Maximum stroke length	10 ft
Number of cylinders	2
Bore through jack	7.06 in.
Operating pressure	3,000 psi

*Data courtesy of Cudd Pressure Control.

The third type of snubbing system is the hydraulic rig-assist unit. **Table 9.3** lists performance characteristics of one model. These units have supplanted mechanical units to the point that many people think of a hydraulic operation when hearing the term "rig-assist." Hydraulic rig-assist units are self-contained, compact, and designed for quick rig up and rig down. They work in conjunction with a drilling or workover rig and save considerable time by using the rig's drawworks and traveling system to handle the workstring and to trip when the pipe is heavy. Hydraulic jacks are used when a pipe-light condition is reached. Though similar to a conventional HWO unit, this system does not have the telescoping guide tube and the allowable stroke length may be limited by the buckling criteria.

9.2.2 Workstring and Downhole Tools. Pipe failure could lead to catastrophic results. Workstring selection is an especially critical part of the job planning. Conventional design factors for a workstring are 1.25 for tension, collapse, and minimum internal yield pressure (MIYP). Alternatively, the pipe is selected or well conditions controlled so that the described loads do not exceed 80% of rating. Compressive loads typically are limited to 70% of what it will take theoretically to buckle the pipe.

Higher strengths and thick-walled tubes are recommended if well conditions and job objectives allow. The presence of H_2S will influence material selection. Proprietary connections incorporating metal-to-metal seal surfaces and redundant seal capabilities are used often. A comprehensive inspection of the tubing and connections should be performed before the job begins and, depending on the service conditions, on a periodic basis thereafter.

Almost any downhole tool used in a conventional drilling or workover operation can be used. However, flow control in the workstring bore is an added consideration when working under pressure. Backpressure valves normally are placed near the bottom of the string if the job procedure includes circulation. Slickline-retrievable plugs are run sometimes in other cases such as running production tubing. Most operators will run at least two backpressure valves or plugs for redundancy and a minimum of two stabbing valves for each connection size/type should be accessible in the work basket. Consideration should be given to installing a landing nipple above the uppermost valve or plug. A pump-down or slickline-set plug can be seated in the nipple and the workstring retrieved if the existing devices fail.

9.2.3 BOP Equipment and Operation. Fig. 9.4 shows a typical BOP stack where a single pipe diameter will be snubbed in a well with working pressures between 5,000 and 10,000 psi. Lower well pressures may not require as many rams and

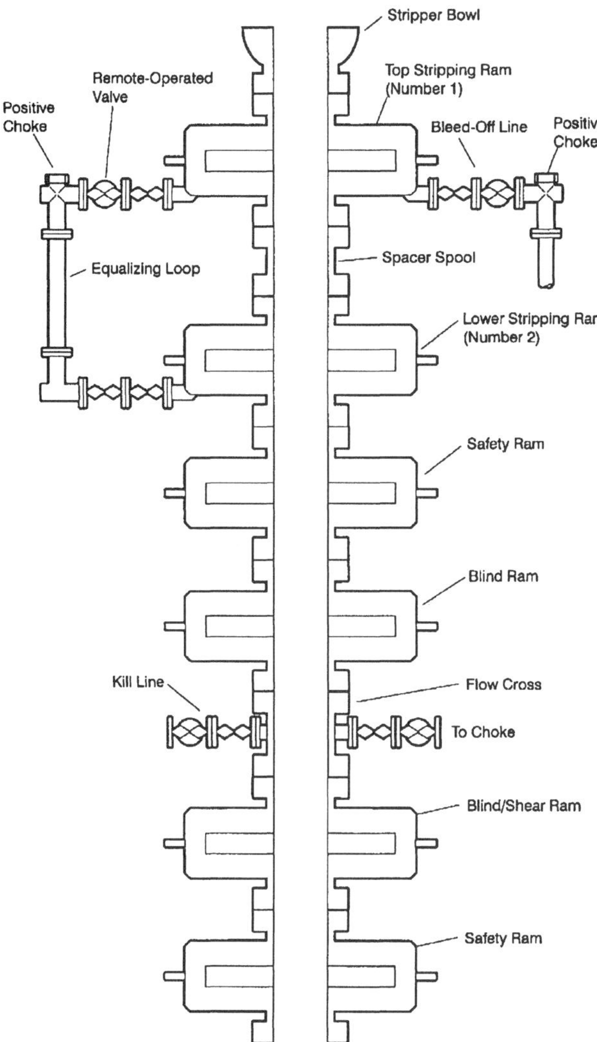

Fig. 9.4—Typical BOP and flow-control equipment for a 10M snubbing operation.

may be single-valved on the side outlets. Pressures greater than 10,00 psi or low-pressure work on a sour well may have added redundancy.

A stripper assembly is located in the bottom part of the window area. The bowl houses a steel-reinforced rubber element that can be used to seal against relatively low pressure. Moving pipe through the stripper rubber is preferred if the conditions allow because the operation is much simpler and faster than a ram-to-ram procedure. The maximum working pressure of a stripper rubber is normally less than 2,000 or 3,000 psi, though a 5,000-psi stripper has been used in the field.[8]

In most designs, packoff is achieved by the well pressure forcing the element against the pipe. Some alternative systems use hydraulic pressure to maintain or supplement the sealing force. The life of any stripper rubber is finite and certain conditions can drastically reduce the time before the operation must be shut down and the element changed. Well pressure, pipe condition, and the type of fluids in the well are factors affecting the stripper-rubber lifespan. The worst case for element wear would be stripping rough pipe through a gas well having surface pressures in excess of 1,500 psi. Lubrication will extend element life under these types of adverse conditions.

Higher pressures, excessive stripper wear, or collared pipe dictate the use of the stripping rams. In a ram-to-ram operation the stripper still serves as a pipe wiper and keeps junk from falling into the well. The needle valve or bull plug between the stripper and top ram must be removed to avoid trapping any pressure below the rubber. Alternatively, the rubber can be replaced with a nonsealing fluted pipe guide.

The top and bottom stripping rams are called the No. 1 and 2 rams, respectively. These are conventional ram preventers with special seal inserts designed to resist wear during pipe movement. A spacer spool long enough to enclose the pipe connections or tools in the bottomhole assembly is required between the top and bottom stripping rams. Buckling failure above the working ram is always a consideration and the pipe should be constrained as much as possible by sizing the BOP and spacer spool to the minimum bore needed to accomplish the job.

The equalizing loop provides the means to equalize pressures above and below the No. 2 ram. The bleed line directs these fluids away from the work site in a controlled manner to the flare or collection pit. The manual valves shown in the equipment schematic are open during the procedure; the helper opens or isolates the two sides by operating the remote-actuated valves from the work basket. A positive choke is needed in the equalizing loop to slow the flow velocity and prevent valve erosion and surge forces on the stack. Normally a positive choke also is included in the bleed line.

One or more conventional pipe rams called "safeties" are located below the lower stripping ram. A safety is used to close in the well whenever the stripper inserts need to be replaced or other repairs made on the overlying components. At least one safety must be included in the stack for each pipe diameter. Also, the stack or tree must contain at least two blind rams or gate valves to secure the well when pipe is out of the hole. As indicated, a blind/shear ram below the flow cross is recommended in a high-pressure operation.

A slip block is a unique type of ram block sometimes used in snubbing. A slip block does not contain any sealing material; its function is to contact and hold the workstring in a stationary position. The gripping teeth are cut in such a way that they will hold either a pipe-light or pipe-heavy workstring.

A means to release well fluids from below the working rams is necessary. As shown on the diagram, a flow cross and choke line below the top blind ram provide this function. A kill line is placed on the opposite side for emergencies or for routine wellbore-fluid replacement. Another access point may be included on a sour gas well where nitrogen or natural gas are pumped to maintain a blanket in the wellbore. This keeps the toxic, corrosive substance away from the personnel and surface equipment during the operation and allows some time for the crews to get away should control be lost.[9]

Snubbing or stripping pipe through the rams is concentrated around getting the tool joints through the stack, an operation that requires coordination between the jack operator (or driller) and the snubbing helper. Take an arbitrary initial condition as depicted in **Fig. 9.5a** where the tool joint or connection is positioned between the two stripping rams. The well pressure, represented by the shading, is contained by the No. 2 stripping ram and the closed valve on the equalizing loop. The procedure for getting the next tool joint in the same position is described in **Table 9.4** and illustrated in **Figs. 9.5b through 9.5e.**

Snubbing with high-pressure gas underneath the stack not only increases safety concerns, but also presents other problems such as difficulties in maintaining a seal, reduced element life, and the potential for hydrate plugging in the stack

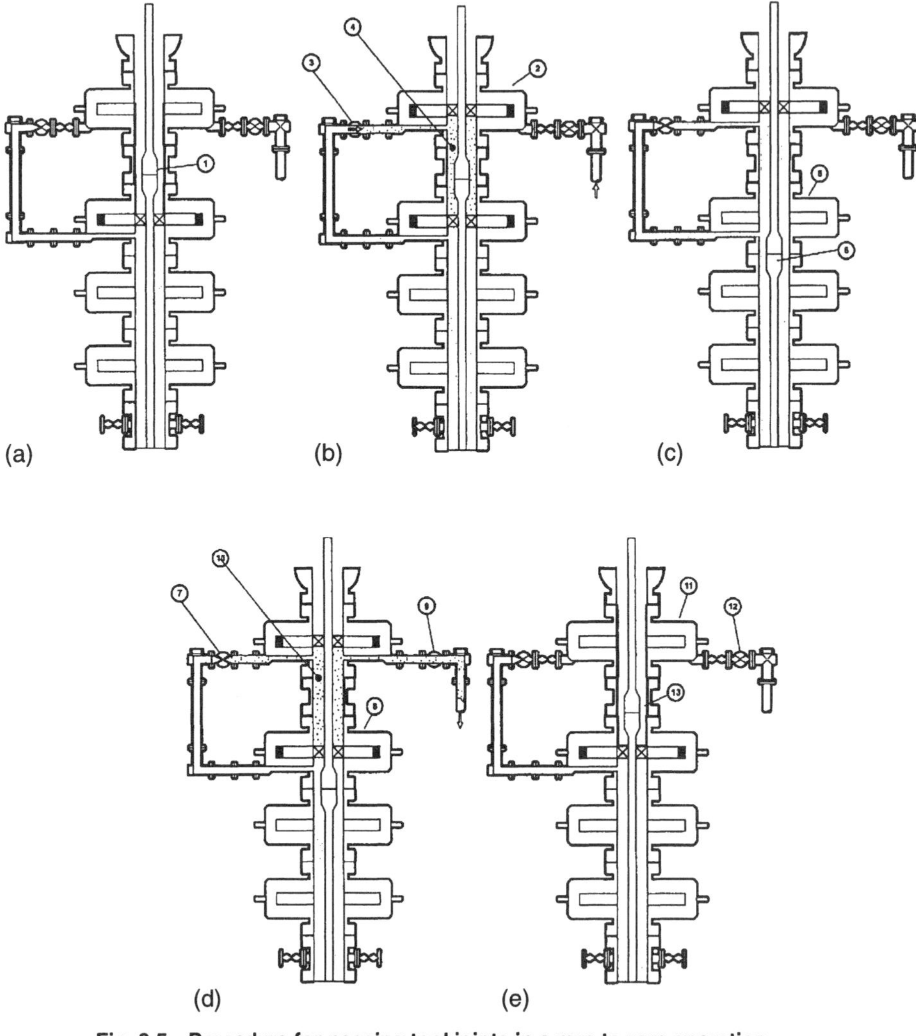

Fig. 9.5—Procedure for passing tool joints in a ram-to-ram operation.

TABLE 9.4—RAM-TO-RAM SNUBBING AND STRIPPING PROCEDURE	
1.	Snub/strip with No. 2 ram closed until the tool joint is about midway between the two stripping rams.
2.	Stop string movement and close the No. 1 ram.
3.	Open the valve on the equalizing loop with the bleed-off line closed.
4.	Allow pressure between the two rams to equalize.
5.	Open the No. 2 ram.
6.	Snub/strip against the No. 1 ram until the tool joint is below the No. 2 ram.
7.	Close the equalizing-loop valve.
8.	Stop string movement and close the No. 2 ram.
9.	Open the bleed-line valve.
10.	Allow pressure between the two rams to dissipate.
11.	Open the No. 1 ram.
12.	Close the bleed line.
13.	Repeat the procedure.

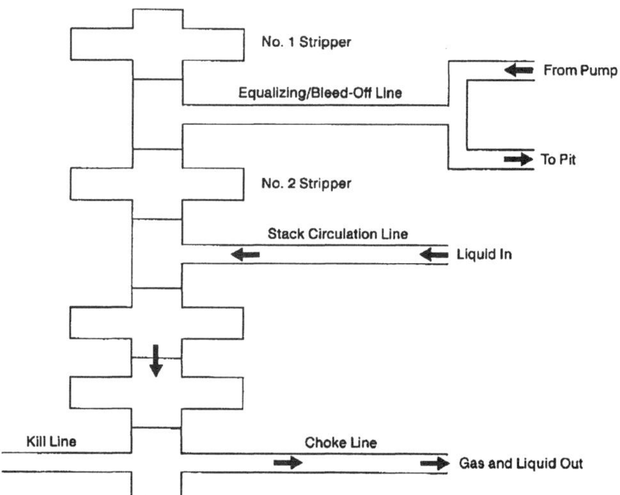

Fig. 9.6—Equipment schematic for a snubbing operation where the stack is circulated continuously.

or lines. Lawson[10] suggested a technique for eliminating these problems by maintaining a water or otherwise "friendly" fluid in the stack at all times. **Fig. 9.6** illustrates the basic flow diagram for the method.

Gas is kept away from the stripping rams by continuously pumping the designated circulation fluid into a line below the No. 2 ram. The fluid and wellbore gas exit through the choke line where the desired surface pressure is maintained by the adjustable choke. Note that the equalizing loop is left out of this arrangement. Instead of equalizing with well fluids, the space between the two strippers is pressured up with a pump. The same line is used to bleed this pressure off before opening the No. 1 ram.

9.2.4 Stripping With Rig Equipment. As in snubbing, a stripping procedure with a drilling or workover rig involves

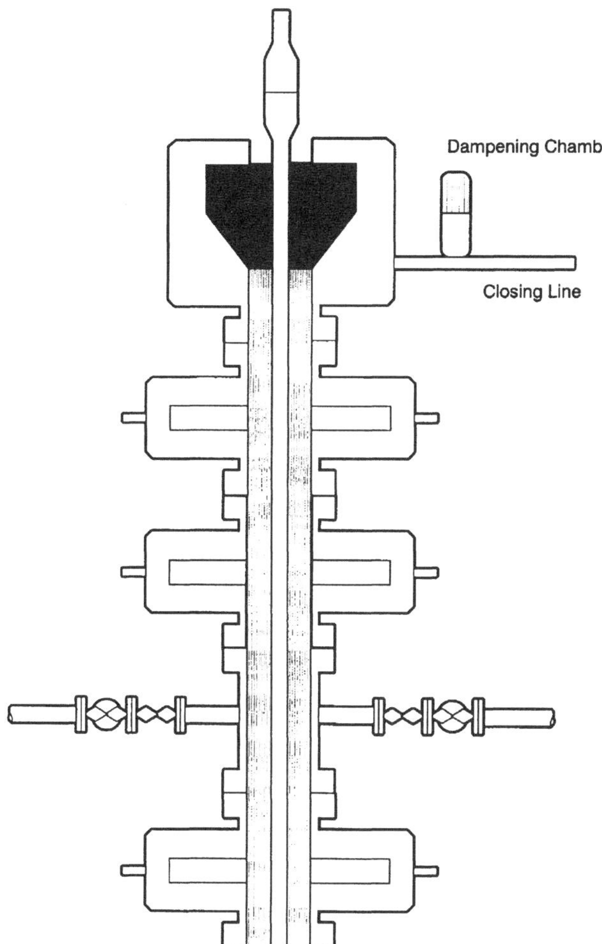

Fig. 9.7—Using a surge chamber to accommodate tool joint expansion in a stripping operation.

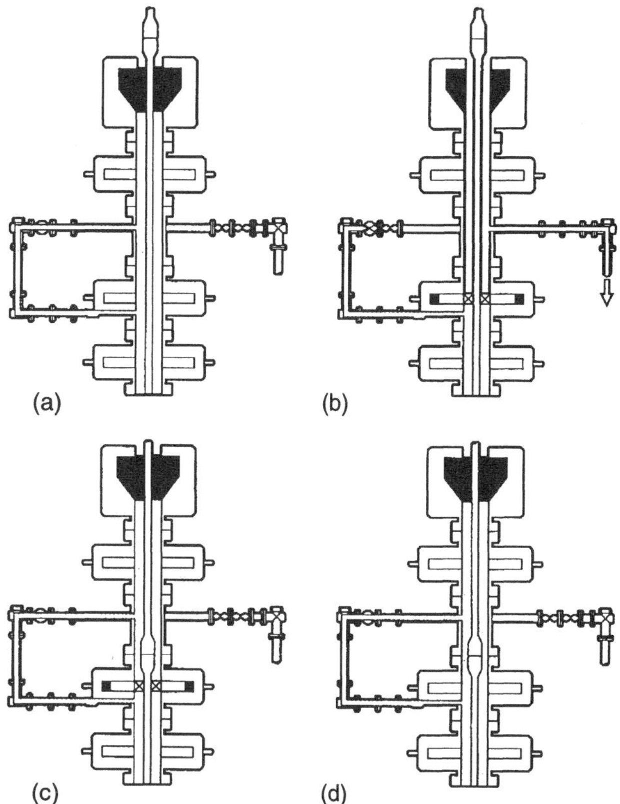

Fig. 9.8—Passing tool joints in an annular-to-ram operation.

getting tool joints in or out of a pressurized wellbore in a controlled manner. If conditions are suitable, stripping through the annular preventer is preferred because tool joints can be stripped into the closed annular. The technique is much simpler than a ram-to-ram procedure, but several factors must be considered.

Preserving the seal element is the major concern. Recommended procedure is to close in the well initially using the minimum closing pressure recommended by the manufacturer. Pressure then can be backed off until a small amount of seepage is observed between the pipe and seal. A rule of thumb holds that 1 to 2 barrels of leakage per 1,000 feet of stripped pipe is sufficient. This fluid should be released into a measuring tank and the closing pressure adjusted during the job so that the target seepage rate is maintained. Placing a gel-water mixture on top of the preventer for added lubrication is recommended when pipe is being stripped into the hole.

Failure to adjust the closing pressure to accommodate packing element expansion across a tool joint will cause the pressure in the closing chamber to increase, resulting in excessive element stress and a greatly curtailed life. A regulator on the closing line set to maintain a constant closing pressure achieves this end. However, sluggish reaction time and closing pressure fluctuations associated with this technique leave much to be desired.

The preferred method is to install a surge or dampening pot on the closing line adjacent to the preventer. **Fig. 9.7** illustrates the concept. The dampening cylinder is simply a precharged accumulator bottle. When a tool joint enters the annular, the relatively incompressible closing fluid enters the chamber, compresses the nitrogen, and results in a small pressure increase rather than a dramatic one.

The string should be run or retrieved at a slow rate to allow for better heat dissipation. The running or hoisting speed should be reduced even more when passing a tool joint. It is important to have an accurate measurement of the distance between the annular preventer and a given reference plane (the rotary table for instance) so that the driller knows when the tool joint is about to enter the preventer.

It may be necessary to use two ram preventers or an annular/ram combination if well pressures are too high to strip with one annular. It must be understood that the packing element in a ram preventer will eventually wear out, particularly if wear-resistant inserts are not installed, and it is imperative that at least one pipe ram below the bottom stripping ram be held in reserve for well closure. The distances between each packing element in the stack must be known and the pipe must be carefully marked and measured to ensure the correct tool-joint placement.

An equalizing loop and bleed-off line must be installed for an annular-to-ram or ram-to-ram operation. So equipped, the sequence for passing a tool joint is similar to the procedure described for the snubbing stack. **Figs. 9.8a through 9.8d** illustrate the steps in an annular-to-ram technique starting with a tool joint above the closed annular. In Fig. 9.5b, the working ram is closed and the pressure between the closed ram and annular is bled to the measuring tank. The next step, shown in Fig. 9.5c, is to open the annular preventer and strip through the ram until the tool joint is below the annular. The annular preventer is then closed and the space between the annular and lower ram is equalized to the well pressure. In Fig. 9.5d,

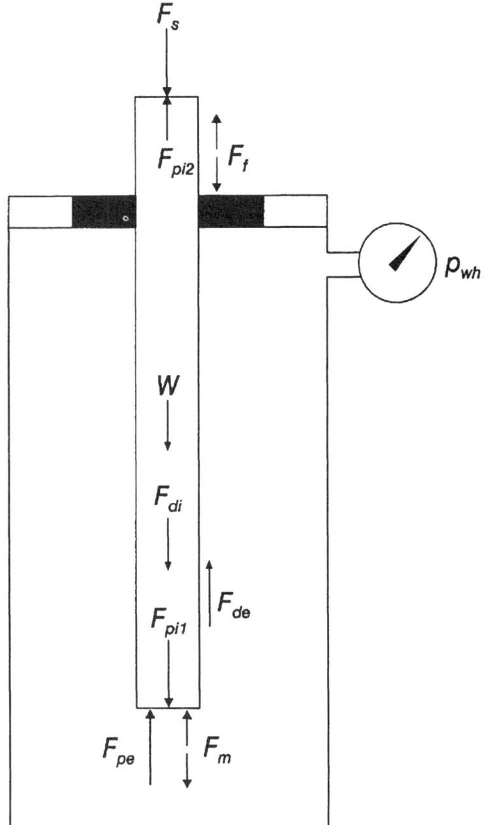

Fig. 9.9—Forces on a snubbed workstring.

the working ram is opened and the tool joint is ready to run below the ram until the next tool joint is positioned as shown in Fig. 9.5a.

9.3 Force and Buckling Calculations

Some engineering predictions are necessary or advisable when planning a snubbing job. These include determining the maximum snubbing force, the potential for pipe buckling, and at what depth the balance point of the workstring will be reached.

9.3.1 Snubbing Force Predictions. Fig. 9.9 is a free-body diagram showing forces that may be encountered when snubbing. For equilibrium, the summation of these forces must equal zero. The externally applied snubbing force is obtained as

$$F_s = F_{pe} - F_{pi1} + F_{pi2} - W \pm F_f \pm F_m \pm F_{de} - F_{di}. \quad \text{(9.6)}$$

A tapered workstring would introduce additional piston forces at the changeover shoulder areas. Eq. 9.6 is a general relationship and several terms may not be present. For example, fluid drag inside the workstring exists only during circulation whereas fluid drag on the outside requires circulation or flow from the formation. Also, a pressure force at the top of the jointed workstring is absent unless pressure is furnished by a pump. The same relation, however, applies to a coiled-tubing operation where well pressure may act on the top of the string.

String weight reduces the snubbing force. Maximum snubbing force normally occurs when the first joint of pipe is shoved into the well or the last joint is restrained from the hole. Most of the force terms drop out when essentially all of the pipe is out of the well and Eq. 9.6 reduces to

$$F_{si} = F_{pe} + F_f, \quad \text{(9.7)}$$

where F_{si} = the initial snubbing force. When the exposed area is close to the pressure gauge,

$$F_{pe} = A_o p_{wh}, \quad \text{(9.8)}$$

where A_o = the cross-sectional area of the outer diameter. Note that A_o is calculated using the connection OD if the pipe is snubbed into a stripper rubber and the tube OD in a ram-to-ram operation.

Two mechanical friction forces may be included in the F_f term; one is the friction between the pipe and packing element whereas the other is the hole drag resulting from contact between the pipe and wellbore. Wellbore friction can be significant in highly inclined or especially crooked holes, but hole drag is obviously not a factor in Eq. 9.7. The resistance to movement through the surface packoff becomes important only when snubbing large pipe and often is not considered in the calculation. Its value depends on many variables, but a rule-of-thumb adopted for coiled-tubing work assumes friction through the stripper is equal to 0.5 lbf for every psi of surface pressure.[11] This estimate may suffice, depending on the required precision. Otherwise, a more reasonable estimate for specific well conditions may be obtained from the snubbing contractor.

Example 9.2. Refer to the data shown in **Table 9.5.**

1. Estimate the initial snubbing force if the job will be done through the stripper rubber.

2. What is the prediction if this is a ram-to-ram operation?

Solution. 1. With the stripper rubber,

$$A_o = \pi(3.210)^2/4 = 8.09 \text{ sq in.}$$

and $F_{si} = (3,200)(8.09) + 1,600 = 27,488$ lbf.

2. In a ram-to-ram arrangement,

$$A_o = \pi(2.875)^2/4 = 6.49 \text{ sq in.}$$

and $F_{si} = (3,200)(6.49) + 1,600 = 22,368$ lbf.

TABLE 9.5—DATA FOR A HYPOTHETICAL SNUBBING OPERATION

Workstring Properties	
Nominal tube OD	2.875 in.
Nominal tube ID	2.441 in.
Unit air weight	6.5 lbf/ft
Connection type	Hydril CS
Connection OD	3.210 in.
Minimum yield strength	80,000 psi
Pipe body tensile yield	144,960 lbf
Minimum internal yield pressure	10,570 psi
Collapse resistance at zero axial stress	11,160 psi
Well Conditions	
Wellhead pressure	3,200 psi
Wellbore fluid	0.6 specific gravity gas
Temperature below the stack	70°F
Circulating fluid density	8.5 lbm/gal
Friction through the packing element	1,600 lbf

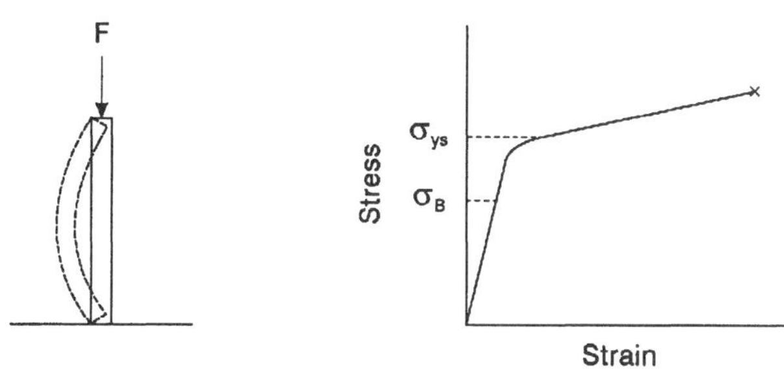

Large Slenderness Ratio - Elastic Buckling

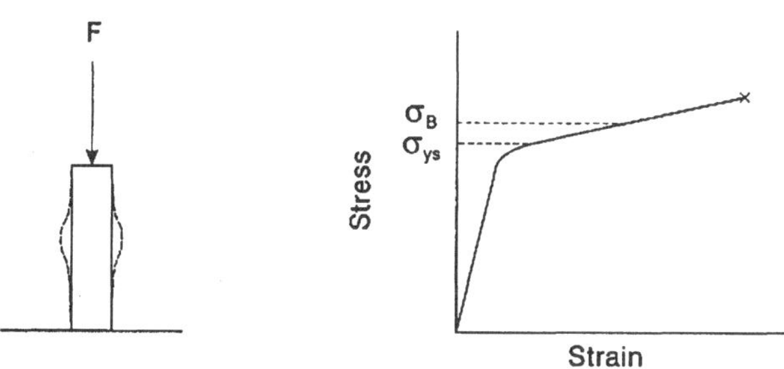

Intermediate Slenderness Ratio - Inelastic Buckling

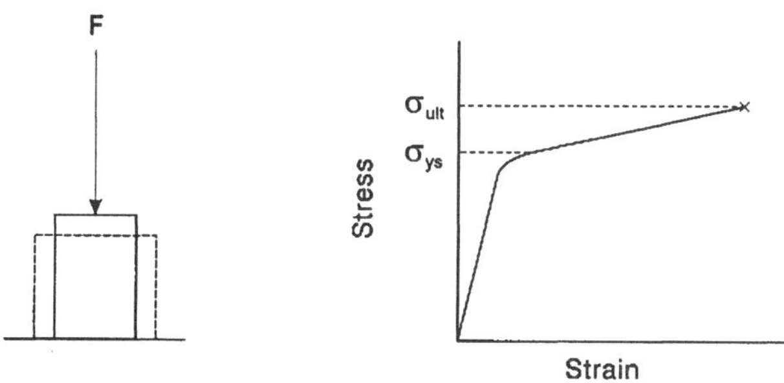

Small Slenderness Ratio - Ultimate Strength Failure

Fig. 9.10—Effect of the slenderness ratio on compressive failure mode.

The maximum snubbing force is not necessarily the initial snubbing force if the workstring diameter changes, wellhead pressures increase, or a large slackoff force (F_m) is necessary during the job procedure. The following example problem demonstrates one of these situations.

Example 9.3. Consider a ram-to-ram operation on the well described in Table 9.5 and Example 9.2. Assume that an obstruction at 1,200 ft requires an 8,000-lbf slackoff force to push through. Determine the snubbing force if the workstring is empty. Assume that the well is not flowing up the annulus (F_{de} = zero).

Solution. Eq. 9.6 reduces to

$$F_s = F_{pe} - W + F_f + F_m.$$

Eq. 1.22 yields a gas density at the wellhead.

$$\rho_g = \frac{(0.6)(3,215)}{(2.77)(0.79)(530)} = 1.66 \text{ lbm/gal}.$$

We choose to simplify the F_{pe} determination by assuming this density does not change betweeen surface and 1,200 ft. Accordingly,

$$F_{pe} = [3,200 + (0.0519)(1.66)(1,200)](6.49)$$
$$= 21,439 \text{ lbf}.$$

The string weight at the obstruction depth is calculated as

$$W = (6.5)(1,200) = 7,800 \text{ lbf}.$$

Substituting terms yields

$$F_s = 21,439 - 7,800 + 1,600 + 8,000$$
$$= 23,239 \text{ lbf}.$$

This discussion has been a simplified introduction to snubbing forces, but the free-body diagram remains valid for more complicated well conditions and the calculation technique can be used for any circumstance.[12] The snubbing contractor may have the software capability to assist in those predictions that involve flow dynamics, hole drag, and different fluid interfaces.

9.3.2 Buckling Predictions. Buckling is different from most other failure modes (except for tubular-collapse resistance) in that the failure load bears no unique relationship to the material stress and strain at failure. In other words, buckling is a stability problem and axial compression can cause failure at any point on the material's compressive stress/strain diagram. For a given strength, a dimensonal parameter called the slenderness ratio controls the axial stress at which compressive failure will occur.

Fig. 9.10 illustrates the important concepts and defines the two distinct buckling modes. A long, slender beam with a "large" slenderness ratio achieves buckling instability at an axial stress less than the proportional limit of the material. The buckling mode is defined as elastic buckling, also designated as Euler or major-axis buckling. A shorter and/or wider column giving an "intermediate" slenderness ratio fails after the material has yielded in compression. This type of buckling is called inelastic or local buckling. Finally, a stocky member with a "small" slenderness ratio will crush at the material's ultimate strength without experiencing any buckling.

The maximum unbraced length above the BOPs is usually our concern. This may occur in an open window or jack, but buckling failure is preventable in these areas by installing appropriately sized guide tubes. On rig-assist units, the pipe is unsupported across the selected stroke length and buckling predictions are advised for determining how much "bite" can be tolerated for a given snubbing force.

The procedure described by Franklin and Abel[13] starts with the maximum unbraced length and then predicts what theoretical force will cause the pipe to buckle. A design factor is applied to the result and compared to the maximum snubbing force. The operator must change conditions if failure is predicted by doing any combination of the following: reducing the snub force, reducing the unbraced length, increasing the strength of the workstring, or increasing the workstring stiffness.

The slenderness ratio for a column is defined by

$$S_u = kL_u/r_g, \quad \text{(9.9)}$$

where S_u = unbraced slenderness ratio, k = end constraint factor, L_u = unbraced length, and r_g = radius of gyration.

The radius of gyration for a tube is given by

$$r_g = (I/A_s)^{0.5}, \quad \text{(9.10)}$$

where I and A_s represent the tube's moment of inertia and wall cross-sectional area, respectively. Recall that for pipe,

$$I = \pi(d_o^4 - d_i^4)/64. \quad \text{(9.11)}$$

The end constraint factor k is a constant that depends on the type of movement allowed at each end of the unbraced length. Right or wrong, k generally is taken to be 1.0, implying that the top and bottom of the unbraced length act like pivots and are free to rotate.

Pipe has an another slenderness ratio controlled by the tube cross section,

$$S_{cs} = (r_{mw}/t)^{0.5}(4.8 + r_{mw}/225t), \quad \text{(9.12)}$$

where t = the wall thickness and r_{mw} = the pipe radius to mid-wall:

$$r_{mw} = 0.5(d_i + t). \quad \text{(9.13)}$$

To be precise, the relation applies only if r_{mw}/t is between zero and 500, but this criterion fits any conceivable workstring. In practice, both slenderness ratios are calculated and the larger of the two is selected as the appropriate value.

The critical slenderness ratio separating the two buckling modes (elastic if $S > S_c$ and inelastic if $S < S_c$) is considered to be

$$S_c = \pi(2E/\sigma_{ys})^{0.5}, \quad \text{(9.14)}$$

where E = the modulus of elasticity and σ_{ys} = the yield strength of the steel. The minimum yield strength for the workstring grade should be used.

The critical snubbing load for elastic buckling is given by

$$F_{sc} = 286 \times 10^6(A_s/S^2). \quad \text{(9.15)}$$

Note that the material strength does not affect the critical buckling force when failure is elastic.

For inelastic buckling,

$$F_{sc} = \sigma_{ys}A_s\left(1 - \frac{S^2}{2S_c^2}\right). \quad \text{(9.16)}$$

The preceding relationships are based on an ideal tube that is initially straight, has a uniform nominal wall thickness, has no ovality, has no residual stresses, and is free from corrosion and handling damage. The importance of the pipe specifications, the inspection efforts, and proper handling practices is obvious. A critical snubbing job, in fact, may prescribe using a new workstring. Regardless of the perceived string condition, a 70% design factor is used when determining the allowable snubbing force.

Example 9.4. Assume the workstring shown in Table 9.5 will be snubbed ram-to-ram in a rig-assist operation and that the initial snubbing force is the maximum snubbing force. Determine the allowable stroke length when the first joint is snubbed into the hole.

Solution. Start with an arbitrary unbraced length of 4 ft and determine the corresponding critical snub force. First the dimensional parameters for a nominal tube are computed:

$$t = (2.875 - 2.441)/2 = 0.217 \text{ in.},$$

$$A_s = \pi(2.875^2 - 2.441^2)/4 = 1.812 \text{ sq in.},$$

$$I = \pi(2.875^4 - 2.441^4)/64 = 1.611 \text{ in.}^4,$$

$$r_g = (1.611/1.812)^{0.5} = 0.943 \text{ in.},$$

and $r_{mw} = 0.5(2.441 + 0.217) = 1.329$ in.

The two slenderness ratios are determined, using Eq. 9.9 for the column. If controlled by the unbraced length,

$$S_u = (1.0)(4.0)(12)/0.943 = 50.90.$$

The slenderness ratio dictated by the tube cross-sectional geometry is determined using Eq. 9.12.

$$S_{cs} = (1.329/0.217)^{0.5}[4.8 + 1.329/(225)(0.217)]$$
$$= 11.95.$$

The former, being larger, is the applicable value. Eq. 9.14 yields the critical slenderness ratio.

$$S_c = \pi[(2)(30)(10^6)/80,000]^{0.5} = 86.04.$$

Inelastic buckling is predicted because the slenderness ratio is less than critical. Eq. 9.16 thus gives

$$F_{sc} = (80,000)(1.812)\left[1 - \frac{50.90^2}{(2)(86.04)^2}\right] = 119,594 \text{ lbf}.$$

The maximum allowable snubbing force is obtained after the design factor is applied.

$$(F_s)_{max} = (119,594)(0.70) = 83,716 \text{ lbf}.$$

A longer stroke length can be tolerated because the result is higher than the initial snubbing force. Other unbraced lengths are selected and the process is repeated until a maximum stroke of approximately 10 ft is obtained. The theoretical and allowable snub forces as function of unbraced length are represented in **Fig. 9.11.** Snubbing contractors publish such charts for common workstrings.

Many snubbing specialists recommend incorporating what may be an additional, somewhat hidden design factor by assuming the pipe has the minimum wall thickness allowed by American Petroleum Institute Specification 5CT.[14] A_s and I are reduced accordingly and a smaller buckling force is predicted. Based on the 12.5% manufacturing tolerance, the minimum ID is calculated as

$$(d_i)_{min} = d_o - 1.75t_{nom}, \quad \text{(9.17)}$$

where t_{nom} = the nominal wall thickness.

The unbraced lengths discussed to this point refer to totally unsupported pipe. Buckling a relatively small workstring in a large-bore riser or BOP stack are two other cases where the pipe may be considered as being effectively unbraced. Buckling is not necessarily a problem if the combined stresses do not yield the pipe. Absent pressure, the hoop and radial stresses are nil and a buckled workstring above the BOPs is subjected to only two axial-stress components. The axial stress resulting from the snub force is consistent throughout the pipe wall and is obtained by dividing this force by A_s. Helical bending produces another axial stress that varies across the wall and around the circumference, going from tension on the long side of the bend to compression on the short side.

Lubinski *et al.*[15] gave the combined axial-stress equation,

$$\sigma_a = \frac{F}{A_s} + \frac{d_o r_c F}{4I}. \quad \text{(9.18)}$$

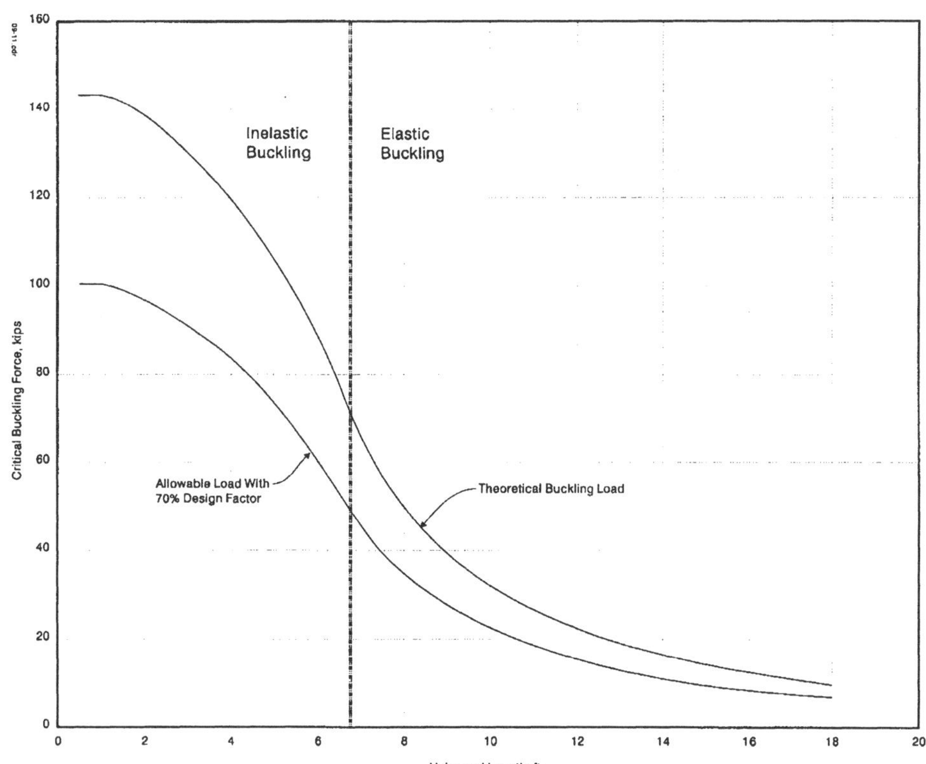

Fig. 9.11—Critical buckling forces and allowable snubbing forces for nominal 2⁷/₈-in. 6.5-lbm/ft N-80 tubing.

The maximum bending stress, given by the second term, occurs at the OD surface. To determine the radial clearance between the tube and its restraint,

$$r_c = 0.5(d_{ic} - d_o), \quad \ldots\ldots (9.19)$$

where d_{ic} = the ID of the confining bore. Setting the stress equal to the minimum yield strength and rearranging gives the snubbing force that could yield the pipe,

$$F_{sy} = \frac{A_s\sigma_{ys}}{\left(1 + \dfrac{A_s d_o r_c}{4I}\right)}. \quad \ldots\ldots (9.20)$$

Example 9.5. Our example workstring will be snubbed into a 15-ft-long riser between the window area and snubbing stack. Will the operation yield the pipe if the riser bore is 12.515 in.?

Solution. Fig. 9.11 shows that the critical buckling load is achieved in the riser. Now the assignment is to determine if this buckling will result in a permanent helical set.

$$r_c = 0.5(12.515 - 2.875) = 4.82 \text{ in.}$$

Eq. 9.20 yields

$$F_{sy} = \frac{(1.812)(80,000)}{\left[1 + \dfrac{(1.812)(2.875)(4.820)}{(4)(1.611)}\right]} = 29,604 \text{ lbf.}$$

F_{sy} is less than F_{si}, indicating that the pipe will not yield.

Conducting the well-work procedure so as to prevent a buckling failure downhole is also necessary. One case in point is what can happen when a macaroni string is run below production tubing and slacked off inside the casing. Above the BOPs, we (normally) consider the snub force as the only parameter contributing to buckling stability. Actually, this approach is a special case of a more general principle in that it does not consider the effect of pressure on buckling stability. A complete discussion of the phenomenon is beyond the scope of this text and the reader is encouraged to review the classic papers on buckling and the neutral point.[16-18] Even today the topic is misunderstood by many experienced drilling and production engineers.

When a workstring is placed into fluid, the axial compression or tension alone does not indicate whether the pipe is buckled or straight. Indeed, pipe in compression can be straight and pipe in tension can be buckled. This seemingly contradictory statement has to do with the fact that the buckling tendency does not correspond directly to the axial force but is derived by integrating the moments over the pipe surface from an arbitrary point along the pipe centerline. A so-called effective force for buckling is derived thereby and has been presented in the literature[19,20] in the form

$$F_e = F_a + A_i p_i - A_o p_o. \quad \ldots\ldots (9.21)$$

The relation adopts the sign convention in which axial compression is positive and tension is negative. Keep in mind that effective force is not a true force; it is a moment with indeterminate arm length.

We can ascertain whether pipe is buckled and the buckling severity at the depth of interest by the sign and magnitude of F_e. A positive value denotes a buckling tendency whereas a negative value indicates the pipe is straight. Internal pressure promotes buckling and external pressure tends to stabilize a tube. The effective force for buckling is the true axial force whenever internal and external pressures are absent.

The slenderness ratio of a pipe string is extremely large and, at low to moderate hole inclinations, the critical buckling force is normally small enough to be ignored. Thus it is common to assume buckling is initiated immediately whenever F_e goes from being negative to being positive, though some minimum force actually is required.

Example 9.6. Consider Example 9.3 and determine the axial force at the bottom of the workstring before the weight was slacked off.

1. Is the pipe buckled?
2. Repeat the exercise for post-slackoff conditions.

Solution. 1. Before slackoff, a free-body diagram at the bottom of the workstring yields

$$F_a = F_{pe} = 21,439 \text{ lbf compression.}$$

The cross-sectional area of the tube ID is

$$A_i = \pi(2.441)^2/4 = 4.68 \text{ sq in.}$$

The pipe has not been filled yet so the internal pressure is essentially zero. The effective force for buckling is obtained from Eq. 9.21 as

$$F_e = 21,439 + (0)(4.68) - (3,303)(6.49) = 0.$$

The string is not buckled.

2. After setting down on the obstruction, the axial force is the sum of the slackoff and pressure-area forces,

$$F_a = 21,439 + 8,000 = 29,439 \text{ lbf compression,}$$

and $F_e = 29,439 + 0 - 21,439 = 8,000 \text{ lbf}.$

Buckling is indicated based on the positive result.

Example 9.6 shows that pressure alone placed the bottom of the tubing in considerable axial compression, yet the pipe remained straight until the mechanical slackoff force was applied.

9.3.3 The Balance Point. String weight will reduce the snubbing force as more pipe is added to a well. A common practice is to snub an empty string until the balance point is approached. Filling the pipe with liquid at or near this depth gives a downward force across the workstring's internal area (F_{pi}) and the operation changes thereafter to a stripping mode if the friction force through the stripper is exceeded. At the balance point of an empty and static workstring, the snubbing by definition is reduced to zero and the force balance is given by $F_{pe} = W$.

In a vertical well, the string weight at the balance-point depth is $W = wD_{bp}$, where w = the unit weight (lbf/ft) of the pipe. The wellbore pressure at D_{bp} = the sum of the surface pressure and the hydrostatic pressure in the workstring annulus. Hence,

$$F_{pe} = \left(p_{wh} + g_{wf}D_{bp}\right)A_o, \quad \text{.................. (9.22)}$$

where g_{wf} = the hydrostatic gradient of the well fluid. Substituting terms and rearranging yields the balance-point depth:

$$D_{bp} = \frac{p_{wh}A_o}{w - g_{wf}A_o}. \quad \text{...................... (9.23)}$$

The term $g_{wf}A_o$ in the denominator is equivalent to the unit weight of the fluid displaced by the workstring. An alternate form of Eq. 9.23 is given by

$$D_{bp} = p_{wh}A_o/w_b, \quad \text{...................... (9.24)}$$

where w_b = the buoyed unit weight of the workstring.

These two relations are not applicable directly to tapered workstrings or to a wellbore where a fluid interface (gas and water, for instance) is present at a depth shallower than the balance point. For these cases we return to the principles of engineering statics. To draw a free-body diagram of the forces acting on the system, sum the forces and equate to zero, and solve for the needed term.

Example 9.7. Estimate the balance point for the workstring described in Table 9.5.

Solution. Determine the gas hydrostatic gradient,

$$g_{wf} = (0.0519)(1.66) = 0.086 \text{ psi/ft.}$$

Eq. 9.23 yields the balance point depth.

$$D_{bp} = \frac{(3,200)(6.49)}{6.5 - (0.086)(6.49)} = 3,249 \text{ ft.}$$

Filling the string or circulating the hole prior to reaching the balance point has other beneficial aspects. For one thing, pipe scale or other debris can plug the backpressure valves and it may be desirable to circulate before reaching this depth. The possibility of collapsing the workstring may be the overriding consideration in some high-pressure applications and it may be necessary to fill the string earlier for this consideration. Finally, a rig-assist operation is faster when the workstring is heavy and an operator may choose to fill the pipe early so that stripping can begin sooner. Whatever the reason, the effect on the surface force (hence slip direction) will need to be calculated beforehand.

9.4 Volumetric Control While Snubbing or Stripping

Assume that a shut-in well acts like a closed container. Inserting a workstring in a snubbing or stripping job will compress the well fluids and drive up the wellbore pressure. Similarly, vacating volume by removing pipe from the hole reduces the pressure. Maintaining a constant bottomhole pressure (BHP) is accomplished by bleeding fluid from or adding fluid to the annulus as pipe is being run or pulled.

The importance of precise volumetric control depends on the job. It may not be especially important if the wellbore is cased and perforations are open to the native or otherwise nondamaging fluids. On the other hand, keeping a relatively constant wellbore pressure may be essential on well-control operations conducted in open hole.

The following fluid conditions may be present during a snubbing or stripping job: the hole is filled with liquid; the hole is filled with gas; a gas cap overlies a liquid column; or an influx is near bottom or rising in the wellbore. Each of these conditions and how they impact volumetric control in a snubbing or stripping operation will be discussed.

9.4.1 Liquid-Filled Holes. Simply bleeding or adding a liquid volume equivalent to the pipe displacement will keep pressures stable if the operation is conducted in a liquid-filled hole. The surface (and wellbore) pressure change caused by a given length of stripped pipe may be estimated using Eq. 9.25.

$$\Delta p_{wh} = L_sC_d/c_lV_l, \quad \text{...................... (9.25)}$$

where L_s = incremental length of the snubbed or stripped pipe, C_d = displacement factor of the plugged string, c_l = liquid compressibility, and V_l = liquid volume in the hole.

The relation ignores wellbore elasticity and thus will overestimate the predicted pressure increase. The liquid compressibility for stratified liquid columns or liquids containing more than one constituent (drilling muds for example) can be determined using

$$c_l = f_w c_w + f_o c_o + f_s c_s, \quad \text{(9.26)}$$

where the f and c subscripts refer to the respective volume fractions and compressibilities of the water, oil, and solid phases.

Example 9.8. 2³/₈-in. tubing will be snubbed into a 10,000-ft well filled with 2% KCl water. The following conditions apply.

Wellhead pressure = 300 psig, casing ID = 4.892 in., casing capacity factor = 0.0232 bbl/ft, tubing displacement factor = 0.0055 bbl/ft, and water compressibility = 3.1×10^{-6} psi^{-1}.

Estimate the pressure increase if 100 ft is inserted without bleeding any fluid. Assume the wellbore is rigid and fill prevents losses to the perforations.

Solution. The initial liquid volume in the well is

$$V_l = (10,000)(0.0232) = 232 \text{ bbl}.$$

Eq. 9.25 yields

$$\Delta p_{wh} = (100)(0.0055)/(3.1)(10^{-6})(232) = 765 \text{ psi}.$$

As shown, running only a short distance into a liquid-filled hole can cause an extreme pressure increase if volume is not released on a semicontinuous basis. Therefore, an accurate volume accounting using a calibrated measuring tank is necessary.

Usually, the preferred method is to control off the wellhead pressure gauge rather than volume. This is not to say that the measuring tank should be bypassed or ignored. Discrepancies between the measured and calculated volumes will indicate activity in the well such as fluid losses or a rising gas bubble. The constant-pressure procedure also will work when removing the workstring if the added fluid density is close to what is in the well.

9.4.2 Gas-Filled Holes. Maintaining a constant BHP while stripping into a gas-filled hole is achieved normally by releasing gas through the choke so as to hold a constant wellhead pressure. Flow from the formation should keep the well charged when pipe is being removed. Eq. 9.25 can be used to estimate the effect of adding or removing volume on the wellbore pressure if gas compressibility is substituted into the relation. Example 9.9 takes an equivalent approach and directly applies the gas law to our hypothetical well.

Example 9.9. The well described in Example 9.8 is filled with a 0.60-specific-gravity gas and has a 4,700-psia shut-in pressure. Determine how much pipe can be snubbed without bleeding any gas if the surface pressure is limited to 5,000 psia. Assume the wellbore temperature at surface is 70°F.

Solution. A simplification will be made for illustration purposes, namely that the properties of the gas at any depth reflect the surface conditions. The gas law gives the gas volume corresponding to the allowable wellhead pressure,

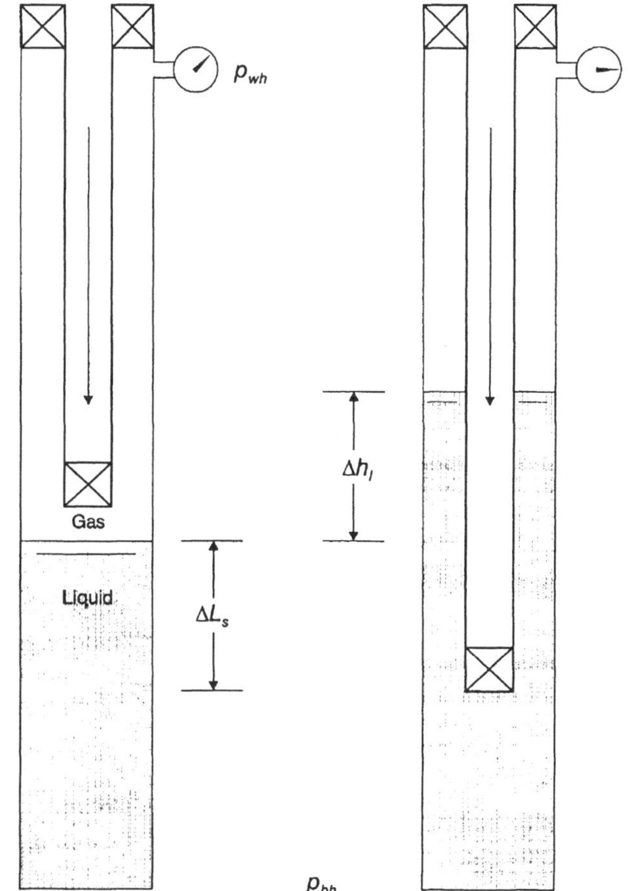

Fig. 9.12—Inserting a workstring into a gas/liquid interface and its effect on wellbore pressure.

$$V_2 = \frac{(4,700)(232)(530)(0.940)}{(5,000)(530)(0.905)} = 226.5 \text{ bbl}.$$

The maximum pipe length is the volume change divided by the displacement factor.

$$L_s = (232 - 226.5)/(0.0055) = 1,000 \text{ ft}.$$

9.4.3 Gas Over Liquid. The task of holding constant BHP becomes more complicated if two or more fluids are in the wellbore. Holding constant wellhead pressure achieves this goal only if column heights of the respective well fluids do not change, hence the technique does not work after the workstring enters the fluid interface.

Fig. 9.12 demonstrates what happens when pipe is run from gas into liquid. For an incremental string length ΔL_s, the liquid column height increases according to the relation,

$$\Delta h_l = \Delta L_s C_d \cos\gamma / C_a,$$

where C_a = the annulus capacity factor opposite the workstring and γ = the hole inclination from vertical. The BHP will increase by the amount,

$$\Delta p_{bh} = g_l \Delta h_l = g_l \Delta L_s C_d \cos\gamma / C_a, \quad \text{(9.27)}$$

where g_l = the hydrostatic gradient of the liquid. The surface pressure must be allowed to fall to account for the changing hydrostatic, otherwise the well will be overpressured.

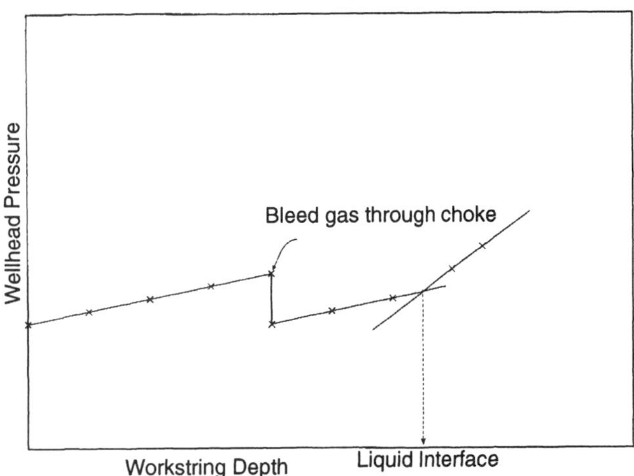

Fig. 9.13—Using surface pressure to identify a gas/liquid interface during a stripping job.

Example 9.10. Assume the example well is vertical and has 1,000 ft of gas on top of an 8.7-lbm/gal brine. The wellhead pressure is 725 psia and the intent is to maintain constant BHP during the snubbing job. Determine the target wellhead pressure when the string gets to 1,500 ft.

Solution. The wellhead pressure must be reduced by an amount corresponding to the hydrostatic pressure increase below 1,500 ft. Use Eq. 9.27 to estimate the surface pressure reduction as

$$\Delta p_{wh} = \Delta p_{bh} = \frac{(8.7)(1,500 - 1,000)(0.0055)(1.0)}{(19.25)(0.0232 - 0.0055)}$$

$$= 70 \text{ psi.}$$

The desired wellhead pressure when the workstring reaches 1,500 ft is then

$$p_{wh} = 725 - 70 = 655 \text{ psia.}$$

Note that we ignored changes in the gas hydrostatic pressure, but the error is only 5 psi for this particular problem.

The operator will not know when to start making pressure adjustments if the liquid level depth is unknown. Measuring this depth by lubricating a wireline tool into the hole or shooting a fluid level may be recommended if precise control is needed. Otherwise, the interface usually can be identified during the job by the rate the surface pressure increases. Refer to Fig. 9.12. Running pipe into gas gives a consistent gas-volume reduction for a specified length and a plot of string depth vs. wellhead pressure will have a constant slope. The gas is compressed much more rapidly after the workstring enters the liquid and, as suggested by **Fig. 9.13,** the interface can be detected by a sharp change in slope.

9.4.4 Stripping With an Influx in the Hole. Example 9.10 was one situation where the wellbore pressure could not be controlled by holding a constant surface pressure, but the problem is largely self-correcting in a cased wellbore with open perforations. This is not the case in an open hole when an off-bottom kick occurs and it becomes necessary to strip the drillstring into the influx. Wellbore pressures throughout the operation must be maintained within the window defined by the pore pressure and frac integrity. Three possible situations are presented.

TABLE 9.6—OFF-BOTTOM KICK DATA FOR EXAMPLES 9.11 THROUGH 9.13	
Vertical well depth	10,000 ft
Estimated kick zone depth	10,000 ft
Bit depth	4,000 ft
Surface Casing Information	
Description	$9^5/_8$-in., 40.0-lbm/ft
Inner diameter	8.835 in.
Setting depth	3,500 ft
Fracture gradient at casing seat	0.81 psi/ft
Assumed openhole diameter	8.5 in.
Drillstring Information	
Drill collar size	$6^1/_4 \times 2^3/_4$ in.
Drill collar section length	600 ft
Drillpipe description	$4^1/_2$-in., 16.60-lbm/ft Grade E NC50
Drillpipe plugged displacement factor	0.02066 bbl/ft
Capacity Factors	
Openhole	0.07019 bbl/ft
Drillpipe × casing	0.05616 bbl/ft
Drillpipe × openhole	0.05052 bbl/ft
Collar × openhole	0.03224 bbl/ft
Mud type	Water Based
Mud density	12.0 lbm/gal
Mud gradient	0.623 psi/ft
Assumed static wellbore temperature	80°F+1.2°F/100ft
Pit gain	23 bbl

Liquid Influx. Consider the situation where the drillstring is a considerable distance from the flowing formation when a well is shut in on a kick. For whatever reason (e.g., area knowledge, the shut-in pressure, no indication of kick migration), the operator suspects a salt water or dead oil kick. One way to deal with the problem is to pump an off-bottom kill and trip either to bottom or to an intermediate point with the well open. The fracture gradient, however, may not tolerate the mud-weight requirement and it may be necessary to strip back to bottom or to a depth where an off-bottom kill-weight mud can be circulated safely. However the task is accomplished, the effect of the string geometry on the influx height must be considered in the plan.

Example 9.11. An operator has swabbed in a kick and is faced with the conditions presented in **Table 9.6.** The shut-in casing pressure is 50 psig and has not changed in over 30 minutes.

1. Estimate the kick-fluid character (gas, oil, or water).
2. Determine the mud weight to pump if the intent is to kill the well from the present depth and keep the well dead as pipe is run back to bottom.
3. What procedure should be followed if the intent is to strip the pipe back to bottom?

Solution. 1. The low shut-in pressure and the fact that the pressure is not changing suggest a liquid kick. However, we do not have the means to read the BHP and cannot be certain until the influx is removed from the well. Assume the mud density exactly balances the pore pressure,

$$p_p = (0.623)(10,000) = 6,230 \text{ psig.}$$

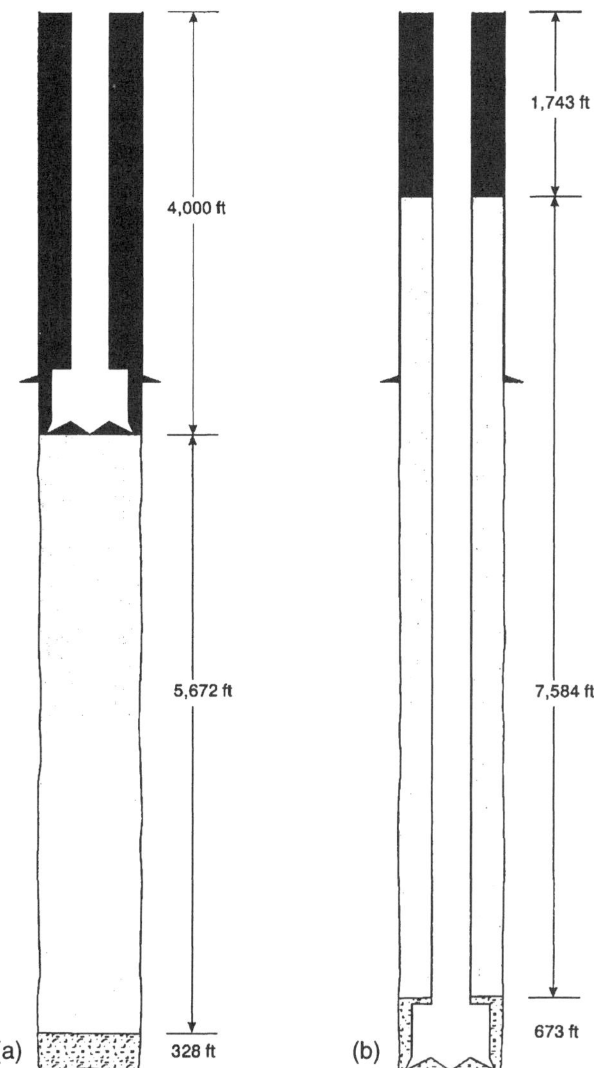

Fig. 9.14—Wellbore fluid positions related to the off-bottom kill discussed in Example 9.11.

The influx height is determined as

$$h_k = 23/0.07019 = 328 \text{ ft}.$$

Thus,

$$50 + (10,000 - 328)(0.623) + 328g_k = 6,230.$$

$$g_k = 0.471 \text{ psi/ft}.$$

The calculation indicates a saltwater kick. Note, however, that our assumed pore pressure is on the high side if the mud overbalanced the formation before the trip. As a result, the kick probably has a lower gradient than predicted and the operator needs to be alert for indications of migrating gas during the subsequent operation.

2. The mud weight must be high enough to counteract the combined effects of the hydrostatic reduction resulting from the elongated influx and heavy-mud displacement out the annulus. The volume opposite the drill collars is

$$(600)(0.03224) = 19.3 \text{ bbl}.$$

The influx will extend into the drillpipe annulus with a final height given by

$$h_{k2} = 600 + (23 - 19.3)/0.05052 = 673 \text{ ft}.$$

The mud volume displaced from the well after pipe reaches bottom is calculated.

$$V_d = (10,000 - 4,000)(0.02066) = 124.0 \text{ bbl}.$$

After pumping the kill, the heavy-weight mud volume in the drillstring annulus will be

$$(3,500)(0.05616) + (4,000 - 3,500)(0.05052)$$

$$= 196.6 + 25.3 = 221.9 \text{ bbl}.$$

This volume ultimately will be reduced by displacement to 97.9 bbl (221.9 – 124.0). The volume will be inside the drillpipe/casing annulus and so its final height is given by

$$97.9/0.05616 = 1,743 \text{ ft}.$$

The 12.0-lbm/gal mud height when the bit reaches bottom is

$$10,000 - 673 - 1,743 = 7,584 \text{ ft}.$$

Figs. 9.14a and 9.14b illustrate the respective fluid positions with the string at 4,000 ft and 10,000 ft. We conservatively assume the predicted saltwater density is correct and determine the mud gradient to pump at 4,000 ft that will kill keep the well dead.

$$1,743g_{km} + (7,584)(0.623) + (673)(0.471) = 6,230,$$

$$g_{km} = 0.682 \text{ psi/ft, and}$$

$$\rho_{km} = (0.682)(19.25) = 13.1 \text{ lbm/gal}.$$

This mud weight will not fracture the shoe and the pipe can be run to bottom after the first kill.

2. Using the choke in attempt to hold 50 psig on the casing will keep BHP constant until the bit reaches the influx. At 9,672 ft, the volume released to the measuring tank should be

$$V_d = (9,672 - 4,000)(0.02066) = 117.2 \text{ bbl}.$$

From this point on, the surface pressure must be allowed to increase by an amount equal to the hydrostatic pressure reduction caused by the lengthened water column,

$$\Delta p_{wh} = (673 - 328)(0.623 - 0.471) = 52 \text{ psi}.$$

The final casing pressure is thus

$$p_{wh2} = 50 + 52 = 104 \text{ psig}.$$

Actually, it would be acceptable to hold this much pressure on the well from the start.

Immobile Gas Influx. Some conditions tend to inhibit gas migration but the potential for migration is always there, even though there may not have been any indication when the well was first shut in. As with a liquid influx, an off-bottom kill may be an option or the pipe can be stripped if the procedure considers the hydrostatic reduction caused by the operation. Well-surveillance measures must be enhanced whenever gas or a gassy influx is suspected. An operator should assume every kick is a gas kick until observations prove otherwise.

Example 9.12. Reconsider Example 9.11 for the case where the initial shut-in casing pressure (SICP) rises to 175 psig and stabilizes.

Solution. 1. The same assumptions apply. The estimated kick fluid gradient is estimated as

$$g_k = [6,230 - 175 - (9,672)(0.623)]/328$$

$$= 0.089 \text{ psi/ft}.$$

TABLE 9.7—BLEED VOLUMES AND WELLBORE PRESSURES FOR EXAMPLE 9.13

Bit Depth (ft)	Casing Pressure (psig)	Expansion Bleed (bbl)	Displacement Bleed (bbl)	Total Bleed (bbl)	Influx Length (ft)	Influx Top (ft)	Mud Hydrostatic (psig)	Influx Hydrostatic (psig)	Bottomhole Pressure (psig)	Shoe Pressure (psig)
Initial conditions.										
4,000	175	0.0	0.0	0.0	328	9,672	6,026	29	6,230	2,356
Strip and bleed for pipe displacement. Allow casing pressure to build by safety and working margins.										
4,040	325	0.0	0.8	0.8	328	9,432	6,026	29	6,380	2,506
Strip and expand gas while maintaining 325 psig on the casing.										
4,555	325	5.6	10.6	16.2	407	5,596	5,976	29	6,330	2,506
Strip and bleed for pipe displacement. Allow casing pressure to build by working margin.										
4,561	375	0.0	0.1	0.1	407	5,555	5,976	29	6,380	2,556
Strip and expand gas while maintaining 375 psig on the casing.										
4,689	375	2.6	2.6	5.2	445	4,689	5,953	29	6,357	2,556
Rapid pressure increase indicates gas is rising above the bit. Stop stripping and allow gas to clear bottom of string. Bleed mud to hold BHP constant.										
4,689	616	1.8	0.0	1.8	871	3,818	5,687	54	6,357	2,797
Allow gas to migrate above the drill collars. Bleed mud to hold BHP constant. Note that gas may be circulated at this point in the operation.										
4,689	536	3.4	0.0	3.4	721	3,368	5,781	40	6,357	2,642
Strip and allow gas to expand while maintaining 536 psig on the casing.										
4,731	536	2.1	0.9	3.0	715	3,072	5,785	38	6,359	2,473
Strip and bleed for pipe displacement. Allow casing pressure to build by working margin.										
4,744	586	0.0	0.3	0.3	706	2,990	5,790	37	6,413	2,475
Strip and allow gas to expand while maintaining 586 psig on the casing.										
4,802	586	4.5	1.2	5.7	766	2,515	5,753	36	6,375	2,325
Continue the process until gas surfaces. Then strip to bottom while maintaining constant casing pressure and bleeding mud for displacement.										

Thus it appears that we have a gas kick.

2. Using the reasoning from the preceding example, the kill-weight mud density to pump at 5,000 feet is

$$1,743g_{km} + (7,584)(0.623) + (673)(0.089) = 6,230;$$

$$g_{km} = 0.829 \text{ psi/ft, and}$$

$$\rho_{km} = (0.829)(19.25) = 16.0 \text{ lbm/gal.}$$

The fracture gradient will not allow killing the well from the present bit depth if the string is run all the way to bottom. As an alternative approach, we will calculate the mud density that will allow the off-bottom kill if the string is run to only 9,000 ft. The heavy-mud displacement will be

$$V_d = (9,000 - 4,000)(0.02066) = 103.3 \text{ bbl.}$$

The heavy-mud volume left in the well is 118.6 bbl (221.9 − 103.3), which gives a column height of

$$118.6/0.05616 = 2,112 \text{ ft.}$$

Hence the 12.0-lbm/gal mud height when the bit reaches 9,000 ft is

$$10,000 - 328 - 2,112 = 7,560 \text{ ft.}$$

The kill-weight mud for the revised plan is

$$2,112g_{km} + (7,560)(0.623) + (328)(0.089) = 6,230;$$

$$g_{km} = 0.706 \text{ psi/ft, and}$$

$$\rho_{km} = (0.706)(19.25) = 13.6 \text{ lbm/gal.}$$

A step-down drillpipe pressure schedule should be followed when filling the drillstring with the 13.6-lbm/gal mud. The well will be controlled from the drillpipe gauge, but the choke operator should monitor the casing pressure during this period. It should not change until the new mud exits the bit whereas a rising annulus pressure indicates gas migration.

Another kill circulation must be pumped after tripping in the hole to 9,000 ft. The density requirement at this depth, considering the longer gas column and small amount of displacement, is 12.8 lbm/gal. To simplify the procedure and maintain system consistency, the operator would elect to pump the 13.1-lbm/gal mud used in the preceding kill.

3. For the stripping option, the backpressure must be allowed to increase before the bit encounters the influx. The final casing pressure is

$$p_{wh2} = 175 + (673)(0.623 - 0.089) = 534 \text{ psig.}$$

A gradient calculation confirms that the shoe will not be risked in the process.

Stripping with a Migrating Gas Influx. When signs of a migrating gas influx are evident, the operator may choose to keep the string where it is and volumetrically control the well until the gas migrates into the drillstring annulus. The gas can be displaced then, a procedure that should kill the well if the mud weight is balanced to the pore pressure. Another option is to strip the drillstring into the hole while the gas rises to meet the bit.

Volumetric control in this situation is complicated because the surface pressure will tend to increase by the rising gas in addition to the drillstring displacement. To maintain constant BHP because of these effects, mud will have to be released and the location of the gas reasonably ascertained. The job will be facilitated if the operator sets up a table to keep track of all changes during the operation. Our final example demonstrates the procedure.

Example 9.13. The initial SICP from the preceding example does not stabilize at 175 psig but continues to increase. Determine a procedure that will maintain a relatively constant BHP if the operator decides to start stripping in the hole immediately.

Solution. **Table 9.7** shows the location of the gas bubble as it migrates up the wellbore, the associated surface and key wellbore pressures, and the volumes of mud the operator must release for both pipe displacement and gas expansion. Calculations are based on a single bubble moving up a rigid, sealed wellbore at 3,600 ft/hr and an average stripping velocity of about 500 ft/hr. Real-world conditions would not be as precise as represented in the table and the operator will not have the means to track influx location exactly.

The volumetric procedures discussed in Chap. 5 still apply here in that mud must be released from the well to expand the gas. In this problem, we arbitrarily use a 100-psi safety margin and a 50-psi working margin. The operator strips and allows the casing pressure to increase by 150 psi while bleeding enough mud to the measuring tank for the pipe displacement. When the casing gauge reads 325 psig, the operator will start bleeding additional mud for the gas expansion. Combining Eq. 5.1 and 5.2 gives the expansion volume when the kick is below the bit.

$$\Delta V_m = (50)(0.07019)/(0.623) = 5.6 \text{ bbl}.$$

After the expansion stage, the casing pressure is allowed to increase by 50 psi while the bled mud volume relates to an additional pipe displacement. The gas is allowed to expand again after these objectives are achieved.

As discussed in Chap. 5, the correct surface pressure and expansion volume to bleed for a given working margin depends on where the influx is located with respect to changes in the downhole geometry. Thus it is important for the operator to recognize when the influx begins to lengthen or shorten because of changes in the hole geometry and then manipulate the surface pressure so that a semiconstant BHP can be realized while these changes are taking place.

A significant dimension change such as going from an openhole section into the drill collar annulus requires that the final surface pressure and expansion bleed during the transition be calculated. This was done in Example 9.13 by specifying the gas pressure at the bottom of the influx needed to hold a constant BHP. The gas law was used in an iterative process to determine the gas volume, which dictates the influx top, which dictates the surface pressure.

The possibility of fracturing the rock must be considered in the predictions and alternatives weighed before the stripping job begins. In the example well described in Table 9.7, the maximum shoe pressure of 2,797 psig was imposed when the drill collar annulus was occupied by gas. This is uncomfortably close to the fracture gradient. Granted, the surface pressure may not need to be this high when one considers the safety margin's cushion and the effect of bubble dispersion on the influx hydrostatic. But an alternative and safer solution would have been to strip out of the hole first so that the drill collars were inside the casing. Volumetrically controlling the well with the bit shallower than 3,500 ft would reduce the maximum shoe pressure because the hydrostatic pressure reduction occurs above the casing seat.

Problems

9.1 How much lift force will be produced if the hydraulic pressure determined in Example 9.1 is applied at the opposite end of the cylinder?

9.2 An HWO unit has two cylinders with cylinder and piston rod diameters of 9.0 and 6.5 in., respectively.

1. Determine the snub-load capacity of the unit at the 3,000-psi system pressure.
2. What lift-pressure setting should be made if the intent is to limit the lift force to no more than 80% of the 261,000-lbf tension rating of the workstring?

TABLE 9.8—SNUBBING DATA FOR A WELL IN THE PROBLEM SET	
Nominal Workstring Properties	
OD	1.315 in.
ID	0.957 in.
Air weight	2.25 lbf/ft
Connection OD	1.600 in.
Minimum yield strength	105,000 psi
Pipe body tensile yield	67,000 lbf
MIYP	25,010 psi
Collapse resistance at zero axial stress	24,690 psi
Well Conditions	
Wellhead pressure	1,500 psi
Wellbore fluid	9.0-lbm/gal brine
Friction through the packing element	600 lbf

9.3 Assume the unit described in the previous problem has a regenerative circuit.

1. What lift-pressure setting should be made if the circuit is activated?
2. Determine the jack lift speed for this case.

9.4 Considering human nature, what is one major weakness to using a crane for evacuating snubbing personnel in an emergency?

9.5 Write a ram-to-ram procedure to retrieve a workstring from a well under pressure.

9.6 List some factors that can affect friction through a stripper rubber or stripping ram.

9.7 Table 9.8 gives conditions related to a through-tubing snubbing job.

1. Determine the initial snubbing force for snubbing ram-to-ram and through the stripper rubber.
2. Which procedure would you recommend? Defend your answer.

9.8 The workstring described in Table 9.8 is being used to circulate and clean out fill at a depth of 7,215 ft. Determine whether the string is light or heavy and by how much if the following conditions apply: slackoff weight = 2,200 lbf, circulating fluid density = 9.0 lbm/gal, fluid drag in workstring = 600 lbf, fluid drag in annulus = 200 lbf, circulating friction in annulus = 150 psi, choke backpressure = 1,400 psi, hole drag = 700 lbf, and packoff friction = 500 lbf.

9.9 Which one of the HWO units described in Table 9.2 would you recommend for the job described in Problems 9.7 and 9.8? Defend your answer.

9.10 In what way does the procedure change or the results differ if the last joint of the workstring described in Example 9.4 is being snubbed out of the hole?

9.11 Consider Table 9.8 and the initial snub force (stripper rubber) determined in Problem 9.7.

1. Will the workstring buckle in an open window if the unbraced length is 36 in.?
2. Regardless of the result, what precaution should be taken in this area?

9.12 Write a spreadsheet program and prepare a set of curves similar to Fig. 9.11 for the workstring described in Table 9.8. Do the same for 7.0-in. casing if the nominal ID and minimum yield strength are 6.094 in. and 110,000 psi, respectively.

9.13 Modify Fig. 9.11 for the case where an API minimum wall is assumed.

9.14 Solve Example 9.4 using the chart created in Problem 9.13.

9.15 Assume that the telescoping guide tube on a conventional HWO unit is 26 ft long at maximum stroke.

1. Will the workstring described in Table 9.8 and Problem 9.7 buckle inside the guide tube if the tube has a 1.94-in. ID?
2. If so, calculate the axial stress including the maximum bending component.

9.16 Derive Eq. 9.17.

9.17 The workstring in Table 9.8 has been snubbed dry to 500 ft with no change in the wellhead pressure or wellbore fluid density.

1. Determine the compressive force at the bottom of the string.
2. Does this compression cause the lower portion of the string to buckle?

9.18 Determine the balance point for the operation in Table 9.8.

9.19 Consider the conditions represented in Table 9.5 and assume that a 3½-in., 9.30-lbm/ft section will be run after placing 2,000 ft of the 2⅞-in. tubing in the well. Determine the balance-point depth.

9.20 Consider the conditions represented in Table 9.5 and assume that the hole will be circulated with a 9.0-lbm/gal brine at 1,000 ft.

1. Will the tubing remain full of brine after the circulation?
2. Determine the balance-point depth if the workstring is not filled again.

9.21 The following information pertains to a proposed through-tubing cleanout: cleanout string = 1.0-in. coiled tubing, tubing OD and ID = 3½ in. × 2.922 in., packer depth = 11,200 ft, production casing ID = 6.184 in., gas zone perforations = 11,290 to 11,350 ft, solid fill depth = 11,250 ft, wellbore fluid = 0.7-specific-gravity gas, shut-in tubing pressure = 1,000 psig, and average temperature = 160°F.

Determine how much coiled tubing can be run before any pressure is bled from the choke if the surface-pressure limitation is 1,500 psig. Assume a rigid wellbore and no leak-off to the perforations.

9.22 For the preceding problem, assume produced water with a density of 9.1 lbm/gal is encountered at 9,700 ft. The wellhead pressure has been maintained at 1,000 psig up until this point. Estimate the surface pressures and BHPs if the coiled tubing is run thereafter to fill depth without bleeding any gas.

9.23 The following information pertains to a trip operation on a drilling rig: well depth = 8,000 ft, gas reservoir depth = 7,900 ft, pore pressure = 3,600 psia, surface casing depth = 2,500 ft, fracture gradient (2,500 ft) = 0.75 psi/ft, casing ID = 8.097 in., average open hole diameter = 8.0 in., mud density = 9.3 lbm/gal, and drillpipe description = 4½-in. 16.60-lbm/ft Grade E NC50.

A flow was noted while rigging up for the logging job. A backpressure valve was installed in the drillpipe and string was run to 3,500 ft before the well was shut-in. Pit gain at this time was 10.0 bbl and the initial SICP was 70 psig.

1. Estimate the gas migration rate if the SICP builds to 200 psig in one hour.
2. What mud volume must be bled from the well strictly for the drillpipe displacement if the string is stripped to bottom?
3. What mud volume must be released for a 100-psi working margin if gas is expanding below the bit?
4. Assume the drillpipe encounters the rising gas bubble at 5,000 ft and that the kick volume at this time is 11 bbl. Your objective is to keep the BHP constant as the bubble moves into the drillpipe annulus. Estimate the incremental mud volume to bleed and the casing pressure when the base of the influx is at the bottom of the string.
5. Your intent is to continue volumetrically controlling the gas as opposed to circulating it out. What mud volume must be released for a 100-psi working margin if gas is expanding in the drillpipe/openhole annulus?

9.24 Estimate the maximum shoe pressure for Example 9.13 if stripping operations are discontinued when the bit reaches 3,500 ft and the well is thereafter controlled volumetrically.

Nomenclature

A_c = cylinder area, in.^{2}
A_o = cross-sectional area of workstring outer diameter, in.^{2}
A_r = rod area, in.^{2}
A_s = cross-sectional area of a pipe wall, in.^{2}
c_l = liquid compressibility, psi^{-1}
c_o = oil compressibility, psi^{-1}
c_s = solids compressibility, psi^{-1}
c_w = water compressibility, psi^{-1}
C_a = annulus capacity factor, bbl/ft
C_d = displacement factor, bbl/ft
d_i = inner diameter, in.
d_{ic} = inner diameter of a restraining tube, in.
d_o = outer diameter, in.
D_{bp} = balance point depth, ft
E = Young's modulus of elasticity, psi
f_o = oil volume fraction, dimensionless
f_s = solids volume fraction, dimensionless
f_w = water volume fraction, dimensionless
F = force, lbf
F_a = axial force, lbf
F_{de} = external fluid drag, lbf
F_{di} = internal fluid drag, lbf
F_e = effective force for buckling, lbf
F_f = friction force, lbf
F_l = lift force, lbf
F_m = mechanically applied force, lbf
F_{pe} = external piston (pressure-area) force, lbf
F_{pi1} = internal piston (pressure-area) force on bottom, lbf
F_{pi2} = internal piston (pressure-area) force on top, lbf
F_s = snubbing force, lbf
F_{sc} = critical snubbing force for buckling, lbf
F_{si} = initial snubbing force, lbf
F_{sy} = snubbing force that could yield a workstring, lbf

g_k = kick-fluid hydrostatic gradient, psi/ft
g_l = liquid hydrostatic gradient, psi/ft
g_{km} = kill-mud hydrostatic gradient, psi/ft
g_{wf} = well-fluid hydrostatic gradient, psi/ft
h_k = kick height, ft
h_l = liquid height, ft
I = moment of inertia, in.^{4}
k = end constraint factor, dimensionless
L_s = length of snubbed or stripped pipe, ft
L_u = unbraced length, in.
N_c = number of active cylinders
p_{bh} = bottomhole pressure, psi
p_h = hydraulic pressure, psi
p_p = pore pressure, psi
p_{wh} = wellhead pressure, psi
q_p = flow rate from the power pack, gal/min
r_c = radial clearance, in.
r_g = radius of gyration, in.
r_{mw} = radius to mid-wall, in.
S = slenderness ratio, dimensionless
S_c = critical slenderness ratio, dimensionless
S_{cs} = slenderness ratio controlled by the cross section, dimensionless
S_u = slenderness ratio controlled by the unbraced length, dimensionless
t = wall thickness, in.
t_{nom} = nominal wall thickness, in.
v_{jl} = lifting jack velocity, ft/min
v_{js} = snubbing jack velocity, ft/min
V_d = displacement volume, bbl
V_l = liquid volume, bbl
V_m = mud volume, bbl
w = unit air weight, lbf/ft
w_b = buoyed unit weight, lbf/ft
W = air weight, lbf
γ = hole inclination from vertical, deg
ρ_g = gas density, lbm/gal
ρ_{km} = kill-mud density, lbm/gal
σ = stress, psi
σ_a = axial stress, psi
σ_B = axial stress at initiation of buckling, psi
σ_{ult} = ultimate strength, psi
σ_{ys} = yield strength, psi

References

1. Joseph, R.A.: "Planning Lessons, Problems Get Benefits of Underbalance," *Oil and Gas J.* (20 March 1995) 86.
2. Lagendyk, R., Loring, G., and Aasen, J.: "Drilling Applications Expand Snubbing Unit Use," *World Oil* (May 1996) 37.
3. Hodgson, R.: "Snubbing Units: A Viable Alternative to Conventional Drilling Rig and Coiled Tubing Technology," paper SPE 30408 presented at the 1995 Offshore Europe Conference, Aberdeen, 5–8 September.
4. Leggett, R.B., Griffith, C.A., and Wesson, H.R. Jr.: "Snubbing Unit Applications in Potentially High-Rate Gas Wells: A Case Study of the Anschutz Ranch East Unit, Summit County, Utah," paper SPE 22824 presented at the 1991 SPE Annual Technical Conference and Exhibition, Dallas, 6–9 October.
5. Maddox, S.: "Hydraulic Rig-Assisted Well Servicing Techniques Can Reduce Formation Damage," paper SPE 23807 presented at the SPE Intl. Symposium on Formation Damage Control, Lafayette, Louisiana, 26–27 February 1992.
6. Robinson, C.E. and Cox, D.C.: "Alternative Methods for Installing ESP's," paper OTC 7035 presented at the 1992 Offshore Technology Conference, Houston, 4–7 May.
7. Maddox, S. and Smith, E.: "An Emergency Escape System for an Elevated Workplace," paper SPE 27258 presented at the 1994 SPE International Conference on Health, Safety & Environment in Oil & Gas Exploration & Production, Jakarta, 25–27 January.
8. Sehnal, Z., Østebø, B., and Rørhuus, K.: "Extending the Limits of Hydraulic Workover Technology," *World Oil* (June 1997) 49.
9. Konopczynski, M.R. and Milligan, M.R.: "Snubbing Workover Operations in Deep Sour Gas Wells a Success," *Oil and Gas J.* (12 February 1996) 35.
10. Lawson, R.: "How to Perform Safer Hydraulic Workovers in Gas Wells," *World Oil* (January 1996) 42.
11. Newman, K.R. and Allcorn, M.G.: "Coiled Tubing in High-Pressure Wells," paper SPE 24793 presented at the 1992 SPE Annual Technical Conference and Exhibition, Washington, DC, 4–7 October.
12. Abel, L.W. "Minimizing Loads During Snubbing Helps Prevent Pipe Failure," *Oil and Gas J.* (12 June 1995) 124.
13. Franklin, R.S. and Abel, L.W.: "Safer Snubbing Depends on Proper Pre-Job Calculations," *World Oil* (October 1988) 85.
14. "Specification for Casing and Tubing," *Spec 5CT*, sixth edition, API, Washington, DC (November 1998).
15. Lubinski, A., Althouse, W.S., and Logan, J.L.: "Helical Buckling of Tubing Sealed in Packers," *JPT* (June 1962) 655; *Trans.*, AIME, **225.**
16. Lubinski, A.: "Influence of Tension and Compression on Straightness and Buckling of Tubular Goods in Oil Wells," *Proc.*, 31st Annual Meeting API, Prod. Sec. IV (1951) 31–56.
17. Klinkenberg, A.: The Neutral Zones in Drill Pipe and Casing and Their Significance in Relation to Buckling and Collapse," *Drill. and Prod. Prac.*, API (1951) 64.
18. Woods, H.B.: Discussion of The Neutral Zones in Drill Pipe and Casing and Their Significance in Relation to Buckling and Collapse," *Drill. and Prod. Prac.*, API (1951) 77.
19. Christman, S.A.: "Casing Stresses Caused by Buckling of Concentric Pipes," paper SPE 6059 presented at the 1976 SPE Annual Technical Conference and Exhibition, New Orleans, 3–6 October.
20. Mitchell, B.J.: *Advanced Oilwell Drilling Engineering Handbook*, ninth edition, SPE, Dallas (September 1992) 30–31.

SI Metric Conversion Factors

bbl	× 1.589 873	E − 01	= m^3
deg	× 1.745 329	E − 02	= rad
°F	(°F − 32)/1.8		= °C
ft	× 3.048*	E − 01	= m
gal	× 3.785 412	E − 03	= m^3
in.	× 2.54*	E + 01	= mm
in.^{2}	× 6.451 6*	E + 02	= mm^2
in.^{4}	× 4.162 314	E + 01	= cm^4
lbf	× 4.448 222	E + 00	= N
lbm	× 4.535 924	E − 01	= kg
psi	× 6.894 757	E + 00	= kPa
psi^{-1}	× 1.450 377	E − 01	= kPa^{-1}
°R	°R/1.8		= K

*Conversion factor is exact.

Chapter 10
Blowout Control

10.1 Introduction

Blowouts may never be eliminated as long as man uses boreholes to tap the earth's energy resources. Situations leading to an uncontrolled flow can occur, even on well-engineered holes equipped with the best control devices available and manned by trained and experienced personnel. Some drilling conditions are uniquely subject to well-control problems. For instance, there is no consensus on how best to prevent or manage shallow gas flows. A uniformity of opinion would imply that the means to prevent shallow blowouts had been discovered, which is not the case. Human error is a major, if not the most important, cause of blowouts. The industry rightly allocates significant resources to prevent blowouts, including well-control training, but the human element will always be a factor.

The detrimental consequences of a blowout can be mitigated, if not entirely eliminated, by planning. Time is crucial when a well is out of control; a plan developed after the fact will be reactive instead of proactive. The financial, environmental, legal, public relations, and safety implications of a blowout will not be addressed adequately unless the operator has a contingency plan in place and ready to activate when that first call to the home office is made.

Emergency response planning for blowouts has received much attention.[1-7] A plan may be general or site-specific, depending on where the operation is located, the consequences of a blowout, logistical difficulties, geologic risk, and numerous other factors. Regardless of the final form, the plan should provide for a designated manager whose sole responsibility lies with the crisis at hand and who is the final authority on all decisions. Teams within the organization may be mobilized to deal with associated tasks such as surface intervention, pollution control, relief wells, and so on. Team leaders and members should not have any distractions until the project has been completed.

Blowouts are not an everyday occurrence. Most operators never experience a blowout and a major oil company may drill for years without having to deal with a wild well. It is to an operator's benefit to obtain the advice and counsel of blowout-control specialists when formulating a contingency plan. Much time and money will be saved and the task will be simplified greatly if the service companies and vendors (firefighters/well cappers, mud company, directional drilling services, etc.) are qualified and selected before the emergency.

10.2 Surface Intervention

In general, time and money will be saved if a surface blowout can be killed using the existing well and its equipment. Surface intervention involves fighting fires, clearing away debris, removing damaged equipment, and restoring flow control. Surface intervention also may be needed to prepare a well for subsurface intervention by means of a kill string. Our intent is to touch briefly on the associated topics. Details are provided in the cited references for those who want to further explore this interesting subject.

10.2.1 Firefighting. Influenced by the news and entertainment media, the public generally perceives a blowout as a massive explosion followed by a conflagration around the well. Those in the petroleum industry realize that fire is not associated with all blowouts, but they may be surprised to know that perhaps 85% or more never catch fire.[8] The main reason is that mobile water from any exposed strata flows into the well, mixes with the hydrocarbons, and thus inhibits if not altogether prevents vapor ignition. But a fire will occur if the oxygen/fuel ratio supports combustion and an ignition source is provided. The Flak and Matthews[9] overview of firefighting principles provided much of the following information.

Temperatures in the core of a blowout can exceed the melting point of steel. Radiant heat proximate to the flame can reduce the material strength to the point that the structure collapses in a matter of minutes. Maintaining a steady stream of water on the rig cools the structure and greatly limits fire damage. Firefighters arrive too late to prevent steel damage, but deluge systems have been incorporated into modern platforms and offshore mobile rigs for this purpose.

A large volume of water is needed before, during, and after the actual extinguishment. Drenching the men and equipment allows work in the vicinity of the blowout; water is used to cool the surrounding area and prevent reignition after the fire has been snuffed. Water is also needed when working around a blowout that is not on fire. It reduces vapor flammability and suppresses the potential for ignition from sparks or static electricity if it is directed into the flow stream and those areas where hydrocarbon accumulations can be expected.

The most popular extinguishment method is to use water if a straight vertical plume is blowing from the well. A powerful water stream displaces the hydrocarbon fuel away from the fire and the steam absorbs heat and displaces oxygen. Typically, three or four 1,000-gal/min monitors direct water flow into the base of the fire while another monitor sprays water along the plume.[10]

The venturi principle has been used to good effect in some cases. Basically, a joint of casing or "chimney" is placed over the plume and water is directed to the bottom opening. The flow velocity tends to suck any small lateral fires up into the chimney along with the water. The system's efficiency is increased and may allow killing a fire using less water. Injecting nitrogen into the base of a chimney reduces the oxygen supply and may be a successful procedure if water supply is limited.[11]

Dynamite has been used to extinguish oil and gas fires since the 1920s and remains an option today if water alone is ineffective. Explosives work because the blast carries the fuel away at a velocity faster than the flame propagates and the explosion briefly uses up the local oxygen. Charge sizes vary and may be as large as 500 lbm for a large fire. The explosives are packed into a small drum and insulated using a silicon-based cloth. A water stream protects the explosives from the fire as the canister is positioned in the flow stream and prevents reignition after the shot has been detonated.

Dry chemical and foam extinguishers also have application, but play a lesser role. Foam has been used to suppress ground fires around a blowing well and may become necessary for extinguishing a large oil fire spreading laterally from a well. Dry chemicals act as a smothering agent and are required if water supply is limited and explosives cannot be used. Some firefighters prefer to use chemical systems instead of explosives.[12]

Some cases, most notably a well spewing lethal H_2S concentrations into the atmosphere, have led operators to ignite a blowout intentionally.[13] An operator also may consider igniting an oil blowout to prevent pollution.[14] It seems paradoxical, but working around a burning well may be safer than exposing workers to the hazards of a potentially explosive vapor cloud. Such conditions as these have led operators to delay putting out a fire until absolutely necessary for the selected capping procedure. Indeed, capping a burning well has become an established technique.

10.2.2 Removing Debris and Wellhead Equipment. The first priority in a surface blowout is to clear access to the wellhead. In the event of a fire, twisted metal and other debris must be cut into pieces and dragged away. Usually, the derrick and upper platform decks will have to be removed before capping equipment can be manipulated over the hole. The blowout preventers (BOPs) and wellhead will be ruined if the well was on fire and must be removed while the well is still blowing. Even without a fire, BOP/wellhead removal often is necessary because of defects, misuse (closing the blinds on pipe for instance), damage, or questionable reliability.

Large bulldozers can be used to drag materials away from the well. Cranes and Athey wagons are other conventional methods for conveying hooks, rakes, and other tools for removing debris. **Fig. 10.1** shows an Athey wagon, which is a tracked vehicle with a long boom extending from the tracks to the work area. A bulldozer is used to maneuver the wagon and the bulldozer's tail winch articulates the boom. A debris hook suspended from the end of the boom or crane can be used to remove junk or, by repeated bending, mechanically fatigue and break up large components.

Other severing techniques described include using reciprocating cables, cutting torches and thermal lances, shaped-charge explosives, pneumatic lathe die cutters, and hydraulic jet cutters.[15-17] Many of these tools have been in the firefighters' arsenal since the early days and technology developments have been a matter of improving the existing techniques. This has been the case particularly with the abrasive cutting tools.

Hydraulic jet cutters use a high-velocity abrasive material carried in a water medium to cut through steel. Their primary advantage is in the speed of the operation as compared to some other methods. For instance, cutting through casing with a reciprocating swab line can take days whereas a jet cutter can accomplish the same task in an hour or two.

The ultra-high-pressure (UHP) cutter shown in **Fig. 10.2** has a small nozzle (less than a millimeter in diameter) that directs a mixture of crushed garnet and clean, distilled water to the cutting surface at nozzle pressures as high as 30,000 psi. The pump rate requirement is less than 5 gal/min, which may be advantageous if water supply is limited. The device is clamped above and below the cut area and the the nozzle is carried on a tractor whose speed and position is controlled from a remote location.

Fig. 10.3 illustrates a schematic of the Hydra-Jet cutter. The device can be attached to an Athey wagon boom and is stabilized by secure contact between the cutter and the target. For the horizontal cut shown in the diagram, a gelled water/sand slurry is directed through two 0.1875-in. nozzles at a combined rate in excess of 150 gal/min and at pump pressures between 8,500 and 12,000 psig. A hydraulic motor advances the cutters along the yoke using a drive screw mechanism and a pivot allows a cut to be made at any angle. One advantage to the Hydra-Jet cutter is that the tool is not manually attached to the target. Hence fire or an otherwise hazardous environment near the well is not an obstacle as long as the device is kept cool with a stream of water. This is a hydraulics-intensive operation, however, and requires more water and pump horsepower than the UHP cutter.

Precise control when cutting the last pipe section may require a pneumatic cutter if the procedure involves stripping away the outer strings to expose the production casing. **Fig. 10.4** shows a common split-type cutter. The device is clamped to the pipe and the die makes a cut as it rotates around the circumference of the tube. The resulting cut is uniform and has a beveled edge.

Damaged or suspect wellheads and blowout prevention equipment (BOPE) components often can be removed by first clamping the two flanges together and removing the bolts. For a controlled procedure, a crane is used to secure the top part of the equipment and snub lines are threaded through the bolt holes or shackles to provide lateral stability.[17] After the clamp is removed. the crane and snub-line winch pull the component off the well. Another method simply uses bulldozers to yank the clamps off and allows the brute force of the well to blow the component free from the wellhead.[19]

10.2.3 Capping Methods. Capping operations may commence after the damaged equipment has been removed and the flow directed vertically. Fires usually are extinguished by this time, although some situations are best managed by installing capping equipment while the well is still on fire. A

Fig. 10.1—An Athey wagon in operation near a burning well. Courtesy of Cudd Pressure Control.

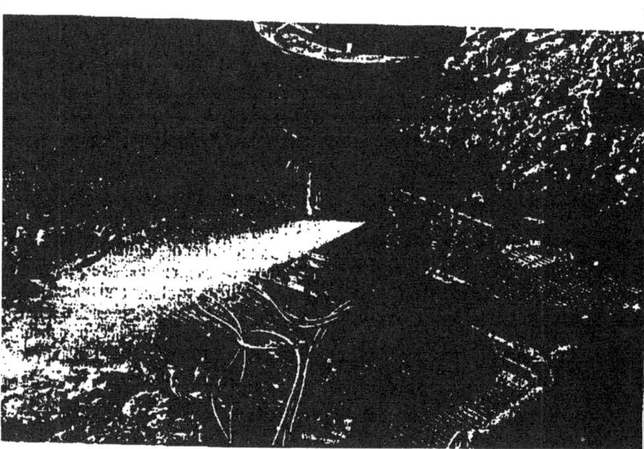

Fig. 10.2—Ultra-high-pressure jet cutter in operation.[18]

flanged capping stack or kill spool can be used if the wellhead is present. Otherwise, a means to attach a capping stack to pipe must be provided.

Figs. 10.5a and 10.5b illustrate a conventional capping and a basic kill spool. Both have a flange on bottom and are secured to a mating flange on the wellhead. The only major differences between the two hookups are the redundancy features and the type of closure device. Both assemblies are maneuvered over the flowing well stream with an open bore and lowered into position so that the flanges can be bolted together. Once a seal has been effected, lines can be installed on the side outlets and the ball valve or lowermost blind ram is closed. Lead may provide an adequate seal if the wellhead is damaged.[15] One method is to mount a lead ring into the capping assembly's ring groove or another groove machined on the flange face inside the existing groove. The lead is deformed when the flanges are mated and fills in any cracks or minor deformations on the wellhead flange. Another type, the "lead smasher" spool has been used with success in cases where the wellhead has been damaged severely, most notably in Kuwait. This spool swallows the entire head and the lead seal is energized when the spool is bolted to the wellhead's bottom flange or to the flange below the damaged head.

Fig. 10.6 illustrates the spin-on method for installing flow-control equipment to a flange. A long bolt stud is placed in the bottom flange of an open-bore stack and a lever arm is attached to the spool. A ring gasket is tack-welded to the groove and a crane or boom maneuvers the assembly into the position shown in the diagram. A long stud dropped into the mating bottom flange acts as a hinge. Personnel push the lever arm, spin the stack over the flow, and install the remaining bolts after the stack is lowered a short distance.

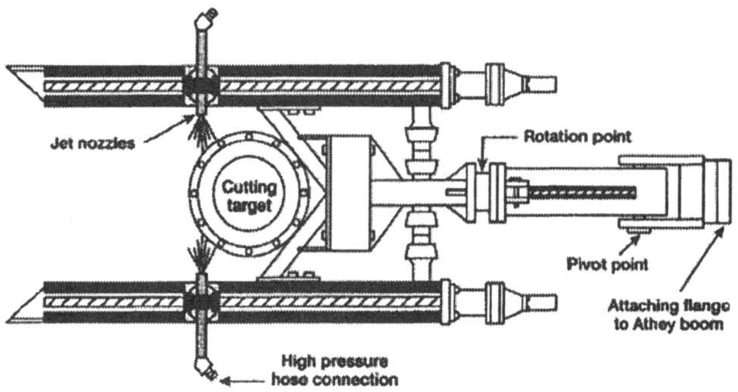

Fig. 10.3—Using the Halliburton Hydra-Jet cutter to sever wellhead bolts.[19]

This technique is not an option if the effluent velocity endangers personnel when flow is diverted sideways or if the well is on fire. The snub-down procedure shown in **Fig. 10.7**

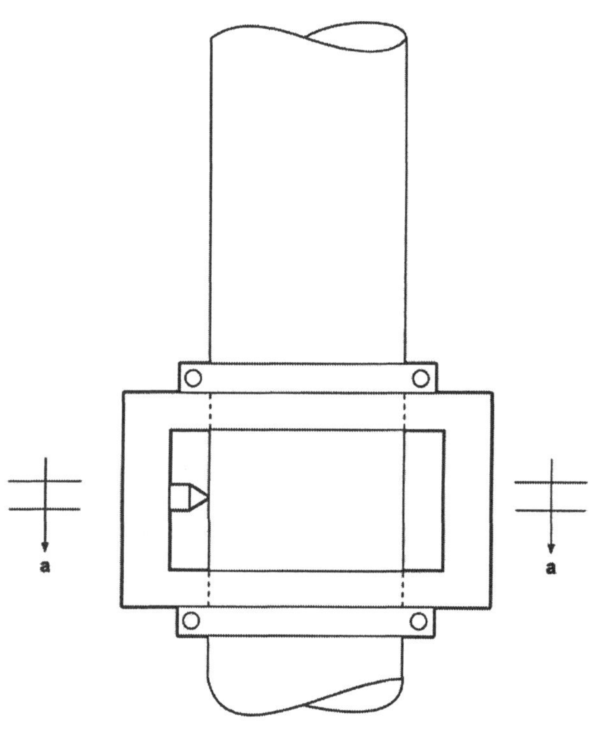

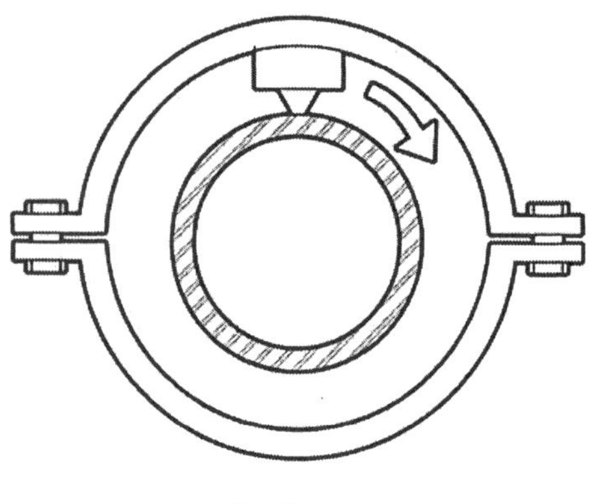

Fig. 10.4—Pneumatic cutter schematic.

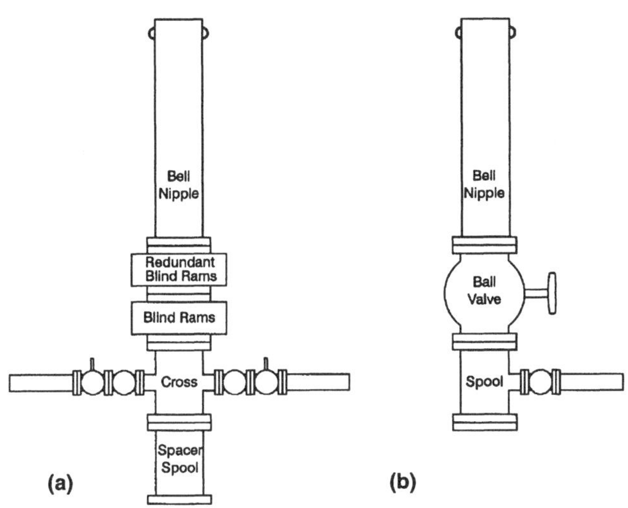

Fig. 10.5—Capping assemblies with a bottom flange.

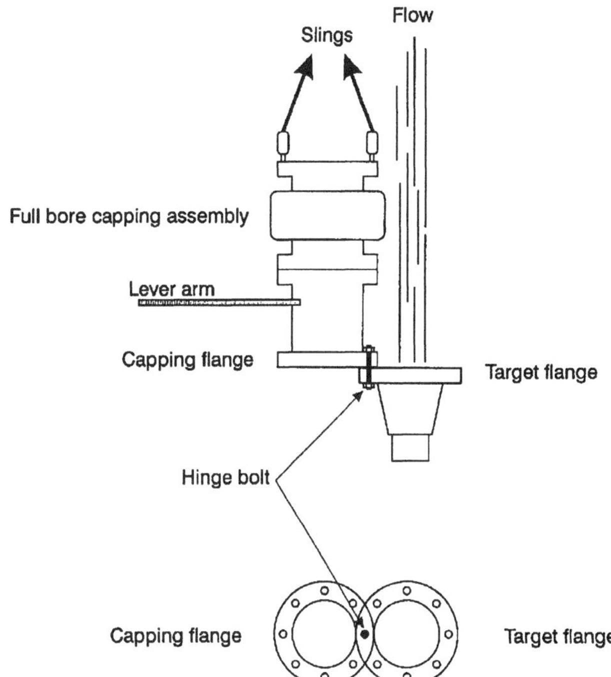

Fig. 10.6—The spin-on technique for installing a capping stack to a flange (after Abel *et al.*[17]).

is necessary in these situations. As in the spin-on method, a ring gasket is tack-welded to the stack's flange. Two snub lines are strung through the bolt holes of the flanges to be mated and a tag line is installed to provide stability along the orthogonal lateral axis. The equipment is pulled over the well when the snub lines are winched in.

Two capping stacks designed to attach to a pipe stub are shown in **Figs. 10.8a and 10.8b.** The installation procedure for Fig. 10.8a is described in Ref. 17 and is represented schematically in **Fig. 10.9.** Note that a clamp has been affixed to the casing and that the snubbing lines in this case are threaded through a pair of sheaves that have been attached to the clamp. The stack assembly is lowered to swallow the casing and another clamp is positioned below the casing spool. Hydraulic jacks are placed between the rigid bottom clamp and the loose upper clamp and conventional casing slips are placed in the bowl of the spool. The seals are energized and the slips are set when the two clamps are jacked apart. The upper clamp is tightened then to maintain the slips in compression before the the jacks are released.

The three-ram assembly illustrated in Fig. 10.8b can be installed on casing, drillpipe, or tubing. Positioning the stack over the pipe is done in the same manner as the previous method. After the stub has been swallowed, the slip rams are closed thereby attaching the stack to the pipe. It should be apparent that severe bending moments could be applied to the pipe if the stack is free to move in the lateral direction. The pipe rams are closed and, being inverted, will seal against shut-in or pumping pressures from above. This assembly with a long riser between the pipe ram and cross was used to attach BOPs to severed casing beneath 95 ft of water.[20]

The post-capping procedure depends on the condition of the well and other factors. Normally, flow from the well is diverted initially to a safe location. The well can be closed in at the choke if the surface- and downhole-pressure limitations are not exceeded. Otherwise, the well may need to be killed from surface using a method that poses less risk. Other options include snubbing a kill string or fishing assembly while

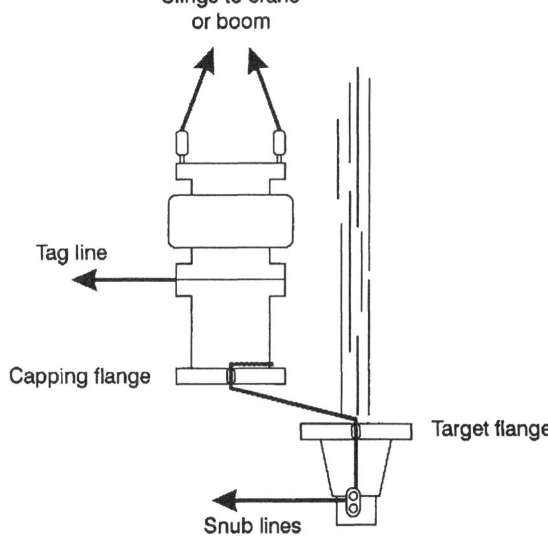

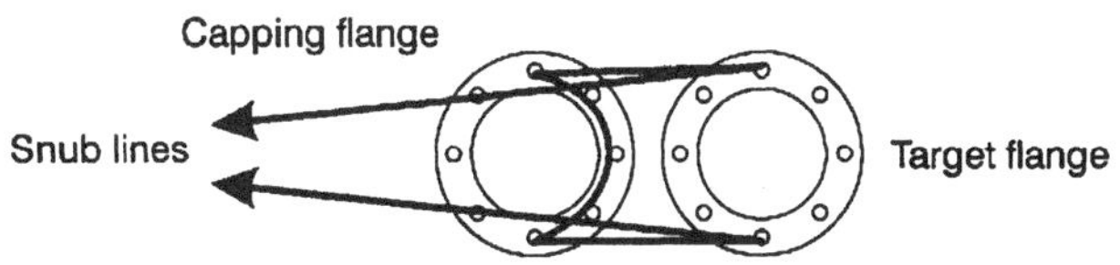

Fig. 10.7—Using snub lines to install a capping stack (after Abel *et al.*[17]).

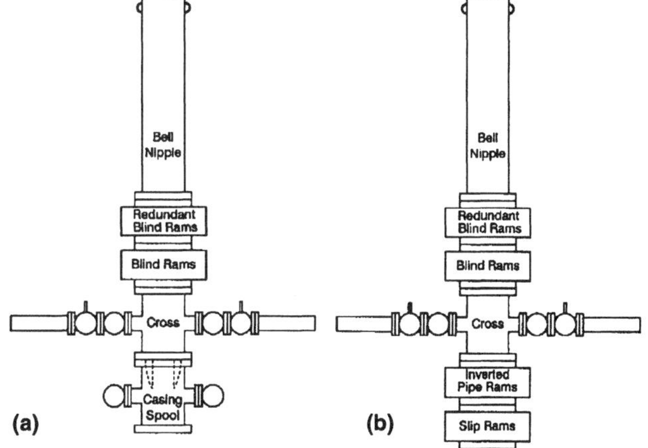

Fig. 10.8—Capping stacks used when a pipe stub is looking up.

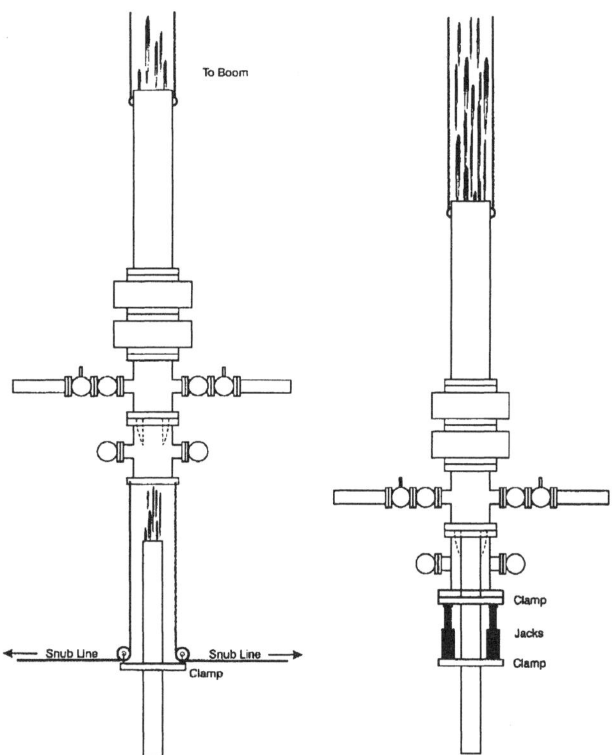

Fig. 10.9—Installing a capping stack on a pipe stub.

Fig. 10.10—Inserting a stinger into a wellhead.[12]

the well is on diverter. Some wells have been diverted into an emergency sales line until reservoir-pressure depletion allows an alternative procedure.

10.2.4 Stingers and Inflatable Packers. A stinger can be used to kill a well without having to remove a damaged wellhead first. The procedure may be an option if the well can be bullheaded and certain other conditions are met. A stinger, shown in **Fig. 10.10,** is an open-bore sub with a taper on the bottom end. The assembly described in Ref. 12 consists of a stinger on bottom followed by a changeover, pumping tee, valve, drill collar, and 45° ell. Changeovers adapt to the stinger size needed for the job and contain lugs for installing snubdown equipment. The tee is tied back to the pumping units and the full-opening valve remains in the open position until the stinger is in place. The drill collar provides weight and stability and is connected to a crane or boom with a saddle clamp. The purpose of the ell is to divert flow away from the suspension equipment.

The tool is inserted into the wellhead cavity with sufficient force to keep the pressure-area force from ejecting the equipment when the well is shut in or at any time during the bullhead procedure. The rigging and tie-downs can handle approximately 35,000 lbf and so the technique is not suited for high-pressure applications.[17] The seal is imperfect, particularly if the bore is oval or cracked. The leak can be plugged by injecting "junk shots" (polypropylene rope sections, ball sealers, or other suitable material) through the stinger until the material bridges and seals.

A stinger-type assembly has been used in Kuwait and other places to convey an inflatable packer a short distance into a blowing well.[21] The objective in these cases is to get the packer below the damaged wellhead equipment and, perhaps, some split casing. Once inflated, the packer isolates the damaged equipment and allows the well to be closed in and bullheaded.

10.2.5 Other Hot-Well Techniques. Hot tapping refers to the process of drilling a hole through the wall of a pressurized vessel. The procedure thus provides a port through which pressure can be relieved or fluids can be pumped. There are numerous routine hot-tap applications and most are not associated with blowout control. Nonetheless, it is an important tool for solving some blowout-related problems as well as other, less serious well-control situations.

Fig. 10.11 illustrates a schematic of a hot-tap tool in operation. The saddle clamp is installed on the pipe, its packoff seal

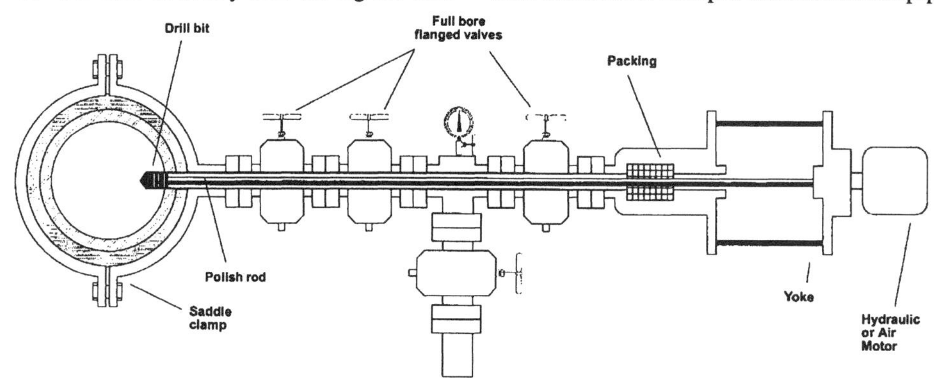

Fig. 10.11—Using a hot tap to drill into pressurized pipe.

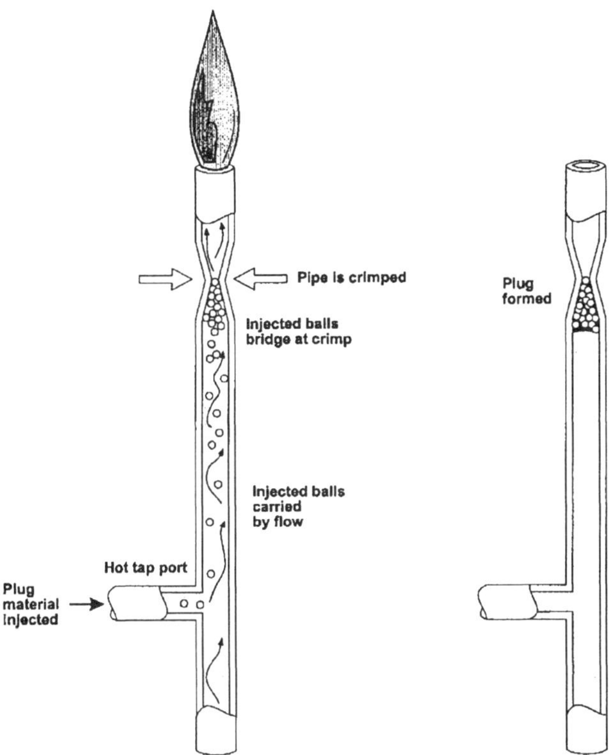

Fig. 10.12—The crimp-and-plug technique (after Taylor[22]).

is energized, and a valve/tee arrangement is placed outboard of the saddle. An undersized bit is installed on the end of the polished rod and a hole is drilled while maintaining a slightly higher pressure in the tool bore than is present in the pipe. Torque may be applied by hand, air, or hydraulics through a threaded system that produces the bit force. A pressure drop indicates the pipe has been penetrated and the bit is withdrawn. The valve is closed, pressure is bled from the tool, and a larger bit (typically 1-in.) is used to ream out the hole. Equipment inboard of the redundant valves serves as the access port after the tapping procedure is completed.

Fig. 10.12 illustrates a hot-tap technique to access an offshore blowout as described by Taylor.[22] The well was blowing out through the tubing string above the rotary table of a jackup rig. A port was hot tapped into the tubing and hydraulic jaws were used to mash the pipe above the tap. Steel balls were injected to create a bridge; a plug was created by injecting ball sealers followed by string rubber.

Similar bridging/plugging materials have been pumped through hot taps and other outlets to seal leaking valves, flanges, and ram preventers. Steel-reinforced rubber and Kevlar® fibers have been used for high-pressure sealing applications.[16] Time is a critical factor because small leaks can erode into big leaks that cannot be plugged with a junk shot.

A variation of hot tapping is drilling through a ball or gate valve that is stuck in the closed position. The only difference between the two techniques is that a saddle is not needed because a flange or threaded connection is available for installing the device. One application from the author's experience was drilling out a kelly cock that failed to open because of an extremely high shut-in drillpipe pressure.

Freezing is a means to create a temporary block in a pressure vessel so that an inoperable valve, BOP, or wellhead component can be removed and either replaced or repaired. The basic technique uses a hot tap or other access port to inject a bentonite/water slurry into the vessel. After placement, a bolt-on split-tub assembly fabricated from a 55-gal drum is packed with dry ice and positioned opposite the slurry. The dry ice draws heat from the pipe and, over time, the slurry freezes into a strong plug capable of containing extremely high pressures from below. Freezing will not work in a flowing well and thus is not strictly applicable to blowout control.

TABLE 10.1—COILED TUBING USE IN BLOWOUT CONTROL

Advantages

1. The operator can move the string under surface pressure while pumping.
2. No threaded connections eliminates the need for personnel to be over the well during most of the operation, allows for a faster operation, and facilitates positioning of the string at a given depth.
3. Remote operation of hydraulic functions enhances safety.
4. Limited equipment requirements over the well reduces the loss exposure.
5. Technological improvements in injector heads, BOP equipment, coiled-tubing strengths, and pipe diameters have made the system more suited for well-control work.

Disadvantages

1. Equipment availability is limited in many areas.
2. Many downhole tools are not designed specifically for use on coiled tubing.
3. Small pipe diameters are plugged more easily with lost circulation material (LCM) and limit pump rates.
4. Flapper valves in the string tend to erode while pumping solid-laden fluids at high rates.
5. Stuck pipe with free point above the release joint poses some operational problems and potential well-control hazards.
6. Rig-up may be difficult if a drilling rig is over the hole.

10.3 Subsurface Intervention

As defined herein, subsurface intervention means that a string of pipe is used in the blowout well to effect the kill. Underground blowouts usually are controlled in this manner if the workstring and surface equipment are accessible and intact. Running pipe into a diverted surface blowout or a post-capped shut-in well may be an option also if one of the surface kill procedures discussed later in the chapter poses a hazard or simply will not work.

Every job is different. The procedures required to get pipe deep enough to effect a kill depend on the present hole conditions. Whatever must be done (fishing, removing restrictions, etc.) to achieve a successful subsurface intervention requires the ability to run and retrieve the workstring under pressure. This usually entails a snubbing and stripping operation with considerations as given in Chap. 9.

Coiled tubing (CT) has been used to intervene with blowing wells and its use as a blowout-control tool will expand. Adams *et al.*[23] discussed the advantages and disadvantages of coiled tubing and **Table 10.1** describes their observations. A diverted blowout in a producing well is one of the more ideal applications, and controlling a blowout in a well drilled with a CT unit has the advantages listed in the table. However, getting the injector head onto a drilling rig floor may not be possible during a blowout and, if it can be done, the head must be connected back in some way to the top flange of the BOP stack.

CT was deployed in a novel way on a propane storage well blowing out through a shallow casing leak.[24] The well was on fire and it was impossible to rig up the injector head in a conventional manner. A stinger with an entry guide on top was shoved into the tubing head and a crane was used to hoist the injector head and an 80-ft-long lubricator above the well. The

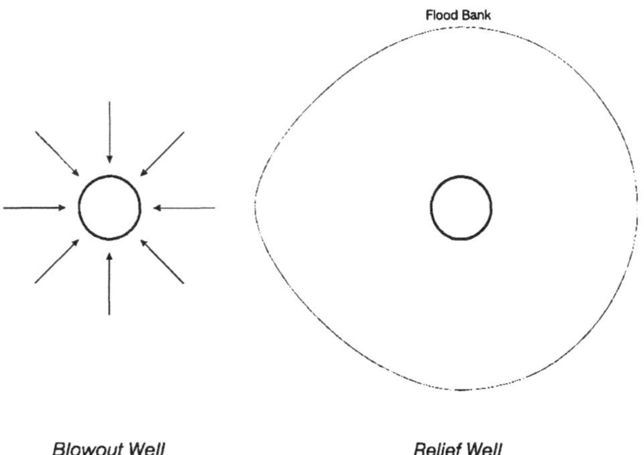

Fig. 10.13—Waterflooding a blowout reservoir from a relief well.

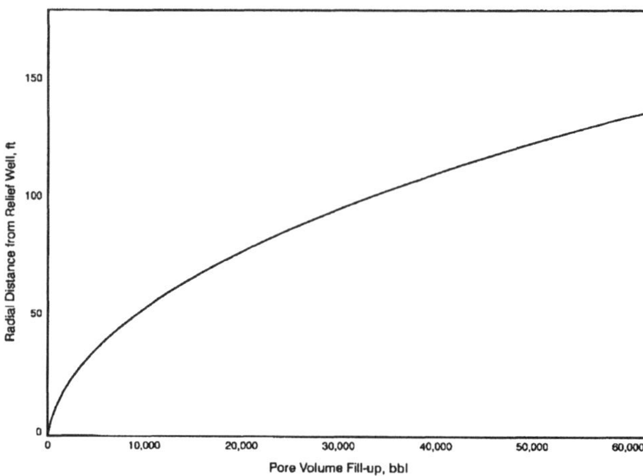

Fig. 10.14—Predicted pore volume fill-up as function of well spacing (after Lewis[32]).

lubricator was stabbed into the stinger and the well was killed ultimately by severing the tubing and setting an inflatable plug in the casing.

Semisubmersible rigs (semis) offer several advantages in offshore well control and it may be possible to re-enter a subsea blowout if a semi can be positioned safely over the hole.[25,26] A blowout in deep water offers the best opportunity for such an intervention procedure.[27] The primary hazard of a blowout from a floater is when gas surfaces in close proximity to the rig. This hazard is reduced greatly in deep water because currents disperse the gas and carry the surface boil a safe distance from the vessel.

10.4 Relief-Well Intervention

Killing a blowout from a relief well may be necessary if no other options are available. Such conditions include inaccessible surface conditions, lack of casing integrity, or a capping procedure that otherwise risks broaching the shoe. The ability to cap a well may be unknown until after much time has been spent in the surface intervention and, for this reason, plans for a relief well are instituted often at the same time the capping operation commences. The operations are conducted in tandem until the surface intervention is successful or the capping effort is abandoned.

Relief wells are no longer considered a last-resort method. Indeed, relief wells compete on equal terms with surface intervention methods in many situations. Technology now provides the ability for a single well to intersect a target directly with a minimum number of sidetracks. The modern capabilities can be attributed to improved ranging tools, better directional surveying methods and procedures, the advent of steerable drilling systems, and the overall expertise gained in planning and drilling relief wells.

10.4.1 Kill Techniques. In the past, relief wells were vertical holes that were drilled around a blowout and produced at high rates to "relieve" the pore pressure in the uncontrolled reservoir. The concept is applied still to vent wells drilled to depressure a zone that has been charged from an underground blowout.[28] But, killing a blowout in this manner could take months to accomplish, involve much waste, and usually is not considered in this day and time.

There have been some exceptions, however. Flak and Goins[29] discussed one application where a dual well was blowing out from two sets of perforations. The wellhead was inaccessible; at least one of the tubing strings had parted; and the flow was lifting a large volume of water from a shallow casing leak. Both zones would have to be killed simultaneously, either from the same relief well or two wells dedicated to each reservoir, if conventional methods were employed. The problem was resolved by positioning a well near the blowout, completing both zones, and producing at high rates. The pressure drawdown at the blowout well combined with static reservoir depletion caused the well to load up with salt water and die.

The year 1933 marks the first time a relief well was drilled directionally to a target bottomhole location (BHL) near a blowout.[30] This well was killed by pumping a large volume of water into the reservoir and destroying the relative permeability to the flowing hydrocarbons. The waterflood technique was the customary approach well into the 1970s and still may be considered as a kill option.[26]

Fig. 10.13 is a conceptual representation of the waterflood technique in a homogenous reservoir. The flood bank grows radially from the relief well until, influenced by the flowing pressure drawdown, it begins to expand preferentially towards the blowout. Reservoir engineering tools are needed to predict how much water must be injected to achieve pore volume fill-up and for optimizing the pump rate.[31,32] **Fig. 10.14** demonstrates that the relationship between fill-up volume and well proximity is not linear. Faster results at less expense will be realized if the relief well is bottomed as close as possible to the blowout. After breakthrough, continued injection should cause the flow to diminish gradually to the point where the well dies.

A flood kill is not applicable to every blowout. The reservoir must have sufficient permeability to allow injecting large water volumes into the rock matrix at reasonably high pump rates. Exceeding the fracture pressure will doom this procedure. Also, a waterflood probably will fail if the static reservoir pressure exceeds the hydrostatic gradient of the water.

In 1971, Shell successfully killed several blowouts on an offshore platform by pumping water into an array of relief wells followed by weighted mud after water breakthrough was observed.[33] Whole mud will not enter a reservoir matrix (absent vugular or gravel-type porosity) and it was posited at the time that mud communication had been established through hydraulic fractures between the relief wells and blowouts.

Consider, however, the likelihood of a fracture propagating tens of feet and intersecting a wellbore with a diameter measured in inches. Granted, the pressure drawdown near a blowout will alter the stress regime, perhaps enough to affect frac-

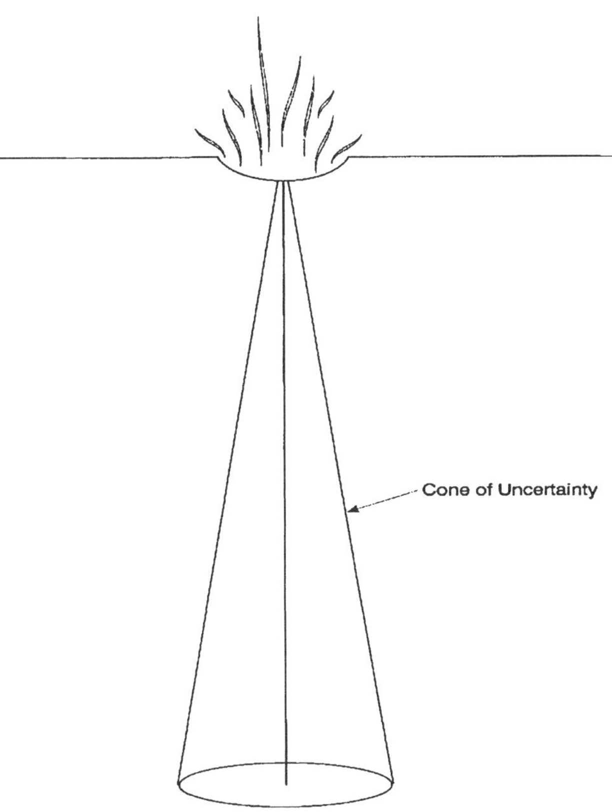

Fig. 10.15—Cone of uncertainty around a blowing vertical well.

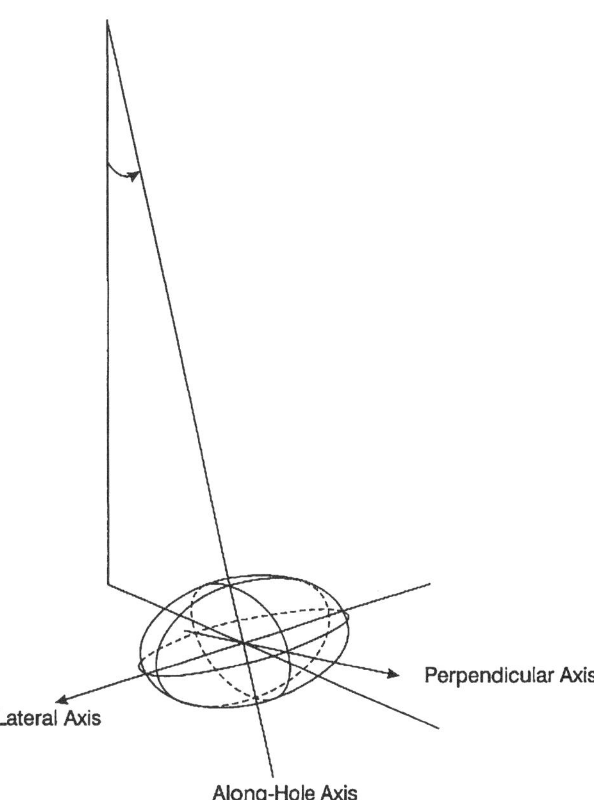

Fig. 10.16—The ellipsoid of uncertainty in a straight, inclined hole (after Wolff and de Wardt[38]).

ture orientation. But a fracture will initiate in a direction governed by the stress state at the relief well and turn only if a stress change is encountered at the fracture tip. Another mechanism is more likely. The wells produced a lot of sand after breakthrough and it is reasonable to conclude that the water eroded flow channels sufficiently large for the mud to traverse.[29]

Another milestone in relief-well technology occurred when the first dynamic kill was pumped to control a massive blowout in Indonesia.[34] The dynamic kill has become the most popular relief-well-kill procedure. We will get into the details later, but one of the most important requirements is to establish an effective hydraulic conduit between the relief well and the blowing wellbore. The referenced blowout was from a vugular carbonate reservoir and communication was not difficult to achieve across the estimated 27-ft distance between the two wells. Even so, between 25 and 35% of the injected water was lost to the rock during the procedure.

Leak-off efficiency is an important factor if the kill fluids must travel a considerable distance through rock and the closer the better if the objective is to establish communication in a reservoir with good permeability. The fracture pressure will inhibit effective communication in tighter formations or rock with relatively low frac integrity. Indeed, obtaining a conduit through the reservoir may be impossible in deep wells regardless of the spacing.[10]

In 1970, a relief well was drilled with the intention of directly intersecting the casing on a blowout in a high-pressured sour-gas blowout in Mississippi.[35] The job was successful because of good planning, skilled execution, and some amount of luck. The casings were in intimate contact for about 100 ft and the two wells were connected when the relief well was perforated between 10,543 and 10,553 ft (about half the depth to the uncontrolled Smackover). The well was killed by pumping a large volume of cement.

State of the art assures that almost any well can be intercepted directly by a relief well. Direct communication eliminates the problems associated with reservoir conduits, allowing for more precision in design and a successful kill with less volume and pump rate. As in its first application, a wellbore intercept also gives the operator the option to pump into the blowing well at any depth.

10.4.2 Position Uncertainty. The need for accurate borehole placement has been discussed, but the means to get there has not. Consider the problem for a cratered blowout in a 10,000-ft "vertical" well. Directional surveys are more uncommon than not when trajectory is not controlled and assume that the only available data are periodic inclination measurements that average ½° from vertical. The maximum horizontal departure at any depth assumes the prior inclinations were in the same direction. At total depth,

$$x_{\max} = (10,000)\sin(0.50) = 87 \text{ ft}.$$

Fig. 10.15 illustrates that the wellbore could be located anywhere within a "cone of uncertainty" that expands with depth. It would be impossible to place a relief well close enough to do a flood kill, much less achieve a direct intersection, based on inclination measurements. Surveying offset wellbores may give the area drift tendencies and thus yield a more educated guess for the blowing well's BHL.[36] But even having actual directional surveys on this well still would lead to positional uncertainty because of inaccuracies inherent to the measurements.

Walstrom *et al.*[37] investigated survey errors, concluded they were random in nature, and introduced the "ellipsoid of uncertainty" concept. At any survey station, the errors

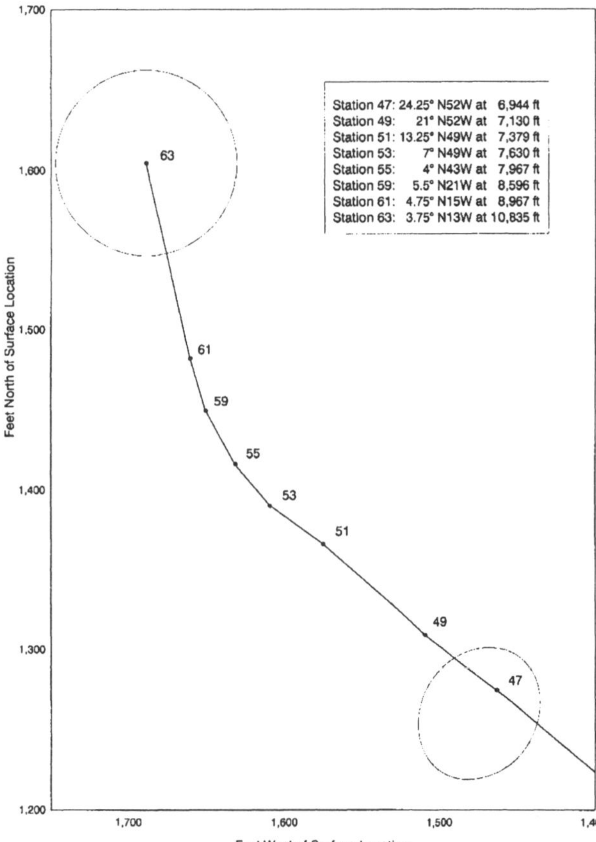

Fig. 10.17—Plan view of a directional well and the ellipses of uncertainty (horizontal plane) at Stations 47 and 63.

associated with wellbore inclination, direction, and depth measurements result in a volume within which the bottom of the hole could be located. As shown in **Fig. 10.16,** the volume is described by an ellipsoid with axes along the hole, lateral to the hole, and perpendicular to the hole.

Later authors[38,39] demonstrated that surveying errors tend to be systematic rather than random. Systematic errors can be attributed primarily to tool misalignment, instrument tolerances, invalid directional reference, magnetic interference from the drillstring as it affects compass readings, inconsistent gyro drift, ineffective gyro gimbaling at high inclinations, the effects of gravity and wireline pull on inclination readings, and imprecise depth measurements. Systematic errors are noncompensating and their cumulative effect increases with depth.

Surveying errors also lead to uncertainties in a relief well's BHL. It is impossible to eliminate errors completely; advances in reducing the overall impact include the introduction of small north-seeking rate-gyro systems for improving accuracy in cased holes, more reliable measurement while drilling (MWD) data and faster transmission times, better understanding and correction of the errors associated with MWD and other methods, and enhanced quality control measures.[30]

Fig. 10.17 shows a portion of the computed well path for example survey data given in *Applied Drilling Engineering*.[40] Position and dimensions of the ellipsoids at 6,944 ft and 10,835 ft were determined using typical error values for a "poor" magnetic survey in the North Sea[38] and the projections onto the horizontal plane are shown on the plan view. Note that there is more uncertainty at total depth than shallower in the slant section. Also, the ellipsoid projection approximates the shape of a circle at low hole inclinations.

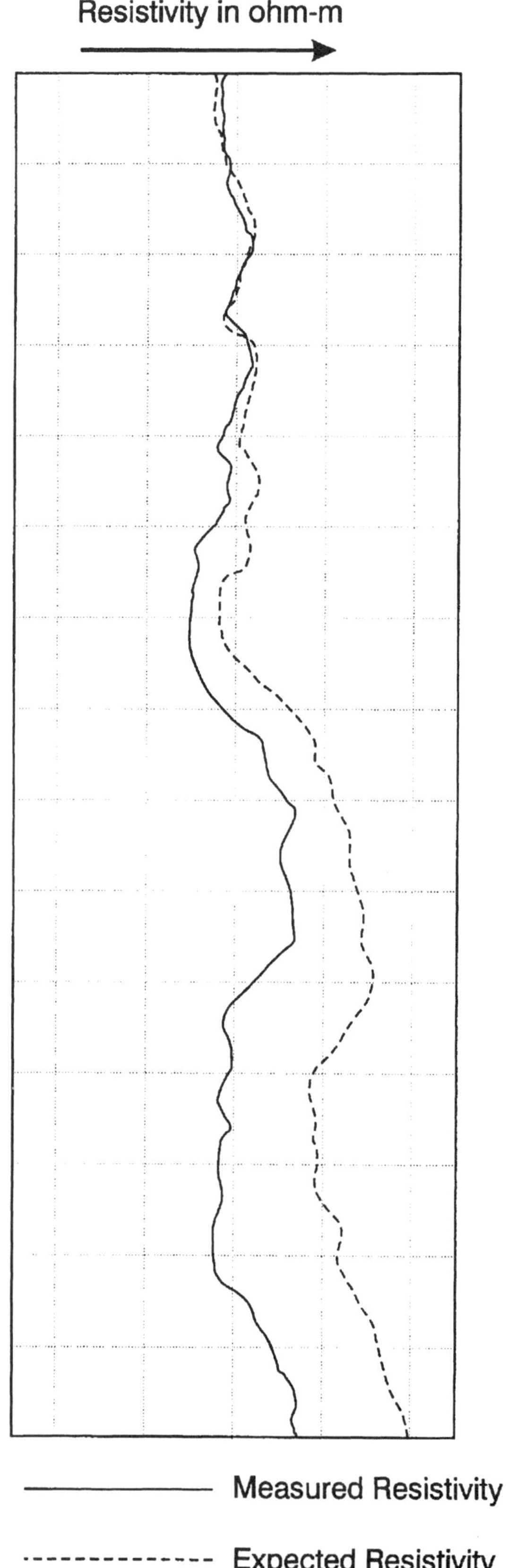

Fig. 10.18—Long-spaced electrical log response when a relief well is approaching a blowout.

Say that this well was blowing out and we needed to direct a relief well to intervene with the blowout near total depth. Despite the best surveying tools and quality control, we cannot assure hitting the target bulls-eye because of positional uncertainties in the relief well. Even if we could, the actual

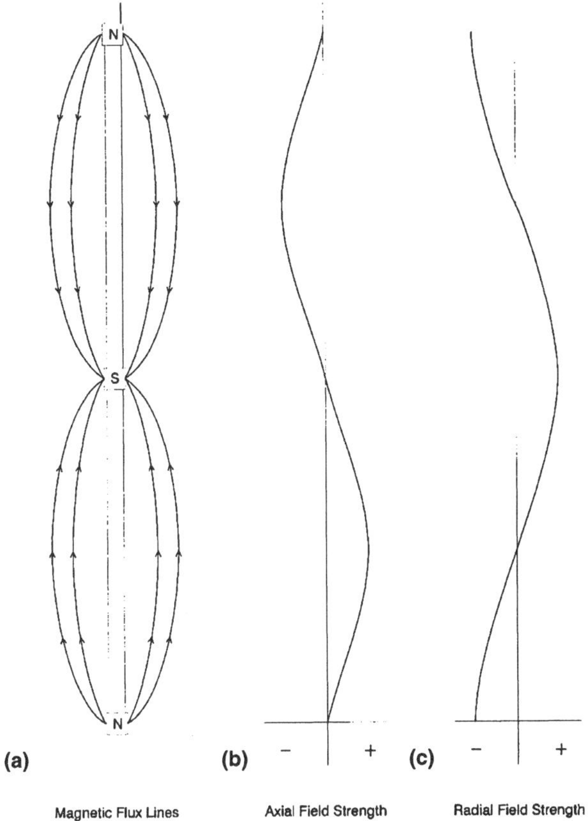

Fig. 10.19—The remanent magnetic field associated with collared pipe.

target could be anywhere in the 56-ft radius circle which defines the blowout well uncertainty. Obviously, some method for sensing the subsurface location of the blowout wellbore must be used if a relief well is to be successful.

10.4.3 Proximity Logging. Logging tools designed to sense the presence of steel through the earth have been devised for detecting a blowout from a relief well. The long-spaced electrical log was one of the earlier methods. The device is a normal-type tool with a long, variable spacing between the A and M electrodes. The spacing allows for resistivity measurements far out into the rock, a capability first used to map salt structures.[41] The tool's first application in a relief well was in 1970.[35,42]

The casing or drillstring in the blowout well short-circuits the current produced at the tool and yields a lower resistivity reading than the adjacent rock bed. The expected resistivity of the formation is modeled using conventional resistivity logs. As shown by **Fig. 10.18,** an increasing divergence between the expected and observed resistivities indicates that the relief well is approaching steel. The ratio of the two resistivities is used to predict distance between the wells.

Theoretically, an efficient normal log system (soft rock absent thin-bed effects) has a radius of investigation equal to twice the electrode spacing. Hence one might expect the ability to detect steel from as far away as 300 ft with a customary 150-ft A/M spacing. The practical detection limit, however, is probably no greater than 100 ft.[39] The inability to predict direction to a blowout and some unavoidable inaccuracies in the resistivity model have limited the method's use since the introduction of the magnetic methods.

Steel tubular goods usually have a residual or remanent magnetism owing to electromagnetic inspection and, to lesser

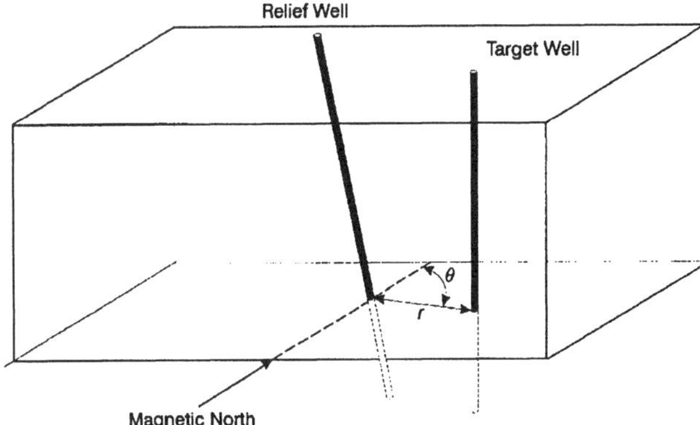

Fig. 10.20—Range and direction from a relief well to a blowout well at the point of closest approach.

extent, induction from the earth's magnetic field. Discontinuities or changes in mass at the tool joints or collars produce a magnetic pole, as does magnetic inspection of the connections. **Fig. 10.19** illustrates the magnetic field associated with collared pipe while the axial and radial components at a specified distance away from the string are shown in Figs. 10.19b and 10.19c, respectively. The strength of the field depends on the degree of remanent magnetism.

The first magnetostatic system was developed in 1970 and one became commercially available a few years later.[43,44] Magnetostatic tools home in on a blowout by detecting the disturbed magnetic field proximate to the well's drillstring or casing. The relative distance and direction between the two wells can be inferred from the data. **Fig. 10.20** depicts a vertical blowout well and an inclined relief well that comes closest to the target at the bottom of the box. The range, designated as r, is the distance of closest approach while θ denotes the direction as referenced from magnetic north.

The magnetostatic tool discussed in Ref. 44 is equipped with sensitive magnetometers that measure the axial and radial components of the magnetic field along the relief wellbore axis. Two axial field magnetometers are set at orthogonal direction in a gradiometric configuration. These sensors are used to establish the small, consistent gradient of the earth's magnetic field above the target zone. Upon approaching the target, the magnetic field emanating from the pipe is indicated by a change in the total gradient. The proximity of the two wells is calculated from the axial sensor measurements.

Note from Figs. 10.19a through 10.19c that each joint is a magnetic dipole and any pipe string is thus a series of connected dipoles. However, it is convenient to consider a dipole as being two impulse monopoles of equal but opposite strengths when modeling the range. For a monopole source, subtracting the earth's contribution to the field along the relief well axis yields the axial component caused by the monopole. As shown in **Fig. 10.21,** its magnitude increases as the relief well enters the altered field, goes to zero at the point of closest approach, and then undergoes a change in polarity. The range is estimated using the relation

$$R = l/(2)^{0.5}, \qquad (10.1)$$

where l = the distance between the maximum and minimum readings along the relief well.

The impulse monopole assumption leads to ranging errors if the distance is less than approximately 20 ft. Jones *et al.*[45] showed that the poles are distributed or "smeared" along the pipe with a strength that decays exponentially with distance from the ends. The authors presented an exponential-pole

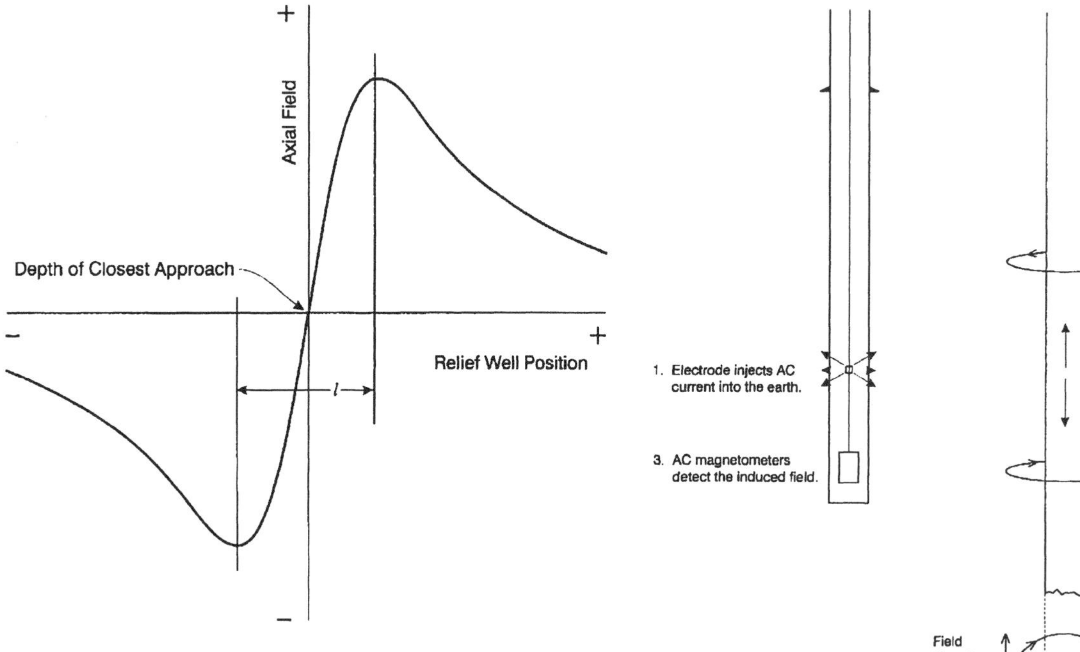

Fig. 10.21—Axial magnetic field of a monopole source as function of relief well position (after Jones *et al.*[45]).

Fig. 10.22—The electromagnetic method for detecting range and direction to a blowout.

model that better approximates a pipe string and thus improves the accuracy of the ranging computations.

Measurements from the radial sensors are used to determine the relative direction and can also be used in the range predictions. The radial field measurements in an openhole section above the target zone are that of the earth, which will not change if adjustments are made for any changes in tool inclination and orientation. Therefore, the radial field components arising from the source can be obtained by subtracting the contribution of the earth from the observed measurements. The direction to the blowout is determined by adding the earth and source vectors and using basic trigonometry to solve for the angle.

The ranging capability of the magnetostatic tools depends on the strength of the source field, which is an unknown parameter in most cases. We can expect a remanent field detection range varying from several feet to, at best, approximately 50 ft. Casing and drillpipe that have been recently inspected present the best targets. Detecting flush-joint pipe may be a problem because the tubes do not have a concentration of metal at the ends.[46] However, flush drill collars usually make good targets because the connections are inspected frequently.

In Ref. 47, de Lange and Darling demonstrated that the magnetostatic detection range could be increased to approximately 100 ft by magnetizing one or more joints of casing with an induction coil during the running procedure. The magnetized joints could be positioned strategically in the string at the optimum depth for a relief well intersection.

The electromagnetic method was invented in 1980 for use in a Louisiana relief well.[48] **Fig. 10.22** illustrates the physical principles of the system. The device is run into the relief well on wireline and an alternating current (AC) electrode emits an electric current into the formation. The current flows symmetrically into the low-conductivity rock and will dissipate rapidly with distance unless it encounters a more conductive material.

The steel pipe in a nearby well short-circuits the current into the pipe producing a magnetic field. According to the right-hand rule, the field associated with the downward current flow is clockwise to an observer looking down. The logging tool, located approximately 300 ft below the electrode, houses two AC magnetometers and a pair of fluxgates. The fluxgate sensors serve as a compass for determining tool orientation and the magnetometers detect the low-frequency field produced at the short circuit. Direction to the pipe is perpendicular to the measured field direction.

Fig. 10.23 is an example plot of the magnetic field strength along the axis of a relief wellbore. The circles represent points where the tool is stopped to measure and process the data. The depth of closest approach corresponds to the maximum amplitude of the signal. The range at this or any other location is determined by a rearranged form of Ampere's law:

$$R = I_p/2\pi H, \quad \text{(10.2)}$$

where I_p = the pipe current and H = the magnetic field strength. Obviously, we cannot directly measure the pipe current; its value must be estimated using assumed conductance properties of the earth and steel.

An electromagnetic tool can detect pipe 200 ft or more away and so the ranging capability is far superior to that of the magnetostatic tools. One drawback to the method has been imprecise range calculations because of the essentially unknown amperage in the pipe, but direct distance measurements independent of the current are now possible up to 30 ft.[30] An electromagnetic tool has been developed that detects the magnetic field along the tool axis.[49] This allows for ranging to a horizontal-well blowout or application in a high-angle approach.

10.4.4 Relief-Well Planning and Execution. The typical path for a relief well designed to intersect a near-vertical well is shown in **Fig. 10.24.** The well is drilled with an "S" trajectory and crosses into the blowout well's cone of uncertainty at a depth where the selected ranging tool can pinpoint the blowout well's position. The well must bypass the casing at a rela-

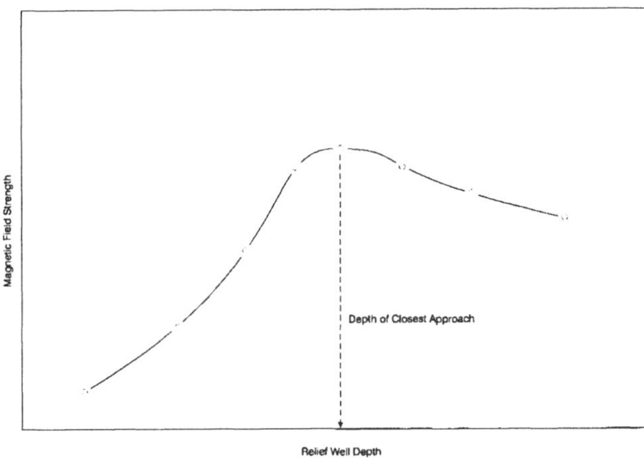

Fig. 10.23—Magnetic field strength along a relief well and the indicated depth of closest approach (after Kuckes *et al.*[48]).

tively shallow depth if the uncertainty is large or the ranging tool has a small radius of investigation. It may be necessary to plugback and sidetrack the relief well after the relative position of the two wells has been triangulated.

Figs. 10.25a through 10.25d show plan views of a hypothetical relief well homing in on casing. The relief well is drilled using MWD surveys and is directed to the southwest of the blowout well's ellipse center. In Fig. 10.25a, drilling is shut down and a north-seeking gyro survey is run in the drillstring to confirm the compass data and more narrowly define the relief well's ellipse. A proximity tool is run and, based on the range and direction tolerances, the casing is indicated to be in the area defined by "Fix 1."

Drilling is resumed and other, progessively smaller, fix areas are defined by the ranging tool as depicted in Figs. 10.25b and 10.25c. The casing has been passed in the diagram illustrated in Fig. 10.25d and another proximity survey is run. The point of closest approach is determined by triangulation and the position of the casing is fixed to within a foot or two. As shown in this series of figures, locating the pipe defines a new cone of uncertainty down to the planned intersection depth.

After the bypass has been made, the hole is directed along a path roughly parallel to the blowout well. Periodic proximity runs are made and the well path is nudged as needed to stay on track until the target depth is approached. More than likely, a rotary assembly can be run across much of this interval because the well path would be expected to follow the same tendencies as the original hole. Ultimately, the hole is kicked off in the direction of the blowout and intersection is achieved.

The first application of a steerable drilling system in a relief well was in 1988 and its use has become an important adjunct to relief-well technology.[52] Basically, a steerable motor/bottomhole assembly combination allows an operator to drill either a straight path or to make a course correction without having to make a trip. This ability greatly facilitates making trajectory adjustments when triangulating well position and intersecting the blowout, and is a significant factor in reducing relief-well costs.[30]

The goal of a relief well is to achieve intersection at a depth sufficient to effect a kill. Surface location is an important consideration for meeting the stated objectives. It is less expensive to spud the relief well as close as possible to the blowout, but safety considerations are in conflict with operational advantages and some tradeoff is necessary. The relief well should be located upwind of the blowout and far enough away

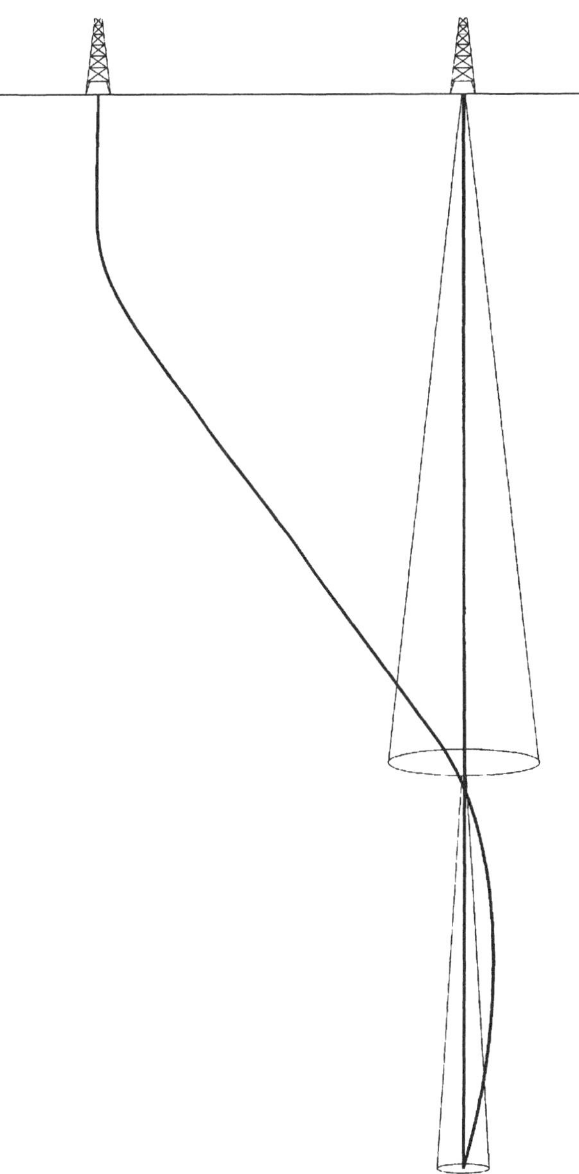

Fig. 10.24—Typical plan for intercepting a blowout with a relief well (after Leraand *et al.*[50]).

so that cratering, fire, or noxious fumes will not be a hazard. An underground blowout poses another risk if there is a chance that the flow is charging a shallow sand. One operator faced with this problem used seismic technology to map the extent of the shallow gas features and spotted the relief well away from the hazard area.[53]

Other factors influence the surface location selection. For instance, the directional plan will be accomplished more easily if the surface location takes advantage of the natural drift and bit-walk tendencies. Magnetostatic and electromagnetic tool measurements are more accurate if the relief well approaches the blowout from the east or west side, and so this may be a consideration.[46]

Low approach angles relative to the blowout well axis are recommended when it comes time to triangulate position.[54] As in all directional wells, the plan should consider the effect of doglegs, especially shallow ones, on hole drag, keyseat formation, and casing wear. However, rock is often softer in the shallower strata and the selected kick-off point may be a tradeoff between having a shallow dogleg and the comparative expense of kicking off in hard, slow-drilling rock. Hole stability is important and so some attention must be focused

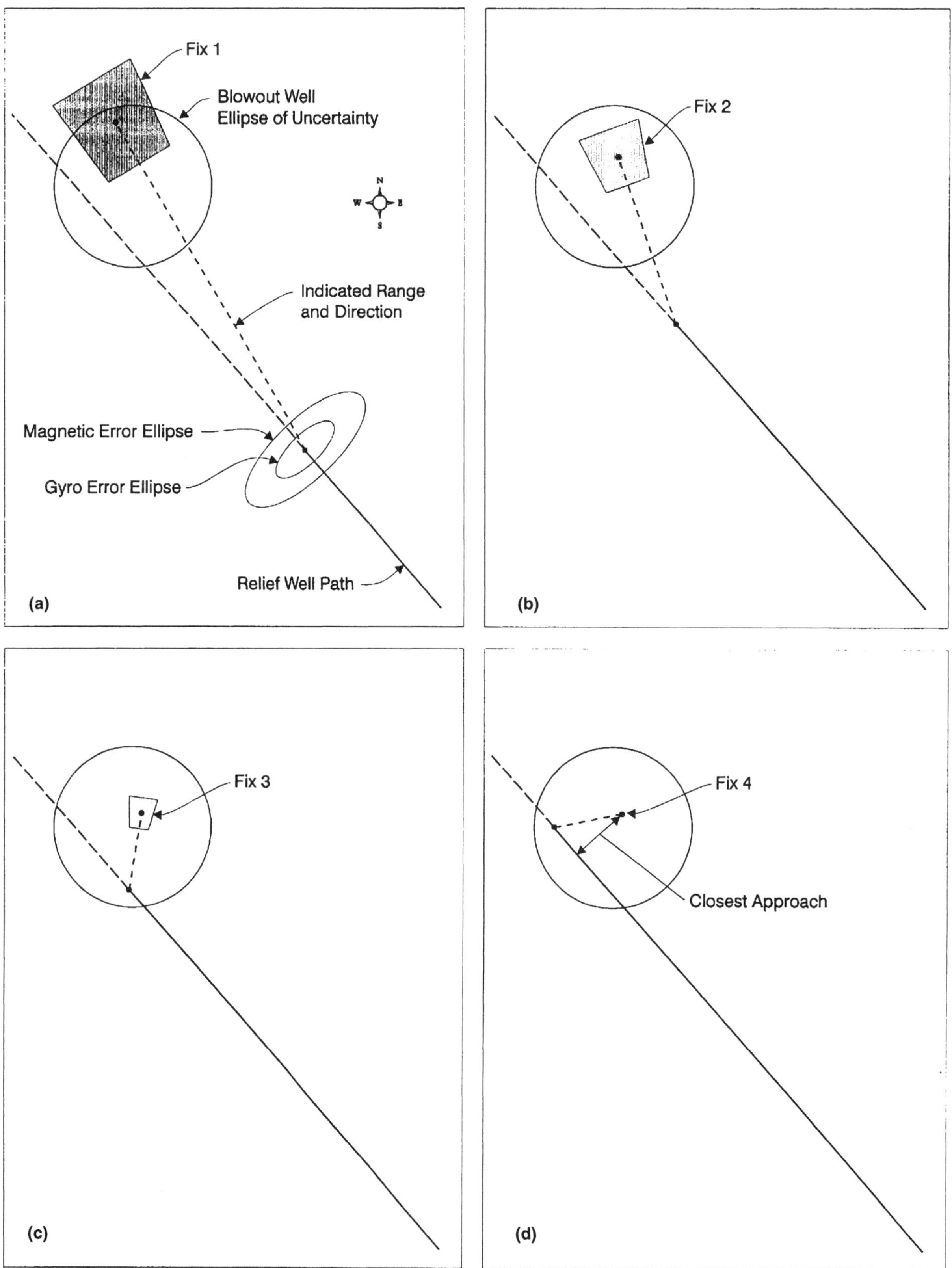

Fig. 10.25—Plan views of a relief well searching for casing (after Voisin *et al.*[51]).

on the mud properties. A gauge hole enhances directional control and improves the accuracy of the surveying and ranging instruments.

Depending on the situation, relief wells drilled for a direct intersection may intercept either open hole or casing. Lost circulation is likely when the relief well breaks into an openhole interval or when a mill penetrates the casing in a blowout well. Casing therefore must be set not far above the planned intersection depth and the operator should be ready to pull up into pipe immediately and start the kill.

It may not be necessary for the bit to drill into the blowing wellbore. One North Sea operator wanted to avoid a potential loss zone located a short distance above the blowout reservoir and selected an overlying impermeable mudstone as the opti-

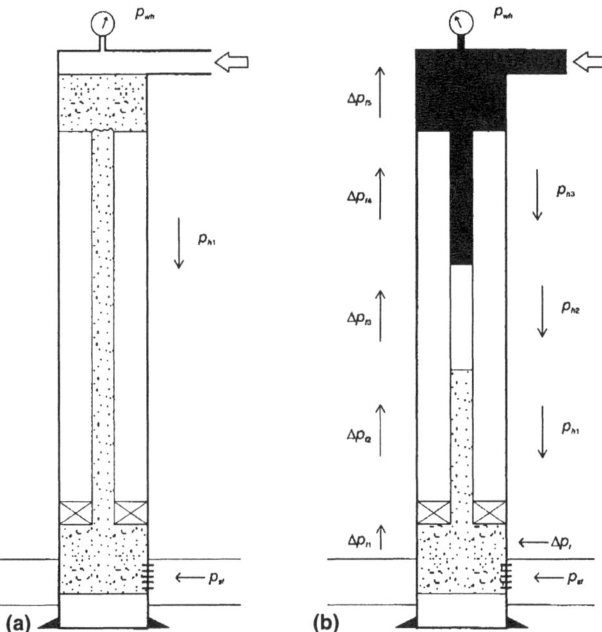

Fig. 10.26—Downhole pressures associated with a bullheading procedure.

mum location for a kill.[50] High compressive stresses are imposed on the walls of a hole when wellbore pressure is relieved by a blowout. Such conditions promote hole collapse and a rock mechanics study suggested the likelihood of breakthrough if the bit penetrated the mudstone within a meter of the blowout.[55]

Casing may be accessed by perforations if pipe has been set in the relief well. The casings need not be in contact for perforations to be effective; a foot or two of separation may be tolerable if the charges are sized large enough. However, better results are achieved with a smaller gap distance. Perforating with a tubing-conveyed gun allows using large charges oriented in the direction of the blowout well's casing.[28]

10.5 Kill Hydraulics

The final kill can be executed after the blowout wellbore has been made accessible from the surface equipment, a subsurface kill string, or a relief well. This section addresses the fluid dynamics principles that must be understood by the design engineer and operations supervisor. Included are the fundamentals related to bullheading, momentum kill procedures, and dynamically killing a well. Other methods including the volumetric control/lubrication procedure have been included elsewhere in the text. Finally, the application of specialty fluids to blowout control will be discussed.

10.5.1 Bullheading. Bullheading is often the easiest, most cost-effective way to kill a well if the conditions are suitable. Its goal is to pump directly into the wellbore, displace the formation fluids back into the rock, and leave a full column of kill-weight fluid in the hole after the task has been completed. Bullheading usually will fracture the rock, especially if there is mud in the hole, and the fluid exit point may be too shallow to kill the well if the blowout zone is a considerable distance below a casing shoe. The technique is best suited to cased holes with perforations or wells that have only a short openhole interval.

Figs. 10.26a and 10.26b show two wellbore schematics. The well illustrated in Fig. 10.26a had been blowing out through a parted tubing string and is left with a hole full of gas after the capping procedure. The operator elects to bullhead the well by pumping a water spacer ahead of the kill-weight fluid. The wellhead pressure when pumping first begins (before any appreciable pump rate is established) is given by

$$p_{wh} = p_{sf} - p_h, \quad \text{(10.3)}$$

where p_{sf} = the sandface pressure and p_h = the hydrostatic pressure of the well fluid.

Later in the job,

$$p_{wh} = p_{sf} + \Delta p_r + \Sigma(\Delta p_f) - \Sigma p_h, \quad \text{(10.4)}$$

where Δp_r = the pressure drop across a wellbore restriction such as perforations. The p_h terms are the hydrostatic pressures furnished by the respective well fluids at this point in the operation and the Δp_f terms are the friction loss components. The sandface pressure either will be the fracture propagation pressure or have a value somewhat higher than the local pore pressure.

The velocity of the pumped fluids must exceed any upward gas migration rates; the minimum kill rate can be determined from the relation,

$$(q_{kr})_{\min} = v_{sl} A_c, \quad \text{(10.5)}$$

where v_{sl} = the gas migration velocity through the largest flow conduit of cross-sectional area A_c.

The planner must arrange an adequate number of pumping units for the job. The hydraulic horsepower requirement at the pump discharge is given by

$$P_h = 0.0245 q_{kr} (p_P)_{\max}, \quad \text{(10.6)}$$

where P_h = the hydraulic horsepower and $(p_P)_{\max}$ = the maximum pump pressure. The constant 0.0245 applies if the pump rate and pressure are expressed in bbl/min and psig, respectively.

The pump pressure is the wellhead pressure plus the friction losses in the lines between the pump manifold and well.

$$p_P = p_{wh} + \Delta p_s. \quad \text{(10.7)}$$

Efficiency losses must be considered before specifying the quantity and size of the pumpers.

Example 10.1. After blowing out, the well described in **Table 10.2** was capped successfully and closed in at the stack with a stabilized shut-in casing pressure (SICP) of 5,930 psig. The decision has been made to bullhead 10 bbl of 9.0-lbm/gal brine followed by a 12.2-lbm/gal mud at 10 bbl/min. The following conditions and assumptions are applicable to the procedure. Friction pressure of the mud = 0.0121 psi/ft; friction pressure of the brine = 0.0044 psi/ft; gas friction pressure = negligible; perforation friction drop = zero; wellbore temperature = geothermal temperature; and borehole and liquid elasticity = zero.

1. Determine the minimum bullhead rate if the assumed gas migration velocity is 4,000 ft/hr.

2. Determine the wellhead pressure when the brine water reaches 6,000 ft if the sandface pressure is equivalent to the fracture-propagation pressure.

3. How much hydraulic horsepower must be furnished at the pump manifold if the maximum wellhead pressure during

TABLE 10.2—SURFACE BLOWOUT INFORMATION	
Wellbore Configuration	
Vertical depth	11,770 ft
Perforation midpoint depth	11,500 ft
Casing description	7-in., 29.0-lbm/ft P-110
Casing nominal inner diameter (ID)	6.184 in.
Casing capacity factor	0.0371 bbl/ft
Perforation quantity and diameter	Fifty × 0.45 in.
Blowout Data	
Formation fluid	Single-Phase Gas
Specific gravity	0.60
Specific heat ratio	1.27
Gas temperature at exit point	120°F
Static pore pressure	12.0-lbm/gal equivalent
Other Known or Assumed Information	
Fracture initiation gradient at perfs	0.82 psi/ft
Fracture propagation gradient at perfs	0.73 psi/ft
Geothermal wellbore temperature	70°F + 1.5°F/100ft
Standard measurement conditions	
Pressure	14.65 psia
Temperature	60°F

the job is 7,450 psig? The loss between the manifold and wellhead is 50 psi.

Solution. 1. Eq. 10.5 gives the minimum bullhead rate to ensure complete gas displacement.

$$(q_{kr})_{\min} = \frac{(4,000\ \text{ft/hr})(\pi)(6.184\ \text{in.})^2}{(60\ \text{min/hr})(4)(144\ \text{sq in./sq ft})(5.6146\ \text{cu ft/bbl})}$$

$$= 2.48\ \text{bbl/min}.$$

Hence the selected 10-bbl/min rate will be adequate.

2. Simplify the problem by assuming the gas is displaced in a piston-like manner (i.e., no gas migrates up into liquid). At this point in the kill, the pressure opposite the perforations is

$$p_{sf} = p_{fp} = (0.73)(11,500) = 8,395\ \text{psig}$$

$$= 8,410\ \text{psia}.$$

We have a 5,500-ft gas column in the well (11,500—6,000) and should consider the variable gas density to compute the pressure at 6,000 ft. The average gas temperature is determined as

$$T = [140 + 0.015(6,000 + 11,500)]/2 = 202°\text{F}$$

$$= 662°\text{R}.$$

A z factor is assumed and Eq. 1.40 is iterated a few times until

$$p_{6,000} = 8,410/e^{\frac{(0.60)(11,500-6,000)}{(53.3)(1.245)(662)}} = 7,801\ \text{psia}$$

$$= 7,786\ \text{psig}.$$

The height of the water pill is calculated as

$$h_w = 10/0.0371 = 270\ \text{ft}.$$

The hydrostatic pressure and friction loss provided by the brine are

$$p_{hw} = (0.0519)(9.0)(270) = 126\ \text{psi and}$$

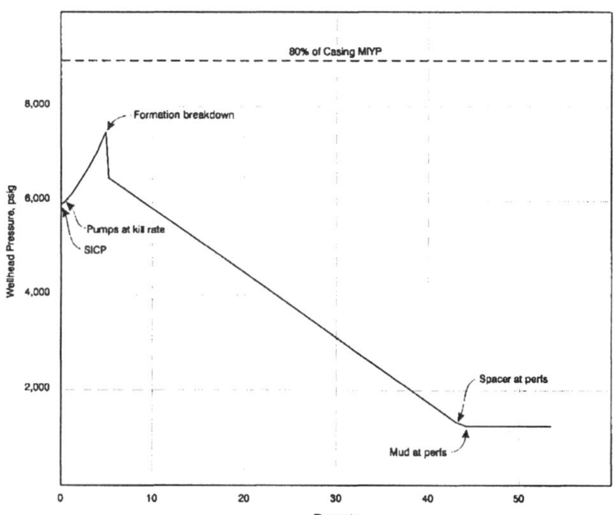

Fig. 10.27—Predicted wellhead pressures for Example 10.1.

$$\Delta p_{fw} = (0.0044)(270) = 1\ \text{psi, respectively.}$$

Similarly for the mud,

$$h_m = 6,000 - 270 = 5,730\ \text{ft};$$

$$p_{hm} = (0.0519)(12.2)(5,730) = 3,628\ \text{psi; and}$$

$$\Delta p_{fm} = (0.0121)(5,730) = 69\ \text{psi}.$$

A modified form of Eq. 10.4 yields the wellhead pressure.

$$p_{wh} = 7,786 + (1 + 69) - (126 + 3,628)$$

$$= 4,102\ \text{psig}.$$

3. Eq. 10.6 gives

$$P_h = (0.0245)(10.0)(7,450 + 50) = 1,838\ \text{hhp}.$$

The wellhead pressures for the example bullhead job were computed as function of pumping time with results as plotted in **Fig. 10.27.** We conservatively assumed no leakoff into the perforations until the fracture initiation pressure was achieved and so the initial pressure increase relates to gas compression. After breakdown, the pressures decline until the gas and brine water are displaced from the well.

For this particular example, the highest wellhead pressure would have been the initial shut-in pressure had the gas and other well fluids been pumped into the formation matrix from the start. Such a condition may be realized for clean fluids at moderate pump rates, but the planner should anticipate otherwise unless he or she has superior knowledge about the rock permeability and nature of the well fluids. The pumps should be slowed before kill-weight fluid hits the perfs if it appears that matrix flow is occurring during the earlier stages of the job.

Bullheading will exert more pressure on a well than any other kill technique and an alternative method must be used if the worst-case predictions approach the pressure ratings of the tubulars or wellhead equipment. Considerations should include the well's age, possible external or internal corrosion, and any effect the blowout may have had on equipment integrity. Friction pressures can override the hydrostatic increase if the pump rate is excessive. A well-control problem will be made much worse if the casing ruptures as result of bullheading a heavy, viscous mud at an extreme pump rate. This has happened more than once.

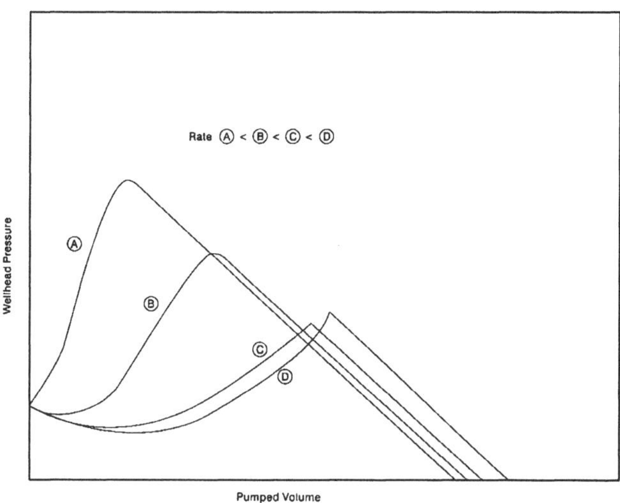

Fig. 10.28—Selecting an optimum bullhead rate based on wellhead pressure limitations (after Abel[56]).

It would seem that the lowest bullheading pressures correspond to the minimum pump rate needed to displace the formation fluids. Abel,[56] however, discussed bullheading applications where surface pressures could be minimized by designing for a higher rate. Say that a blowout well is on diverter and that the intent is to avoid shutting in the well before starting kill fluids downhole. It will take some time for the sandface pressure to build up to the average reservoir pressure and a faster pump rate can reduce the maximum wellhead pressure by furnishing hydrostatic pressure faster than the buildup rate.

Fig. 10.28 illustrates this concept. Curves A through D are simulated wellhead pressures for four pump rates with $q_A < q_B < q_C < q_D$. Note that the formation is assumed to take fluids from the start and that the pressure increase is driven solely by the transient pore-pressure buildup. The lowest wellhead pressure is realized when $q_{kr} = q_C$. Friction pressures begin to impact the minimum wellhead pressure if pump rates get any higher and q_C would be considered the optimum rate.

10.5.2 The Momentum Kill. The momentum kill was introduced in 1976[57] and the method has been applied successfully to a number of wild wells, many of which would have otherwise required a relief well. The objective is to counteract the flowing momentum of the formation fluids with enough fluid density and pump rate to arrest the flow and force these fluids to reverse direction back into the rock.[57-60] **Fig. 10.29** illustrates this simple concept. Sufficient hydrostatic pressure must be stacked in the hole to keep the well dead after the operation.

The theoretical basis of the method is the impulse-momentum principle, which holds that the force exerted by a flowing fluid is equal to the rate of change in the fluid's momentum. In equation form,

$$F = \frac{1}{g_c}\frac{\mathrm{d}(mv)}{\mathrm{d}t}, \qquad (10.8)$$

where the mass/velocity product mv = the momentum. The relation is another way of stating Newton's second law of motion. The principle has several other oilfield applications, including force predictions when the velocity vector (magnitude or direction) changes in pipe flow and is the basis for the jet impact force equation used in bit hydraulics.

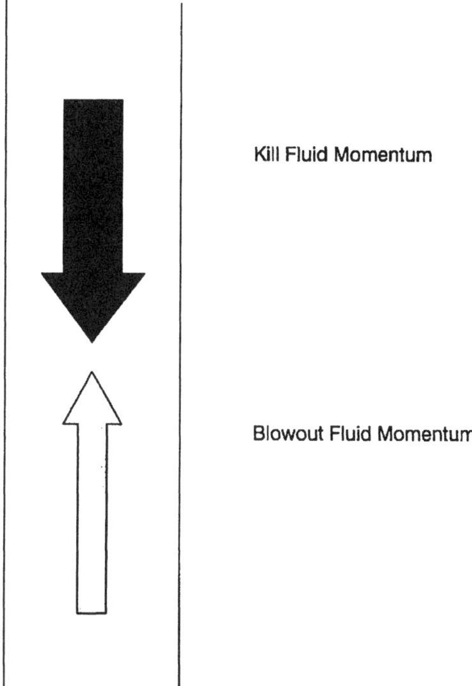

Fig. 10.29—Force balance concept for a momentum kill.

The objective is to provide enough kill-fluid momentum to reduce the blowout fluid velocity to zero. For steady-state conditions (constant mass flow), Eq. 10.8 can be rewritten as

$$F = m(0 - v)/g_c\Delta t = -mv/g_c\Delta t. \qquad (10.8a)$$

The negative sign indicates that the countering force is applied in a direction opposite to flow. The mass rate can be expressed in terms of the fluid density ρ and volume flow rate q,

$$m/\Delta t = \rho q.$$

Also, $v = q/A_c$ and $A_c = \pi d_c^2/4$,

where A_c and d_c are the cross-sectional area and diameter of the flow conduit, respectively. Substituting terms yields the basic relation for the momentum force F_{mv},

$$F_{mv} = -4\rho q^2/\pi g_c d_c^2. \qquad (10.9)$$

Applying conversion constants leads to the equation for an incompressible fluid.

$$F_{mv}(\text{lbf}) =$$

$$-\frac{\rho(\text{lbm/gal})(7.48\ \text{gal/cu ft})q^2(\text{bbl/min})^2(5.6146\ \text{cu ft/bbl})^2(144\ \text{sq in./sq ft})}{(32.17\ \text{lbm-ft/lbf-s}^2)(\pi/4)d_c^2(\text{sq in.})(60\ \text{s/min})^2}.$$

Finally,

$$F_{mv} = -\rho q^2/2.68 d_c^2. \qquad (10.10)$$

Example 10.2. A well is blowing salt water at a rate of 40,000 bbl/d up casing. The casing ID is 8.921 in. and the brine weighs 9.6 lbm/gal. Determine the minimum pump rate that will effect a momentum kill with 18.0-lbm/gal mud.

Solution. Enough momentum force must be furnished to counter that of the salt water. This force is obtained from Eq. 10.10 as

$$-F_{mv} = (9.6)(40,000)^2/(2.68)(8.921)^2(1,440)^2$$
$$= 34.7 \text{ lbf}.$$

Rearranging this equation gives the required pump rate.

$$q_{kr} = \left[(2.68)(8.921)^2(34.7)/(18.0)\right]^{0.5} = 20 \text{ bbl/min}.$$

The momentum of an incompressible fluid is constant for a given conduit diameter. A compressible fluid like gas, however, introduces variability to the density and velocity terms. It follows that the gas momentum depends on the pressure and, hence, on location. We use the gas law to relate the flow rate at p and T to standard conditions.

$$q_g = q_{gsc}p_{sc}zT/z_{sc}T_{sc}p.$$

The gas density in lbm/ft^3 is

$$\rho_g = 2.70p\gamma_g/zT.$$

Use 14.65 psia and 520°R as standard conditions, substitute terms into Eq. 10.9, and apply conversion constants:

$$F_{mv} = -\frac{(2.70)\gamma_g(14.65)^2q_{gsc}^2(10^6)^2zT(4)(12)^2}{(32.17)p(1.0)^2(520)^2\pi d_c^2(86,400)^2},$$

when q_{gsc} is expressed in MMcf/D. Arithmetic yields the momentum force relation for gas,

$$F_{mv} = -1.64\gamma_g q_{gsc}^2 zT/pd_c^2. \quad \text{.............. (10.11)}$$

At the exit point of a blowing well, changes in the outlet pressure travel upstream at the speed of sound. Hence fluids cannot travel through a constant-area conduit any faster than the critical (sonic) velocity as this would require the pressure wave to be transmitted back to the source at supersonic speed. Also, critical velocities cannot be achieved anywhere in the wellbore other than at the outlet. The critical velocity for gas is given by

$$v_g^* = 41.4(\kappa zT/\gamma_g)^{0.5}, \quad \text{.................... (10.12)}$$

where v_g^* = the critical velocity (ft/sec) and κ = the specific heat ratio of the gas. The critical flow rate is given by

$$q_{gsc}^* = \frac{v_g^* d_c^2 p_e}{59.8z_eT_e}, \quad \text{....................... (10.13)}$$

where q_{gsc}^* is in MMcf/D and the e subscripts refer to the exit conditions at the well outlet.

Eq. 10.14 gives the speed of sound in a liquid.

$$v_l^* = 24.9(1/\rho_l c_l)^{0.5}, \quad \text{.................... (10.14)}$$

where ρ_l and c_l are the respective density (lbm/gal) and compressibility (1/psi) of the liquid.

Space does not allow for an in-depth discussion of sonic conditions for two-phase flow and the reader is encouraged to review the literature related to the subject.[61-63] Suffice it to say that the critical velocity of a mixture depends on the volume fractions of the respective components and usually is lower than the individual critical velocities.

Example 10.3. Refer to the information given in Table 10.2.

1. Determine the critical velocity and flow rate of the gas if the fluid is to exit the casing at an atmospheric pressure of 14.7 psia.

2. Determine the critical velocity of an 8.7-lbm/gal salt water if the brine's compressibility at the outlet is 2.82×10^{-6} psi^{-1}.

Solution. 1. The critical velocity and flow rate for the gas are determined using Eq. 10.12 and 10.13, respectively.

$$v_g^* = 41.4[(1.27)(1.0)(580)/(0.60)]^{0.5} = 1,451 \text{ ft/s}$$

and $$q_{gsc}^* = \frac{(1,451)(6.184)^2(14.7)}{(59.7)(1.0)(5809)} = 23.6 \text{ MMcf/D}.$$

2. Eq. 10.14 yields the critical liquid velocity.

$$v_l^* = 24.9[1/(8.7)(2.82 \times 10^{-6})]^{0.5} = 5,027 \text{ ft/s}.$$

The flowing pressure (and to lesser extent, temperature) at the point where the fluids interact must be reasonably approximated before we can attempt to design the kill-fluid density or pump rate. Determining this pressure may not be as straightforward as it might seem.

Consider **Figs. 10.30a through 10.30c** which represent a well blowing out through a piece of pipe. The exit pressure will be atmospheric if the discharge is subsonic. Gas or two-phase flow rates greater than the critical (sonic) rate determined at atmospheric pressure imply that the well outlet is behaving like a venturi nozzle. Supersonic speeds are realized downstream of the outlet and the exit pressure will be greater than atmospheric pressure.

Fig. 10.30b illustrates one possible configuration for a momentum kill. Flow is routed through a diverter hookup and a kill string is hot-snubbed into the well to intercept the blowout below the diverter side outlet. Assume the fluids achieve critical rate at the point where the fluids exit the diverter line. The discharge pressure will be greater than atmospheric and the pressure used to predict the momentum-kill parameters will be the exit pressure plus the acceleration and friction losses through the diverter equipment.

Friction loss for a dry gas in turbulent flow can be estimated using the Weymouth[64] formula, which has been modified to include the compressibility factor.

$$p_{wh}^2 - p_e^2 = q_{gsc}^2 T\gamma_g L/1.24d_c^{5.333}. \quad \text{........... (10.15)}$$

The presence of oil, water, or mud requires using a two-phase correlation for horizontal flow.[65]

Significant expansion occurs in a diverter that is flowing gas at a high rate and the acceleration effects cannot be ignored. The pressure associated with a change in a fluid's kinetic energy is given by

$$\Delta p_{ac} = \rho(v_2^2 - v_1^2)/2g_c. \quad \text{.................. (10.16)}$$

Gas densities at the line inlet and outlet can be averaged when applied to diverter flow. Adiabatic expansion will cool the gas. Nonetheless, we choose to simplify matters by assuming that the gas temperature at the wellhead is the same as the exit temperature. Also, diverter pressures are usually low enough that the error will be acceptable if we set the z factors equal to unity. Substitution yields

$$\Delta p_{ac} = \frac{\gamma_g q_{sc}^2 T_e(p_e + p_{wh})}{1.92d_c^4}\left(\frac{1}{p_e^2} - \frac{1}{p_{wh}^2}\right). \quad \text{...... (10.17)}$$

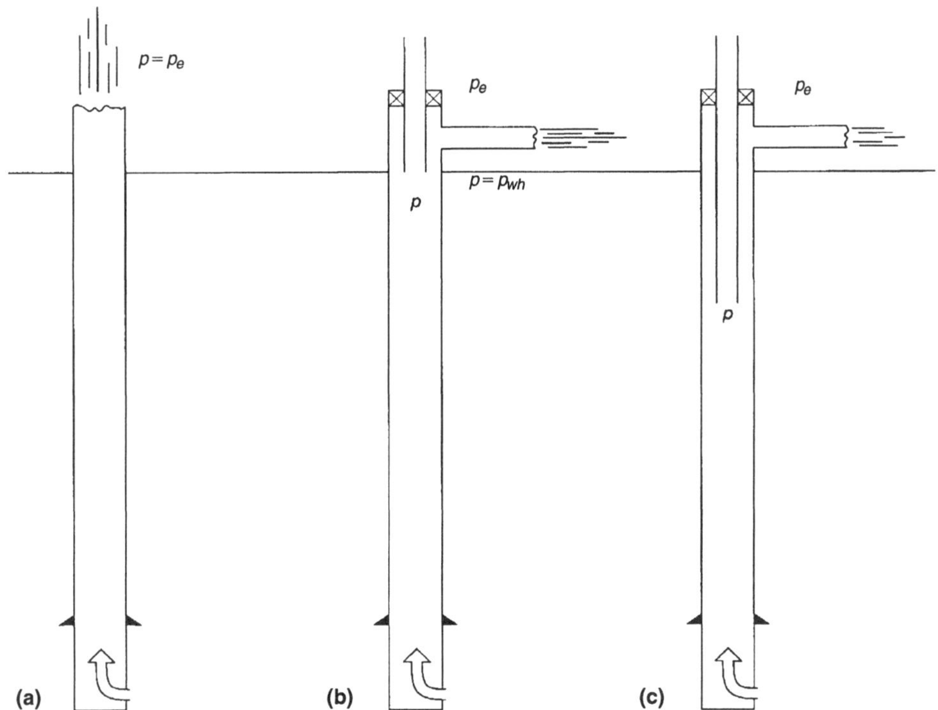

Fig. 10.30—Exit and intercept pressures in a blowing well.

Finally, a rather unwieldy relationship for the wellhead pressure is obtained by combining Eq. 10.15 and 10.17:

$$p_{wh} = \left(\frac{q_{sc}^2 T_e \gamma_g L}{1.24 d_c^{5.333}} + p_e^2 \right)^{0.5}$$

$$+ \frac{q_{sc}^2 T_e \gamma_g (p_e + p_{wh})}{1.92 d_c^4} \left(\frac{1}{p_e^2} - \frac{1}{p_{wh}^2} \right) \quad \text{........ (10.18)}$$

Iteration will be necessary to solve for p_{wh} because the term appears on both sides of the equation.

Example 10.4. The well described in Table 10.2 is blowing an estimated 100 MMcf/D through 200 ft of 8-in. diverter line. Determine the pump rate that will effect a momentum kill if a kill string is hot-snubbed below the diverter side outlet. The kill-fluid density is 20.0 lbm/gal.

Solution. Intuitively we expect that flow at the diverter terminus is critical and rearrange Eq. 10.13 to calculate the exit pressure.

$$p_e = \frac{(59.7)(1.0)(580)(100)}{(1,451)(8.0)^2} = 37 \text{ psia.}$$

Eq. 10.17 is used to estimate the wellhead pressure by first assuming $p_{wh} = 50$ psia.

$$p_{wh} = \left[\frac{(100)^2(580)(0.60)(200)}{(1.24)(8.0)^{5.333}} + 37^2 \right]^{0.5}$$

$$+ \frac{(100)^2(580)(0.60)(37 + 50)}{(1.92)(8.0)^4} \left(\frac{1}{37^2} - \frac{1}{50^2} \right),$$

$$p_{wh} = 112 \text{ psia.}$$

Another iteration using $p_{wh} = 112$ psia on the right side of the relation yields $p_{wh} = 143$ psia. The process is repeated and, ultimately, the result $p_{wh} = 160$ psia is revealed. The kill fluid will encounter the flowing gas stream in the 7-in. casing immediately below the diverter outlet. Ignore the minor expansion effect between the casing and diverter and calculate the momentum force of the flowing gas using Eq. 10.11.

$$-F_{mv} = \frac{(1.64)(0.60)(100)^2(0.982)(580)}{(160)(6.184)^2} = 916 \text{ lbf.}$$

Eq. 10.10 yields the required pump rate.

$$q_{kr} = \left[(2.68)(6.184)^2(916)/(20.0) \right]^{0.5} = 69 \text{ bbl/min.}$$

The example problem suggests that extreme pump rates may be necessary to kill a blowout from a prolific gas reservoir. A heavier kill fluid may be an option. For instance, a 35-lbm/gal density can be realized by using galena as the weighting agent. This would have reduced the example's pump rate requirement to 52 bbl/min, but these ultra-heavy fluids are expensive, abrasive, and difficult to maintain.

A better way to reduce the fluid-momentum requirement is to take the intercept point deeper in the well where the gas momentum is lower. The procedure is illustrated in Fig. 10.30c. The kill string has been run to a predetermined depth where the wellbore pressure is high enough to allow a kill using the available momentum. This pressure and string depth can be predicted based on an equation used to determine the bottomhole pressure (BHP) in a flowing gas well.

One of these equations (and there have been several) was presented by Katz and Lee[66] in the form,

$$p^2 - e^s p_{wh}^2 = 667 q_{gsc}^2 T^2 z^2 f_M (e^s - 1)/d_c^5, \quad \text{.... (10.19)}$$

where the intermediate variable s is given by

$$s = 0.0375 \gamma_g D/Tz.$$

D = the depth of interest and f_M = the Moody friction factor. (The Moody friction factor is four times the Fanning factor.) The temperature and z factor are averaged over depth. Turbulent flow is assumed and acceleration effect is ignored, which may lead to significant error if the flowing wellhead pressure is low.

Friction factors level off and remain constant at high Reynolds numbers. The authors suggested using the following empirical relations if this condition is met.

$$f_M = 0.01750/d_c^{0.224}, \quad \text{.................... (10.20a)}$$

if $d_c \leq 4.277$ in. and

$$f_M = 0.01603/d_c^{0.164} \quad \text{.................... (10.20b)}$$

for larger diameters. Substitution yields

$$p^2 - e^s p_{wh}^2 = 11.67 q_{gsc}^2 T^2 z^2 f_M (e^s - 1)/d_c^{5.224} \quad \text{................ (10.21a)}$$

and

$$p^2 - e^s p_{wh}^2 = 10.69 q_{gsc}^2 T^2 z^2 f_M (e^s - 1)/d_c^{5.164}, \quad \text{................ (10.21b)}$$

respectively.

The depth where a given flowing wellbore pressure is realized is obtained by rearranging these two equations. Accordingly,

$$D_{ks} = 26.67 q_{gsc} T z \gamma_g^{-1} \ln\left(\frac{p^2 + a}{p_{wg}^2 + a}\right). \quad \text{........ (10.22)}$$

D_{ks} in this case will be the minimum kill-string depth for the available kill-fluid momentum. The intermediate variable a depends on the conduit ID. Use

$$a = 11.67\left(q_{gsc} T z\right)^2 d_c^{-5.224} \quad \text{................ (10.23a)}$$

if $d_c \leq 4.277$ in. and

$$a = 10.69\left(q_{gsc} T z\right)^2 d_c^{-5.164} \quad \text{................ (10.23b)}$$

when $d_c > 4.277$ in.

Example 10.5. Assume that the operator of the well described in Table 10.2 wants to pump 22.0-lbm/gal mud at a rate no faster than 40 bbl/min. Determine the minimum depth at which the kill string must be snubbed to effect a momentum kill. Ignore the choking effect of flowing friction in the kill string annulus.

Solution. The available momentum force of the kill fluid is

$$F_{mv} = (22.0)(40)^2/(2.68)(6.184)^2 = 343.5 \text{ lbf}.$$

The available force is equated to that of the gas.

$$\frac{(1.64)(0.6)(100)^2 Tz}{(6.184)^2 p} = 343.5$$

$$p/Tz = 0.7492.$$

We must iterate to solve this problem. First assume that the temperature and z factor at the kill-string depth correspond to the wellhead conditions.

$$p = (0.7492)(5809)(0.982) = 427 \text{ psia}.$$

Now calculate the minimum kill-string depth using Eqs. 10.23b and 10.22:

$$a = 10.69[(100)(580)(0.982)]^2(6.184)^{-5.164}$$
$$= 2.844 \times 10^6 \text{ psi}^2.$$

Therefore, $D_{ks} = (26.67)(580)(0.982)(0.6)^{-1} \ln$

$$\left(\frac{427^2 + 2.844 \times 10^6}{160^2 + 2.844 \times 10^6}\right)$$
$$= 1,346 \text{ ft}.$$

Assume the flowing temperature at the perforations is geothermal and the wellbore gradient is linear. The temperature at 1,346 ft is determined then as 134°F and the z factor at 427 psia is 0.961. Repeat the iteration procedure using these parameters:

$$p = (0.7492)(594)(0.961) = 428 \text{ psia};$$

$$a = 10.69[(100)(594)(0.961)]^2(6.184)^{-5.164}$$
$$= 2.857 \times 10^6 \text{ psi}^2;$$

and $D_{ks} = (26.67)(594)(0.961)(0.6)^{-1} \ln$

$$\left(\frac{425^2 + 2.857 \times 10^6}{160^2 + 2.857 \times 10^6}\right)$$
$$= 1,329 \text{ ft}.$$

No further iterations are necessary.

Some basic criteria must be met for a momentum-kill procedure to be successful. First, the means to introduce the kill fluid directly into flow conduit somewhere below the exit port must be available. Snubbing a kill string into a diverted blowout is one possibility. Pumping down the annulus of a well blowing out a string of tubing or drillpipe is another.

High fluid densities and pump rates are required often, particularly when killing a gas flow. Wellbore pressures below the intercept depth are not usually a problem, but the pressure rating of the kill string may be a consideration. In a backside kill, the pressure ratings of the wellhead and casing will be a factor in the design.

A reasonable estimate of the production rate is necessary and the nature and physical properties of the produced fluids should be known to some extent. Finally, the static pore pressure of the producing reservoir must be known. Some momentum kills have succeeded initially, but failed to keep the hole dead because the pore-pressure buildup subsequently caused the hole to unload again.

10.5.3 Dynamic Kills. In 1978, Blount and Soeiinah[34] were involved in the design of the first dynamic kill, which was applied successfully to a a prolific gas blowout in Indonesia. The technique has been used numerous times to bring wild wells under control. It is the most common relief-well kill procedure and can be used sometimes to kill a well by circulation within the blowout wellbore. Basically, a dynamic kill uses friction pressure furnished by the pumps to supplement the hydrostatic pressure of the initial kill fluid. The initial kill-fluid density may be either underbalanced to the static pore pressure or of magnitude such that a clean (non-gas-cut) fluid at zero pump rate is sufficient to keep the well dead.

A dynamic kill often is accomplished initially with a fresh or salt water that is underbalanced to the shut-in pore pres-

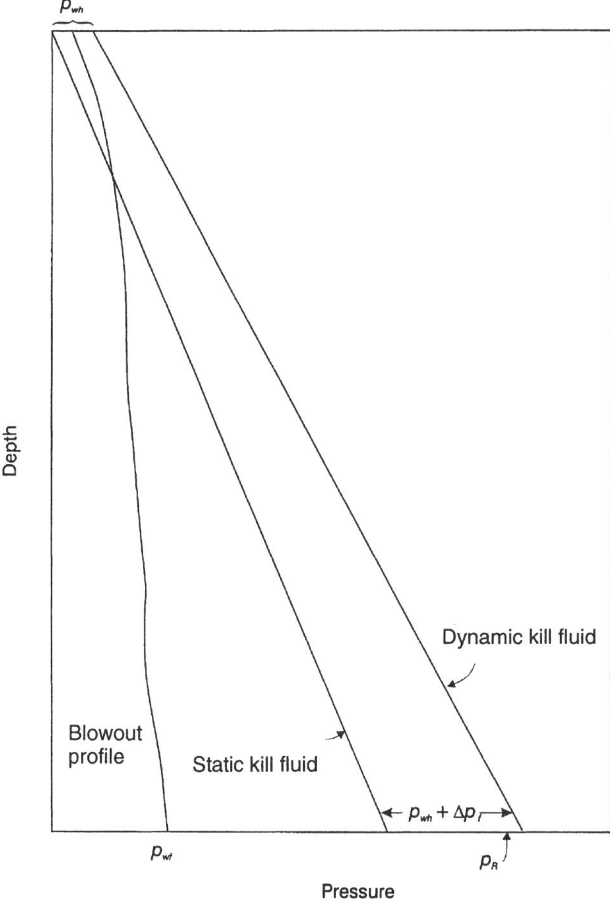

Fig. 10.31—Pressure profiles associated with a dynamic kill.

sure. **Fig. 10.31** illustrates certain concepts for such a case. The flowing wellbore pressures for an uncontrolled gas well are represented by the blowout pressure profile on the left side of the diagram. The flowing pressure is shown as p_{wf} whereas the average static reservoir pressure is p_R. The selected kill fluid has a hydrostatic pressure less than p_R and will not prevent flow into the well under static conditions. Given enough pump rate, however, the circulating friction along with, perhaps, some pressure at the wellhead adds enough backpressure to subdue and ultimately stop the flow.

Fig. 10.32 represents the gas flow rate and flowing BHP as functions of the injection rate. The gas is flowing at the maximum, blowout rate at zero injection rate. Pumping kill fluid into the flow stream introduces a liquid fraction, which increases the hydrostatic backpressure, and gives additional friction pressure. Hence p_{wf} increases and the gas rate diminishes because of the smaller differential pressure. One design objective is to determine what liquid flow rate will produce a backpressure equivalent to p_R.

The fundamental energy equation describes the flowing pressure in a wellbore.

$$p_{wf} = p_{wh} + \frac{g}{g_c}\rho h + \frac{\rho\left(v_{wh}^2 - v^2\right)}{2g_c} + \frac{\rho f_M v^2 L}{2g_c d}. \qquad (10.24)$$

The first term is the outlet pressure at the wellhead. It could be a choke backpressure, exit pressure to ambient conditions, or equivalent to the sea water hydrostatic in the case of a subsea blowout. The second term is the hydrostatic pressure of the wellbore fluids having a density ρ across vertical height

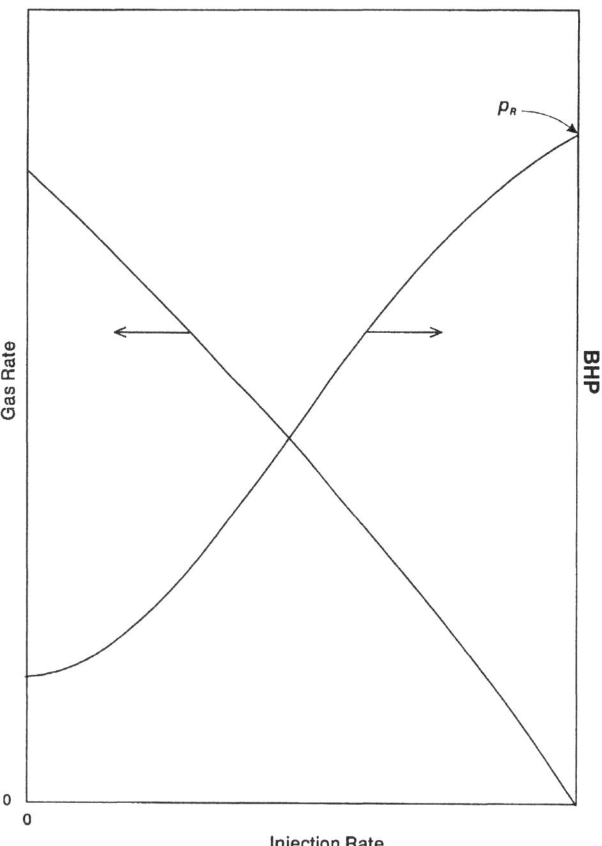

Fig. 10.32—BHP and gas flow rate as functions of injection rate (after Warriner and Cassity[26]).

h. The kinetic energy term represents the acceleration or deceleration changes associated with going from velocity v to the wellhead velocity v_{wh}.

The final term in Eq. 10.24 is the friction loss. Eq. 10.24 is obtained after applying conversion constants.

$$\Delta p_f(\text{psi}) = \frac{\rho(\text{lbm/gal}) f_M v^2(\text{ft/s})^2 L(\text{ft})}{103.2 d(\text{in.})}. \qquad (10.25)$$

The Jain[67] friction factor correlation can be used for Newtonian fluids in fully developed turbulent flow:

$$f_M = \left[1.14 - 2\log\left(\frac{\epsilon}{d} + \frac{21.25}{N_{\text{Re}}^{0.9}}\right)\right]^{-2}, \qquad (10.26)$$

where ϵ = the pipe roughness and N_{Re} = the Reynold's number. N_{Re} is determined using

$$N_{\text{Re}} = 928 dv\rho/\mu, \qquad (10.27)$$

where μ = the fluid viscosity.

In this text, d = the hydraulic diameter. For flow in round pipe,

$$d = d_c. \qquad (10.28)$$

In annular flow we assume

$$d = d_h - d_o, \qquad (10.29)$$

where d_h and d_o are the hole and pipe diameters, respectively. For flow in bbl/min, the respective average velocities in pipe and the annulus are obtained as

$$v = 17.16 q/d_c^2 \qquad (10.30)$$

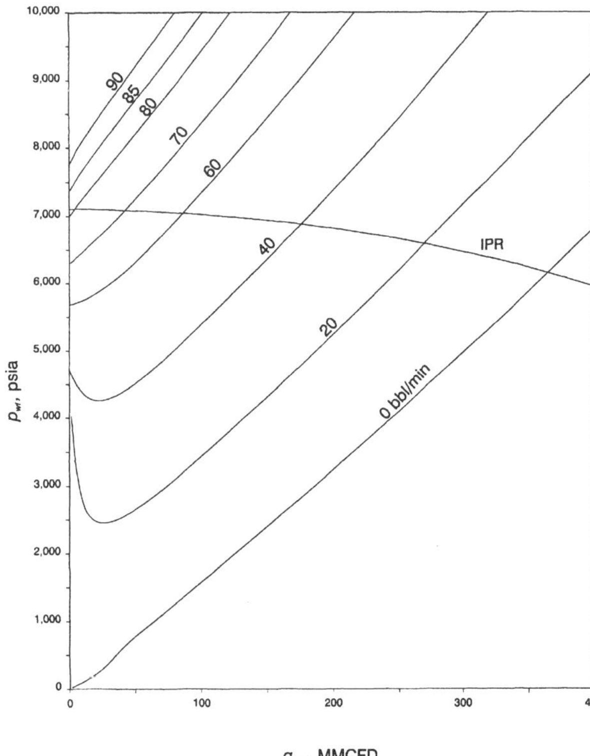

Fig. 10.33—IPR and system-intake curves at several water injection rates as predicted for a high-rate blowout in Indonesia (after Kouba *et al.*[68]).

and $v = 17.16q/(d_h^2 - d_o^2)$. .... (10.31)

Note that Eq. 10.24 applies to the flowing pressure at formation depth only if the wellbore fluid is consistent and incompressible. For gas and multiphase flow, we can apply the equation to a discrete segment from the wellhead to a shallow depth of interest and determine p_{wf} at the prevailing pressure and temperature state. Numerical integration across successive segments is necessary then to determine the BHP. The task is accomplished best by using a good multiphase flow simulator.

Reservoir deliverability can be described by the inflow performance relationship (IPR), which gives the flow rate at any backpressure. The IPR for a gas well in pseudosteady state flow can be expressed as

$$q_{gsc} = J\left(p_R^2 - p_{wf}^2\right)^n, \quad \text{.....................} (10.32)$$

where J = the productivity index and includes the remaining variables from Darcy's law. The exponent n accounts for turbulence and theoretically ranges from 0.5 to 1.0.

For a given liquid rate, the multiphase solutions to Eq. 10.24 for various blowout rates describes a system-intake relationship. As shown in **Fig. 10.33,** several system-intake curves can be graphed and compared to the IPR. A successful dynamic kill requires that the system-intake curve remain above the IPR at all blowout rates. Otherwise the kill fluids will be gas-lifted from the hole without achieving the needed backpressure. For this particular analysis, the minimum kill fluid rate in the blowout wellbore is between 80 and 85 bbl/min.

Note from the diagram that we do not need to consider the complications of multiphase flow or to know the formation's IPR when the gas rate is zero. In other words, a liquid rate that gives a BHP equal to the static reservoir pressure can be calculated based on the single-phase properties of the kill fluid. The equation for the minimum liquid rate is derived from Eq. 10.24 by setting p_{wf} equal to p_R, ignoring any change in kinetic energy, and rearranging terms. Eq. 10.33 is thereby obtained for pipe flow.

$$(q_l)_{\min} = 0.592 d_c^{2.5}\left(\frac{p_R - p_{wh} - p_{hl}}{\rho_l f_M L}\right)^{0.5}. \quad \text{.....} (10.33)$$

The annular flow expression is given by

$$(q_l)_{\min} = 0.592\left(d_h^2 - d_o^2\right)\left[\frac{(p_R - p_{wh} - p_{hl})(d_h - d_o)}{\rho_l f_M L}\right]^{0.5}. \quad \text{.................} (10.34)$$

Fig. 10.33 illustrates a characteristic dip in some of the system-intake curves when the gas flow rate approaches zero. Designing an injection rate based on Eq. 10.34 will not effect a kill if the system-intake curve drops below the static pore pressure. Kouba *et al.*[68] defined an upper limit to the minimum liquid rate requirement called the zero-derivative solution that accounts for this behavior.

They started with Eq. 10.24 and eliminated the kinetic energy term, reasoning that acceleration would be neglible at the low gas rates where the pressure sags. They took the partial derivative of the flowing pressure with respect to gas flow rate, set the derivative equal to zero, and solved for the liquid rate. Eq. 10.35 is thus obtained for pipe flow.

$$(q_l)_{\min} = 0.135\left[\frac{(\rho_l - \rho_g) d_c^5 h_l}{(\rho_l + \rho_g) f_M L}\right]^{0.5}. \quad \text{.........} (10.35)$$

In the annulus,

$$(q_l)_{\min} = 0.135\left[\frac{(\rho_l - \rho_g)\left(d_h^2 - d_o^2\right)^2 (d_h - d_o) h_l}{(\rho_l + \rho_g) f_M L}\right]^{0.5}. \quad \text{.................} (10.36)$$

Iteration will be necessary to solve for the minimum liquid rate because the friction factor depends on the Reynold's number, which in turn depends on the desired result. In practice, liquid rates are computed using both the static-pore-pressure and zero-derivative equations; the larger of the two is considered to be the design requirement. The following example demonstrates the application to our hypothetical blowout.

Example 10.6. The example well in Table 10.2 has been intercepted by a relief well close to the perforation midpoint depth. Determine the mimimum kill-fluid rate if 8.5-lbm/gal water is selected as the initial fluid. Use 0.00065 in. as the absolute pipe roughness and, to be conservative, assume the wellhead pressure is only 15 psia during the kill.

Solution. The friction factor in the blowout well's casing is estimated initially by assuming an infinitely large Reynolds number. Eq. 10.26 yields

$$f_M = \left[1.14 - 2\log(0.00065/6.184)\right]^{-2} = 0.01208.$$

The job must be designed to balance the average static reservoir pressure, which is

$$p_R = (0.0519)(12.0)(11,500) + 15 = 7,177 \text{ psia}.$$

The hydrostatic pressure furnished by the initial kill fluid is

$$p_{hl} = (0.0519)(8.5)(11,500) + 15 = 5,088 \text{ psia}.$$

The minimum liquid rate to sustain a kill after all the gas has been cleared from the hole can be calculated now using Eq. 10.33.

$$(q_l)_{\min} = 0.592(6.184)^{2.5}\left[\frac{7,177 - 15 - 5,088}{(8.5)(0.01208)(11,500)}\right]^{0.5}$$

$$= 74.6 \text{ bbl/min}.$$

Determine the Reynolds number to verify the friction factor assumption.

$$v = (17.16)(74.6)/(6.184)^2 = 33.5 \text{ ft/s, so}$$

$$N_{Re} = (928)(6.184)(33.5)(8.5)/1.0 = 1,634,000.$$

Use Eq. 10.26 again and obtain

$$f_M = \left[1.14 - 2\log\left(\frac{0.00065}{6.184} + \frac{21.25}{1,634,000^{0.9}}\right)\right]^{-2}$$

$$= 0.01311.$$

The minimum liquid rate is recalculated as

$$(q_l)_{\min} = 0.592(6.184)^{2.5}\left[\frac{7,177 - 15 - 5,088}{(8.5)(0.01311)(11,500)}\right]^{0.5}$$

$$= 71.6 \text{ bbl/min}.$$

No further iterations are necessary.

Now we must obtain the zero-derivative solution. Estimate the average gas density in the well by assuming a relatively cool temperature and an average wellbore pressure equal to half the static reservoir pressure.

$$\rho_g = (0.60)(3,589)/(2.77)(0.830)(540) = 1.7 \text{ lbm/gal}.$$

The zero-derivative rate is obtained from Eq. 10.35 as

$$(q_l)_{\min} = 0.135\left[\frac{(8.5 - 1.7)(6.184)^5(11,500)}{(8.5 + 1.7)(0.01311)(11,500)}\right]^{0.5}$$

$$= 91.6 \text{ bbl/min}.$$

The friction factor at the higher pump rate is calculated to be 0.01296. Another iteration yields the result 92.1 bbl/min, which is therefore the minimum design rate.

Similar principles are applicable when the kill fluid enters the flow stream a considerable distance above the blowing formation. This may be necessary when a relief well purposefully or inadvertently intersects the blowout well at a relatively shallow depth or when a blowout ensues with an off-bottom workstring. In their simplest form, the minimum-kill-rate equations are modified only slightly by determining the upper and lower limits corresponding to having the space below the point of injection completely occupied by either the kill fluid or formation fluids. More sophisticated techniques model the effect of counterflow on BHP.[68,69] Counterflow occurs when gas velocity diminishes to the point that kill-fluid droplets begin to fall and accumulate below the point of injection.

In a relief-well kill, the liquid rate in the blowout wellbore will not be the same as the injection rate if any rock is exposed to the injected fluids. In these situations, fluid leak-off to the rock matrix must be estimated and factored into the design. The dynamic kill rate is thus

$$q_{kr} = q_l/E_{lo}, \quad \text{.......................} \quad (10.37)$$

where q_l = the desired liquid rate in the blowout well and E_{lo} = the fluid's leak-off efficiency. The fluid efficiency depends on the surface area of the exposed rock, differential pressure, rock properties, and characteristics of the pore and injected fluids. A direct intercept will result in much less leak-off than a communication channel through several feet of rock.

Consider the system shown in **Fig. 10.34.** The relief well on the right has achieved hydraulic communication with the blowing well on the left and a dynamic kill is underway. In the relief well, a dead string, say drillpipe or tubing, is equipped with a pressure gauge. The dead string gives a semidirect reading of the BHPs because friction losses in the relief well are not a component to the gauge reading. A dead string allows the operator to monitor progress off the kill and to adjust the pump rate as needed to achieve the desired pressure in the blowout well. It also affords the means to tell when the BHP is approaching the fracture pressure.

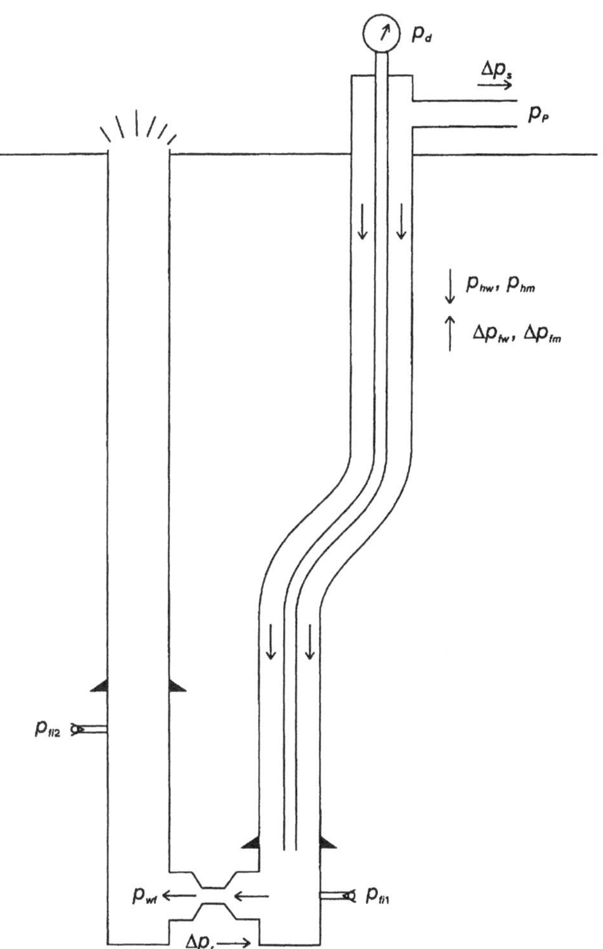

Fig. 10.34—Hydraulic system associated with a dynamic kill from a relief well (after Warriner and Cassity[26]).

The pressure gauge reading during the pumping operation is given by

$$p_d = p_{wf} + \Delta p_r - p_{hd}, \qquad (10.38)$$

where Δp_r = the pressure drop across any restriction between the two wells and p_{hd} = the hydrostatic pressure of the fluid in the dead string. The restriction can be viewed as a downhole positive choke. The term represents the resistance to flow offered by the rock matrix if communication is achieved through the blowout reservoir or perforations if the two wells are connected in this manner. The restriction loss is essentially zero if a direct intervention is achieved with a bit or mill.

Once flow has ceased, the pumps cannot be slowed down until the hydrostatic component starts to replace the friction component, which means that kill-weight fluid must replace the lighter fluid before the operation is complete. How this fluid changeover is accomplished depends on several factors, the most important being the pipe pressure ratings and fracture gradient. These limitations seldom allow pumping kill-weight mud to surface at the original kill rate. Some step-down schedule, guided by the drillpipe gauge measurements, is necessary. Staging the density may be needed if the original kill fluid is significantly underbalanced.

Before injection begins, the flowing BHP in the blowout well can be obtained directly from the pressure gauge if the hole will stand full. This observation will not be possible if the well goes on vacuum after communication has been established. A positive drillpipe pressure, on the other hand, allows measurement of the restriction loss as it will be the difference in the static- and dynamic-pressure gauge readings,

The fracture initiation pressure, indicated as pressure relief valves in the diagram, controls the maximum pressure limit for the operation. In the planning stage, the maximum pump rate can be predicted by replacing p_R with p_{fi} and incorporating Δp_r into Eq. 10.33 or Eq. 10.34. Doing so yields the respective relations for pipe and annular flow:

$$(q_l)_{max} = 0.592 d_c^{2.5}\left(\frac{p_{fi} - p_{wh} - p_{hl} - \Delta p_r}{\rho_l f_M L}\right)^{0.5} \qquad (10.39)$$

and

$$(q_l)_{max} = 0.592(d_h^2 - d_o^2)\left[\frac{(p_{fi} - p_{wh} - p_{hl} - \Delta p_r)(d_h - d_o)}{\rho_l f_M L}\right]^{0.5}. \qquad (10.40)$$

The fracture initiation pressure in these two relations refers to that of the relief well. The limitation at the shoe or other weak point in the blowout well is another, separate design consideration.

The casing pressure at any point in the operation is given by

$$p_c = p_{wf} + \Delta p_r + \Sigma(\Delta p_f) - \Sigma p_h, \qquad (10.41)$$

where the hydrostatic pressures and friction losses are summed in the relief well. The maximum casing pressure controls the pipe design. The pump pressure, given by Eq. 10.42, is needed to quantify the hydraulic power requirement.

$$p_p = p_c + \Delta p_s. \qquad (10.42)$$

Example 10.7. The relief well described in Example 10.6 is equipped with 9⅝-in. 53.50-lbm/ft casing and a 3½-in. dead string. The hole was displaced with 8.5-lbm/gal brine before perforating into the blowout well at a midpoint depth of 11,950 ft measured depth (MD). The blowout will be killed dynamically by injecting the water at 100 bbl/min down the relief well casing.

1. Determine the initial pump pressure if the flowing BHP in the blowout well is 450 psia and the brine friction pressure in the dead string annulus is 0.178 psi/ft. Other losses include 200 psi across the perfs and 150 psi in the surface equipment. Assume the pumps are brought to kill rate instantaneously.

2. Determine the pump pressure when dynamic kill is achieved. Assume that access through the perforations gives 100% leak-off efficiency.

3. Determine the maximum water injection rate if the fracture initiation gradient is 0.85 psi/ft.

4. Determine the pump pressure when the mud reaches the perfs if 12.1-lbm/gal kill-weight mud follows the water. The pump rate cannot change until after mud starts stacking up in the blowout well. Friction pressure has increased to 0.359 psi/ft in the drillpipe annulus and 400 psi in the surface lines.

5. At the initial kill rate, the friction loss through the perfs increases to 280 psi when mud replaces the water and the mud friction loss in the 7-in. casing is 0.477 psi/ft. Will the BHP limitation allow pumping kill-weight mud to surface in the blowout well without reducing injection rate?

6. Determine the number of pumps if each unit can deliver 400 hhp under long-term (longer than two hours) service. Use a 50% design factor as recommended by Abel.[70]

Solution. 1. Eqs. 10.41 and 10.42, respectively, yield

$$p_p = 450 + 200 + (0.1782)(11,950) - (0.442)(11,500) + 150,$$

$$p_p = 450 + 200 + 2,127 - 5,088 + 150 = -2,161 \text{ psia}.$$

The result indicates that the water in the relief well will be drawn into the blowout without any impetus from the pumps for a considerable period.

2. With no leak-off, q_l in the blowout well is equal to q_{kr}. The BHP is higher than the average static reservoir pressure because the selected pump rate is significantly higher than the result obtained from Eq. 10.35. We obtain the friction loss in the blowout well's casing as follows:

$$v = (17.16)(100)/(6.184)^2 = 44.9 \text{ ft/s};$$

$$N_{Re} = (928)(6.184)(44.9)(8.5)/1.0 = 2,190,000;$$

$$f_M = \left[1.14 - 2\log\left(\frac{0.00065}{6.184} + \frac{21.25}{2,190,000^{0.9}}\right)\right]^{-2} = 0.01290; \text{ and}$$

$$\Delta p_f = \frac{(8.5)(0.01290)(44.9)^2(11,500)}{(103.2)(6.184)} = 3,983 \text{ psi}.$$

Add the hydrostatic and outlet pressures,

$$p_{wf} = 3,983 + 5,088 + 15 = 9,086 \text{ psia}.$$

Use Eq. 10.41 and 10.42:

$$p_p = 9,086 + 200 + 2,127 - 5,088 + 150$$

$$= 6,475 \text{ psia} = 6,460 \text{ psig.}$$

3. Eq. 10.39 yields the maximum pump rate.

$$(q_l)_{max} = 0.592(6.184)^{2.5}$$

$$\left[\frac{(0.81)(11,500) - 15 - 5,088 - 200}{(8.5)(0.01290)(11,500)}\right]^{0.5}$$

$$= 106.0 \text{ bbl/min.}$$

Another iteration using a slightly lower friction factor corresponding to the higher rate leads to essentially the same result.

4. The predicted pump pressure at this stage of the operation is

$$p_p = 9,086 + 200 + (0.359)(11,950)$$

$$- (0.629)(11,500) + 400,$$

$$p_p = 9,086 + 200 + 4,290 - 7,234 + 400$$

$$= 6,742 \text{ psia} = 6,727 \text{ psig.}$$

5. The BHP in the relief well is now

$$p_{bh} = (0.477)(11,500) + 7,234 + 15 + 280$$

$$= 13,015 \text{ psia;}$$

that gives a gradient of

$$g_{bh} = (13,015 - 15)/11,500 = 1.13 \text{ psi/ft.}$$

The rock will fracture long before the mud surfaces.

6. Eq. 10.6 gives the hydraulic horsepower requirement. Choose the maximum pressure from Part 4 and obtain

$$P_h = (0.0245)(100)(6,727) = 16,481 \text{ hhp.}$$

The number of pumps is, therefore,

$$1.5(16,481/400) = 62 \text{ pumps.}$$

The excess power provides for higher-than-expected power needs and reserve pump capacity when units start to fail under the rigors of the operation.

The anticipated pump rate controls the relief-well casing program and pressure rating of the last string. Thus it is extremely important to quantify the requirement accurately long before spud. One of the largest unknowns is the leak-off efficiency. A severe over- or under-prediction of the parameter could spell doom for the effort or lead to a grossly overdesigned relief well. A high leak-off efficiency, probably no lower than 90%, is one of the major advantages to a direct intercept.

Achieving a dynamic kill with water usually demands a great amount of hydraulic energy and power. The job logistics and space requirements for such an endeavor may be difficult to achieve in remote and offshore locations. And a large volume of water, perhaps thousands of barrels, is required with the consequent concerns related to surface storage capacity. Water, however, is abundant and cheap in an offshore dynamic-kill operation.

A two-phase pipe flow simulator was developed in 1989 to assist in the planning of a dynamic kill in the North Sea and commercial simulators have been available since 1992.[70,30] Current practice uses simulation when designing a relief well or kill procedure. The hand calculations presented in this chapter are valid, but by their nature yield conservative kill-rate predictions. The ability to model the kill process is a valuable tool that can remove many computational uncertainties, especially with regard to the zero-derivative solution.

The advent of software models has changed fundamentally the way many dynamic kills are performed. Modeling and field experience have shown that weighted mud is generally a more cost-effective initial kill fluid than water if the formation pressure equivalent is much greater than that of the water.[56,71] The friction component becomes less important than the hydrostatic pressure, which can result in a kill at substantial savings in fluid volume and horsepower requirements. Simulation is needed to design one of these so-called "mass-flow-rate" kills. Regardless of the selected kill fluid, computer simulation using reasonable estimates of the IPR and other input data allows the design engineer to consider numerous options and thus to optimize fluid density, kill rate, and relief-well design based on the available resources.[72]

Though our focus has been relief wells, the same principles apply to a dynamic kill pumped through a workstring. This application, however, has been limited because it is often difficult or impossible to pump down a small-bore workstring at a rate necessary to achieve adequate friction in the annulus. Drag-reducing polymers have been successful in reducing pipe friction losses to the point where the kill-rate requirement could be met.[73] But a significant reduction in annulus friction removes an important component to the kill and drag reducers should be used with care and only after modeling the overall effect.

Relying on friction pressure to help kill a well is not strictly a blowout-control concept. By necessity, the dynamic kill is also an important secondary control measure on certain slimhole projects. Recall from Chap. 4 that friction losses in the annulus are considered as a small safety margin in a Driller's Method or Wait-and-Weight procedure. The annular friction losses in a slimhole well, however, dominate the circulating pressure and cannot be ignored. Coupled with the annulus friction is the effect of a moderate kick volume in a small annulus on the shut-in and circulating casing pressures. Following a conventional shut-in and kick displacement procedure may exceed the wellbore pressure limit in these types of wells.

A slimhole procedure recommended by Bode *et al.*[74] is to start immediately circulating at a maximum tolerable rate as soon as a kick is detected. Positioning the tool joint, closing the preventer, and placing returns through the choke line are all accomplished while circulating at the dynamic-kill rate. The friction backpressure afforded by the procedure can be estimated on a routine basis by shutting down the drilling or continuous coring operation, circulating (and perhaps rotating) at a designated rate, and measuring the drillpipe pressure. The annulus friction is obtained by subtracting the computed string losses from the observed pressure. The procedure is repeated at different rates and, in the event of a kick, the desired equivalent circulating density (ECD) can be determined from a graph of ECD *vs.* pump rate.

One major disadvantage to the practice is that the static pore pressure is an unknown and the well will continue to flow if the BHP does not overbalance the rock. This may be an acceptable risk if the pore pressure gradient can be quantified as a range and the ECD is based on the maximum. Large kicks also pose a problem as less friction can be generated by the mud; some predetermined kick volume may dictate the use of a more conventional kill.

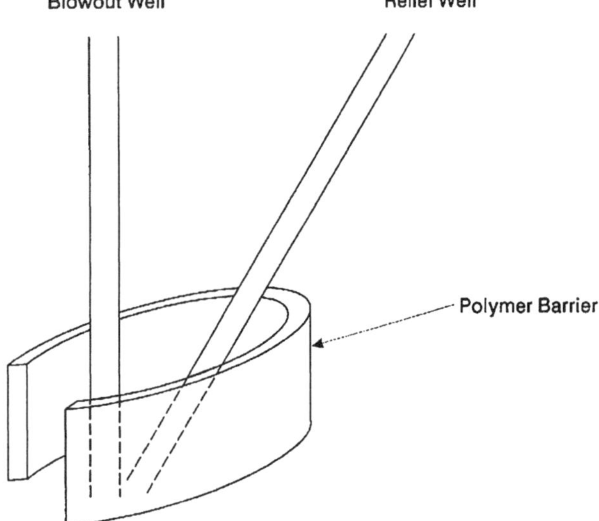

Fig. 10.35—Creating a reservoir permeability barrier using a delayed-viscosity polymer (after Ely and Holditch[77]).

Prince and Cowell[75] recommended shutting a slimhole well in on a kick and obtaining the mud-weight requirement, but using a low-choke method (see Chap. 5) when displacement begins. The annular-friction losses at the designated kill rate are predicted using the described technique before the well-control problem. Following a kick, the casing pressure is not maintained constant as the pump is brought up to speed, but rather is adjusted to a value corresponding to the initial SICP less the annulus friction plus some safety margin. As in a conventional procedure, the drillpipe gauge is used then to control the well for the remainder of the kill. The primary disadvantage to this approach is that slimhole geometries may not allow shutting the well in except on very small kicks. But prudent well design, state-of-the-art detection sytems, and diligent, trained crews can do much to lessen the hazard.

10.5.4 Specialty Fluids in Blowout Control. Fluids designed for a specific function have been used in blowout control for many years. For example, crosslinked polymers were pumped into a relief well on one job from the 1970s.[76] The objective in that case was to fill up leached-out salt cavities in the blowout wellbore so that the subsequent kill fluid could develop hydrostatic pressure more rapidly instead of being wasted in the enlarged hole sections.

Another application described by Ely and Holditch[77] involves injecting a polymer slug from a relief well into a high-permeability blowout reservoir. Delayed viscosity characteristics allow the polymer slug to flow radially from the relief well and towards the pressure sink caused by the blowout before setting up. Ideally, a "wall" is created in the reservoir similar to the one illustrated in **Fig. 10.35.** The permeability barrier accomplishes two purposes: it forms a flow restriction to the hydrocarbons and it channels kill fluids more directly into the blowout well. The technique, of course, is not usable for a direct intercept.

More recent attention has been focused on fluids or slurries that thicken and either create a solid plug in the blowout conduit or, by generating high friction pressure, assist a dynamic-kill effort. These materials remain fluid and thus pumpable until they mix downhole with a reactant that triggers the setting action. Reactive plugs have been used since the early 1960s to combat lost circulation. Most well-control applications thus far have been in the control of underground blowouts.

Flak[78] discussed using the reactive plug formulations listed in **Table 10.3.** The so-called gunk plugs thicken by a hydration or swelling mechanism. For example, the bentonite in a BDO slurry hydrates when contacted by fresh or moderately saline (less than 50,000 mg/L chlorides) water and almost instantaneously creates a plug. Salt clays can be substituted in the BDO or guar gum formulations to react with saltier brines. Similarly, organophilic clays swell in the presence of oil and cause an invert gunk to form a solid mass. Sodium silicate solution, on the other hand, reacts chemically with $CaCl_2$ brine, cement, or another source of dissolved calcium to precipitate calcium silicate.

Gunk plugs have limited strength and their ability to remain competent at a high differential pressure is questionable unless cement is used in conjunction with the material. Dry cement can be mixed directly into oil-based gunks at concentrations up to 150 lbm/bbl or the water in a cement slurry can be used as a reactant. Wet cement, upon contacting sodium silicate, forms a hard plug very quickly.

A means to inject the reactive material into the flow stream is necessary and the reactant must be present at the right place and in the correct mixing ratio to institute the solidification process. A single injection path may be used to pump a gunk squeeze into a blowout that is producing a large volume of oil or water. Sampling the produced fluid and pilot testing is recommended for determining the optimum gunk/fluid ratio, but one problem is knowing how fast to mix and pump the slurry to achieve this optimum ratio. Too much or too little slurry will not accomplish the desired result. An accurate estimate of the production rate is needed.

Two separate flow paths (pipe and annulus, relief well and pipe, etc.) give more options and allow for more precise control of the downhole reaction. The plug is pumped down one conduit while the reactant is injected down the other and the respective injection rates are controlled to yield the desired mixing ratio at the point where the fluids come in contact.

TABLE 10.3—REACTIVE MATERIALS USED IN BLOWOUT CONTROL

Description	Formulation	Reactant
Bentonite-diesel oil (BDO) gunk	≤350 lbm/bbl bentonite in diesel	Fresh water
Guar gum gunk	≤400 lbm/bbl guar gum in diesel	Brine
Invert gunk	≤300 lbm/bbl amine-treated clay in water	Oil
Acid-soluble gunk	≤300 lbm/bbl polymer,	Fresh or brine water
	≤200 lbm/bbl $CaCO_3$, and	
	≤10 lbm/bbl viscosifier in diesel or oil	
Nonpolluting gunk	≤300 lbm/bbl polymer and	Fresh or brine water
	≤100 lbm/bbl organic fiber in nontoxic oil	
Sodium silicate	40% dissolved sodium silicate in water	$CaCl_2$ or cement

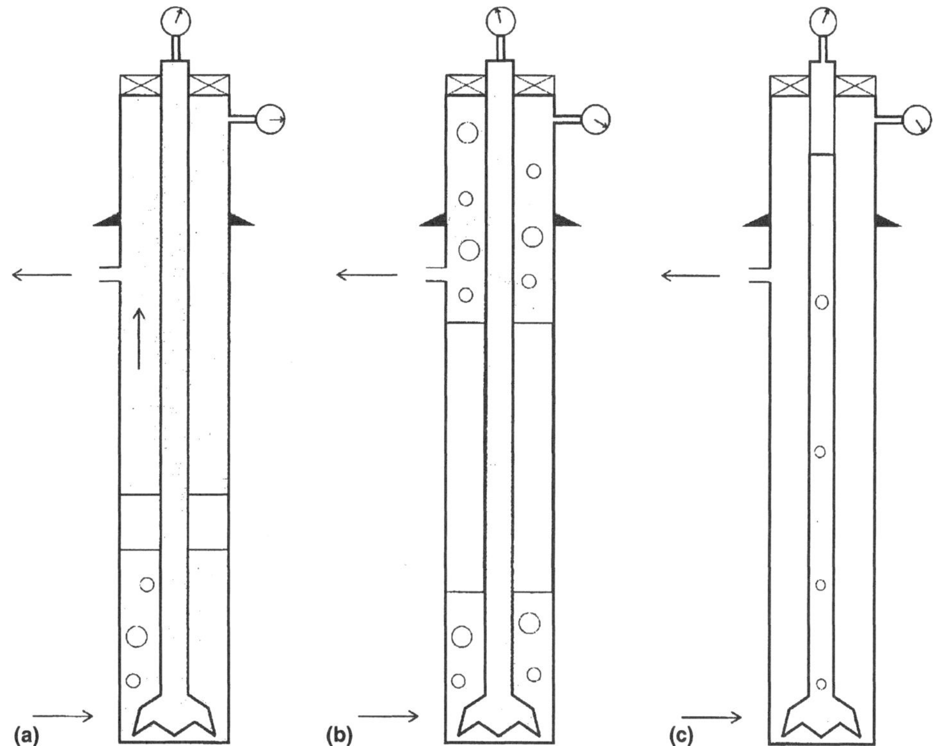

Fig. 10.36—Three stages in the development of a typical underground blowout.

Operational concerns when using a reactive fluid include taking steps to assure that the plug is kept away from anything that will cause a premature set in the surface equipment or well tubulars. Mixing equipment and tanks should be cleaned thoroughly and at least 10 bbl of spacer should be used ahead of and behind the slurry. Continuous agitation with a batch mixer is necessary to keep any solid constituents in suspension until the slurry can be started downhole.

Finally, the operator needs to consider the consequences if the plug sets partially but fails to stop the blowout. Stuck pipe is a possibility and access to the blowout zone may be restricted, both of which may limit subsequent control options.

10.6 Underground Blowouts

An underground blowout is the uncontrolled flow of formation fluids from one zone to another. While not as spectacular as a surface blowout, an underground blowout can be just as difficult and costly to control. Most underground blowouts result from a combination of factors and event sequences. In a drilling operation, underground blowouts usually occur when there is a kick in the hole and may transpire when the well is first shut in or at some other point in the control procedure. In cased holes, underground blowouts often result when internal pressure ruptures the production casing. The outer strings and any exposed rock are subjected to this pressure and also fail.

The most effective way to control underground blowouts is to prevent them from happening. The effort starts when the drilling engineer is handed the well prognosis and does not finish until the well is plugged and abandoned. Proper casing point selection, adequate pipe designs, measuring fracture gradients, predicting pore pressures, crew training, and prudent tripping practices are all important steps to achieving this goal.

10.6.1 Diagnostics. The first order of business is recognizing when an underground blowout occurs. Consider a drilling well with a kick in the hole. Recall the U-tube analogy from Chap. 4. The pressure gauges on the drillpipe and casing reflect the pressure at the bottom of the string less the hydrostatic pressure of the overlying fluids. Knowing what these fluids are and their respective heights, one can use either side of the U-tube to determine the BHP. Also, changing the surface pressure on one side should be noted, after some lag time, by a corresponding change at the other gauge.

The U-tube analogy is appropriate only when the wellbore is a closed system. A well becomes an open system when the rock is taking fluid and then these principles no longer apply. Direct pressure communication between the two gauges is no longer present; the pressure may change on the drillpipe with no corresponding change on the casing and vice versa. Thus the absence of pressure transmission between the surface gauges is one indication of a downhole crossflow.

Refer to the wellbore schematics shown in **Figs. 10.36a through 10.36c.** On a shut-in well, the drillpipe and casing pressures should increase or stabilize. A pressure reduction, however, indicates that the fracture integrity has been exceeded and that flow has commenced into the rock. Fig. 10.36a depicts a shut-in well not long after a zone below the shoe fractures. The casing pressure has dropped to a value corresponding to the fracture-propagation pressure less the hydrostatic pressure of the mud down to the loss depth. The drillpipe pressure has fallen also and the kick zone, being underbalanced, is flowing more gas into the hole.

As the blowout proceeds, more mud is displaced into the fractured rock and less pressure is applied to the flowing zone. The drillpipe-gauge reading continues to fall as the flowing BHP is reduced. On the casing side, gas is percolating above the loss zone, accumulating below the preventers, and causing the surface pressure to increase. More mud is forced into the formation.

Absent any action by the operator, the situation depicted in Fig. 10.36c could develop. The backside is essentially full of gas and the casing pressure has reached its maximum level.

What happens on the drillpipe gauge depends on other factors. Gas percolation through the drillstring will cause the pressure to increase and displace more mud into the annulus. Absent gas, the drillpipe pressure plus the mud hydrostatic will be the flowing BHP. The gauge will read zero if the hydrostatic exceeds this flowing pressure.

The schematic series illustrates one of many possible scenarios. Interpreting what is going on downhole by the pressure-gauge readings may be a difficult if not impossible task. Casing pressures, in fact, may decline during an underground blowout.[79] Also, the formation-fluid composition can change and a rich gas or volatile oil will undergo a phase change in response to the changing wellbore pressure. The formation can collapse around the drillstring and prevent or restrict the ability to transmit pressure. Pressure depletion can occur if the reservoir is small or the fracture-propagation pressure may drop. The last situation suggests that the path of least resistance is not confined to the fractured zone and, hence, the possibilty that fluids will ultimately broach to surface.

The location of the loss zone must be known before any attempt can be made to control the well. This may be obvious in some cases, for example when the bit drills into severe lost circulation and an underground blowout ensues from the shallower rock. Most situations, however, require running a production log of some sort to identify where the fluids are going.

Estimating the loss-zone depth may be possible using only surface observations. Normally, an operator will pump water down the annulus to keep the formation fluids below the loss zone and suppress casing pressure. The casing pressure with a full column of water down to the receptive formation is given by

$$p_c = p_{fp} - g_w D_{lz}, \quad \text{(10.43)}$$

where g_w = the hydrostatic gradient of the water and D_{lz} = the depth to the loss zone. This relation assumes the pump rate is slow enough that friction losses are negligible. Similarly,

$$p_c = p_{fp} - g_g D_{lz}, \quad \text{(10.44)}$$

if the backside is full of gas. Combining these two equations yields

$$D_{lz} = \Delta p_c / (g_w - g_g). \quad \text{(10.45)}$$

Grace[80] discussed an offshore underground blowout where the result of this calculation technique may have prevented a disaster. The drillpipe and casing pressures had stabilized over a long period and indications were that the losses were well below the 2,500-ft casing shoe. Then both sides suddenly went on a decline to lower stabilized readings, which indicated a significant change in the downhole conditions. The depth to the loss zone was calculated at 1,500 ft, which was 1,000 ft above the shoe. Either the casing had failed or the flow had worked its way up to the shallower zone. In any event, working conditions were deemed unsafe and the rig was abandoned. The flow erupted at the sea floor the very next day.

Example 10.8. An underground blowout occurred when a well was closed in on a kick. SICP increased and ultimately stabilized at 2,550 psig. Seawater weighing 8.6 lbm/gal is pumped into the annulus and, over time, the casing pressure decreases to 940 psig. Surface casing is set at 3,100 ft. Estimate the depth to the loss zone.

Solution. We have to make some assumptions regarding the gas to get an average gradient for the fluid. Assume that the loss is occurring directly below the casing shoe and that the wellbore temperature is given by T = 70°F + 1°F/100 ft. The average gas temperature from surface to 3,000 ft is

$$T = (70 + 100)/2 = 85°\text{F} = 545°\text{R}.$$

Assume the gas specific gravity (SG) = 0.60 and iterate Eq. 1.44 to calculate the pressure at 3,000 ft.

$$p_{3,000} = (2,564)\exp\left[\frac{(0.60)(3,000)}{(53.3)(0.795)(545)}\right] = 2,772 \text{ psia}.$$

Therefore, the average gas gradient to the loss zone is approximately

$$g_g = (2,772 - 2,564)/3,000 = 0.070 \text{ psi/ft}.$$

Eq. 10.45 gives an approximate loss-zone depth of

$$D_{lz} = (2,550 - 940)/(0.447 - 0.070) = 4,271 \text{ ft}.$$

The most common way to locate a loss zone is to run a temperature log through the drillpipe or tubing. **Fig. 10.37** represents a typical log response when the loss zone is being charged from a deeper formation. The formation fluids are hotter than the adjacent strata and so the stored heat is manifested on the log as an anomaly in the wellbore temperature profile. The effect can be subtle and a differential track, which records the gradient rate-of-change, is used to assist in the interpretation.

A radioactive tracer survey can pinpoint the loss zone without the ambiguities so often associated with temperature logs. The tool is run into the workstring and the pump is engaged

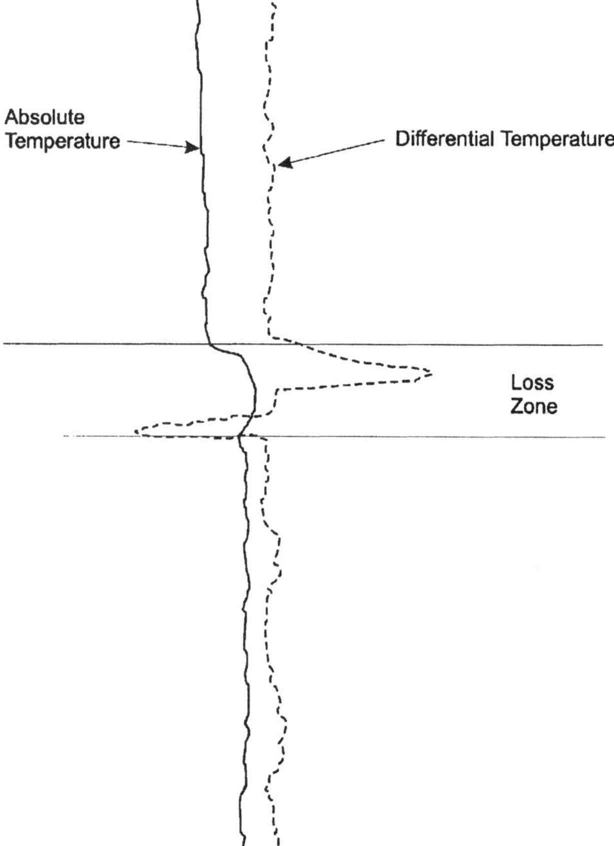

Fig. 10.37—Typical temperature log response near a loss zone.

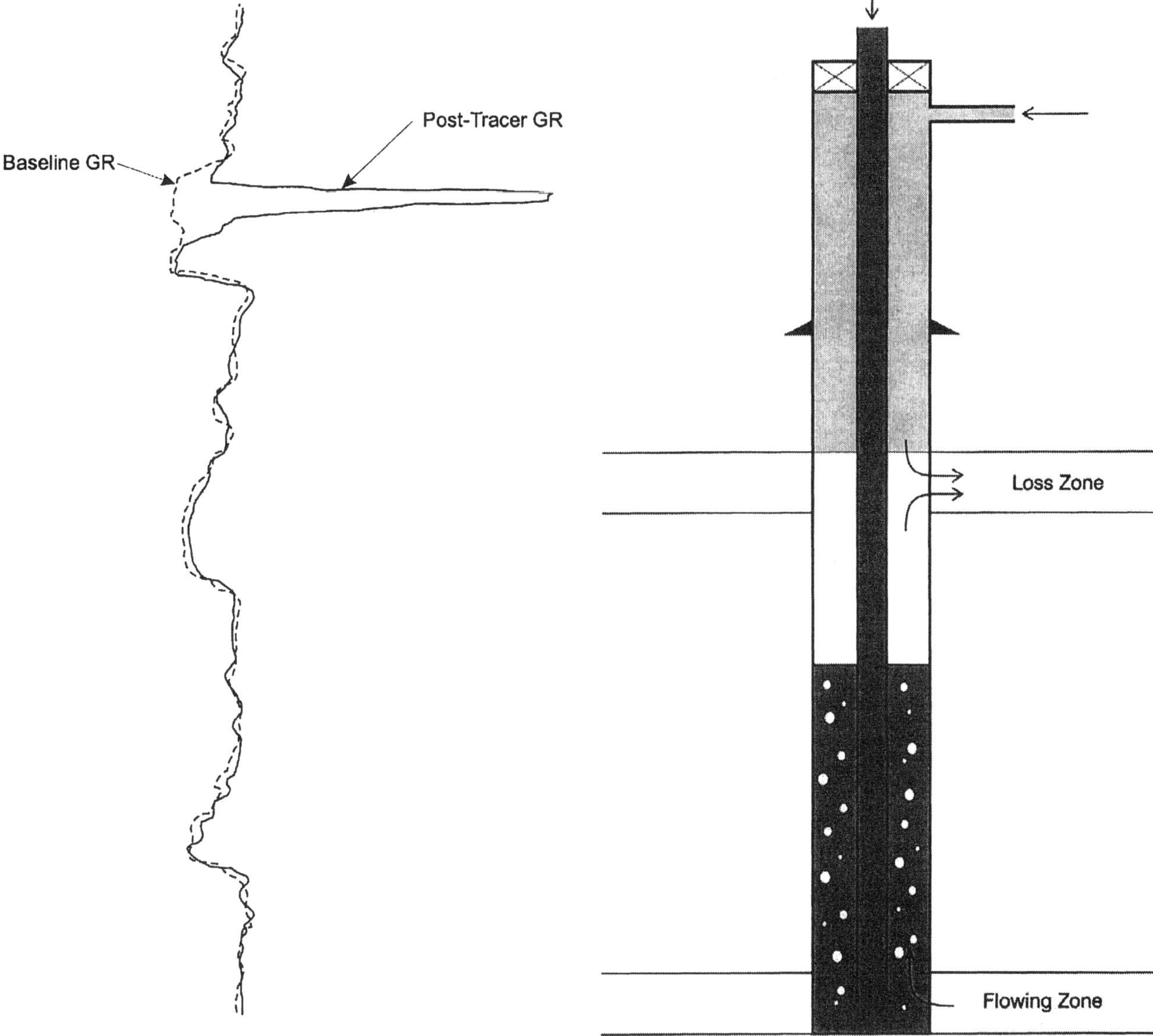

Fig. 10.38—Using a radioactive tracer survey to identify a loss zone.

Fig. 10.39—Killing an underground blowout with heavy mud.

at a slow rate. A small slug of radioactive (RA) material is released from the tool and is conveyed into the annulus by the mud. In the annulus, the RA material is swept up by the flow stream and is deposited in the loss zone. As shown in **Fig 10.38,** the tool's gamma ray detector identifies a hot spot where the RA material left the wellbore.

Noise logs are run sometimes, but normally in conjunction with some other diagnostic tool. As the name implies, the tool detects sound. Ideally, a static region above the loss zone will be relatively silent whereas flow will create noise. Hence the loss zone is inferred at the depth where the noise level increases.

10.6.2 Control Measures. It has been said that controlling an underground blowout is more of an art than a science. This may be true to some extent, but only because the problem cannot be observed directly and some assumptions must be made with regard to flow rate, fluid compositions, wellbore conditions, downhole pressures, reservoir IPR, and so on. Many potential remedies may be implemented before one ultimately does the trick. Or the problem may cure itself if the hole bridges off or the blowing formation depletes. Nonetheless, one can take what information is available and use sound engineering judgement to rule out those control measures having little merit and thereby concentrate on those having the best chance of success. Whatever steps are taken, there needs to be some reasonable assurance that the potential cure will not cause the problem to deteriorate or limit future options if the procedure should fail.

Be it cased or open hole, the most common flow circuit is from a deeper formation into a shallower zone. The other type of underground blowout is flow from an upper zone into a lower zone, usually a result of drilling into a severe lost circulation zone or failing to set casing soon enough when pore pressures undergo a reverse transition. The remainder of this chapter discusses control measures that have proved successful in both flow circuits. A relief well will be necessary if these methods are not an option or fail to do the job; the well does not kill itself; or cratering is a distinct possibility.

Killing a Bottom-to-Top Flow. It may be possible to kill an underground blowout by pumping heavy mud at a fast rate. **Fig. 10.39** illustrates this procedure. Heavy mud leaves the bottom of the string and commingles with the formation fluids at a mixture ratio that depends on the blowout and pump rates. The mixture gains height in the annulus and increases the pressure at the bottom of the hole, which reduces the influx rate and increases the volume fraction of the mud. The well can be killed if the BHP ever exceeds the average static reservoir pressure.

Assume for now that the zone is producing gas. It will prove useful to express the downhole volumetric flow rate in terms consistent with the pump rate. The gas law gives

$$q_g = q_{gsc}zT/287{,}000p_{wf}, \quad \ldots\ldots\ldots\ldots\ldots\ldots (10.46)$$

when q_g is expressed in bbl/min. Assume the gas does not achieve turbulent flow in the reservoir ($n = 1.0$). Substituting Eq. 10.45 into Eq. 4.1 and dividing by the drawdown ($p_R - p_{wf}$) then gives the productivity index at sandface conditions.

$$J^* = \frac{k_g h_f(p_R + p_{wf})}{408\mu_g(\ln r_e/r_w - 0.75)p_{wf}}. \quad \ldots\ldots\ldots\ldots (10.47)$$

The units of J^* are bbl/(min-psi).

Wessel and Tarr[81,82] derived a set of equations that indicate the "killability" of a well with an underground flow. Several simplifying assumptions were made to make the solutions more tractable and these are listed in **Table 10.4.** The pump rate that will kill the well when the heavy mud first reaches the loss zone is presented here as

$$(q_{kr})_{OH} = \frac{J^*\left(p_{hk} + p_{fp} - p_R\right)^2}{2\left(p_{hm} + p_{fp} - p_R\right)}, \quad \ldots\ldots\ldots\ldots (10.48)$$

where p_{hk} and p_{hm} are the hydrostatic pressures of the kick fluid and mud between the two depths of interest, respectively. The former can be calculated using Eq. 10.49.

$$p_{hk} = \frac{gV_a\rho_k\cos\alpha}{g_cA_a}. \quad \ldots\ldots\ldots\ldots\ldots\ldots\ldots\ldots (10.49)$$

Note that p_{hm} can be expressed as

$$p_{hm} = p_{hk}\rho_m/\rho_k. \quad \ldots\ldots\ldots\ldots\ldots\ldots\ldots\ldots (10.50)$$

Injecting the kill fluid at a rate lower than that determined by Eq. 10.47 will require more volume to effect a kill. A system-intake curve must exceed the IPR at all formation flow rates before a well can be killed dynamically. The same holds true for this method and the authors derived Eq. 10.51 for the pump rate that causes the system-intake curve to touch the linear IPR.

TABLE 10.4—ASSUMPTIONS USED BY WESSEL AND TARR[81,82] TO DERIVE EQ. 10.48 AND 10.51	
1.	The drillstring and hole have a constant diameter.
2.	The drillstring and flowing zone are at the bottom of the hole.
3.	Friction losses in the annulus are negligible.
4.	The kill mud and formation fluids are incompressible and will mix readily at downhole conditions.
5.	The IPR is a linear function of the drawdown pressure.
6.	The wellbore pressure opposite the loss zone is independent of the flow rate into the zone.
7.	The fluid above the loss zone is incompressible.
8.	The kill mud is pumped at constant rate with the annulus closed in.

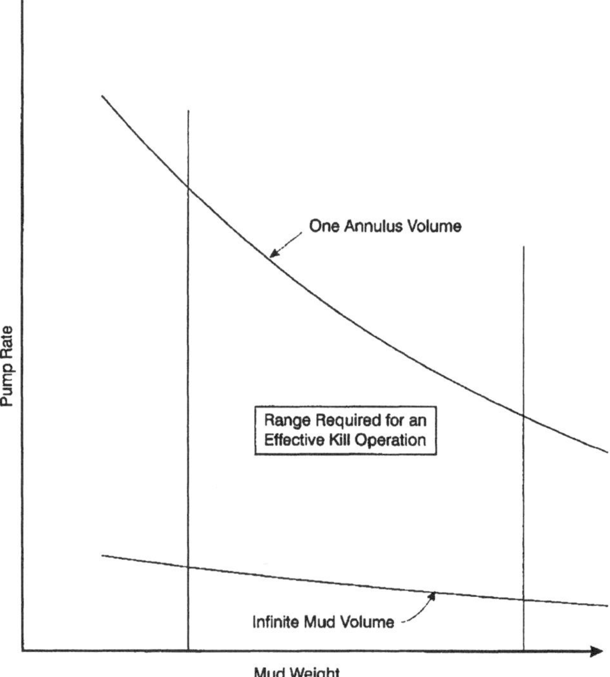

Fig. 10.40—Pump rate and mud density combinations needed to kill an underground blowout (after Wessel and Tarr[81]).

$$(q_{kr})_\infty = J^* \left\{\left(p_{fp} - p_R + 2p_{hm} - p_{hk}\right) - 2\left[\left(p_{hm} + p_{fp} - p_R\right)\left(p_m - p_{hk}\right)\right]^{0.5}\right\}. \quad \ldots\ldots\ldots\ldots\ldots\ldots (10.51)$$

An infinitely large mud volume is needed to meet this criterion, hence the reason for the "∞" subscript.

As indicated in **Fig. 10.40,** the pump rates that will kill an underground flow fall between the two extremes and a numerical integration procedure is necessary to solve for the time (hence mud volume) to kill a flow at the intermediate kill rates. This involves using a small time increment to determine the flowing BHP; calculating flow from the formation based on the IPR; using this result to determine the mixture density; and repeating the process until the flowing BHP equals the static reservoir pressure. A plot of the simulated BHP as a function of time or volume pumped will be similar in appearance to **Fig. 10.41.**

Costs will be reduced proportionally with increasing pump rate because the time and mud volume requirements are lessened. If possible, the workstring should be open-ended and have a large ID. For one thing, it may not be possible to pump

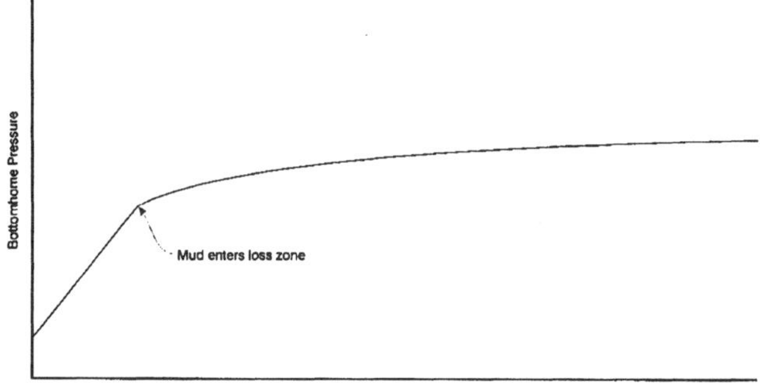

Fig. 10.41—Typical BHP behavior when an underground flow is killed with heavy mud (after Wessel and Tarr[82]).

TABLE 10.5—INFORMATION RELATED TO AN UNDERGROUND BLOWOUT

Wellbore Configuration	
Vertical depth	9,475 ft
Measured depth	9,832 ft
Kickoff point (hole is vertical above)	4,198 ft
Surface casing depth	3,420 ft
Loss zone depth from temperature survey	3,605 ft
Bit diameter	12¼ in.
Drillstring OD	5 in.
Annulus capacity factor	0.12149 bbl/ft
Annulus cross-sectional area	98.224 in.^{2}
Assumed or Calculated Reservoir and Fluid Properties	
Formation fluid density	1.4 lbm/gal
Static reservoir pressure	6,254 psig
Flowing BHP	3,260 psig
Permeability to gas	100 md
Gas viscosity	0.03 cp
Drainage radius	1,000 ft
Exposed reservoir thickness	5 ft
Surface Observations	
Water density above loss zone	9.0 lbm/gal
Casing pressure during water Injection	1,240 psig

through the existing string at the desired rate because of friction-loss limitations. Another benefit not considered in the calculations is the added friction backpressure corresponding to a fast pump rate in a relatively narrow annulus.

The main problem with the procedure is that there are inaccuracies inherent to the calculation method. Reasonable estimates can be obtained for most of the input variables. But one important parameter, formation permeability, is probably unknown to any degree of precision and may easily differ by an order of magnitude from the estimated value. Local knowledge may be a reliable guide for ranges to consider.

Example 10.9. Table 10.5 lists wellbore and formation characteristics for a hypothetical underground blowout.

1. Determine the minimum mud weight if the intent is to balance the static formation pressure after all of the influx has been displaced into the fracture and annulus pressure is released.

2. Determine the maximum allowable mud weight if the intent is to keep the pressure gradient at all points in the well less than the loss zone's fracture propagation gradient.

3. Estimate the kill rate range for both of these muds.

Solution. 1. First determine the fracture propagation pressure based on the casing pressure and salt water hydrostatic.

$$p_{fp} = 1{,}240 + (0.0519)(9.0)(3{,}605),$$

$$p_{fp} = 1{,}240 + 1{,}684 = 2{,}924 \text{ psig, and}$$

$$g_{fp} = 2{,}924/3{,}605 = 0.811 \text{ psi/ft.}$$

Water will remain above the loss zone as long as pump pressure on the annulus exceeds 1,240 psig. The casing pressure will be released when the kill is completed, the fracture closes, and clean mud resides below the loss zone. The final mud-density requirement below the loss zone is thus

$$1{,}684 + g_m(9{,}475 - 3{,}605) = 6{,}254;$$

$$g_m = 0.779 \text{ psi/ft} = 15.0 \text{ lbm/gal.}$$

The mud gradient is less than the fracture-propagation gradient, so it would be safe to go ahead and completely displace the hole with this mud.

2. Using the frac gradient as the limiting criterion, the maximum allowable pressure at total depth is

$$(0.811)(9{,}475) = 7{,}684 \text{ psig.}$$

The maximum mud density is therefore

$$(g_m)_{max} = (7{,}684 - 1{,}684)/5{,}870 = 1.022 \text{ psi/ft}$$

$$= 19.7 \text{ lbm/gal.}$$

This mud *cannot* be circulated to surface and would have to be removed later in stages.

3. Eq. 10.47 is used to estimate the productivity index at bottomhole pressure and temperature.

$$J^* = \frac{(0.100)(5)(6{,}254 + 3{,}260)}{(408)(0.03)\left[\ln(1{,}000/0.51) - 0.75\right](3{,}260)}$$

$$= 0.01745 \text{ bbl/(min-psi).}$$

The average hole inclination across the interval between the loss zone and the reservoir is determined as

$$\alpha = \cos^{-1}\left(\frac{9{,}475 - 3{,}605}{9{,}832 - 3{,}605}\right) = 19.5°.$$

The annular volume between the two formations is

$$V_a = (0.12149)(42)(9{,}832 - 3{,}605) = 31{,}774 \text{ gal.}$$

Eq. 10.50 yields the gas hydrostatic pressure across this hole section.

$$p_{hk} = (31{,}774)(1.40)\cos(19.50)/98.224 = 427 \text{ psi.}$$

Eq. 10.49 gives the contribution of the 15.0-lbm/gal mud to the hydrostatic pressure.

$$p_{hm} = (15.0)(427)/1.40 = 4{,}575 \text{ psi.}$$

Eq. 10.48 gives the mimium pump rate for the upper limit.

$$(q_m)_{OH} = \frac{(0.01745)(427 + 2{,}924 - 6{,}254)^2}{2(4{,}575 + 2{,}924 - 6{,}254)}$$

$$= 59.1 \text{ bbl/min.}$$

Now use Eq. 10.51 to determine the kill pump rate for an infinite volume of 15.0-lbm/gal mud.

$$(q_m)_\infty =$$

$$(0.01745)\{[2{,}924 - 6{,}254 + (2)(4{,}575) - 427]$$

$$- 2[(4{,}575 + 2{,}924 - 6{,}254)(4{,}575 - 427)]^{0.5}\}$$

$$= 14.8 \text{ bbl/min.}$$

The requirements for the 19.7-lbm/gal mud are determined in similar fashion.

$$p_{hm} = (19.7)(427)/1.40 = 6{,}008 \text{ psi,}$$

$$(q_m)_{OH} = \frac{(0.01745)(427 + 2{,}924 - 6{,}254)^2}{2(6{,}008 + 2{,}924 - 6{,}254)}$$

$$= 27.5 \text{ bbl/min,}$$

and $(q_m)_\infty =$

$$(0.01745)\{[2,924 - 6,254 + (2)(6,008) - 427]$$

$$- 2[(6,008 + 2,924 - 6,254)(6,008 - 427)]^{0.5}\}$$

$$= 9.2 \text{ bbl/min.}$$

This well appears to be a candidate for the procedure.

Another method for solving an underground flow problem is to deal directly with the loss zone by attempting to seal the fracture or matrix with lost circulation material (LCM), gunk plug, or other reactive material. The loss zone may or may not be totally curable, but it may be possible to retard the flow if not stop it entirely. Large LCM concentrations are required usually and open-ended pipe is desirable to prevent plugging. It may be necessary to strip/snub from the hole to remove a bottomhole assembly or at least to remove the bit jets with a primer-cord shot. Stripping up to a point above the loss zone is recommended to prevent annulus bridging and for enhanced effectiveness. Cement should be used only as a last resort for controlling lost circulation.

Another option may be to isolate the loss zone from the blowout reservoir. The reactive plugs discussed in the last section have been used for this purpose. Barite plugs also have a proven track record in underground flow control. A barite plug is basically a mixture of water and barite. Before mixing barite, the water pH is adjusted to approximately 10.0 using caustic soda and a small amount of suitable thinner such as lignosulfonate or sodium acid pyrophosphate (SAPP) is added to the water. The thinner reduces friction loss of the plug slurry during placement and enhances settling once the plug is in place. Barite is mixed in sufficient concentration to yield a plug density between 18 and 22 lbm/gal.

In practice, the drillstring is positioned above the flowing zone and the plug is mixed with a volume sufficient to yield a height of at least 500 ft.[83] The barite will settle quickly in water and so the blend must be agitated vigorously during the mixing process. Once target density is achieved, the plug is started downhole and kept moving until placed on the bottom. If everything goes according to plan, the barite settles out into a tough, solid plug. Pressure buildup below the plug further compacts the mass and enhances its sealing ability.

The key to getting the barite to form a plug is to have the material settle faster than the upward flow rate from the reservoir. Otherwise the material will be washed away before it can set. It may be necessary in a high-rate situation to pump a heavy pill ahead of the plug in an attempt to minimize the flow and give relatively quiescent conditions below the pill.

Barite may not settle into a plug if the formation is flowing salt water at high rate. A reactive plug or cement may be the best control options in these situations. Cement, either by itself or in combination with a reactive material, is usually a last-resort tactic. This decision should be made only after considering all of the potential difficulties.

Isolating the flow with a packer or bridge plug may be a solution, but there are many drawbacks and a failed attempt can make a bad situation worse. All but the harder formations will erode in an underground blowout and finding a packer seat may not be possible. Also, sticking the workstring across the loss zone is a possibility with any control procedure. Snubbing a large-diameter tool across the loss zone only increases the sticking risk.

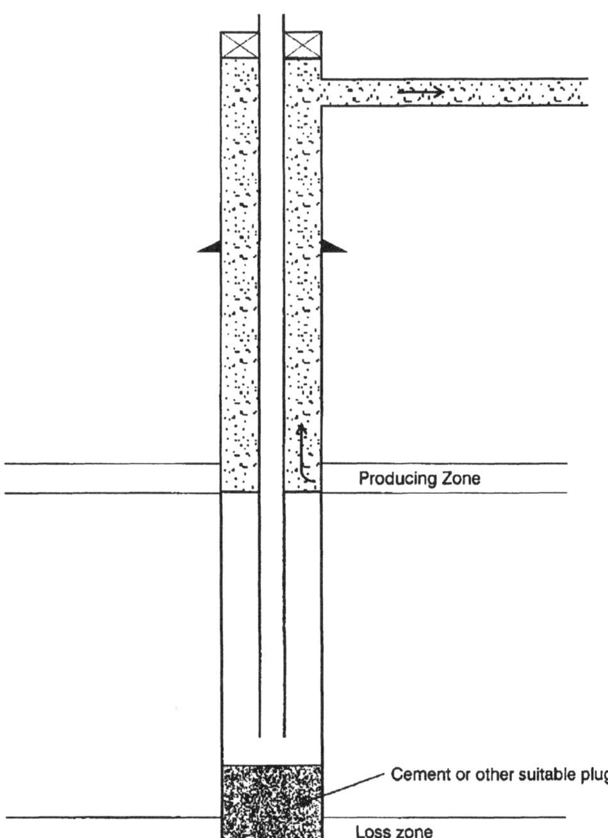

Fig. 10.42—Diverting flow to surface as an intermediate step in controlling an underground blowout.

Another major concern is getting the packer or plug to the objective depth without having the elements cut by flow erosion and particulate abrasion. Inflatable tools would reduce this effect, but setting them is difficult under flowing conditions. An operator should be reasonably sure of success before attempting to use a mechanical device to isolate a dynamic well.

Killing a Top-to-Bottom Flow. Underground flow from a shallower zone into a deeper formation is the most difficult condition to control. About the only advantage to this flow circuit is the minimal risk of flow broaching to the surface. The list of control options is limited to curing the lost circulation and isolating the producing formation from the deeper loss zone.

As before, pumping high concentrations of LCM into the loss zone or squeezing the zone with a reactive material may be successful in regaining full or partial circulation. Alternatively, it may be possible to isolate the two zones using one of the described techniques and achieve circulation as result. Regardless of the method, the ability to circulate kill-weight fluid allows control to be restored so that casing can be set.

Fig. 10.42 illustrates how producing the well at the surface may effect zone isolation. The objective is to bring all the flow to the surface so that a plug can be set in the bottom of the hole under static conditions. The well can be killed and casing set above the plug after the task is accomplished.

Flowing a well wide-open through a vent line or diverter is an inherently hazardous undertaking and should be done only after studied consideration of the failure potential. All that is needed is to reduce the pressure opposite the loss zone to something less than its pore pressure (matrix flow) or fracture gradient (fracture flow). Normally, this can be done by producing the well in controlled fashion through the choke.

Problems

10.1 You are planning to kill a well by shoving a stinger into a damaged tubing head and bullheading kill-weight mud. The mimimum bore through the head is 6.766 in. and the expected shut-in pressure is 500 psig.

1. How much hold-down force must be available when the valve is closed?
2. Determine the maximum bullheading pressure if the hold-down force is limited to 35,000 lbf.

10.2 A well is bullheaded at a slow rate because of pump pressure limitations. The wellhead pressure, however, is increasing instead of falling. What might be an explanation for this behavior?

10.3 Use the real gas law to verify the maximum wellhead pressure seen in Fig. 10.27.

10.4 Consider the well described in Table 10.2 and Example 10.1. Determine the wellhead pressures when the gas is displaced to 1,000 ft, 6,000 ft, and 11,000 ft if the problem is modified as follows.

1. The bullhead kill rate is increased to 70 bbl/min. Calculate friction pressures based on Newtonian behavior for the brine. For the mud, use the power law relations discussed in Appendix A and assume the viscometer dial readings at 600 and 300 rpm are 62 and 40 lbf/100 ft^2, respectively.
2. Injection pressure at the sandface is 500 psi higher than the static pore pressure.
3. Mud density is increased to 14.0 lbm/gal and the kill rate is increased to 70 bbl/min. Assume the same viscometer readings as above.
4. Injection is down a string of 3½-in. 9.30-lbm/ft tubing that is landed in a packer at 11,450 ft. Use a bullheading rate of 10 bbl/min.
5. The well is flowing immediately before the job starts and the local pore pressure builds up according to the curve shown in **Fig. 10.43.** Assume the injection pressure at the sandface exceeds the formation pressure by 500 psi.

10.5 Derive the constant for Eq. 10.10 expressed in SI metric units.

10.6 Reconsider Example 10.2 and determine the required kill rate if the well is blowing out through 3½-in., 9.30-lbm/ft tubing.

10.7 Derive the SI metric relationship for the pressure drop resulting from gas expansion in a diverter line. Show the units in each step of the derivation.

10.8 Grace[58] discussed the successful application of the momentum-kill method to a blowout in Wyoming. Drill collars were above the rotary table when the well blew in. A shallow zone is producing an estimated 30 MMcf/D gas and entrained water through the collar bore. The objective is to pump 12.0-lbm/gal mud down the annulus with sufficient momentum to counter the blowout momentum at the bottom of the collars, which is approximately 180 ft below the rig floor. 13⅜-in., 54.5-lbm/ft casing is set at 400 ft. A gauge on the annulus is installed and the measurement is used to predict a pressure of 835 psia at 180 ft. The estimated gas temperature at this depth is 60°F and a 0.60 SG gas is assumed. Determine the pump rate requirement if the effect of the formation water is to double the momentum calculated for dry gas.

10.9 Derive Eq. 10.21a and 10.21b. Show your work.

10.10 A gas well is blowing an estimated 55 MMcf/D dry gas through 150 ft of 6-in. diverter piping. Other known or assumed data follows. Wellbore conduit diameter = 4.276 in.; depth of the flowing zone = 9,900 ft; gas SG = 0.70; gas specific heat ratio = 1.25; exit temperature = 610°R; and wellbore temperature = 150°F + 0.3°F/100 ft.

1. Estimate the wellhead pressure.
2. Determine the pump rate that will effect a momentum kill at the wellhead if the kill-fluid density is 19.0 lbm/gal.

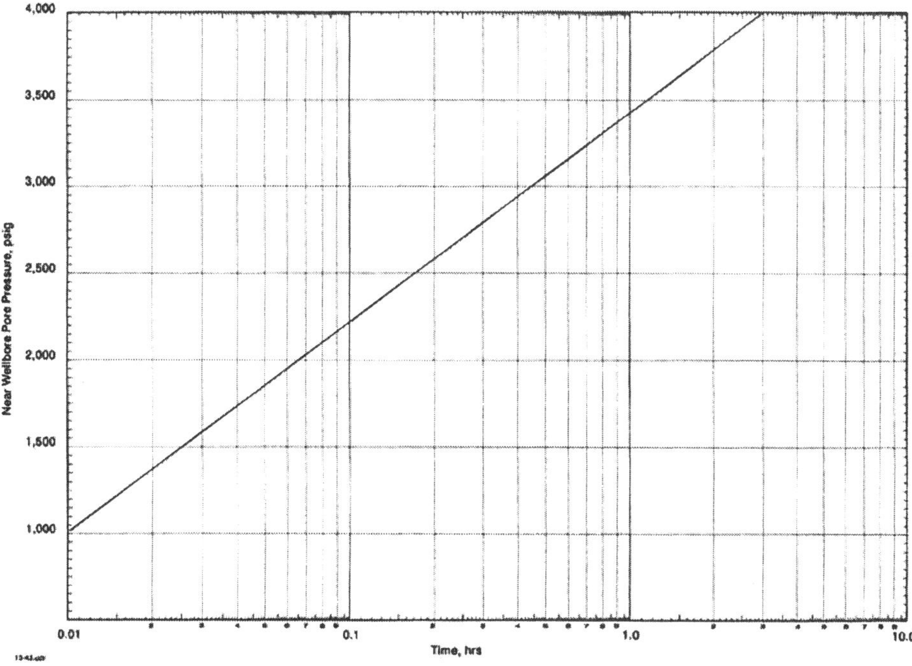

Fig. 10.43—Pore pressure buildup function for Problem 10.4.

TABLE 10.6—INFORMATION RELATED TO A SUBSEA BLOWOUT

Blowout Well Configuration	
Blowout reservoir depth	4,300 ft
Openhole diameter	12¼ in.
Last casing depth	4,200 ft
Casing nominal ID	12.347 in.
Drillstring description	
Top section	4,280 ft of 5-in. drillpipe and HWDP
Bottom section	120 ft of 7-in. drill collars
Relief Well Configuration	
Openhole intercept depth	5,240 ft measured depth (MD), 4,250 ft total vertical depth (TVD)
Openhole diameter	12¼ in.
Drilling liner description	
Depth	5,031 ft MD, 4,047 ft TVD
Liner top	3,993 ft MD, 3,724 ft TVD
Nominal ID	8.681 in.
Casing description	
Depth	4,293 ft MD, 3,936 ft TVD
Nominal ID	12.347 in.
Drillstring description	
Top section	4,940 ft of 3½-in. drillpipe and HWDP
Bottom section	91 ft of 6½-in. drill collars
Injection riser description	
Length	1,075 ft
Nominal ID	8.681 in.
Blowout Data	
Formation fluid	Single phase gas
Gas SG	0.60
Flow rate	1,850 MMCFD
Gas temperature at the mud line	60°F
Gas temperature at formation	136°F
Static pore pressure	2,250 psia
Other Information	
Fracture initiation pressure at the shoe	2,850 psig
Water depth	1,000 ft
Seawater density	8.6 lbm/gal

3. Determine the minimum depth to insert a kill string if the desired pump rate is 30 bbl/min. Ignore the choking effect of flowing friction in the kill string annulus.

4. Determine the minimum fluid density that follows the heavy mud if the static pore pressure is 5,200 psig.

10.11 Data shown in **Table 10.6** were presented by Warriner and Cassity.[26] The well was being drilled from a semisubmersible rig, blew out in a prolific gas reservoir, and is now venting gas at the mud line. The drillstring has a float and the most likely flow path is in the annulus. The plan is to intersect the blowout well, pull up into casing, close the preventers, and kill the well dynamically with 8.6-lbm/gal seawater. The specially-designed injection riser is parallel to the drilling riser. All depths are referenced from the rotary kelly bushing, which is 75 ft above sea level for both wells.

1. Determine the minimum kill rate if the leak-off efficiency is 90%.
2. Determine the casing pressure and pump pressure when dynamic kill is achieved. The pressure loss between the pump manifold and injection riser is 100 psi. There are no downhole restriction losses.
3. Determine the maximum water-injection rate.
4. Determine the final kill-fluid density.
5. Determine the pump pressure when the final kill fluid starts to enter the blowout well. Include an adjusted surface equipment loss. Also, assume that a heavy brine is the selected fluid and that the brine behaves as a Newtonian fluid.
6. Determine the number of pumps if each unit can deliver 400 hhp. Use a 50% safety factor.

TABLE 10.7—INFORMATION RELATED TO A SLIMHOLE CIRCULATION TEST[74]

Depth		7,010 ft
Bit diameter		3.06 in.
Last casing depth		3,219 ft
Casing ID		3.27 in.
Mud density		7.5 lbm/gal
Pump Rate (gal/min)	**Pump Pressure (psig)**	**Calculated String Loss (psi)**
11	121	15
13	164	19
16	241	27
19	331	37
23	471	51
27	634	68
31	820	88
35.5	1,055	108
40	1,323	147

10.12 You are drilling a slimhole well and shut down the operation to measure the pump pressures at various circulation rates. Details related to the well and the test results are shown in **Table 10.7.** Estimate the ECDs at each pump rate and plot as a function of circulation rate.

10.13 You took a kick while drilling and the well has been closed in at the preventers. Total depth is 9,000 ft, casing is set at 4,000 ft, and the mud density is 10.0 lbm/gal. You ask the driller to record the shut in drillpipe pressure (SIDPP) and SICP values every five minutes while you drive into town and call your supervisor. Upon returning an hour later, he hands you the following table.

Time, min	SIDPP, psig	SICP, psig
0	500	600
5	600	700
10	650	750
15	700	800
20	750	850
25	800	900
30	850	950
35	900	1,000
40	800	950
45	650	850
50	600	850
55	550	850
60	550	900

1. What is going on here?
2. What instructions should you have left with the driller?
3. Estimate the pore pressure of the kicking formation.
4. Estimate the fracture gradient if the weakest point is at the shoe.

10.14 After several hours, the casing pressure in the previous problem increased to 2,960 psig and stabilized. An 8.4-lbm/gal mud is injected into the annulus, causing the casing pressure to decline to 1,170 psig. Estimate the depth to the loss zone.

TABLE 10.8—DATA FOR PROBLEM 10.16	
Measured depth of the flowing zone	9,711 ft
Measured surface casing depth	5,000 ft
True vertical casing depth	4,833 ft
Average inclination in open hole	38.9°
Annulus area in open hole	98.224 in.^{2}
Annular volume in open hole	24,050 gal
Static reservoir pressure	7,482 psig
Fracture pressure in loss zone	4,348 psig
Kill-mud density	19.0 lbm/gal
Formation-fluid density	2.0 lbm/gal
Downhole productivity index	0.02260 bbl/(min-psi).

10.15 Write a spreadsheet program and plot the range of pump rates that will kill the blowout described in Table 10.5 for mud densities between 12.7 and 19.7 lbm/gal.

10.16 Wessel and Tarr[82] presented the data listed in **Table 10.8** related to an underground blowout. The loss zone is assumed to be immediately below the shoe.

1. Determine the minimum mud weight if the intent is to balance the static formation pressure after all of the influx has been displaced into the fracture and annulus pressure is released.
2. Determine the maximum allowable mud weight if the intent is to keep the pressure gradient at all points in the well less than the loss zone's fracture propagation gradient.
3. Estimate the kill-rate range for both of these muds.

Nomenclature

a = variable in the equation used to calculate kill-string depth, psi^2
A_a = annulus cross-sectional area, in.^{2}
A_c = flow conduit area, in.^{2}
c_l = liquid compressibility, psi^{-1}
d = diameter, in.
d_c = flow conduit diameter, in.
d_h = hole diameter, in.
d_o = pipe outer diameter, in.
D = vertical depth, ft
D_{ks} = kill string depth, ft
D_{lz} = vertical depth to a loss zone, ft
E_{lo} = leak-off efficiency, dimensionless
f_M = Moody friction factor, dimensionless
F = force, lbf
F_{mv} = momentum force, lbf
g = acceleration of gravity, ft/s
g_{bh} = bottomhole pressure gradient, psi/ft
g_c = gravitational system conversion constant, 32.17 (lbm-ft)/(lbf-s)
g_g = gas hydrostatic gradient, psi/ft
g_m = mud hydrostatic gradient, psi/ft
g_w = water hydrostatic gradient, psi/ft
h = vertical height, ft
h_f = formation thickness, ft
h_l = liquid height, ft
h_m = mud height, ft
h_w = water height, ft
H = magnetic field strength, A/m
I_p = pipe current, A
J = productivity index at standard flow rate, MMcf/(D-psi)
J^* = productivity index at sandface flow rate, bbl/(min-psi)
k_g = permeability to gas, darcy
l = distance between the maximum and minimum axial field readings, ft
L = length, ft
m = mass, lbm
n = turbulence exponent, dimensionless
N_{Re} = Reynold's number, dimensionless
p = pressure, psi
p_{bh} = bottomhole pressure, psi
p_c = casing pressure, psi
p_e = exit pressure, psi
p_{fi} = fracture initiation pressure, psi
p_{fp} = fracture propagation pressure, psi
p_h = hydrostatic pressure, psi
p_{hd} = hydrostatic pressure of fluid in dead string, psi
p_{hk} = kick fluid hydrostatic pressure, psi
p_{hl} = liquid hydrostatic pressure, psi
p_{hm} = mud hydrostatic pressure, psi
p_{hw} = water hydrostatic pressure, psi
p_p = pore pressure, psi
p_P = pump pressure, psi
p_R = average static reservoir pressure, psi
p_{sc} = standard pressure
p_{sf} = sand-face pressure, psi
p_{wf} = flowing wellbore pressure, psi
p_{wh} = wellhead pressure, psi
p_{ac} = pressure change resulting from change in a fluid's kinetic energy, psi
p_f = friction pressure, psi
p_{fm} = friction pressure of mud, psi
p_{fw} = friction pressure of water, psi
p_r = pressure drop across restriction, psi
p_s = pressure drop in the surface equipment, psi
P_h = hydraulic power, hhp
q = flow rate, bbl/min
q_g = gas flow rate at sandface conditions, bbl/min
q_{gsc} = gas flow rate at standard conditions, MMcf/D
q^*_{gsc} = critical gas flow rate at standard conditions, MMcf/D
q_{kr} = kill rate, bbl/min
$(q_{kr})_{OH}$ = underground flow kill rate for one openhole annulus mud volume, bbl/min
$(q_{kr})_4$ = underground flow kill rate for an infinite mud volume, bbl/min
q_l = liquid flow rate, bbl/min
R = range, ft
r_e = drainage radius, ft
r_w = wellbore radius, ft
s = variable in the relation used to determine flowing bottomhole pressure
t = time, s
T = temperature, °F or °R
T_e = exit temperature, °F or °R
T_{sc} = standard temperature, °F or °R
v = velocity, ft/s
v_g = gas velocity, ft/s
v^*_g = critical gas velocity, ft/s
v^*_l = critical liquid velocity, ft/s
v_{wh} = velocity at wellhead conditions, ft/s
V_a = annular volume between flowing and loss zones, gal
x = horizontal departure, ft
z = compressibility factor, dimensionless

z_e = compressibility factor at exit conditions, dimensionless
z_{sc} = compressibility factor at standard conditions, dimensionless
z_{wh} = compressibility factor at wellhead conditions, dimensionless
α = average hole inclination, deg
α_g = gas specific gravity, dimensionless
ϵ = pipe roughness, in.
θ = direction, deg
γ = specific heat ratio, dimensionless
μ = viscosity, cp
μ_g = gas viscosity, cp
ρ = density, lbm/gal
ρ_g = gas density, lbm/ft
ρ_k = kick fluid density, lbm/gal
ρ_l = liquid density, lbm/gal
ρ_m = mud density, lbm/gal

References

1. Flak, L.H., Wright, J.W., and Ely, J.W.: "Blowout Control: Response, Intervention and Management: Part 1—Strategy and Planning," *World Oil* (November 1993) 71–78.
2. Bell, S. and Wright, R.: "*New* Well Control Companies Stress Planning, Engineering," *Pet. Eng. Intl.* (April 1994) 39–44.
3. Cudd, B. and Goodman, E.: "Well Control Management Moving Beyond Blowouts," *Offshore* (January 1995) 50–52.
4. Abel, L.W.: "Blowout Contingency Plans Can Cut Firefighting and Capping Risks," *Oil and Gas J.* (1 May 1995) 88–96.
5. Oberlender, G.D. and Abel, L.W.: "Project Management Improves Well Control Events," *Oil and Gas J.* (10 July 1995) 56–63.
6. Abel, L.W.: "Blowout Contingency Plans Can Cut Firefighting and Capping Risks," *Oil and Gas J.* (1 May 1995) 88–96.
7. Eby, D.: "Precautions in Planning HTHP Well Control," *Offshore* (January 1997) 80.
8. Westergaard, R.H.: *All About Blowout*, Norwegian Oil Review Ltd., Oslo (1987) 40–41.
9. Flak, L.H. and Matthews, C.: "Blowout Control: Response, Intervention and Management: Part 9—Firefighting," *World Oil* (October 1994) 101–108.
10. Grace, R.D.: *Advanced Blowout & Well Control*, Gulf Publishing Co., Houston (1994) 367–94.
11. Littleton, J.: "Proven Methods Thrive in Kuwait Well Control Success," *Pet. Eng. Intl.* (January 1992) 31–37.
12. "Stinging Procedures Allow Rapid Well Control," *World Oil* (May 1992) 80–81.
13. Abel, L.W.: "Planning, Training, Equipment All Crucial in HS Blowout," *Oil and Gas J.* (5 June 1995) 56–60.
14. Flak, L.H., Kelly, M.J., and Tuppen, J.: "Case History of Timbalier Blowout Shows Necessity of Capping While Burning," *Offshore* (June 1995) 39–44.
15. Flak, L. "How Well Control Techniques Were Refined in Kuwait," *World Oil* (May 1992) 72–75.
16. Flak, L.H. and Kelly, M.J. Jr.: "Blowout Control: Response, Intervention and Management: Part 10—Blowout Surface Intervention Methods," *World Oil* (November 1994) 85–93.
17. Abel, L.W., Campbell, P.J., and Bowden, J.R. Sr.: "High-Pressure Jet Cutters Improve Capping Operations," *Oil and Gas J.* (8 May 1995) 60–69.
18. "Ultrahigh Pressure Hydrocutter Cuts Kill Time for Kuwait Wells," *Oil and Gas J.* (30 September 1991) 30.
19. "India Gas Well Blowout Capped and Killed in 17 Days," *World Oil* (June 1995) 39.
20. "Well Control Team Caps Underwater Blowout: Part 2—Execution Phase," *Pet. Eng. Intl.* (December 1987) 32–38.
21. Miller, D. and Conover, G.: "Inflatable 'Kill' Packers Used in Working Over Kuwaiti Wells," *Oil and Gas J.* (9 March 1992) 74–76.
22. Taylor, D.M.: "A Better Way to Kill Blowouts," *Ocean Industry* (April 1971) 69–72.
23. Adams, N.J. *et al.*: "Coiled-Tubing Applications for Blowout-Control Operations," *JPT* (May 1996) 398–405.
24. Gebhardt, F., Eby, D., and Barnett, D.: "Utilizing Coiled Tubing Technology to Control a Liquid Propane Storage Well Fire," IADC Well Control Conference for Europe, Aberdeen, 22–24 May 1996.
25. Adams, N.J. and Kuhlman, L.G.: "Case History Analyses of Shallow Gas Blowouts," paper SPE 19917 presented at the 1990 SPE/IADC Drilling Conference, Houston, 27 February–2 March.
26. Warriner, R.A. and Cassity, T.G.: "Relief-Well Requirements to Kill a High-Rate Gas Blowout From a Deepwater Reservoir," *JPT* (December 1988) 1602–608.
27. Furlow, W.: "Deepwater Well Control: Where's the Fire?" *Offshore* (September 1997) 46.
28. Cruz, H.U.J. and Rodriguez, F.: "Frequent Surveys Guide Relief Well to Underground Blowout," *Oil and Gas J.* (2 September 1991) 80–84.
29. Flak, L.H. and Goins, W.C. Jr.: "New Techniques, Equipment Improve Relief Well Success," *World Oil* (January 1984) 113–18.
30. Wright, J.W. and Flak, L.H.: "Blowout Control: Response, Intervention and Management: Part 11—Relief Wells," *World Oil* (December 1994) 59–64.
31. Miller, R.T. and Clements, R.L.: "Reservoir Engineering Techniques Used to Predict Blowout Control During the Bay Marchand Fire," *JPT* (March 1972) 234–40.
32. Lewis, J.B. Jr.: "New Uses of Existing Technology for Controlling Blowouts: Chronology of a Blowout Offshore Louisiana," *JPT* (October 1978) 1473–80.
33. Nelson, R.F.: "The Bay Marchand Fire," *JPT* (March 1972) 225–33.
34. Blount, E.M. and Soeiinah, E.: "Dynamic Kill: Controlling Wild Wells a New Way," *World Oil* (October 1981) 109–26.
35. Bruist, E.H.: "A New Approach in Relief Well Drilling," paper SPE 3511 presented at the 1971 SPE Annual Meeting, New Orleans, 3–6 October.
36. Grace, R.D. and Storts, P.D.: "In Search of the Apache Key," paper SPE 12624 presented at the 1984 SPE Deep Drilling and Production Symposium, Amarillo, Texas, 1–3 April.
37. Walstrom, J.E., Brown, A.A., and Harvey, R.P.: "An Analysis of Uncertainty in Directional Surveying," *JPT* (April 1969) 515–23; *Trans.*, AIME, **246.**
38. Wolff, C.J.M. and de Wardt, J.P.: "Borehole Position Uncertainty—Analysis of Measuring Methods and Derivation of Systematic Error Model," *JPT* (December 1981) 2339–50.
39. Warren, T.M.: "Directional Survey and Proximity Log Analysis of a Downhole Well Intersection," *JPT* (December 1981) 2351–62.
40. Bourgoyne, A.T. Jr. *et al.*: *Applied Drilling Engineering*, Textbook Series, SPE, Richardson, Texas (1991) 459.
41. Runge, R.J., Worthington, A.E., and Lucas, D.R.: "Ultra-Long Spaced Electric Log (ULSEL)," *The Log Analyst* (September–October 1969).
42. Mitchell, F.R. *et al.*: "Using Resistivity Measurements to Determine Distance Between Wells," *JPT* (June 1972) 723–40; *Trans.*, AIME, **253.**
43. Robinson, J.D. and Vogiatzis, J.P.: "Magnetostatic Methods for Estimating Distance and Direction from a Relief Well to a Cased Wellbore," *JPT* (June 1972) 741–49; *Trans.*, AIME, **253.**
44. Morris, F.J., Waters, R.L., and Costa. J.P.: "A New Method of Determining Range and Direction from a Relief Well to a

Blowout Well," paper SPE 6782 presented at the 1977 SPE Annual Meeting, Denver, 9–12 October.

45. Jones, D.L., Hoehn, G.L., and Kuckes, A.F.: "Improved Magnetic Model for Determination of Range and Direction to a Blowout Well," *SPEDE* (December 1987) 316–22.
46. Flak, L.H. and Goins, W.C. Jr.: "New Relief Well Technology is Improving Blowout Control," *World Oil* (December 1983) 57–61.
47. de Lange, J.I. and Darling, T.J.: "Improved Detectability of Blowing Wells," *SPEDE* (March 1990) 34–38.
48. Kuckes, A.F. *et al.*: "An Electromagnetic Survey Method for Directionally Drilling a Relief Well Into a Blown Out Oil or Gas Well," *SPEJ* (June 1984) 269–73.
49. Tarr, B.A., Kuckes, A.F., and Ac, M.V.: "Use of a New Ranging Tool to Position a Vertical Well Adjacent to a Horizontal Well," *SPEDE* (June 1992) 93–99.
50. Leraand, F. *et al.*: "Relief-Well Planning and Drilling for a North Sea Underground Blowout," *JPT* (March 1992) 266–73.
51. Voisin, J. *et al.*: "Deep Relief Well Parallels Capping Efforts," *Pet. Eng. Intl.* (March 1988) 28–33.
52. Maduro, W.P. and Reynolds, J.: "Enchova Blowout: Record Relief Time," paper SPE 18717 presented at the 1989 SPE/IADC Drilling Conference, New Orleans, 28 February–3 March.
53. Booth, J.E.: "Use of Shallow Seismic Data in Relief Well Planning," *World Oil* (May 1990) 39–42.
54. Patton, B.J. and Foster, M.: "Discussion of Relief-Well Requirements to Kill a High-Rate Gas Blowout from a Deepwater Reservoir," *JPT* (July 1989) 740.
55. Aadnøy, B.S. and Bakøy, P.: "Relief Well Breakthrough at the Problem Well 2/4-14 in the North Sea, paper SPE 20915 presented at the 1990 SPE European Petroleum Conference, The Hague, 22–24 October.
56. Abel, L.W.: "Kill Operation Requires Thorough Analysis," *Oil and Gas J.* (15 May 1995) 32–38.
57. Grace, R.D. "Practical Considerations in Pressure Control Procedures in Field Drilling Operations," *JPT* (August 1977) 1031–35.
58. Grace, R.D.: "Fluid Dynamics Kill Wyoming Icicle," *World Oil* (April 1987) 45–53.
59. Grace, R.D. and Cudd, B.: "Fluid Dynamics Used to Kill South Louisiana Blowout," *World Oil* (April 1989) 47–50.
60. Watson, W.D. and Moore, P.L.: "Momentum Kill Procedure Can Quickly Control Blowouts," *Oil and Gas J.* (30 August 1993) 74–77.
61. Wallis, G.B.: *One-Dimensional Two-Phase Flow*, McGraw Hill Book Co. Inc., New York City (1969) 265.
62. Clark, A.R. and Perkins, T.K.: "Wellbore and Near-Surface Hydraulics of a Blown-Out Oil Well," *JPT* (November 1981) 2181–88.
63. Beck, F.E., Langlinais, J.P., and Bourgoyne, A.T. Jr.: "An Analysis of the Design Loads Placed on a Well by a Diverter System," paper SPE 16129 presented at the 1987 SPE/IADC Drilling Conference, New Orleans, 15–18 March.
64. Weymouth, T.R.: "Problems in Natural Gas Engineering," *Trans.*, ASME (1912) **34.**
65. Dukler, A.E., *et al.*: "Gas-Liquid Flow in Pipelines, I. Research Results," AGA-API Project NX-28 (May 1969).
66. Katz, D.L. and Lee, R.L.: *Natural Gas Engineering: Production and Storage*, McGraw-Hill Publishing Co., New York City (1990) 239–43.
67. Jain, A.K.: "An Accurate Explicit Equation for Friction Factor," *J. Hydraulics Division* (May 1976) **102,** 694.
68. Kouba, G.E., MacDougall, G.R., and Schumacher, B.W.: "Advancements in Dynamic Kill Calculations for Blowout Wells," *SPEDC* (September 1993) 189–94.
69. Gillespie, J.D., Morgan, R.F., and Perkins, T.K.: "Study of the Potential for an Off-Bottom Kill of a Gas Well Having an Underground Blowout," *SPEDE* (September 1990) 215–19.
70. Rygg, O.B. and Gilhuus, T.: "Use of a Dynamic Two-Phase Pipe Flow Simulator in Blowout Kill Planning," paper SPE 20433 presented at the 1990 SPE Annual Technical Conference and Exhibition, New Orleans, 23–26 September.
71. Abel, L.W.: "Planning a Dynamic Kill," *JPT* (May 1996) 422–26.
72. Smestad, P., Rygg, O.B., and Wright, J.H.: "Blowout Control: Response, Intervention and Management: Part 5—Hydraulics Modeling," *World Oil* (April 1994) 75–80.
73. Lynch, R.D. *et al.*: "Dynamic Kill of an Uncontrolled CO Well," *JPT* (July 1985) 1267–75.
74. Bode, D.J., Noffke, R.B., and Nickens, H.V.: "Well-Control Methods and Practices in Small-Diameter Wellbores," *JPT* (November 1991) 1380–86.
75. Prince, P.K. and Cowell, E.E.: "Well Control Equations Modified for Slim-Hole Kill Operations," *Oil and Gas J.* (27 September 1993) 71–77.
76. Arnwine, L.C. and Ely, J.W.: "Polymer Use in Blowout Control" *JPT* (May 1978) 705–11.
77. Ely, J.W. and Holditch, S.A.: "Polymers Used to Direct Kill Fluids in a Blowout," *Oil and Gas J.* (1 August 1988) 44–48.
78. Flak, L.H. "Reactive Materials Can Quickly Form Plugs for Blowout Control," *Oil and Gas J.* (17 April 1995) 63–68.
79. Grace, R.D.: "Analyzing and Understanding the Underground Blowout," paper SPE 27501 presented at the 1994 SPE/IADC Drilling Conference, Dallas, 15–18 February.
80. Grace, R.D.: *Advanced Blowout & Well Control*, Gulf Publishing Co., Houston (1994) 349–52.
81. Wessel, M. and Tarr, B.A.: "Underground Flow Well Control: The Key to Drilling Low-Kick-Tolerance Wells Safely and Economically," *SPEDE* (December 1991) 250–56.
82. Wessel, M. and Tarr, B.A.: "Supplement to SPE 22217, Underground Flow Well Control: The Key to Drilling Low-Kick-Tolerance Wells Safely and Economically," paper SPE 23764.
83. Barnhill, C.C. and Adams, N.J.: "Underground Blowouts in Deep Well Drilling," paper SPE 7855 presented at the 1979 SPE Deep Drilling and Production Symposium, Amarillo, Texas, 1–3 April.

SI Metric Conversion Factors

bbl	× 1.589 873	E − 01	= m
bbl/min	× 2.649 788	E + 00	= L/s
bbl/(min-psi)	× 3.843 193	E + 02	= L/(sAMPa)
cp	× 1.0*	E + 03	= Pa·s
ft^3	× 2.831 685	E − 02	= m^3
darcy	× 9.869 233	E − 01	= μm^3
deg	× 1.745 329	E − 02	= rad
ft	× 3.048*	E − 01	= m
ft/hr	× 8.466 667	E − 02	= mm/s
°F	(°F − 32)/1.8		= °C
°F/100 ft	× 1.822 689	E + 01	= mK/m
gal	× 3.785 412	E + 00	= L
hhp	× 7.460 43	E − 01	= kW
in.	× 2.54*	E + 01	= mm
$in.^2$	× 6.451 6*	E + 02	= mm^2
lbf	× 4.448 222	E + 00	= N
lbm	× 4.535 924	E − 01	= kg
psi	× 6.894 757	E − 03	= MPa
psi^{-1}	× 1.450 377	E − 01	= kPa^{-1}
°R	°R/1.8		= K

*Conversion factor is exact.

Chapter 11
Casing Seat Selection

11.1 Introduction

Of the many reasons for setting casing, some of the more important are completion purposes, regulatory compliance, isolating troublesome hole sections or worn casing, reducing torque and drag, and well control. The ability to manage a pressure-control problem successfully is related directly to how deep the casing is set and casing point selection as it relates to well control is the focus of our final chapter.

The objective of conventional design practices is to have casing set deep enough that the pore pressures are compatible with the frac gradients in the next hole section. Ideally, all wells would be designed to contain any conceivable pressure condition. But there is a practical limitation to the number of casing strings that can be set in a well. Casing seat selection for purposes of well control is a compromise relative to cost, safety considerations, and ability to reach objective depth with a usable hole diameter. A well designer may not be able to preclude the possibility of an underground blowout under all worst-case scenarios. An example well will be used to demonstrate why this is so.

Fig. 11.1 is a plot of the predicted pore pressure and fracture gradients (in mud weight equivalent) for a prospect located offshore Louisiana.[1] Normal pore pressures are anticipated down to about 8,300 ft. The variable transition is expected then to increase the pore pressure to a 17.4-lbm/gal equivalent at 11,000 ft. Fracture gradients were computed using Eaton's[2] method.

To be completely safe, the planner of this well might conclude that the maximum wellbore pressure would be experienced if a gas kick completely unloaded the mud from the hole before the well could be closed in by the preventers. This approach is adopted often when designing casing for burst loads. The engineer in this case uses the same reasoning and plans to set casing deep enough so that the maximum shut-in pressure will not fracture the rock.

Pore pressures and fracture pressures are plotted against depth as shown in **Fig. 11.2.** Starting at total depth, the assumption is made that the well unloads and is shut in with a dry gas gradient of 0.1 psi/ft. The next casing point is obtained easily from the graph by drawing a line with the gas gradient slope from the pore pressure to the frac pressure curve. The recommended intermediate casing point is determined at the intersection depth of 10,400 ft.

The process is repeated. The gas gradient line is projected from the pore pressure at the intermediate point to the frac pressure curve and a drilling liner depth is obtained at 9,500 ft. Ultimately, a hole program is submitted that calls for surface casing at 2,750 ft, three intermediate strings, a drilling liner, and a production liner. There are eight casing strings in this well when we include the drive pipe and conductor casing.

This program would be difficult to accomplish, expensive, and may inhibit the well's productivity because of the small conduit required of the tubing. In addition, designing the shallower strings may be a problem because the internal and external pressure ratings decrease as pipe diameter increases. The need is apparent for a compromise approach based on more realistic criteria and more accepting of some risk. After all, it is unlikely that the deeper hole sections will ever be evacuated if the crews are trained and the equipment is in working order. Furthermore, some risk may be acceptable if the worst thing likely to happen is an underground blowout that stays underground.

The first design method discussed in this chapter is based on setting up the hole program so that the mud densities fit within an allowable window dictated by the highest pore pressure and lowest fracture pressure. The second design uses a so-called kick tolerance to establish where casing must be set if the wellbore is to maintain integrity during a well-control problem. Finally, we discuss some considerations that should be given to the shallower hole sections.

11.2 Designs Based on Mud Density

Minimum mud weight may be controlled by borehole collapse, but it is more common to base the minimum on controlling the pore pressure. With the important exception of underbalanced drilling, the hydrostatic pressure of the drilling fluid provides enough overbalance to counteract the unavoidable pressure losses associated with a trip. A 0.5-lbm/gal trip margin is common and adequate for most well conditions, though the actual margin should consider such factors as the depth, mud properties, and downhole geometry. A higher margin

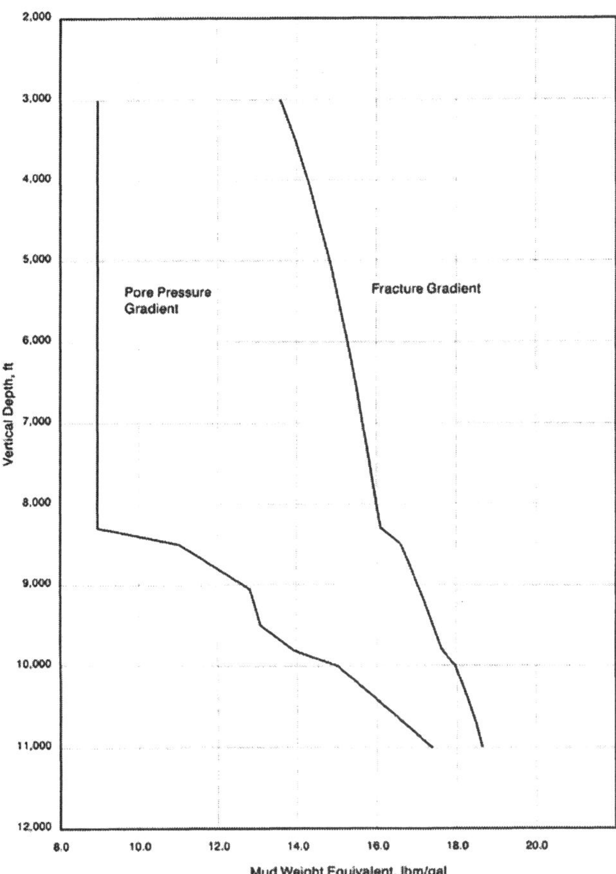

Fig. 11.1—Pore pressure and fracture gradients for a well located offshore Louisiana (from Ref. 1).

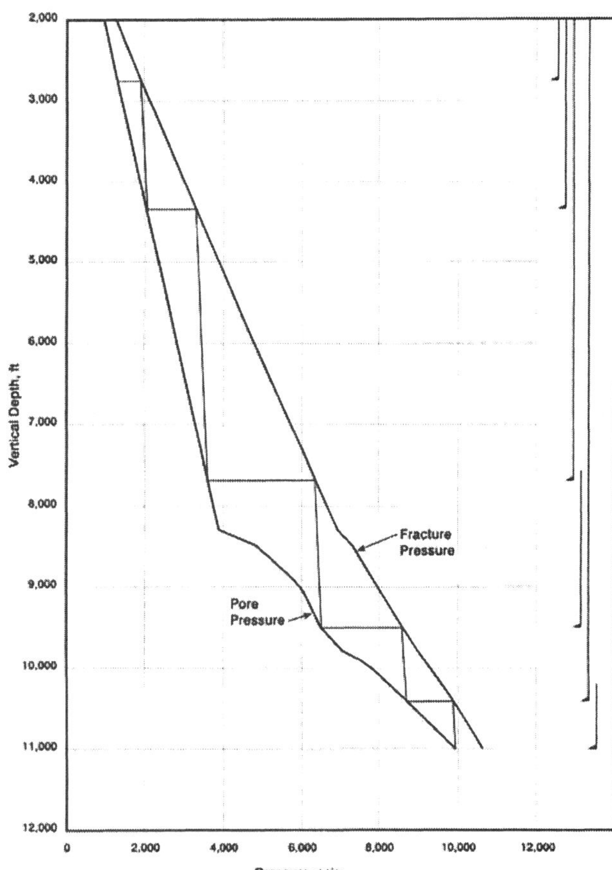

Fig. 11.2—Well design based on shutting in a hole section on dry gas without breaking down the last casing shoe.

may be needed in shallow hole sections, especially where trips are being made with tight hole clearances and thick mud. Less may be acceptable in relatively deep hole sections.

On the high end, circulating friction losses in the annulus, surge pressures, and kick control demand that the maximum mud gradient be something less than the fracture gradient. The allowable pressure difference is specified normally as a density equivalent and has been defined as the kick margin.[3] This is a somewhat misleading term in that a 0.5-lbm/gal kick margin does not imply that a 0.5-lbm/gal kick can be contained and controlled without breaking down the shoe and we prefer to call the factor the frac margin.

The method is straightforward: casing seats are selected so that the minimum mud density does not exceed the maximum allowable density. In the planning phase, reasonably accurate pore-pressure and fracture-gradient predictions are essential. One or two contingency strings should be planned if this knowledge is lacking.

Example 11.1. Use anticipated mud weights to select casing seats for the well shown in Fig. 11.1. The designated trip and frac margins are 0.5 lbm/gal.

Solution. Two curves that offset the pore pressure equivalent and fracture gradient by 0.5 lbm/gal are constructed as shown in **Fig. 11.3.** The mimimum and maximum mud densities for the well are thus defined. The technique is illustrated in **Fig. 11.4.** We start with the minimum mud weight at total depth (17.9 lbm/gal) and graphically determine the depth at which the minimum starts to exceed what is allowed by the frac margin. Intermediate casing must be set at 10,600 ft to meet the criterion. In similar fashion, the remaining setting depths are selected and the following hole program is recommended by the drilling engineer: 13$^3/_8$-in. surface casing at 4,250 ft; 9$^5/_8$-in. intermediate casing at 9,600 ft; 7-in. intermediate casing at 10,600 ft; and 4½-in. production liner at 11,000 ft.

The program thus established is a guide because the assumed conditions undoubtedly will vary from the actual data obtained at the rigsite. Casing points will shift up or down depending on the leak-off tests and pore-pressure predictions made while drilling. Again, assure that the drilling, completion, and production objectives can still be met if the differences turn out to be considerable.

There is often a problem designing casing points in extremely overpressured sediments because the spread between frac and pore pressures is reduced to a narrow window. Refer to Fig. 11.3 again and note that the allowable mud weights are converging and would intersect for a deeper well if the pressure transition continued its present course. The operator would have no choice then but to reduce one or both of the safety margins. Accordingly, greater emphasis has to be placed on keeping the hole filled on trips, reducing swab and surge pressures, and keeping annulus friction losses to a minimum.

11.3 Casing Seat Selection Based on Minimum Kick Tolerance

A well designed using the preceding method may not have enough frac integrity to control a kick. The maximum pressure experienced at the shoe or other weak section of the exposed open hole is experienced when the well is shut in initially or when the bulk of an expanding gas influx reaches the

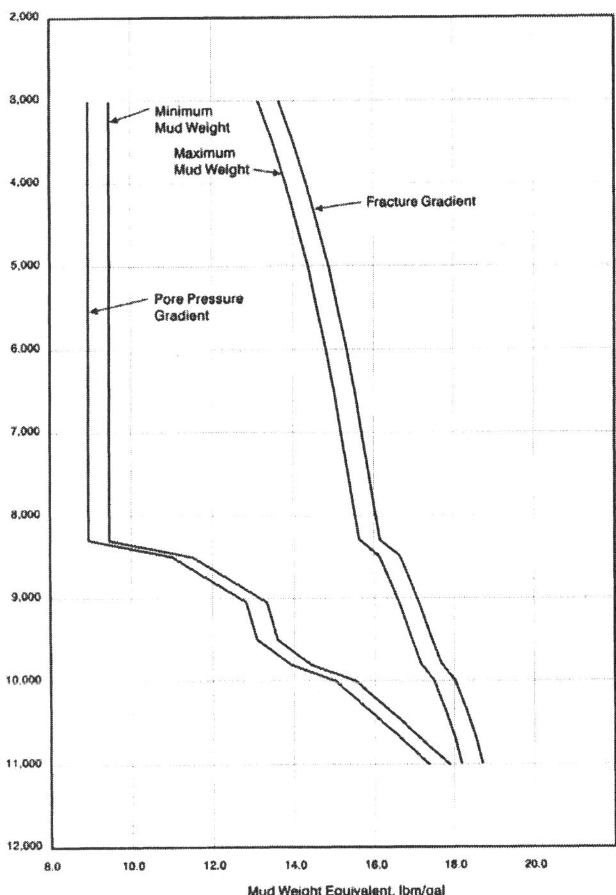

Fig. 11.3—Minimum and maximum mud weights for the offshore Louisiana well.

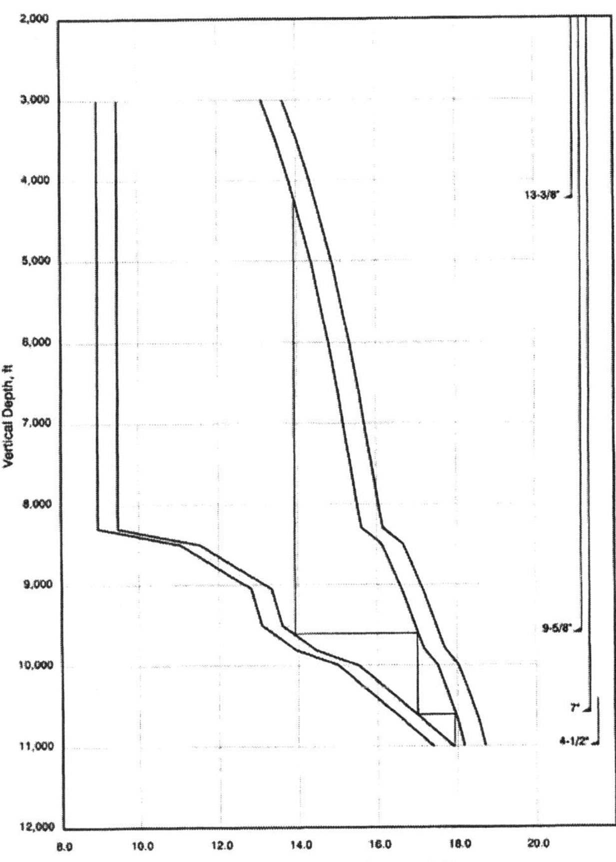

Fig. 11.4—Well design based on allowable mud density.

critical depth. The preceding method makes no direct provision for these effects and a well design based on mud weight alone may not incorporate enough frac integrity to control a kick. An example demonstrates the nature of the problem for our offshore well.

Example 11.2. The operator has almost reached the 9⁵/₈-in. casing point and is drilling with a mud weight exactly balanced to the predicted 13.4-lbm/gal pore pressure. A permeable sand is encountered with an unexpected 13.9-lbm/gal pressure and the zone kicks. Determine if the shoe can withstand the shut-in pressure if the frac gradient immediately below the surface casing is 0.749 psi/ft. For this problem, ignore the effect of the kick fluids on the hydrostatic pressure and assume the annulus is filled with mud.

Solution. The pore pressure at 9,600 ft is

$$p_p = (0.0519)(13.9)(9,600) = 6,926 \text{ psig}.$$

The shut-in pressure at the surface casing shoe is therefore

$$p_{sh} = 6,926 - (0.0519)(13.4)(9,600 - 4,250)$$

$$= 3,205 \text{ psig}.$$

This gives a gradient of

$$g_{sh} = 3,205/4,250 = 0.754 \text{ psi/ft},$$

and fracturing is predicted.

The problem is more pronounced when casing is set relatively shallow with a long openhole section. For instance, the same underbalance conditions when applied to the next hole section will not break down the 9⁵/₈-in. shoe if the well is shut-in on a "zero-gain" (i.e., no kick fluids in the well) kick at 10,600 ft.

11.3.1 Static Kick Tolerance. Before deriving the static kick tolerance equation, the basic concepts will be presented for the conditions described in **Table 11.1.** Say that a well is drilling this strata at 10,000 ft with a mud density of 1.0 lbm/gal underbalanced to the pore pressure when a gas zone kicks. Furthermore, assume the gas enters as a slug, has a 0.1-psi/ft hydrostatic gradient, and occupies 500 ft in the drillstring annulus. The pore pressures, fracture pressures, and shut-in annulus pressure profile are plotted as shown in **Fig. 11.5.** As indicated, lost returns would be experienced if intermediate casing is set any shallower than about 8,700 ft.

Fig. 11.6 illustrates the effect of kick height for the same kick intensity and formation fluid whereas **Fig. 11.7** retains the same mud weight and kick height, but changes the effective gradient within the kick region. The depth at which the annulus pressure intersects the fracture line depends on the pit gain and how much the hydrostatic pressure in the annulus is reduced by the formation fluids. A casing program based on managing a shut-in kick must anticipate and include these factors.

Fig. 11.8 considers a gas zone underbalanced by 1.0-lbm/gal that kicks at 9,000 ft instead of 10,000 ft. The concept is less intuitive perhaps, but the chart demonstrates that casing must be set deeper as drilled depth increases if the objective is to avoid lost circulation for a specified kick intensity.

Maintaining an adequate kick tolerance simply means that casing is set deep enough so that a given size kick with a given

TABLE 11.1—EXAMPLE PORE PRESSURE AND FRAC GRADIENT DATA				
Depth	Pore Pressure Gradient		Fracture Gradient	
(ft)	(psi/ft)	(lbm/gal)	(psi/ft)	(lbm/gal)
2,000 to 8,000	0.450	8.66	0.686	13.21
8,200	0.475	9.14	0.700	13.48
8,400	0.500	9.63	0.714	13.74
8,600	0.525	10.22	0.729	14.03
8,800	0.550	10.59	0.743	14.30
9,000	0.575	11.07	0.757	14.57
9,200	0.600	11.55	0.771	14.84
9,400	0.625	12.03	0.786	15.13
9,600	0.650	12.51	0.800	15.40
9,800	0.675	12.99	0.814	15.67
10,000	0.700	13.48	0.829	15.96

mud weight in the hole will not fracture the formation. Its use as an operational and planning tool has been addressed in the literature,[4,5] but our discussion takes a somewhat different tack. The end result, however, is equivalent and we hope to extend the previously published work and clarify any misconceptions as to how the kick tolerance should be used.

Kick tolerance can be defined as the maximum kick intensity that a well can tolerate before lost circulation is experienced at the last casing seat. Expressing the kick intensity as a gradient gives

$$g_{kt} = g_p - g_m, \quad \text{(11.1)}$$

where g_{kt} = the kick tolerance gradient, g_p = the pore pressure gradient, and g_m = the hydrostatic gradient of the existing mud.

Fig. 11.9 shows the parameters used to derive the kick tolerance equation when a well is closed in with an influx at the bottom of the hole. The pressure at the shoe is given by

$$p_{sh} = p_p - g_m(D - h_k - D_{sh}) - g_k h_k. \quad \text{(11.2)}$$

The kick tolerance requires that the fracture pressure below the casing seat be equal to or greater than the pressure exerted at this depth if the well is shut in on a kick:

$$p_{fi} = p_{sh},$$

where p_{fi} = the fracture-initiation pressure at the casing seat.

The fracture pressure and pore pressure can be expressed as

$$p_{fi} = g_{fi} D_{sh}$$

and $p_p = g_p D$, respectively. Substituting terms into Eq. 11.2 gives

$$g_{fi} D_{sh} = g_p D - g_m(D - h_k - D_{sh}) - g_k h_k. \quad \text{(11.3)}$$

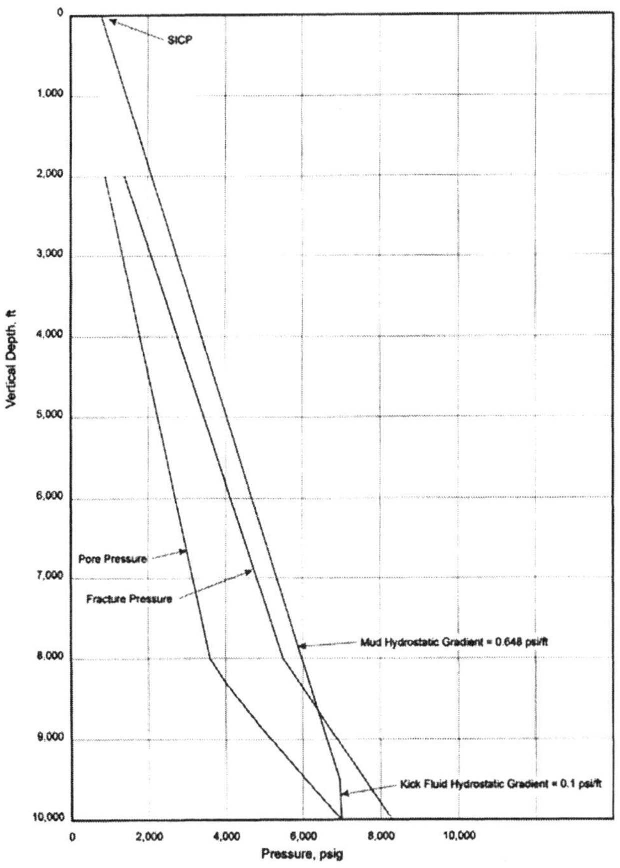

Fig. 11.5—Pore pressures, fracture pressures, and shut-in wellbore pressures for a hypothetical kick.

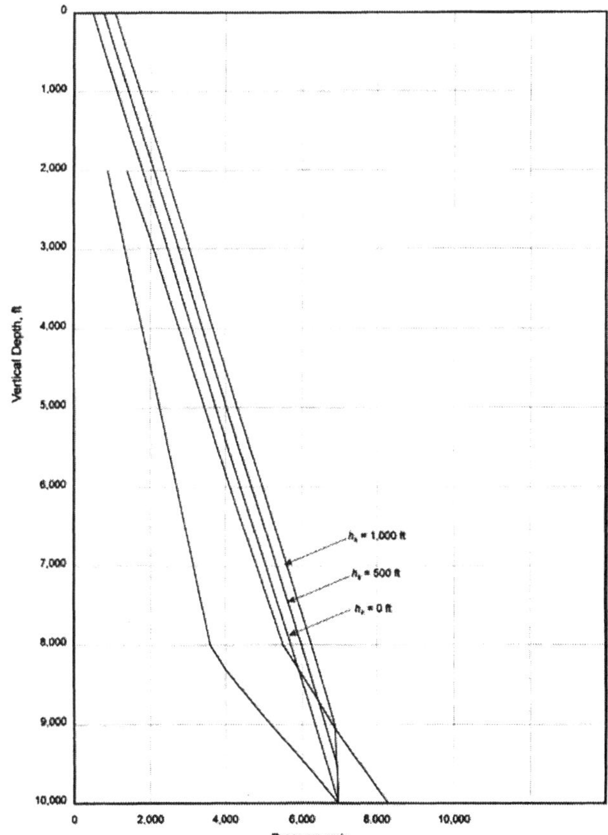

Fig. 11.6—Effect of kick height on shut-in wellbore pressures.

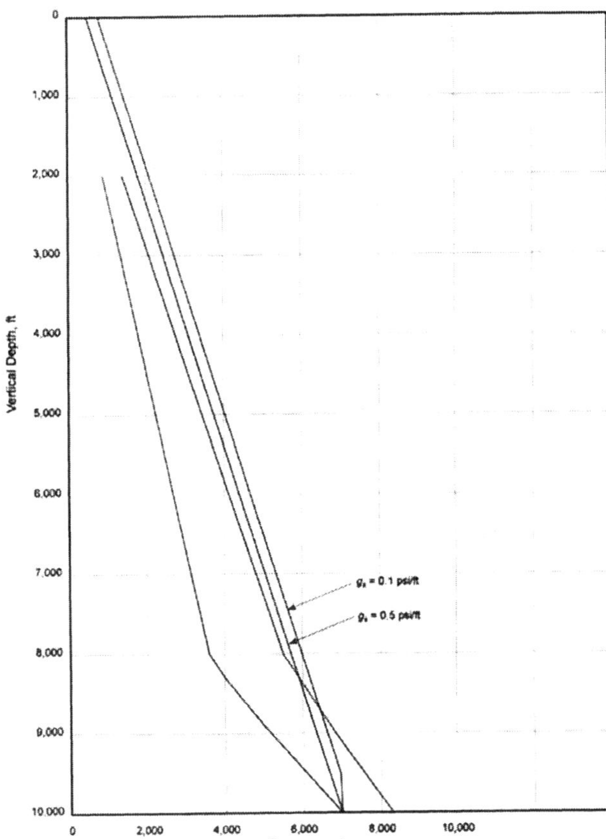

Fig. 11.7—Effect of kick-fluid gradient on shut-in wellbore pressures.

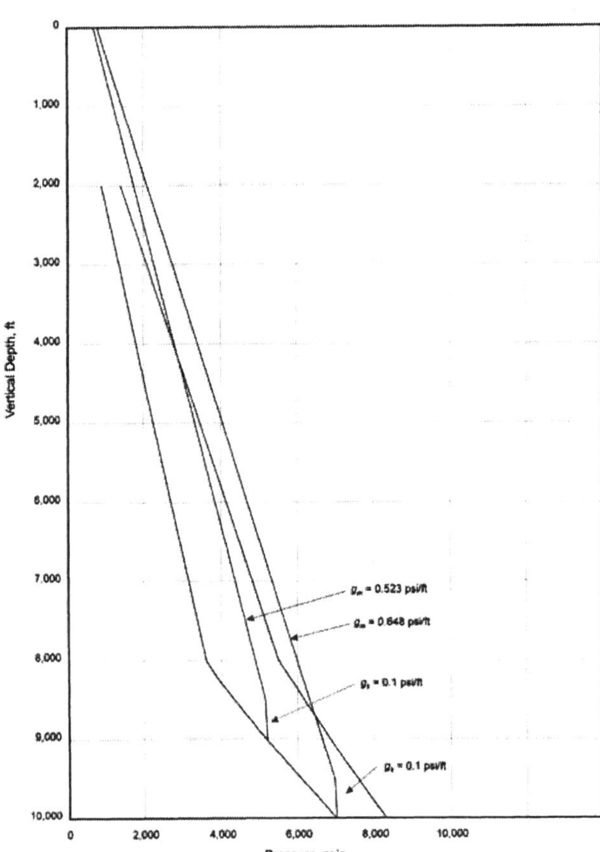

Fig. 11.8—Effect of drilled depth on shut-in wellbore pressures for a specified kick intensity.

Rearranging yields

$$g_p - g_m = \frac{D_{sh}}{D}\left(g_{fi} - g_m\right) - \frac{h_k}{D}(g_m - g_k). \quad \ldots (11.3a)$$

The left side of the equation is the kick tolerance. The gradients can be replaced with density terms and the kick tolerance in density equivalent is given thus by

$$\rho_{kt} = \frac{D_{sh}}{D}\left(\rho_{fi} - \rho_m\right) - \frac{h_k}{D}(\rho_m - \rho_k). \quad \ldots\ldots\ldots (11.4)$$

As discussed, the relation shows that the kick tolerance for a given shoe depth (or other critical depth) and frac gradient is reduced when drilled depth and pit gain (i.e., kick height) are increased and when the density of the kick region is reduced.

Note that Eq. 11.4 also indicates that the kick tolerance is reduced whenever the mud weight is increased. This poses an apparent paradox because the pore-pressure gradient of a well is fixed and increasing the mud weight reduces the kick intensity of any underbalanced zones ahead of the bit. Nonetheless, increasing the mud weight also diminishes the spread between the mud hydrostatic and fracture gradient, which means that the allowable shut-in casing pressure (hence kick tolerance) must be reduced. Keep in mind that the use of kick tolerance for selecting casing points is most applicable in those areas where the pore-pressure gradient, though fixed, is not known with any precision.

The maximum depth to drill without exceeding the specified kick tolerance is obtained by rearranging Eq. 11.4.

$$D_{max} = \frac{D_{sh}}{\rho_{kt}}\left(\rho_{fi} - \rho_m\right) - \frac{h_k}{\rho_{kt}}(\rho_m - \rho_k). \quad \ldots\ldots\ldots (11.5)$$

Example 11.3. An operator is drilling in a pore-pressure transition at 11,300 ft with a 9.8-lbm/gal mud. A 13⅜-in. surface casing is set at 2,350 ft, where the leak-off test established a fracture gradient of 13.8-lbm/gal. Crew reaction times suggest a maximum expected kick volume of 20 bbl, which translates to 289 ft in the drillstring annulus opposite the 12¼-in. hole. Assume an influx enters the hole as a slug and weighs 1.8 lbm/gal.

1. Determine the current kick tolerance.

2. Determine the intermediate casing depth if a 0.5-lbm/gal kick tolerance has been specified. Assume the mud weight does not change prior to this depth.

Solution. 1. Eq. 11.4 yields the kick tolerance at present bit depth.

$$\rho_{kt} = \frac{2,350}{11,300}(13.7 - 9.8) - \frac{289}{11,300}(9.8 - 1.8)$$

$$= 0.61 \text{ lbm/gal.}$$

2. Eq. 11.5 gives the casing depth required to meet the 0.5-lbm/gal kick tolerance.

$$D_{max} = \frac{2,350}{0.50}(13.7 - 9.8) - \frac{289}{0.50}(9.8 - 1.8)$$

$$= 13,706 \text{ ft.}$$

The value will change if mud weight is increased in response to the pore pressure indicators. For instance, a 10.0-lbm/gal mud reduces the intermediate casing depth to 12,650 ft.

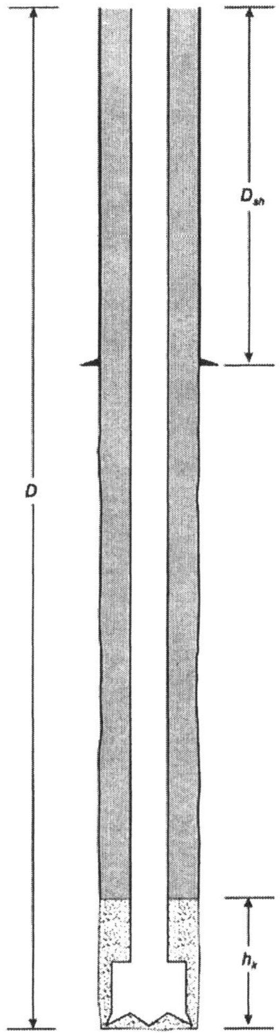

Fig. 11.9—Schematic of a well shut in on a gas kick.

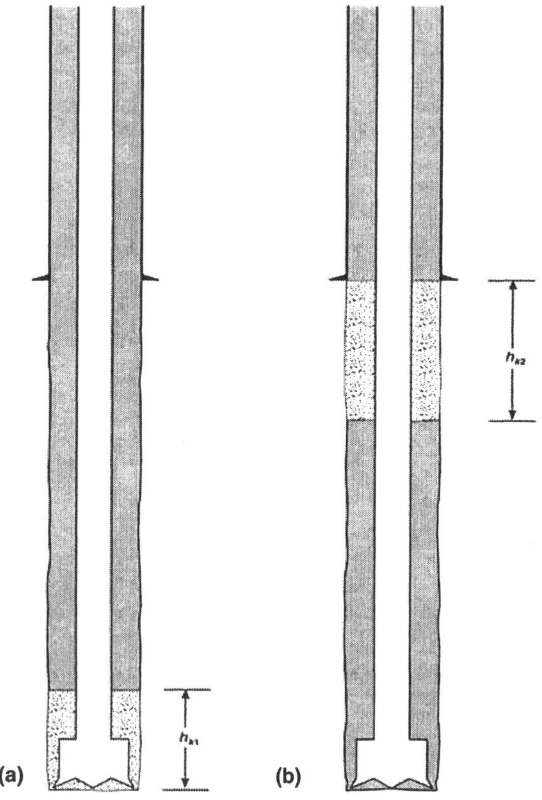

Fig. 11.10—Kick height comparison between initial shut-in condition and when gas is displaced to the shoe in a Driller's Method kill.

11.3.2 Dynamic Kick Tolerance. Fig. 11.10 shows an ideal bubble displaced to the casing shoe in a Driller's Method procedure. The maximum shoe pressure may be experienced at this time if the influx has more height than when the well was closed in and it stands to reason that casing points selected by the kick-tolerance method should consider this contingency. Solving for the dynamic kick tolerance is an iterative procedure that, if we assume the influx is a gas bubble, uses the annular-pressure prediction equations discussed in Chap. 4. **Table 11.2** describes a method for calculating the dynamic-kick tolerance. **Table 11.3** gives a procedure for determining the casing setting depth to achieve a specified kick tolerance.

TABLE 11.2—PROCEDURE TO CALCULATE THE DYNAMIC KICK TOLERANCE	
1.	Assume a kick intensity at present bit depth.
2.	Determine the pore pressure for the assumed kick intensity.
3.	Estimate temperature and compressibility factor at present bit depth.
4.	Calculate maximum pressure at the last casing seat or other critical depth for the selected kill procedure (Driller's or Wait and Weight Method).
5.	Determine equivalent density at the casing shoe and compare to the fracture gradient.
6.	Adjust the assumed kick intensity based on the result.
7.	Repeat Steps 2 through 6 until the equivalent density at the casing shoe equals the fracture density.

Example 11.4. Determine the dynamic-kick tolerance for the previous example if the Driller's Method is the preferred kill procedure. Assume a 0.60-specific gravity (SG) gas kick and use a circulating temperature given by T = 80°F + 1.2°F/100 ft. The assumed hole diameter is 12¼ in. and the drillstring consists of 270 ft of 9-in. collars, 90 ft of 7-in. collars, and 10,840 ft of 5-in. drillpipe and heavy-wall drillpipe.

Solution. The static-kick tolerance is the controlling factor if the maximum dynamic pressure at the shoe is less than the frac gradient. Start with this basis and compute the pore pressure as

TABLE 11.3—PROCEDURE TO DETERMINE CASING DEPTH USING DYNAMIC KICK TOLERANCE	
1.	Assume a casing setting depth and the mud density at this depth.
2.	Add the specified kick tolerance to the mud density at the assumed depth.
3.	Determine pore pressure at the assumed depth and specified kick tolerance.
4.	Estimate the temperature and compressibility factor at the assumed depth.
5.	Calculate maximum pressure at the last casing seat or other critical depth for the selected kill procedure (Driller's or Wait and Weight Method).
6.	Determine the equivalent density at the casing shoe and compare to the fracture density.
7.	Adjust the casing setting depth based on the result.
8.	Repeat Steps 2 through 7 until the equivalent density at the casing shoe equals the fracture density.

$$p_p = (0.0519)(9.8 + 0.5)(11,300) = 6,041 \text{ psig}$$
$$= 6,056 \text{ psia}.$$

The bottomhole temperature is 676°R and the compressibility factor at the bottomhole conditions is 1.091. The initial pit gain translates to a 289-ft kick height and so the effective capacity factor opposite the drill collar annuli is

$$C_{ai} = 20/289 = 0.06920 \text{ bbl/ft}.$$

Eq. 4.30 yields the initial hydrostatic pressure of the gas.

$$p_{hgi} = \frac{(0.60)(6,056)(289)}{(53.29)(1.091)(676)} = 27 \text{ psi}.$$

The gas will be in the drillpipe annulus when the influx reaches the shoe. Its hydrostatic pressure at this depth will be

$$p_{hg} = (27)(0.06920)/(0.12149) = 15 \text{ psi}.$$

Solve for the parameter X using Eq. 4.37.

$$X = 6,056 - (0.509)(11,300 - 2,350) - 15$$
$$= 1,485 \text{ psia}.$$

The circulating temperature at the casing shoe is 568°R. We assume a z factor of 1.00 and calculate the pressure at 2,350 ft when the gas reaches this depth using Eq. 4.36.

$$p_{2,350} = \frac{1,485}{2} + \left[\frac{1,485^2}{4} + \frac{(0.509)(6,056)(289)(1.000)(568)(0.06920)}{(1.091)(676)(0.12149)}\right]^{0.5}$$

$$p_{2,350} = 1,711 \text{ psia}.$$

At this pressure, $z_{2,350} = 0.843$ and another iteration yields

$$p_{2,350} = \frac{1,485}{2} + \left[\frac{1,485^2}{4} + \frac{(0.509)(6,056)(289)(0.843)(568)(0.06920)}{(1.091)(676)(0.12149)}\right]^{0.5}$$

$$p_{2,350} = 1,679 \text{ psia}.$$

The new pressure yields the same z factor and 1,679 psia is the solution. The gradient at the shoe is therefore

$$g_{sh} = (1,679 - 15)/2,350 = 0.708 \text{ psi/ft}$$
$$= 13.62 \text{ lbm/gal}.$$

The result is larger than the 13.15-lbm/gal frac gradient so the dynamic kick tolerance is less than the initial, static tolerance. Assume another kick tolerance, say 0.40 lbm/gal, and repeat the preceding exercise. The pore pressure, z factor at total depth, and gas hydrostatic pressures also change.

$$p_p = (0.0519)(9.8 + 0.4)(11,300) = 5,982 \text{ psig}$$
$$= 5,997 \text{ psia};$$

$$z_i = 1.086;$$

$$p_{hgi} = \frac{(0.60)(5,997)(289)}{(53.29)(1.086)(676)} = 27 \text{ psi};$$

and $p_{hgi} = (27)(0.06920)/(0.12149) = 15$ psi.

X is determined as

$$X = 5,997 - (0.509)(11,300 - 2,350) - 15$$
$$= 1,441 \text{ psia}.$$

We assume what happens to be the correct z factor and compute the maximum pressure at the shoe,

$$p_{2,350} = \frac{1,441}{2} + \left[\frac{1,441^2}{4} + \frac{(0.509)(5,997)(289)(0.847)(568)(0.06920)}{(1.086)(676)(0.12149)}\right]^{0.5}$$

$$p_{2,350} = 1,642 \text{ psia}.$$

Thus

$$g_{sh} = (1,642 - 15)/2,350 = 0.692 \text{ psi/ft}$$
$$= 13.33 \text{ lbm/gal}.$$

We are not there yet. A 0.38-lbm/gal tolerance is used; that yielding a maximum shoe pressure equivalent to the fracture gradient. Hence the dynamic kick tolerance is 0.12 lbm/gal less than the static tolerance.

The selected kill method is another variable that affects kick tolerance when we consider the dynamic case. A Wait and Weight procedure will increase the kick tolerance as long as kill-weight mud enters the annulus before gas is displaced or migrates to the shoe. Whatever the chosen kill method, dynamic kick tolerance computations are time consuming if done by hand and are an unwieldy tool for selecting casing points unless assisted by a computer or programmable calculator.

Conditions continually change in the process of drilling a well and it is difficult to forecast casing points using either the static or dynamic kick tolerance when a well is in the planning stage. In practice, kick tolerances are computed using actual data as the bit generates more hole and casing is set once the value is reduced to the specified tolerance.

A 0.5-lbm/gal tolerance is common, but how much is acceptable depends on many factors. Regulatory agencies may specify the minimum in some areas. Otherwise, the operator should consider such things as the reliability of the pore-pressure predictions, formation deliverability characteristics, expected pore fluids, and the consequences of losing returns with a kick in the hole. Many wells are drilled in known areas with little tolerance for taking a kick whereas 0.5 lbm/gal may be deemed insufficient in less-understood environments.

Well costs will increase as the kick-tolerance variables become more stringent. An operator should provide the best available input and do a risk/cost assessment when specifying the unknown parameters. Again, our single-bubble kick model will yield conservative results in the vast majority of cases. Simulators that consider the effects of kick dispersion, gas solubility, and gas migration are available and offer a better way to select casing points.[6] Ideally, a rigsite system would be used with current inputs for the vertical depth, penetration rate, and pump rate. User-specified data would include the leak-off test gradient, formation permeability, and crew reaction time.

Very little tolerance may be available when drilling a transition into extreme overpressures because the spread between the minimum mud weight and fracture gradient becomes so narrow. The offshore Louisiana well shown in Fig. 11.1 is a good example of this. Kick tolerances may have to be reduced in other cases if the well is to be drilled at all.

A risk-management tool for helping the operator decide what hole sections pose lesser risks with the lower tolerance was presented by Wessel and Tarr[7] and implemented on a low-tolerance well offshore Indonesia.[8] The permeability-thickness product defines the flow capacity of a formation. An underground blowout from rock having a low flow capacity often can be stopped by using the rig equipment to pump weighted mud at a fast rate. On the other hand, controlling the well may be impossible absent a relief well or other extraordinary measures if the flow capacity is high enough. Basically, the technique quantifies the flow capacity of a rock at present bit depth that can be killed without much effort. A lower kick tolerance may be acceptable if the result is high compared to the known rock characteristics in the same hole section.

11.4 Shallow Casing Seat Considerations

The first string of pipe set in a well is called either structural casing, drive pipe, or conductor pipe, depending on the operator's terminology and whether the well is located on land or offshore. This string usually is set from as shallow as 50 ft on land jobs to no more than a few hundred feet below the ground surface or mud line. The next hole section almost always is drilled with a diverter.

The second string is called the conductor pipe or surface casing, again depending on local usage and the operator's preference. To be consistent and avoid any confusion, we call the second casing string the conductor casing and the section drilled below conductor is the surface hole. Well flows in the surface hole are diverted in the majority of cases because the risks of broaching the shoe to surface are deemed more unacceptable than the risk of a failed diverter. The following discussion, however, considers those well designs where the intent is to control surface hole kicks rather than let the well blow uncontrolled.

The objective is the same as for the deeper hole sections—i.e., to be able to shut-in the well and pump a controlled kill without exceeding the wellbore pressure limitations at or near the conductor shoe. But the consequences of an inadequate design may not be as forgiving as in a deep underground blowout where the normal worst-case scenario does not risk life or limb. Broaching concerns are real and contingencies must be addressed in the planning and execution.

One of the designer's problems is that the region matrix stress correlations do not extrapolate very well into the shallower sediments. For various reasons, there tends to be a great deal of scatter in the data when people start compiling shallow leak-off data. Aadnoy *et al.*[9] discussed some of these reasons including measurement inaccuracies and inconsistencies in the technique; different lithologies (sand instead of shale for instance); the effect of tensile strength if the rock is intact; differing mud properties; and a poor cement job across the shoe. We believe the last is a significant reason. Shallow leak-off tests often approach or correspond to a curve that describes the overburden stress gradient. This behavior has been observed in the North Sea, Gulf of Mexico, Canada, and other areas.[9-13]

Another consideration is whether to set casing based on the kick-tolerance concept, gas completely unloading the mud from the hole, or other criteria. The prolific nature of the shallow gas sands in many areas may suggest the most conservative approach, though assuming a large pit gain is acceptable if justified by the hole volume and local knowledge.

Finally, the designer must incorporate the pore pressure in the setting point calculations. Normal compaction usually prevails in shallow rock, but a thick gas sand may be abnormally pressured at the penetration depth whereas the pressure in a thin sand will be closer to a normal gradient. Stratigraphic information from the area should be the guide.

Consider the situation where the surface casing depth has been determined and the next step in the design phase is to determine the conductor pipe depth that will allow for kick control. An evacuated surface hole might be the design as could be a straightforward application of the kick-tolerance method. An alternate procedure discussed by Aadnoy *et al.*[9] is similar to the kick-tolerance method and uses an allowable pit gain to determine the conductor depth. In other words, we determine where casing must be set if the shoe is to retain integrity for a given kick height.

Rearranging Eq. 11.3 to solve for the kick height yields

$$h_k = \frac{D_{sh}\left(g_{fi} - g_m\right) + D\left(g_m - g_p\right)}{g_m - g_k}. \qquad (11.6)$$

The solution involves taking an arbitrary shoe depth, the fracture gradient at this depth, and using Eq. 11.6 to calculate the minimum kick height that will result in a fractured shoe. If the bottomhole assembly is known, the kick height can be translated to a pit gain. The pit gain and shoe depth represent a point on a graph and the process is repeated until a curve is defined. The casing depth for the specified pit gain then can be obtained directly from the curve.

Gas does not weigh very much at low pressure and it is reasonable to set g_k equal to zero. Also, assume for now that the risk of encountering abnormal pressures is low. The kick is swabbed into the hole and a conservative approach is to set the pore pressure equal to the mud hydrostatic. Eq. 11.6 is simplified thus.

$$h_k = D_{sh}\left(g_{fi} - g_m\right)/g_m \qquad (11.7)$$

or its equivalent,

$$h_k = D_{sh}\left(\rho_{fi} - \rho_m\right)/\rho_m. \qquad (11.8)$$

These relations should be used only if the potential kick zones are thin and buoyancy-assisted overpressures are unlikely. The technique and effect of the assumed fracture gradient is demonstrated in the next example.

Example 11.5. The fracture-gradient curve shown on the left side of **Fig. 11.11** was determined using Eaton's[2] Poisson's ratio and overburden-gradient correlations for the Gulf Coast. Poisson's ratios shallower than 1,000 ft were extrapolated from Eaton's data. A well is to be drilled in the area and 13³/₈-in. surface casing will be set in a 17½-in. hole at 3,000 ft. Abnormal pore pressures are not anticipated based on offset information and the expected mud density at the casing point is 9.6 lbm/gal. The drillstring will have 500 ft of 9-in. collars and 5-in. drillpipe.

1. Determine the conductor-casing depth for a maximum pit gain of 150 bbl.

2. For the same kick volume, determine the conductor-casing depth if the fracture gradient corresponds to the overburden.

Solution. 1. The problem is solved by iteration. First, assume a conductor casing depth of 1,000 ft and obtain a

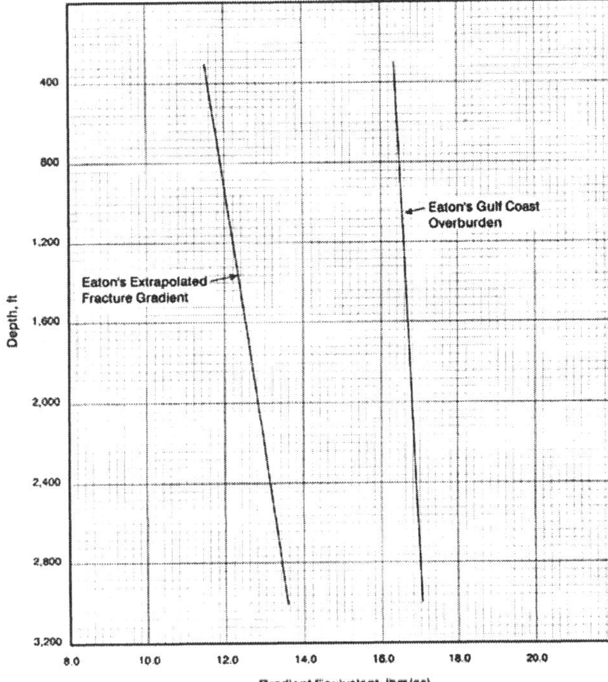

Fig. 11.11—Shallow fracture gradients extrapolated from Eaton's Gulf Coast correlations compared to Eaton's[2] overburden gradient for the area.

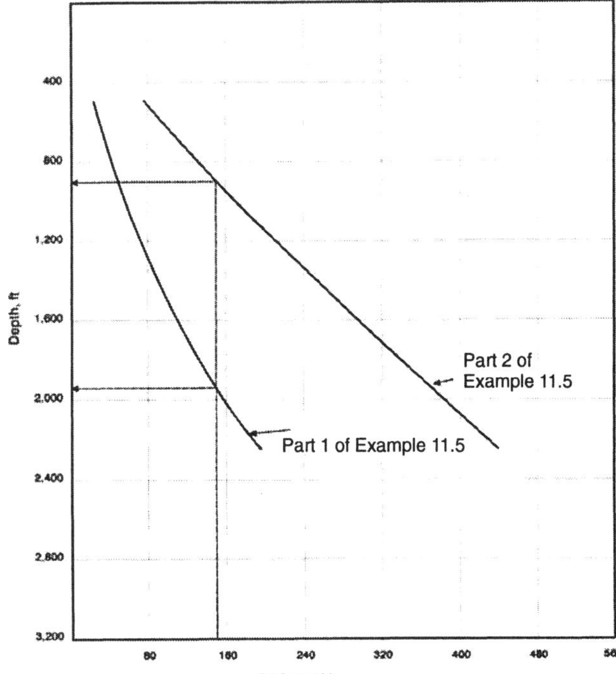

Fig. 11.12—Conductor casing seat selection based on tolerable pit gain in the surface hole.

12.1-lbm/gal fracture gradient from Fig. 11.11. Eq. 11.8 yields the kick height at which the wellbore will fail if conductor is set at 1,000 ft.

$$h_k = (1,000)(12.1 - 9.6)/9.6 = 260 \text{ ft}.$$

The kick does not extend up into the drillpipe annulus and the pit gain is

$$G = (260)(0.21882) = 57 \text{ bbl}.$$

The kick volume is less than 150 bbl and so casing must be set deeper than 1,000 ft. At 2,000 ft,

$$h_k = (2,000)(12.9 - 9.6)/9.6 = 688 \text{ ft}.$$

Part of the influx is in the drillpipe annulus, which has a 0.27322-bbl/ft capacity factor. The volume opposite the drill collars is

$$(500)(0.21882) = 109 \text{ bbl}.$$

The pit gain for the 688-ft kick is therefore

$$G = 109 + (688 - 500)(0.27322) = 160 \text{ bbl}.$$

Hence casing can be set somewhat shallower than 2,000 ft. The allowable pit gains for other depths are computed and the curve shown on the left side of **Fig. 11.12** is constructed. A conductor setting depth at 1,950 ft is obtained directly from the plot.

2. The allowable pit gain for the higher fracture gradient at 1,000 ft determined:

$$h_k = (1,000)(16.6 - 9.6)/9.6 = 729 \text{ ft}$$

$$\text{and } G = 109 + (729 - 500)(0.27322) = 172 \text{ bbl}.$$

Other points are computed, the curve shown on the right side of the diagram is constructed, and the casing point is found at 900 ft.

One problem with the technique is that it does not necessarily allow the kick to be displaced to the shoe, but a similar method could be used for the dynamic case. Nonetheless, the example shows that the key to shallow casing-seat selection is having good fracture-pressure data.

Consider the advantages of having conductor set at 900 ft instead of 1,950 ft. Savings on pipe and cement are the most obvious benefits, but the well-control risks are much lower with the shallower casing point if the seat is competent and the well can be closed in and controlled. In the example, the exposed hole below the structural casing is reduced by 1,050 ft when conductor is set at 900 ft. A shorter conductor hole infers less chance of encountering and having to divert a shallow gas sand.

Say that an operator is committed to eliminating the use of diverters and designs wells so that every hole section below structural casing can be shut-in on a kick. Furthermore, the well plan is established on the basis of a swabbed gas kick displacing all the mud before shut-in. The effect of the fracture-gradient assumption is demonstrated when we compare the hole programs illustrated in **Figs. 11.13 and 11.14.** Both start with structural casing set at 300 ft below the mud line (or ground surface), but Fig. 11.13 is based on Eaton's extrapolated frac gradients whereas Fig. 11.14 predicts failure when the wellbore pressure matches the overburden.

The more stringent design would demand eight casing strings before 3,000 ft is reached. Obviously such a hole program would be impractical and impossible to accomplish except for the shallowest of normally pressured wells. The same design criteria result in only four casing points across the section if the fracture pressure equals the overburden and involves only two or three more strings than would be run in a conventional program.

The design practice is by no means customary, but could be achieved given good leak-off test data at each shoe depth and enough faith in the results to close the well in. The task would

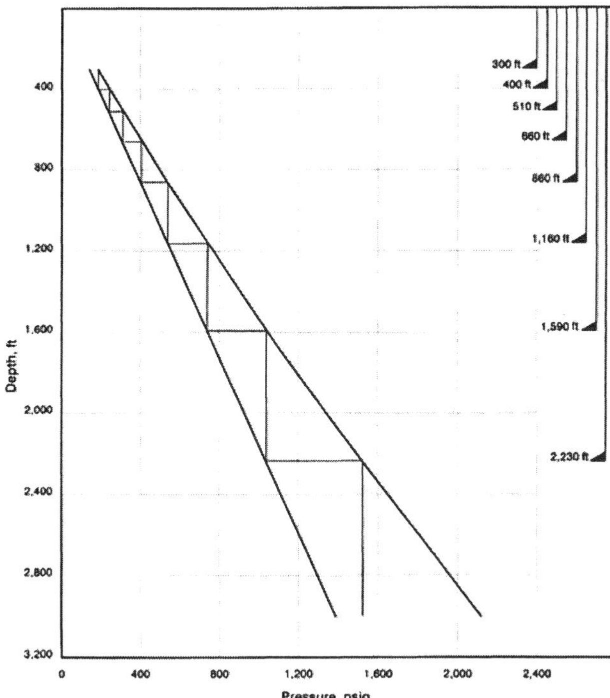

Fig. 11.13—Shallow casing program based on shutting a normally pressured well in on gas with frac gradients extrapolated from Eaton's[2] correlations.

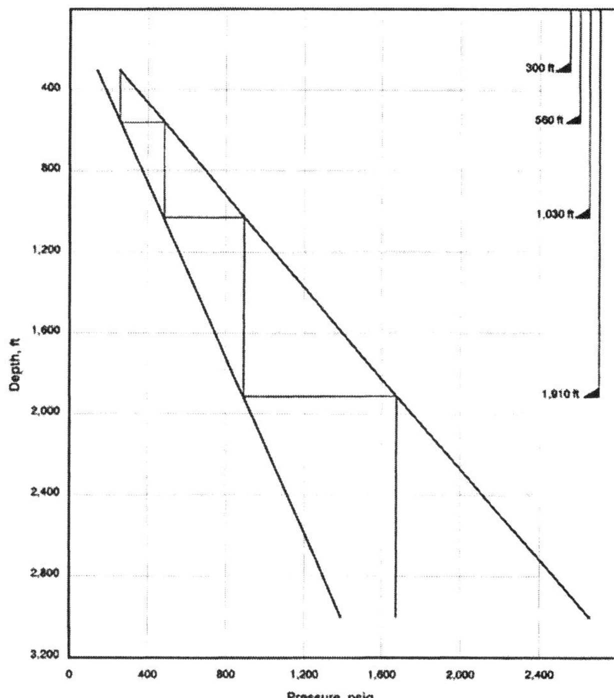

Fig. 11.14—Shallow casing program based on shutting a normally pressured well in on gas with frac gradients equivalent to Eaton's[2] overburden.

TABLE 11.4—PORE PRESSURE AND FRACTURE GRADIENTS FOR THE WELL IN PROBLEM 11.2

Depth (ft)	Pore Pressure (lbm/gal)	Fracture Pressure (lbm/gal)
2,000	7.7	11.0
7,000	7.7	11.0
7,600	9.0	12.4
8,700	11.3	14.0
10,200	11.3	14.0
10,300	9.5	12.9
11,500	9.5	12.9

not be easy, involving higher well costs and the need for larger hole sizes and casings in the shallower sections. But the option may pose the least risk and provide the lowest long-term costs unless diverter systems can be made 100% reliable.

There are no industry standards when it comes to selecting casing depths for well control. This is particularly true for the shallow strings. All the concepts presented here are subject to debate and may not be universally applicable. We hope the reader has gained an appreciation of the basic dilemma posed by the question: Is it safer and more cost-effective to divert shallow flows or to plan our wells to control kicks? For the latter, remaining issues to address include what worst-case scenarios to consider and the validity of the design assumptions.

Much work remains to be done on both sides of the question. Diverter designs have improved but are not yet 100% reliable. The industry needs to devote more study to shallow broaching modes and operators need to focus more on measuring, compiling, and distributing shallow leak-off data.

Problems

11.1 Company policy requires that mud density overbalance the pore pressure by 300 psi. Determine the trip margin in density equivalent for a zone at 7,000 ft vertical depth.

11.2 You are writing the prognosis for a land drilling operation. Projected total depth of the well is at 11,500 ft and the objective reservoir is a permeable limestone with expected top at 11,400 ft. Stuck pipe can be expected if the differential pressure exceeds 500 psi. The required minimum hole size at total depth is 6.0 in. and surface casing must be set at a depth no greater than 2,000 ft to meet regulatory requirements. **Table 11.4** lists anticipated pore pressure and fracture gradients.

Design the casing program for this well based on 0.5-lbm/gal frac and trip margins.

11.3 The pore pressure gradients shown in **Table 11.5** were predicted by applying Eaton's[14] sonic log relation to a well on the Texas Gulf Coast.[15] The fracture gradients were computed using Eaton's[2] overburden and Poisson's ratio correlations for the area.

1. Plot the data against depth on coordinate graph paper.
2. Determine casing points for the well based on the allowable mud densities if 0.5-lbm/gal frac and trip margins are used.

11.4 Write a FORTRAN program to compute the dynamic kick tolerance for a Driller's Method kill.

11.5 Surface casing on your well is set at 3,515 ft and you are drilling a 9½-in. hole at 7,200 ft with a 9.4-lbm/gal mud. The shoe tested to a 14.1-lbm/gal equivalent, but a formation at 5,300 ft is known to breakdown if the mud density exceeds 11.5 lbm/gal.

1. Determine the static kick tolerance for a 30-bbl pit gain if 5-in. drillpipe and 450 ft of 7-in. collars comprise the drillstring. Assume the kick is gas and weighs 1.8 lbm/gal.
2. Determine the static kick tolerance if fluid mixing results in a 60-bbl kick region of gas-cut mud weighing an average 5.6 lbm/gal.

TABLE 11.5—PREDICTED PORE PRESSURE AND FRACTURE GRADIENTS FOR A WELL IN JEFFERSON COUNTY, TEXAS

Depth (ft)	Pore Pressure Gradient (lbm/gal)	Fracture Gradient (lbm/gal)
2,820	8.95	13.49
3,210	8.95	13.76
4,000	8.95	14.26
4,170	8.95	14.36
4,520	8.95	14.56
5,210	8.95	14.92
6,000	8.95	15.28
6,210	8.95	15.37
6,970	8.95	15.65
7,500	8.95	15.82
7,810	8.95	15.92
8,000	8.95	15.97
8,320	8.95	16.06
8,410	8.95	16.09
9,000	8.95	16.24
9,220	8.95	16.29
9,300	8.95	16.31
9,390	9.60	16.46
9,410	10.23	16.63
9,580	10.23	16.63
9,620	11.00	16.79
9,710	11.72	16.95
9,810	12.24	17.07
9,900	12.94	17.22
10,010	13.24	17.30
10,100	13.71	17.40
10,200	14.05	17.48
10,370	14.07	17.51
10,540	14.10	17.54
11,190	14.50	17.72
11,380	14.50	17.74
11,720	14.89	17.86
12,300	14.90	17.94
13,000	15.01	18.06

3. Mud weight was increased to 10.0 lbm/gal to combat sloughing shales. Determine the static kick tolerance if the present bit depth is 8,150 ft. Estimate the hydrostatic pressure of the kick by assuming a 0.7 gas SG and using a circulating temperature given by $T = 70°F + 1.0°F/100$ ft.

4. Determine the dynamic kick tolerance for the preceding conditions if the Driller's Method is the preferred kill procedure.

11.6 Intermediate casing has been set in the slant section of a directional well at 13,200 ft TVD. You intend to drill to the next casing point with a 13.0-lbm/gal mud. The following conditions apply: leak-off test pressure = 15.6 lbm/gal, hole diameter = 8½ in., drillpipe outer diameter (OD) = 4½ in., drill collar OD = 6¼ in., drill collar length = 540 ft, hole inclination = 30°, design pit gain = 25 bbl, gas SG = 0.60, and wellbore temperature = 90°F + 1.5°F/100 ft.

1. Starting at the casing shoe, plot the static kick tolerance with vertical depth. Estimate the kick hydrostatic using the given gas and temperature parameters.

2. Where must 7-in. casing be set if the static kick tolerance is 0.5 lbm/gal?

3. Determine the dynamic kick tolerance for a Driller's Method kill at this casing point.

11.7 Refer to Example 11.4 and determine the dynamic kick tolerance for a Wait and Weight kill procedure.

11.8 Refer to Fig. 11.11. Say that a leak-off test data base for the prospect area suggests fracture gradients that fall midway between the two curves.

1. Determine the conductor setting depth for the conditions given in Example 11.5.

2. Determine the conductor depth if a 0.465-psi/ft pore-pressure gradient is assumed.

3. Determine the conductor depth if the design engineer anticipates thick gas sands and assumes a 10.5-lbm/gal pore pressure.

11.9 Consider the 900-ft conductor depth determined in part 2 of Example 11.5.

1. Will the shoe retain integrity if a 150-bbl gas kick is displaced using the Driller's Method? Assume the gas behaves as a bubble and make your own assumption with regard to the gas density.

2. If not, where must the casing be set?

11.10 Refer to the fracture-gradient profile given in Problem 11.8. Determine casing points on the well to allow every hole section below structural pipe to be closed in on dry gas. The first string is set at 300 ft and the pore-pressure gradient is 0.465 psi/ft.

Nomenclature

C_{ai} = annulus capacity factor at initial conditions, bbl/ft
D = vertical depth, ft
D_{sh} = casing shoe depth, ft
g_{fi} = fracture initiation gradient, psi/ft
g_k = kick fluid hydrostatic gradient, psi/ft
g_{kt} = kick tolerance gradient, psi/ft
g_m = mud hydrostatic gradient, psi/ft
g_p = pore pressure gradient, psi/ft
g_{sh} = pressure gradient at the casing shoe, psi/ft
G = pit gain, bbl
h_k = kick height, ft
p_{hg} = gas hydrostatic pressure, psi
p_{hgi} = gas hydrostatic pressure at initial conditions, psi
p_{fi} = fracture initiation pressure, psi
p_p = pore pressure, psi
p_{sh} = pressure at the casing shoe, psi
T = temperature, °F or °R
X = variable in the pressure-prediction relation where mud density is constant, psi
z = compressibility factor, dimensionless
z_i = compressibility factor at initial conditions, dimensionless
ρ_{fi} = fracture gradient in density equivalent, lbm/gal
ρ_k = kick fluid density, lbm/gal
ρ_{kt} = kick tolerance density, lbm/gal
ρ_m = mud density, lbm/gal

References

1. Eaton, B.A.: "How to Drill Offshore with Maximum Control," *Offshore Handbook Vol. 2, World Oil* Reprint Series 8 (1971) 57–61.
2. Eaton, B.A: "Fracture Gradient Prediction and its Application in Oilfield Operation," *JPT* (October 1969) 1353.

3. Bourgoyne, A.T. Jr. *et al.*: *Applied Drilling Engineering*, Textbook Series, SPE, Richardson, Texas (1991) 330.
4. Pilkington, P.E. and Niehaus, H.A.: "Exploding the Myths About Kick Tolerance," *World Oil* (June 1975) 59.
5. Redmann, K.P. Jr.: "Understanding Kick Tolerance and Its Significance in Drilling Planning and Execution," *SPEDE* (December 1991) 245.
6. Leach, C.P. and Quentin, K.M.: "How to Design for Well Control Events," *World Oil* (June 1995) 43.
7. Wessel, M. and Tarr, B.A.: "Underground Flow Well Control: The Key to Drilling Low-Kick-Tolerance Wells Safely and Economically," *SPEDE* (December 1991) 250.
8. Quitzau, R. and Muchtar, J.B.: "Drilling Safely at Well Design Limits: A Critical Well Design Case History," paper SPE/IADC 23930 presented at the 1992 SPE/IADC Drilling Conference, New Orleans, 18–21 February.
9. Aadnoy, B.S., Soteland, T., and Ellingesen, B.: "Casing Point Selection at Shallow Depth," paper SPE/IADC 18718 presented at the 1989 SPE/IADC Drilling Conference, New Orleans, 28 February–3 March.
10. Bourgoyne, A.T., Kelly, O.A., and Sandoz, C.L.: "New Ideas for Shallow Gas Well Control," *World Oil* (June 1996) 50.
11. Baron, S. and Skarstol, S.: "New Method Determines Optimum Surface Casing Depth," *Oil and Gas J.* (7 February 1994) 51.
12. Black, D. and Laurie, A.M.: "Control of Shallow Gas Kicks: A Case History," presented at the 1995 Asia Pacific Well Control Conference, Jakarta, Indonesia, 29 November.
13. Rocha, L.A. and Bourgoyne, A.T.: "A New Simple Method to Estimate Fracture Pressure Gradient," paper SPE 28710 presented at the SPE International Petroleum Conference and Exhibition, Veracruz, Mexico, 10–13 October 1994.
14. Eaton, B.A.: "Graphical Method Predicts Geopressures Worldwide," *World Oil* (July 1976) 100.
15. Hottman, C.E. and Johnson, R.K.: "Estimation of Formation Pressure from Log-Derived Shale Properties," *JPT* (June 1965) 717.

SI Metric Conversion Factors

bbl	× 1.589 873	E − 01	= m^3
deg	× 1.745 329	E − 02	= rad
ft	× 3.048*	E − 01	= m
°F	(°F − 32)/1.8		= °C
°F/100 ft	1.822 689	E + 01	= mk/m
gal	× 3.785 412	E − 03	= m^3
in.	× 2.54*	E + 01	= mm
lbm	× 4.535 924	E − 01	= kg
psi	× 6.894 757	E − 03	= MPa

* Conversion factor is exact.

Appendix A
Circulating Pressure Losses

Introduction

The pump pressure required to overcome flow resistance in a wellbore is the sum of the losses in each hole segment as defined by changes in geometry or rheological properties of the circulating medium. In a drilling well, consistent mud properties are normally assumed and the system losses can be characterized as referenced from the standpipe pressure gauge.

$$p_{dc} = \Delta p_s + (\Delta p_d)_1 + (\Delta p_d)_2 + \Delta p_c + \Delta p_b + \Delta p_{ca} + (\Delta p_{da})_1 + (\Delta p_{da})_2, \quad \text{(A-1)}$$

where p_{dc} = circulating drillpipe pressure, Δp_s = pressure drop in the surface equipment downstream of the gauge, $(\Delta p_d)_1$ = pressure drop in first drillpipe segment, $(\Delta p_d)_2$ = pressure drop in second drillpipe segment, Δp_c = pressure drop in the collars, Δp_b = pressure drop across the bit, Δp_{ca} = pressure drop in the drill collar annulus, $(\Delta p_{da})_1$ = pressure drop in first drillpipe annulus, and $(\Delta p_{da})_2$ = pressure drop in second drillpipe annulus.

The system thus described has two drillpipe sections, drill collars of consistent dimension, and two hole sizes opposite the drillpipe. The pressure drop relationships offered in *API RP13D*[1] for power law fluids are presented in the following discussion. More sophisticated methods are available, but we find these simple procedures to be reasonably accurate given the limitations and questionable data inherent to most drilling-related problems.

Friction Losses in Pipe

It is assumed that rheological data from a six-speed Fann viscometer are available. Furthermore, the shear rates found in the drillstring are assumed to correspond to the power-law relationships of the 300- and 600-rev/min readings. Inside a selected pipe segment, the flow behavior index, n_p, and fluid consistency index, K_p, are thus described by

$$n_p = 3.32 \log(\theta_{600}/\theta_{300}) \quad \text{(A-2)}$$

and $$K_p = 5.11\theta_{300}/511^{n_p}, \quad \text{(A-3)}$$

where θ_{300} and θ_{600} = the dial readings at the 300- and 600-rev/min rotor speeds.

The average velocity in the pipe bore is

$$v_p = 0.408q/d_i^2. \quad \text{(A-4)}$$

Having the velocity, the effective viscosity (centipoise) in the pipe bore is determined with Eq. A-5.

$$\mu_{pe} = 100K_p\left(\frac{96v_p}{d_i}\right)^{n_p-1}\left(\frac{3n_p+1}{4n_p}\right)^{n_p}. \quad \text{(A-5)}$$

Ref. 1 presented charts for adjusting the effective viscosity of water- and oil-based muds for downhole temperature. The effect of pressure is considered also in the oil-based mud correlations.

The next step is to determine the Reynold's number.

$$N_{\text{Re}p} = 928v_p d_i \rho_m/\mu_{pe}, \quad \text{(A-6)}$$

where the mud density, ρ_m, is in lbm/gal. Laminar flow is assumed if the Reynold's number is less than or equal to 2,100 and the corresponding Fanning friction factor is

$$f_{pl} = 16/N_{\text{Re}p}. \quad \text{(A-7)}$$

At higher Reynold's numbers, turbulence is assumed and the friction factor is described by

$$f_{pt} = a/N_{\text{Re}p}^b, \quad \text{(A-8)}$$

where the intermediate variables are

$$a = (\log n + 3.93)/50 \quad \text{(A-9)}$$

and $$b = (1.75 - \log n)/7. \quad \text{(A-10)}$$

Finally, the friction loss in the selected pipe segment, Δp_p, is determined with Eq. A-11.

$$\Delta p_p = f_p v_p^2 \rho_m L_p/25.81d_i, \quad \text{(A-11)}$$

where L_p = segment length in feet.

TABLE A-1—DRILLPIPE EQUIVALENT DIMENSIONS FOR TURBULENT FLOW THROUGH TYPICAL SURFACE CONFIGURATIONS

	Configuration							
	Number 1		Number 2		Number 3		Number 4	
Component	ID (in.)	L (ft)	ID (in.)	L (ft)	ID (in.)	L (ft)	ID (in.)	L (ft)
Standpipe	3	40	3½	40	4	45	4	45
Kelly hose	2	45	2½	55	3	55	3	55
Swivel washpipe and gooseneck	2	4	2½	5	2½	5	3	6
Kelly	2¼	40	3¼	40	3¼	40	4	40
Drillpipe equivalent ID	Drillpipe Equivalent Length (ft)							
2.764 in. (3½ in. 13.30 lbm/ft)	437		161					
3.826 in. (4½ in. 16.60 lbm/ft)			761		479		340	
4.276 in. (5 in. 19.50 lbm/ft)					816		579	

Annular Friction Losses

The shear rates in the annulus are usually much lower than in the pipe bore. The 3- and 100-rev/min readings are assumed to match conditions in the annulus best, which leads to the following expressions for the fluid behavior and consistency indices.

$$n_a = 0.657\log(\theta_{100}/\theta_3) \quad \text{(A-12)}$$

and $K_a = 5.11\theta_{100}/170.2^{n_a}$. ..... (A-13)

The average velocity in the annulus is

$$v_a = 0.408q/\left(d_h^2 - d_o^2\right), \quad \text{(A-14)}$$

where d_h and d_o = hole diameter and pipe outer diameter (OD), respectively. The effective viscosity at this average velocity is

$$\mu_{ae} = 100K_a\left(\frac{144v_a}{d_h - d_o}\right)^{n_a - 1}\left(\frac{2n_a + 1}{3n_a}\right)^{n_a}. \quad \text{(A-15)}$$

Annulus expressions for the Reynold's number, Fanning friction factors, and pressure loss follow:

$$N_{\text{Rea}} = 928v_a(d_h - d_o)\rho_m/\mu_{ae}, \quad \text{(A-16)}$$

$$f_{al} = 24/N_{\text{Rea}}, \quad \text{(A-17)}$$

$$f_{at} = a/N_{\text{Rea}}^b, \quad \text{(A-18)}$$

and

$$\Delta p_a = f_a v_a^2 \rho_m L_a/25.81(d_h - d_o). \quad \text{(A-19)}$$

Surface Equipment Losses

Estimating the pressure drop through the equipment between the standpipe gauge and top joint of drillpipe is accomplished most easily by using an equivalent pipe diameter and length for the system comprising the standpipe, kelly hose, swivel, and kelly. The equivalent dimensions then are substituted in the relations previously given for pipe. **Table A-1** gives published data for four surface equipment combinations.[2]

Pressure Drop Through a Bit

The relation for the bit pressure drop is derived from the fundamental conservation of energy equation. For a 0.95 discharge coefficient,

$$\Delta p_b = \left(9.209 \times 10^{-5}\rho_m q^2\right)/A_n^2, \quad \text{(A-20)}$$

where A_n = bit total flow area (TFA) in square inches.

Example A-1. The following conditions apply to a drilling well.

Depth = 12,000 ft
Casing setting depth = 9,000 ft
Casing inner diameter = 8.921 in.
Average openhole diameter = 8.50 in.
Bit TFA (11-11-12) = 0.296 in.^{2}
Mud type = water based
Mud density = 15.0 lbm/gal
Pump rate = 350 gal/min
Average drillstring temperature = 110°F
Average annulus temperature = 180°F

The drillstring consists of surface equipment corresponding to Type No. 4; 10,410 ft of 4½-in. OD × 3.826-in. ID drillpipe; 1,500 ft of 4½-in. × 2.75-in. heavy-wall drillpipe (HWDP), and 90 ft of 6¼-in. × 2.25-in. collars. A heat cup is used to approximate the average downhole temperature and a Fann multispeed viscometer yields the following readings.

θ_{600} = 70 lbf/100 ft^2
θ_{300} = 43 lbf/100 ft^2
θ_{200} = 32 lbf/100 ft^2
θ_{100} = 22 lbf/100 ft^2
θ_6 = 4.4 lbf/100 ft^2
θ_3 = 3.5 lbf/100 ft^2

Estimate the (1) standpipe pressure and (2) equivalent circulating density (ECD) at the last casing seat.

Solution. 1. First, we apply Eqs. A-2 and A-3 and obtain the rheological parameters in the pipe bore.

$$n_p = 3.32\log(70/43) = 0.703$$

and $K_p = (5.11)(43)/511^{(0.703)} = 2.741$ dyne-s$^{0.703}$/cm^2.

For the surface equipment, Table A-1 gives an effective length of 4½-in. drillpipe as 340 ft. Hence, the combined drillpipe length to use in the flow relations is 340 + 10,410, or 10,750 ft. The average velocity in the drillpipe bore is

$$v_p = (0.408)(350)/(3.826)^2 = 9.76 \text{ ft/s}.$$

Eq. A-5 gives the effective viscosity at this velocity.

$$\mu_{pe} = (100)(2.741)\left[\frac{(96)(9.76)}{3.826}\right]^{(0.703-1)}$$

$$\left[\frac{(3)(0.703)+1}{(4)(0.703)}\right]^{0.703} = 57.4 \text{ cp}.$$

The Reynolds number is computed with Eq. A-6.

$$N_{\text{Rep}} = (928)(9.76)(3.826)(15.0)/57.4 = 9,056.$$

Hence flow is turbulent and the friction factor must be determined with Eq. A-8. Eqs. A-9 and A-10 yield the intermediate variables.

$$a = [\log(0.703) + 3.93]/50 = 0.0755$$

and $b = [1.75 - \log(0.703)]/7 = 0.2719$.

Thus,

$$f_{pt} = 0.0755/9,056^{0.2719} = 0.0063.$$

Eq. A-11 yields the pressure loss in the surface equipment and 4½-in. drillpipe.

$$\Delta p_s = (0.0063)(9.76)^2(15.0)(10,750)/(25.81)(3.826)$$
$$= 980 \text{ psi}.$$

In similar fashion pressure drops of 610 psi and 89 psi are predicted in the HWDP and drill collars, respectively.

Eq. A-20 gives the pressure drop through the bit jets.

$$\Delta p_b = (9.209 \times 10^{-5})(15.0)(350)^2/0.296^2 = 1,931 \text{ psi}.$$

Now our attention turns to the annulus where Eq. A-12 and A-13 yield

$$n_a = 0.657 \log(22/3.5) = 0.525$$

and

$$K_a = (5.11)(22)/170.2^{(0.525)} = 7.579 \text{ dyne-s}^{0.525}/\text{cm}^2.$$

Eq. A-14 gives the average velocity in the drill collar annulus.

$$v_a = (0.408)(350)/(8.5^2 - 6.25^2) = 4.30 \text{ ft/s}.$$

Eqs. A-15 and A-16 yield an effective viscosity and Reynold's number of

$$\mu_{ae} = (100)(7.579)\left[\frac{(144)(4.30)}{8.5 - 6.25}\right]^{(0.525-1)}$$

$$\left[\frac{(2)(0.525)+1}{(3)(0.525)}\right]^{0.525} = 60.4 \text{ cp}$$

and

$$N_{\text{Rea}} = (928)(4.30)(8.50 - 6.25)(15.0)/60.4 = 2,224.$$

Turbulent flow is therefore predicted and the friction factor is obtained from Eq. A-18 as

$$f_{at} = 0.0755/2,230^{0.2719} = 0.0093.$$

Eq. A-19 gives the friction loss in the drill-collar annulus,

$$\Delta p_{ca} = (0.0093)(4.30)^2(15.0)(90)/(25.81)(8.50 - 6.25)$$
$$= 4 \text{ psi}.$$

In the drillpipe x open hole section,

$$v_a = (0.408)(350)/(8.5^2 - 4.5^2) = 2.75 \text{ ft/s},$$

$$\mu_{ae} = (100)(7.579)\left[\frac{(144)(2.75)}{8.5 - 4.5}\right]^{(0.525-1)}$$

$$\left[\frac{(2)(0.525)+1}{(3)(0.525)}\right]^{0.525} = 98.1 \text{ cp},$$

and

$$N_{\text{Rea}} = (928)(2.75)(8.5 - 4.5)(15.0)/98.1 = 1,561.$$

Laminar flow is predicted for this (and the next) hole section. Eq. A-17 yields the corresponding friction factor.

$$f_{al} = 24/1,561 = 0.0154.$$

Thus

$$(\Delta p_{da})_1 = (0.0154)(2.75)^2(15.0)(12,000 - 90 - 9,000)$$
$$/(25.81)(8.5 - 4.5) = 49 \text{ psi}.$$

The procedure is repeated for the cased-hole interval and the pressure drop $(\Delta p_{da})_2 = 122$ psi is computed.

Finally, Eq. A-1 yields the expected standpipe pressure.

$$p_{dc} = 980 + 610 + 89 + 1,931 + 4 + 49 + 122$$
$$= 3,785 \text{ psig}.$$

2. The ECD at the 9⅝-in. casing point includes the hydrostatic and friction loss components above the depth of interest.

$$\rho_{eq} = [(19.25)(122)/9,000] + 15.0 = 15.26 \text{ lbm/gal}.$$

In well control, our concern is normally with the pressure applied in the annulus at either the weakest point in the wellbore or at the bottom of the hole. Unfortunately, the annulus is where our pressure assumptions are most suspect and, thus, most subject to computation inaccuracy. A pressure-while-drilling tool can eliminate the uncertainties associated with ECD computations, at least at the bottom of the hole, but this technology is applied only to a fraction of the holes drilled.

Nomenclature

a = intermediate variable in the turbulent friction factor relation, dimensionless
b = intermediate variable in the turbulent friction factor relation, dimensionless
A_n = total flow area through a bit, in.^{2}
d_h = hole diameter, in.
d_i = pipe inner diameter, in.
d_o = pipe outer diameter, in.
D_{sh} = shoe depth, ft
f_{al} = Fanning friction factor in the annulus for laminar flow, dimensionless
f_{at} = Fanning friction factor in the annulus for turbulent flow, dimensionless
f_{pl} = Fanning friction factor in pipe for laminar flow, dimensionless
f_{pt} = Fanning friction factor in pipe for turbulent flow, dimensionless
K_a = fluid consistency index in the annulus, dyne-s^n/cm^2
K_p = fluid consistency index in pipe, dyne-s^n/cm^2
L_a = length of an annular segment, ft

L_p = length of a pipe segment, ft
n_a = flow behavior index in the annulus, dimensionless
n_p = flow behavior index in pipe, dimensionless
N_{Rea} = Reynold's number in the annulus, dimensionless
N_{Rep} = Reynold's number in pipe, dimensionless
p_{dc} = circulating drillpipe pressure, psi
Δp_a = pressure drop in an annulus segment, psi
Δp_b = pressure drop across the bit, psi
Δp_c = pressure drop in collar bore, psi
Δp_{ca} = pressure drop in collar annulus, psi
Δp_d = pressure drop in drillpipe bore, psi
Δp_{da} = pressure drop in drillpipe annulus, psi
Δp_p = pressure drop in a pipe segment, psi
Δp_s = pressure drop in the surface equipment, psi
q = volumetric flow rate, gal/min
v_a = average velocity in the annulus, ft/sec
v_p = velocity in the pipe bore, ft/sec
θ_3 = Fann viscometer reading at 3 rev/min, lbf/100 ft^2
θ_6 = Fann viscometer reading at 6 rev/min, lbf/100 ft^2
θ_{100} = Fann viscometer reading at 100 rev/min, lbf/100 ft^2
θ_{200} = Fann viscometer reading at 200 rev/min, lbf/100 ft^2
θ_{300} = Fann viscometer reading at 300 rev/min, lbf/100 ft^2
θ_{600} = Fann viscometer reading at 600 rev/min, lbf/100 ft^2
μ_{ae} = effective viscosity in the annulus, cp
μ_{pe} = effective viscosity in pipe, cp
ρ_m = mud density, lbm/gal
ρ_{eq} = equivalent density, lbm/gal

References

1. "Recommended Practice on the Rheology and Hydraulics of Oil-Well Drilling Fluids," *Recommended Practice 13D*, third edition, API, Washington, DC (1 June 1995).
2. Craft, B.C., Holden, W.R., and Graves, E.D. Jr.: *Well Design: Drilling and Production*, Prentice-Hall Inc., Englewood Cliffs, New Jersey (1962) 51.

SI Metric Conversion Factors

cp	× 1.0*	E − 03 = Pa·s
ft	× 3.048*	E − 01 = m
gal	× 3.785 412	E − 03 = m^3
in.	× 2.54*	E + 01 = mm
in.^{2}	× 6.451 6*	E + 02 = mm^2
lbm	× 4.535 924	E − 01 = kg
psi	× 6.894 757	E + 00 = kPa

* Conversion factor is exact.

Appendix B
Surge and Swab Pressure

Pressure to Break Gel Strength

The pressure surge when pipe movement begins is given by

$$\Delta p_g = \pm \frac{\tau_g L}{300(d_h - d_o)}. \quad \text{(B-1)}$$

A positive sign is selected if a surge is created by downward pipe movement while a negative sign corresponds to an upward movement and swabbing condition. Note that Eq. B-1 is identical to the relationship for the pump pressure required to break circulation.

Steady-Flow Friction Pressure

We must first estimate the annular velocity of the mud displaced by the constant velocity pipe string. If the pipe bottom is closed, the mean annular velocity for Newtonian fluid displacement is given by

$$v_a = \frac{d_o^2 v_s}{d_h^2 - d_o^2}, \quad \text{(B-2)}$$

where v_a and v_s = the mean annular velocity and string velocity, respectively. If the tube is of consistent inner diameter and completely open on bottom, v_a is given by

$$v_a = v_s \left[\frac{3d_i^2 - 4d_o^2(d_h - d_o)^2}{-6d_i^4 - 4(d_h - d_o)^2(d_h^2 - d_o^2)} \right], \quad \text{(B-3)}$$

where d_i = the pipe inner diameter.

Pipe velocities in the field are measured typically by clocking the time it takes to run a joint or stand. However, we are interested in the maximum rather than average running or hoisting speed and Moore[1] recommended a time correction factor such that

$$(v_s)_{\max} = 1.5\bar{v}_s, \quad \text{(B-4)}$$

where $\bar{v}_s$ = the average pipe velocity measured over the stem length. An alternate procedure for 90-ft stands is to time only the middle joint.

Eq. B-3 is applicable only if the relative velocity of the mud is in laminar flow and the fluid is Newtonian. Complexities are introduced if the pipe bottom is restricted (say by bit nozzles) or the string has inconsistent geometry. If this is the case, the reader is referred to *Applied Drilling Engineering*[2] or another appropriate text for the solution methods. These, however, involve lengthy, iterative equations best performed on a computer. A worst-case approach often used in hand calculations is to assume the string is plugged on bottom.

For non-Newtonian fluids, Burkhardt[3] suggested the relation

$$v_{ae} = v_a + Cv_s, \quad \text{(B-5)}$$

where v_{ae} = the effective mean annular velocity and C = a parameter called the mud-clinging constant. Burkhardt developed laminar and turbulent flow correlations for Bingham plastic fluids as a function of the d_o/d_h ratio and presented the two curves shown in **Fig. B-1.** Also shown in the chart is the Schuh[4] correlation for power law fluids.

Having a mud velocity relative to the pipe (v_a or v_{ae}), we simply substitute the parameter into the selected Newtonian, Bingham plastic, or power law friction-pressure relationship and obtain the steady-flow surge pressure. Swab pressure is obtained simply by taking the calculated result to be negative. Unfortunately, a Reynold's number transition between laminar and turbulent flow has not been defined for these relative velocities: standard practice is to compute the pressure both ways and adopt the higher answer.

Inertial Pressure

A moving pipe string may experience several acceleration and deceleration cycles over the length of a stand. Eq. B-6 quantifies the pressure change resulting from accelerating or decelerating a closed-end tube through an incompressible liquid.

$$\Delta p_{ac} = \pm \frac{0.00162 \rho_m a_s d_o^2 L}{d_h^2 - d_o^2}. \quad \text{(B-6)}$$

Inertial effects with open-ended pipe usually are considered insignificant.

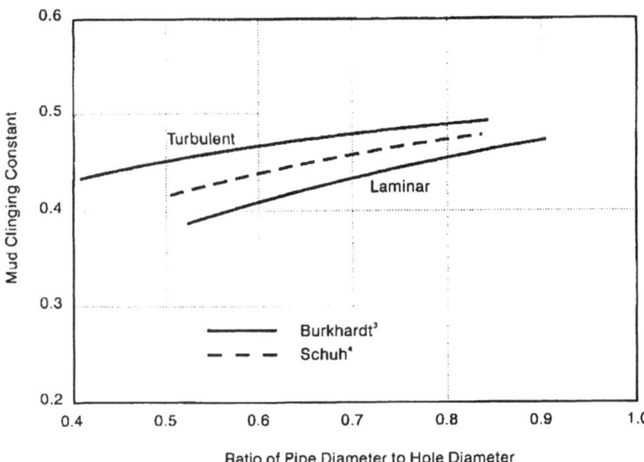

Fig. B-1—Mud-clinging constant correlations for Bingham plastic and power law fluids.

Example B-1. The following conditions apply to a drilling liner job on a deep well.

Present depth = 16,000 ft
Last casing depth = 12,100 ft
Last casing inner diameter = 8.835 in.
Liner outer diameter = 7.625 in.
Drillpipe outer diameter = 5.0 in.
Liner length = 4,300 ft
Mud density = 15.8 lbm/gal
Average running speed = one minute for a 90-ft stand, and
Maximum acceleration = 0.60 ft/s^2.

Assume the mud has developed an average gel strength of 30 lbf/100 ft^2 and use the following Fann multispeed viscometer data.

θ_{600} = 65 lbf/100 ft^2
θ_{300} = 39 lbf/100 ft^2
θ_{200} = 27 lbf/100 ft^2
θ_{100} = 17 lbf/100 ft^2
θ_{6} = 5 lbf/100 ft^2
θ_{3} = 4 lbf/100 ft^2

Estimate the maximum equivalent mud density experienced at the last casing seat.

Solution. When dealing with tapered string geometries (liner strings, drilling assemblies, etc.), the maximum surge or swab pressure usually is experienced when the bottom of the string reaches the depth of interest. Accordingly, we will determine a surge pressure at 12,100 ft for each of the three effects and calculate the equivalent density based on the highest value.

1. Eq. B-1 yields the pressure required to break the gel strength:

$$\Delta p_g = \frac{(30)(4,300)}{300(8.835 - 7.625)} + \frac{(30)(12,100 - 4,300)}{300(8.835 - 5.0)}$$

$$\Delta p_g = 355 + 203 = 588 \text{ psi}.$$

2. We estimate the maximum string velocity using Eq. B-4:

$$(v_s)_{max} = 1.5(90/60) = 2.25 \text{ ft/s}.$$

The relative velocities of a Newtonian fluid in the liner and drillpipe annuli are determined using Eq. B-2.

$$(v_a)_1 = (7.625)^2(2.25)/(8.835^2 - 7.625^2) = 6.57 \text{ ft/s}.$$

$$(v_a)_2 = (5.0)^2(2.25)/(8.835^2 - 5.0^2) = 1.06 \text{ ft/s}.$$

The liner/casing clearance expressed as a ratio is (7.625/8.835) or 0.863. Assuming power law behavior, Schuh's extrapolated mud-clinging constant is approximately 0.48. Similarly, the constant in the drillpipe annulus is obtained as 0.43. Eq. B-5 gives the effective annular velocities in the two annuli.

$$(v_{ae})_1 = 6.57 + (0.48)(2.25) = 7.65 \text{ ft/s}.$$

$$(v_{ae})_2 = 1.06 + (0.43)(2.25) = 2.03 \text{ ft/s}.$$

We go now to the procedure given in Appendix A for estimating annular friction losses. Eqs. A-12 and A-13 yield the fluid behavior and consistency indices, respectively, in the annulus:

$$n_a = 0.657 \log(17/4) = 0.413.$$

$$K_a = (5.11)(17)/170.2^{(0.413)} = 10.411 \text{ dyne-s}^{0.525}/\text{cm}^2.$$

Eq. A-15 gives the effective viscosity in the liner and drillpipe annuli.

$$(\mu_{ae})_1 = (100)(10.411)\left[\frac{(144)(7.65)}{8.835 - 7.625}\right]^{(0.413-1)}$$

$$\left[\frac{(2)(0.413) + 1}{(3)(0.413)}\right]^{0.413} = 22.4 \text{ cp}.$$

$$(\mu_{ae})_2 = (100)(10.411)\left[\frac{(144)(2.03)}{8.835 - 5.0}\right]^{(0.413-1)}$$

$$\left[\frac{(2)(0.413) + 1}{(3)(0.413)}\right]^{0.413} = 96.0 \text{ cp}.$$

The two Reynold's numbers are calculated using Eq. A-16.

$$(N_{Rea})_1 = (928)(7.65)(8.835 - 7.625)(15.8)/22.4$$
$$= 6,059.$$

$$(N_{Rea})_2 = (928)(2.03)(8.835 - 5.0)(15.8)/96.0$$
$$= 1,189.$$

Friction losses will be considered for both turbulent and laminar flow and we must determine the Fanning friction factors for each flow regime. Eq. A-17 yields the results for laminar flow.

$$(f_{al})_1 = 24/6,059 = 0.0040.$$

$$(f_{al})_2 = 24/1,189 = 0.0202.$$

The intermediate variables used in the turbulent relation are computed.

$$a = [\log(0.413) + 3.93]/50 = 0.0709.$$

$$b = [1.75 - \log(0.413)]/7 = 0.3049.$$

The turbulent friction factors are thus

$$(f_{at})_1 = 0.0709/6,059^{0.3049} = 0.0050.$$

$$(f_{at})_2 = 0.0709/1,189^{0.3049} = 0.0082.$$

Considering the liner annulus, the turbulent friction factor is higher than the laminar factor. Substituting this value in Eq. A-19 gives

$$(\Delta p_a)_1 = (0.0050)(7.65)^2(15.8)(4,300)/(25.81)$$

$$(8.835 - 7.725) = 637 \text{ psi.}$$

Now we substitute 0.0202 for the friction factor in the drill-pipe annulus and obtain

$$(\Delta p_a)_2 = (0.0202)(2.03)^2(15.8)(7,800)/(25.81)$$

$$(8.835 - 50) = 104 \text{ psi.}$$

The total surge pressure at 12,100 ft is therefore

$$\Delta p_a = 637 + 104 = 741 \text{ psi.}$$

3. Finally, Eq. B-6 yields the pressure increases caused by the pipe acceleration.

$$(\Delta p_{ac})_1 = \frac{(0.00162)(15.8)(0.6)(7.625)^2(4,300)}{8.835^2 - 7.625^2} = 193 \text{ psi.}$$

$$(\Delta p_{ac})_2 = \frac{(0.00162)(15.8)(0.6)(50)^2(7,800)}{8.835^2 - 5.0^2} = 56 \text{ psi.}$$

Thus

$$\Delta p_{ac} = 193 + 56 = 249 \text{ psi.}$$

The steady-flow condition is the most significant of the three effects and the maximum equivalent density imposed at the last casing seat is

$$\rho_{eq} = 15.8 + 19.25(741/12,100) = 17.0 \text{ lbm/gal.}$$

Nomenclature

a = intermediate variable in the turbulent friction factor relation, dimensionless
a_s = acceleration, ft/s^2
b = intermediate variable in the turbulent friction factor relation, dimensionless
C = clinging constant, dimensionless
d_h = hole diameter, in.
d_i = pipe inner diameter, in.
d_o = pipe outer diameter, in.
f_{al} = Fanning friction factor in the annulus for laminar flow, dimensionless
f_{at} = Fanning friction factor in the annulus for turbulent flow, dimensionless
K_a = fluid consistency index in the annulus, dyne-s^n/cm^2
L = length, ft
n_a = flow behavior index in the annulus, dimensionless
N_{Rea} = Reynold's number in the annulus, dimensionless
Δp_a = pressure drop in an annulus segment, psi
Δp_{ac} = pressure change caused by pipe acceleration or deceleration, psi
Δp_g = pressure surge caused by gel strength, psi
v_a = average velocity in the annulus, ft/s
v_{ae} = effective average velocity in the annulus, ft/s
v_s = string velocity, ft/s
θ_3 = Fann viscometer reading at 3 rev/min, lbf/100 ft^2
θ_6 = Fann viscometer reading at 6 rev/min, lbf/100 ft^2
θ_{100} = Fann viscometer reading at 100 rev/min, lbf/100 ft^2
θ_{200} = Fann viscometer reading at 200 rev/min, lbf/100 ft^2
θ_{300} = Fann viscometer reading at 300 rev/min, lbf/100 ft^2
θ_{600} = Fann viscometer reading at 600 rev/min, lbf/100 ft^2
μ_{ae} = effective viscosity in the annulus, cp
ρ_m = mud density, lbm/gal
ρ_{eq} = equivalent density, lbm/gal
τ_g = gel strength, lbf/100 ft^2

References

1. Moore, P.L.: *Drilling Practices Manual*, second edition, PennWell Publishing Co., Tulsa (1986) 285.
2. Bourgoyne, A.T. Jr., Chenevert, M.E., Millheim, K.K., and Young, F.S. Jr.: *Applied Drilling Engineering*, second printing, Textbook Series, SPE, Richardson, TX (1991) 164–72.
3. Burkhardt, J.A.: "Wellbore Pressure Surges Produced by Pipe Movement," *JPT* (June 1961) 595–605.
4. Schuh, F.J.: "Computer Makes Surge Pressure Calculations Useful," *Oil & Gas J.* (3 August 1964) 96–104.

SI Metric Conversion Factors

cp	× 1.0*	E − 03 = Pa·s
ft	× 3.048*	E − 01 = m
gal	× 3.785 412	E − 03 = m^3
in.	× 2.54*	E + 01 = mm
lbm	× 4.535 924	E − 01 = kg
psi	× 6.894 757	E + 00 = kPa

* Conversion factor is exact

Author Index

Subject Index

C

D

E

F

G

H

I–K

L

M

N

O

P–Q

R

S

T

U–V

W–Z